THE SAMPLING AND ASSAY

OF THE

PRECIOUS METALS

COMPRISING GOLD, SILVER, PLATINUM, AND THE PLATINUM GROUP METALS IN ORES, BULLION, AND PRODUCTS

by

ERNEST A. SMITH, Assoc. R.S.M.

FORMERLY OF THE ROYAL SCHOOL OF MINES, LONDON: MEMBER OF THE INSTITUTE OF METALS, AND PAST MEMBER OF THE INSTITUTION OF MINING AND METALLURGY.

SECOND EDITION

REVISED AND RE-SET

With 167 Illustrations

Published By

Met-Chem Research

from the 2nd edition

originally published by

Charles Griffin & Company Ltd.

1st Edition 1913
2nd Edition 1947

2nd Edition Reprinted 1987
2nd Edition Reprinted 2003 by
Met-Chem Research
P.O. Box 67347
Albuquerque, New Mexico 87193-7347 USA

ISBN 0-931913-23-3

Dedicated

TO THE MEMORY OF

RICHARD SMITH,

ONE OF THE ORIGINAL STAFF OF THE ROYAL SCHOOL OF MINES,

ON ITS FORMATION IN 1851,

BY HIS SON.

PREFACE TO SECOND EDITION

As stated in the preface to the first edition, this book forms one of the series of treatises on metallurgy written by Associates of the Royal School of Mines, London, inaugurated by the late Sir W. C. Roberts-Austen, F.R.S., on whose staff the author served for some years.

The work is primarily intended for students, having been originally written at the request of the author's old students at the Royal School of Mines; hence more detail has been given than would be necessary in a work intended only for the use of experienced assayers. The subject-matter includes the author's varied experience in the assay of precious metals in industrial practice.

In preparing the revised edition, it has been considered advantageous to make important alterations in several places, especially in the chapter on the assay of the platinum-group metals, which in recent years have increased considerably in industrial importance. Nevertheless, it has been the author's endeavour to conserve as much as possible of the original, and all alterations have been carefully considered.

Special acknowledgments are due to Messrs. Schoeller and Powell for permission to quote from the platinum metals section of their well-known work, *The Analysis of Minerals and Ores of the Rarer Elements*; and also to the Gas, Light and Coke Company, London, for the illustration of their gas assay muffle furnace.

The author is indebted to A. E. W. Smith, B.Sc., Ph.D., for assistance given in proof-reading.

ERNEST A. SMITH.

December, 1946.

CONTENTS

CHAPTER XVII

BULLION: VALUATION AND SAMPLING

CHAPTER XVIII

THE ASSAY OF GOLD BULLION

CHAPTER XIX

THE ASSAY OF SILVER BULLION

APPENDIX

TABLES

THE SAMPLING AND ASSAY OF THE PRECIOUS METALS

CHAPTER I

INTRODUCTORY

"Precious Metals" Defined.—The term "precious metals" (or "noble metals") has long been applied to gold and silver on account of their remarkable properties and costliness, and in more recent years it has been extended to include the six comparatively rare metals known as the *platinum group*, which are of considerable industrial value. Of this group platinum is the best known and was the first discovered (in 1735). The other five, in order of their importance to the jewellery and other industries, are palladium, iridium, rhodium, ruthenium, and osmium. Since all of these latter metals are now finding increasing application in industry, it is essential to include the methods in use for detecting and determining them in a book dealing with precious-metal assaying. One of the distinguishing features of the group is that none of these metals oxidizes at ordinary temperatures, and they are therefore permanent under ordinary conditions of wear. Other characteristic features of individual metals of the platinum group are dealt with later in the text.

The assay of the precious metals, especially those of the platinum group, belongs to a specialized branch of analytical chemistry and requires considerable chemical knowledge of the metals involved, as well as experience in detecting and separating them.

Assaying Defined.—Originally, "assaying" was the term applied to the art of determining the quantity of metal contained in an ore, in alloys, or in metallurgical products by the dry or "fire" method. At the present time the term assay has a much wider meaning, being applied to the determination of metal content by whatever method the operations are carried out. Assaying is said to be either "dry" or "wet" according as the agency of either fluxes and fire or of liquid solvents is employed. Formerly, in order to obtain the result with the greatest expedition, the dry method (also called "furnace method") was chosen for producing the chemical reaction; but as this was frequently done at the expense of accuracy in the result of the assay, the wet method is also used in modern times. It has not, however, entirely displaced the dry method, for the latter is employed in all cases where sufficiently accurate results can be reached more quickly, as in assaying lead, tin, gold and silver ores, or where suitable wet-assay methods cannot conveniently be substituted for it.

Furnace methods, in combination with wet methods, may frequently be used with advantage. It must be admitted, however, that except in a few cases such as those given, the tendency of modern assay practice is to discard purely furnace methods in favour of wet methods.

During the last few years rapid strides have been made in the application of quick and reliable chemical methods to the assay of ores, metals, etc., and assaying is also becoming wider in its scope; consequently the line of demarcation between it and chemical analysis has almost disappeared. Indeed, assaying has now become so nearly identified with the latter that the two terms are often used synonymously, and even the determination of an *organic* constituent of a mixture is not infrequently called an "assay."

History of Assaying.—Assaying was the earliest-known branch of chemistry, and in fact the branch which afterwards, through the medium of the various investigations of alchemy, drew attention to theoretical chemistry, and thus founded that science. The origin of the art of assaying is somewhat obscure, and dates back to a remote period. To get to the root of a number of assay operations, it is necessary to consult the records of the alchemists who did so much to develop them.

The first explicit records of assaying are contained in a series of anonymous booklets published in German early in the sixteenth century under the title *Probierbüchlein*. In these, the art of assaying is disclosed as being well advanced towards maturity so far as concerns gold and silver, and some notes on lead and copper are given. The booklets form a collection of disconnected recipes, lacking in systematic arrangement and constituting, apparently, a heritage of wisdom passed from father to son or apprentice over many generations. Brief notes on assay methods appear in the treatise on metallurgy by Vannocio Biringuccio, published at Venice in 1540 under the title of *De la Pirotechnia.* Georgius Agricola, who published his *De Re Metallica* in 1556, was, however, the first writer to deal systematically with the art of assaying minerals and metals.[1] Much valuable information on early methods of assay is also found in the well-known work on assaying by Lazarus Ercker, published at Frankfurt in 1574.[2]

It appears very probable that the art of assaying had its origin in the testing of gold to detect the presence of adulterants, notably copper, which, when polished, more closely resembles gold in colour than any other metal and was undoubtedly palmed off as gold on the uninitiated, and also alloyed with gold with fraudulent intent at a very early period. This is evident from one of the Tel-el-Amarna tablets, dated about 1380 B.C., which records the fact that when a bar of "gold" received from a merchant was tested by melting, it proved to be largely dross, and left only a small amount of pure gold after the operation.

It would appear, therefore, that the earliest method of testing the purity of gold was by "fire," that is, by melting it and observing the surface appearance of the melted metal, which in the case of pure gold (or silver) would be clean and bright, whilst the presence of copper or other impurity would give rise to the production of a certain amount of dross. This testing by fire was doubtless the forerunner of the "cupellation" process for refining the precious metals and for assay purposes. Refining silver by fire, as in cupellation, is graphically described by Jeremiah in the Old Testament records; and a cupellation process for gold is mentioned by Diodorus Siculus as being in use in the second century B.C.

Pliny records that the Roman metallurgists used lead for the purification of gold and silver, and the refining of these metals is described by Geber, the Arabian and greatest of the early alchemists, who died A.D. 777.

[1] *De Re Metallica*, libri XII, Basle, 1556.

[2] L. Ercker, *Allerfurnemisten mineralischen Eerzt u. Bergwerks Arten*, Frankfurt, 1574, translated by Sir John Pettus, 1683 (*The Laws of Art and Nature in Assaying Metals*).

The exact period at which the cupellation process was first used for assaying is not accurately known, but conducting assays by the cupel (or "coppels," as the word is given in old treatises on assaying) on what would at the present day be considered very large samples of metal, seems to have been held necessary in the twelfth century, for in certain official trials of coin in the time of Henry II (1154-89) the "Miles Argentarius," or Assayer of Silver, is instructed to make tests by cupellation. Thus the cupellation test was already recognised in this country in the latter part of the twelfth century, while in France the first official mention of the process occurs about the year 1314.

Another method of valuing gold has survived from very ancient times, and is usually known as the "touch." It was certainly in use by the Greeks as early as 500 B.C., and was probably devised by the Lydians for testing the quality of their gold coinage. Other names given to this test are "touchstone," "streak" or "stripe test." It was originally made on a black stone known as "Lydian stone," found in that district of Asia Minor, and the earliest written account of the test and the kind of stone employed is that given by Theophrastus in the third century B.C. It is a simple physical test, and consists in rubbing the gold to be tested on a piece of hard black stone, and comparing the colour of the streak thus produced with similar streaks made with a series of standard gold alloys of known composition, termed "touch-needles." The "touchstone" was well known to the Romans; it was in use in England in the twelfth century and probably much earlier; and Prof. Gowland[1] has stated that this method of testing the quality of gold was also used in the first Government Mint established in Japan in the sixteenth century.

The only other physical method, as distinguished from chemical methods, that has been widely adopted, is that of ascertaining the purity of metals by their density; this was originally devised by Archimedes (212 B.C.) for ascertaining whether the crown of Hiero was or was not genuine. It consists in comparing the weight of the metal with that of an equal volume of water, and is still sometimes resorted to when the metal to be examined must be preserved intact.

Pliny states that in his time a method was in use for estimating the amount of silver in an alloy of silver and copper by the degree of discoloration or blackening which attends the heating of the alloy in air. This method, long practised in France, and known by the name of *essai à la raclure*, is described by Rochon,[2] who says that it was generally recognized in the Roman mints in the time of Marius Gratidianus, "Triumvir of the Money"; but recent experiments have shown that it ceases to be useful for alloys of silver and copper which contain less than 80 per cent. of silver, as such an alloy becomes quite black when heated in air, and in such alloys as are poorer in silver, variations in tint cannot be detected.[3]

The "parting assay," in which gold is separated from an alloy of gold and silver by the action of nitric acid, was certainly known to Geber and the early alchemists, but the first official mention of its use appears to be in a decree of Philippe de Valois in the year 1343, confirming its adoption in the French Mint.

The methods of procedure in the seventeenth century have been briefly described by Savot[4] and by Reynolds,[5] and more fully in Glaser's *Compleat*

[1] *J. Inst. Metals*, 1910, **4**, 11.
[2] *Essais sur les Monnaies*, 1792, p. 17.
[3] See Chaudet, *L'Art de l'Essayeur*, Paris, 1835, p. 77; also Percy, *Metallurgy of Silver*, p. 157.
[4] *Discours sur les Médailles Antiques*, Paris, 1627, p. 72.
[5] *A New Touchstone for Gold and Silver Wares*, London, 1679, p. 362.

Chymist (1677). "In 1666 Pepys saw the parting assay being practised at the Mint in the Tower of London, and from his description it is clear that the method then employed bears a surprisingly strong resemblance to that of the present day" (Rose).

With regard to the application of the early methods of assaying to the determination of metals in ores, very little appears to have been recorded. It is not, however, unreasonable to suppose that the ancient process of cupellation originally suggested the idea of fusing precious-metal ores with lead or lead compounds to collect and concentrate the gold and silver as a preliminary to cupellation proper. It would appear that the art of assaying, as applied to ores, received attention in Germany at a very early date. The methods used by German assayers of the sixteenth century are fully described by Ercker in the work already referred to.

The method of assay by "scorification" is mentioned, and was preferred to fusion in a crucible, which is also described. When fusion in a crucible was recommended, it was only as a preliminary to scorification.[1] Ercker describes many of the methods and precautions which are generally adopted at the present time, such as the treatment of cupriferous precious-metal ores with nitric acid and the use of common salt as a cover to the charge in the crucible; while the use of "proof-centners," the old German system of weights, corresponds to the present-day "assay-ton" system. It is also of interest to note that he mentions the forerunner of the buck-board for the fine pulverizing of ores; and he instances assay-offices in which 200 assays of ores were made in a week. These references are sufficient to indicate the advanced state of the knowledge of assaying at that period.

With the advance of metallurgical science and the extraction of the common metals by smelting processes on a large scale, the need must have been felt for methods of assaying the ores of these metals also. A study of the dry methods of assay, many of which were in use until quite recent years, shows that they may be regarded as smelting processes in miniature, as the older methods of dry assaying frequently follow, on a small scale, the operations of smelting. These methods, as previously stated, are now to a very large extent obsolete in good assay practice for determining with accuracy the value of ores, etc., and have been generally replaced, when possible, by the more accurate and usually more rapid methods of chemical analysis. The dry methods of assaying iron, copper, zinc, nickel, and cobalt are now taught only as exercises in furnace manipulation or to instruct the student in the principles of metallurgy; occasionally they are used in the works laboratory to test the smelting qualities of ores.

To-day the metallurgical chemist relies very largely on volumetric methods for the analytical determinations required. Many volumetric assays can be performed in a shorter time than gravimetric determinations for corresponding constituents, and this is an important item where much work has to be done. The results yielded are either very accurate or at least sufficiently exact for controlling processes in metallurgical work; and these assays are in most cases less expensive, though they require special apparatus of accurate construction.

Of the few dry methods of assay still in use, those employed for the determination of the precious metals are by far the most important. The quantity of gold and of silver contained in ores is too minute to allow of its being accurately determined by wet methods. An impression prevails that the results obtained by furnace methods are less accurate and reliable than those

[1] Ercker, *loc. cit.*, Book 1, Chap. X; Book 2, Chap. VIII.

obtained by other methods. Although this is true of certain dry methods that are now obsolete, the accuracy with which small amounts of gold and silver can be determined by the fire-assay probably exceeds that of any known method of analysis for any other metal.

The assayer is frequently called upon to determine the gold in samples that yield only one-twentieth of an ounce to the ton, or even less.

Qualifications of Assayers.—It may be well here to make a brief reference to the qualifications of assayers. Before a student enters a course on assaying he should have gone through a course of practical chemistry in which he has been taught to work with the greatest accuracy, and has acquired that degree of confidence in his work which every operator ought to possess, so that he may be prepared to commence training in the special methods necessary for the guidance of metallurgical operations as well as for the valuation of ores and metallurgical products.

It would be impossible to insist too strongly upon the importance of an assayer receiving sound instruction in the specific department of analytical chemistry which relates to assaying. Occasionally a large sum of money may depend upon an assayer possessing a profound knowledge of certain departments of analytical chemistry. Thus, material submitted for assay might contain small quantities of some of the metals of the platinum group, which may not be detected by an assayer possessing only a limited knowledge of chemistry. An assayer of the precious metals may be called upon to determine the presence or absence of iridium and of other metals also in gold, which, for their satisfactory detection, require no ordinary skill, and involve some of the most difficult operations of analytical chemistry. But although a knowledge of chemistry is absolutely necessary, this of itself would be a very insufficient qualification for the man who has to perform assay operations: no one can attain to any degree of eminence in assaying unless he has had some amount of practice in the laboratory. Theoretical knowledge counts for much, but skill counts for more; and we cannot get skill ready made. It can come only as the result of long and steady practice. Without experience an assayer could not decide whether the requisite temperature was produced in any of the furnaces, or whether any of the various operations of assaying had been properly performed. It is well, however, to bear in mind that while scientific knowledge alone will not qualify a man to take charge of an assay laboratory, so neither is empirical knowledge the only qualification desirable in such cases: it is clearly the combination of scientific with practical knowledge that will render the head assayer of a large metallurgical works in the highest degree competent for his responsible position.

A man once brought his son to the Royal School of Mines, London, with the request that he might be taught to "do copper." He did not want his boy to "waste his time learning about oxygen and hydrogen and all that," but he wished him simply to learn to "do copper."[1] Although seldom expressed with such refreshing candour, the desire to "do assaying" without learning more than the minimum amount of chemistry is still prevalent; and, unfortunately, assaying is a subject which may be, and frequently is, taught and practised in such a manner as to degrade it to the level of a purely mechanical and often quite unintelligible series of rule-of-thumb operations.

The author fully endorses the late Prof. J. O. Arnold's [2] statement that students "cannot be too strongly urged to remember the fact that an assayer or metallurgisal chemist and an 'analytical machine' who turns out so many

[1] Quoted by G. S. Newth, in his preface to *Manual of Chemical Analysis*.

[2] Preface to *Steel Works Analysis*, J. O. Arnold, of the University of Sheffield.

estimations per day are two very different personalities. Assayers deficient in a thorough knowledge of at least elementary chemistry, physics, and mathematics, and the principles of qualitative and quantitative analysis, can claim to rank only with skilled artisans: in both assaying and chemical analysis the head and the hands should always work together. Even with students really well grounded in pure chemistry the possession of knowledge and the ability to apply it are two widely different things. After the preliminary training, all that can be recommended, in other respects, to one who desires to perfect himself in assaying, is the most scrupulous cleanliness, order, and precision as regards his assays and implements, and the most unwearying adroitness and attention in performing the various manipulations required. Students should bear in mind that the whole aim of assaying is to get the most accurate result with economy of time and material, and, when these factors are incompatible, to sacrifice everything to the most accurate result."

Some assayers, after a few years' experience, hold the view that there is nothing to be learnt in assaying, and that the work is humdrum and monotonous. But, as one writer has said, "a man's work in any call of life is what he makes it, and the art of assaying especially calls for the exercise of the highest qualifications of the chemist—irreproachable honesty and perseverance under difficulties within and without the office. It must not be permitted to sink to the rule-of-thumb state, and the interest evoked by indefatigable attempts to attain excellency and to improve on existing methods is unending. There is no real truth in the common assumption that the assayer cannot err, but it is open to him to endeavour to live up to the assumption, which can only be accomplished by an intelligent application and use of all his faculties."

It is generally recognized that assaying has a very important bearing on the mining and metallurgical departments of mineral enterprises, but it must be admitted that the work of the assayer is seldom valued as it should be. The operations of mill and smelter are being more and more directed according to the results obtained in the laboratory, and the metallurgical chemist is now required to make daily a number of determinations that would have appalled his predecessor of even a few years ago. The time allowed for making individual determinations is also being steadily reduced. Under these conditions it is imperative that the man placed in charge of the assay department of a large works or mill should be thoroughly qualified for the important position he holds, and his qualifications and position should be duly recognized.

It is the chief function of the assayer and metallurgical chemist to *interpret* results, and it is he alone who can be trusted with the practical carrying out of his deductions. He should, therefore, be valued chiefly in his advisory capacity, and should be accorded a position and power commensurate with the importance of his functions. If it be worth while to take aid from chemical science at all, then this is cheapest and best done by employing a competent analyst and assayer provided with the best and most convenient instruments and apparatus that money can procure. It is evident, in many cases in metallurgical works, that this view is as yet not acted upon. As pointed out some years ago, by Sir Philip Magnus, "Few manufacturers in this country yet realize the economy and industrial advantage of attaching to their works intelligence departments staffed with scientific experts."

The importance of the mine-assay department has, however, long been recognized among practical mining men, and in South Africa the demand for good assayers and a fully-equipped assay office has increased in recent years owing to the close attention being paid to the lower-grade ores and to the demands of the reduction works.

CHAPTER II

THE DESIGN AND EQUIPMENT OF ASSAY OFFICES

In assay offices where quick and accurate work is required, the general arrangement and convenience of the different rooms is of fundamental importance. The assayer has usually to adopt the quickest methods consistent with the degree of accuracy required, and such details as space and apparatus have to be carefully considered, and are considerably influenced by the design and equipment of the assay office.

It may be well here to remind the student who has received his training in a well-equipped laboratory, supplied with every convenience, that his success as an assayer will be largely dependent on his ability to adapt himself to new and less favourable conditions of work such as would exist in a small laboratory unprovided with many of the appliances which he has been in the habit of using, and regards as indispensable.

The general arrangement and equipment of an assay office will necessarily vary somewhat according to the amount and class of work to be done, and also according to the individual preferences of the assayer. The city assayer seldom has much opportunity of planning his office to suit his particular requirements: usually he has to adapt existing buildings as best he can. When, however, no such restrictions exist, it is true economy of time and labour, and also conducive to accurate working, to have the assay-office buildings especially planned to meet the requirements of the work to be done. No design of assay office can be given that is suitable for all requirements, but a few general remarks may be made.

The most important consideration in planning an assay office is so to economize space, by the careful arrangement of the positions of the several rooms and of the appliances and apparatus, that no unnecessary steps will be required in moving from one room to another and from one piece of apparatus to another. If possible, all the rooms should be on one level.

A well-equipped assay office for general assaying should consist of at least six rooms, which are allotted as follows:—

(1) Grinding and sampling room.
(2) Furnace room.
(3) Laboratory for wet work.
(4) Weighing or balance room.
(5) Store room.
(6) General office.

One or more of these rooms may of course be omitted according to circumstances.

In many cases the whole of the assay work has to be performed in two rooms, or even one room. Unless, however, circumstances render it unavoidable, the several operations should be performed as far as possible in separate rooms, so as to avoid the possibility of the contamination of samples from dust resulting from crushing and grinding and from the manipulation of the furnaces. When a large number of samples have to be assayed by furnace methods, a separate room is sometimes kept solely for "fluxing" or preparing assays for fusion, but as a general rule this is performed in the furnace room.

The appliances for the equipment of assay offices are described subsequently in Chapters III and IV.

The size of the various rooms will depend on the amount and character of the work to be done in them, but the furnace room and the laboratory for wet work should be of ample proportions, well lighted and ventilated, and 15 or 16 ft. high. Hot-water pipes or gas or electric fires are best for heating the rooms, because they create no dust.

(1) **Grinding and Sampling Room.**—This room should contain all the appliances for crushing, grinding, and sampling. It should be divided by a screen, and the crushing and grinding machinery kept in one section, so that the dust produced by these operations shall not contaminate material that is being sampled. Electric or other power should be available for driving the machinery. The floor of this room should be of concrete or similar material. Ample bench room with cupboards should be provided; but shelves on the walls should be dispensed with as far as possible, as they retain considerable amounts of fine ore dust.

The sampling room is frequently absent, but it will amply justify its existence in an assay office, unless the samples are always received in a condition suitable for assay purposes.

(2) **Furnace Room.**—The furnace room, as of primary importance in many cases, should be as large as possible, considering the sedentary and unhealthy nature of the work, especially in hot climates. The ground floor of a building is the most convenient, so that a good height of chimney or stack may be obtained for the furnaces. The floor near the furnaces should be of stone or concrete; all other parts may be boarded, or if a concrete floor extends all over the furnace room, it is advisable to have wooden stages in front of the work-benches so as to avoid damp. The work-benches should be of thoroughly seasoned non-resinous wood, and stoutly built. They should be fitted with drawers and cupboards, with shelves above for reagents and fluxes. The tops of the benches require to be made of hardwood such as teak or American walnut.

The furnaces usually needed, various types of which are described subsequently, are one or more assay-melting furnaces and one or more muffle furnaces for cupellation and scorification, etc. When furnaces burning coke or coal are used, it is advantageous to have facilities for firing and for clearing the ash-pit from the back of the furnace, when this can be conveniently arranged, so as to keep as much dust and dirt out of the furnace room as possible. For the same reason, even when the furnaces are fired from the front, the coke- or coal-bin should be placed if possible outside the furnace room and behind the furnaces, but communicating with the furnace room by means of a trap or shoot, through which the fuel is drawn with a shovel as required. The furnaces should not be placed where they will be subjected to direct sunlight, as this will occasion error in judging the temperature. As a general rule, the main light in the furnace room, except in hot climates, is best derived from a large top-light.

An iron table on which to stand the moulds when pouring assays is frequently desirable, especially when furnaces with sloping tops are used. It is made of boiler-plate $\frac{3}{8}$-inch thick, and, say, $2\frac{1}{2}$ feet wide and 4 to 6 feet long. It is supported on a frame of angle iron or on brick pillars about 30 inches high.

A hot-plate, described below, should be provided in the furnace room for drying samples, etc., and is best contained in a cupboard or recess surmounted by a hood to protect it from draughts. A good sink is also very necessary. Provision should be made near the furnaces for drying cupels, crucibles, scorifiers, etc.

(3) **Laboratory for Wet Work.**—A room of about 30 by 20 ft., and 15 or 16 ft. high, is large enough for most work. A ceiling sheathed with wood is very desirable in the laboratory, as this prevents the possible falling of pieces of whitewash or plaster loosened by the action of acids. The windows should be all on one side, as cross-lights are objectionable.

The working benches are made of well-seasoned and well-oiled wood as described for the furnace room, and should be placed close against a side-light in preference to a central position in the room, where the light is not so good.

A good draught-cupboard should be provided to carry off acid fumes, etc. This is lined with slate, white tiles, or white opal glass, and fitted with balanced glazed sashes.

Heating or "hot"-plates are almost indispensable if much wet work has to be done. These should be at least $\frac{3}{8}$-inch thick, otherwise they are very liable to buckle with constant heating. The plates, made in any convenient size, are supported on an iron frame or on bricks, and heated by several ring or other burners placed underneath them, or electrically. It is advisable to lay a sheet of $\frac{1}{4}$-inch asbestos millboard or cloth on the top of the plate. The plate should be at a convenient level from the floor, say $2\frac{1}{2}$ feet.

Sand-baths are used by many assayers in preference to hot-plates; they can be made of any convenient size to accommodate the number of assays to be made. In a large laboratory sand-baths may be 3 by 2 ft., and 3 to 4 inches high. Hot-plates and sand-baths are best contained in a good draught-cupboard.

A good earthenware sink is required, and an additional tap connected with a good head of water will be needed if a filter-pump is to be used.

Ventilation may be effected preferably through louvres on the ridge of the roof or by means of swinging upper windows.

(4) **Weighing or Balance Room.**—The balance is the most important part of the assayer's outfit, and in all properly equipped assay offices a balance room is provided. The sensitive balances used in assaying are very liable to be injured by fine dust and acid fumes, so that a separate room for them is necessary. This room, if possible, should be quite apart from the laboratory used for wet work. The north aspect, if obtainable, is the best for the windows of the balance room; direct sunlight coming into the room is very inconvenient. The room should be as far away as possible from the crushing and grinding machinery, so that the balances may be placed where they will be subject to the least possible vibration. The bench on which the balances are placed should rest on concrete or brick pillars, say 18 inches square at the bottom and 12 inches at the top, set in the ground underneath the floor, and projecting a few inches above the floor, which is cut away from the pillars by a space of about half an inch. By this arrangement the balance bench is free from all vibrations of the building. The bench should be placed parallel to the window, so that the balances face the window with the light behind the operator. The top of the balance bench is best made of hardwood, well seasoned; slate is sometimes recommended, but is objectionable as it is very cold to the hands for continuous weighing, and is very liable to cause breakage of the bottles, etc. into which substances are being weighed if these are accidentally knocked over or are placed hurriedly on the bench.

The walls of the balance room are best painted some light colour, or tiled with cream-coloured tiles.

(5) **Store Room.**—This is best attached to the main working rooms, and not situated in a basement, where the damp is always objectionable. It is desirable to have it divided by a partition, so that stock, apparatus and reagents, etc. can be kept quite apart from samples that require to be stored.

(6) **General Office.**—This should be arranged near the main entrance to the assay office and be readily accessible. It should be provided with a writing-table or desk, cabinets for filing correspondence, typewriter, safe, and a book-case to contain standard works of reference, etc. Suitable arrangement should be made in the lobby for accommodating coats and hats.

Mill Assay Office.—The accompanying plan (Fig. 1) illustrates the arrangement suggested by A. Whitby,[1] slightly modified by E. J. Laschinger, for an assay office for a 200-stamp cyanide mill on the Rand, South Africa. The plan will serve as a guide in the building and fitting up of new laboratories.

Where convenient, the office building should have foundations quite distinct from those of the battery, and removed as far as possible from it in order to avoid vibration. It should be in as central a position as possible, and not attached to any reduction plant. With the exception of the balance room, each working room in the structure should be under observation from the adjoining rooms. Whitby considers this most important, as a greater number of assays can then be controlled with a smaller staff (native). The assay office shown in the plan consists of six rooms contained within the main building, viz. :—

(1) Drying and sampling room.
(2) Fluxing room.
(3) Furnace room.
(4) Laboratory for wet work.
(5) Balance room and general office.
(6) Store.

In the lean-to outside the main building, provision is made for the crushing and grinding appliances. The building is of brick, with concrete or granolithic floor throughout, excepting the balance room, which has a wood floor and a wood ceiling. The furnace and sampling rooms are not ceiled, but it is of the utmost importance to have a dust-proof ceiling for the balance room, if not quite so necessary for the wet laboratory and fluxing rooms. The roof trusses are of steel. The following details are given :—

(1) *Drying and Sampling Room.*—This should be provided with an ample drying hearth at one end for drying such samples as slimes, residues, etc. A long bench is also provided for mixing and quartering samples. It communicates with an outside bin as shown, for the reception of waste or rejected ore from sampling.

The crushing appliances should consist of a small, fine crusher with motor or other convenient power. An auxiliary hand machine for crushing should also be supplied.

(2) *Fluxing Room.*—This room is used for " fluxing " the samples and preparing them for fusion. Under the window which overlooks the sampling room is a small trap, through which the samples, after thorough mixing, are passed to the counter behind the fluxer, who works at the bench on the opposite side of the room.

[1] *J. Chem. Met. Min. Soc. S. Africa,* March, 1906, p. 266. The plan is reproduced by kind permission of the Council of the Chemical, Metallurgical, and Mining Society of South Africa, and of Mr. E. J. Laschinger. See also plans for small and large assay office by A. M'Arthur Johnston, *Rand Metallurgical Practice,* 3rd edn., vol. I, 1926, pp. 287-288 (Griffin & Co.).

The fluxing room, in addition to a balance, fluxing bench, and a few drawers, is provided with one or more small benches or shelves to take the day's work

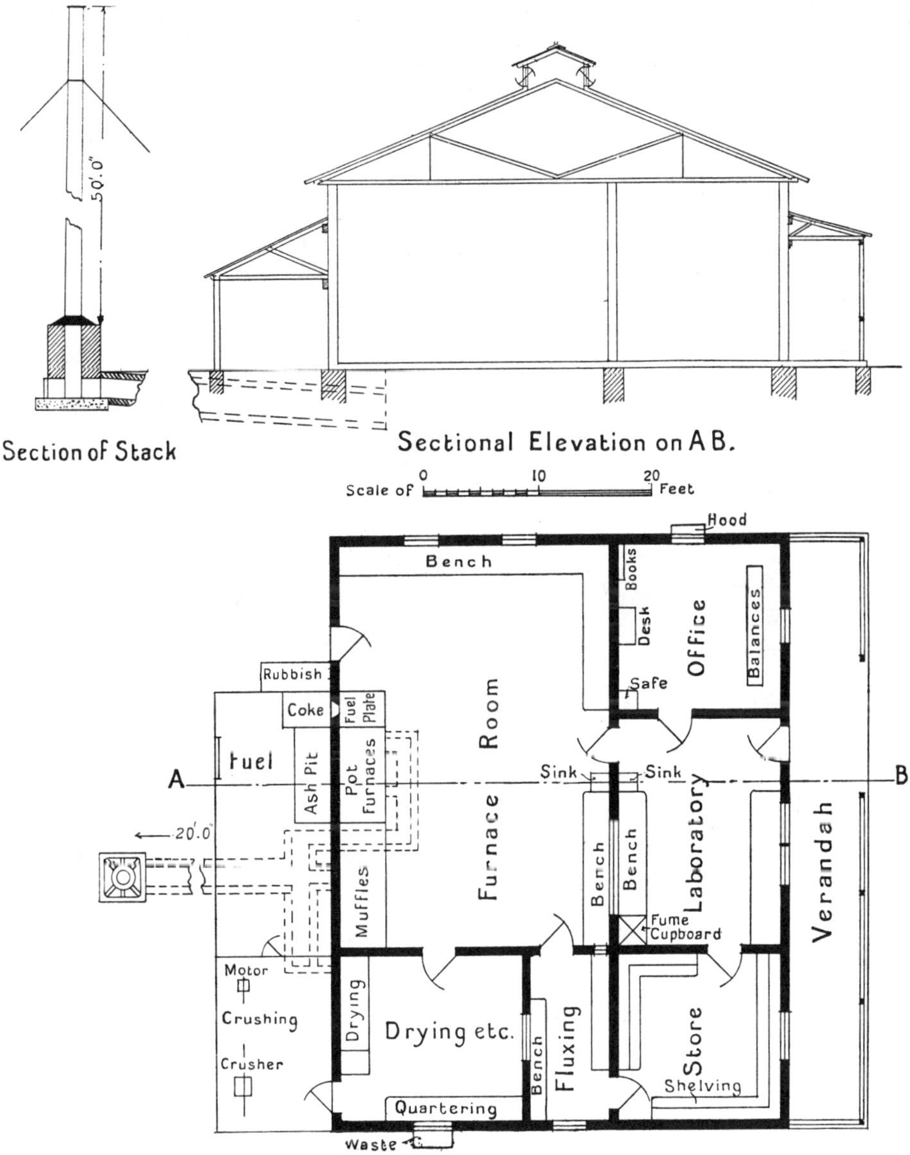

FIG. 1.—PLAN OF MILL ASSAY OFFICE (*A. Whitby*).

only, *i.e.* samples not yet reported on. After the fluxer has added all his ingredients to the samples, the tins into which they are placed are passed through a small trap into the furnace room, where the contents are mixed and transferred to the crucibles.

(3) *Furnace Room.*—The furnace room, being of primary importance, should be given ample proportions. It is provided with wind and muffle furnaces, which are connected with a stack 50 ft. high, separate from the building and connected, as shown, by an underground flue, into which all the flues from the furnaces are led. The stack is high enough to give a strong draught without the aid of fans. The fuel for the furnaces is stored outside the main building in a shed, which communicates with the furnace room and supplies fuel to the fuel plate as required. The ashpits to the furnaces are cleared from the outside, so as to keep dirt out of the furnace room. A cast-iron plate floor is provided in front of the furnaces. Ample bench accommodation and a sink are provided. Anvil blocks are supported on solid timber to the foundations. Machinery, such as the crushing rolls etc., is better placed in the furnace room than in the wet room, where it would be affected by the acid fumes, etc.

(4) *Laboratory.*—No special comment is needed regarding this room. It has good benches with cupboards for apparatus, and is fitted with a sink and fume-cupboard with good draught.

(5) *Balance Room.*—As previously stated, it is of the utmost importance to have a dust-proof ceiling in this room. The balance bench should be away from the walls and supported on solid pillars resting on the foundations. With these precautions any kind of top may be used. The use of slate tops is, as mentioned above, deprecated by Whitby. The light for the balances should be ample, and come from behind the operator, but no direct sunlight should ever fall on the cases.

Three balances are necessary: (i) a chemical balance with a minimum sensibility of $\frac{1}{5}$ milligramme, used for weighing out by-product charges; (ii) an assay balance, sensible to $\frac{1}{20}$ mg., for bullions and mine samples; and (iii) another sensible to $\frac{1}{200}$ mg., for surface or reduction work.

The balance room is also used as a general office, in which all the assay books, etc. are kept.

(6) *Store Room.*—The samples, when reported on, are stored away in numbered bags on shelves allotted to them in the store room. In this room the fluxes are also kept. The various materials used by the fluxer are stored in small bins which communicate with the fluxing bench in the adjoining room.

Custom Assay Offices.—In many of the mining districts of America and Australia there have been erected mills that undertake to crush, sample, and treat parcels of ore for the general public. These mills are known as *Custom Mills*, and the assay offices connected with them are known as *Custom Assay Offices.*

Assay-Office Records.—A careful record should be made of all samples received for assay, and of the results obtained. The record must be clear and neat, so that reference, even after an interval of years, may be certain and easy. One method should be adopted and adhered to. In some assay offices the card-index system and other modern methods of keeping records are adopted. Where a large number of samples have to be dealt with daily, three books are required, viz.: (i) sample book, (ii) laboratory book, and (iii) assay book. These books will necessarily vary in detail in every office according to the character of the samples to be dealt with, but the following examples will serve to illustrate the method of record:—

Sample Book.—The sample book contains particulars of the samples (marks, etc.), which are entered by the office clerk as they arrive. The clerk at the same time puts on each sample the distinguishing number against which it has been entered, and by which it is known in its course through the various assay operations. The marks on the samples very frequently consist of the initial letters of the name of the mine or smelting company, and the name of the vessel in which the ore was shipped.

SPECIMEN PAGE OF SAMPLE BOOK.

Date	No.	Description of Sample.	Remarks.
1943.			
Nov. 8	95	R. J. T., argentiferous lead ore	For Pb and Ag.
,, 9	96	Silver ore, ex " Sorata "	For Ag and Au.
,, 9	97	Silver ore, ex " Potosi "	For Ag.
,, 9	98	A. M. Co., concentrates	For Au.
,, 10	99	R. J. T., argentiferous lead ore	For Pb and Ag.
,, 10	...	...	Report Ag per { ton of ore and ton of lead.
,, 11	100	T. Min. Co., gold ore, X.	For Au.
,, 11	101	T. Min. Co., gold ore, XX.	For Au.
,, 11	102	S. S. Co., burnt ore, Spain	For Ag, Cu, and S.
,, 12	103	Argentiferous copper ore, ex " Liguria "	For Ag and Cu. Test for Au.

Laboratory Book.—This is the assayer's note-book, in which he enters clearly the particulars of his work—the results obtained, and details as to how these results were arrived at. The calculations should be made on scrap-paper, and only sufficient detail entered in the laboratory book to enable the results to be verified should this become necessary. The results of each assay should be recorded as soon as obtained, and not left to memory for later entry.

SPECIMEN PAGE OF LABORATORY BOOK.

No.

95. Argentiferous lead ore.
Clay-pot fusion : 25 grammes gave 15·3 grammes Pb
= 61·2 per cent. Pb.
0·0615 gramme Ag
= 80·36 oz. per ton.

96. Silver ore.
Pot fusion : 25 grammes gave 0·159 gramme Ag = 207·76 oz. per ton.
0·002 gramme Au = 2·613 ,, ,,
(Moisture, 1·3 per cent.)

97. Silver ore.
Scorification : 3 grammes gave 0·0352 gramme Ag
= 383·29 oz. per ton.
(Moisture, 2·2 per cent.)
Metallics from 100 grammes = 0·0051 gramme = 1·666 oz. per ton.

100. Gold ore (quartzose).
50 grammes gave Au + Ag 0·0043 gramme.
Au 0·0027 gramme = 1·764 oz. per ton.
Ag 0·0016 ,, = 1·045 ,, ,,

101. Gold ore (pyritic).
50 grammes gave Au + Ag 0·0018 gramme.
Au 0·0015 gramme = 0·9800 oz. per ton.
Ag 0·0003 ,, = 0·1960 ,, ,,

Assay Book.—The assay book is the official book for the regular entry of all assay results, and is a combination of the sample and laboratory books. It corresponds to the report form given below. The assay book should contain sufficient detail to distinguish each sample, but excessive detail should be avoided. When only a small number of assays have to be dealt with, the sample book may be omitted, and the entries made directly in the assay book as the samples arrive.

SPECIMEN PAGE OF ASSAY BOOK.

Date.	Date Reported.	Description of Sample.	Reference No.	Gold oz. per ton.	Silver oz. per ton.	Percentage.			Remarks.
						Copper.	Lead.	Other Metals.*	
1943. Nov. 8	9	R. J. T., argentiferous lead ore	95	...	80·36	...	61·2	...	
" 9	14	Silver ore, ex "Sorata"	96	2·613	207·76	...	...	...	Moisture, 1·3 °/₀
" 9	14	Silver ore, ex "Potosi"	97	...	383·29	...	...	...	Moisture, 2·2 °/₀. Metallics, 1·666 oz.
" 9	18	A. M. Co., concentrates	98	52·15	10·69	...	...	...	Contained Cu.
" 10	12	R. J. T., argentiferous lead ore	99	...	105·40	...	50·0	...	Ag per ton of lead, 210·8 oz.
" 11	14	T. Min. Co., gold ore, X.	100	1·764	1·045				
" 11	14	T. Min. Co., gold ore, XX	101	0·980	0·196				
" 11	20	S. S. Co., burnt ore, Spain	102	...	1·630	0·13	...	...	Sulphur, 0·15 °/₀.
" 12	20	Argentiferous copper ore, ex "Liguria"	103	...	15·190	5·3	...	...	Au trace only.

* Separate columns are added for other metals or constituents, such as S, As, FeO, SiO_2, etc., according to requirements.

Report Forms or Assay Certificates.—These are printed forms on which the result of the assay is reported to the person from whom the sample was received. They should entail as little writing as possible in making out the report. Report forms differ considerably according to the character of the samples to be reported upon. Almost every assayer has his own particular form of assay certificate; but so long as the certificate states clearly the results of his work, any little differences of detail are unimportant.

The two accompanying report forms may be taken as examples of those generally used by assayers.

Assay Office, London, E.C.

Reference No.................. ....19......

Richard Smith & Son,
Consulting Metallurgists and Assayers, etc.

CERTIFICATE OF ASSAY.

To M..............................

....

The sample of.., marked as under, has been assayed, with the following result:

Mark of Sample:

.... ex " ...."

(Signed)..

Assay Office, London, E.C.
.....................19......

Reference No.............

To M...
....

CERTIFICATE OF ASSAY.

I hereby certify that the samples of... herein described have been assayed, with the following results :

Sample Number and Mark.	Description of Sample.	Gold,		Silver,		Total Value per ton.	Percentage,		Remarks.
		Oz. per ton, 2240 lb.	Value at ...s. per oz.	Oz. per ton, 2240 lb.	Value at ...d. per oz.		Copper.	Lead.	

Assayer.................................

CHAPTER III

ASSAY FURNACES

Two kinds of furnaces are used in assaying, viz. (i) Crucible or "wind" furnaces, worked with natural or with forced draught, and (ii) Muffle furnaces, which consist of a fireclay chamber enclosed in a suitable casing and heated externally.

They vary in design according to the fuel used in heating them, and may conveniently be classified as follows :—

(A) Furnaces employing solid fuel.
(B) Furnaces employing gaseous fuel.
(C) Furnaces employing liquid fuel.

The choice of furnace is usually dependent on the nature of the fuel that is available, and this varies with different localities. They may be permanent or portable, according to requirements.

The furnaces described below may be taken as representative of the different classes.

(A) Furnaces employing Solid Fuel.

The furnaces of this class are perhaps more generally used than those in which gaseous or liquid fuel is employed. The solid fuels used are coal, coke, wood, and charcoal.

The furnaces are built of good bricks, solidly cemented with clay, and tightly bound with iron hoop and rods. Firebricks are used for the interior, which is exposed to the greater heat. The furnaces are connected with a tall chimney to provide a good draught.

Crucible or Wind Furnace.—Fig. 2 is a vertical section through the middle of the fire grate of a permanent wind furnace ordinarily used for assaying. It consists of the fireplace, 8 inches square by 12 inches deep, and the ashpit, provided with two sliding doors for the regulation of the draught. The doors are made of sheet iron, running in a stout wrought-iron frame built into the brickwork. By means of the sliding doors the size of the opening may be varied and the air supply regulated as required. The capacity of the ashpit should be at least equal to that of the fireplace. A shallow iron tray may be placed in the ashpit for receiving and removing the ashes. The fire bars, made of bar-iron 1 inch square, are 21 inches long, and project a short distance beyond the front of the furnace so as to be easily removable for drawing the fire ; their ends rest loosely on iron supports. The short flue from the fireplace communicates with a main flue, connected with a chimney not less than 30 feet high, with which other similar furnaces may be connected as described later. The interior of the furnace is of firebrick ; the exterior and the part below the grate are built of ordinary brick of good quality. The whole of the brickwork is kept firmly bound together by means of cast-iron plates and wrought-iron tie rods. The furnace mouth can be closed by means of two firebricks, each of which is clamped with a piece of flat bar-iron firmly wedged at one end : these bricks are of two sizes, as shown, and the larger one only need be removed when the crucible is taken out of the furnace. The furnaces are sometimes provided with a sloping top closed with a sliding iron plate, some assayers preferring this arrangement to a flat top. The draught may be regulated not only by the sliding doors, but also by means of a damper in the main flue or by placing a piece of firebrick in the short horizontal flue above the fireplace. The damper itself may consist of an iron plate sliding in grooves as shown in Fig. 2, or of a fireclay tile as shown in Fig. 6, page 22. When an iron damper is used, the handle should be hinged so that it drops when the damper is fully open. With these arrangements perfect control may be obtained over the temperature of the furnace ; it can be kept below a dull red heat, or increased sufficiently to melt iron or nickel if so desired.

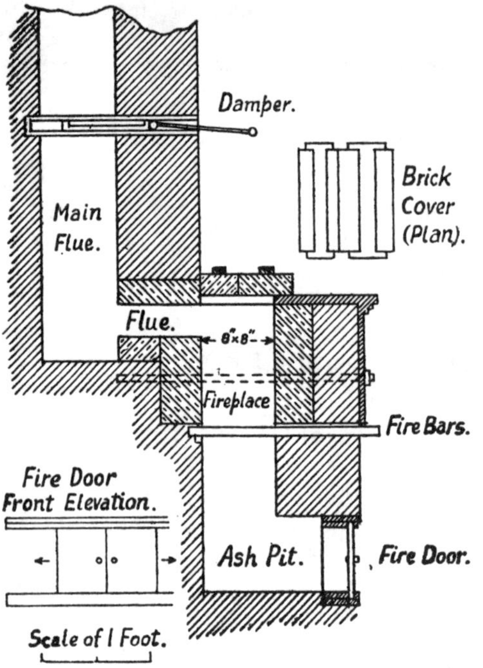

FIG. 2.—SECTION THROUGH FIRE-GRATE OF NATURAL-DRAUGHT CRUCIBLE FURNACE.

The fuel employed in these furnaces is almost invariably coke. Charcoal may be used, but it is more costly, and does not give such a high temperature as coke.

Free-burning (flaming) coal is sometimes employed in places where good coke is dear and difficult to obtain, and in this case the crucible should always be supported on a piece of firebrick about 3 or 4 inches high, resting on the firegrate, so that it is completely surrounded by the flame.

In many cases an old crucible inverted will serve as a convenient support.

Wind furnaces are generally square, but they may be made rectangular to accommodate a larger number of crucibles. In assay offices where a large amount of furnace work has to be done, the furnaces are frequently made 12 inches square, with a flue, say, 4 by 5 inches, so that a number of fusions can be simultaneously made in one furnace. For instructional work in mining schools, however, the smaller sizes are preferable, as each student should have a furnace to himself, so that he alone is responsible for the temperature employed in his assays.

Beginners should not attempt more than one fusion at a time, so as to learn the exact conditions of the fusion.

Furnace Manipulation.—Wind furnaces require to be properly manipulated to obtain satisfactory results. The temperature possible to attain depends on the depth of the furnace, *i.e.* the distance between grate and flue, the height of the chimney, and the quality of fuel used. " A very essential condition in obtaining the maximum heating effect of a furnace, the importance of which can alone be appreciated by experience, is to choose pieces of fuel of a suitable size. If, on the one hand, a shovelful of coke be taken at random, it generally contains the dust and dirt found in most fuel, which, by filling the interstices, prevents the air from passing as required, and consequently retards the combustion. On the other hand, if a furnace be filled with large pieces, considerable spaces are left between them, so that only a comparatively small surface is exposed to the action of the atmospheric oxygen, and a correspondingly small quantity of fuel is consumed in a given time, so that the maximum heat can never be obtained.

" In order to produce the desired result, it is necessary that the pieces shall have a certain uniform size, and experience has proved that pieces about 1 to $1\frac{1}{2}$ inches in diameter produce the best effect.

" These may be selected by sifting the coke through two strong wire screens, one of which has meshes about $1\frac{1}{2}$ inches square, and the other about 1 inch square. The coke which passes through the larger one, but will not go through the smaller screen, will be the right size for use in the furnace." (Mitchell.)

The coke used should be of good quality and not too dense ; it should be broken into lumps of a uniform size before being brought into the furnace room.

In lighting a furnace, a start is made with wood, as this readily ignites and aids the combustion of the coke, which of itself does not ignite readily. A little charcoal, when obtainable, is a great help at the start. Before commencing to use the furnace the fuel should be well packed by poking, raising the coke and not ramming it, and it should be uniformly heated, not hot below and cold above.

Crucibles must be placed in the middle of the fire, at equal distances from the sides and bottom of the furnace, and must be completely surrounded with the fuel, so that they are uniformly heated. The fuel should be packed sufficiently to give a good solid foundation on which the crucible can be placed.

If the fire has not been well packed, the fuel, in burning away, allows the crucible to tilt, and may cause the loss of the contents. A plan adopted by many assayers is to make a hollow in the fire, place an old crucible in it, and pack round with coke, so that the surface slopes upwards from the mouth of the crucible to the sides of the furnace. When the empty crucible is uniformly heated, it is taken out, and the crucible containing the assay substituted. It is rarely advisable to have a very hot fire at first, for the reasons stated on p. 139, where fuller details of the fusion of assays are given.

The highest temperature is found at about 2 inches above the grate, a fact which should be taken into consideration in placing the crucibles in the furnace.

The temperature is regulated in the manner indicated above, and, if necessary, fuel is added from time to time ; but before this is done, the glowing coke must be poked down to do away with empty spaces.

In commencing a second assay immediately in the same furnace, certain precautions must be taken to ensure success. All ash and clinker must be removed from the grate by means of a poker ; the fire must be well poked to pack the fuel, and fresh fuel added.

A great saving of fuel is effected by working off the assays continuously, so that the furnace is not allowed time to get too cold.

Carr's Crucible Furnace.—Carr's patent crucible furnace[1] (Fig. 3), or some similar portable furnace burning coke and with natural draught, is used in many assay offices and mining schools.

The outer case of the furnace consists of wrought-iron plates securely bolted to the angle framework, and is lined internally with fireclay slabs. For convenience in transport it is supplied in sections, and can readily be fitted up.

[1] Morgan Crucible Company, London.

The fire-bars are placed several inches lower than the bottom of the casing, thus permitting a current of air to enter above the fire-bars on all sides of the furnace. By thus admitting air to the fuel, a perfect and regular combustion of the coke takes place, and a high uniform heat of the required intensity completely surrounds the crucible. The air passing through the fire-bars is reduced to a minimum, and this prevents rapid consumption of the fuel bed, so that the crucible remains in its proper position and is not chilled at the base by the incoming cold air.

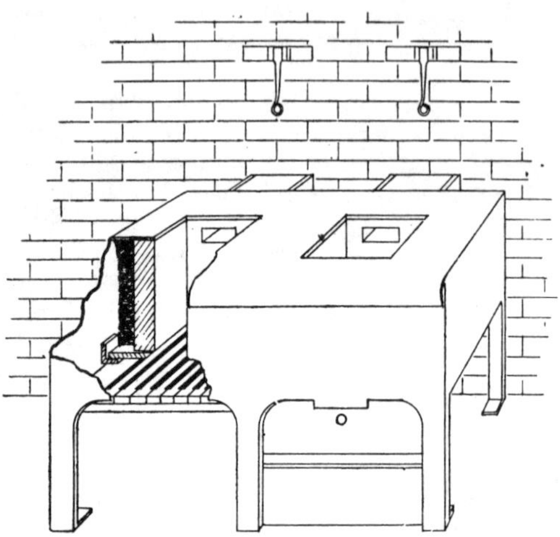

FIG. 3.—CARR'S CRUCIBLE FURNACE.

The solid lining rests on the bottom plates, and is kept in a central position by a backing of non-conducting material, and the heat is thereby effectively retained within the furnace and prevented from destroying the metal casing, which will last for many years.

Two fires in one frame are shown in the illustration. The furnaces are connected with a chimney shaft, and the draught is regulated by means of a damper in the flue in the ordinary way.

Muffle Furnaces.—The principal part of these furnaces is the muffle (Fig. 4), which is a ∩-shaped chamber of fireclay open in front and closed at the back. It serves for the reception of the assays, and is heated externally either by direct contact of glowing fuel such as coke or by the flame of burning fuel such as bituminous coal. These furnaces are absolutely necessary for cupellation and other oxidizing processes described subsequently.

Muffles are made in varying sizes and shapes. The best shape for general use is one of nearly rectangular cross-section, with a slightly arched top, as shown in Fig. 103. This shape is invariably used when crucible fusions are made in a muffle. The largest muffles ordinarily used in coal-fired furnaces are 19 inches long, 12 to 14·5 inches wide, and 7·75 inches high (outside dimensions.).

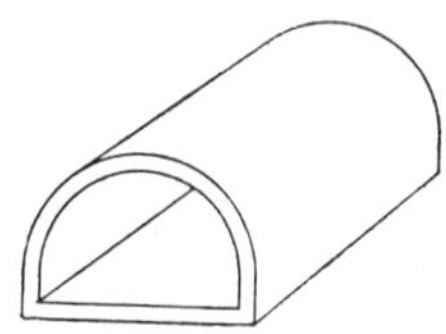

FIG. 4.—MUFFLE.

The main requirements in bullion assaying are uniformity of temperature and air-supply throughout the muffle, in order to keep the absorption of gold and silver, and slight volatilization losses, as small and regular as possible.

In most assay muffles the air-supply enters through the door at the front, but not infrequently the muffle is provided with a slit or a number of small holes about $\frac{3}{8}$-inch diameter at the back, in order that a current of air may be kept constantly passing through it. The position, number, and size of the holes in the muffle should be carefully abjusted according to the draught of the furnace and other conditions.

Instead of holes, the muffle is sometimes provided with a fireclay or graphite tube at the back or sides, through which the air is withdrawn, as illustrated in Fig. 5 and also in the gas muffles in Fig. 18.

The mouth of the muffle is closed during working by means of a thin firebrick standing on edge, or some similar form of door.

The essential dimensions in a muffle furnace are (i) the area of the firegrate; (ii) the depth of the fireplace, *i.e.* the distance from the fire-bars to the bottom of the muffle; and (iii) the "fire-space," *i.e.* the distance between the muffle and the walls of the furnace. These dimensions are dependent on the nature of the fuel used. The solid fuels usually employed for heating muffle furnaces are bituminous coal (*i.e.* long-flame coal), anthracite, and coke. The fuel preferred by many assayers, and usually the one most easily obtained, is bituminous coal, giving a fairly long flame. Anthracite and coke, either alone or a mixture of the two, are very frequently used, but as they burn without flame they must surround the muffle.

The stoking of muffle furnaces is done from the top or from the back. With the latter arrangement much dust and dirt are kept out of the furnace room.

Coal-fired Muffle.—A common form of muffle furnace for coal with a stokehole at the back is shown in cross-section in Fig. 6 A, and in longitudinal section in Fig. 6 B.

The external dimensions of the muffle are 19 inches long, 12 inches wide, and 8 inches high. In front the muffle rests upon the front wall of the furnace, and at the back or at the bottom it rests on two or more firebrick supports as shown. It should be clayed well round the mouth, so that no air may enter except through the mouth itself.

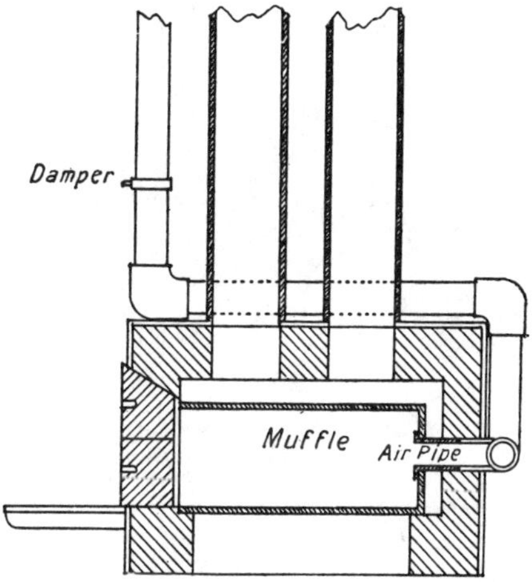

FIG. 5.—TUBE-MUFFLE IN NATURAL DRAUGHT FURNACE.

The luting material used for this purpose consists of a mixture of 1 part of raw fireclay and 3 parts of burnt fireclay, *i.e.* firebrick or crucibles crushed to pass a sieve of 8- to 10-mesh. Sufficient water is added to make a plastic mass, but the mixture should be used as dry as possible. Raw fireclay has too great a shrinkage when drying to be used alone.

The dimensions of coal-fired muffles vary somewhat with the nature of the coal employed, *i.e.* whether long- or short-flame, but the fireplace is always deeper than in the case of muffles heated by coke. The distance from the fire-grate to the bottom of the muffle varies from about 12 to 18 inches: in Fig. 6 it is 15 inches. The fire-space round the sides and top of the muffle is $2\frac{1}{2}$ inches. The fire-space above the muffle is slightly arched.

The flue is best placed over one end of the muffle as shown, but it may also be placed over the middle of the muffle. The flue-area in muffle furnaces is usually from $\frac{1}{6}$ to $\frac{1}{8}$ of the grate-area. The three middle fire-bars should be made of square bar-iron, and should be removable, to facilitate discharge of the fuel when closing down the furnaces.

The ashpit is closed with two sliding doors, as described for the wind furnace, which are used for adjusting the draught. The draught is also regulated by the damper, consisting of a movable fireclay tile inserted in the flue or an iron damper. The walls of the furnace should be about 10 inches thick to prevent loss of heat by radiation. A similar type of furnace is used for heating two muffles placed one above the other, or three muffles, two below placed side by side and one above.

Coke-fired Muffle.—The method of firing constitutes the chief difference between coke-fired and coal-fired muffles. In the case of the coke the firing is somewhat more difficult, as it has to be done from the top of the furnace, and the fire requires careful attention.

A coke-fired muffle is shown in front elevation in Fig. 7 A, in vertical section on line AB in Fig. 7 B, and in vertical section through the middle of the furnace and the axis of the muffle in Fig. 7 C. The muffle is about 12 in. long, 8 in. wide, and 7 in. high; the distance from the bottom of the muffle to the fire-grate is 6 in., and the fire-space at the sides of the muffle 3 in.

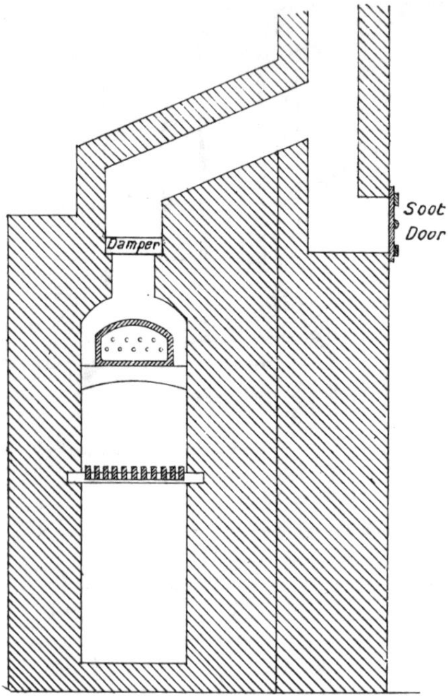

Section on line A-B.

FIG. 6 A.—Cross-section.

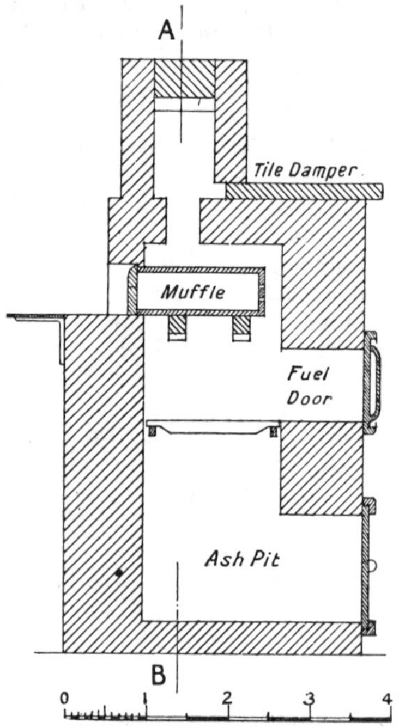

FIG. 6 B.—Longitudinal Section.

BACK-FIRED COAL MUFFLE FURNACE.

The fuel is introduced through an opening in the top of the furnace, which is closed with a fireclay slab bound with iron and fitted with a handle. The fuel is carefully packed round the sides and the top of the muffle. The coke should be broken to pieces not larger than 1½ inches in diameter, otherwise the fuel is liable to get wedged at the side of the muffle and tend to break it during stoking. The fire is stoked through the opening in the front wall of the furnace immediately below the muffle door, which is closed with a loosely-fitting tile: it is also stoked through the opening in the top. The draught is regulated by the iron damper in the flue and the sliding doors in the ashpit. Some of the fire-bars are removable for the reason already given.

Coal- and coke-fired furnaces are in very general use for assaying purposes. When the amount of furnace work to be done is not large, the coke-fired wind furnace is usually preferred for fusions, and a coal- or coke-fired muffle furnace

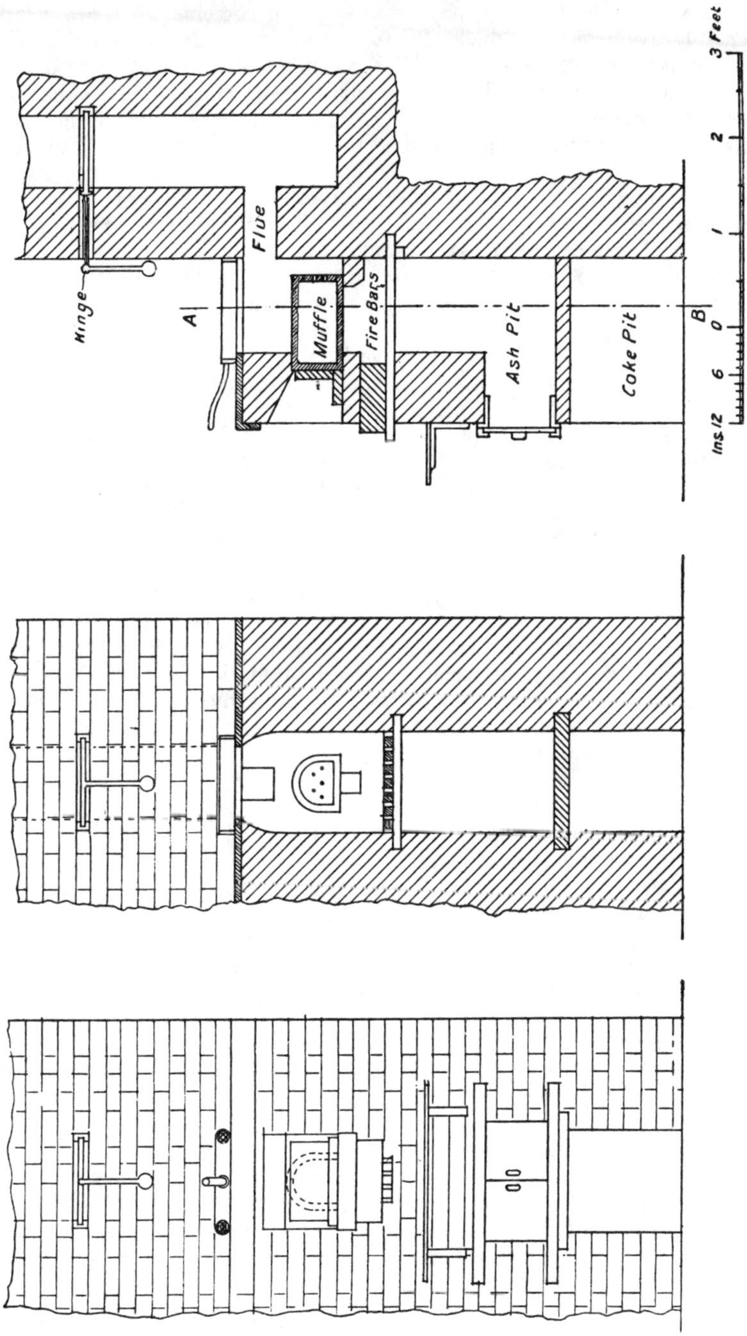

FIG. 7 A.—Front Elevation.

FIG. 7 B.—Cross-section.

COKE MUFFLE FURNACE.

FIG. 7 C.—Longitudinal Section.

for muffle work. The furnaces may be separate or combined.[1] When a large number of assays have to be made, the fusions are generally performed in wind furnaces of large dimensions or in a large muffle. For reasons of economy, however, coke-fired fusion and muffle furnaces are gradually being eliminated in mine-assay offices in South Africa, and coal furnaces, chiefly of the reverberatory type, substituted. The coal used in these is a long-flaming, bituminous coal.

The advantages of furnaces of the reverberatory type are: considerable reduction in cost of fuel, less wear and tear of muffles, and the heating of a large number of crucibles at the same time.

The furnaces can be fitted into the side wall of the building with the working door opening into the room, and outside firing adopted, thus adding considerably to comfort in working.

Three types of coal-fired furnaces have been introduced on the Witwatersrand, viz.: (i) the Tennant furnaces, (ii) the Rival furnace, and (iii) the Rusden furnace. They may be briefly described as follows[2] :—

Tennant Furnaces.—In the Tennant pot furnace the crucibles are placed on bars immediately above the coal fire, whilst another set can be introduced above these, also on bars. The crucibles are thus heated from below, the flame playing round each pot.

In the Tennant muffle furnace the muffles are arranged in a large vertical flue down which the combustion gases pass after being deflected by a bend in the flue. The muffles (two or more) are placed at different intervals to permit of uniform heating as far as possible, but the one nearest the fire-grate gets by far the strongest heat. This constitutes one of the drawbacks to the furnace, as it frequently happens that when the lower muffle is ready for cupellation the top muffle is too hot and its use has to be temporarily discontinued. Before entering the main chimney the gases from the downcast pass beneath a drying plate, which can be used for drying wet slime, etc.

Rival Furnace.—This is a combined pot and muffle furnace. The reverberatory chamber containing the muffles is placed above that holding the crucibles, whilst by-passes allow of the waste gases from the latter being used to heat the muffles initially or supplementarily.

Rusden Furnace.—This furnace differs from the others mainly in that the roof of the crucible reverberatory forms the door for charging and withdrawing the crucibles. The crucibles are easy of access, so that fusions can readily be examined, and on this account the Rusden furnace is preferred by many assayers. The door consists of two firebrick slabs. The muffle furnaces are on either side, with a by-pass to divert the flame to them when the crucible-furnace door is open.

Arrangement of Furnace Flues.—The proper construction of the furnaces is a matter of first importance, as a slight defect in the arrangement of the flues will considerably affect the draught, and prevent the temperature requisite for assay operations from being attained.

The short flue of an assay furnace is horizontal, and is generally connected with a larger main flue, which will vary in length according to the distance of the stack from the furnace. The size or cross-section of the flue has a great influence on the working of the furnace; if it is too narrow, the draught will be slow as a result of friction of the air current on the walls.

[1] A combined crucible and muffle furnace built of ordinary firebrick has been described by G. T. Holloway, *Trans. Inst. Min. Met.*, 1907, pp. 341-349. Drawings and specifications are given.

[2] For fuller details, see *Rand Metallurgical Practice*, 3rd edn., vol. I., pp. 297-299.

For an active and strong draught the flue must be wide and the chimney large and high. As a general rule the section of the flue is from $\frac{1}{6}$ to $\frac{1}{8}$ the area of the fire-grate. When several furnaces are built side by side so that all the flues open into a common conduit or main flue, special attention must be paid to the arrangement of the flues, otherwise defective draught will result. For example, if the flues from several furnaces are connected with one horizontal conduit and each opens into it at right angles, the stack being at one end of the conduit, the draught will be defective owing to the air-currents from the several flues interfering with one another. The furnace near the stack will draw well, while the draught in those farthest from the stack will be so feeble that only a low temperature will be maintained.

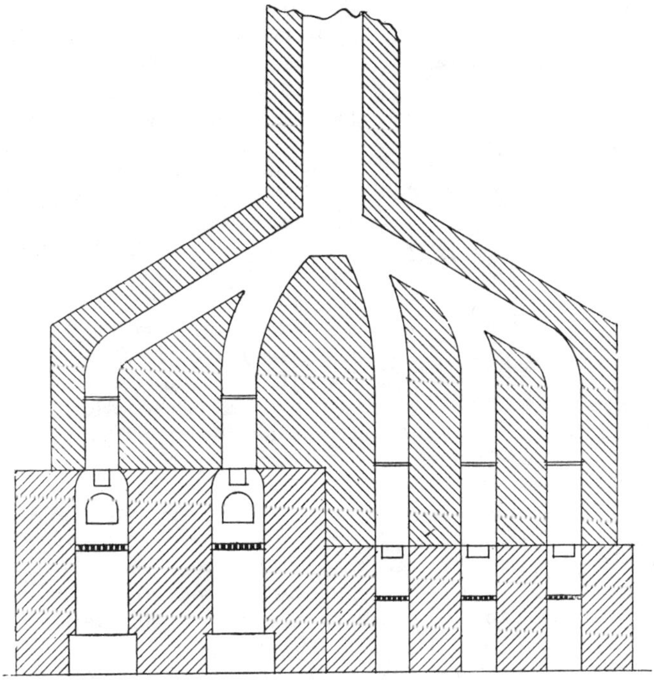

FIG. 8.—ARRANGEMENT OF FLUES FOR SEVERAL FURNACES.

For producing an equally high temperature in each of a series of similar furnaces, so that each furnace will act as well as if it were connected with a separate chimney, the arrangement of the flues should be such that the separate currents are diverted into one uniform stream. The arrangement shown in section in Fig. 8 for two muffle furnaces and three wind furnaces placed side by side has been found to be satisfactory.

Another very essential structure connected with a wind furnace is the stack or chimney, which is used for creating a draught and for carrying off the products of combustion. It is on the height and size of the stack that the draught depends, and, in consequence, the temperature produced within the furnace. In general, a draught sufficiently strong to produce a temperature high enough for all assay operations can be obtained with a chimney from 20 to 30 feet high, but higher chimneys are sometimes used with advantage. The height of the stack will depend on the position of the furnace. A height of 20 feet is usually regarded as the minimum that can be allowed in order to ensure a good draught, but, according to Sir T. K. Rose, satisfactory results can be obtained with a stack only 16 feet high.

Chimneys are generally made square or rectangular, and in the case of small wind furnaces have interiorly the same dimensions as the body of the furnace in which the crucibles are placed.

When the chimney is connected with several furnaces having an arrangement of flues as shown in Fig. 8, so as to produce a current in one direction, a single opening will suffice; but if furnaces are arranged on opposite sides of the stack, and the flues open into it at right angles, the chimney shaft must be divided for a certain distance, so as to divert the two streams into one vertical direction, as shown in the accompanying Fig. 9, otherwise they will interfere with each other and check the draught.

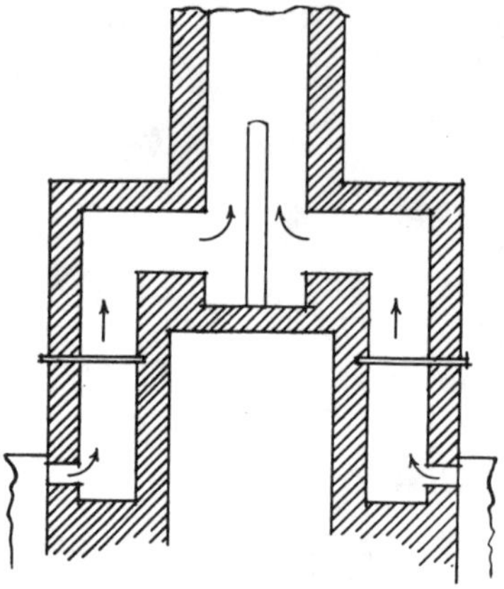

FIG. 9.

To attain the maximum temperature in wind furnaces the air supply must be allowed to pass into the furnace as directly as possible; and if the furnace is in a confined space with little access of air, it is necessary to lead air to it from outside the building through an air channel or pipe under the floor, with an opening, covered by an iron grating, immediately in front of the furnace. The channel should be furnished with a valve or door at the outer end, so that the quantity of air admitted may be regulated as desired.

Portable Furnaces for Solid Fuel.—These furnaces consist of a casing of stout sheet-iron lined with firebrick, or are made entirely of fireclay, strapped with iron bands.

A portable prospector's furnace made by the Morgan Crucible Co. is shown in Fig. 10. It is a crucible and muffle furnace combined. The casing is of stout sheet-iron with removable fittings, lined with strong fireclay tiles, which are kept in position by a top plate firmly screwed down. The tiles have rebated edges to ensure tight joints. The grate consists of nine wrought-iron bars fitted diagonally into a cast-iron frame. The bars can readily be removed to clean the grate, and, being of ordinary size and section, they can be replaced without difficulty. A small iron plate can be supported on the projecting ends of the fire-bars, to form a convenient shelf on which to place scorifiers, etc. Cast-iron draught doors, fitted with regulators, are provided for adjusting the air supply. A chimney collar is fitted to the back of the furnace body, so that a chimney of suitable length can be added.

The top of the furnace is closed with a fireclay cover, provided with a sight-hole, through which the contents of the crucibles can be inspected from time to time, thus obviating the necessity of removing the cover. The sight-hole is closed by a fireclay stopper. The muffle (12 by 6 by 4 inches) is placed in the middle of the furnace, and rests on suitable supports back and front. The muffle door is provided with a mica sight-hole, through which the contents of the muffle can be watched. The furnace can be used for crucibles without removing the muffle.

The total height of the furnace is 28 inches ; depth, back to front, 14 inches ; and width, 16 inches ; diameter of chimney collar, 5 inches. Total weight of furnace complete, 130 lbs.

Portable furnaces are also made with a circular body of stout sheet-iron lined with three solid fireclay rings 2 inches thick, which cannot fall in pieces when fired, as frequently happens to the linings made of firebricks or tiles cemented together by a clay lute such as are used in square furnaces.

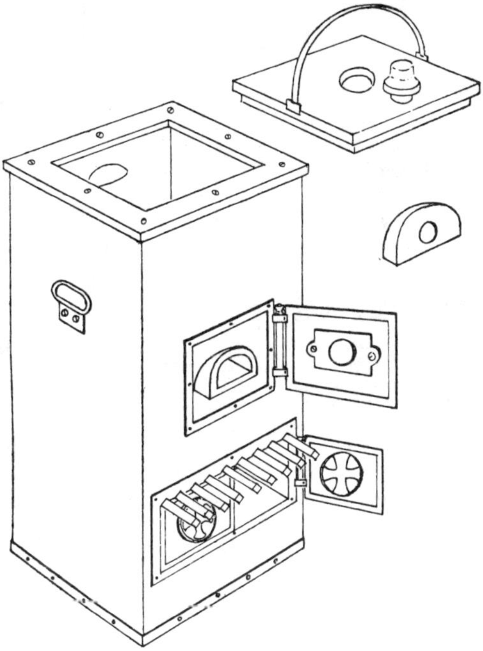

FIG. 10.—PORTABLE ASSAY FURNACE.

To reline these circular furnaces a mould of wood is placed in the middle, and the open space between the surface of the mould and the old lining is filled with a stiff paste of refractory clay, each addition being well rammed down. When the space is filled, the case is withdrawn, and the clay lining dried very slowly. Any cracks that result from the drying must be filled up.

A good mixture for the lining consists of 1 part of refractory clay and 3 or 4 parts of sifted quartz sand. The old lining should be wetted before the new lining is added.

The wind furnace ordinarily used for assay purposes, being of simple construction, can be built by any bricklayer acting under instructions, and it is advantageous to erect a brick furnace in all cases where the assayer is likely to be more or less permanently settled in any locality. No mortar is used in its construction, the bricks being cemented with fireclay, mixed with an equal bulk of sand. The stack must be of brick for a distance of 2 or 3 feet from the furnace, but may be of wrought-iron piping in its upper part.

(B) Furnaces employing Gaseous Fuel.

Where coal-gas or other gaseous fuel is available, assay furnaces heated by gas are in general use. They are easily manipulated, and clean to work.

The use of gas instead of solid fuel has many advantages to recommend it, the chief drawback being the increased cost of working where coal gas is burned, as gas-furnaces are useless for assay purposes without a good gas pressure and a large supply-pipe. When these are secured, the gas-fired muffle furnaces will be found very convenient for assay-office use. For crucible assays, however, the ordinary wind furnace burning coke is preferable to a gas-fired crucible furnace, especially where a large number of fusions have to be made.

When coal-gas is used as a source of heat, the combustion should be as complete as possible, so as to obtain the maximum temperature and prevent a non-smoky flame. This is secured by using a burner constructed on the Bunsen principle, and surrounding the flame with some refractory and non-conducting material such as fireclay. The gas-furnaces described below are constructed on this principle.

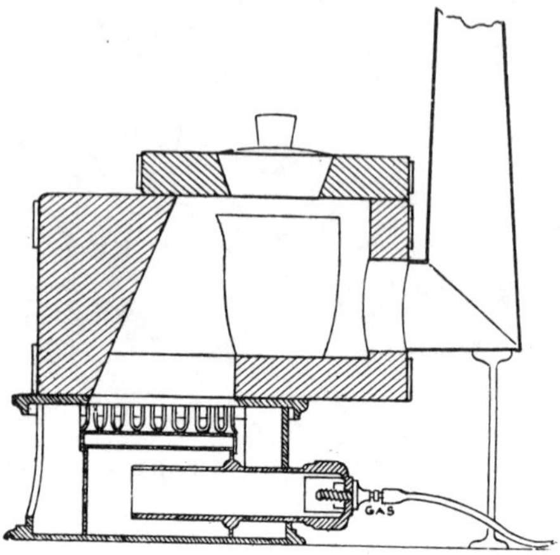

FIG. 11.—FLETCHER DRAUGHT FURNACE.

For the successful working of gas-fired furnaces the gas-supply tap and pipe must be large and clear, so as to give as great a pressure of gas as possible at the burners. It is best to connect the furnaces throughout with iron gas piping, but if flexible metal or rubber tubing is used it must of necessity be perfectly smooth inside.

Gas - fired Crucible Furnaces.—Fig. 11 represents, in section, a Fletcher draught crucible furnace taking crucibles not exceeding 4 in. by 3·5 in. It consists of a circular fireclay body bound with sheet-iron hoops, and a burner. The furnace is covered with a lid which can be lifted by a handle, or can be pushed sideways sufficiently to enable the crucible to

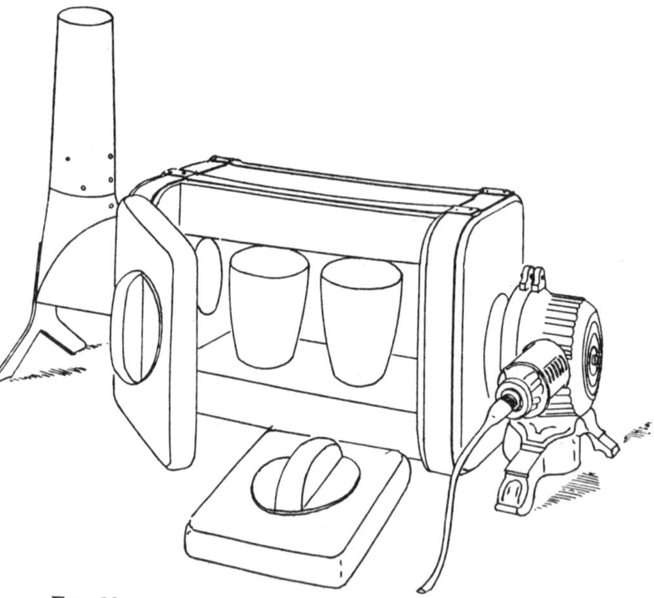

FIG. 12.—FLETCHER GAS-FIRED REVERBERATORY FURNACE.

be lifted out. During fusion the contents of the crucible can be inspected by removing the small fireclay plug in the middle of the cover. The flame from the burner surrounds the crucible and then passes up the sheet-iron chimney.

The gas-supply required is 60 cubic feet per hour, equal to $\frac{1}{2}$-inch clear-bore gas pipe and tap. When used for ordinary crucible assays the weight of the ore taken for assay is limited to 25 to 35 grammes, as this is the largest charge, with the necessary fluxes, that can be accommodated conveniently in the crucibles employed.

Gas-fired Reverberatory Furnace (Fletcher).—This furnace (Fig. 12) consists of a rectangular fireclay chamber with a draught burner at one end and a chimney at the other. The floor-space in the furnace is 14 inches long, 6 inches wide, and 7·5 inches high, and will permit of several crucibles being heated at the same time. The furnace can be placed with the opening either at the side or the top, as it works equally well in both positions. The opening is closed with two fireclay doors. The furnace can be used for crucible fusions only, or can be made to take one muffle and two or three crucibles, or two muffles, at the same time. The muffles are 4 by 7 by 4¼ inches, and are placed across the furnace after removing the doors. With a chimney draught these furnaces give a temperature about equal to the fusing-point of silver, but with an air-blast cast iron can be melted.

The burner is at one end, out of the way of injury in case of accident to a crucible : it should be placed close up to the furnace when in use.

When the draught burner is used, the blue cones of flame should be clearly seen on the burner : if they disappear, the gas-supply must be increased, or the slide over the burner air-tube closed until they reappear. In the latter case the furnace works with a smaller gas-supply at a lower temperature, and by closing this slide and reducing the gas-supply any temperature required can be obtained. If the gas and air are not properly adjusted, the burner grid becomes red hot, and is quickly rendered useless. When properly used, the grids will last for some years.

These furnaces are very convenient for general assay-office work when the amount of furnace work is not large and is intermittent.

Gas-fired Muffle Furnaces.—These furnaces are now very generally adopted for bullion assays, as they are capable of giving very uniform temperatures when proper attention is given to the adjustment of the burners. Fig. 13 illustrates a type of Fletcher gas-muffle furnace used in assaying. It is made in several sizes : the No. 8 size, with a muffle 8¼ by 5¼ by 14 inches, clear inside working-space, is very suitable for general assay work. The furnace casing is made of sheet-iron, lined with refractory material, and standing on an iron frame with four legs. The top of the furnace is removable to permit of a new muffle being fitted when necessary. In some designs of gas-muffle furnaces the muffle is introduced through an opening or door at the back of the furnace instead of through the top. The muffle is heated by means of eleven large burners of the Bunsen type, mounted on an iron frame and arranged in a row beneath the muffle. A single burner is represented on a larger scale in Fig. 14.

The flame from the burners passes through a rectangular hole in the bottom of the furnace, heats the muffle on all sides by direct contact, and passes finally into the chimneys at the top.

The furnace is provided with two chimneys 2 feet 6 inches in height, each having a cast-iron foot to enable it to stand steadily, and a damper by means of which the draught can be regulated. The muffle door is made in two parts, each provided with a slot for handling by means of the holder shown in Fig. 15. The tang end of an old file fitted to a handle makes a very suitable holder for the purpose.

For exact temperatures a pressure governor is necessary, and it is very advisable to use a quadrant gas tap, with arm and pointer, for exact adjustment, as shown in Fig. 18. The most uniform temperature is obtained by working with some of the middle burners turned down slightly, and two or three burners at each end left full on: the burners are soon adjusted after a few trials. Each burner is provided with a slide for the regulation of the air-supply, the slides being fastened to a wire handle as shown (Fig. 13), so that the air-supply may be adjusted or cut off from all the burners simultaneously.

When lighting the furnace, the air openings are covered with the slides, a lighted taper is put to the burners, and the gas is turned on slowly. When the gas is full on, the air openings are uncovered. It is very important that a light should be put to the burners before turning on the gas, or an explosion will take place, which may break the thin muffle or otherwise damage the

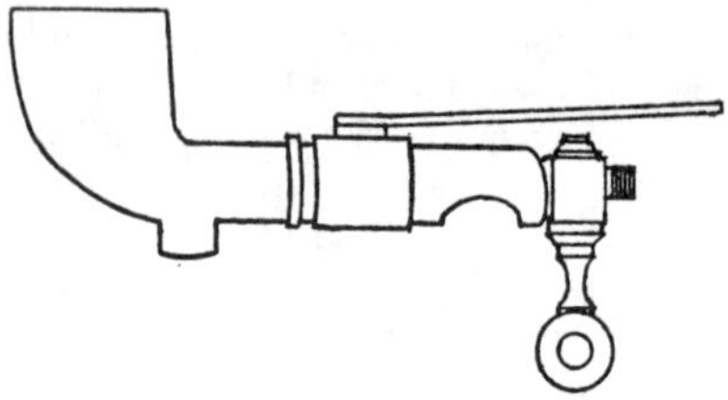

FIG. 14.—ENLARGED VIEW OF BURNER.

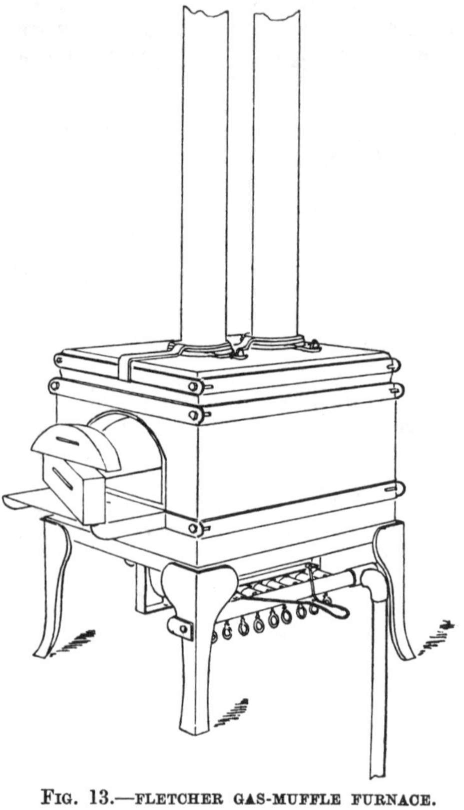

FIG. 13.—FLETCHER GAS-MUFFLE FURNACE.

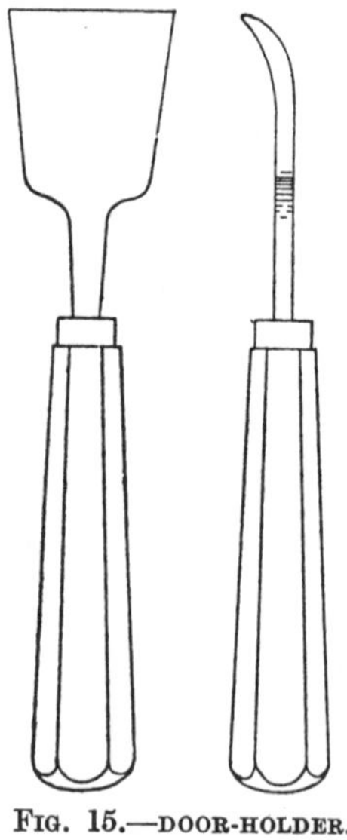

FIG. 15.—DOOR-HOLDER.

furnace. The burners must be kept clean: care should therefore be taken to cover them when a new muffle is being fixed in the furnace.

When the furnace is working at its full power, the flame must be just visible at the top of the chimney. Muffles always require great care in heating and cooling, or they very soon crack and become useless. When the furnace is cooling down after turning off the gas, the dampers should be shut to prevent a current of cold air from cooling the muffle too quickly, and the doors should not be closed too tightly, otherwise they are very liable to stick fast owing to the contraction of the muffle.

Another form of gas muffle for assay purposes is that designed by the Gas, Light, and Coke Company, London, for operations carried out at temperatures between 700° and 1100° C.; a cross-section and longitudinal section of this are shown in Fig. 16. It consists of a lining of refractory brickwork of low

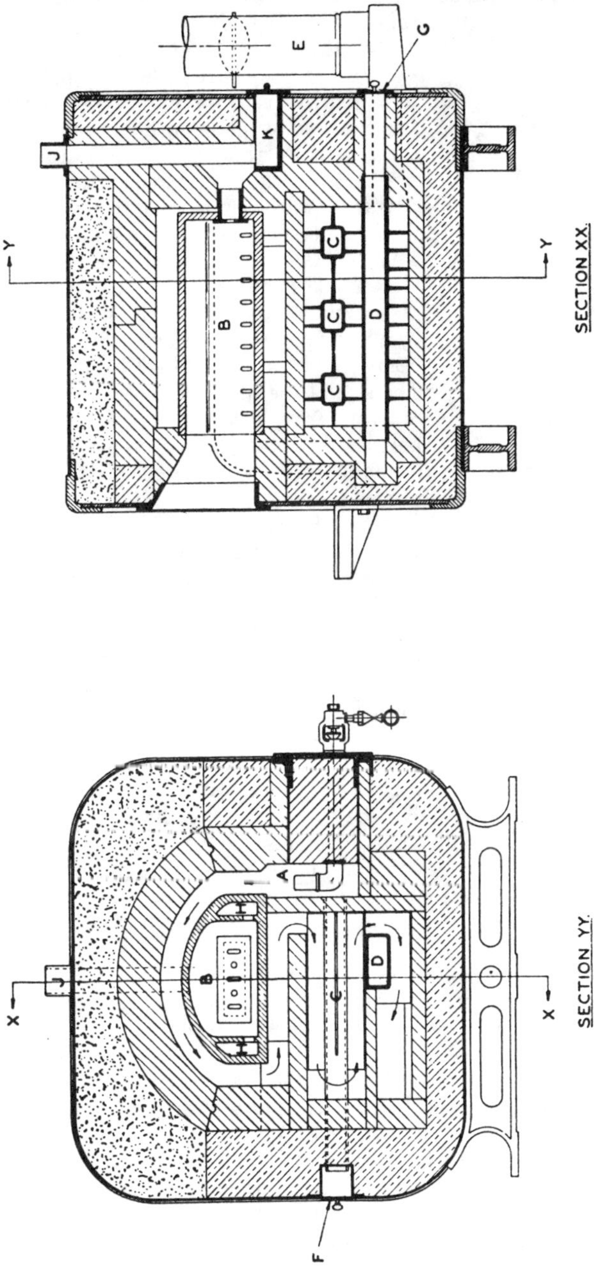

FIG. 16.—G.L.C. ASSAY FURNACE.

thermal capacity and conductivity, enclosed in a polished aluminium casing to reduce loss by radiation. The fireclay muffle *B* is heated by six burners *A*, three on each side, the secondary air-supply for which enters through adjustable shutters *F* in the side of the furnace, and is pre-heated before being fed to the burners by passing through the three-finned heat-resisting steel recuperator tubes *C*. The flames from the burners pass round the muffle and recuperators, as shown by the arrows, and finally to the flue *E*. A regulated air-supply to the muffle interior for oxidizing purposes enters through an adjustable shutter *G* at the back of the furnace, and is pre-heated in the lower recuperator tube *D*. It then passes upwards near the front of the furnace, finally entering the inner working space through small ports or slots on each side of the muffle as shown. The muffle fumes flow towards the back of the muffle and are discharged into a small flue *J*, which is taken up through the top of the furnace casing, and can be closed by a sliding damper. The size of the muffle is $15\frac{1}{4} \times 11\frac{1}{2} \times 6$ inches. Starting from cold, the muffle reaches a temperature of 1000° C. in 75 min. This temperature is maintained by a gas consumption of 65 cubic feet per hour.

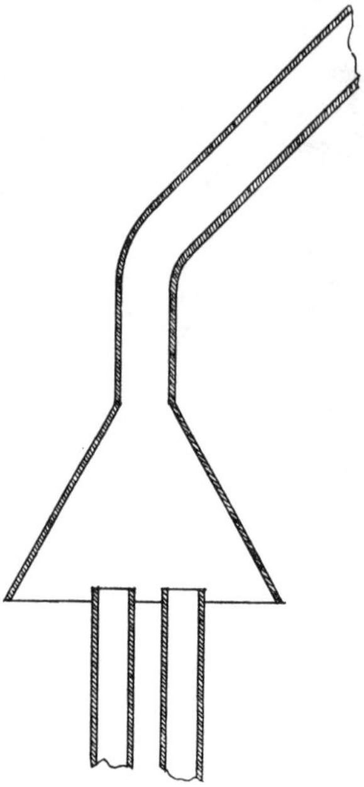

FIG. 17.—HOOD FOR MUFFLE FURNACE.

Gas furnaces should be worked under a suitable hood connected with a flue to draw off the products of combustion, etc. A sheet-iron hood for a two-chimney muffle furnace is shown in Fig. 17. Instead of a hood, however, the furnace may be connected up direct with a flue in the wall by means of sheet-iron elbow-joints.

A convenient arrangement for a series of Fletcher gas muffles is illustrated in Fig. 18. The furnace bench is of iron plates or of fireclay or stone slabs resting on brick supports, the plates or slabs being placed as shown so as to give an open air-channel under the furnaces, extending the whole length of the bench and open at both ends. By this arrangement more space is given below the muffles, and the gas taps are more readily accessible for adjustment: it also allows of the gas-supply pipes being placed under the front part of the bench, where they are out of the way, thus leaving the working part of the bench free.

The hood is of sheet-iron and is supported on brackets, the gases and fumes being drawn through three flues in the wall, the flues being connected with a chimney stack. To shield the operator from the heat radiated from the furnaces, sheet-iron screens are suspended from the hood. These are hinged so that they can be thrown up out of the way when a new muffle is being fitted. A small movable sheet-iron screen is also placed in front of each furnace to keep the heat off the working bench: one of these screens resting against the legs of the furnace frame is shown in the end muffle on the left. A quadrant gas tap, with arm and pointer for exact adjustment, is fitted to each furnace, and is almost indispensable in cases where the muffle has to be slowly cooled to prevent the "spurting" of silver assays.

When a new muffle has been inserted, the best point for heating up and for cooling down the furnace should be carefully ascertained by a few trials, and then recorded in some convenient place for reference.

(C) Furnaces employing Liquid Fuel.

Furnaces heated by petrol or other liquid fuel are used when coal, coke, and gas are difficult to obtain. Furnaces of this type find little application in Great Britain, but in the U.S.A. they are in more or less common use in assay offices where the amount of furnace work to be done is comparatively small.

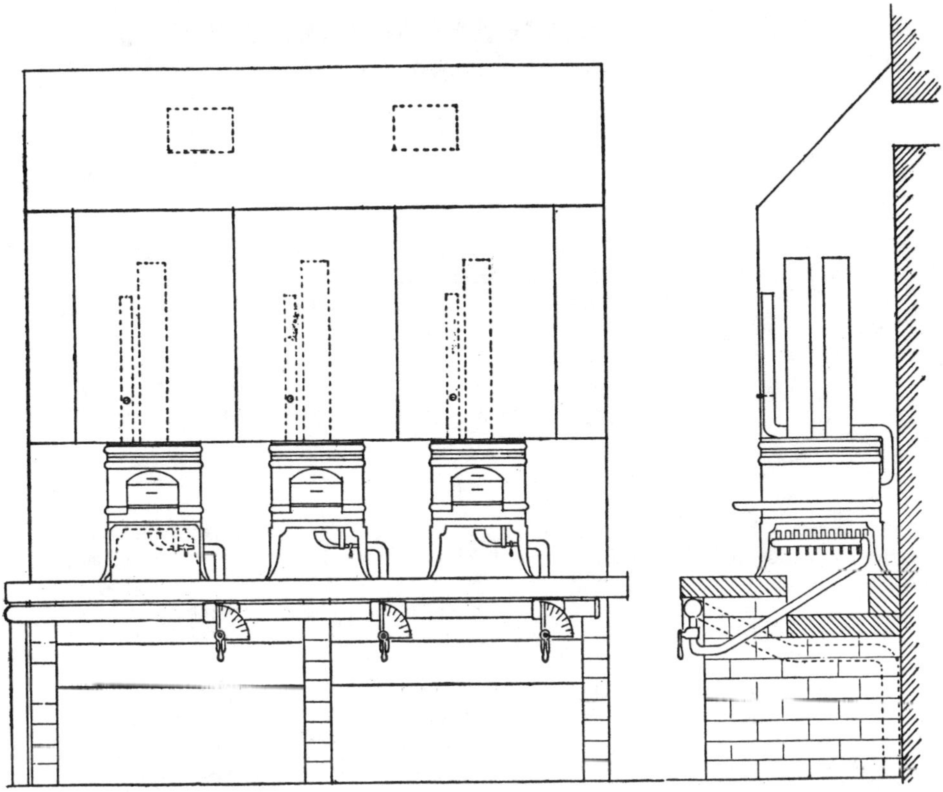

FIG. 18.—FLETCHER GAS MUFFLES.

The furnaces are similar in construction to those used for gaseous fuel, and are made of fireclay bound with sheet-iron. The chief difference is in the special burner which is so designed that it vaporizes the petrol, which then mixes with the requisite quantity of air to ensure complete combustion and the maximum heat inside the furnace.

A little experience is necessary before these furnaces can be successfully manipulated.

CHAPTER IV

FURNACE IMPLEMENTS AND APPARATUS USED IN ASSAY OPERATIONS

I. FURNACE IMPLEMENTS

COMPARATIVELY few tools are required for assay furnaces: the more important are pokers for stoking the fire, tongs for handling crucibles, etc., and iron moulds in which to pour assays, etc.

Pokers.—These are made of stout round bar-iron ½-inch in diameter, and are used for stirring the fuel and clearing the fire-bars when choked with clinkers and ashes. Two straight pokers, one 30 to 36 in. long and a shorter one 18 in. long, are very suitable. For clearing the fire-bars from underneath a 36-in. poker, bent at the end, is advisable (Fig. 19).

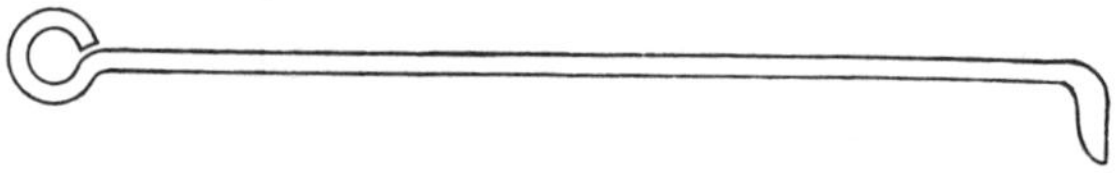

FIG. 19.—BENT POKER.

Tongs.—Tongs should be as light as possible, consistent with strength, and capable of grasping the crucible, etc. firmly without danger of slipping. Many of the tongs sold for assay purposes are heavy and clumsy and unsuitable; it is usually more satisfactory to have them made by a good blacksmith. The following different kinds of tongs are used: crucible tongs, scorifier tongs, and cupel tongs.

Crucible Tongs.—Several forms of crucible tongs are required. A useful length for these is 24 in. A very convenient pair of crucible tongs to grasp the sides of the crucible, and operating in little space, is shown in Fig. 20. This form of tongs is the one commonly used for furnace work.

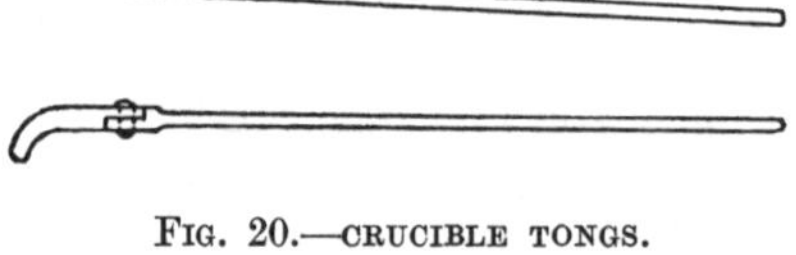

FIG. 20.—CRUCIBLE TONGS.

Two forms of crucible tongs designed to grasp the body of the crucible are shown in Figs. 21 and 22. These tongs are sometimes useful for pouring the contents of a crucible, but owing to the space they take up in opening they cannot be conveniently manipulated when there are several crucibles close together in a large wind furnace or in a muffle.

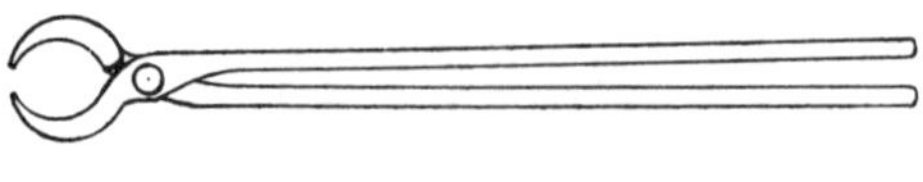

FIG. 21.

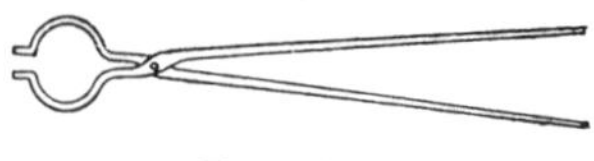

FIG. 22.

For lifting large crucibles the basket tongs (Fig. 23) are used.

Fig. 24 illustrates a convenient form of crucible tongs for handling the "Colorado"-shaped crucibles used for the fusion of assays in the muffle (see Fig. 59, page 42). They are 24 to 30 inches long.

The scissor tongs shown in Fig. 25 are very useful for packing coke round a crucible and for handling small crucibles. A convenient size is 20 inches.

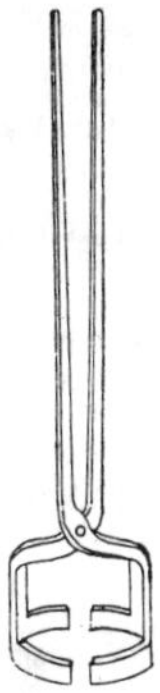

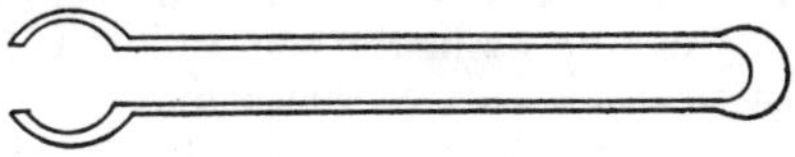

FIG. 24.

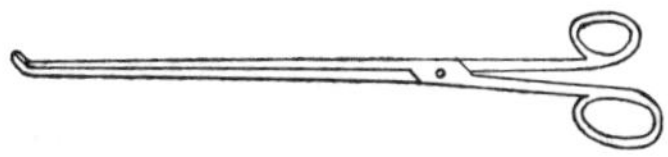

FIG. 23.

FIG. 25.

Scorifier Tongs.—These are used for handling the scorifiers described on page 43. For this purpose the lower arm of the tongs is divided near the end into two prongs, while the upper arm is made straight throughout. The lower arm fits the bottom of the scorifier, the long upper arm extending across the top.

Scorifier tongs are of different shapes and sizes: those represented in Fig. 26 will be found generally suitable. The length is usually about 24 to 28 inches.

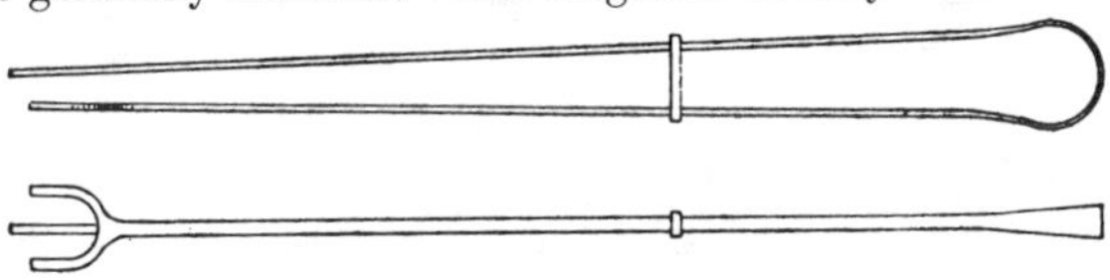

FIG. 26.—SCORIFIER TONGS.

The guide in the middle of the tongs should be so constructed that the arm moves *inside* the guide as shown. Scorifier tongs and cupel tongs (described below) are frequently made with a guide running through a slit or hole in one arm of the tongs; this arrangement is unsatisfactory, as the guide tends to stick owing to rusting or through being bent accidentally. Two or three pairs of scorifier tongs of different sizes are generally required to accommodate the various sizes of scorifiers. Special forms of scorifier tongs have been introduced from time to time, but are not in very extensive use.

Cupel Tongs.—These are used for manipulating cupels (see page 43). Two common forms of cupel tongs are shown in Figs. 27 and 28. They are made of different sizes, but tongs about 28 inches long will be found generally suitable.

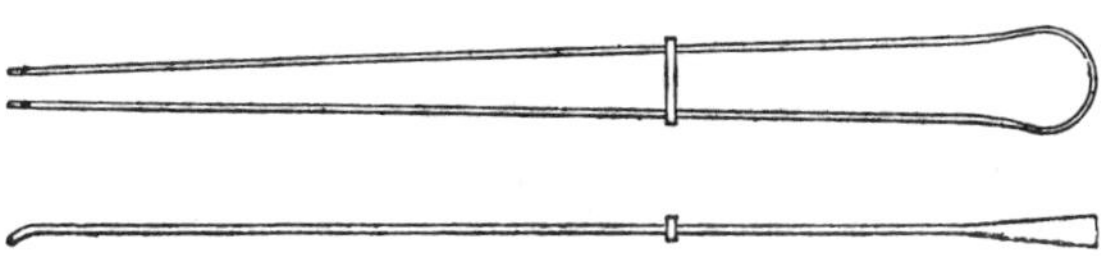

FIG. 27.

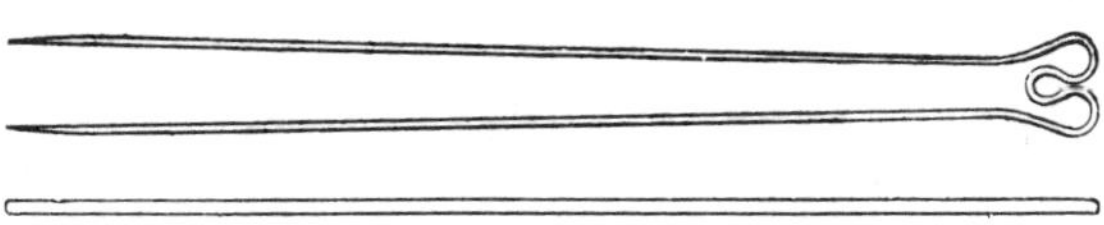

FIG. 28.—TWO FORMS OF CUPEL TONGS.

Some assayers prefer cupel tongs that are curved at the ends so that they fit round the body of the cupel.

Stirring-Rod.—Stirrers are used in roasting operations. A round rod of iron about $\frac{1}{4}$-inch in diameter and 24 inches long, flattened out at one end, is generally used as a stirrer. When required for roasting in a muffle, the flattened end is bent at right angles for the distance of about an inch. Two forms of stirrers are shown in Figs. 29 and 30.

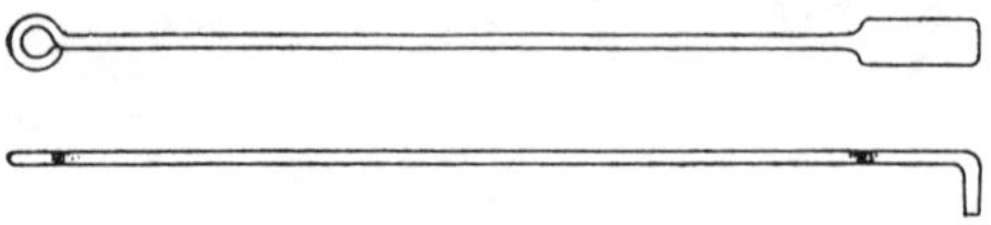

FIG. 29.

FIG. 30.—STIRRERS.

Muffle Scraper. — Two forms of scraper in general use are shown in Figs. 31 and 32. They are 24 to 30 inches long, and consist of a small iron plate riveted to an iron-rod handle. They are used for introducing bone-ash or sand into the muffle and for smoothing down the surface of the floor of the muffle. The scraper with the end bent at right angles is employed for cleaning the floor of the muffle when necessary.

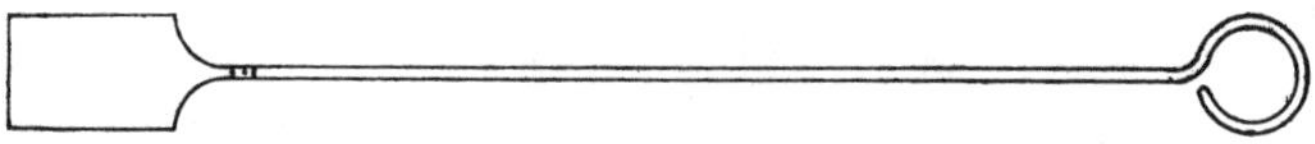

FIG. 31.

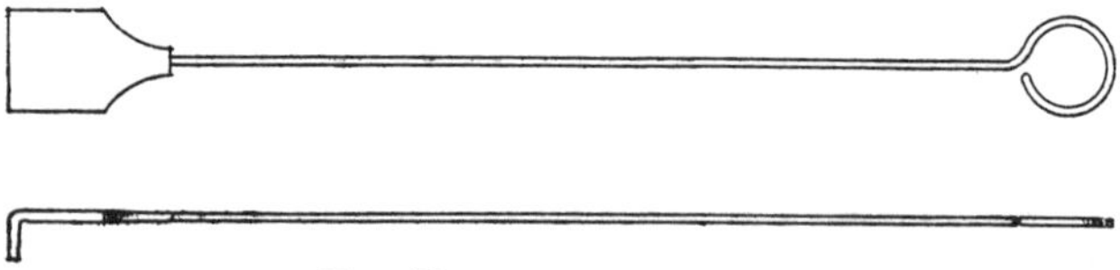

FIG. 32.—MUFFLE SCRAPERS.

Scoop.—This is generally made of copper, and is employed chiefly for charging crucibles; it is shown in Fig. 33. A scoop about 9 to 12 in. long, not including the handle, is a convenient size. It should be kept clean so that the assay charge may flow freely into the crucible.

Cupel Trays.—These are used to hold cupels when carrying them to or from the muffle. Cupel trays are generally made of sheet-iron with or without partitions. Two different forms are are shown in Figs. 34 and 35.

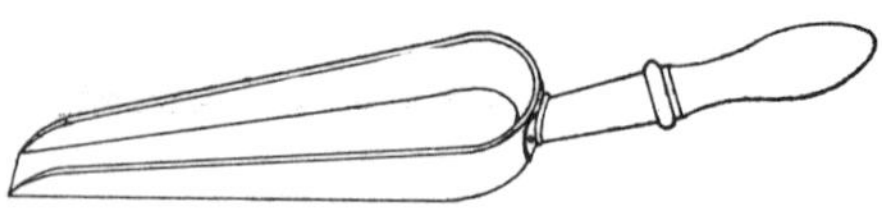

FIG. 33.

Moulds.—Moulds for receiving the molten contents of crucibles, scorifiers, etc. are of various shapes and sizes. They are either open or closed, and are generally made of cast iron: open moulds have either hemispherical or conical cavities, as shown in Figs. 36 and 37. When large numbers of assays have to be made, a mould of thick sheet-iron, containing 9 or 12 or more hemispherical cavities, formed by stamping, is frequently used (Fig. 38). A common form of ingot mould for receiving comparatively large quantities of molten metal is shown in Fig. 39. When thin flat ingots of metal are required, a closed mould (Fig. 40) is used. The various parts are detachable, and can be adjusted so as to give ingots of different sizes.

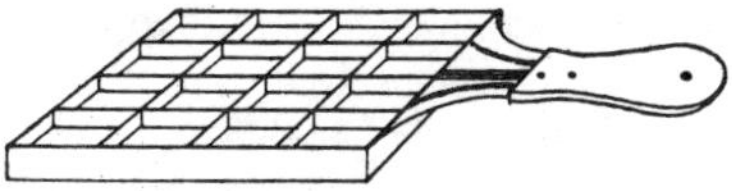

Fig. 34.

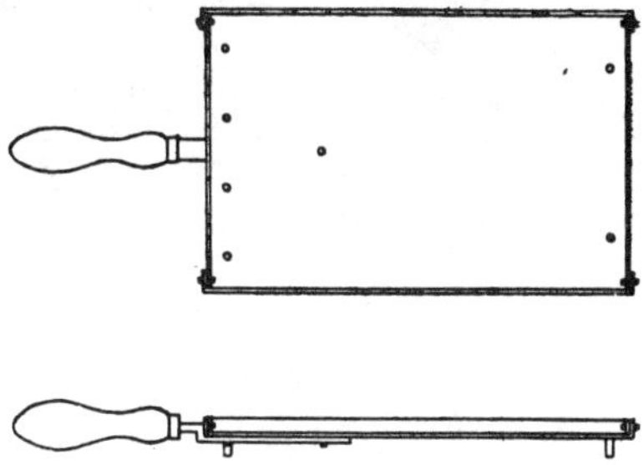

Fig. 35.

CUPEL TRAYS.

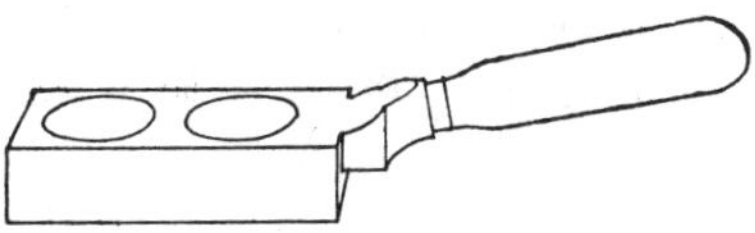

Fig. 36.

Fig. 37.

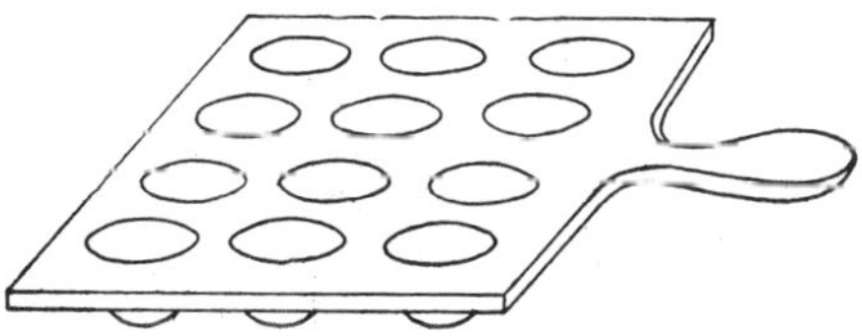

Fig. 38.

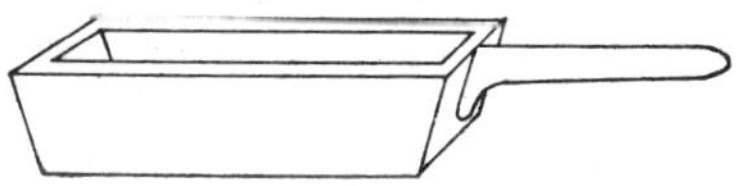

Fig. 39.

FOUR TYPES OF MOULDS.

All moulds should be cleaned thoroughly inside before being used. Spherical moulds are usually cleaned by rubbing with plumbago powder. This is best done by means of a "plug" consisting of a piece of stick rounded at one end and covered with a piece of leather or cloth like a drum-stick. The plug is dipped into plumbago powder and then rotated several times in the mould. Care must be taken to remove all powder from the mould before use.

Ingot moulds may be blackleaded inside although it is more usual to prepare these by wiping with an oily rag. Small ingot moulds may be prepared by holding them over a gas or oil flame so that the inside becomes coated with a thin deposit of "soot" or "lampblack."

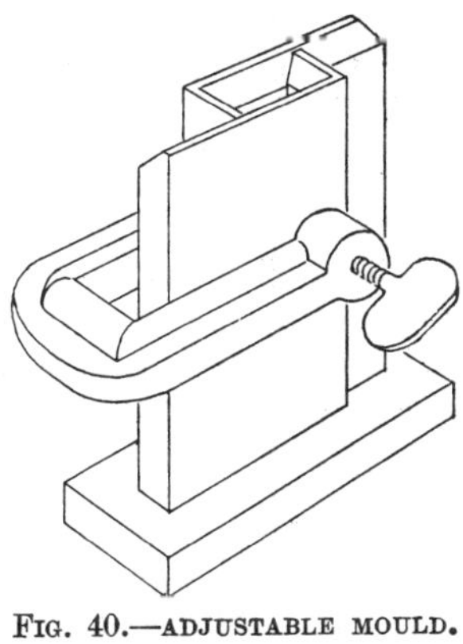

Fig. 40.—ADJUSTABLE MOULD.

II. MISCELLANEOUS APPLIANCES.

Anvils.—Two "anvils" of hardened steel are usually required, one for hammering the lead buttons from assays to free them from slag, etc., and the other for flattening assay beads of gold and silver. An anvil or stake for cleaning lead buttons is shown in Fig. 41. It has a hard smooth surface 4 or 5 inches square and a fang for fixing it securely into a wood block. The anvil should be mounted on a strong bench in the furnace room so arranged that the slag, etc. when hammered off is collected and not allowed to fall on to the floor and become a nuisance. The anvil is best mounted on a solid block of wood which extends to the floor. A suitable bench may be made of strong sheet-iron, supported on angle iron. The top should measure about 4 ft. by 2 ft. 6 in., and should be turned up at right angles at both ends and at the back to form a flange $2\frac{1}{2}$ to 3 inches high. The legs are quite firm if made of $1\frac{1}{4}$-inch angle iron braced up with $\frac{3}{4}$- by $\frac{1}{8}$-inch iron strip riveted on.

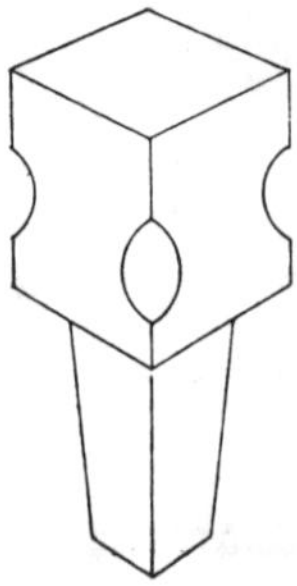

FIG. 41.—ANVIL.

Two forms of anvil or "slagging" bench are shown in cross-section in Figs. 42 and 43. In Fig. 42 the anvil is fixed in the middle of an oblong trough placed in front of the bench as shown, so that the slag as it is hammered off falls into the trough. The top of the bench in this case may conveniently measure 4 ft. by 1 ft. 6 in. A small trap or door is provided at the bottom of the trough to permit the ready removal of the slag that accumulates from the day's work.

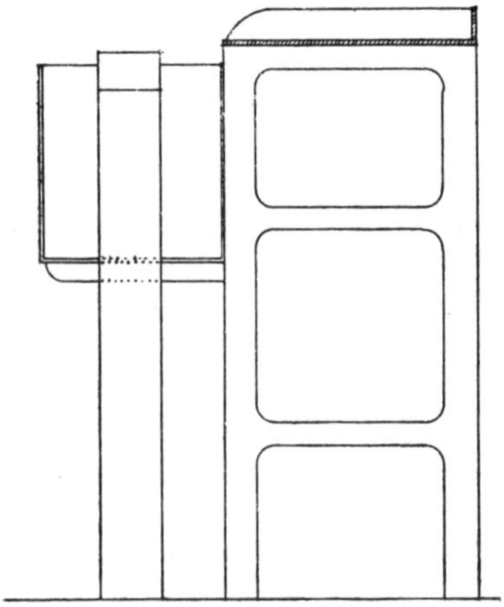

FIG. 42.

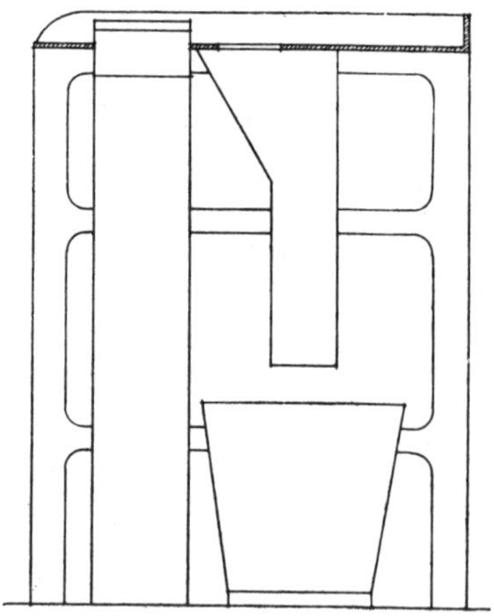

FIG. 43.

TWO FORMS OF ANVIL BENCH.

The anvil shown in Fig. 43 should be from $6\frac{1}{2}$ to 7 inches square and fixed into the end of a 10-inch square timber standing on the floor, so that it projects through the bench top about $1\frac{1}{2}$ to 2 inches. As a preventive against slag flying about, a flange about $\frac{3}{8}$-inch thick and $\frac{1}{2}$-inch high should be cast on three sides of the anvil except about 3 inches along the back, measuring from one corner, to allow of the broken slag being readily brushed away. Behind the anvil, about 2 inches away from it and in direct line with the opening in the flange, a hole, $2\frac{1}{2}$ or 3 inches square, is cut in the bench top, and through this the slag is brushed so that it falls into a bucket or box placed to receive it under the bench.

The anvil (Fig. 44) used for flattening large beads, etc. of gold or silver should be of hardened steel with a face about 6 to 10 in. square. It should be firmly fixed in a block of wood. The face should be highly polished, and kept bright by rubbing with an oiled rag and by being covered with a suitable wood cover when not in use.

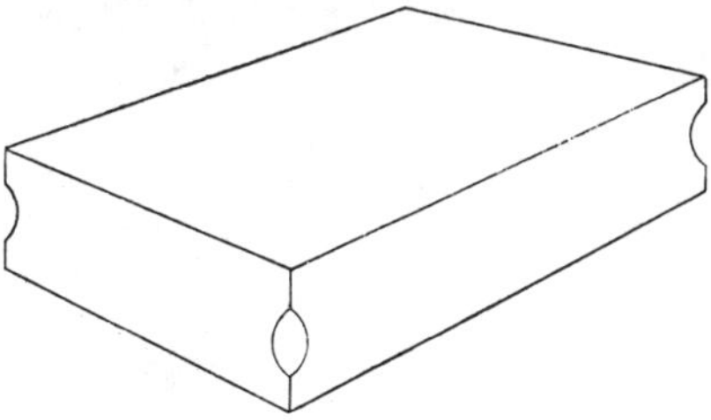

FIG. 44.

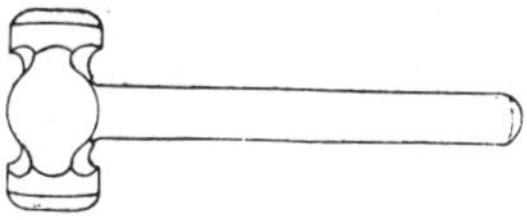

FIG. 45.

Hammers.—For use with this anvil it is well to have two hammers of the shape shown in Fig. 45, the heads of which weigh 7 and 11 lbs. respectively. The faces of the hammers should be highly polished like the anvil and kept bright: it is a good plan to have an anvil cover sufficiently large to allow of the hammers being kept under it with the anvil.

To prevent loss when hammering brittle metal it is advisable to have an iron ring about 8 inches in diameter and 1¼ inches high to stand on the anvil as occasion requires.

Shears.—A strong pair of shears about 9 inches long with a good cutting edge is generally useful for cutting metals, such as sheet gold, silver, and lead.

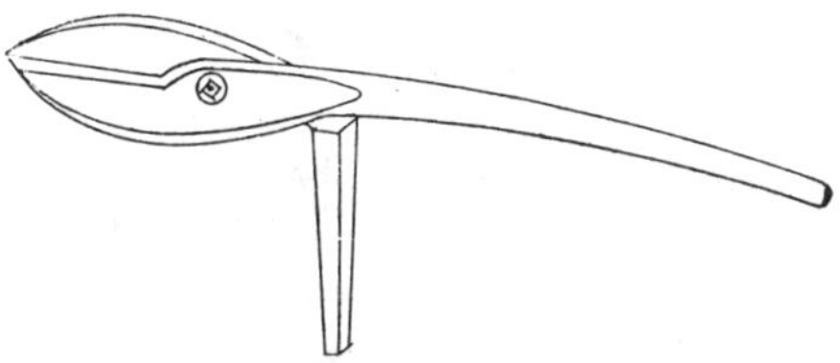

FIG. 46.—SPIKED SHEARS.

For cutting hoop-iron, required in certain assays, a large pair of shears with a spike for mounting on a wood block as shown in Fig. 46 is desirable.

Chisels.—The cold chisel (Fig. 47) is employed for cutting off portions of metal for assay, etc. It should be 5 or 6 inches long, and about ½ to ¾-inch wide, which is the best size for general use. However, for some purposes, as when cutting very tough metals, it is convenient to have a chisel only ¼ or ⅜-inch wide, as these metals are so much more difficult to cut, and the small chisel meets with less resistance.

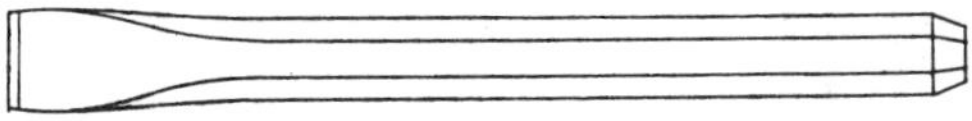

FIG. 47.

Pliers.—Several forms are used for removing gold and silver buttons from cupels and squeezing them to loosen adhering bone-ash. Half-round-nose pliers (Fig. 48) are useful for small buttons, while flat-nose pliers (Fig. 49) are generally used for large buttons. The nose of these pliers may be conveniently filed to the curvature of the surface of the cupels used.

Cutting Nippers.—These are employed for cutting metal that cannot be cut conveniently with shears. The form of cutters shown in Fig. 50 is commonly used. The spring cutting nipper (Fig. 51) is a very suitable form of cutter, as it puts less strain upon the hand. The steel jaws are held in position by screws, and can be adjusted for narrow opening or for wide opening, as illustrated. The jaws can be easily replaced when necessary.

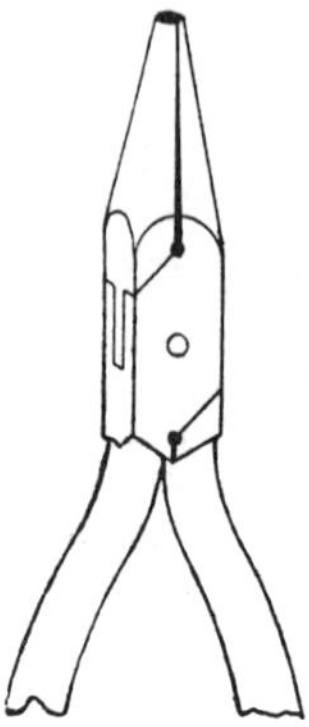

FIG. 48.

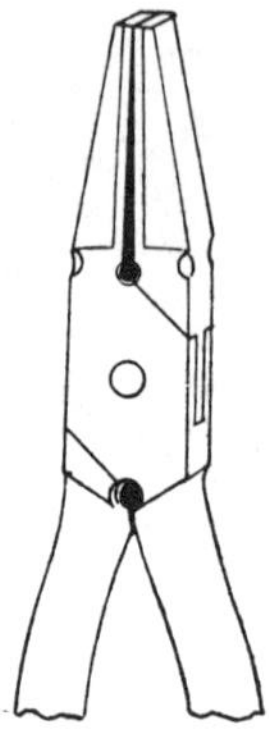

FIG. 49.

FORMS OF PLIERS.

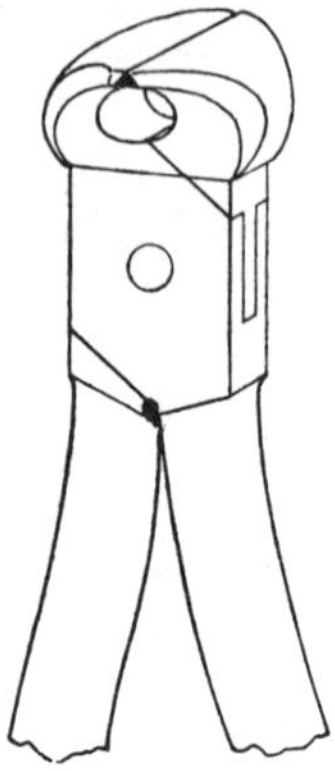

FIG. 50.

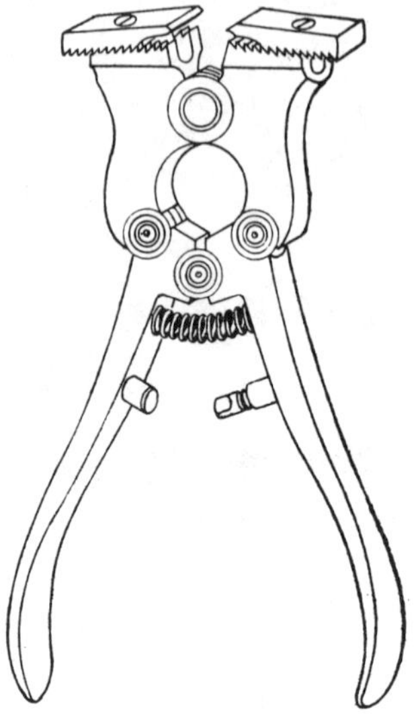

FIG. 51.

FORMS OF CUTTING NIPPERS.

Brushes.—Hard stiff brushes are used for cleaning silver and gold buttons from adhering bone-ash. For small buttons a hard tooth-brush is frequently used. A form of brush known as the Freiberg pattern consists of a bundle of stiff bristles fitted into a metal tube (Fig. 52). Another form of brush (Fig. 53) is made of hog bristles set at a slight angle as shown at *A*. By mounting the bristles in this way the brush retains its stiffness until worn out, and does not spread out as is usual with brushes in which the bristles are mounted in the ordinary way.

FIG. 52.

A FIG. 53.

Punches and Dies.—Good-quality steel punches are used for stamping bullion. They should comprise the numerals from 0 to 9 and the letters of the alphabet, and may vary in size from $\frac{1}{32}$ inch to $\frac{5}{8}$ inch. In addition to these it is usual to have certain word-punches in one piece, such as "Gold," "Silver," "Fine," "Value," "Total," "No.," "Oz.," and "$." Special punches with name or mark of the assayer, mine, or company are also in use and can be procured as desired.

III. APPARATUS USED IN ASSAY FURNACE OPERATIONS.

The principal apparatus used in assaying by furnace methods consists of crucibles, scorifiers, roasting-dishes, and cupels. The materials from which these are made vary according to the objects for which they are to be used.

Crucibles.—These are vessels made of some refractory or heat-resisting material, for the fusion of assays, melting of metals, etc.

The essential qualities of crucibles to be used for assay purposes are as follows: (i) they must be able to resist a high temperature without softening; (ii) they must not be friable when hot; (iii) they must be capable of withstanding sudden changes of temperature without cracking, as, for example, when a white-hot crucible is brought out of a furnace into cold air; (iv) they are required to resist the corrosive action of metallic oxides that may be contained in the material being fused.

Crucibles are made of fireclay mixed with burnt clay (grog), sand, plumbago, or other infusible material, in order to counteract the tendency to shrink when drying which raw clay possesses. The various substances thus mixed with the raw clay do not contract on drying; by their addition, therefore, this tendency is considerably reduced.

The crucibles are moulded, dried, and then burnt in kilns. All crucibles should be carefully annealed before use by placing them in an *inverted* position over the furnace, otherwise they are liable to crack when placed in a red-hot fire.

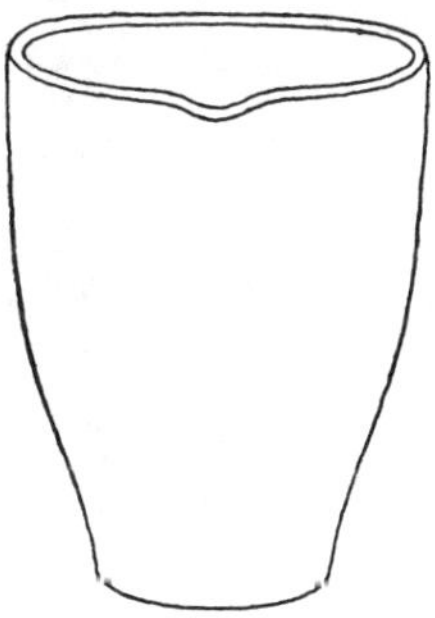

FIG. 54.

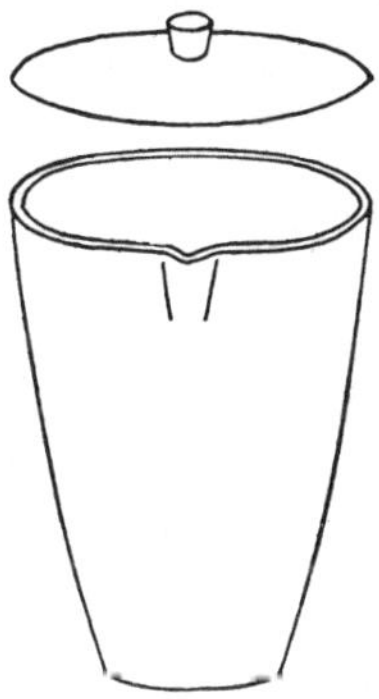

FIG. 55.

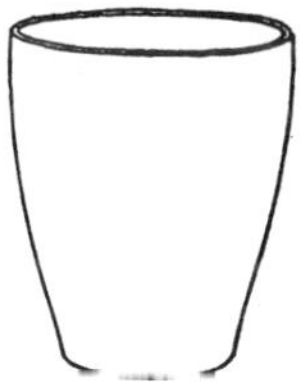

FIG. 56.

TYPES OF CRUCIBLES.

The crucibles best known in commerce are the varieties designated London, Battersea, Cornish, Hessian, and French. The sizes and shape of crucible best suited for assay purposes are indicated when giving the charges for crucible work.

London Crucibles (Fig. 54).—These are round in shape, and are preferred by many assayers on account of their very refractory nature. They resist the corrosive action of fused metallic oxides better than most clay crucibles.

Battersea Crucibles (Fig. 55).—These are similar in shape to the London crucible, but a little taller. They are fairly refractory, and are largely used for the fusion of gold and silver ores. For silver-ore assays the "Battersea Round," E or F sizes, are generally used, while for gold ores the G size is most suitable.

Cornish Crucibles (Fig. 56).—These are round, short, and somewhat open crucibles used sometimes for the roasting of ores. They are fairly refractory, though liable to be corroded by metallic oxides. The usual size employed is No. 79, which is $3\frac{3}{8}$ inches high and $3\frac{1}{4}$ inches in diameter at the top.

Hessian Crucibles ("London Triangles").—These are triangular in shape, and have three spouts, from any of which the contents can be poured. For this reason they are sometimes preferred, but they are not largely employed by assayers. They are fairly refractory, and resist the corrosive action of metallic oxides very well. They are very liable, however, to crack with sudden changes of temperature, and to break when held with the tongs. A crucible known as the "Battersea London Triangle" is another and somewhat more convenient form of triangular crucible (Fig. 57).

French Crucibles.—Fig. 58 shows the form of the French clay or "fluxing pot." These crucibles are very carefully made, being smooth, close in grain, and of good quality. They are somewhat brittle, but will withstand high temperatures without softening, and resist the corrosive action of metallic oxides remarkably well. In the assay office they are used occasionally for melting small quantities of gold, silver, and other metals.

Colorado Crucibles (Fig. 59).—These are special-shaped crucibles used for the fusion of gold and silver ores in the muffle. They are made of the same material as Battersea Round Crucibles. The various sizes are specified by their gramme capacity, *i.e.* by the number of grammes of ore, with the proper amount of fluxes necessary for fusion, which the crucible will hold. The chief

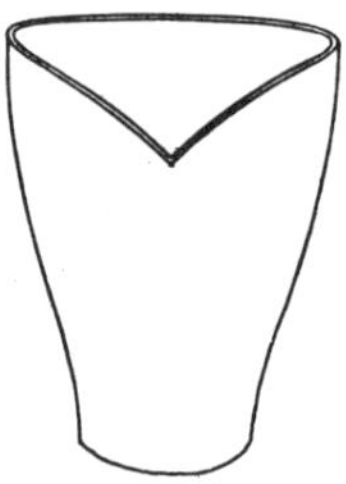
FIG. 57.

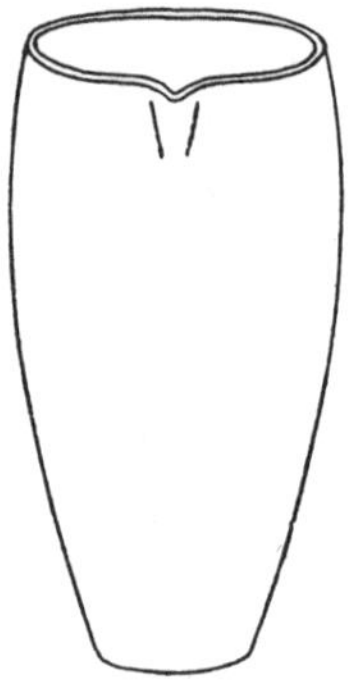
FIG. 58.

FIG. 59.

FIG. 60.

TYPES OF CRUCIBLES.

sizes have capacities of 5, 10, 12·5, 15, 20, 30, and 40 grammes. The sizes most frequently used are the 20-gramme crucible for 0·5 assay-ton charges, and the 30-gramme for 1·0 assay-ton charges.

Plumbago Crucibles.—The so-called plumbago crucibles are made of raw fireclay and graphite, in the proportion of 51 of the former to 49 of the latter. The usual shape of these crucibles is shown in Fig. 60. Good plumbago crucibles, after a preliminary careful annealing, withstand the greatest changes of temperature without cracking, bear the highest heat without softening, and may be used repeatedly. They are the least corroded by metallic oxides, but, owing to the reducing property of the carbon they contain, they must not be employed when oxidizing actions are required.

The careful annealing of plumbago crucibles is very essential, as they have a very great tendency to split when plunged into a hot fire without proper annealing.

In order to prevent absorption of moisture and salts—as, for instance, in orders for shipment—and to enable rapid alterations of temperature to be better withstood, the finished crucibles are in some cases dipped in milk of clay, dried, glazed, and then dipped in tar. Crucibles made in this way by the Morgan Crucible Company are known as "Salamander" crucibles; they will stand forty meltings of gold without sensible deterioration (Roberts-Austen).

Plumbago crucibles possess many advantages over clay crucibles, and they have almost entirely superseded the latter for melting gold and silver. Their surface is very smooth, and permits of the contents being poured clean.

Scorifiers.—Scorifiers are shallow, circular, cup-like vessels made of fireclay, in which the substances to be subjected to scorification are placed.

Fig. 61 shows a section of the ordinary type of scorifier. They are made in various sizes designated by their outside diameters. The sizes ordinarily made are 1·5, 2, 2·5, 3, and 3·5 inches. The 2·5-inch scorifier is the one most frequently used.

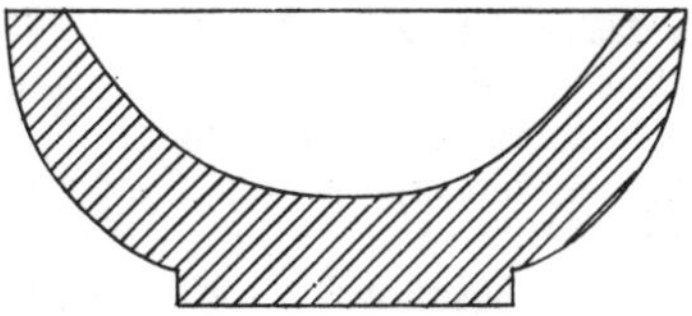

FIG. 61.—SECTION OF SCORIFIER.

The uses of both scorifiers and cupels in assaying are given in Chapter XII.

Good-quality scorifiers are very essential for assay work. A scorifier should resist as much as possible the corrosive action, at high temperatures, of molten lead oxide, either by itself or in admixture with other highly corrosive substances, such as copper or iron oxides. It should not crack on exposure to great and sudden changes of temperature. It should be wide across the top, so that a large surface of the materials operated upon may be exposed to oxidation; and so deep in the centre that a considerable proportion of the lead used in the operation may be kept well covered and preserved from oxidation.

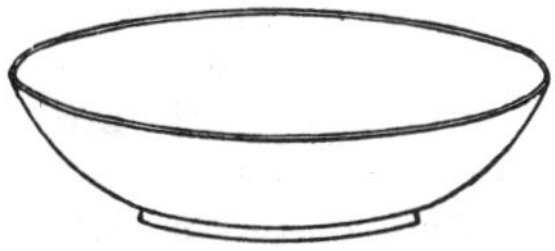

FIG. 62.—ROASTING-DISH.

Scorifiers should be well annealed before use.

Roasting-Dishes.—Roasting-dishes are flat, shallow, circular vessels of fireclay (Fig. 62). They are sometimes employed for the roasting of ores, etc. containing sulphides or arsenides, an operation which may be conducted in a muffle, or by placing the roasting-dish in front of the flue of an air-furnace. Roasting-dishes 4 and 5 inches in diameter are the sizes most generally useful.

Cupels.—These are cup-like vessels in which the operation termed *cupellation* is carried on. In old treatises on assaying they are described as "copples." Both "cupel" and "copple" are derived from the French *coupelle*, which means a little vessel in the form of a cup. Cupels are generally round, and more or less tapered towards the base, but square cupels are also used. A cupel of the usual form is represented in section in Fig. 63. These cupels must be made of substances that are not acted upon by certain fused metallic oxides, such as those of lead and copper; and their texture has to be sufficiently loose to allow of the molten oxides penetrating their substance readily, and yet be sufficiently strong to bear handling without breaking.

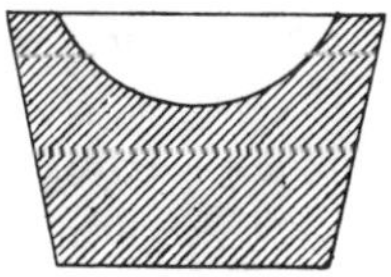

FIG. 63.—CUPEL.

There are several materials of which cupels are made which fulfil all these conditions, but only two are in general use, viz. the ash of burnt bones (bone-ash), which consists principally of calcium phosphate, and magnesia (MgO). Mixtures of bone-ash and magnesia are frequently used.

Preparation of Bone-Ash.—The best bones for the preparation of bone-ash are stated to be those of sheep and horses, and next, those of calves and oxen. They are calcined with free access of air, but before burning it is advisable to boil them repeatedly in water to dissolve organic matter. If the bones are not quite white after the first ignition, but contain a little carbon, they should be ground up, moulded into small balls, and burned again. The reheating should not be too strong, otherwise the bone-ash when compressed will have a smooth glassy surface, and will not be sufficiently absorbent to make good cupels. When the bones are burnt white throughout, they must be finely ground, sifted, and washed several times with distilled water till all soluble salts (sodium carbonate and sodium chloride) are removed.

According to Bettel,[1] the bone-ash should be especially treated by washing with an aqueous solution of ammonium chloride, this salt being used to the extent of 2 per cent. of the weight of the bone-ash to be treated. The bone-ash, thus purified, is dried, again pounded, and passed through a sieve of 40-mesh. The fineness of division of the bone-ash is of much importance. If it is too coarse, the cupel will be too porous and will permit a comparatively large absorption of silver and gold. If the bone-ash is too fine, the cupels made from it will be too dense, and the cupellation will be prolonged and tend to increase the loss by volatilization. Cupels too fine in grain are, moreover, very liable to crack when heated, and so cause mechanical loss of silver.

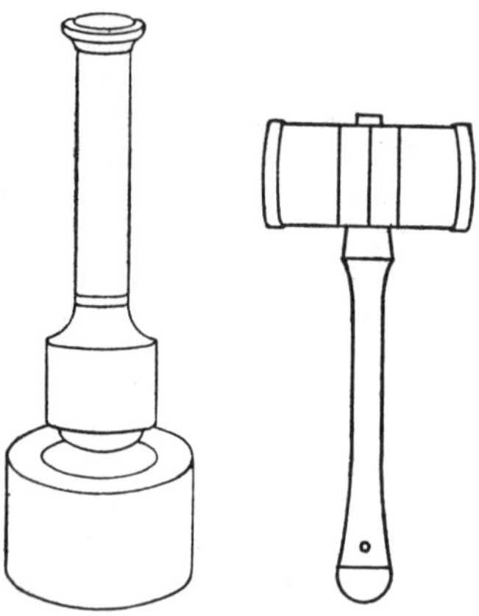

FIG. 64. FIG. 65.

CUPEL MOULD, AND MALLET.

Bone-ash cupels do not bear transport well, hence the assayer has generally to make them, or to supervise their making; they can, however, be purchased. When small quantities only are required they are usually made in moulds by hand, but as cupels should be as uniform as possible as regards density, they are best made by machine, since by machinery a constant pressure may be obtained. When large quantities are required, a machine is generally employed in preference to hand moulds.

Manufacture of Cupels.—The process is conducted as follows:—Finely-ground bone-ash is moistened with water and thoroughly kneaded, so that the mass may be of uniform consistence throughout, and quite free from lumps. The material is best worked up by rubbing between the hands. The bone-ash is sufficiently wet when it coheres by gently squeezing it in the hand. If too much water has been used, the cupel will be too dense and compact; if too little, it will be tender and friable. As a general rule, bone-ash requires the addition of about one-twelfth its weight of water.

It is very important, for the reasons stated on p. 145, that cupels should not contain carbonaceous matter or carbonates; and therefore, in making them, the bone-ash should not, as sometimes recommended, be mixed with stale beer, or water containing adhesive substances such as molasses, or with alkaline carbonates, etc. Nothing but pure water is necessary, and when the correct proportion is added, the results are perfectly satisfactory.

Cupel moulds (Fig. 64) are generally made of cast iron or gun-metal. They consist of three parts: (i) a hollow cylinder, (ii) a disc of metal, and (iii) a plunger or plug for compressing the bone-ash and shaping the top of the cupel. The disc (not shown in Fig. 64) forms a false bottom for the mould.

[1] W. Bettel *Proc. Chem. Met. Min. Soc. S. Africa*, **11**, 599.

In forming the cupel, the disc is put in its place, and the cylinder filled with the moistened bone-ash. The plunger is then placed in position, and the bone-ash compressed by three or four smart blows from a wooden mallet (Fig. 65). Before removing the plunger it should be turned half-way round upon its axis, so as to free it and to smooth the face of the cupel. The cupel is then carefully pushed out by pressing up the disc of metal forming the false bottom.

The removal of the cupels is more easily effected if the mould is slightly tapered, and not cylindrical, in form. The thickness of the metal disc is varied according to the desired thickness of the cupel. When cupels with a fine smooth surface are required, a thin layer of very fine bone-ash is placed in the mould on top of the coarser bone-ash before inserting the plug. In this way the face of the cupel becomes coated with a layer of finer material.

The cupels are placed on iron or wooden trays and well air-dried by storing in a warm place. If possible they should be allowed to dry for two or three months before using. It is important to preserve cupels from the access of nitrous or other acid fumes. These are absorbed by the bone-ash and given out in the furnace, and may cause "spurting." The mould should be oiled after use to preserve the smooth surface from rusting.

Cupels vary in size according to requirements: those upon which buttons of 15, 30, and 40 grammes of lead respectively can be cupelled are most generally useful. A cupel should be capable of absorbing its own weight of molten litharge: this will serve as a guide in the selection of cupels of suitable size.

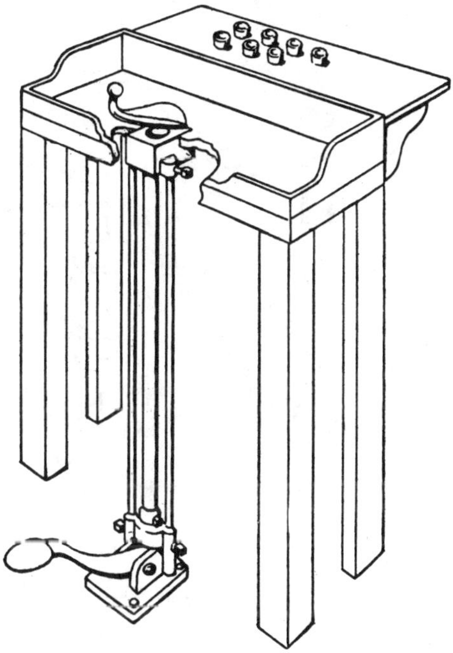

FIG. 66.—CUPEL MACHINE.

The weight of lead absorbed by different sized cupels of the shape shown in Fig. 63 is as follows:

Diameter at top	$\frac{3}{4}$, $\frac{7}{8}$, 1, $1\frac{1}{8}$, $1\frac{1}{4}$, $1\frac{3}{8}$, $1\frac{1}{2}$, $1\frac{5}{8}$ inches.
Absorption in grammes ...	3, 5, 8, 10, 16, 20, 28, 40.

A simple form of cupel machine is shown in Fig. 66. It consists essentially of an ordinary cupel mould as described above, placed in an inverted position with the plug attached to a long rod connected with a pedal fixed to the floor. To make the cupel the mould is filled with bone-ash, then closed by means of the sliding cover shown, and pressure is applied to the plug by means of the pedal, thus forming the cupel in an inverted position. When made, the cupel is released by sliding the cover on one side and pushing the cupel out by a slight pressure of the pedal. Considerable pressure may be used with this machine, and the cupel made quite firm.

Another type of machine is Calkin's automatic machine, illustrated in Fig. 67. This machine consists of a hopper to hold the moistened bone-ash, and a revolving disc with a number of holes which are automatically filled and brought into position under the plunger. To operate the machine, the moistened bone-ash is placed in the hopper, in which there is a small wheel with a rubber rim that keeps the material stirred up and automatically fills the moulds. The lever handle is then raised, the handle on the lower disc being grasped at the same time, and the filled mould is brought beneath the plunger. The lever handle is then forced down to compress the cupel; by pulling the disc handle in a reverse direction to that formerly given it, the opening in the lower disc is brought beneath the cupel, and further pressure on the lever handle brings a new system of levers into action, ejecting the cupel, which may be caught in the hand. An automatic attachment stops the disc at the proper

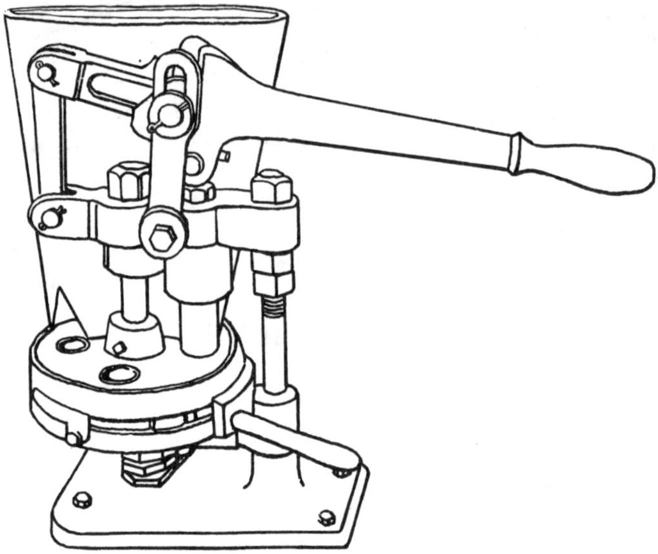

FIG. 67.—AUTOMATIC CUPEL MACHINE.

points, and an adjusting device is fitted for giving different degrees of compression.

Cupels of various sizes may be made by using interchangeable discs and dies, which are easily adjusted to the machine. Perfect edges and a homogeneous cupel can always be obtained. It is claimed that 600 cupels an hour can be made with this machine.

A smaller machine to be bolted to a wall, post, or other convenient support can also be obtained. The same principle of leverage is applied, but it is necessary to feed the bone-ash by hand.

Square cupels are used in some assay offices. The cupels in use at the Royal Mint, London, are square blocks of bone-ash, a little less than 2 inches across, and a little more than $\frac{3}{4}$-inch deep. Each block carries four circular cavities about 0·7 inch in diameter and 0·3 inch deep. The muffles in use will take six rows of these blocks placed three abreast, thus making provision for seventy-two assays at one time.

A mould for making such cupels is represented in Fig. 69. It will be seen that the mould is in three pieces, and the cupel blocks are made with the cavities downwards. To obtain the necessary pressure the mould is placed in the press shown in Fig. 68 and the lever pulled down. A mould for making a block with one large cavity only is also shown.

Magnesia Cupels.—Magnesia was introduced, some years ago, as a substitute for bone-ash, and a large number of brands of so-called "patent" cupels made with a magnesia base are now on the market. The magnesia (MgO) is produced by strongly calcining crude mineral magnesite (magnesium carbonate, $MgCO_3$). The cupels require to be made under high pressure, and should be baked at a high temperature before use. They are then very hard and firm and light-brown in colour, resembling fireclay. Owing to the high pressure required (usually hydraulic), magnesia cupels cannot be made satisfactorily in the laboratory. The exact composition of these cupels is generally a trade secret; several

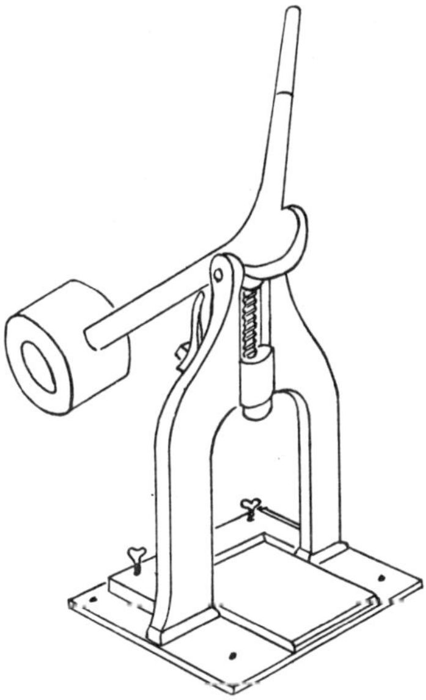

FIG. 68.—PRESS FOR MAKING SQUARE CUPELS.

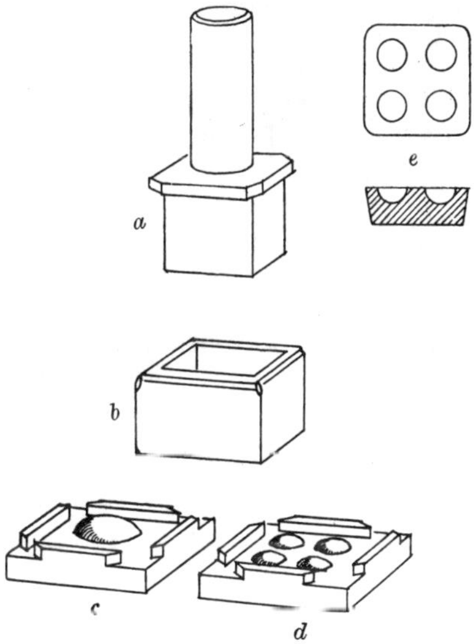

FIG. 69.—CUPEL MOULD IN THREE PIECES. *a*, *b*, and *c*, for cupel with one cavity; *d*, for cupel with four cavities; *e*, cupel with four cavities (plan and vertical section).

brands contain a certain proportion of bone-ash and other ingredients. Magnesia cupels bear transport well, and on that account have come into wide use in South Africa and other mining centres. They absorb and hold less litharge than cupels made of bone-ash of the same size. Their properties are discussed further on p. 153.

Portland-cement Cupels.—According to Holt and Christensen,[1] ordinary Portland cement forms a convenient substitute for bone-ash in making cupels, provided the amount of water added is carefully adjusted. This should be between 6 and 10 per cent. of the weight of the cement: 8 per cent. is the most suitable amount. If less than 5 per cent. of water is used the cupels are too fragile; with 20 per cent. of water the cement cannot be readily moulded in the cupel machine.

[1] "Experiments with Portland Cement Cupels," T. P. Holt and N. C. Christensen, *Eng. Min. J.*, 1910, **90**, 560-561; also "Cement *versus* Bone-ash Cupels," J. W. Merritt, *Min. Sci. Press*, 1911, **100**, 649.

Cupels with less than 5 per cent. and more than 15 per cent. of water crack at the edges when heated. Cement cupels are strong, and will absorb their own weight of litharge; the loss of silver by absorption is stated to be about the same as with bone-ash. The cupels should be thoroughly dried before use, otherwise they will develop cracks during heating.

Cupels made of a mixture of equal parts of cement and of bone-ash were found to be in some respects superior to either cement or bone-ash cupels. Compared with bone-ash, cement forms a very cheap substitute for making cupels and is frequently more easily procured. The cupels may be made entirely of cement, or the lower portion only, from one-half to three-quarters, made of cement and the upper portion of bone-ash or some magnesia composition. In this case the required quantity of moistened cement is put into the cupel mould, and bone-ash or other material added to fill the mould; the whole is then compressed in the ordinary way.

A very smooth surface can be given to the cupels by dusting them over with very finely powdered bone-ash or some magnesia material (about 200 mesh) just before compression.

CHAPTER V

BALANCES AND WEIGHTS: WEIGHING

THE delicate balances used in precious-metal assaying are the most important part of the assayer's equipment, and should be of reliable make and used with care. In practice, differences between assay results are not infrequently traced to the use of worn weights or to balances that have lost their sensitiveness through improper handling. Consequently the provision of a balance room not subject to vibration, and free from contamination with acid fumes and dust, is very essential for accurate work.

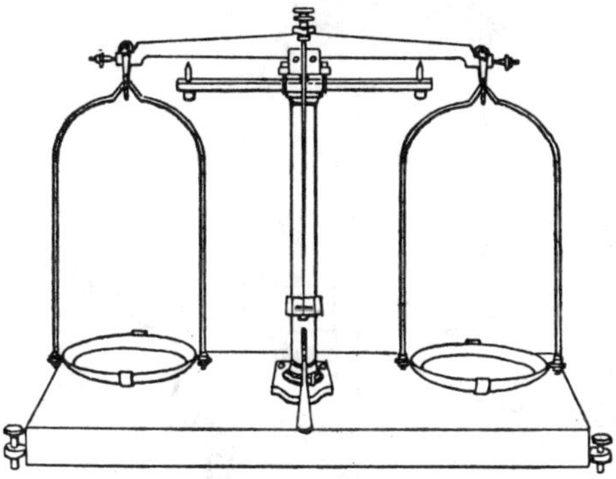

FIG. 70.—PULP BALANCE.

At least two balances are employed for the assay of gold, silver, and platinum, etc., one for weighing fluxes and finely pulverized ore or *pulp*, and the other for weighing the small *buttons* of precious metal obtained from the assay. This latter—the assay balance proper—is extremely sensitive. In assay offices where general assays are made, a third balance, more sensitive than the pulp balance, is usually necessary: in this case some form of chemical balance is employed.

(1) **Pulp Balance.**—The pulp balance, for weighing the charges of ore or products and the fluxes for assay, should be capable of carrying at least 250 grammes in each pan, and be sensitive to 2 milligrammes. A convenient form of pulp balance is shown in Fig. 70.

(2) **Assay or "Button" Balance.**—For all precious-metal assaying, balances of the highest precision are necessary. In purchasing an assay balance it is true economy to buy a first-class instrument by some reliable maker. It should be capable of carrying 1 or 2 grammes in each pan, and, for ordinary work, should be sensitive to $\frac{1}{20}$ milligramme or less. For work requiring extreme accuracy balances sensitive to $\frac{1}{100}$ or even $\frac{1}{200}$ milligramme are in use. In cases where a large amount of general gold and silver assaying has to be done it is usual to employ two button balances, one for the silver buttons and the other for the gold, as the finely-adjusted balances used for gold are too slow

for rapid silver weighing, and are liable to lose their extreme sensitiveness if subjected continually to the strain of weighing large silver buttons.

The accompanying illustration (Fig. 71) shows a typical assay balance by Messrs. L. Oertling, Ltd., with a 6-in. horizontal beam to carry 2 grammes

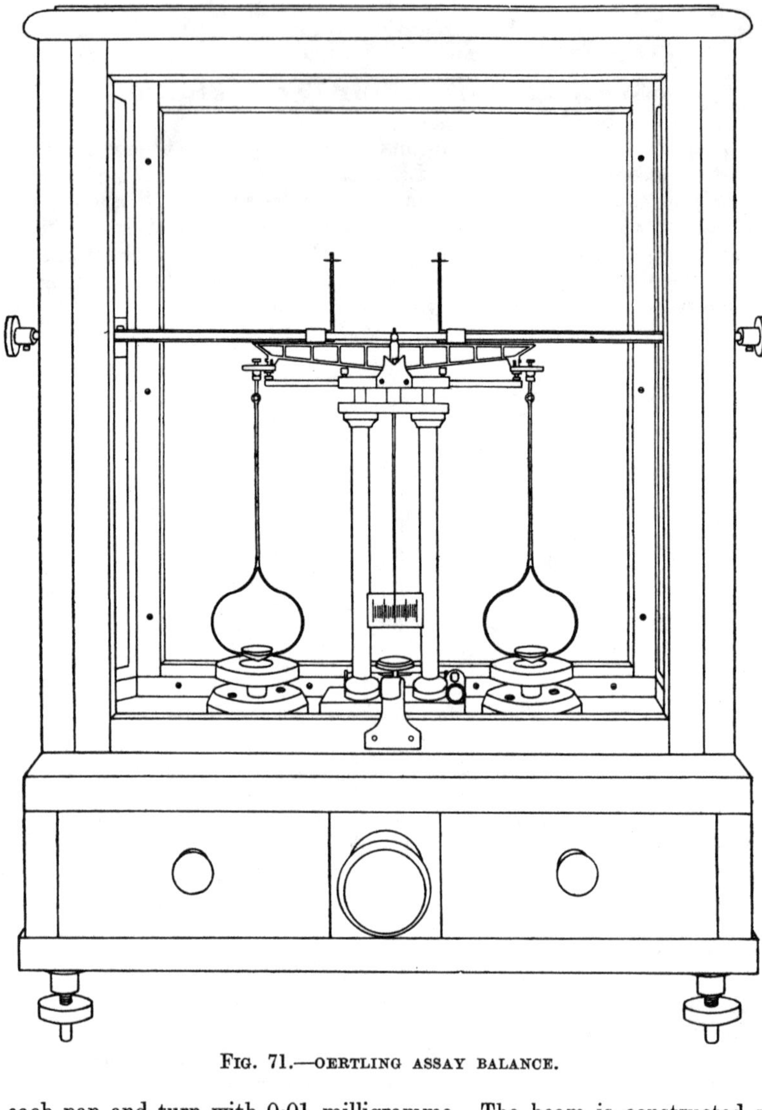

FIG. 71.—OERTLING ASSAY BALANCE.

in each pan and turn with 0·01 milligramme. The beam is constructed with agate knife-edges working on agate planes, and divided into 100 parts.

Another type of button balance (by Oertling) for very exact work is shown in Fig. 72. In this "short-beam" type the pan suspenders are decreased in length, thus permitting of very short brass supports for the beam. This arrangement tends to concentrate the movable mass near its central axis, thus giving great stability of poise, without diminishing the sensitiveness. The beam is

4 inches long and is divided into 100 parts. The pointer extends upwards, and the scale is above the beam. The pointer makes one complete oscillation in 10 seconds, and with 0·1 milligramme shows 10 divisions on the scale. Lenses, which are removable, are provided for reading the divisions on the beam and the scale: they are shown in position in Fig. 72. The capacity of the balance is 1 gramme and the sensitiveness $\frac{1}{200}$ milligramme.

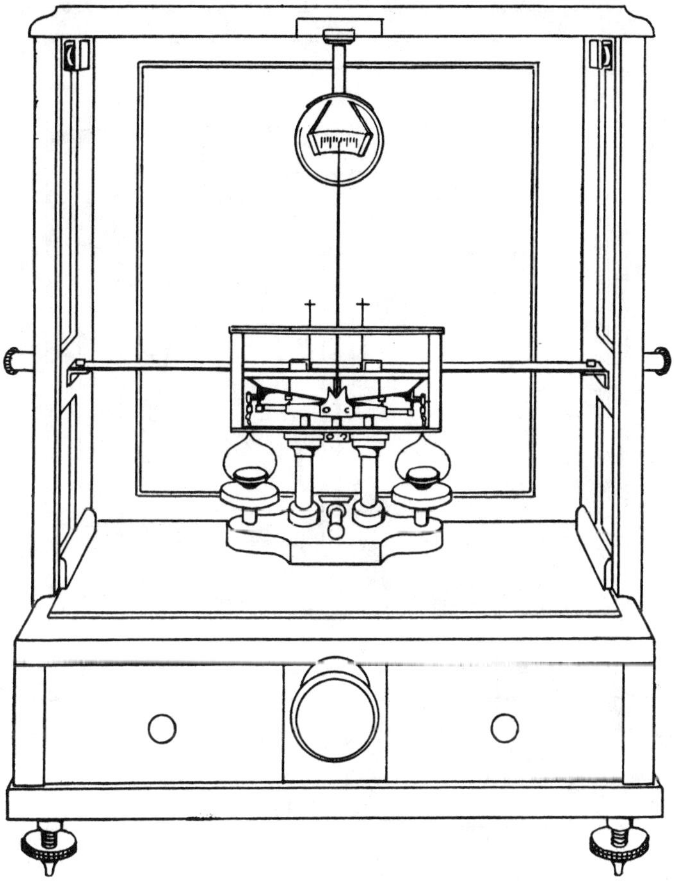

FIG. 72.—OERTLING BALANCE (SHORT-BEAM TYPE).

In some designs of button balances the pointer is fixed at the end of the beam in a horizontal position and the scale is vertical.

A portable lens or telescope to stand inside the balance case is very frequently employed for reading the divisions on the scale and beam of assay and chemical balances. A simple arrangement of lenses for this purpose is that devised by G. T. Holloway.[1] The usual telescopic or portable lens arrangements are replaced by lenses cemented on the rising glass front of the balance case. Plano-convex lenses, cemented with Seccotine on the side of the glass, are preferred. A 3-inch lens is a convenient size for the pointer reading, but a 2-inch lens is sufficient for reading the beam graduations. The latter is cemented to face the

[1] *The Analyst*, February, 1906.

right-hand side of the beam, but a slight movement of the head suffices to bring any part of the beam into the field. The focal length of the lenses is of little importance, but they must not be powerful enough to produce distortion, or thick enough to prevent the rise and fall of the balance-front on which they are fixed. There is always sufficient clearance for any ordinary lens, and as the lenses are fixed inside, they remain clean and require no attention or adjustment.

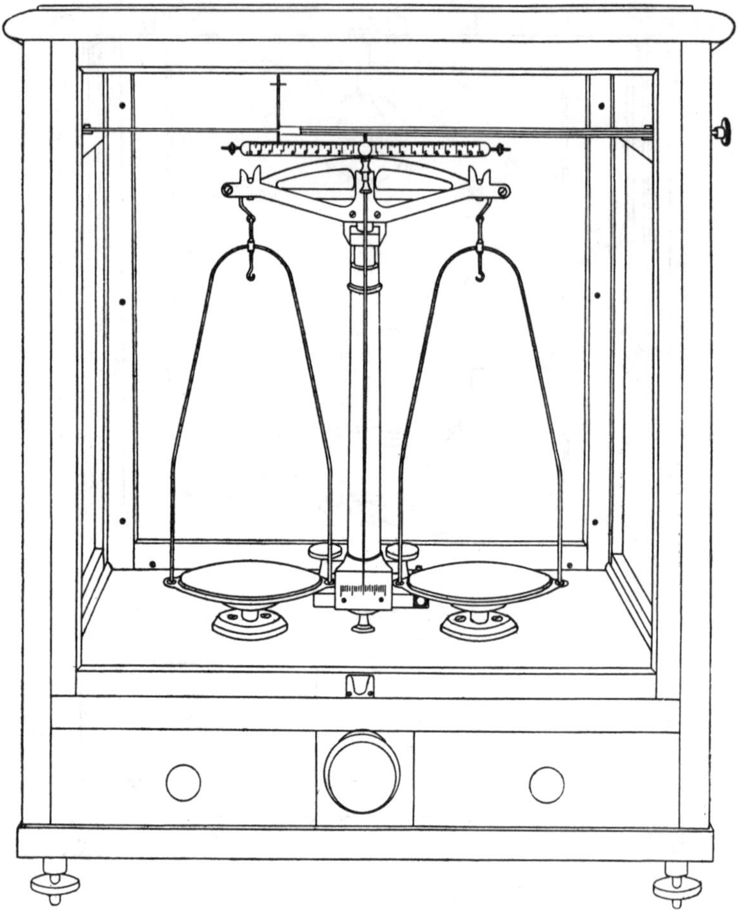

FIG. 73.—SHORT-BEAM CHEMICAL BALANCE.

The latest type of balance, introduced by Messrs. Oertling, is fitted with a prismatic reflecting device giving direct scale-readings without the use of small weights or riders.

On account of the great accuracy required in precious-metal assaying, it is very desirable that the assayer should make himself thoroughly acquainted with the capabilities of the balance which he employs.

The degree of sensitiveness may be determined by accurately noting the deviations of the pointer from the zero-point which overweights of a milligramme and decimals of a milligramme effect under varying conditions, as follows: (i) when the balance is unloaded, (ii) when carrying 1 gramme on each pan, and (iii) when carrying 2 grammes on each pan. The variation in the rates of oscillation as the load increases should also be observed at the same time. When the balance is approaching equilibrium it is possible, after a little experience, readily to estimate the weight required to establish perfect equilibrium from the extent and rapidity of the oscillations as indicated by the pointer. This knowledge is very useful when a large number of weighings have to be made.

Auxiliary Assay Balance.[1]—A balance of special design is in use at the Melbourne Mint in order to lessen the time of weighing when dealing with a large number of gold cornets of widely differing weights. It resembles somewhat the balance commonly used for weighing letters. The cornet is placed in a pan suspended at the end of a short counterpoised beam, supported at an angle of 60° from the horizontal on a central fulcrum, to which is attached a long vertical pointer.

The pointer moves over a scale through an angle of 45°, and its position when it comes to rest denotes the approximate weight of the cornet. The cornet is then transferred to the pan of a delicate button balance, and the necessary weights, already determined by the auxiliary balance within narrow limits, are placed on the other pan and adjusted as necessary to obtain the exact weight.

The advantages of an auxiliary balance are said to be not only a saving of time, but also a saving of wear on the more sensitive assay balance and weights. Commercial modifications of the auxiliary balance are available.

(3) Chemical Balance.—Fig. 73 represents a modern form of short-beam chemical balance suitable for weighing rich ores and products which require to be more accurately weighed than is possible with the ordinary pulp balances. This balance will carry 200 grammes in each pan and turn with $\frac{1}{10}$ milligramme. The length of the beam is 6 inches.

Setting-up Balances.—The balance must stand in a position free from vibration: otherwise it is liable to be frequently thrown out of adjustment. In ordinary buildings it is often difficult to obtain a firm support for the balance: if placed on a table which stands directly on the floor the vibration caused by persons walking in the room is quite sufficient to affect a very sensitive balance. The best foundations are stone or concrete piers set a few feet into the ground and projecting upwards through the floor but without touching it. A simple method of support consists of brackets firmly attached to the wall. This will usually be sufficient to prevent vibrations from the floor. Where slight vibrations occur they may be counteracted by placing a rubber block under each levelling screw of the balance. A suitable rubber block ($1\frac{1}{4}$ inches in diameter) mounted between two brass caps is shown in Fig. 74. A shallow cavity is drilled in the top cap to receive the levelling screw and keep the block in position. An ordinary rubber bung is frequently used, a hole being bored in the bung sufficiently large to admit the leg of the levelling screw, so that the milled plate rests on the bung, which is placed with the wide end resting on the bench.

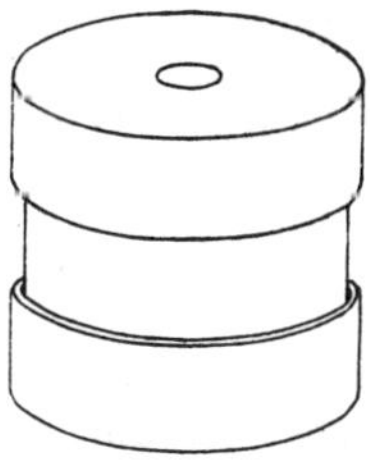

FIG. 74.

The bench top on which the balance stands is best made of some hard, well-seasoned wood: slate is sometimes recommended, but is not desirable. The balance should not stand in a position where it may be exposed to direct sunlight. It should be so arranged in the balance room that in weighing the light falls over the left shoulder of the operator, or from overhead.

[1] R. Law, *J. Chem. Soc.*, 1896, 526; *Mint Report*, 1913, **44**, 140.

Facing a strong light while manipulating a sensitive balance is very trying to the eyes. On this account the north aspect, if obtainable, is best for the balance room. The balances should be arranged on the bench with ample room allowed on each side for samples and for the bottles, etc. into which the samples are to be placed after being weighed. It is to be regretted that more attention is not paid to the position of assay balances in mining schools.

To economize space it is sometimes convenient to place the balance bench under the window so that the operator faces a direct light, and the disadvantage of this position may not be felt with the intermittent weighing done by students. This position should, however, be avoided wherever possible, otherwise students may get the impression that it is the most suitable for weighing, while experience proves that it is most injurious to the eyes. In all assay offices where a large amount of continuous weighing has to be done and where the welfare of the staff is considered, the balances are arranged so as to cause as little eye-strain as possible.

Care of Balances.—A balance suffers more from imperfect preservation than from proper usage. When not in use it should be provided with a well fitting baize or linen cover to keep out dust. A balance in constant use will require cleaning about every three or four months, but this must be attempted only by a person thoroughly acquainted with the construction. The instrument should be carefully taken to pieces, and the various parts rubbed with a soft leather or brushed with a camel-hair brush. The movements should be cleaned and oiled with good watch-oil. Occasionally the agate planes will require repolishing by the maker, as the constant working of the knife-edges wears a minute groove in them, easily perceptible by a lens.

The systematic examination of balances should on no account be neglected if they are to retain the high degree of accuracy expected from them in precious-metal assaying.

The practice of keeping lime or other substance in the balance case as a precaution against damp is to be deprecated, and is not advised by balance makers.

Artificial Lighting of Balances.—The method adopted for the artificial lighting of balances is dependent on the source of light available. Electric light is to be preferred, and is the most convenient and most suitable. In arranging the illumination of a balance, in addition to securing satisfactory lighting of the beam, etc., it is necessary to avoid any unequal expansion of the arms due to heat, or disturbance of equilibrium arising from air currents. The source of light is usually placed either on top of the balance case, or inside the case, facing the beam and parallel to it. If placed on top of the case, the electric lamp should be in the best position to reduce heat by radiation to a minimum, and to distribute uniformly such heat as is unavoidable. Excessive heating is avoided by suspending the electric lamp a short distance above the balance case and not resting on it, thus permitting air currents to pass between the source of light and the case.

Many assayers prefer to use the type of electric lamp in which the source of light consists of a metallic filament suspended along a narrow tube, as it appears to be free from the defects of the ordinary type of bulb lamp, since the heat evolved is slight and largely dissipated; and also, by reason of its shape, the lamp can easily be arranged so that any slight heating effect will be uniformly distributed over the beam of the balance. Porritt [1] recommends a metallic-filament lamp of 16 candle-power, 8½ in. long and enclosed in a semi-circular aluminium reflector, which serves both to direct the light through

[1] B. D. Porritt, *J. Soc. Chem. Ind.*, 1918, **37**, 85T.

the glass top into the balance case and to screen the light-source from the view of the person weighing.

In order to ensure the lamp and reflector always being placed symmetrically to, and slightly in front of, the balance beam, two upright brass springs are attached to the woodwork of the balance case, and into circular countersunk holes in these are sprung the bosses at each end of the reflector. Both these fittings are standard articles in the electrical trade.

Similar filament lamps are sometimes fixed inside the balance case, but unequal expansion of the balance arms is more difficult to overcome by this arrangement.

Other methods of lighting consist of suitably placed gas- or electric-light balanced pendants suspended in front of the balance and behind the weigher, so that the light is reflected over his shoulder on to the balance. A convenient arrangement consists of a double-swing bracket with ball-sockets fixed on the wall behind the balance at a convenient height to the right or left of the balance. Ball sockets are used for the joints so that the light can be moved into any desired position and the light thrown from behind the weigher. An onion-shaped lamp is fitted into a ball socket at the end of the bracket. This arrangement satisfactorily avoids any unequal heating of the beam, and permits of a stronger light being used. In this connection, the prismatic reflecting device referred to above, giving direct scale-readings without the use of small weights or riders, has proved to be a considerable advance in balance precision.

Assayer's Weights.—The system of weights now very generally adopted by assayers is the French or metric system, in which the gramme (15·43236 grains) is taken as the unit. The different denominations in weight of the system are related to the gramme by multiples and submultiples of ten, the fractional weight most frequently referred to being the milligramme, written "mg." The denominations are:

Gramme ...	... 1·0	
Decigramme	... 0·1	or $\frac{1}{10}$ part of a gramme.
Centigramme	... 0·01	or $\frac{1}{100}$ " "
Milligramme	... 0·001	or $\frac{1}{1000}$ " "

A set of weights usually extends from 50 grammes to a milligramme, and should contain weights of the following denominations:—

Grammes.	Grammes.	Gramme.	Gramme.	Gramme.
50	5	0·5	0·05	0·005
20	2	0·2	0·02	0·002
10	1	0·1	0·01	0·001
10	1	0·1	0·01	0·001

with several 0·01-g. (centigramme) riders. For assay work it is convenient to have a 100-gramme weight included in the set.

The weights from 1 gramme upwards are usually made of brass, and the smaller weights, from 0·5 gramme downwards, of platinum or of aluminium. Although weights of less than 0·01 gramme are generally contained in the set, in practice it is usual to employ only the deci- and centigramme weights (*i.e.* the first and second places of decimals), the milligrammes and fractions of a milligramme (the third and fourth decimals) being determined far more conveniently and with equal if not greater accuracy by means of the sliding weight or *rider* than by the use of the small weights themselves. According to Rose, the small weights often differ from their stated value by 5 per cent., or even more, and for this reason riders should be used.

The rider consists of a piece of platinum, gold, or aluminium wire weighing 1 centigramme (0·01 gramme), and bent so that it can be placed astride the beam and moved to any desired position on the graduated beam of the balance. It must fit the beam, so that when placed in position it always rests in a plane perpendicular to it. The effective weight of the rider decreases as it travels from the end to the centre of the beam.

Each arm of the balance (*i.e.* the distance from the central knife-edge to the extremity of the beam) is divided into ten equal parts, and when a centigramme rider is used, each part, counting from the centre outwards, is equivalent to 0·001 gramme or 1 milligramme. If each of these tenths be further subdivided, fractions of a milligramme are obtained. The divisions of the beam, therefore, represent milligrammes and fractions of a milligramme, and thus give figures in the fourth place of decimals.

When a rider (weighing 1 centigramme) is placed on the tenth main graduation at the end of the beam, its effect is the same as if it were placed in the scale-pan, *i.e.* it is equal to 1 centigramme or 10 milligrammes; if placed on the fifth main graduation it exerts only half this effect, and is equivalent to 5 milligrammes; and so on for the other main graduations. The ten main graduations are (generally) subdivided into five equal parts, each of which is equivalent, with a centigramme rider, to $\frac{10}{50}$ or $\frac{1}{5}$ milligramme (0·2 mg. or 0·0002 g.). A rider placed midway between two main divisions, say between 6 and 7, would therefore be equivalent to 6·5 milligrammes or 0·0065 gramme; and if placed on the fourth subdivision between the two main divisions 6 and 7, the weight would be 6·8 milligrammes or 0·0068 gramme.

It will be evident that when the rider is placed in the middle of one of the subdivisions it is equivalent to $\frac{1}{10}$ milligramme or 0·0001 gramme. If a rider weighing only 5 milligrammes is used, each subdivision will obviously equal $\frac{5}{50}$ or $\frac{1}{10}$ milligramme, and with a 1-milligramme rider each subdivision will be equivalent to $\frac{1}{50}$ milligramme. A thorough knowledge of the use of the rider is very essential to the assayer, as it is almost invariably employed in weighing the very small buttons of precious metal obtained from assays. Its use is not only attended with much economy of time and trouble but also with greater accuracy, as the rider allows the minutest variations to be determined. Riders weighing 5 milligrammes, 1 milligramme, and in special cases 0·5 milligramme are generally used with the most sensitive assay balances, their employment enabling the fractions of a milligramme to be determined with greater accuracy than is possible with a centigramme rider. Needless to say, the balance must always be brought to rest before altering the position of the rider. Further details on the use of riders for weighing gold and silver assay buttons will be given later.

It is very necessary to point out that riders are frequently sold which are not of correct weight, and it is essential to test them carefully before using. The errors on riders are reduced according to their position on the beam. At a distance of one-tenth of the half-beam length from the centre the error of the rider is reduced to one-tenth of its full amount. For ordinary work the riders will usually be sufficiently accurate, but when great accuracy is required they must, if necessary, be carefully adjusted by the assayer.

"Platinum riders, if too light, may be gilded in a cyanide bath, but light aluminium riders must be rejected. Any rider may be reduced in weight by light rubbing on a sheet of ground glass, or by other mechanical means."[1] The error on a rider can in this way be made less than 0·005 milligramme

[1] *Metallurgy of Gold*, Sir T. K. Rose and W. A. C. Newman, 7th edn., 1937, p. 504 (Griffin & Co.).

without difficulty; but, if greater accuracy is desired, it saves time to determine the error of the weight and to apply it as a correction to the result of each assay.

A good set of weights will be found to be practically unchanged even after many years' use provided they are used with due care. The weights must never be touched with the fingers, but handled always with ivory-tipped or brass forceps: steel forceps should not be used. When not in actual use, the box should be kept closed, and the greatest care taken to protect the weights from the action of acid fume. A new set of weights should always be tested before being taken into general use to see if they agree among themselves, and the testing should be repeated from time to time, particularly when the weights begin to be tarnished.[1] It is very desirable that an assayer should be able to compare his weights with a standard set, and in many assay offices a set of carefully standardized weights is kept exclusively for the purpose.

In addition to an ordinary box of decimal weights, the following special weights are also used by assayers:—

The Assay-Ton System.—In reporting the result of an assay for gold or for silver, the number of ounces troy, per ton avoirdupois, is always returned. In England and Australia the result is reported upon the English or *long ton* of 2240 lbs. (avoirdupois), and in North America and Africa upon the *short ton* of 2000 lbs. (avoir.). To facilitate calculation a system of weights, known as the *assay-ton system* (abbreviated A.T.) has been devised, which is a combination of the troy and avoirdupois systems with the French metrical system. The unit of the system is the "assay ton," which for the English system (2240 lbs. to the ton) is 32·6667 grammes. It is derived as follows:—

There are 2240 lbs. in an English ton, and 7000 grains in 1 lb. avoirdupois; therefore there are

$$2240 \times 7000 = 15{,}680{,}000 \text{ grains in a ton.}$$

There are 480 grains in 1 ounce troy; consequently there are

$$\frac{15{,}680{,}000}{480} = 32{,}666{\cdot}7 \text{ troy oz. in 1 ton avoir.}$$

Now employing the French system: If we take 32,666·7 grammes of ore for an assay, each gramme of precious metal obtained would indicate 1 ounce troy per ton avoirdupois; or, if we take 32·6667 grammes of ore, each milligramme of precious metal obtained will indicate 1 ounce per ton; this quantity, 32·6667 grammes, is the English assay ton: it contains the same number of milligrammes as there are troy ounces in the ton avoirdupois. As an example of its use, suppose the weight of the resulting "button" of precious metal from one assay ton of ore is 1·2 milligrammes, then the ore contains 1·2 ounces of gold (or of silver) per statute ton.

The "assay ton" for the American system (2000 lbs. to the ton) is 29·166 grammes, and is derived in the same way. Sets of assay-ton weights for both the English and American systems ranging from $\frac{1}{5}$ A.T. to 4 A.T. can be purchased, or a set can be made up from an ordinary box of gramme weights and its subdivisions. Although in some cases it may be advantageous to employ assay-ton weights, in general assay practice it is more convenient to work on a purely decimal system, and convert when required into ounces per ton, etc., either by actual calculation or by reference to a set of tables.

Assay Weights.—In assaying silver and gold bullion a special decimal series of weights may be employed in which the unit for *silver assay weights* is 1 gramme,

[1] For the method of testing weights, see Appendix.

which is stamped "1000," and for *gold assay weights* is ½ gramme, which also is stamped "1000." The decimal subsidiary weights are stamped as follows:—

Silver Assay Weights.		Gold Assay Weights.	
Number stamped on Weights.	Actual weight, gramme.	Number stamped on Weights.	Actual Weight, gramme.
1000	1·0	1000	0·5
500	0·5	500	0·25
200	0·2	200	0·1
100	0·1	100	0·05
100	0·1	100	0·05
50	0·05	50	0·025
20	0·02	20	0·01
10	0·01	10	0·005
10	0·01	10	0·005
5	0·005	5	0·0025
2	0·002	2	0·001
2	0·002	2	0·001
1	0·001	1	0·0005
Riders 0·01 gramme (10 milligrammes)		Riders 0·005 gramme (5 milligrammes)	

The figures stamped on the weights in the silver series denote the number of milligrammes contained in the weight, and in the gold series the number of ½-milligrammes (termed *milliemes*) contained in the weight. The advantage of these special "assay weights" is that the report is at once indicated by the figures on the weights without further calculation, provided 1 gramme of silver or 0·5 gramme of gold is used for the assay respectively. They are in use in most bullion assay offices dealing with a large number of assays per diem.

Methods of Weighing.[1]—There are three methods of determining the weight of a substance by means of the balance: (1) the *direct* method, (2) the method of *substitution*, and (3) the method of *reversal*. Before describing these methods it is well to point out that, whatever method of weighing is adopted, it is best to work with the balance *on the swing*, as it is expressed, that is, the state of adjustment of the balance is indicated by the extent to which the pointer swings or is deflected over the ivory scale. If the balance is in equilibrium, the pointer will swing approximately an equal number of divisions on each side of the zero, losing, however, half a division (or thereabouts) on each swing, the small loss being due to a gradual return of the balance to rest, caused by friction.

Equilibrium is established when the number of divisions traversed by the pointer to the left is the mean of the preceding and succeeding number of divisions traversed to the right. For example, if the pointer swings 7 divisions to the right on one swing and 6 divisions on the next, the intermediate swing to the left should have been 6·5 divisions. When the swings are thus approximately equal, the balance is in equilibrium, and if any substance is being weighed, the weight would be at once recorded without spending the time necessary to bring the pointer to rest at the zero-point.

The most sensitive balances in coming to rest do not usually lose as much as half a division on each swing, and are consequently slower in coming to rest.

[1] The author has availed himself of the excellent description of the balance and methods of weighing given in *Quantitative Chemical Analysis*, by T. E. Thorpe.

The first few swings, after releasing the beam, are always disregarded; but if afterwards the pointer swings further in one direction than in the other, either the balance is not level, or is out of adjustment.

In using this method to ascertain the state of adjustment of the balance it is advisable not to let the deflection of the pointer exceed five or six divisions each way. The method of determining the position of rest or actual zero-point of the balance is fully described below.

Some assayers work by the method of *no deflection* or *no swing*: that is, when equilibrium is established and the beam is lowered gently on its knife-edge, no deflection of the pointer takes place. "This method, however, is not recommended, as it disregards friction and inertia, and for small weights gives inaccurate results" (Fulton). If carefully conducted, the method of working by noting the amount of deflection of the pointer to the right and the left of the zero, as described above, is more accurate and rapid than the method of bringing the balance to complete rest, and is the method in very general use.

(1) **Direct Weighing.**—To determine the weight of a substance by direct weighing, the balance is first adjusted to equilibrium, then the substance is placed in one pan, and weights are added to the other pan until the substance is exactly counterpoised. As this method is the most expeditious, it is the one in common use. If, however, the balance is not in perfect equilibrium, or the arms of the beam are of unequal length (though this is rare in good balances), the true weight is not registered by this method. This is of no consequence in ordinary operations, since in any one series of weighings the *relative* weights of the different substances only are required. These will be correctly given, provided the different substances are placed always upon the same scale-pan and the conditions of the balance remain unaltered between the successive weighings of the series.[1]

If the arms are of equal length, and the balance is in a perfect state of equilibrium, it is indifferent upon which pan the substance to be weighed is placed. The right-hand pan is the most convenient for manipulation, and in practice this pan is invariably used for manipulating weights when ascertaining the weight of a substance, but when weighing out a definite quantity of a substance it is usual to put the weight on the left-hand pan and manipulate the substance on the right-hand pan.

(2) **Weighing by substitution** gives both the relative and the true weights of a substance, and the results are absolutely accurate, even when the balance is not in perfect equilibrium or the arms are not of exactly equal lengths. In this method the substance to be weighed is first accurately counterpoised; it is then removed, and weights are substituted for it until equilibrium is again restored. It is obvious that the weights thus substituted for the substance will represent the true weight of the substance. "In practice this method of weighing may be facilitated by using one of the larger weights, heavier than the object to be weighed, as a counterpoise, and adding weights to the object until equilibrium is established. The object is then removed, and weights are substituted until the balance is again in equilibrium. The substituted weights express the real weight of the object" (Thorpe). Weighing by substitution is in use in some bullion assay offices where very great accuracy is required.

(3) **The Method of Reversal.**—In this method "the substance is weighed first in one pan, and then in the other. If the weights are identical, the true weight of the object is at once given. If the weights are unequal, their geometric mean will be the true weight: this is found by multiplying the apparent weights together and taking their square root. Practically the common arithmetic mean

[1] Clowes and Coleman, *Quantitative Chemical Analysis.*

of the two weights will be found sufficiently accurate, unless their difference is considerable" (Thorpe).

The Method of Swings.—Within certain limits the deviation from the horizontal in a balance is proportional to the weight causing it. Advantage may be taken of this fact to determine the weight of minute quantities (*i.e.* fractions of a milligramme) of gold or silver with great accuracy. The position of rest of the balance, *i.e.* the true zero or equilibrium point, is first found, and then the value of one division of the ivory scale is ascertained. When the balance is very nearly in equilibrium it is caused to oscillate, and the position of the pointer when at rest is determined from successive observations of the extreme points reached by the pointer when swinging. Disregarding the first swings, so soon as the deflections of the pointer fall within a certain limit their extent, as stated above, commences to decrease at a regular and uniform rate. Let A_1, A_2, A_3, A_4 be the extreme points (end of swing) consecutively reached by the pointer in its deflections, A_1 and A_3 being on one side of the middle of the ivory scale and A_2 and A_4 on the other. Then the position of rest or equilibrium X of the balance is given by the formula—

$$X = \frac{A_1 + 2A_2 + A_3}{4}$$

or with greater exactness by

$$\frac{A_1 + 3A_2 + 3A_3 + A_4}{8}.$$

The following is an example :—

Position of Rest of Empty Balance.—The balance, having been adjusted as nearly as possible, is made to oscillate, and the extreme positions of the pointer in its deflections are observed, counting swings to the left as minus and to the right as plus. (Plus divisions are always reckoned on the side where the weights are placed, minus divisions on the side where the gold or silver is added.) The observations should be made through a lens or telescope. Suppose the swings, after the first two, are in succession as follows: $+ 4{\cdot}4$, $- 3{\cdot}8$, $+ 4{\cdot}2$. The position of rest or true zero is

$$\tfrac{1}{2}\left(\frac{4{\cdot}4 + 4{\cdot}2}{2} - 3{\cdot}8\right) = +0{\cdot}25 \text{ division, or one-fourth of a division to the right.}$$

In the actual determination of the position of rest X an odd number (conveniently five or seven) of successive readings are made so soon as the pointer reaches division 6 on the scale: the arithmetical mean of the half-difference between consecutive pairs of observations gives the position of rest of the pointer along the graduated scale as shown above.

Value of One Division.—If we know the weight corresponding to a given deviation from the zero, the estimation of the minute fraction required for exact equilibrium becomes a simple matter. We have first to determine the values of one division of the graduated scale (*i.e.* the weight required to make the pointer deviate one division) for varying loads. "Suppose a balance at rest in perfect equilibrium, with the pointer exactly over the middle point of the scale. Let the scale be a series of points at equal distances along a horizontal line; then, if a small weight be placed on one pan, the pointer will deviate from its vertical position and come to rest opposite some definite part of the scale, which will depend upon the magnitude of the weight added. The law determining this position is a very simple one; the deviation as measured along the points of the scale varies directly as the weight added. For example, with an ordinarily sensitive balance, such as is used for general purposes, ½ milligramme will move the pointer along, say, four divisions of the scale; then 1 mg. will move it eight divisions; $\frac{1}{4}$ mg. two divisions; and so on. Of course with a more sensitive balance the deviations will be greater" (Beringer[1]).

To determine the value of one division a weight of, say, 0·3 mg. is placed (by means of a rider) on the side used for weights, and the number of divisions traversed by the pointer are noted as described above. Suppose the swings are $-6{\cdot}7$, $+0{\cdot}7$, $-6{\cdot}3$; then the new position of rest or zero is

$$\tfrac{1}{2}\left(+0{\cdot}7 - \frac{6{\cdot}7 + 6{\cdot}3}{2}\right) = -2{\cdot}9 \text{ divisions.}$$

The deviation from the middle point due to the overweight of 0·3 mg. is therefore 2·9 divisions, and one division of the scale would be equivalent to $0{\cdot}3 \div 2{\cdot}9 = 0{\cdot}103$ mg. This value is practically constant from day to day.

If the pointer is not exactly over the middle point when the balance is empty, the slight deviation must be allowed for. Suppose, for instance, that the true zero of the balance was one-fourth of a division to the right (*i.e.* $+0{\cdot}25$), as in the first example, then the deviation due to the overweight of 0·3 milligramme would be $2{\cdot}9 - 0{\cdot}25 = 2{\cdot}65$ divisions: or one division would be equivalent to 0·113 mg. A weight of 0·6 mg. ought to produce double the amount of deviation caused by 0·3 mg.: if any difference is observed, the mean of the two results is taken as representing the true value of one division. In these determinations the greatest care is necessary to preserve the balance under perfectly uniform conditions. The value of one scale division on a very sensitive balance is invariably greater in summer than in winter.

Weight of Minute Quantity of Gold.—The method of determining minute weights by swings or deflections affords a simple means of weighing the minute gold and silver beads obtained from assays, and is sometimes employed by assayers. Suppose the gold is found to require about 1 mg. to counterbalance it, and the swings are now $-4{\cdot}6$, $+0{\cdot}2$, $-4{\cdot}4$. The new position of rest is

$$\tfrac{1}{2}\left(0{\cdot}2 - \frac{4{\cdot}6 + 4{\cdot}4}{2}\right) = -2{\cdot}15.$$

Assuming the position of rest of the empty balance to be $+0{\cdot}25$ and the value of one division to be 0·113 mg., the gold weighs

$$1 - \{(2{\cdot}15 + 0{\cdot}25) \times 0{\cdot}113\} = 0{\cdot}73 \text{ mg.}$$

In the case of very minute gold beads which, without the addition of weights, deflect the needle only a few divisions, the weight of the bead is the difference in deflection between the two points of rest (*i.e.* of empty balance and balance with bead) multiplied by the value of one division. This method of working by swings or deflections adds very considerably to the power of a balance in distinguishing small quantities. According to Sir T. K. Rose,[2] by weighing in this way results correct to 0·002 mg. are easily obtained by using the best assay balances, in which the value of the scale division is 0·005 mg.: and still greater accuracy is attainable by weighing the gold in each pan in succession and by taking the means of a number of observations. Unless, however, very sensitive balances are available, the method of weighing by swings is not generally to be recommended: fractional parts of a milligramme should be determined with the rider.

In working by the method of swings, it is a good rule always to place the button to be weighed on the left-hand pan of the balance, and the weights on the right; count the divisions of the scale from the middle to right and to left, marking the former *plus* and the latter *minus* as indicated above.

[1] C., J. J., and H. R. Beringer, *A Text-book of Assaying*, 15th edn., 1921, p. 132 (Griffin & Co.).

[2] Rose & Newman, *op. cit.*, p. 503.

CHAPTER VI

PRECIOUS-METAL ORES : VALUATION OF ORES

Ores.—" The term *ore* is applied by the metallurgist only to those minerals from which, on a large scale, metals may be obtained with profit."

Ores are always accompanied by more or less extraneous earthy matter, to which the terms *gangue*, *vein stuff*, or *matrix* are applied. The gangue may consist of one or more of the following substances : silica, in the form of quartz ; various silicates, such as felspar and mica, in granite, hornblende, clay-slate, etc. ; carbonates of lime and magnesia, barytes, fluor-spar, iron pyrites, iron oxides, etc.

One of the most striking differences between the ores of the precious metals and those of almost all other metals lies in the extremely small proportion which the gold or silver bears to the worthless gangue with which it is accompanied. For example : Copper ores are regarded as " rich " when they contain 15 per cent. of metal, whilst ores yielding only a few ounces of silver or gold per ton are of commercial value. The average value of the ore now treated on the Rand in South Africa is well under 7 dwt. of gold to the ton, but ore of half that value is treated profitably. The gold and silver in ores are usually referred to as the " values." When ores or metallurgical products of other metals contain gold they are said to be *auriferous*, and when they contain silver they are described as *argentiferous*. Thus a copper ore carrying gold may be described as auriferous copper ore, and one carrying silver as an argentiferous copper ore.

Gold Ores.—Gold occurs very widely distributed in nature, and almost always, but not invariably, in the native, *i.e.* metallic state, usually in veins or lodes in rock formations and in alluvial deposits formed by the disintegration of auriferous rocks. Native gold is never found pure, but sometimes contains less than 1 per cent. of impurities. It is always alloyed with a varying proportion of silver and smaller quantities of copper and other base metals. It is sometimes associated with platinum and metals of the platinum group ; thus gold from Mexico was found to contain as much as 34 to 43 per cent. of rhodium. The quality of native gold and silver is always expressed in parts per 1000, pure gold and pure silver being described as " 1000 fine." Thus an alloy containing 90 per cent. of gold or of silver and 10 per cent. of base metal would be described as " 900 fine."

The following analyses will serve to show the general composition of native gold :—

	Australia.	California.	West Africa.	U.S.S.R.[1]	Wales.
Gold	94·64	89·10	97·81	98·96	89·83
Silver	4·95	10·50	2·19	0·16	9·24
Copper	...	...	...	0·05	...
Iron	0·41	0·20	...	0·35	...
	100·00	99·80	100·00	99·52	99·07

[1] Formerly obtained at Ekaterineburg (Sverdlovsk) in the Urals.

The purest native gold yet found is that from the Pikes Peak Mine at Cripple Creek, Colorado, which contained 99·9 per cent. of pure gold.[1]

The average fineness of gold from different countries is as follows :—

	Gold.	Silver.	Base.
South Africa (banket ore)[2] ...	875	125	...
Western Australia	831	107	62·0
Australia[3]	921	39·0	40·0
„	868	89·0	43·0
South America (Andes) ...	600–700	...	...
Colorado	800	200	...
California	880	...	...
Canada	850–900	...	...

Native gold generally occurs in the form of scales, threads, or grains disseminated through the ore; nuggets are rare. It is sometimes visible to the eye without magnification, but is more often in too fine a state of division to be seen. Gold also occurs in copper and iron pyrites and with other metallic sulphides and arsenides of base metals, most probably in the metallic state, not as a sulphide. Gold is also met with in combination with tellurium as telluride. The tellurides of gold, containing the compound $AuTe_2$, are very common.[4] They occur in considerable quantities in Western Australia, Colorado, and Transylvania, and have been reported from many other localities. Various mineralogical names have been assigned to the tellurides from different localities, the best known being *calaverite*, which has the composition $AuTe_2$; *sylvanite* or graphic tellurium, which appears to be a variable mixture of $AuTe_2$ and Ag_3Te_4; *Petzite*, Ag_2Te, in which some silver is replaced by gold; and *nagyagite*, or foliated tellurium, which contains a considerable percentage of lead. The tellurides are for the most part dark grey or black in colour, rarely silver-grey. They are often mixed with metallic gold, which sometimes gives them a brassy-yellow colour.

The greater part of the rock material or " gangue " forming the veins or lodes in which gold is found is usually quartz, but various silicates also occur, and sulphides, arsenides, or antimonides of base metals are almost invariably present. The upper portions of these veins near the surface of the ground are often found to have undergone oxidation from the influence of the atmosphere, so that the sulphides of iron, etc., have been converted into oxides. These oxidized portions are termed " gossans," and consist of a loose mixture of silica and iron oxide, and the finely divided gold contained in them is often readily separated.

On the Witwatersrand (Rand), South Africa, the gold ores consist almost entirely of " banket " formation, a conglomerate consisting of water-worn pebbles of translucent quartz set in a matrix composed mainly of iron oxides and silica, known as "puddingstone." The gold exists in the cement, the pebbles being generally barren. The ore contains on an average about 7 dwt. gold per ton. Early assays of ore from the reef discovered at Odendaalsrust in the Orange Free State in 1946 proved it to be of unprecedented richness.

Veins of iron and copper pyrites, galena, argentite, etc., traversing crystalline rocks are frequently sufficiently auriferous to permit of the profitable extraction

[1] Furman in *Colliery Manager and Metal Mines*, October, 1896, p. 89.
[2] R. B. Young, *Proc. Geol. Soc. of S. Africa*, 1911.
[3] " Gold deposited at Sydney and Melbourne Mints," from *Mint Report*, 1910.
[4] Sir T. K. Rose, *The Precious Metals.*

of the gold by smelting processes, although the gold may be present in insufficient quantity to render it visible in the ore.

Auriferous Black Sands.—The heavy sands derived from gold-bearing gravel deposits are generally known as black sands by reason of the preponderance of dark-coloured iron minerals they contain; The black sand from some districts consists largely of titaniferous iron ore as well as chromite. These sands are occasionally rich in gold and to a less extent in platinum. Most of the value is found in the particles below ⅛ inch diameter.

Minerals resembling Gold.—Several minerals, when disseminated through rock material, resemble native gold in appearance and are sometimes mistaken for gold by those not familiar with minerals.

Yellow mica, which is found in some sands on the western coast of America and in other places, looks remarkably like gold ; but when the sand is washed in an iron pan, the mica, being very light, floats off, even before the grains of quartz that accompany it, and can thus be easily distinguished from gold, which on account of its density always remains on the pan. Copper pyrites and iron pyrites also resemble gold, especially when they are slightly oxidized on the surface. When these sulphides are heated they are decomposed and converted into oxides and are thus readily distinguished from gold, which retains its colour on heating and is not altered.

Silver Ores.—Silver occurs in the native state, but unlike gold it is generally combined with sulphur, chlorine, bromine, and iodine in its ores. It occurs also in combination with other metals, as arsenic, antimony, etc., and with mercury as a native amalgam, whilst also frequently occurring in ores of other metals, such as lead, copper, zinc, etc., constituting argentiferous ores of these metals. The latter class of ores frequently contains sufficient silver to render its extraction profitable, in which case such minerals are also considered as ores of silver. The silver occurs in veins traversing gneiss, granite, clay-slate, quartz, calc-spar, mica-schist, etc., in ores which are usually associated with ores of lead and copper, blende (ZnS), spathic iron ore, brown hæmatite, pyrites, ores of nickel and cobalt, heavy-spar, quartz, earthy carbonates, etc. The ores of silver are more or less abundantly distributed in all parts of the world.

Native silver occurs in sufficient quantity to constitute one of the chief ores of the metal. Native silver is rarely found perfectly pure, but usually contains also gold, copper, platinum, or other metals in larger or smaller proportions, and is itself always present in small quantities in native gold.

Native silver usually occurs in laminated masses or capillaries and in grains minutely disseminated through silver ores and other minerals. Large masses are rare.

Amongst the more important natural compounds of silver are the following :—

Argentite, or silver-glance, Ag_2S, is one of the most important ores of silver. It occurs in black soft sectile crystals and masses, which, when pure, contain 87·1 per cent. of silver, but it rarely occurs pure as an ore, being often associated with the sulphides of lead, copper, iron, zinc, antimony, arsenic, and tin, or with the ores of nickel and cobalt : so that the ore containing this mineral as its source of silver yields much smaller proportions of silver than the pure sulphide. Silver-glance constitutes one of the richest, most abundant, and usual forms of occurrence of silver found in veins, and sometimes in large masses. It occurs in great quantities at the Comstock Lode, Nevada, Mexico, and in many other localities. Other widely-distributed silver sulphides are :—

Ruby Silver Ores.—Proustite, $3Ag_2S + As_2S_3$; pyrargyrite, $3Ag_2S + Sb_2S_3$; and stephanite or brittle silver ore, $5Ag_2S + Sb_2S_3$.

Fahl ore, or grey silver, another important ore, is a mixture of antimony and arsenic sulphides, with sulphides of silver, mercury, copper, zinc, and iron. Its formula is $4(Ag_2Hg,CuFeZn)S(SbAs)_2S_3$. It may contain any proportion of silver up to 31 per cent., arsenical grey silver being poorer than the antimonial varieties. Ores consisting of fahl ore are sometimes very rich in silver, containing occasionally as much as 1400 oz. of silver per ton of ore, but the more usual content of these ores is from 2 per cent. to 4 per cent. of silver. Fahl ore occurs in Germany, the Tyrol, Bolivia, Nevada, Mexico, Colorado, etc.

Polybasite, $9(Ag_2Cu)S + (SbAs)_2S_3$, is also a commonly occurring mineral, especially in America.

Stromeyerite is an argentiferous copper-glance, sulphide of silver and copper, Ag_2S,Cu_2S. The pure mineral would contain 53·08 per cent. of silver, but it is usually more or less mixed with copper-glance (Cu_2S), and hence yields only from 3 per cent. to 30 per cent. of silver. Other sulphide minerals containing silver also occur, but they are less important than those given above.

In combination with tellurium it occurs as *hessite*, Ag_2Te; and *petzite*, a telluride of gold and silver; and combined with chlorine as *horn silver* or kerargyrite, AgCl; and *embolite*, Ag(ClBr), all of which are of some importance as ores. Horn silver occurs as a soft, horny-looking mass and accompanies other silver ores somewhat abundantly in Peru, Chili, and Mexico.

Silver also occurs in the ores of lead, copper, zinc, nickel. Galena, PbS, often contains considerable amounts of silver (up to 5 per cent.), and zinc blende, ZnS, chalcopyrite, CuS,FeS,FeS_2, bournonite, and other minerals also yield more or less silver.

Owing to the comparatively high value of gold and of silver, and the facility with which they are extracted from ores or minerals containing them, their separation is rendered profitable, although the ores contain only an exceedingly small percentage of the metal. Thus gold ore containing only 4 dwt. of gold per ton is profitably treated in South Africa; and in the case of silver, while certain of the silver minerals in their pure state contain a very high percentage of the metal, yet the ores usually treated in the smelting works of Europe and America contain only a comparatively small proportion of silver.

Argentiferous galena is extensively treated for silver, the facility with which the silver may be concentrated in the lead and subsequently separated by the process of cupellation rendering its extraction profitable if it be present to the extent of only 2 or 3 oz. per ton of lead, and a similar remark extends to the extraction of silver from certain copper ores. The amount of gold and silver in precious-metal ores varies considerably, and no exact statement can be made as to what amount of precious metal will constitute a rich or high-grade ore and a poor or low-grade ore. An ore that is considered rich in one mining district may be regarded as poor in another district. For assay purposes, however, a gold ore is as a rule considered to be rich if it assays over 3 oz. of gold per ton, and poor if the gold is less than 10 dwt. per ton: between these amounts the ore would be considered to be of average quality.

In the case of silver, an ore containing over 200 oz. of silver to the ton would as a general rule be regarded as rich, while an ore with any amount under 50 oz. would be classed as poor. Average ores would contain amounts between these.

Valuation and Purchase of Ores.[1]—The following information has reference to pre-war conditions, but may be taken as representing the general procedure

[1] Consult also "Notes on the Valuation of Ores, Concentrates and Smelter Products," L. C. Stuckey, *Trans. Inst. Min. Met.*, March, 1922.—H. K. Picard, *Costs and Marketing of Copper* (Common Commodities of Commerce Series, Copper), 1916.—C. H. Fulton, *Buying and Selling of Ores and Metallurgical Products*, 1915, Bureau of Mines, Washington, Technical paper No. 83.

5

in normal times. Changes however will unquestionably be made, especially in prices, in the post-war years and will be duly noted in the mining press. Changes in industrial procedure have also to be allowed for ; *e.g.*, the tendency in recent years has been to remit the fine formerly imposed for undesirable constituents in a smelting ore.

It is very necessary that the assayer should be familiar with the principles of ore valuation. The value of an ore depends upon the quantity and nature of the metal it contains and the ease with which this can be extracted. The *gross* value of an ore, based on the market price of the metals contained, is termed the " assay value," as it is calculated from the result of the assay. The " assay value " merely indicates the money value of the ore if the precious metal could be extracted *without cost*. The " assay value " must, therefore, be distinguished from the real value of the ore, which is the gross value less the total cost of extraction ; in other words, the net return to the proprietors. It is necessary to take note of the distinction, as the term " value " is frequently used indiscriminately as a synonym for " assay value " to indicate the *gross* value of an ore, thus leading to confusion. For instance,[1] an ore that contains 7·6 dwt. gold per ton has not the " value " of 7·6 dwt., which is about 32s., for from that amount must be subtracted the cost of treatment and realization. In some cases the " assay value " does not even roughly indicate the actual value of the ore. Assay values of gold and silver ores and products are usually represented in ounces and decimals or in pennyweight and decimals per ton ; the method of representing the assay values in ounces, pennyweight, and grains per ton has been discontinued. The values are expressed in terms of fine gold and fine silver respectively, not as " bullion." In the U.S.A. the assay values of gold and silver ores, etc., are represented in money value in dollars per ton according to the prevailing market price of the metals.

Owing to the higher price of gold and silver compared with other metals, the valuation of precious-metal ores and products is carried to a higher degree of accuracy than in the case of ores of lead, copper, zinc, etc. In the sale of ores, both the gold and silver are always paid for when present above a certain amount, which varies with the different classes of ore and in different mining centres. As a general rule, gold is not paid for if below $\frac{1}{20}$ (·05) ounce per ton, and silver if below 2 ounces per ton. The minimum amount of silver paid for varies, however, with the richness of the ore or product. In the purchase of ores in the United States for smelting, the gold is usually priced at 20·0 dollars per ounce, as this is more convenient for calculation and is the price adopted by smelters.[2]

The market price of silver fluctuates considerably, and the value of the silver in ores, etc., is always determined by the prevailing market price of fine silver. The price of fine silver may be taken at 2d. or 2½d. above the London market quotation for " Standard " silver, which contains 925 parts of silver and 75 parts of copper per 1000.

As indicated above, the value of precious-metal ores is not merely dependent on the amount of gold and silver they contain, but also on the cost of treatment, which varies with the nature of the ore and method employed. A deduction from the gross value of the ore is always made to cover the cost of treatment.

In the case of ores treated by smelting processes the general rule is for the smelting company to buy the ores from the various mines and smelt them in

[1] Consult note on " Values," *Mining Magazine*, 1911, **4**, pp. 251–252.

[2] See " Table of Assay and Coinage Values for Gold," by R. J. Holland, *Min. Mag.*, **11**, 1905, 525–530.

consideration of a smelting charge which is deducted from the gross value of the ores when paying for the same to the mines. The decline in the price of silver and other metals within recent years has brought down the net value of ores to the miner, but, on the other hand, the cost of mining and smelting has been reduced by improved methods of treatment, better transport, cheaper labour, etc., enabling ores that were formerly worthless to be sold and treated at a profit.

Ores in which the precious metals exist in such a state that they can be recovered readily by pulverizing and amalgamation are said to be *free-milling*, while ores that require some preliminary treatment before the precious metals can be extracted by ordinary methods of amalgamation, or that resist the action of heat and chemical reagents and are therefore difficult to treat, are said to be *refractory*. Ores that can be smelted readily are sometimes termed "docile" ores.

When the precious metals are to be extracted by smelting processes the value of the ore is considerably affected by the nature of the gangue material, as this largely decides the cost of treatment. In the large majority of cases the predominating gangue material is *quartz* (*silica*), which alone fuses with difficulty and for smelting purposes requires the addition, in suitable proportions, of fluxes such as iron oxide and lime (calcium oxide), to convert it into a fusible compound or slag; consequently, from the smelter's point of view, the proportion of silica in an ore very largely determines its value. The value of an ore is also influenced by the presence of metalliferous minerals. *Lead* and *copper*, if present in any considerable amount, are always paid for by the smelter, as they are recovered in the smelting operation, and in addition to the value of the metals themselves one or the other is indispensable to the operation, as they are utilized as agents for the concentration of the precious metals disseminated through the furnace charge. When the ores are smelted, the lead or copper they contain is reduced to metal, which sinks slowly through the melting material in the furnace and alloys with the gold and silver, thus collecting and concentrating these in the mass of lead or copper which accumulates at the bottom of the furnace, forming what is termed "base bullion."

Zinc is frequently present in precious-metal ores, but is not paid for, as it is not recovered in smelting. When present to the extent of 10 per cent. or more it depreciates the value of the ore, as it gives trouble in the smelting operation, as described below under "Treatment Charges."

The following system of valuation employed on the American continent, where the extraction of silver and gold from their ores by smelting processes is conducted on such a large scale, is, with slight differences in detail, almost universally adopted for the valuation of precious-metal ores to be treated by smelting processes.[1] The gross value of the ore is calculated from the amount of the different metals present, and the prevailing market price of the respective metals. The assay value of the gold, silver, and lead is always determined by the dry or fire assay. Copper when present is almost invariably determined by volumetric or electrolytic methods.

Silver is always paid for on the assay value at a given rate per ounce (troy) based on New York quotations, with a reduction of 5 to 10 per cent. to cover loss in smelting. When the price of the metal is high, only 90 per cent. of the contents of medium and low-grade ores is paid for; but when low, 95 per cent. is almost invariably paid, as in the case of high-grade ores. Assays are reported to $\frac{1}{10}$ ounce, and payment is made accordingly.

[1] H. F. Collins, "Smelting Processes," *Proc. Inst. Civil Eng.*, 1893, 112.

Gold also, when amounting to $\frac{1}{20}$ ounce and upwards per ton of ore, is paid for at 95 per cent. of its full market value.

Lead.—The manner of paying for the lead content is variable, the principle being to pay for what can be recovered, less refining charges and freight to the refinery; but a sort of sliding scale is adopted according to the richness of the ore, so as to cover increased loss in working poor ores. Thus, for example, the lead in poor ores containing between 10 per cent. and 20 per cent. will be paid for at, say, 20 cents to 30 cents per unit (*i.e.* 1 per cent. of a short ton, or 20 lbs.), whilst between 30 per cent. and 40 per cent. it will be paid for at 40 cents to 45 cents per unit, the price of course varying with the market price of the metal. Under 5 per cent. of lead is not paid for.

Copper is not paid for in lead ores (and this is one of the sources of profit), but when the ores are bought comparatively rich in copper and poor in lead, the former metal is paid for at a settled rate per unit according to the market price of copper, in which case usually nothing is paid for lead, whatever its amount, the practice of not paying for both metals together always leaving a good margin for profit in refining. A sliding scale is usually adopted for the payment of copper, as in the case of lead. For example, a low price is paid per unit for copper under 5 per cent., a little higher price per unit for copper from 5 to 10 per cent., and a still higher price when over 10 per cent., and so on.

Under favourable circumstances, and if worked on a sufficiently large scale, ore containing as little as 2 per cent. of copper may be worked at a profit: but in such cases the small amounts of silver and gold which often accompany the copper in the ore, and are recovered therewith, are usually the determining factors of commercial success.[1] This is equally true of the ores of other metals.

Abbreviations used in Ore Sales.—In the sale of ores, etc. certain abbreviations are used with which the student should be familiar. Thus, for export business the price quoted is usually "c.i.f.," that is, cost, insurance,and freight to some European or other port. Another abbreviation is "f.o.b.," free on board. "R/C" is the "returning charge" or smelting expense, which will vary according to the richness of the ore; the lower the produce or yield of metal, the higher will be the returning charge.

Charges for Ore Treatment.—To cover the cost of treatment, the smelter makes certain charges which are deducted from the gross value of the ore. The losses of gold and silver in treatment are usually assumed to be 5 per cent. The treatment charges vary enormously in different smelting centres according to the class of ore to be dealt with and the quantity and quality. They are also affected by local conditions such as fuel-supply, cost of labour, and freight. The smelting rate is chiefly fixed according to the fluxing qualities of the ores, and this, as previously stated, is regulated principally by the percentage of silica. Ores containing excess of fluxing constituents such as iron oxide, lime, etc., are termed *fluxing ores*. In districts where fluxing ores are not very plentiful they are in great demand as fluxes for the much more plentiful siliceous ores, and to procure them, smelters will make only a nominal charge for treatment; whilst a very siliceous ore may have an almost prohibitory smelting charge put upon it or perhaps be refused altogether. In general it may be stated that when lead ores and "fluxing ores" are plentiful the treatment charges on them rise, whilst those on siliceous and "dry ores" fall, and *vice versa*. As a rule, the charges for smelting are arranged on a sliding scale according to the amount of silica present and other constituents in the ore. As a basis for calculation a price is from time

[1] H. K. Picard, *Costs and Marketing of Copper*, p. 9.

to time fixed for an assumed ore with neutral or self-fluxing gangue; that is, one in which the slag-forming components balance, leaving no excess of either silica or iron oxide, etc. An extra charge or penalty is made for each unit or per cent. of silica in excess of the neutral basis, and an allowance granted per unit when it is below the iron oxide, lime, etc. Manganese if present is reckoned as iron. A small allowance is made for all lime, magnesia, and baryta present. Ores containing zinc are very intractable, as part of the zinc enters the slag as oxide, and renders it both more pasty and more infusible. Smelters therefore always make a deduction (in America usually 50 cents) from the value of the ore for every unit of zinc in excess of a certain amount, usually 10 per cent., sometimes only 8 per cent. Special deductions (about 5 per cent.) are sometimes made in the case of arsenic. The necessity for determining the silica, iron, etc., is judged of according to the apparent mineralogical composition of the ore. From the foregoing remarks it will be evident that whilst only gold, silver, lead, and copper are paid for or mentioned in the bid of a smelting company for any particular lot of ore, every element in its chemical composition is taken into consideration in calculating its value, and extra charges (penalties) made or allowances granted on a sliding scale based upon the fixed smelting charge for the assumed neutral ore as stated above.

The method of calculating the value of an ore is well illustrated in the following Tables I and II, given by Brunton[1] for the valuation of ores in the large smelting district of Denver or Pueblo, Colorado. The rates, of course, are changed from time to time, and the actual prices paid for silver and lead vary daily with quotations from the New York markets, but the figures serve to illustrate the method. The debit column includes penalties charged for objectionable constituents. The credit column includes the value of the metals present and the allowances granted for "fluxing" constituents. The fixed smelting charge for the assumed neutral ore is 6 dollars per ton.

TABLE I.—TABLE OF PRICES AND RATES (DENVER, COL.).

Ore.	Debit.	Credit.
	Dollars.	Dollars.
Au per oz.	...	19·00
Ag ,, (per cent. N.Y. quotation)	...	...
Pb per unit (with N.Y. quotation 4 cents per lb.) ...	...	0·50
Fe ,, ...	...	0·15
CaO ,, ...	...	0·10
Mn ,, ...	...	0·15
SiO_2 ,, ...	0·13	...
$BaSO_4$,, (small percentages occasionally disregraded)	0·05	...
Zn ,, ,, ,, ,, ,,	0·30	...
S ,, ,, ,, ,, ,,	0·15	...

As an example, an ore of the following composition,

Au	Ag	Pb	SiO_2	CaO	Fe	S	Zn	$BaSO_4$	Mn
Oz.	Oz.	per cent.	per cent.	per cent.	per cent.	per cent.	per cent.	per cent.	per cent.
0·3	30	5	14	8	6	12	7	24	4

would be calculated by the table given above in the following manner, showing

[1] In discussion on paper by H. F. Collins, *ibid.*

that after deducting the total charges, amounting to \$12·92, the ore would have a net value in Denver of \$14·68 per ton :—

TABLE II—Valuation of an Ore by Method given in Table I.

Prices :—Silver, 60 cents per oz. ; lead, \$4·00 per 100 lb.

Ore.	Debit.	Credit.
	Dollars.	Dollars.
Au, 0·3 oz. at \$19·00	...	5·70
Ag 30 oz. at 95 per cent. of 60 cents (\$0·570) ...	...	17·10
Pb, 5 per cent. at 50 cents	...	2·50
SiO_2, 14 ,, at 13 ,, ...	1·82	...
CaO, 8 ,, at 10 ,, ...	...	0·80
Fe, 6 ,, at 15 ,, ...	...	0·90
S, 12 ,, at 15 ,, ...	1·80	...
Zn, 7 ,, at 30 ,, ...	2·10	...
$BaSO_4$, 24 ,, at 5 ,, ...	1·20	...
Mn, 4 ,, at 15 ,, ...	...	0·60
Smelting charges per ton	6·00	...
Net value per ton in Denver or Pueblo	14·68	...
	27·60	27·60

The precious-metal ores smelted in England are exclusively confined to imported ores. Imported silver ores average about 250 to 300 oz. per ton, but run up to 3000 or 4000 oz. and down to 50 oz. per ton. The valuation of the ores and deductions for smelting charges are made in a similar manner to that already described.

Terms of Settlement in Ore Deals.—The terms of settlement for the purchase of ores, etc. are always arranged between the purchaser and the seller before a lot is dealt with. The terms vary somewhat according to the nature of the ore, size of the lot, and other conditions, but the following description may be taken as representing the general practice. In the ordinary method of assaying ores that are sold, the ore is sampled and the final sample thoroughly mixed in the presence of representatives of the buyer and seller. Each representative then usually selects four or more portions (sometimes less), which are each wrapped in a separate packet and sealed. The details of the lot, number and weight, etc., are written on each packet. Some of the packets are delivered to the assayer acting for the purchaser and some to the assayer acting for the seller. A certain number of the packets, agreed upon by buyer and seller, are held under seal in case of dispute, and this is called the " referee " or " umpire " sample.

In Custom assay offices (see p. 12) in the U.S.A., the final sample, weighing about 2 to 3 lb., is usually subdivided into four portions, each of which is put into a paper bag and sealed. Two portions are assayed by the seller, one by the purchaser, while the fourth portion is retained under the seal for an " umpire " assay, should such become necessary. In America the assays for the purchase and sale of ores are called " control assays " or simply " controls." The control assay, both by the purchaser and seller, is made in duplicate or triplicate and an average of the assays in either case is taken for comparison. If the results of purchaser and seller agree within a certain limit, depending on the value of the ore, settlement is made on the purchaser's assay, or sometimes on the average of the two assays. If these assays do not agree within specified limits, it is the practice of the purchaser and seller to re-assay their own samples, or to exchange samples and re-assay. If the results of the seller and the purchaser do not agree when the samples are re-assayed, the "referee" (or umpire) sample is sent to an assayer mutually agreed upon, and his result is decisive, both parties abiding by his result, on which settlement is made. The party whose result is farthest away from the umpire's result has to pay for the assay. The charge made for referee or umpire assays is usually a fee and a half or twice the ordinary fee, and sometimes more, according to the professional position of the referee.

As a matter of fact, in the buying and selling of ores the reference to an umpire is not frequent, and usually the seller and the purchaser agree as to the values within prescribed limits on the first attempt. The assays made by assayers representing the buyer and seller of an ore are expected to agree within 0·03 to 0·04 oz. per ton in the case of gold, each 0·01 oz. being termed a *point*, and usually about 2·0 oz. in the case of silver. The amount of the difference to be allowed is mutually agreed between purchaser and seller.

For copper they are expected to agree to within half a unit, that is, 0·5 per cent.

In some cases it is previously agreed that if the umpire's result is outside the assays of the purchaser and seller it shall be rejected, and the average of the buyer's and seller's results taken as the basis of settlement, or the lot may be resampled and the same process repeated.

It is usual for the sellers to appoint some firm of assayers to look after their interests, whose experienced representative attends the works of the smelter on the arrival of the ore, superintends the weighing of the parcel and its sampling, and forwards an accurate sealed sample to his principals for assay on behalf of the seller. Upon an appointed day the buyer and seller exchange the results of their respective analyses and agree as to a "settled" assay. Usually, it is stipulated in a copper-ore contract that assays are to be made by the electrolytic method, and 1·3 units deducted to arrive at the "dry" assay on which payment is made. The "dry" or smelting assay is presumed to indicate more nearly the amount of copper recoverable by smelting methods. The deduction of 1·3 units is assumed to correspond with the amount of irrecoverable copper.[1]

The bona-fides of the seller being beyond question, payment is in many cases arranged as to 80 per cent. of the approximate value on or against the delivery of the ore, the balance becoming due on "settlement of assays," etc.

In most cases it is arranged that the purchase is for copper ore said to contain about — per cent. copper, and the R/C basis is fixed accordingly. Should the settled assay turn out below the contract figure, an increased R/C would be demanded by the buyer; it is therefore necessary that the assay of the bulk-ore shall be an accepted figure, arrived at by exchange of assay figures, according to definite trade custom. It is also understood that the copper ore offered for sale shall be of good quality and contain no deleterious elements —if such be present, the buyers will reject the parcel or exact a penalty for the impurity.

[1] H. K. Picard, *Copper*, p. 118.

CHAPTER VII

SAMPLING PRECIOUS-METAL ORES AND MATERIALS

Definition.—" By the word ' sample ' is meant a small portion of a mass which accurately represents the mean composition of the whole of the mass " (Percy).

Introductory.—It is to be regretted that the word " sample " is frequently used in a loose manner as a general term for a small portion of material. It would avoid confusion if its use were restricted to such small portions as are truly representative of the bulk. Moreover the terms " sample" and " specimen " are often applied indiscriminately to a part of some material that has been selected to show the quality, etc. of the whole, but there is a very special difference in the meaning of the terms from the assayer's point of view. Thus a few lumps of copper ore may be selected haphazard as representative specimens of the nature and general quality of the ore, but such lumps cannot be regarded as a sample from which the true percentage of metal in a ton of the ore can be determined.

It may be well at the outset to mention that there are very few assay offices which do not occasionally receive samples which are not truly representative of the ore mass from which they are taken. They consist usually of pieces of ore, selected because they are worse or better than the average of what they are meant to represent.

Assays of such individual " specimens " may be accurately performed, but the results are of little value if the assayer has not operated upon an *average* sample. For example, it may be stated, in a prospectus, that the assay of a portion of quartz from a certain mine is 50 oz. of gold to the ton, and this may be certified by an assayer whose name would be sufficient guarantee for a bona-fide *assay*. Yet this report is generally of no practical use as an indication of the value of the property, simply because there is no guarantee that the portion of quartz is an average of what there is in the mine: in other words, the portion is not representative, and cannot therefore be truly labelled a " sample."

An assayer who values his reputation, and wishes his report to be above doubt and suspicion, will not put his name to any report without having first ascertained on good evidence that the samples submitted to him are really average samples of the ore.

The object of an assay is to ascertain the true value of the material represented by the sample ; it is therefore absolutely necessary that the small quantity operated upon should accurately represent the average composition of the ore-heap, etc., from which it is taken ; in other words, it must be a *true* sample, otherwise the careful work of the assayer is worse than useless, because the assay result is misleading.

The selection of a small but truly representative portion may be regarded as the first step in determining the composition and value of a large quantity of material. To select a bona-fide sample may seem an easy matter, but those who have had experience testify that it is not so easy as it looks, and generally regard the selection of the sample as a more difficult matter than the actual assay of it. The work of sampling is too frequently entrusted to subordinates ; but although the manual labour may be left to them, the principles on which it is conducted and the precautions which need to be taken are not unworthy the study of experts, and the work should be conducted under experienced supervision.

The extreme importance of correct sampling, and of the utmost care in assaying, are emphasized when it is borne in mind that the sample submitted for assay may represent many tons of rich ore, etc., and large sums of money may be involved in the settlement made on the assay results ; consequently, a want of care in the selection of the sample or in conducting the assay may give rise to errors which may mean considerable monetary loss to the buyer or the seller.

Sampling and assaying are closely related, but they are usually performed by different persons. The selection of the sample is done by men termed "samplers," whose special work it is to select a portion which is truly representative of the mass. An assayer is not as a rule responsible for the accuracy of the sample submitted to him, his duty being simply to ascertain and report its value. But, as pointed out by Beringer, "although 'sampling' is thus distinct from 'assaying,' the assayer should be familiar with the principles of sampling, and rigorous in the application of these principles in obtaining, from the sample sent to him, that smaller portion upon which he performs his operations."

It is also to the advantage of the assayer to have some knowledge of the methods in general use for sampling large lots of ore.

Since the introduction of mechanical sampling into many smelting-works, and the establishment of public sampling-mills in most of the great mining centres where the accuracy of the sample is guaranteed by the works, the samples representing large lots of material are in many cases handed to the assayer in sealed paper bags, and need very little preparation for assay purposes.

But in general assay-office practice the samples submitted for assay are more often received in small lots, weighing from 10 to 100 lb., either in the state of coarse powder or as lumps of various sizes, and the assayer requires a knowledge of sampling to enable him satisfactorily to reduce the bulk of the sample to a convenient size for assay.

Principles of Sampling.[1]—The fundamental principle of good sampling may be said to be "a gradual reduction in bulk simultaneously with a gradual reduction in the size of the pieces" ; or, in other words, the bulk should not be reduced a second time without first crushing to a degree of greater fineness. This principle is well illustrated by Beringer[2] in the following example :—

Taking a heap of ore, A, and selecting, say, every twentieth shovelful (*i.e.* 5 per cent.) as the ore is shovelled into the ore-bins, there is obtained a second heap, B, containing $\frac{1}{20}$ of the heap A. If the ore in the heap B is then crushed until it contains approximately the same number of pieces that A did—which means crushing every piece in B into about twenty pieces—B will become the counterpart of A. Selecting in the same manner $\frac{1}{20}$ of B, there is obtained a third heap C, which in turn is crushed and $\frac{1}{20}$ selected as before. This alternate

[1] The information on the principles of Sampling is mainly derived from papers on "Sampling" by T. Clarkson, A.R.S.M. ; see *J. Iron and Steel Inst.*, 1893, **44**, 131 ; *J. Soc. Chem. Ind.*, 1894, **13**, 214.

[2] *Text-Book of Assaying*, 15th edn., p. 2.

reduction and pulverizing is carried on until a sample of suitable size is obtained. This may be expressed very clearly thus :—

A, 1000 tons of ore in lumps.
B, 50 tons of large pieces, $\frac{1}{20}$ of A or 5 per cent. of A.
C, 2·5 tons of small pieces, $\frac{1}{20}$ of B or ·25 per cent. of A.
D, 0·125 tons of coarse powder, $\frac{1}{20}$ of C or ·0125 per cent. of A.

The conditions affecting and governing the selection of a true average sample from a large bulk of material are summarized by Clarkson as follows :—

(1) The uniform composition or thorough mixing of the bulk.
(2) The fineness of division, or the size of the pieces forming the bulk.
(3) The mode of occurrence of the metallic constituents.
(4) The ratio of weight of sample to the fineness of division.
(5) The ratio of weight of first sample to total weight of bulk.
(6) The method of cutting out the sample.

These conditions are mutually related and interdependent, and are, of course quite independent of the question of the relative merits of hand and machine sampling.

(1) Uniformity of Composition.—It is evident that, if the bulk of material is thoroughly uniform, a portion taken from any part at random would be truly representative of the whole, quite irrespective of all other considerations. This is the ideal simple case, which seldom, or probably never, occurs in practice.

It is common knowledge to all acquainted with the constitution of ores that the valuable metallic constituents are not uniformly disseminated throughout the whole mass, but generally occur in separate and distinct materials that are associated in a greater or lesser degree with non-metallic and valueless minerals, such as quartz, etc. Thus, in a heap of ore, one piece may contain absolutely no metal, another piece may have 60 per cent., and another may be made up partly of valuable mineral and partly of waste rock, so that the average percentage may differ very considerably from that of any particular portion.

To obtain a true sample of such a heap, so as to ascertain how much metal is actually present, it is necessary to take some of the richest, some of the poorest, and some of the intermediate value ; and further, the quantity of each kind taken to make up the sample must be carefully regulated so that the ratio between the different kinds is the same as in the original bulk. For example, if there are twice as many rich pieces as poor ones in the bulk, there must be twice as many in the sample, and so for the pieces of other values.

In general, the composition of ore-heaps is far from uniform throughout the bulk ; and in some cases, owing to the difference in the size and specific gravity of the various constituents there is a great tendency for them to separate, the fine and heavy material going to the bottom.

The difference in the size of the pieces forming the bulk is also an important factor, as the fine material is almost invariably different in value from the coarse ; sometimes one is richer, and sometimes the other ; hence it is very necessary, in selecting a sample, to keep the fine and the coarse in the same proportion as in the bulk.

To ensure uniformity of composition as far as possible, it is very important to mix the bulk thoroughly, especially in the case of " spotted ores " (*i.e.* ores of very irregular composition), for unless the mass is well mixed so that the fine and coarse, and heavy and light, particles are kept together in their proper proportion and relation to each other, it will require a larger percentage to be cut out to obtain a truly representative sample.

Much difficulty is frequently experienced in thoroughly mixing ores that contain the highest values in the finest portions, and to overcome this difficulty it is the practice in some sampling works to damp the ore at the beginning so that the fine and coarse will adhere and thus prevent the coarse from separating out by rolling to the bottom. This practice has been found to give very satisfactory results with rich material.[1]

(2) **The proper fineness of division,** or the size of the single pieces forming the bulk before a cut for sample can safely be made, is controlled by the size of the bulk and (3) the **mode of occurrence of metallic constituents.** Small lots should always be crushed fine before cutting.

If the total bulk is very large and uniform in composition, as in a cargo of pyrites or iron ore, the pieces may be proportionately large, provided that the first sample selected is of fair size, say from 2 to 10 per cent. of the whole, according to the material ; but in the most difficult cases of sampling, where the metal is carried by only a comparatively few rich pieces in a large mass of barren material, it is absolutely imperative to reduce everything to small size before taking even the first cut. This has more distinct reference to the ores of the precious metals, especially silver ores, which require great care in sampling, owing to the presence of minerals rich in silver and very irregularly distributed. After the first sample has been obtained, it is, according to the aforementioned principle of gradual and simultaneous reduction, generally necessary to crush it finer before making a second cut, unless the original material was fine. In that case, two or more cuts may be made before re-grinding, according to the original degree of fineness.

The importance of particle size in samples of certain metallurgical materials has been emphasized in the report of a special investigation by Scherrer and Lundell[2], in which it is pointed out that the sampling of such materials gives rise to particles that differ not only in size but also in composition. The extent of the differences is shown, and the procedures that must be followed in selecting the sample for analysis are outlined.

The necessity of first crushing to a finer state of division before making a second cut, in cases where the pieces are large, is illustrated by the following example. " Let it be assumed that a lot of ore has been crushed to cubes of 1 inch average size, and that to obtain a correct sample at this size it is necessary to cut out, say, 25 per cent. If the sample, representing 25 per cent. of the original mass, is reduced in bulk by one-half, *i.e.* 50 per cent., without first recrushing, it is obvious that it simply amounts to taking out $\frac{1}{8}$ or 12·5 per cent. of the original mass in two cuts instead of one ; and this proportion, as pointed out, will be 50 per cent. too small to give a correct sample with this particular ore if crushed only to 1-inch cubes

" It follows, therefore, that if 50,000 lb. (*i.e.* 25 per cent.) are taken as the first cut in sampling a 100 short-ton lot of 1-inch average cubes, and this quantity is then crushed to $\frac{1}{4}$-inch average cubes, a weight of 3125 lb. will give as accurate a sample at $\frac{1}{4}$ inch, as 50,000 lb. at 1 inch ; or, if the original mass is all crushed to $\frac{1}{4}$ inch, 3125 lb. will do for the first cut ; and further, that on a reduction to $\frac{1}{16}$ inch, 195 lb. bear the same ratio between size of cube and weight of sample as the 50,000 lb. did to 1-inch cubes, and hence will give a correct sample " (Argall).

[1] See " Sampling of Ores from Cobalt Mines," by F. W. Pugsley, *Eng. and Min. Journ.*, **1911, 91,** p. 776.

[2] J. A. Scherrer and G. E. F. Lundell (U.S. Bur. Stand. J. Research, 1930, **5,** 891–895 : Research Paper No. 237).

This leads up to the important question of

(4) Ratio of Weight of Sample to Fineness of Division.—Experience has proved that for successful sampling a definite ratio must exist between the weight of the sample and the size of the ore particles, and, before further reducing the weight of samples, the size of the pieces must be reduced upon the basis that there must be a certain minimum number of particles in a sample below which it is unsafe to go if the representative character of the sample is not to be endangered. What this minimum number is, must be determined experimentally for each class of materials. "If the ore particles are large (10 to 12 inches diameter,) a large sample must be taken; if the particles are small (0·10 to 0·20 inch diameter), a small sample will, if properly taken, accurately represent the lot of ore."[1]

As an illustration of this principle, let it be assumed that there are

50	bricks of a material designated by		A.
25	,, another material	,,	B.
15	,, ,,	,,	C.
10	,, ,,	,,	D.

Dividing these numbers by their greatest common measure, namely, 5, there would be 10 of A, 5 of B, 3 of C, and 2 of D. These bricks, if taken together, would be a representative sample of the original lot, and would be one-fifth of its weight. To attempt to reduce the bulk of the sample below this point without subdividing the pieces would clearly result in destroying the sample. This also illustrates the difficulty of properly sampling a large bulk of waste material containing only a few particles of value, as, for example, in the case of some gold ores, consisting of a large amount of barren quartz containing a few small nuggets or coarse particles of metal. In such a case it is imperative to crush it all very fine and sift out any flattened nuggets that will not crush. The sifted material may then be sampled and the coarse metallic residue reported separately.

Within recent years considerable attention has been given to the subject of the "safe" weight, *i.e.* the minimum quantity which shall be truly representative of the whole of the ore-mass, and there is general agreement amongst sampling authorities as to the safe weight to be taken for certain classes of ore; but the minimum allowable quantity for any particular ore will depend on the nature of the ore and can only be determined by experiment.

Much time has been devoted to the subject by H. A. Vezin of Denver, Colorado, and as the result of experiments he considers it a safe rule to take 1 troy ounce (31·1 grammes) as the minimum weight to which a sample should be reduced from material passing a sieve of 1-millimetre (·04 inch) mesh; and, using this as the basis of calculation, he states that the weight to be taken for sample, for material of other sizes (coarser or finer), may be found by "making the minimum weight of the sample proportionate to the cube of the linear dimensions of the average particles." Thus, if the material has passed a sieve of 2-millimetre mesh, by this rule the sample should not be cut down below 8 ounces (249 grammes) without the material being first ground finer. This is equivalent to stating that the "number of particles in a sample should never be below a certain minimum, whatever their size may be."

The number of particles of all sizes in 1 ounce of material (with a specific gravity of, say, 4) that has passed through a 20-mesh sieve is at least 50,000, and on this basis the weight of the smallest sample of the same material passed only through a 3-inch mesh would be about 8 tons. Vezin's experiments were

[1] Brunton, "The Theory and Practice of Ore Stamping," *Trans. Amer. Inst. Min. Eng.*, **25**, 826, quoted by Fulton & Sharwood, *Manual of Fire Assaying.*

made on the pyritic ores of Gilpin County, Colorado, carrying from 1 to 4 ounces of gold per ton, and the "safe weights" for coarser samples of the same ore, obtained by calculations, are indicated in the following table (the rule is in use in Colorado as the basis of calculation) :—

TABLE III—"SAFE WEIGHTS" FOR COARSE SAMPLES OF ORE (DENVER, COL.).

Diameter of Largest Particles (inches).	Minimum Weight of Samples (lbs.).
0·04 or $\frac{1}{25}$	0·0625 or $\frac{1}{16}$ (1 oz.)
0·08 ,, $\frac{1}{12}$	0·50 ,, $\frac{1}{2}$
0·16 ,, $\frac{1}{6}$	4·00
0·32 ,, $\frac{5}{16}$	32·00
0·64 ,, $\frac{5}{8}$	256·00
1·25 ,, $1\frac{1}{4}$	2048·00 or 1 short ton (approx.)
2·50 ,, $2\frac{1}{2}$	16348·00 ,, 8 ,, tons (approx.)

With regard to this rule, Vezin has remarked[1] that in giving the safe minimum weight of a sample at 1 ounce for material of 1 millimetre diameter, he refers to ores containing the precious metals in which these are finely disseminated and not in large quantities. With rich ores in which the precious metals are very irregularly distributed, a larger sample must be taken, while for ores uniform in composition, such as iron ores, much smaller samples will suffice.

Wright has pointed[2] out that Vezin's rule implies that the ratio between the number of particles of mineral and the number of particles of gangue remains the same through all the stages of the crushing of the ore, and that during the grinding, the mineral and gangue are crushed at the same rate, whereas in the large majority of commercial ores the valuable mineral—owing to its greater brittleness, etc.—is more readily broken down than the gangue.

From experiments made by Wright upon two samples of the same ore, taken respectively at the $\frac{1}{8}$ inch and the 100-mesh stages of the grinding, it was shown that the ratio of the safe weights—for this ore at least—was very much smaller than that required by Vezin's rule. From these remarks it will be seen that the weight of the sample is dependent not only on the size of the ore particles but also on the nature of the ore.

As already stated, in cases where the valuable constituents of the ore are uniformly distributed, smaller samples will suffice than are necessary for "spotted" ores in which the values are irregularly distributed. For example, the rich telluride ores of Colorado, assaying 10 to 15 ounces of gold per ton, are usually very difficult to sample, especially by hand sampling, chiefly on account of the great difficulty in properly mixing the fine and coarse particles. In sampling such ores with automatic samplers, Argall found the following ratio between the average size of the ore cubes and proportional weight of the sample to give accurate results[3] :—

Average size of ore cubes, in inches ...	1·0	0·25	0·0625 (8-mesh)	0·0171 (30-mesh)
Weight of sample (for 100-ton lot) in lb.	40,000	2,500	157	10
,, ,, per cent.	20	1·25	0·0785	0·005

[1] In discussion on Clarkson's paper, *J. Iron and Steel Inst.*, 1893, **44**, 139.
[2] L. T. Wright, *Mining Magazine*, 1910, **3**, 353–358.
[3] *Trans. Inst. Min. and Met.*, 1902, **10**, 234–273.

With reference to these figures, Argall remarks that in practical work larger quantities of the finer-sized material would be taken simply as a matter of extra precaution, more especially so in mills where all the ore is ultimately ground fine.

Vezin's rule frequently demands finer crushing than practice indicates to be necessary, and in many sampling works, to save the excessive cost of crushing, sampling rules based on experience, and deviating from Vezin's rule, have been made which are sufficiently accurate for the purposes of buying and selling.

Prof. R. H. Richards of the Massachusetts Institute of Technology has collected a large amount of data based on actual sampling practice, and as the results of his investigations he has given the following rule for ascertaining the weight of sample to be taken for any given ore :—"The weight taken for sample shall be proportional to the square of the diameter of the largest particles."[1]

By adopting this rule, Prof. Richards has drawn up the set of figures given in the accompanying Table IV, which have been found to give results that conform to good practice.

TABLE IV—WEIGHTS TO BE TAKEN IN SAMPLING ORE (*R. H. Richards*).

Weight.		Diameter of Largest Particle (in millimetres).[2]					
		Very low-grade or very uniform Ores.	Low-grade or uniform Ores.	Medium Ores.		Rich or "Spotted" Ores.	Very rich or excessively "Spotted" Ores.
		A.	B.	C.	D.	E.	F.
grammes.	lbs.	mm.	mm.	mm.	mm.	mm.	mm.
...	20,000·000	207·00	114·00	76·20	50·80	31·60	5·40
...	10,000·000	147·00	80·30	53·90	35·90	22·40	3·80
...	5,000·000	104·00	56·80	38·10	25·40	15·80	2·70
...	2,000·000	65·60	35·90	24·10	16·10	10·00	1·70
...	1,000·000	46·40	25·40	17·00	11·40	7·10	1·20
...	500·000	32·80	18·00	12·00	8·00	5·00	0·85
...	200·000	20·70	11·40	7·60	5·10	3·20	0·54
...	100·000	14·70	8·00	5·40	3·60	2·20	0·38
...	50·000	10·40	5·70	3·80	2·50	1·60	0·27
...	20·000	6·60	3·60	2·40	1·60	1·00	0·17
...	10·000	4·60	2·50	1·70	1·10	0·71	0·12
...	5·000	3·30	1·80	1·20	0·80	0·50	
...	2·000	2·10	1·10	0·76	0·51	0·32	
...	1·000	1·50	0·80	0·54	0·36	0·22	
...	0·500	1·00	0·57	0·38	0·25	0·16	
90·0	0·200	0·66	0·36	0·24	0·16	0·10	
45·0	0·100	0·46	0·25	0·17	0·11		
22·5	0·050	0·33	0·18	0·12			
9·0	0·020	0·21	0·11				
4·5	0·010	0·15					
2·25	0·005	0·10					

The figures in each column are based on data obtained in practice, full details of which are given by Prof. Richards in his *Ore Dressing*. The equivalents in inches and " mesh " may be seen by reference to Table VI, p. 107.

[1] R. H. Richards, *Ore Dressing*, vol. II, p. 850.

[2] 25·4 mm. (millimetres) = 1 inch ; 1 ton = 2000 lbs.

The figures given in columns *A* and *B* are applicable to poor or low-grade ores and to ores in which the valuable constituent is very evenly distributed, while the figures in columns *C* and *D* are applicable to ores of similar character to *A* and *B* but richer or higher-grade. Most of the ores of the common metals would come under the heads given in these columns.

Gold ores may also be included under the head of low-grade ores when the gold is combined with pyritic material which is more or less evenly disseminated through the ore.

The figures in columns *E* and *F* mainly apply to ores of the precious metals in which the values, as already stated, are very frequently contained in a few particles of metal (or nuggets), or a few lumps of rich mineral, distributed very irregularly through a large mass of barren material. Exceptionally rich ores of the common metals are also included under the heading of "Spotted Ores."

With reference to the table, Richards points out that it is not intended that the reduction in size shall proceed by all the successive stages shown in any one column, but whenever any weight given in the first column is taken as a sample, the ore should first be crushed to the corresponding size shown in one of the other columns.

For example, with a 20,000-lb. (10-ton) lot of low-grade ore, the entire lot might be crushed to 25·4 millimetres (1 inch) and 1000 lbs. taken for a sample. If this were all reduced to 2·5 millimetre ($\frac{1}{10}$-inch) size, the quantity could be reduced to 10 lbs.; and then by grinding to 0·80 millimetre (about 20-mesh) the weight could safely be reduced to 1 lb.

(5) **The ratio of weight of first sample to total weight of bulk,** or the smallest percentage to be cut out as a sample at one operation, will depend upon the uniformity of composition and fineness of division of the bulk, and also upon the method of cutting out the sample. In some cases 1 per cent. or even less would be safe, whereas in others 20 per cent. or even more would be desirable. The percentage to be taken for sample for any particular ore is usually agreed upon by buyer and seller before sampling. In general sampling practice it is usual to cut out for first sample from $\frac{1}{5}$ (20 per cent.) to $\frac{1}{20}$ (5 per cent.) of the ore bulk, according to the nature of the ore and size of the pieces, experience having shown that if the same quantity of ore is taken each time and at frequent and regular intervals the sample so obtained will truly represent the bulk from which it is taken. Where the ores are of low grade, or very uniform in composition, a small percentage will suffice for first cut. Take iron ore, for example, or ores to be used as fluxes for blast-furnace work, where it is important to keep the ore as coarse as possible: 10 per cent. of ore as coarse as 6-inch cubes can be taken if necessary and, reduced proportionately, will give an accurate sample.

In the case of copper ores sampled automatically, Peters states[1] that the proportion of the ore stream deflected into the sample bin in the first instance may vary from 10 to 50 per cent., the latter amount only being required for coarse ores of enormous and very variable richness, while for ordinary lump ores from 10 to 20 per cent. is the maximum required. Ores of the precious metals that are to be treated in stamps or roller mills, and reduced ultimately to a state of fine division, should preferably be crushed finer for the first cut, and as a matter of precaution, say, 20 per cent. taken out for the first sample.[2]

When the nature and fineness of the ore permit of a very small percentage being taken for first sample, it is generally better to obtain this in two cuts.

[1] Peters, *Modern Copper Smelting*.

[2] Argall, *loc. cit.*

For example, a 1 per cent. sample is obtained by cutting 10 per cent. of 10 per cent. instead of cutting 1 per cent direct, or a 2 per cent. sample is taken by cutting 10 per cent. of 20 per cent.

(6) Method of cutting out Sample.—This is obviously of the utmost importance, so that a fair average proportion of all the different qualities and sizes may be obtained. The method to be adopted is dependent on the quantity and nature of the ore and the metallurgical treatment to which it is subsequently to be subjected. "In milling and concentrating plants where fine crushing is a necessary preliminary to the metallurgical treatment of the ore, sampling is a comparatively easy and reasonably accurate operation, but where the ore is to be treated by blast-furnace smelting, crushing of any kind is objectionable and fine sub-division is prohibited"[1]; under these conditions sampling becomes more difficult.

When a very large lot has to be dealt with, the chances of error are lessened by dividing it into several smaller lots of about 50 or 100 tons each and sampling each of these lots seperately. The commercial value of the ore or product to be sampled is also an important consideration in determining what the allowable limit of error in sampling should be, and this will affect the method of sampling and the number of samples to be taken. In the case of metallurgical products, an intimate knowledge of their nature and the process by which they are obtained is of great assistance to the sampler in deciding the method of sampling to be adopted. The tendency of modern practice in ore sampling is to employ methods which are as far as possible automatic and independent of the will or judgment of the sampler, and in public sampling-works machine methods are very generally employed. Accurate sampling can, however, be performed by hand, and the methods of hand sampling which have been in use for many years are still largely employed for mine sampling and smelting works, as they avoid crushing a large proportion of the ore. When dealing with large lots of ore, the first sample is invariably taken by some method of fractional selection, that is, selecting a certain proportion while the ore is being dumped or deposited in a heap; or, if the ore has already been dumped, trenches are cut through the heap and portions reserved for sample, as described below.

If, for instance, an entire cargo of ore has to be sampled, there would be set aside at the time of unloading, according to the importance of the mass or the intrinsic value of the ore, every fifth, tenth, or twentieth skipload or sackful, etc., for an especial heap, which may be considered to have the same composition as the whole mass. The quantity thus set aside is then reduced by successive crushing and division until about 8 oz. is finally obtained, which is crushed to pass a very fine sieve and divided into several portions for assay purposes.

The degree of fineness to which an ore is crushed in the successive stages of the sampling will of course vary with the class of ore and other considerations. Ores direct from the mines usually range in size from 12-inch to 16-inch cubes down to the finest particles, and contain from 1 to 2 per cent. of moisture in hard ores up to in some cases as much as 25 to 30 per cent. moisture in clayey or talcy ores (*i.e.* ores containing the mineral talc—$(MgO)_6(SiO_2)_4+H_2O$).

The first operation is, therefore, weighing the ore and reducing it to a suitable size for sampling.

If the ore is destined for the blast furnace, it is crushed as little as possible and a large sample cut out, not less than 20 per cent., depending of course largely on its value; this sample is then coarsely crushed and reduced as stated above. As an example of the amount of crushing necessary, the following course of procedure is recommended by Argall in the machine sampling of 8-ounce gold

[1] D. W. Brunton, *Trans. Amer. Inst. Min. Eng.*, 1909, 837–855.

ores to be treated by smelting. The ore is crushed to 3-inch cubes and passed through a Vezin sampler (see page 87), which can be made to cut out pieces as large as 3 inches, the sampler taking, say, 33 tons out of 100. The sample is next crushed to about $\frac{3}{4}$-inch, and 4 tons are cut out for the sample ; these 4 tons, or 4 per cent. of the original weight of the ore, is all that would, in this case, be reduced to a fine state of division in the sampling process, leaving 96 per cent. of the ore in good shape for blast-furnace operations.

The proportion of size of cubes and weight of ore taken as sample in this case is, according to Argall, quite safe for 8-ounce gold ores to be treated by smelting, as already stated.

If the ore is to be ultimately ground fine for chemical processes, it is usually crushed to $\frac{1}{2}$-inch or $\frac{3}{4}$-inch before sampling, and in such cases a smaller sample will answer the requirements of commercial accuracy.

In sampling ores that are to be sold, it is the usual practice to reserve a separate portion, say 200 to 300 lbs., of the finer crushed material obtained towards the end of the sampling, and to utilize this quantity for resampling should this become necessary. By adopting this plan the resampling of the entire lot is avoided.

For crushing the ore, rock-breakers of the Gates or Blake type are invariably used to reduce large rocks to cubes of $1\frac{1}{2}$-inch to 2-inch, but below 2-inch cube rolls are preferable to rock-breakers as a medium of reduction. Rolls are used for any size, from 2-inch to 30- or 40-mesh ; below that, some form of laboratory machine for fine grinding is employed.

Moisture Sample (see also page 111).—When it is necessary to determine the amount of moisture in the ore, a separate " moisture " sample, weighing usually from $\frac{1}{2}$ to 2 lbs., is taken when the ore has been crushed to about 10 mesh (0·05 inch). In sampling-mills, larger moisture samples, weighing from 15 to 20 lbs., are frequently taken. The whole of the sample is dried, and the percentage of moisture determined from the loss in weight as described on page 112. If the sample taken for moisture is too large, it is reduced in bulk by quartering or by any of the methods given later.

The importance of determining the moisture as soon as possible after the sample is taken will be obvious.

METHODS OF SAMPLING.

The methods of sampling are conveniently classified under two heads: (1) hand sampling, and (2) machine sampling.

I.—Hand Sampling.

The following are the methods of hand sampling in general use :—

(*a*) Grab Sampling.
(*b*) Trench and Pit Sampling.
(*c*) Coning and Quartering.
(*d*) Fractional Selection or Shovel Sampling.
(*e*) Split Shovels.

(a) Grab Sampling.—This method consists of taking proportionate amounts of the lumps and of the fine ore, at equal distances over the pile, after first removing the surface ore which has been dried by the atmosphere. In sampling the lumps the whole piece should be broken and chips taken from both the outside and the inside ; no piece should be larger than about $\frac{1}{2}$-inch cube.

The sample thus selected is crushed finer, well mixed, and then reduced in bulk by coning and quartering or other method.

The method is imperfect, as the sample only represents the upper part of the pile. The accuracy of the sample is largely dependent on the discretion of the sampler: great care and judgment are needed to take the coarse and fine parts in the correct proportions. The method is used chiefly for obtaining an approximate idea of the value or composition of an ore pile, as it is quick and inexpensive.

(b) Trench Sampling.—In this method the ore is formed into a flattened square or rectangular pile from 2 to 3 feet deep for lots of about 100 tons, and 1 to 2 feet deep for smaller lots.

Trenches about 1 foot wide and crossing each other at right angles are then dug through the centre of the pile with a shovel, as shown in Fig. 75, thus converting the large pile into four or more smaller piles. The workman, as he advances in cutting the trench, reserves for sample either all the ore he removes,

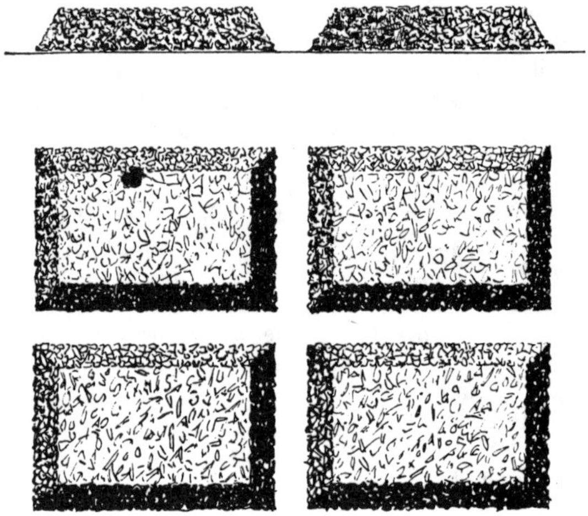

FIG. 75.—TRENCH SAMPLING.

or a certain proportion, every fifth, tenth, or twentieth shovelful, and throws the rejected portion alternately right and left on to the pile.

The sample obtained in this way may, after crushing, be formed into a new rectangle and reduced in bulk by again trenching or by coning and quartering. Instead of reserving an aliquot portion for sample as just described, it is also the practice when cutting the trenches to reject all the ore removed, each shovelful being thrown to the right and left as before, and when the trenches are completed the sample is taken from the sides of each of the small piles into which the original pile has been divided.

In this case portions are removed from each of the freshly exposed sides of the piles by carefully "rising" the shovel from the bottom to the top, the total quantity taken in this way being about 10 per cent. of the original pile.

Pit Sampling.—Another method, less expensive than trench-cutting, consists in digging a series of holes or pits at equal distances over the surface of the pile and reserving either all or equal amounts, usually a few shovelfuls, of the material excavated from each hole. The reserved portions are then well mixed and reduced in bulk as before.

In digging the pits care must be taken to go right through the layer, otherwise a fair proportion of the finer material, which generally gets to the bottom, will not be obtained.

(c) Coning and Quartering.—This old Cornish method of hand sampling has been in use for many years, and is still largely employed for reducing the bulk of large samples or for sampling small lots.

It consists in piling the ore into the form of a cone which is quartered (hence the term quartering), two opposite quarters being reserved for sample and the other two rejected. The various stages of the coning and quartering method of sampling are illustrated in Fig. 76. When the coning and quartering are

Cone.

Partly reduced cone (elevation).

Partly reduced cone (plan).

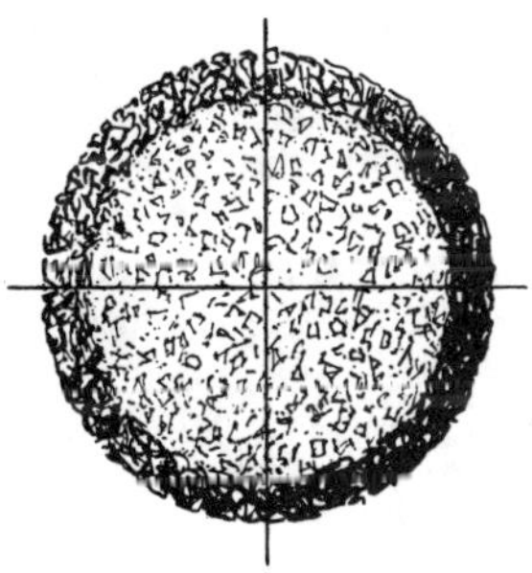

Plan of reduced cone ready for dividing.

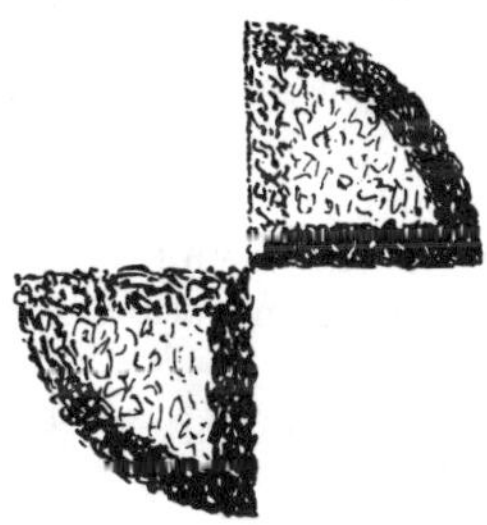

Plan of reduced cone after dividing.

FIG. 76.—CONING AND QUARTERING.

skilfully performed the results are accurate. Before applying this method, and to save unnecessary labour, it is usual with large lots to take an aliquot portion of the ore lot, say every tenth or twentieth truck-load, etc. as previously described; but however the first sample may be selected, the subsequent treatment of it to obtain the final sample is the same.

If the selected material is in lumps it must be crushed down to, say, $1\frac{1}{2}$-inch to 2-inch cubes before the bulk is reduced.

After crushing, the ore is well mixed by turning it over several times with a shovel and then piled into the form of a cone, special care being taken that each shovelful is thrown upon the apex of the cone. To do this successfully requires some skill; the ore must not be thrown as in ordinary shovel work, but must be tossed vertically into the air, so that it will drop directly upon the apex and fall evenly down the side of the cone.

The even distribution of the material is absolutely necessary in order to obtain a correct sample, and this can only be attained by keeping the cone truly vertical. Dishonest samplers perform the operation of what is known as "drawing the cone," *i.e.* deliberately building up the cone away from the centre of the heap and thus altering the relative proportions of lump and fines in their own favour. In building up the cone the fine material always forms the apex, while the coarse particles fall towards the base, and as the fine ore is usually richer than the coarse it will be evident that if the cone is not truly vertical when it is divided vertically into four sections, the section towards which the apex leans will be richer than the other sections. On subsequently levelling and quartering the cone this local richness may be partly corrected or may be intensified, according to whether less or more of the fine ore forming the apex is shovelled into one of the sections and not evenly distributed between all four sections.

To ensure the cone being vertical, it is a common practice to drive a rod into the ground as a guide to fix the apex.

The object in forming the cone is to mix the ore thoroughly, and when it is completed it is "pulled" down by the men walking around it and shovelling the ore from the apex to the base so that it falls outward, as shown in Fig. 76. The lower portion of the cone is not disturbed, and when this process is finished there remains a truncated cone of ore from 6 to 12 in. deep which should be uniform in thickness and truly circular in form.

This is divided by diametrical lines into four equal sections by pressing a board or iron plate edgewise into the pile, or by placing a cross, formed of iron plates riveted together, on the pile and driving it into the ore to mark the divisions.

Two opposite quarters are then cut out with the shovel and removed to the ore-bins, and the other two, which represent the sample, are formed into a fresh cone which is drawn down and quartered as before. This process is repeated until the sample requires recrushing, after which the coning and quartering are continued until the sample has been reduced to a few pounds in weight.

The sample is finally prepared for assay purposes by grinding it sufficiently to pass a fine-mesh sieve, but to facilitate this fine grinding it is frequently necessary to dry the sample.

The details of this practice will vary in different works, and the number of recrushings necessary will depend on the quantity and nature of the ore and the size of the pieces forming the original bulk.

In some of the works in Colorado it is the practice first to dump the ore into the form of a ring, leaving a clear space inside of about 5 to 10 feet in diameter according to the quantity of the ore. The men then move around the outer circumference of the ring and shovel the ore into the centre of the clear space to form a cone, which is dealt with as already described.

The coning and quartering method requires considerable floor-space, a 100-ton lot requiring at least 2000 sq. ft.,[1] and when large lots have to be sampled it is somewhat slow and laborious; but the results are trustworthy, and as good as can be obtained by hand sampling, provided proper care is taken and sufficient time allowed. Insufficient crushing and inefficient mixing during the progress of sampling are the chief sources of error.

[1] Bridgman, *Trans. Amer. Inst. Min. Eng.*, 1891.

(d) Fractional selection or shovel sampling[1] consists in reserving for sample a certain proportion of the ore as it is being shovelled into the stock-bins or otherwise handled.

For example, if one-fifth is to be cut out for sample, four shovelfuls will be thrown direct into the bins and every fifth shovelful reserved. The proportion to be reserved is dependent on the nature and size of the ore pieces. The selected portion, after crushing, is coned for the purpose of mixing, and then every fifth shovelful again reserved. This is coned as before, and the process repeated until it becomes necessary to recrush, after which the reduction in bulk proceeds in the same way.

To conduct this method successfully, properly shaped shovels must be used, and the size of the shovel must accord in capacity with the size of the bulk to be handled, so that smaller shovels must be used towards the end of the sampling.

Sampling by shovel is considered by many to be a more convenient and accurate method of hand-sampling a pile of crushed ore than coning and quartering, but the latter method is more generally preferred. The advantages claimed for shovel sampling are: (1) it is cheaper in operation; (2) it is quicker, the ore-bulk being reduced at a much faster rate; (3) that the sample is more accurate for a given reduction as a large number of cuts are made. This is illustrated by Argall[2] as follows: "Let it be assumed that a heap of 10,000 lbs. has to be divided for sample and reject; that the shovel used holds 10 lbs., and that the labourer places alternate shovelfuls in the stock-bin for the rejected portion, and on the floor for the sample. In disposing of this heap of ore he will handle 1000 shovelfuls or make 1000 cuts in the heap as against two cuts in the 'coning and quartering' method, a manifest advantage in favour of the shovel sampling.

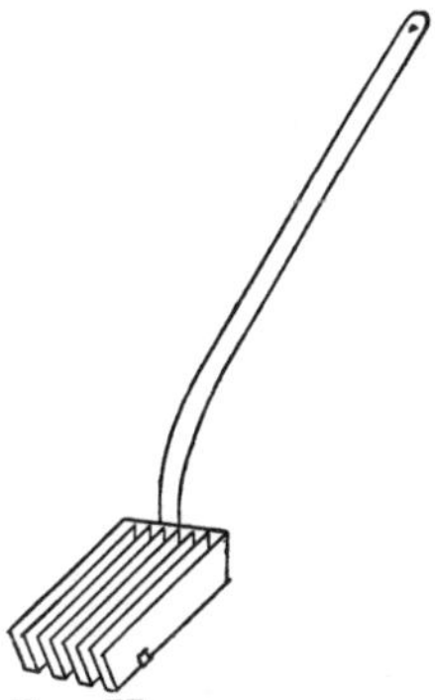

FIG. 77.—SPLIT SHOVEL FOR SAMPLING.

"Should the ore, however, be sufficiently fine to warrant a reduction to, say, one-fourth, then every fourth shovelful would go to form the sample and three to the stock-bin. In this case the sample is made up of 250 cuts and reduced to 25 per cent. of the original ore in the operation, whereas the tedious process of coning and quartering would have to be twice performed to give the same reduction, and even then the sample is the result of but four cuts."

An objection to the shovel method is that the selection of the portions to be reserved for sample is left entirely to the discretion of the operator.

(e) Split Shovel.—This shovel consists of four long parallel troughs, each 12 in. long and 2 in. wide, with open spaces between them, the spaces being the same width as the troughs (Fig. 77). It is sometimes used for sampling small lots of ore weighing not more than about half a ton. The ore must be crushed to ½-inch diameter. The ore is dug from the pile with an ordinary shovel and then spread over the split shovel, the sampler moving his shovel backwards and forwards across the troughs so that the ore as it slides off is evenly distributed. Half the ore remains in the troughs and the other half drops through the spaces. When the troughs are full the split shovel is lifted and the contents reserved as the sample. When all the ore-heap has been dealt with, the portion reserved for sample can be reduced in bulk by again passing it over the split shovel, thus reducing the bulk by one-half.

Pipe Sampler or Sampling Iron.—Heaps of finely-crushed material, such as dry tailings or concentrates, are sometimes sampled by driving in iron pipes at

[1] Also called the "alternate shovel method" when every second shovelful is reserved.
[2] *Trans. Inst. Min. and Met.*, 1902, **10**, 237.

regular intervals and mixing the samples withdrawn in the pipes. A form of sampling iron (or cheese-taster sampler) for very dry material is described by Richards.[1]

It consists of an iron pipe A, Fig. 78, with nearly half its circumference cut away, except for a few inches at the upper end. The pipe is fitted with a T-piece to hold the handle, by means of which the pipe can be rotated. This pipe fits inside a slightly larger pipe B which is similarly cut away, and has its lower end sharpened or drawn out to a point and its upper end provided with a handle. To obtain the sample, the pipes are driven into the ore with their openings arranged as shown in the sectional plan. The inner pipe is rotated until it reaches the position shown in B′, and moved backwards and forwards until it is filled with ore. It is then replaced in its original position A′ and both pipes withdrawn with the sample enclosed.

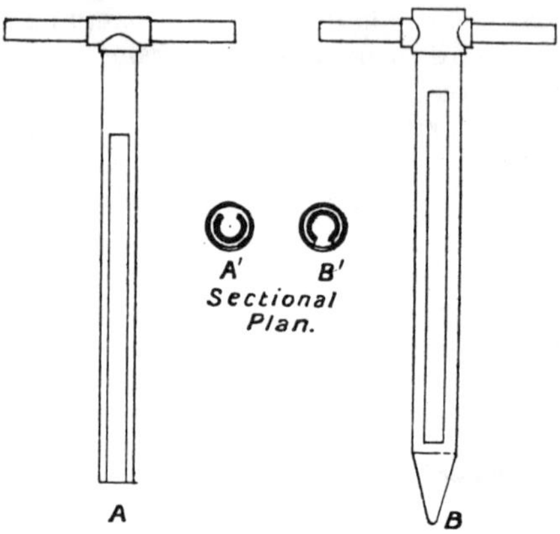

FIG. 78.—PIPE SAMPLER.

In order to obtain correct samples they should be taken systematically from all parts of the heap and throughout the entire depth of the heap, so as to prevent an undue proportion of the upper layers from being included. In the case of tailings the samples are usually taken from the charge in the vat or from the ore-trucks, and it is also usual to sample the lowest 6 inches to 18 inches separately.

Running Samples.—The term running sample is applied to the sample of any running stream of material, such as the pulp suspended in water, issuing from the screens in a stamp battery used for crushing gold ores, etc.

Dipper or Bucket Samplers.—Running samples are generally taken by catching the whole stream in a dipper or bucket at stated intervals, for example, every fifteen or thirty minutes, for some definite period, usually a few seconds. When a bucketful is collected the sample is allowed to settle for several hours, the clear water is decanted, then the sample dried, well mixed, and reduced in bulk for assay purposes.

Automatic dipper and bucket samplers are now being adopted in many gold stamp mills.[2]

"Salted" Samples.—Samples which have been intentionally or accidentally contaminated with either richer or poorer material so as to alter the true value of their contents are said to be "salted." The intentional salting of samples for deceptive purposes is now seldom attempted.

II.—Machine Sampling.

Mechanical sampling is based upon the principle of intercepting or cutting out a portion of a running stream of ore. This may be done by taking either—

(1) Part of the ore stream continuously, or
(2) The whole of the ore stream intermittently.

[1] *Ore Dressing*, 1903, **11**, 844.

[2] Text-books on mining should be consulted for detail of the methods of taking running samples.

Machines for taking a definite portion, usually one-sixteenth, of a falling or sliding stream of ore continuously, were constructed on the assumption that the values are evenly distributed across the ore stream; but experience has proved that this is not the case, and machines based on this principle have been largely displaced by those which automatically deflect the entire ore stream into the sample division for a varying portion of the time, usually one second in each five seconds. In many machines now in use a narrow scoop passes steadily across the stream of ore at regular intervals of time, and this is considered to be equivalent to the method used in the machines of class (2).

Brunton[1] has pointed out that "it is not practicable to produce a stream of ore which shall be continuous in value through every part of its length, any more than it is possible to produce a stream of ore that is constant in value throughout its width; but by taking a small sample entirely across a falling stream at very short intervals it is found that, while no single cut would give an exact representation of the composition of the entire lot, the average of a large number of these small samples is so nearly correct that results can be duplicated within very narrow margins, or, in other words, that individual errors are balanced." The accuracy of the sample obtained in this way, provided it is taken sufficiently often, is now well established and recognised.

"Numerous different machines working on this principle, such as the Snyder, Vezin, and Brunton samplers, described below, are now in use; and these types of sampling plants have been perfected to such a degree that where the hopper ore-cars which are now coming into general use are employed, ore may be unloaded, crushed, sampled, and reloaded into the outgoing cars, and the ground samples delivered in a locked steel box without ever having been handled—the entire chain of operations being performed automatically."[2]

The chief drawbacks to machine sampling are the necessity of fine crushing and the difficulty of cleaning the machine.

The following are types of the sampling machines now in general use:—

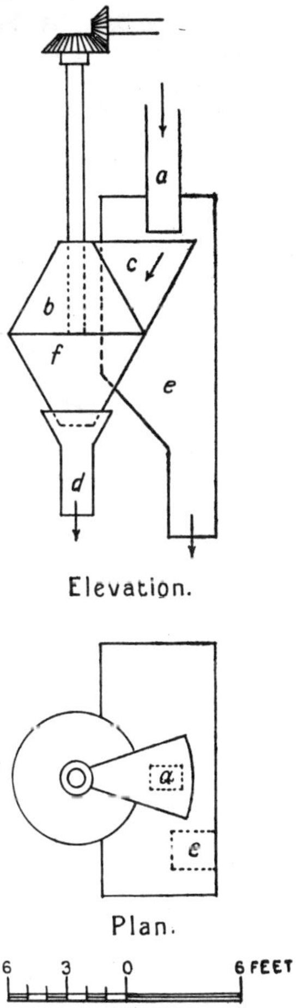

FIG. 79.—VEZIN SAMPLER.

Vezin Sampler (Fig. 79).—This sampler is accurate, simple of operation, and suitable to handle all grades of ore. It is coming into very general use and has been widely employed in Colorado.[3]

It consists of two truncated cones *b*, *f*, of sheet-steel, with their bases bolted together. A scoop *c*, made of sheet-steel, with a sector-shaped opening at the top, is riveted to the upper cone *b*. The angle of the sector may subtend any desired portion of the circumference of a circle, such as one-tenth (36°) or one-sixteenth ($22\frac{1}{2}$°). Both cones and the scoop are rotated at about 25 or 30 revolutions per minute by bevelled gearing. The stream of ore falls from a spout *a*, and when the scoop *c* passes below the spout, the whole stream of ore falls into

[1] *Trans. Amer. Inst. Min. Eng.*, 1909, 837–855. [2] Brunton, *loc. cit.*
[3] *Trans. Inst. Min. and Met.*, 1902, **10**, 240–241, with working drawings.

the scoop and is delivered into the interior of the lower cone *f* and thence through *d* to the sample bin or to the rolls, if it is to be further crushed and reduced in bulk. In this way a sample is taken about every two seconds. The main portion of the ore stream falls into the hopper *e* and passes through the shoot to the storage bins.

A Vezin sampler of about 3 feet in diameter, and requiring a fall of about 6 feet, is trifling in cost, and treats 30 to 40 tons an hour (Rose). There is no patent on the sampler.

Collom Sampler (Fig. 80).—In this machine one or more diverting scoops *b* are mounted on a horizontal arm *a* carried by a revolving vertical shaft. As a scoop passes under the stream of ore from the spout *C* it cuts out a sample and delivers it through a vertical spout *d* into the sample bin. When the scoop

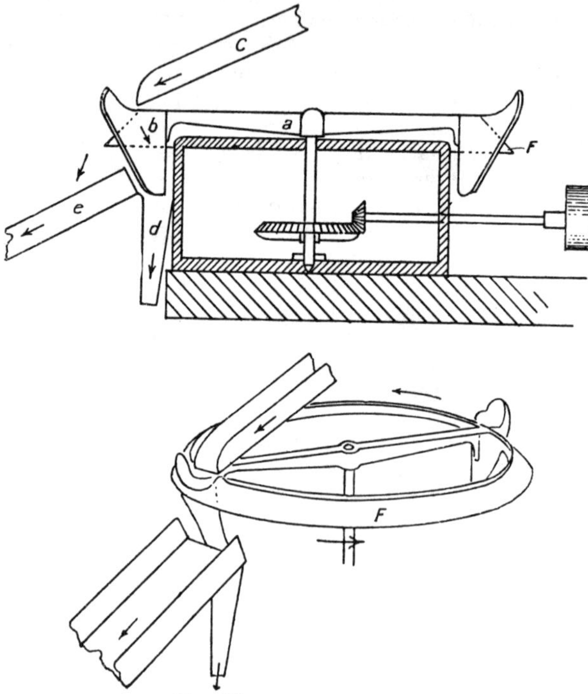

FIG. 80.—COLLOM SAMPLER.

has passed, the ore stream falls on to the shoot *e* and thence to the storage bin. To prevent any ore from falling into the sample spout, except when the scoop is passing, the machine is fitted with a sloping metal flange *F*. This machine is driven at a constant speed in one direction by bevelled gear and pinion.

Snyder Sampler (Fig. 81).—This sampler has the form of a circular pan *c* with flaring sides, set on edge on a revolving horizontal shaft. A spout *b* projects through the flaring side, and passes through the ore stream from the feed-spout *a* at each revolution of the machine, thus cutting out a sample which is diverted to the sample spout *d*. During the rest of the revolution the ore falls on the flaring sides and is diverted into the spout *e*.

Brunton Sampler (Fig. 82).—This sampler consists of a sheet-iron deflector *A*, with a sloping side and a scoop *B* attached to it, slanting in the opposite direction, as shown in section, Fig. 82 A. The deflector is fastened to a horizontal shaft connected with machinery which causes it to swing backwards and forwards beneath the end of the feed-spout *C*. As the deflector oscillates the scoop cuts completely across the stream of falling ore at each oscillation, so that a sample is thus deflected to the left by the scoop *B* and the rest of the ore deflected to

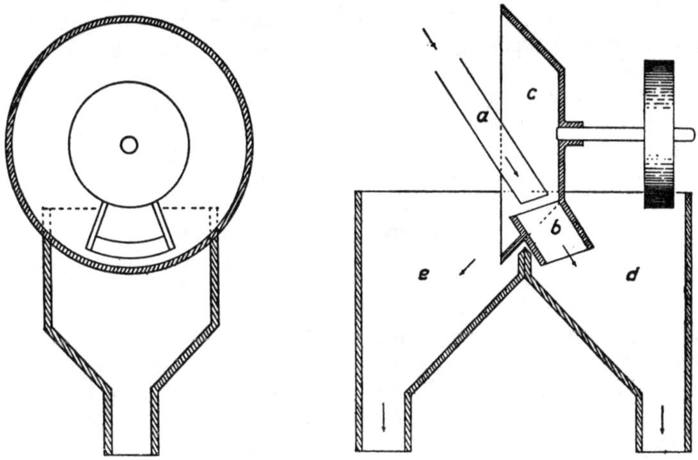

FIG. 81.—SNYDER SAMPLER.

the right by the sloping face *A*. The ratio of rejected ore to sample is governed by the width of the scoop and the width of the ore stream through which the deflector travels. A common ratio is 5 to 1.

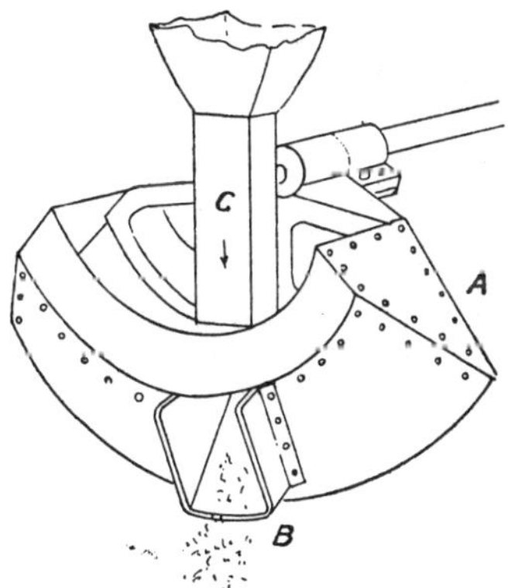

FIG. 82.

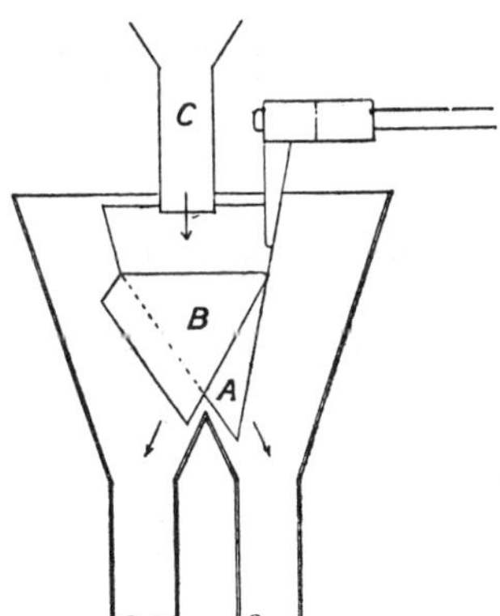

FIG. 82 A.

BRUNTON SAMPLER.

Sampling-Works or Mills.—The general arrangement of a works especially designed for sampling varies according to the quantity and nature of the ore to be sampled and the method of sampling to be employed.

Hand and machine sampling are sometimes combined. In the case of ore of high value and of which it is desired to obtain very accurate samples, it is generally desirable to put the entire lot through a sampling machine.

In cases, however, where it is desired to preserve the ore in lump form, rather than to obtain greater accuracy of sample, only such fractional part of the lot as may have been predetermined is reserved as the first sample and passed to the sampling machinery, while the main bulk of the lot is thrown into the storage bins. In this latter case, hand sampling is usually employed in taking the first sample, as it is advisable to limit the maximum size of ore passing through automatic samplers to not more than 2-inch cubes, although ore of larger size is sometimes sampled.

The plant in a sampling works consists of a series of large storage bins for keeping the different lots of ore separate, scales for weighing the ore before sampling, crushers or rock-breakers, crushing rolls or other machinery for fine crushing, one or more automatic sampling machines, elevators for elevating the ore up into bins of sufficient height to discharge into cars or storage bins, etc. If hand sampling is employed, sufficient floor-space must be allowed for manipulating large lots of ore. Various devices are also in use for sacking the ore when such is necessary.

In some cases revolving screens are used after the first crushing to remove the fine ore from the coarse as soon as possible, and thus prevent the production of a large proportion of very fine ore. The fine ore thus removed is mixed with the finely-crushed ore that results from the recrushing of the coarse ore.

As a general rule, the sampling machine is placed below the crushing machinery, so that the stream of ore from the crusher is fed directly into the sampler. When this cannot be arranged it is necessary to use an elevator to feed the sampler. The feed-spout above the sampler, whether delivering the ore from the crusher or from the elevator, should be short and should not exceed 3½ inches in diameter. The ore is fed to the sampler in a regular stream while the sampler is running.

The discard spout from the sampler delivers directly into a car, bin, or elevator as required. When, for the reasons stated on page 79, it is desired to pass the ore through two samplers in succession without intermediate recrushing, a bin is interposed between the samplers, in which the ore from the first sampler is allowed to accumulate until there is sufficient to ensure a continuous feed to the second sampler. Sampling mills are best built on a hill-side when this can be conveniently arranged, as this avoids the necessity of using elevators.

Examples of Sampling-Mill Practice.—The following brief descriptions and accompanying flow-sheets will serve to illustrate the arrangement of sampling-works and show how the principles of sampling are carried out in practice.

Sampling Mill of Standard Smelting Company, Rapid City, South Dakota.[1]—The method of sampling generally employed consists of shovel (fractional selection) and riffle sampling (see p. 108). The first sample is cut out with the shovel. Assuming that a 100-ton lot is taken as a unit and every tenth shovelful reserved, a sample of 10 tons is obtained.

This is passed to a Blake crusher with a 9- by 15-inch mouth opening, discharging on a Λ-divider which divides the sample into two parts. One half is fed directly to a pair of 24- by 12-inch rolls, and the discharge from these is automatically halved as before, giving a sample of 2·5 tons with no particle larger than 0·375 inch in diameter.

The rolls discharge the crushed ore directly upon a plate-iron floor, where it is reshovelled, every one-fifth or one-tenth (or other fraction) shovelful being reserved as a sample, which now amounts to 500 or 1000 lbs., according to the fraction reserved. This is put through a pair of 12- by 12-inch sampling rolls and crushed fine, and then sampled by a large Jones riffle sampler, which takes halves,

[1] Fulton & Sharwood. *Manual of Fire Assaying.*

the operation being repeated until finally a sample of between 15 and 20 lbs. is obtained. This is put through a small cone-grinding mill, and after a determination of moisture, is sent to the assay office. Here it is cut down to about 2 lb. by a small Jones sampler, and then crushed on a buck-board to pass a 120-mesh screen, furnishing the assay sample. This sample is supposed to contain no moisture, but as all settlements are made on dry samples, this final sample is

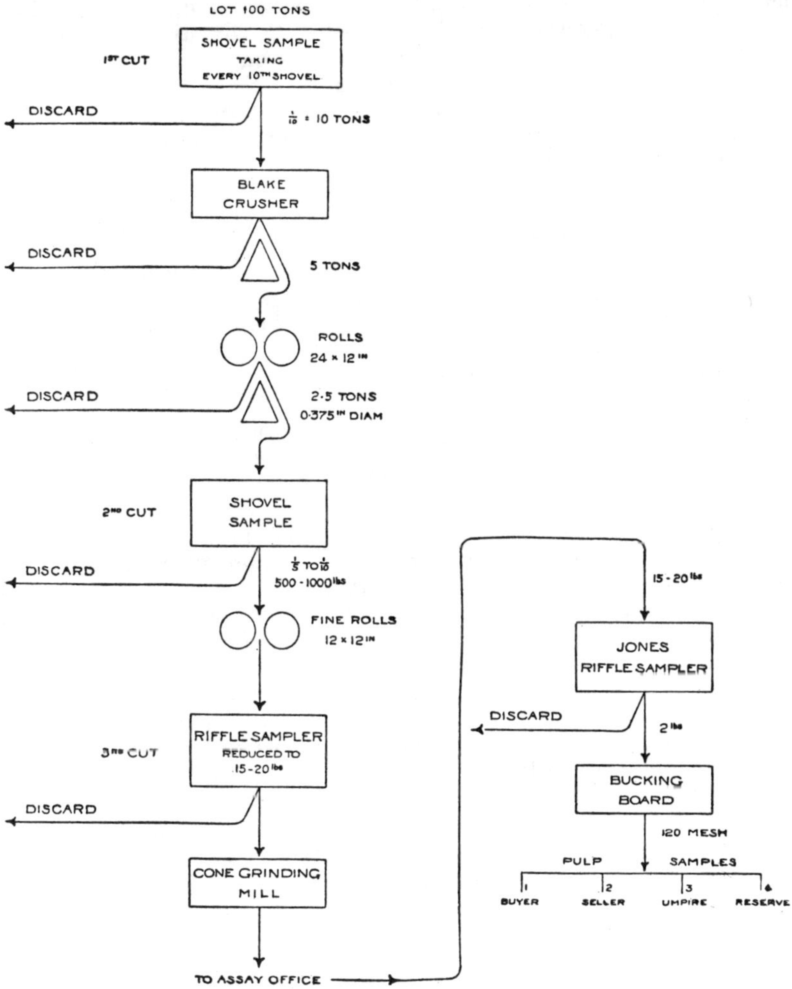

FIG. 83.

again heated at 120° C. for some time in order to expel any moisture which the sample may have absorbed in its passage from the sampling works to the assay office.

The assay sample is divided into four parts and put into paper sacks. One part is assayed by the seller of the ore or product; one part by the purchaser; a third part is kept for emergency; and a fourth part is laid aside for an umpire assay, if such becomes necessary.

The method is illustrated in the flow-sheet given above (Fig. 83).

Sampling Plant for Cripple Creek Ores, Colorado[1].—The following course of procedure is recommended by Argall in machine-sampling ordinary quartzose gold ores assaying from 2 to 6 oz. of gold per ton. For lots of 50 to 100 short tons in large lumps :—Crush in 20- by 12-inch crushers to a maximum size of 1·70 inch, and crush this in 36- by 16-inch rolls (at 35 revolutions per minute) to pass a 0·75-inch mesh screen. The product of the rolls is passed over a Vezin sampler, and 25 per cent. taken for the sample, which should be deflected to a 26- by 15-inch rolls to be crushed to 0·20-inch mesh. At this size about 6 per cent. is ample for a correct sample, but to be quite safe one-tenth is cut out ; if the ore does not contain over 5 per cent. of moisture, the sample, $2\frac{1}{2}$ per cent. of the original, can be passed direct to a fine grinder, reducing it to 0·0525 inch, and again cut to one-tenth or one-quarter per cent. of the original volume ; if damp, the sample is then dried and ground, then passed over a small Vezin sampler, or riffled, as found most desirable, being finally reduced by successive crushing and division until 8 oz. is finally crushed to pass a 120-mesh sieve for assay purposes. The flow-sheet is given in Fig. 84.

In sampling rich telluride ores assaying 10 to 15 oz. of gold per ton, Argall recommends the following method, taking 100 short tons as unit :—With these richer and " spotty " ores finer preliminary crushing is necessary. The ore is crushed to an average of about 1-inch cubes, and 20 tons (*i.e.* 20 per cent.) taken as the first sample ; this sample is then crushed to $\frac{1}{4}$ inch, and 2 tons (10 per cent.) cut out by the second sampler. Crush this to 8 mesh ($\frac{1}{16}$ inch) and reduce by " riffling " to 250 lbs. Dry and crush the sample to 30-mesh ($\frac{1}{80}$ inch), and riffle down to 15 lbs. Pass this through a sample grinder to 90-100 mesh, and riffle down to 1 lb. This is ground on the buck-board to pass 120-mesh ($\frac{1}{280}$ inch), and divided for assay.

Sampling at Cananea, Mexico.[2]—The following plant has for some years been in use at Cananea for sampling the domestic smelting ore, which contains a small percentage of copper, with silver and a little gold. The ores are comparatively uniform, soft, and lean. The ore is brought from the mines in train-loads, the cars holding about 30 tons each. These are discharged into receiving bins at the reduction works. One or more train-loads constitute a lot, say 100 tons. The ore is fed on to a screen made of steel rails with a 4-inch clearance. The under-size drops on to a moving belt, and the over-size passes through two 24- by 36-inch Blake crushers set with 4-inch maximum opening, and the ore after crushing immediately joins the under-size on the belt. This delivers the ore to a Vezin sampler, which cuts out one-tenth for sample. This is conveyed to a Blake crusher set to crush to 1-inch maximum size. After passing the latter crusher, another sample is cut out automatically by a Vezin machine, giving $\frac{1}{100}$th or 1 per cent. of the original lot as a sample. This passes to a small Gates crusher set to crush very fine, and the stream is again cut, $\frac{3}{2000}$th of the original lot being taken. This passes to fine rolls, where it is crushed, after which it is again cut and about $\frac{1}{4000}$th remains as the sample.

This sample, now under 8-mesh maximum size and weighing from 100 to 200 lbs., according to the weight of the lot, is then thoroughly mixed and split in a Jones riffle sampler, until about 4-lb. weight is obtained. The 4-lb. sample is dried and ground in a disc-grinder, split to about 2 lbs., screened and ground to pass a 100-mesh screen. The pulp thus obtained is thoroughly rolled in a mixer and divided into four pulps of about 8 ounces each, which are used for the assays. This process, which is shown in the accompanying

[1] Argall, " Sampling and Dry Crushing in Colorado," *Trans. Inst. Min. Met.*, 1902, **10**, 234–299.

[2] L. D. Ricketts, *Min. Mag.*, 1911, **4**, 128.

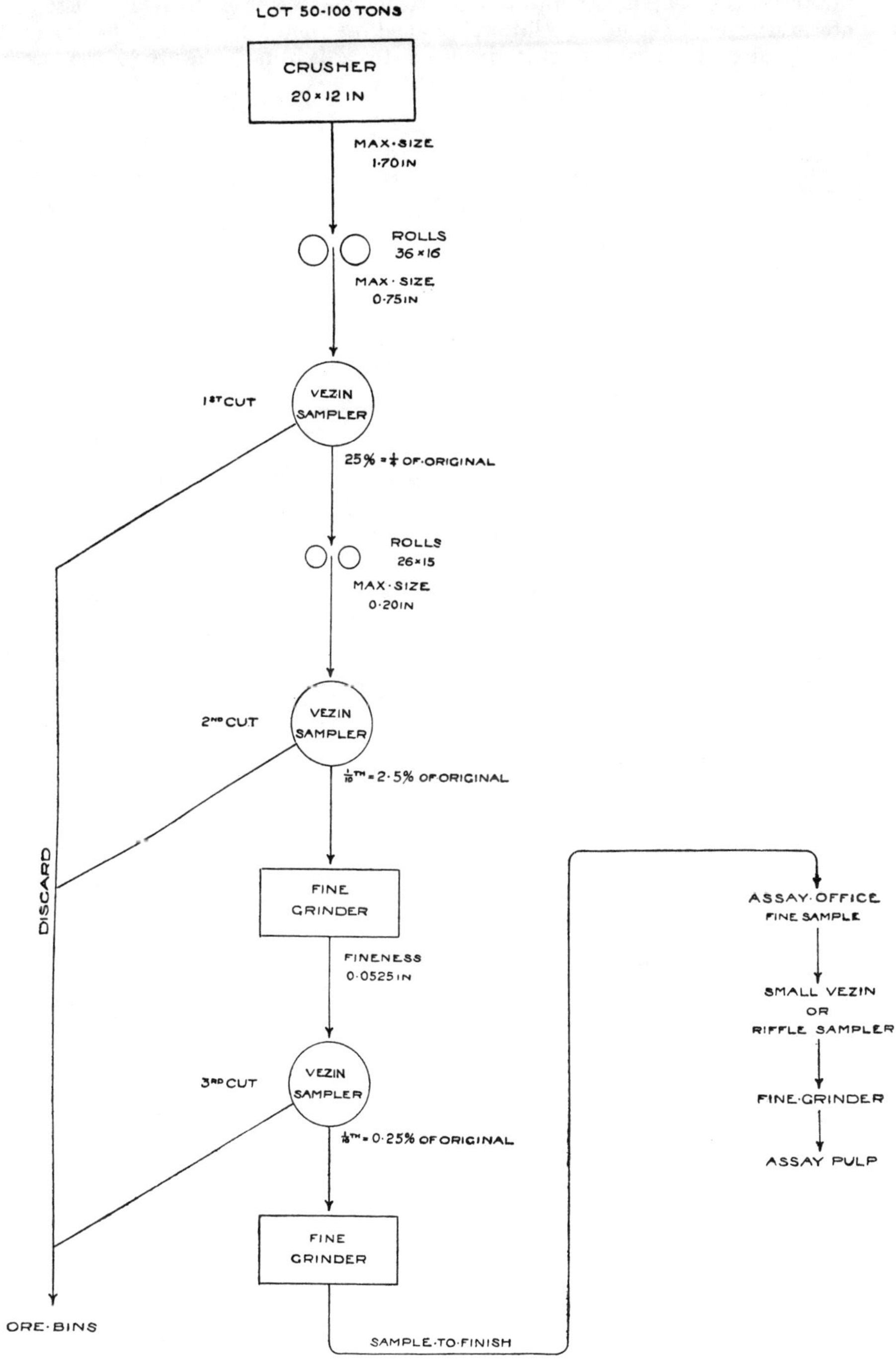
LOT 50-100 TONS
CRUSHER
20 × 12 IN
MAX·SIZE
1·70 IN
ROLLS
36 × 16
MAX·SIZE
0·75 IN
1ST CUT
VEZIN
SAMPLER
25% = ¼ OF·ORIGINAL
ROLLS
26 × 15
MAX·SIZE
0·20 IN
2ND CUT
VEZIN
SAMPLER
1/10TH = 2·5% OF ORIGINAL
FINE
GRINDER
FINENESS
0·0525 IN
3RD CUT
VEZIN
SAMPLER
1/10TH = 0·25% OF ORIGINAL
FINE
GRINDER
DISCARD
ORE·BINS
SAMPLE·TO·FINISH
ASSAY·OFFICE
FINE SAMPLE
SMALL VEZIN
OR
RIFFLE SAMPLER
FINE·GRINDER
ASSAY PULP

Fig. 84.

flow-sheet (Fig. 85), is stated to be accurate for the class of ore treated. Custom ore is sampled differently. All the discarded ore from this sampling passes to an elevator, from which it is delivered to a belt and thence to the ore beds.

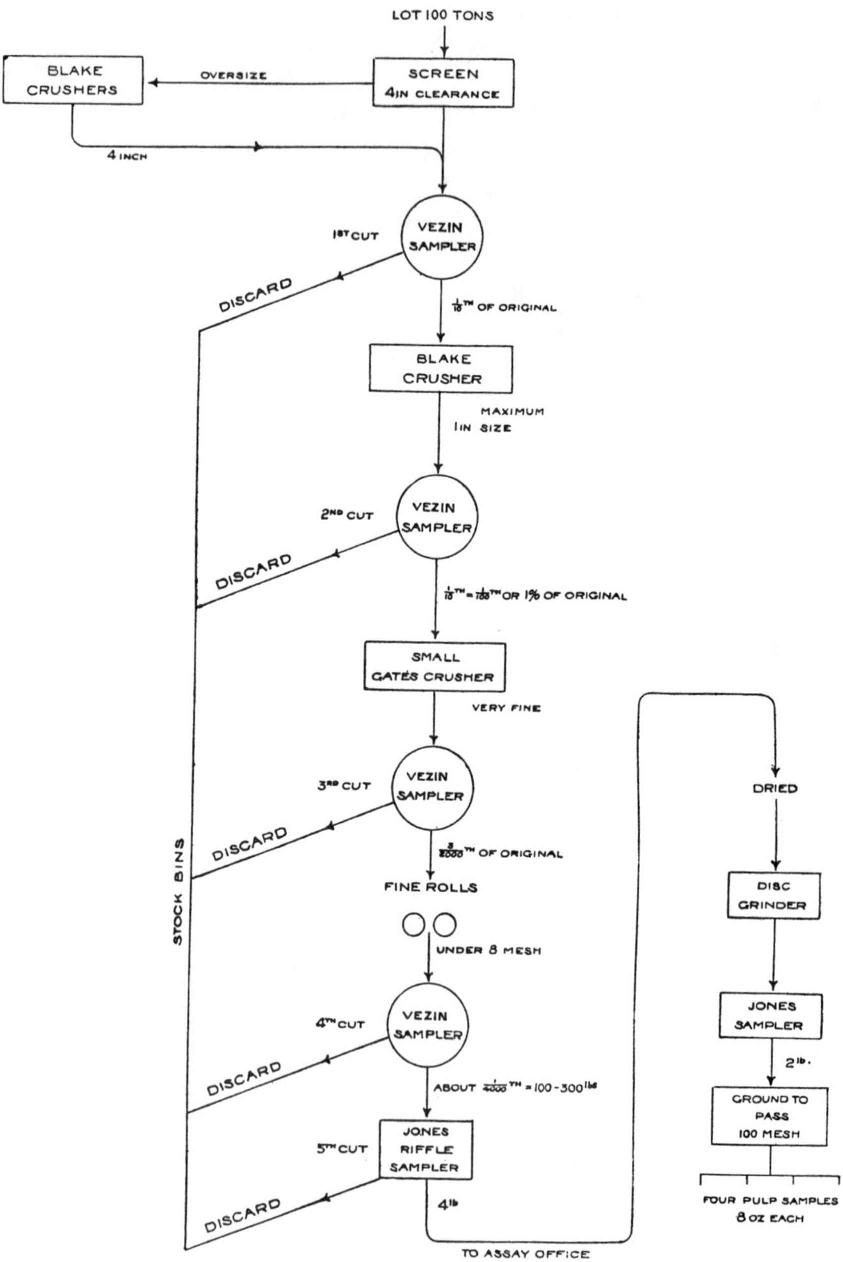

Fig. 85.

Sampling of Silver Ores containing Cobalt, Nickel, and Arsenic.—The following details of the sampling of ores containing the arsenides of cobalt, nickel, and silver, and also metallic silver, are given by Handy.[1] The ores are mined in the Cobalt district, Ontario, and are usually very rich in silver, carrying sometimes over 2000 oz. per ton.

Having first been sampled when being unloaded and a determination made of the moisture, the low-grade ore is passed through a jaw-crusher to reduce it to $1\frac{1}{2}$ inches or smaller, and is then divided by shovelling two parts to the storage bin and one to the sampling floor, from which it is again divided, one shovel to bin and one to sample. It is then reduced to $\frac{1}{2}$-inch size and quartered to not less than 1700 lbs. The ore is then crushed again to pass a $\frac{1}{4}$-inch screen, turned over twice, quartered to not less than 300 lbs., crushed to pass a 10-mesh screen, mixed and quartered to not less than 60 lbs., when it is ground to pass a 20-mesh screen and cut down by a Jones sampler to about 4 kilos, which is ground in a pebble mill to pass a 120-mesh screen. The high-grade ore containing over 500 oz. of silver per ton is crushed to 1-inch size, and alternately shovelled to the storage bin and to the rolls, which reduce to $\frac{1}{2}$-inch size. The " metallics " on the screen are collected, weighed, melted, and cast into bars which are assayed. The $\frac{1}{2}$-inch ore is coned and quartered to about 2400 lbs., crushed to pass a $\frac{1}{4}$-inch screen, again coned three times, and quartered to about 400 lbs. It is then crushed in rolls to pass a $\frac{1}{10}$-inch screen, and cut down by a Jones sampler to about 75 lbs.; again ground to pass a $\frac{1}{20}$-inch screen, and reduced to about 5 lbs. It is then finely ground, and screened on a 120-mesh screen to give " metallics " and " pulp."

The samples thus obtained are then assayed by the methods described on p. 223.

[1] J. O. Handy, Eighth Int. Cong. Appl. Chem., 1912, Sect. IIIA, abstract, *J. Soc. Chem. Ind.*, **31**, 880.

CHAPTER VIII

PREPARATION OF SAMPLES FOR ASSAY

General Remarks.—When received by the assayer, samples of ore and metallurgical products are seldom in a fit state for assay purposes: they are generally too large in bulk and too coarse. To prepare them for assay they have to be crushed finer and reduced systematically to small bulk. If the sample is in lumps, these are broken down to the size of $\frac{1}{4}$-inch cubes or less before reducing the bulk, and preparatory to fine grinding. If the original bulk is small, the entire sample may be pulverized to a fine state and then reduced to the required bulk.

In the assay office the reduction of samples is usually effected by coning and quartering, or by means of a riffle or an automatic machine such as those described later; but whatever method is employed, the sample is alternately crushed finer and reduced in bulk, until finally the mass is reduced to about 1 or 2 lbs. in weight, the whole of which is pulverized until it is fine enough to pass a sieve of at least 80 holes to the linear inch (*i.e.* 80-mesh). This finely pulverized sample is termed the *pulp*, and is used for the assay. It is transferred to a sample bag, box, or bottle and carefully labelled.

The number of crushings and reductions to which a sample has to be subjected is dependent on the nature and size of the ore particles and the weight of the sample, and the assayer, in obtaining the portion he requires for assay, must rigorously apply the principles of sampling already considered.

The preparation of assay samples has been studied by Huntoon,[1] who states that "with ores containing the values in brittle minerals the rich ore is almost entirely separated from the poor portions, if several passes are required in pulverizing and the fine material is screened out of each pass. As it is extremely difficult to properly mix pulps varying largely in value, such a procedure often leads to discrepancies in the assay results. Screens should be avoided, the pulverizing machinery being adjusted, whenever possible, to deliver a product of the desired degree of fineness, without over-size. The following weights are sufficient to ensure reliable samples:—

Size of Ore Particle.	Weight of Sample.
$\frac{1}{2}$ inch	50 lbs.
$\frac{1}{4}$,,	25 ,,
10 mesh.	6 ,,
20 ,,	3 ,,
40 ,,	24 oz.
60 ,,	5 to 6 ,,

Pulverizing to 100-mesh assists the fusion, but it is not necessary for accuracy of sampling." Experiments on Cripple Creek gold ores are quoted to bear out the statements.

It is very important that the final reduction of samples in the assay laboratory should be done with the utmost care, as there is little doubt that it

[1] L. D. Huntoon, "Preparation of Assay Samples," *Eng. and Min. Journ.*, 1911, **91**, 1249.

is in this last stage of the sampling that errors very frequently occur. The samples received at the assay offices attached to sampling-mills are usually in a state of coarse powder from the crushing-rolls or grinding-machines, and weigh from 12 to 15 lbs. They are reduced in bulk to about 2 lbs., and finally pulverized, then well mixed and divided into four or more portions of about 8 oz. each. These are put into separate sample bags or bottles and sealed. Some of these are assayed by the different assayers representing the buyer and the seller, and some are reserved for a " reference " or " umpire " assay in the case of dispute, as previously described on page 70.

"Metallics" or "Scales."—During the pulverizing of both gold and silver ores a residue of small particles is frequently left upon the sieve, and an inexperienced person, not understanding that they may be the most valuable portion of the sample is inclined to throw them aside. These particles, which are termed *metallics* or *scales*, generally consist of metallic gold or of metallic silver, associated sometimes with silver chloride, bromide, or iodide, and native copper. Owing to their malleability the " metallics " cannot be readily pounded to a fine state of division, and although a part always passes through the sieve, some of the larger particles fail to do so. The particles left on the sieve must be very carefully collected, weighed and put aside to be separately assayed (see page 190.)

The sifted sample (without the metallics) is weighed, well mixed, and assayed. The total value of the ore is found by adding the results of these two assays, as shown in the example on page 203.

" In pulverizing ores and products containing ' metallics ' it is of great importance to powder the whole of the selected sample without loss during the process, and of even greater importance to mix well the sifted portion, of which the last particles to come through the sieve are apt to be more than ordinarily rich through the grinding down of some of the metallic particles " (Beringer).

In some assay offices it is the practice to throw the metallics back into the mortar or on to the bucking-plate with a small quantity of the pulverized ore, and to continue the grinding until everything passes through the sieve. This practice is open to grave objection, as it is impossible to ensure the even distribution of the metallic particles throughout the sample; and, moreover, experience has proved that it gives rise to low results owing to the metallic particles being flattened out and left adhering to the surface of the sample grinder. For example, a sample of rich gold ore in which the gold was present almost entirely as metallic flakes, tested by F. A. Thompson[1] by both methods, gave an average result of 11·71 oz. per ton for the sample ground until all the metallics passed through a 100-mesh sieve, and an average result of 17·92 oz. per ton for the sample in which the metallics were assayed separately and allowed for in the final result.

Rich ores in which the precious metals are present largely in the metallic state are especially difficult to sample and, as indicated above, require very careful attention during the final stage of the sampling.[2] To check any carelessness in the preparation of samples, they should be weighed before and after grinding. In sifting the ore after grinding, the material left on the sieve, which usually consists of the hard gangue, must be again pulverized until the whole of it, with the exception of any metallics, passes the sieve: it is obvious that if this hard portion is rejected the sample will be vitiated, as the sifted portion is thus made proportionately richer.

[1] F. A. Thompson, *Western Chemist and Metallurgist*, 1907, **3.**

[2] Consult F. White, " Notes on Tests made to ascertain the Effect on Assays of unusually Large Particles of Gold in Ore Samples," *Trans. Inst. Min. Met.*, 1912–1913, **22.**

Crushing and Pulverizing.—The implements employed for crushing and pulverizing samples differ according to the number of samples to be dealt with. In assay offices the crushing and reduction of samples is generally done by hand: large lumps are broken down in an iron mortar or by a hammer, and the fine grinding done on a bucking-plate or in some form of laboratory grinding-machine such as those described later. At smelting-works and sampling-mills where large numbers of samples are dealt with daily, small rock-crushers with reciprocating jaws worked by hand or by steam or other power are used for breaking down large lumps; while for fine pulverization, small, steel-faced, high-speed rolls are very generally employed. In selecting a crusher or grinder for laboratory use, preference should be given to those machines that can be easily and thoroughly cleaned, and in which the wearing parts can be readily replaced. The wearing parts of all crushing and grinding implements should be made of some tough and hard material such as manganese steel or chromium steel.

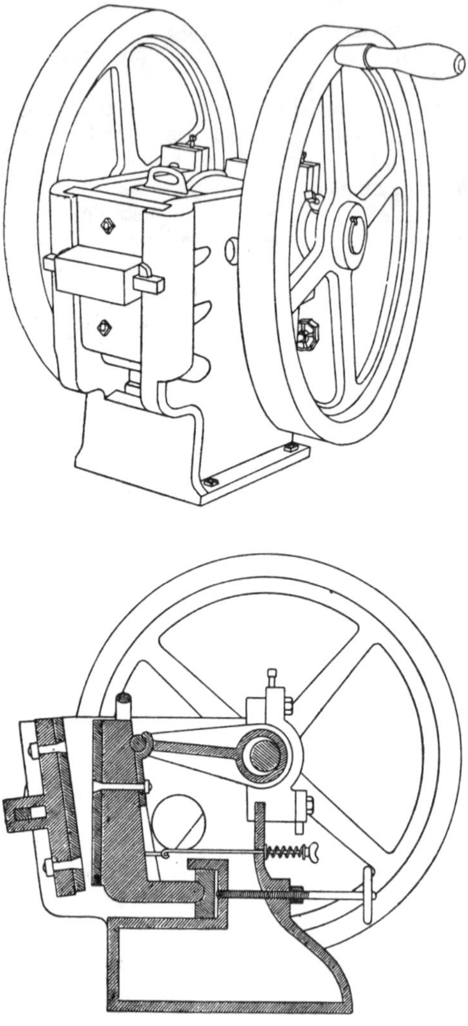

FIG. 86.—SIMPLEX HAND ROCK-CRUSHER.

When made of grey cast-iron, notable amounts of iron are ground off and dilute the finely-ground ore. This is especially noticeable in the fine grinding of exceedingly hard and tough ores, which may become diluted with iron to the extent of 2 or 3 per cent, The iron should be removed by passing a good magnet through the pulp several times. With soft ores the wear on the grinding surfaces is comparatively small. Whatever form of crushing or grinding implement is used, a supply of barren rock or sand should be kept at hand and a small quantity crushed or ground in the apparatus after each sample has been dealt with in order to clean the grinding surfaces thoroughly before starting a fresh sample. This is very important when dealing with samples of rich ores and products. In assay offices where both rich and poor samples have to be dealt with, it is a wise plan to have two grinding-machines, one for grinding rich ores and the other for poor ones; the danger of "salting" (*i.e.* contaminating) the poor sample is thus greatly diminished.

Crushing Implements.—Iron pestles and mortars are still largely used in the laboratory for breaking down lumps when the sample is of small bulk. To prevent fragments of ore flying out during the crushing, the mortar should be covered with a cloth.

Rock-Crushers.—The small crushers used in the laboratory for breaking down the lumps in comparatively large samples are very similar to those employed on a large scale.

The Simplex Crusher shown in elevation and in section in Fig. 86 may be take as a type of hand-crusher for laboratory use. In this machine the crushing is effected by a motion imparted to the vibratory jaw from the eccentric on the shaft, and as the heel of this jaw is stationary, the lower portion moves through an arc of a circle, crushing rapidly and ensuring the discharge of the product. Two adjusting screws pressing on the brass box which engages the heel of the

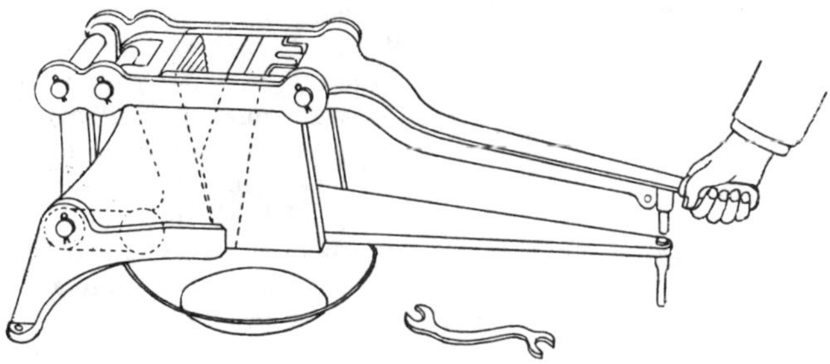

FIG. 87.—TAYLOR HAND ROCK-CRUSHER.

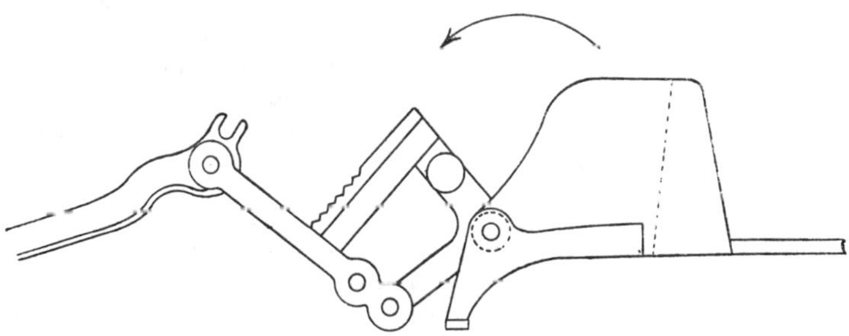

FIG. 88.—METHOD OF OPENING THE ABOVE FOR CLEANING.

vibratory jaw regulate the size of the discharge outlet, and material may be crushed from 2-inch to $\frac{1}{4}$-inch or to a 10-mesh powder.

The crusher is cleaned after pulling out the wedge holding the stationary jaw; removing this jaw gives sufficient access to the interior. By loosening the thumb-screw on the spring attachment the vibratory jaw may be thrown back, exposing the entire interior.

When very fine crushing is desired time will be saved by running the ore twice through the machine, set the first time as widely as possible, and the second time at its maximum fineness.

Great care must be taken to clean the crushing jaws before putting a fresh ore through the machine. If necessary, a quantity of barren quartz should be crushed to remove any particles of rich material that may be left adhering to the jaws. For cleaning purposes a hand bellows may be used in place of a brush.

Taylor Rock-Crusher (Figs. 87 and 88).—This is a small hand-machine of the "Blake" type. The construction of this crusher is such that the size of the discharge opening does not vary when the machine is operated; consequently, when the jaws are set close together, a product of uniform fineness is obtained. It is possible to pulverize an entire sample to a comparatively fine mesh with one feeding through this machine. The adjustment for a coarse or fine product is instantly made by turning the adjustable pin shown directly under the hand (see Fig. 87).

The straps are made of steel, and the stationary and vibratory jaw-plates are of chilled iron, the frame and hand lever are made of cast iron; all parts of the crusher are sufficiently strong to withstand hard usage.

The method of opening the machine for cleaning, by throwing the handle back, is illustrated diagrammatically in Fig. 88.

Each machine has a cover (not shown in the illustration) to prevent pieces of ore from flying out.

Grinding Implements.—After coarse crushing, the sample is finely pulverized on the buck-board or in a grinding-machine. Before the fine pulverizing can be done it is frequently necessary to dry the ore, in which case the percentage of moisture is usually determined at the same time by weighing the sample both before and after it is dried, as described on page 111. The iron pestle and mortar is sometimes used for the finer grinding of small samples, but although useful for breaking down the ore to a coarse sand, the mortar is far inferior to the buck-board for grinding material to pass through a very fine sieve. If a mortar is used for this purpose, it should be borne in mind that pulverization is rendered much easier by operating on a small quantity at a time and removing it at frequent intervals for sifting. The quantity operated upon at one time is regulated by the hardness and friability of the ore; the harder it is, the smaller the quantity taken.

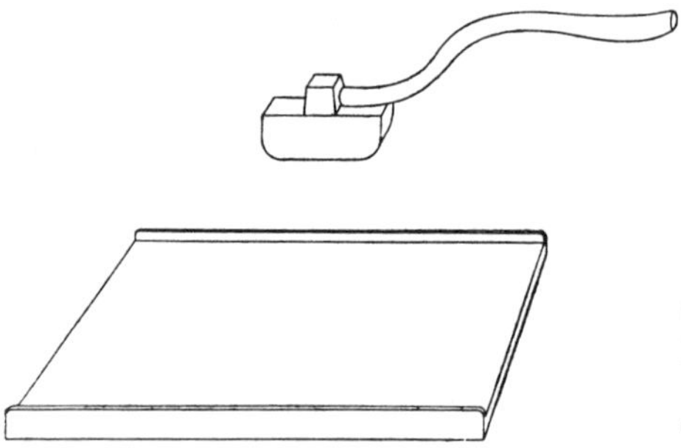

FIG. 89.—BUCKING-PLATE AND MULLER.

Bucking-plate.—For general assay-office work, the bucking-plate (Fig. 89) is most suitable for fine grinding. It is a smooth iron plate about 18 by 24 inches in size, and 1 inch thick, with a 1-inch rim on two or three sides. In the latter case the back corners of the plate are usually rounded to facilitate the removal of the fine material. The " muller " or grinder to be used with the plate is a heavy piece of iron 4 to 6 inches wide with a smooth, curved surface, weighing about 15 to 20 lbs., and fitted with a handle. To grind the coarsely crushed ore, it is spread on the plate and the muller placed upon it. The muller is then held with both hands, one holding the handle and the other firmly pressing the iron block downwards, the curved surface being on the ore. It is then moved backwards and forwards while an oscillating movement is imparted by the handle. After a little practice the method of grinding is soon acquired.

Circular buck-plates from 3 to 4 feet in diameter are also in use. They are usually placed on a low table so situated that the operator can walk round it and thus alter the position of the muller as he works it backwards and forwards. By working in this way the grinding takes place in different directions across the plate, and the wear on the surface of the plate is much more even than in the case of rectangular plates, on the surface of which grooves are liable to be formed owing to the fact that the grinding is always in one direction. The buck-plate is very effective, and saves considerable time and labour compared with fine grinding in a mortar. It is in very general use for the final pulverizing of ores. The pulp is removed from the plate and the plate cleaned by means of metal scrapers and the stiff brushes shown in Figs. 90 A and 90 B.

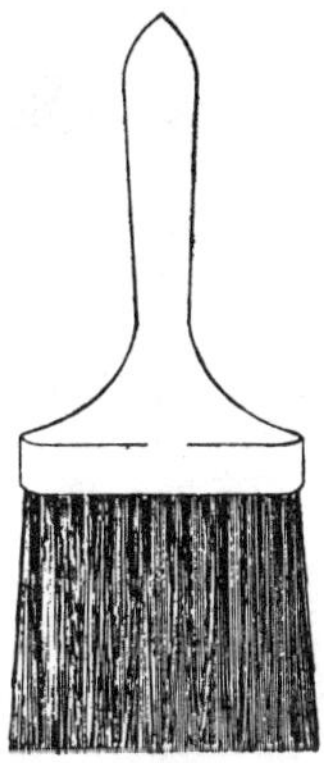

FIG. 90 A.

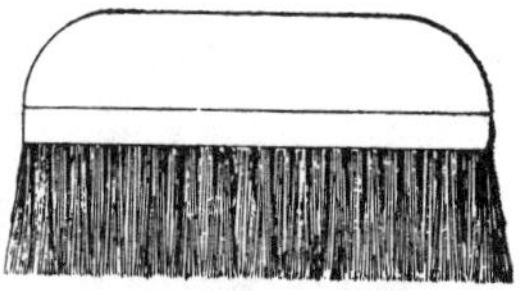

FIG. 90 B.

Grinding-Machines.—The following may be taken as types of laboratory grinding-mills, etc., that are used more or less generally as substitutes for the pestle and mortar and the buck-plate for reducing ores and other material to fine pulp for assay purposes.

End-Runner Mills.—These mills are essentially large mortars and pestles worked by hand or other power. The mills vary in size and in detail of construction, but they are all worked on the same principle. Fig. 91 shows a usual type of end-runner mill for either hand- or other power. The mill consists of an iron mortar and a "runner" or pestle, the runner being turned true to fit the inside of the mortar. The mill is shown in position for operating. The mortar is mounted either on a table which runs on ball-bearings, power being applied to the runner; or the table is made to revolve by hand or by gearing below the mortar, the runner being held down by two heavy weights as shown in the illustration. A scraper is attached to the runner frame and so arranged that it removes any material that may adhere to the surface of the runner. The runner is hinged at *A* so that it can be swung back easily for cleaning purposes and the mortar lifted off and emptied.

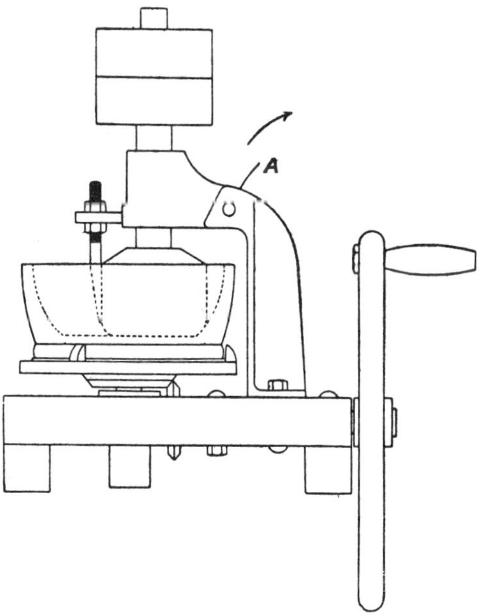

FIG. 91.—END-RUNNER MILL IN WORKING POSITION.

When the runner is raised, it rests directly over the mortar, and may be cleaned with a brush and palette knife. All ore that is cleaned off the runner falls directly into the mortar. The usual size of the mortar is $15\frac{1}{2}$ inches diameter and 5 inches deep. The size of the runner or pestle is $7\frac{1}{2}$ by 6 inches. The mill is run at a speed of 90 to 120 revolutions per minute. These mills are very suitable for grinding ore samples which have previously been coarsely crushed in a rock-breaker.

In some of the smaller machines of this type attempts have been made to make the motions of the machine identical with those given to the pestle when worked by hand, and in addition the mortar is, in some cases, made to rotate with the table to which it is attached.

Radial Bucking-Plate (Calkins).—This machine has been devised as a substitute for the buck-plate. The motions of the machine are very similar to those given to the muller when working a buck-plate by hand. Fig. 92 A shows a perspective view. Fig. 92 B is a vertical section showing the automatic refeeding of the material. Fig. 92 C shows the machine open for cleaning.

The appliance is designed to reduce crushed ore samples quickly to a pulp of any desired fineness. It consists of a semicircular iron trough the thick bottom of which forms the bucking-plate, and to the sides of which is journalled an axle carrying a shoe or muller which is pressed against the plate by two spiral springs. More or less compression is obtained by tightening or loosening these springs, and both ends of the plate depart somewhat tangentially from the true radial line of the grinding surface, which prevents the material being thrown out while in operation. At one end a screen of any desired mesh may be inserted and from time to time the muller should be swung clear of the plate as shown in Fig. 92 C and the material brushed against the screen, causing that which is fine enough to drop through. Both ends of the shoe terminate in a sharp edge which gathers up the coarse particles as it works, and as it approaches the extremity of the plate these particles roll or slide over the top of the shoe or muller and enter the cross-wise aperture in the centre of the shoe, and are thus automatically re-fed between the grinding surfaces, as shown in Fig. 92 B.

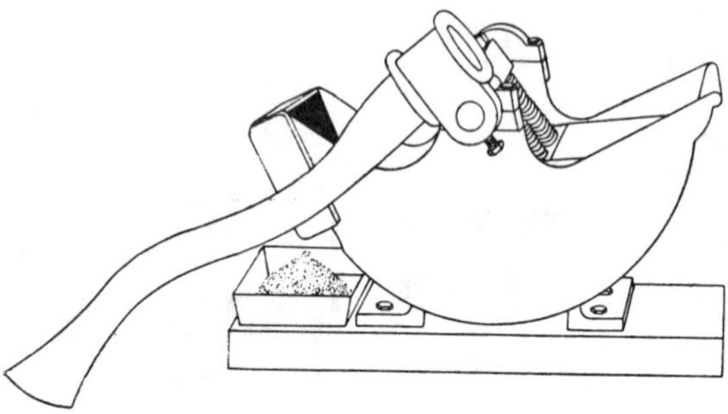

FIG. 92 A.

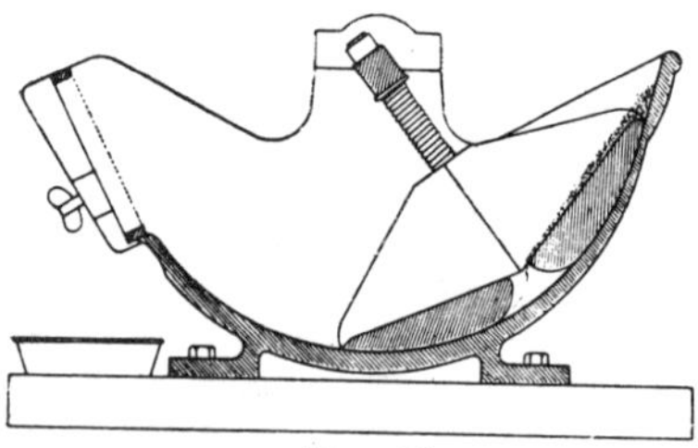

FIG. 92 B.

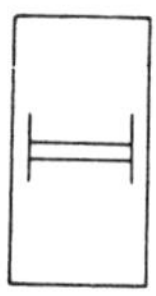

FIG. 92 D.

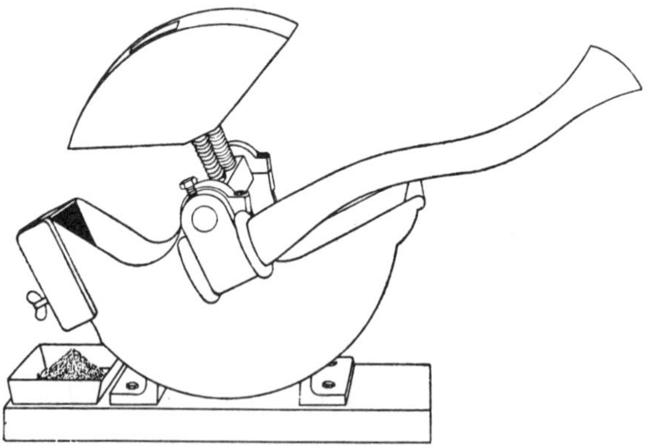

FIG. 92 C.

RADIAL BUCKING-PLATE IN WORKING AND OPEN POSITION.

The face of the shoe with aperture is shown in the illustration (Fig. 92 D) on a reduced scale.

The importance of this construction is apparent, as by gathering up the coarser particles it prevents their being surrounded and protected by the fine powder. The shoe and muller also wear to place, and this prevents their becoming ineffective through use. The screen is held in place by means of a thumb-screw, and as it may be removed easily, various meshes may be used.

The length of the machine is 18 in. and the width 6 in.

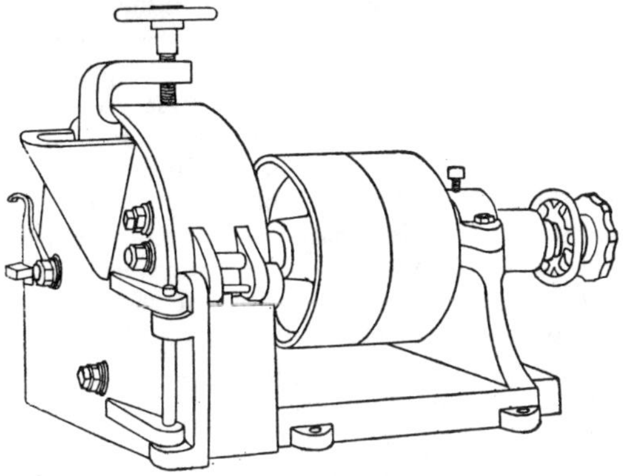

FIG. 93.—STURTEVANT DISC-GRINDER.

FIG. 94.—THE ABOVE MACHINE IN OPEN POSITION.

Disc Grinders.—This type of grinding-machine was introduced some years ago and has gradually gained favour. Figs. 93 and 94 show a grinder of this type made by the Sturtevant Engineering Co., Ltd. The construction of the machine is shown clearly in the illustrations. The grinding is done between two hard steel discs or grinding plates, one of which is stationary and the other revolves at a speed of 750 to 800 revolutions per minute. The machine is fed through the spout or hopper in the door to which the stationary disc is attached. The door and cover are hinged, and can be swung open as shown in Fig. 94 to give full access to the interior for cleaning. This arrangement enables all the pulp to be brushed into a pan, for which provision is made when the machine is mounted.

The space between the discs is regulated by the hand-wheel shown, so that the machine will deliver a coarse or fine product as desired. The stationary disc is so arranged that it can be moved frequently from its centre, thus changing the circle of the grinding surfaces. By this means any cuts or scores made on the surfaces when coarse or very hard-cutting ores are being treated will grind themselves out. This simple arrangement also enables the plates, when grinding most ores, to be kept true, so that the output does not become uneven in character.

To give the best results, the hand-wheel must be turned slightly as the grinding proceeds, and very frequently when the ore is cutting.

The capacity of pulverizers of this type depends on the fineness to which the material has been previously crushed. If the product fed into the pulverizer is ¼-mesh, it will be slower than if this product were crushed finer before being put through the machine. If large quantities are to be pulverized, it will usually be accomplished in less time if the product is first ground to about 10-mesh and then re-fed through the machine and ground to 100-mesh or other desired mesh. The life of the grinding-plates is naturally dependent on the character of the material treated, but when worn out they are easily replaceable. As a general rule, one set of discs is capable of grinding about 2000 to 3000 lbs. of quartz ore of medium size to a pulp of very fine mesh. The grinding-plates in machines of this class must be made of steel of very hard quality, otherwise the product will be contaminated with notable amounts of iron. With very fine pulverizing of hard and tough ores the product may contain as much as 2 or 3 per cent. of iron, even when grinding-plates of good quality are used.[1] This fact constitutes one of the chief drawbacks of disc grinding-machines. With softer ores the dilution with metallic iron may be nearly or quite negligible. The amount of metallic iron in a sample may be determined by digesting a weighed portion of the pulp overnight in a solution of copper sulphate and determining the iron dissolved in the solution ; or the ore may be treated with dilute sulphuric acid and the iron that passes into solution determined by titration with permanganate or dichromate.

The grinding surfaces are inclined to clog with soft ores containing much limestone or iron oxide.

When cleaning these machines before starting a new sample a hand bellows may be used with advantage in place of a brush. The machine described requires about 3 H.P. to run to full capacity, but smaller machines requiring only 1 H.P. are obtainable.

Cone-and-Disc Grinder.—This form of laboratory grinder has come into favour in recent years and has given good results. A typical example of this class is the Braun Grinder illustrated in Fig. 95.

This machine consists of two especially designed grinding-discs of hard steel, one of which is stationary whilst the other is made to revolve at a comparatively high speed.

The upper or stationary disc has a corrugated surface, and is screwed to the base of the hopper into which the ore is introduced. The lower grinding surface consists of a second disc with a coned centre, which is corrugated and secured to a vertical shaft and is usually operated at a speed of about 200 revolutions per minute. The vertical shaft runs on ball- and roller-bearings to reduce friction to a minimum.

The ore is fed into the hopper and falls between the grinding surfaces, where it is speedily reduced to very fine pulp. When completely pulverized, the pulp falls directly into a small bucket, preferably of copper or brass with concave bottom, suspended from two screws and fitting closely so that it is dust-proof, The hopper is also provided with a readily adjustable sheet-metal cover to prevent

[1] Consult V. Lenher, "Contamination of Laboratory Samples by Iron derived from Crushing Machinery," *Chem. News*, **106**, October 11, 1912.

loss by dusting. The distance between the grinding surfaces is readily adjusted by means of the milled screw at the top of the revolving shaft, so that any desired degree of fineness can be obtained.

The machine will readily pulverize 4 oz. of ordinary quartzose material from coarse powder to 60-mesh in 2 minutes, to 80-mesh in 2½ minutes, and to 100-mesh in 3 minutes.

To facilitate cleaning, the revolving disc is lowered by simply releasing the hinged adjusting-screw at the top of the vertical shaft, as shown in the illustration. By this means the grinding-plate is lowered sufficiently to render all parts readily accessible for thorough cleaning with a brush.

The gears are enclosed to prevent clogging with dust and to obviate all danger. The machine may be operated by hand or by ½ H.P. motor. The total

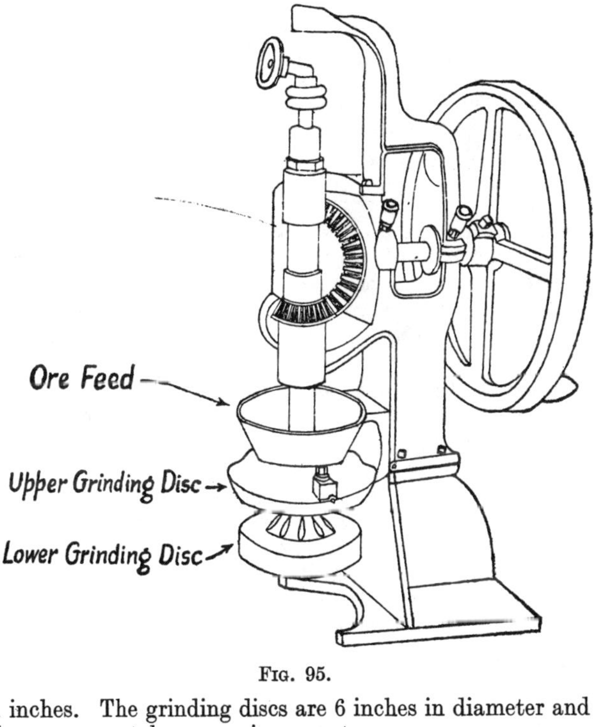

FIG. 95.

height is 31 inches. The grinding discs are 6 inches in diameter and are readily replaced when worn out by removing a nut.

Fineness of Samples.—Opinions differ as to the degree of fineness to which assay samples should be pulverized, but recent experience has proved that better results are generally obtained when the samples are crushed finer than was formerly thought to be necessary. There is now general agreement amongst assayers that samples should never be coarser than 80-mesh, and many assayers recommend 100-mesh and some 200-mesh. The author's experience is in agreement with that of many other assayers that " as a general rule all ores require crushing to 100-mesh, and that finer crushing may be dispensed with in the case of simple ores, but that complex ores, such as tellurides, should be crushed as finely as possible." The want of close agreement in assay results may frequently be traced to the use of samples that have not been ground sufficiently fine.

The influence of fine crushing on the assay value in the case of South African gold ores is shown in the following Table V, which gives the results of experiments by A. Whitby.[1] The figures show that very finely crushed material gives higher results on assay by fusion than does the same material when less finely crushed. Whitby recommends the 100-linear mesh as an approximate standard.

TABLE V.—SHOWING THE INFLUENCE OF FINE CRUSHING ON THE ASSAY VALUE (*A. Whitby*).

Assay No.	Fineness of pulp.			
	30–60-mesh.	60–90-mesh.	150-mesh.	150-mesh.
	oz. per ton.	oz. per ton.	oz. per ton.	oz. per ton.
1	4·10	4·20	4·52	4·88
2	5·00	5·00	5·55	5·70
3	5·10	4·40	5·64	5·28
Average Result	4·73	4·53	5·23	5·28

In mine-assay offices where the assaying is largely confined to one class of ore, a few tests should be made on the ore to ascertain what degree of fineness is necessary to give accurate results: if 80-mesh gives correct results, finer grinding is obviously unnecessary. However in the case of ores with which the assayer is not familiar, it is better not to omit grinding the ore sufficiently fine to pass a 100-mesh.

Screens or Sieves.—To ensure, so far as possible, uniformity in the size of the particles, the finely pulverized ore must be sifted. For this purpose sieves of various sizes and different degrees of fineness are necessary. Two kinds of sieves are in general use, open or " drum " sieves and closed or " box " sieves. The *open sieve* is a round wooden frame with a meshwork of brass or iron wire gauze stretched over one end. For finely pulverized gold and silver ores, where it is very desirable to prevent loss, the *box sieve* of tinned iron is frequently employed. It consists of three parts: the sieve, a cover which fits tightly on the top to retain the dust, and a tightly-fitting box or receptacle to receive the sifted pulp. Iron wire gauze is used for sieves coarser than, say, 20-mesh, while brass wire cloth is invariably used in the laboratory sieves of finer mesh. The size to which the ore is pulverized is generally stated in terms of *mesh*, that is, the number of holes per linear inch measuring from centre to centre of the wires. Thus a 40-mesh sieve would have 40 holes to the linear inch, or 1600 to the square inch; and a 90-mesh sieve 90 holes to the linear inch, or 8100 to the square inch.

Wire Screen Mesh.—In view of the tendency of modern assaying practice towards very fine grinding, it may be advisable here to make brief reference to wire screens. To those familiar with laboratory wire screens, the mesh number, such as 40-, 80-, or 100-mesh, represents a fairly definite maximum size of ore particle. It is, however, generally known that the number of meshes to a linear inch does not clearly define the size of opening through which the ore particle passes, as this is dependent on the thickness of the wire used. For example, a 26-mesh sieve made from No. 26 wire (·0181 inch) gives an opening or mesh

[1] " Gold Assaying: Note on the Influence of Fine Crushing on the Assay Value," A. Whitby, *J. Chem. Met. Min. Soc. S. Africa*, 1904, **5**, 95.

of ·0203 inch, while a 26-mesh sieve made from No. 36 wire (·0090 inch) gives an opening of ·0294 inch, a variation of over 30 per cent.[1] With coarse sieves the error is not great and often unimportant, but with the finer sieves the wire itself occupies so much space that the size of particle passed by the sieve may be anything from about one-half to about four-fifths of the size indicated by the word "mesh." It is well therefore, in using fine sieves for the preparation of assay samples, to bear in mind that the stated mesh is not always an indication of the degree of fineness of the pulp obtained. A sieve of 120-mesh, if made with No. 43 gauge wire, will pass particles not exceeding $\frac{1}{200}$ inch in diameter; but this size particle will also pass a 130-mesh sieve if made of No. 45 wire, and a 140-mesh sieve if made from No. 46 wire.

Owing to great variations in the thickness of wire in different screens it is always necessary, in referring to wire-screen mesh (if anything approaching accurate definition is required), to give either the gauge of the wire or the actual measurement of the opening between the wires.

The importance of knowing the actual dimensions of sifted ore particles, both in laboratory work and in making grading tests with dressing-plant, etc., has long been recognized, and has resulted in the adoption of several series of standard screens. As a result of careful investigation by a special committee, the following Table VI, of standard laboratory screens, was established, and adopted in 1907 by the Institute of Mining and Metallurgy, London.[2] It is intended for use in making grading tests and for the correlation of screens used in commercial or other work. In this series, which has since been very generally adopted, the diameter of the wire is equivalent to the linear width of the opening, *i.e.* the open spaces total 25 per cent. of the entire screen surface.

TABLE VI.—I.M.M. Standard Laboratory Screens.[3]

Mesh or Apertures per linear inch.	Diameter of Wire.		Aperture.		Screening Area.
	in.	mm.	in.	mm.	per cent.
5	0·1	2·540	0·1	2·540	25·00
8	0·063	1·600	0·062	1·574	24·60
10	0·05	1·270	0·05	1·270	25·00
12	0·0417	1·059	0·0416	1·056	24·92
16	0·0313	0·795	0·0312	0·792	24·92
20	0·025	0·635	0·025	0·635	25·00
30	0·0167	0·424	0·0166	0·421	24·80
40	0·0125	0·317	0·0125	0·317	25·00
50	0·01	0·254	0·01	0·254	25·00
60	0·0083	0·211	0·0083	0·211	24·80
70	0·0071	0·180	0·0071	0·180	24·70
80	0·0063	0·160	0·0062	0·157	24·60
90	0·0055	0·139	0·0055	0·139	24·50
100	0·005	0·127	0·005	0·127	25·00
120	0·0041	0·104	0·0042	0·107	25·40
150	0·0033	0·084	0·0033	0·084	24·50
200	0·0025	0·063	0·0025	0·063	25·00

[1] Argall, *Trans. Inst. Min. Met.*, 1902, **10**, 244.

[2] *Bulletin No.* 38, Institute of Mining and Metallurgy, Sept., 1907. For other standard series and bibliography of the subject, see paper by T. J. Hoover on "A Standard Series of Screens for Laboratory Testing," *Trans. Inst. Min. Met.*, 1910, **19**, 486.

[3] Sole manufacturers, N. Greening & Sons, Ltd., Warrington, England.

Reducing the Bulk of Samples.—In the laboratory, large samples are reduced in bulk by the method of coning and quartering, fully described on p. 83, or one of the following methods may be employed.

Riffle Sampler (or sampling tin).—This consists of a frame containing a number of narrow troughs or "riffles" placed parallel to one another, with spaces between, equal in width to the width of the riffles (see Fig. 96). In operation, an even stream of ore is allowed to fall on to the riffles from a scoop the same width as the series of riffles. Half the ore falls into the riffles, and is reserved for sample, and the other half falls through the open spaces and is rejected or reserved as a duplicate sample. A repetition of the process, with recrushing when necessary, reduces the sample to any required extent. The width of the riffles varies according to the size of the ore particles, but they should never be less than $3\frac{1}{2}$ times the diameter of the largest ore particle to be passed through them.

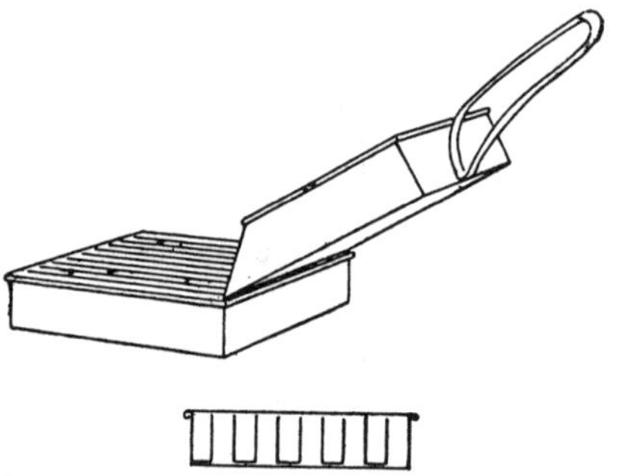

FIG. 96.—RIFFLE ORE SAMPLER AND SCOOP.

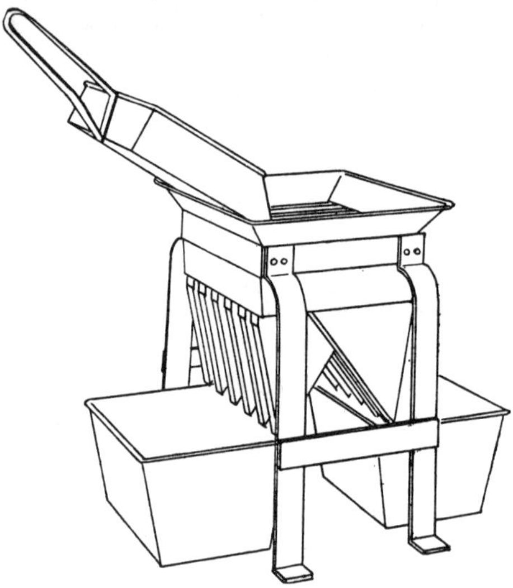

FIG. 97.—JONES RIFFLE SAMPLER.

For material with particles not more than a $\frac{1}{4}$ to $\frac{1}{3}$ inch in diameter, a sampler with riffles 1 inch wide is suitable, while for finely-ground materials riffles about $\frac{3}{8}$ inch wide are convenient. From six to twelve riffles are usually fitted to each sampler. Care must be taken not to overload the riffles, otherwise the samples will not be accurate—one half will contain a larger percentage of coarse material, and the other half a larger percentage of fine material.

To prevent errors from overloading, riffle samplers of the Jones type, shown in Fig. 97, were introduced. In samplers of this type, instead of alternate riffles and spaces, two sets of riffles with inclined bottoms sloping in opposite directions are placed side by side, and so arranged that the stream of ore, as it falls upon the riffles, is deflected alternately to the right and to the left, thus yielding two samples which fall into separate pans as shown. For large samples this type is preferable to the flat-bottomed riffles. To obtain accurate results, the riffle samplers of both types must have an even number of divisions, and the scoop used in feeding the ore on to the riffles must be the same width as the sampler, so that the ore is evenly distributed over the whole series of riffles.[1] Careful experiments have proved that, under these conditions, each half of the sample is a representation of the whole. Riffle samplers are clean, convenient, and rapid in operation, and are now extensively used for reducing the bulk of laboratory samples.

Umpire Sampler. — A sectional view of this machine is shown in Fig. 98. The pulp is charged into a hopper, which is continually agitated, causing the material to fall in a steady stream into two buckets, which are revolved in opposite directions by means of gear-wheels. Each bucket is divided into four divisions, two being closed and two open; therefore, if 16 lbs. of ore be fed into the machine from the hopper, 8 lbs. will remain in the upper bucket, 4 lbs. in the lower bucket, and 4 lbs., or one-quarter of the original amount, will pass into the receptacle at the bottom. By repeating the operation, the original sample may be reduced to $\frac{1}{16}$, $\frac{1}{64}$, $\frac{1}{256}$, etc. as desired.

A cam on the crankshaft agitates the hopper by striking a strap, which is attached to the hopper with a coiled spring, and by means of an eccentric lever the force of the blows may be varied at will. The upper and lower buckets may be quickly removed, and all portions are readily accessible for cleaning, making it impossible for subsequent samples to be contaminated.

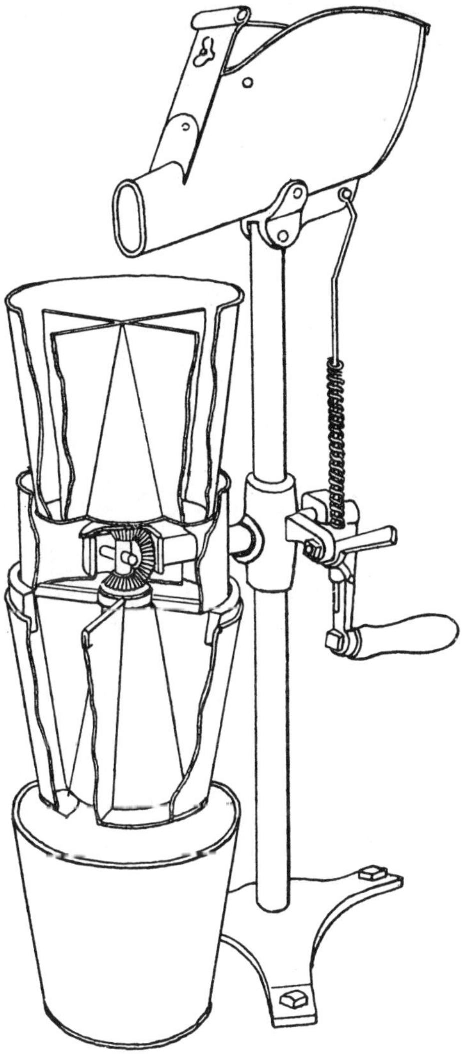

FIG. 98.—UMPIRE SAMPLER, SHOWING REVOLVING BUCKETS.

Sample Dividers.—After the pulp sample has been obtained, it usually happens that several persons wish to have a portion of it for separate assay; thus the sample may be divided into three parts—one for the buyer, another for the seller, and a third for the referee. Sometimes six or more distinct samples are wanted, and in order that the results of the separate assays may agree, it is clearly essential that all the samples should be identical—that each should have a fair proportion of the fine and the coarse, the rich and the poor, the heavy and the light particles. A very usual method of dividing small lots of pulverized

[1] See "Accuracy of Riffle Ore Samplers," L. D. Huntoon, *Eng. and Min. Journ.*, 1910, **90**, 62–65. A description, with working drawings, of a complete set of riffle samplers arranged by W. A. Brown for reducing the bulk of mine and mill samples is given in *Eng. and Min. Journ.*, 1907, **87**, 232.

ore in the laboratory consists in spreading the pulp out in a thin layer with a spatula on a sheet of glazed paper or oilcloth and digging out small portions at regular intervals. A small scoop made of sheet brass is preferable to a spatula for digging out the portions. Before spreading the ore, it must be thoroughly mixed by drawing up first one corner and then another of the paper or oilcloth so as to roll the ore over on itself. After one sample has been taken as described, other samples may be obtained by repeating the process. The pulp may also be divided by quartering in the usual way after being well mixed. In assay offices where a large number of samples have to be dealt with, special " sample dividers " are frequently used.

Dividing Machine (Clarkson).—The divider is shown in elevation in Fig. 99 A, and in section in Fig. 99 B. It consists of a rotating hopper *A*, a cone *B*, and a dividing-cup *C*. The material to be divided is formed into a rotating annular stream by means of the rotating hopper and cone, and at the same time thoroughly mixed by three arms in the upper part of the hopper. The stream of ore splits on the cone *B*, and is divided into six samples by means of the

FIG. 99 A.

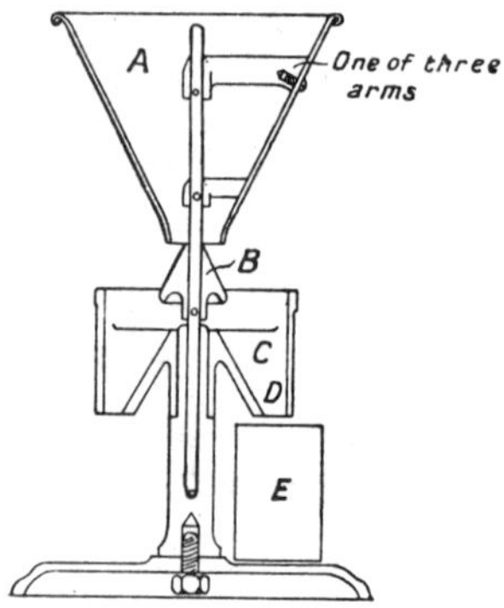

FIG. 99 B.

SAMPLE DIVIDING MACHINE.

dividing-cup *C*, which is made with six equal segments, delivering the samples by the spouts *D* into the tins (or bottles) *E*, one of which only is shown. To use the divider, the hopper is set spinning by a touch of the hand, then the sample is thrown in, and it is divided into six equal and exactly similar portions in less than the same number of seconds. The speed of the hopper should be from 60 to 70 revolutions per minute. It must be set turning before the material is charged in. The height of the dividing-cup can be adjusted to suit the size of bottle or tin, and the hopper and cup can be removed in two or three seconds for thorough cleansing. The material put through the divider should not be coarser than $\frac{1}{8}$-inch. This machine may be also used as a mixer for mixing fluxes, etc. The machine can be arranged for more or less than six samples according to requirements.

Mixer and Divider (Bridgman's).—These are used for the final mixing and distribution of assay samples. The pulverized material is introduced into the large covered funnel or mixer (Fig. 100), the outlet being first closed by the thumb or finger as may be most convenient. The mixer and contents are then well shaken for a few moments, and then, with opened outlet, passed to and fro over the set of distributing funnels (divider) and bottles as shown. With very finely pulverized or very light material the flow may be assisted bv a slight shaking or tapping with the hand. The little skill necessary to use the apparatus successfully is readily acquired. The mixer will also be found very useful for the rapid and thorough mixing of crucible assay charges, and all other work of similar character.

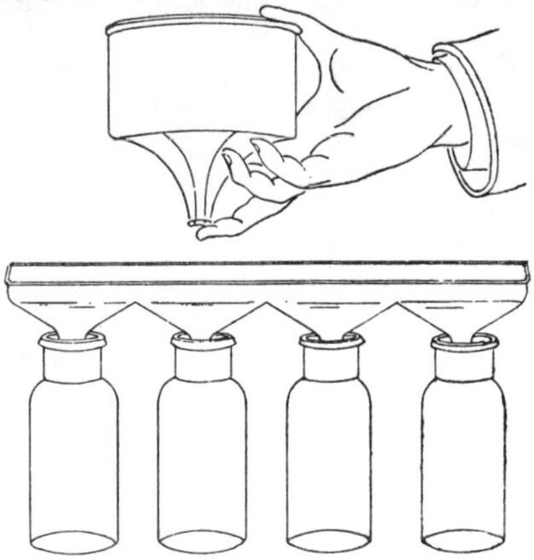

FIG. 100.—SAMPLE MIXER (*above*) AND DISTRIBUTING FUNNELS.

Labelling Samples.—The prepared pulp samples are stored in proper sample bottles, bags, or boxes, and clearly labelled by numbers for easy reference. Many assayers and smelting works use linen or strong manilla paper bags, printed with the name of the assayer or company, and provided with a tag for convenience in noting details of the samples or in sending them by post.

Drying: Determination of Moisture.—The term "moisture" is applied to all water lost on heating at 100° C., and ore which has been dried at this temperature is termed "dry ore." As settlements in the sale and purchase of ores are made on the "dry ore," it is very necessary that the samples received by the assayer should be carefully dried before being assayed, as they frequently contain more or less moisture, even though they may have been previously dried. "The reason for reporting assay results on the 'dry ore' will be apparent when it is borne in mind that 'dry ore' has a constant composition, and the results of all assays of it will be the same, no matter when made. On the other hand, the percentage of moisture may vary from day to day, and it is best to limit this variability of the moisture by considering it separately and thus avoid having the percentage of the gold or silver rising and falling under the influence of the weather" (Beringer). The determination of the moisture present in any given sample of ore is quite as important as the determination of the metallic contents, for the moisture sample decides what percentage of the weight is to be excluded in the settlement between buyer and seller. This apparently simple process must, therefore, be conducted with care. In practice the ore is generally dried during sampling, and the moisture determined at the same time by the samplers, the percentage of moisture found being stated on the label attached to the sample sent to the assay office. It is frequently assumed that the ore dried by the sampler contains no moisture: this may or may not be the case, according to the nature of the ore. If the sample contains hygroscopic minerals it may reabsorb moisture, especially if a considerable time has elapsed between the drying of the ore and the assay of it. It is a safe rule, therefore, to dry all ores received for assay and to expel any moisture which may have been absorbed during their passage from the sampling floor to the assay office, unless the assayer knows from the source or nature of the sample that this step is unnecessary. The method usually adopted is to take from one-half to one pound of the coarsely-crushed ore and dry it in a shallow iron pan, Fig. 101 (such as a frying-pan), heated over a gas burner or in an ordinary oven, with frequent stirring, until no deposition of moisture takes place when a cold piece of mirror is held over it. The ore is

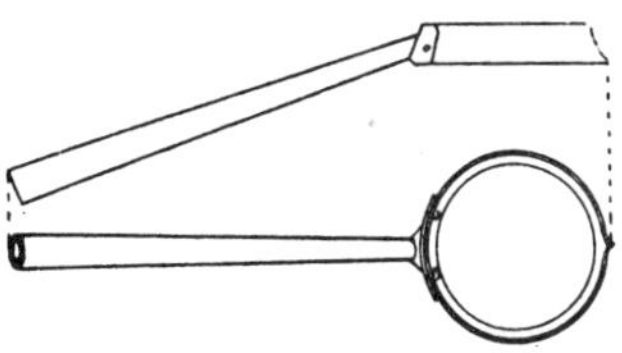

FIG. 101.—ORE-DRYING PAN.

then weighed, heated again for about thirty minutes, and re-weighed: when two successive weighings agree, the ore is considered dry. The percentage of moisture is calculated from the loss of weight. Thus the difference between the weight of the original ore and the dried ore is the loss by driving off the water; this difference, divided by the amount of ore taken, and multiplied by 100, gives the percentage of moisture in the ore. The heating should be carefully conducted, and the temperature kept as near 100° C. as possible, especially in the case of ores or products that oxidize readily, such as those containing finely-divided sulphides, etc. Samples of ore and products that are hygroscopic should only be dried shortly before the assay is made. In the case of small samples the whole sample is dried, and frequently the drying is done in a porcelain basin heated over a sand-bath or water-bath; but if smaller quantities than those stated above are used, extra precaution must be taken, especially in the weighing.

CHAPTER IX

FLUXES, ETC., USED IN ASSAYING

THE dry, or furnace, method of assay is almost invariably employed for the determination of the precious metals in ores, etc., the ore being brought into a liquid state by fusion at a high temperature with suitable reagents or "fluxes." Very few of the constituents of ores are fusible when heated alone at the temperature attainable in an assay furnace, but if certain solid reagents are mixed with the pulverized ore, compounds are formed which fuse readily at a moderate temperature, producing a molten mass sufficiently liquid to enable a heavy metal such as lead, that may be diffused through it in the form of small globules, to sink and unite into one mass at the bottom of the crucible. The reagents added for the purpose of forming fusible compounds with infusible or difficultly fusible bodies are termed *fluxes.*

Slag.—The fusible compounds resulting from the fusion of fluxes with infusible or difficultly fusible bodies are termed *slags.* For example, in auriferous quartz the gold exists in the metallic state and is diffused through the mass in particles and in extremely small proportion in comparison with the quartz particles. When such powdered quartz is heated to a temperature far above the melting-point of gold (1064° C.), no separation of the metal takes place, owing to the infusibility of the quartz. If, however, sodium carbonate is mixed with the quartz, a fusible sodium silicate will be formed, and the melted gold, by virtue of its high density, will then separate out easily. The sodium carbonate in this case would be designated a flux, and the resulting sodium silicate a slag. The quantity of gold (and silver) present in ores is generally so small that it tends to remain diffused through the molten mass, and in assaying, it is necessary to collect it by means of metallic lead, which alloys with the gold and silver and sinks to the bottom of the crucible. The metallic lead is obtained by the reduction of lead oxide added for the purpose.

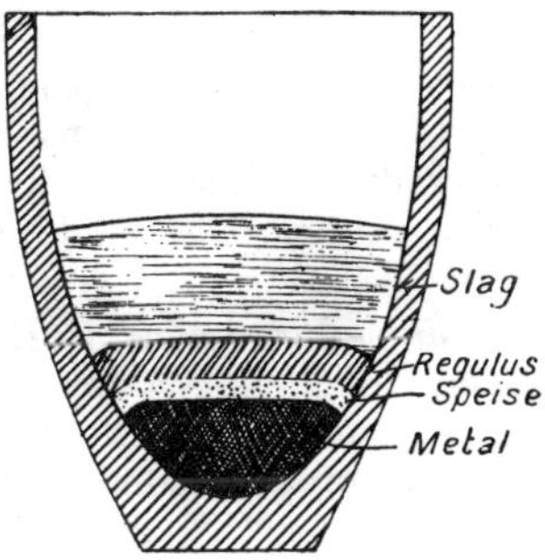

FIG. 102.—PRODUCTS OF ORE FUSION.

In addition to slag and metallic lead, other products may be obtained as the result of the fusion of an ore, according to the constituents present. If metallic sulphides are present, an artificial sulphide or *regulus* (sometimes termed *matte*) may be formed; and if the ore contains arsenical minerals, a compound of a metal or metals with arsenic, termed a *speise*, may result. Assuming all these substances to be present in a charge, they would separate according to their densities, when the fused mass solidified, in the order shown in Fig. 102. At the bottom of the crucible a button of lead would be found; above this a thin layer of speise; then a regulus, next a slag, and, in special cases, on the top of this a layer of more fluid slag, consisting usually of fusible alkaline chlorides and sulphates. A regulus, which is a compound of one or more metals with sulphur, is a yellowish-grey metallic-looking mass, usually brittle and often crystalline. A speise is usually hard, brittle, and greyish-white in appearance.

The correct use of fluxing materials is a subject of considerable importance to the assayer, and ought to be regulated according to chemical principles. It will be conducive to the thorough appreciation of the principles which should underlie the fluxing of ores, etc. if reference is first made to the action of the various reagents usually employed for this purpose. The principal fluxes employed in assaying are as follows :—

Basic Fluxes :

Sodium carbonate (Na_2CO_3).
Sodium bicarbonate ($NaHCO_3$).
Litharge (lead monoxide) (PbO).
Red lead (lead oxide) (Pb_3O_4), or (PbO,Pb_2O_3).
Hœmatite (iron oxide) (Fe_3O_4).
Lime (CaO).

Acid Fluxes :

Borax (sodium pyroborate) ($Na_2B_4O_7$. $10H_2O$).
Borax glass ($Na_2B_4O_7$).
Silica (SiO_2).
Glass.

Neutral Fluxes :

Fluor-spar (calcium fluoride).
Common salt (sodium chloride).

Sodium Carbonate (commonly called soda), Na_2CO_3.—The powdered anhydrous sodium carbonate is the most convenient for fluxing purposes. It is a readily fusible basic flux, and when heated with silica (quartz) forms a fusible sodium silicate and liberates carbon dioxide (carbonic acid).

$$Na_2CO_3 + SiO_2 = Na_2SiO_3 + CO_2.$$

Sodium Bicarbonate, $NaHCO_3$, may be used, but it is much less convenient, as the same bulk contains a smaller quantity of soda, the fluxing constituent. When heated, water and carbonic acid are given off at a comparatively low temperature, and sodium carbonate left.

$$2NaHCO_3 = Na_2CO_3 + CO_2 + H_2O.$$

The fluxing power of the bicarbonate is about two-thirds that of the carbonate. The fusion of charges containing sodium carbonate or bicarbonate must not be performed rapidly, otherwise the escaping gas will cause frothing and loss through the boiling-over of the charge. This is very liable to occur when bicarbonate is used, and on that account its employment is to be deprecated. Potassium carbonate may be wholly or partially substituted for sodium carbonate, but it has the disadvantage of deliquescing and of being more costly. A mixture of equal parts of potassium and sodium carbonates forms a very suitable flux and is sometimes employed. Owing to their great fusibility, the alkaline-carbonates can retain in suspension, without losing their fluidity, a large proportion of pulverized infusible materials such as an earth, charcoal, graphite, etc.

Lead Oxides.—These are valuable basic fluxes, used in the form of red lead (PbO,Pb_2O_3) or of litharge (PbO). With silica, lead monoxide forms a very fusible lead silicate :—

$$PbO + SiO_2 = PbSiO_3.$$

With silicates, also, it forms very fusible double silicates, and in the presence of silicates or of borax it acts as a flux for lime and magnesia. Metallic oxides that are infusible, or difficultly fusible alone, are easily dissolved by lead oxide, with which they form a strongly basic slag which has a very corrosive action on the crucible. This is particularly the case with fused mixtures of the oxides of lead and copper, which readily unite with the silica of the crucible. Red lead, when fused, is decomposed into litharge and oxygen, so that whether the lead oxide used as a flux is added in the form of red lead or of litharge the lead will exist in the slag as monoxide (litharge) :—

$$2Pb_3O_4 = 6PbO + O_2.$$

The excess of oxygen in red lead is thus available for oxidizing purposes, and on that account red lead is in some cases to be preferred to litharge. Red lead of good quality is often more readily obtained than good-quality litharge, and is therefore more frequently used than litharge. Lead monoxide, in addition to its use as a flux, is used for obtaining the lead necessary for collecting the precious metals. For this purpose it is mixed with a reducing agent such as charcoal or flour, whereby metallic lead is obtained, as follows :—

$$2PbO + C = 2Pb + CO_2.$$

As red lead and litharge always contain silver in sensible quantity, this must be determined as described on p. 131, and taken into account in returning the assay result. The following more or less frequent constituents of ores also act as fluxes when present :—

Iron Oxides are of frequent occurrence in ores, either as hæmatite (natural ferric oxide, Fe_2O_3) or more commonly as limonite (hydrated ferric oxide, $Fe_2O_3 . H_2O$). Ferric oxide is very infusible, but when converted by reducing agents into the lower oxide it unites readily with silica to form ferrous silicate, which is easily fusible. The reactions are expressed as follows :—

$$2Fe_2O_3 + C = 4FeO + CO_2,$$
$$2FeO + SiO_2 = 2FeO,SiO_2.$$

Manganese oxides, which are also found in ores, act as a flux in a similar manner.

Lime, present in ores as limestone (calcium carbonate, $CaCO_3$), acts as a powerful base for fluxing silica. When heated, limestone is converted into calcium oxide and the carbon dioxide driven off. Lime is very infusible, but in combination with silica it forms a fusible silicate, the fusibility of which is increased by the addition of another base.

"Carbonate of lime" also occurs in ores in combination with magnesium in the mineral dolomite, a double carbonate of lime and magnesia ($CaO,CO_2 + MgO,CO_2$). As a flux, magnesia acts in a similar manner to lime.

Alumina (Al_2O_3) is often found in ores, usually in combination with silica. It acts as a basic flux. It is a common constituent of rocks and materials such as clays and slates.

Borax, $Na_2B_4O_7 . 10H_2O$.—This may also be written $Na_2O,2B_2O_3 . 10H_2O$. Borax is a hydrated sodium biborate containing nearly half its weight of water. When heated, the water is expelled, causing the borax to swell and finally fuse, at 760° C., into a clear glass known as "borax glass." The swelling-up may cause loss in the fusion of an assay by forcing part of the charge out of the crucible. On this account, the borax should be freed from the water by carefully heating it at a temperature a little above the boiling-point of water, or it may be fused in a crucible and the resulting glass poured upon a clean surface, and then powdered for use.

Borax acts as an acid flux, and is the chief acid flux employed by assayers. It contains an excess of boron trioxide (B_2O_3) which is capable of dissolving many metallic oxides to form very fluid slags. It is valuable in fluxing infusible substances such as lime, zinc oxide, etc. Being very fluid, borax increases the fluidity of almost any charge, and on that account is frequently added. In the form of borax glass it is largely used as a "cover" for scorification assays.

Silica, SiO_2.—This is sometimes used, in the form of very finely pulverized quartz or siliceous sand, as a flux for extremely basic ores, to lessen the corrosive action on the crucible. It combines with the metallic oxides or bases present in the charge to form a slag mainly composed of silicates. It is only very occasionally that an ore requires to be fluxed with silica, as most basic ores contain a sufficient amount of silica for fluxing purposes. Powdered glass is a very convenient substitute for silica: broken beakers, etc., cleaned, dried, and powdered, are very suitable for the purpose. Glass is a mixture of sodium and potassium silicates with small amounts of other silicates. It contains generally from 60 to 80 per cent. of silica.

Fluor-spar (calcium fluoride, CaF) should be selected as free as possible from other minerals (such as galena) and powdered. It is occasionally used in assaying, as it is a suitable flux for infusible substances like calcium phosphate (bone-ash) and infusible silicates. It also fuses easily with barytes and gypsum. Fluor-spar requires a comparatively high temperature to fuse it (900° C.), and melts without decomposition. When melted it is very fluid, and is capable of holding in suspension unfused particles of mineral matter without seriously diminishing the fluidity of the slag. A great part of the fluor-spar is found unchanged in the slag merely present as mechanical mixture. Its extreme fluidity constitutes its chief value as a flux; it facilitates the fusion, and increases the fluidity of the mass. Slags containing fluor-spar have a stony appearance.

Salt (sodium chloride, NaCl).—Ordinary table salt. Fuses easily into a very thin fluid and is not decomposed during fusion. It volatilizes at a red heat. It is occasionally used as a "cover" on the top of assay charges, as it helps to preserve certain slags from oxidation, and also, by reason of its fluidity, washes down particles of metal adhering to the sides of the crucible. Sodium chloride decrepitates when heated, it should therefore be dried and powdered before being used as a flux.

Use of Metals.—The following metals are employed in dry assaying:—

Iron.—Wrought-iron, in the form of hoop-iron, rods about half an inch in diameter, or nails, is used in the crucible assay of ores containing sulphur, arsenic, and lead. It acts as a desulphurizing agent, as later described. Wrought-iron crucibles are employed with advantage for ores consisting largely of galena (PbS).

Metallic Lead.—This is used in assaying in the form of (*a*) sheet-lead, or (*b*) granulated lead. Ordinary commercial lead generally contains too much silver to be suitable for assay purposes; lead refined by desilverization processes and sold as assay lead or test-lead is therefore employed.

(*a*) *Sheet-Lead.*—Sheet-lead is chiefly used in the direct cupellation of pieces of metal, which are wrapped up in the lead. The thickness of the sheet should be about equal to that of an ordinary visiting card. It is usually purchased, but may be prepared if necessary by casting the lead in the mould shown in Fig. 40 to obtain a thin flat ingot, which is rolled out to the desired thickness. Very pure assay sheet lead may be obtained by the reduction of white lead with charcoal: 100 parts of white lead require about 3 parts of charcoal powder to reduce it to metallic lead.

(*b*) *Granulated Lead.*—Granulated lead is used for assays by the scorification method. It may be prepared by pouring the molten metal into a strong wooden box and rapidly shaking the box until the lead granules have solidified. A rectangular oak box about 13 in. by 9 in. by 9 in., provided with handles at the ends, is very suitable for the purpose. The inside is blackleaded, and the lead, when just on the point of solidifying, is quickly poured in and subjected to a vigorous shaking by which means the metal is broken up into a fine state of division. The temperature of the molten metal may be tested by inserting a piece of stick, and when this is no longer sensibly charred the lead is ready for pouring. If the lead is poured too hot, the box is burnt and the resulting fume causes inconvenience in shaking. After being granulated, the lead is thrown on to a sieve of about 13-mesh, and any metal that fails to pass the sieve is remelted and again granulated. Small quantities of granulated lead may be prepared by pouring the molten metal into a large crucible, throwing a little charcoal powder upon it, and stirring the mixture rapidly with a stout stick or iron rod until the lead has solidified. The charcoal powder is afterwards removed by winnowing or by washing.

Lead-shot is sometimes used as a substitute for granulated lead, but its use is to be deprecated, since it contains arsenic.

Reducing and Oxidizing Agents.

In assaying ores by the dry method, reactions frequently take place which involve reduction or oxidation, and it is essential that the assayer should thoroughly understand these reactions. When a metal is separated or set free from a state of chemical combination it is said to be reduced, and the process of separation is termed *reduction*. Also, when a compound is changed from a higher to a lower state of oxidation, it is said to be reduced.

Reducing Agents.—The agent by which reduction is effected is termed the reducing agent. In ores the metals exist, in many cases, in the state of oxides or of sulphides, and the term reducing agent is most frequently used to signify " a substance that removes oxygen or sulphur (or analogous element) from a compound and sets free the metal " or " that changes a compound from a higher to a lower state of oxidation." Thus when lead oxide is heated with carbon, carbon dioxide is formed and the lead is reduced to the metallic state ; or when iron is heated with lead sulphide, iron sulphide is formed and the lead set free in the metallic state. Also, when ferric oxide (the higher oxide) is heated with a suitable proportion of carbon, carbon dioxide is formed and the iron reduced to ferrous oxide (the lower oxide).

In these cases the carbon and iron are reducing agents. As a general rule, reducing agents consist mainly of carbon and hydrogen, or their compounds, but in some instances the common metals and their compounds are employed. Although the above definition of a reducing agent is generally applicable, it must be remembered that under suitable conditions sulphur itself, either alone or in combination, will act as a reducing agent. For example, when sulphur is heated with lead oxide, sulphur dioxide is formed, and the lead is reduced to the metallic state, as in the case with carbon : thus, $S + 2PbO = 2Pb + SO_2$. Also, when the sulphides and oxides of certain metals are heated together, complete reduction takes place, and in such cases the sulphides are regarded as reducing agents. The following reactions between the sulphides and oxides of lead and of copper are examples :—

$$PbS + 2PbO = 3Pb + SO_2$$
$$CuS + 2CuO = 3Cu + SO_2.$$

A reducing agent that removes oxygen from a compound is also termed a *deoxidizer*, and one that removes sulphur is termed a *desulphurizer*.

From the assayer's standpoint the subject of reduction and of reducing agents is mainly confined to the consideration of the substances that will reduce lead oxide to metallic lead. For this purpose the assayer chiefly employs some form of carbonaceous matter such as charcoal, argol, or flour, but he also makes use of the reducing power of the sulphur present in the metallic sulphides frequently found in ores. He also employs iron as a reducing agent for the reduction of certain metallic sulphides such as galena (lead sulphide), as stated above. The chief reducing agents are the following :—

Charcoal.—Powdered wood charcoal is used. It contains about 90 per cent. carbon, the rest being mainly ash and moisture. The advantage of carbon as a reducing agent consists in its great affinity for oxygen, and at a red heat it surpasses most other substances in this respect. When heated with metallic oxides, carbon removes the oxygen and liberates the metal with the production of, at low temperatures, carbon dioxide, and at higher temperatures carbon monoxide. The reducing effect of charcoal varies with the quality and the conditions under which it is used ; the actual reducing effect of any given sample should therefore be determined under the conditions in which it is to be employed. Finely-powdered coal or coke may be used as substitutes for charcoal, but are not so suitable. Coal, especially soft coal, swells on heating and is inconvenient, but as the quantity required in an assay is usually small, it is sometimes employed. Coke is not very suitable as a reducing agent, as it burns very slowly. It should never be used when it is possible to avoid it. When employed, it must be very finely powdered.

Flour.—Ordinary wheat flour is frequently employed as a substitute for charcoal and has many advantages. It can usually be readily obtained, and being in such a fine state of division it mixes very intimately with the substance to be reduced. It contains moisture which varies in amount with different samples. When heated, flour gives off inflammable reducing gases and leaves a residue of finely-divided carbon. The reducing effect of flour is about one-half that of charcoal. When flour is employed it is useful to remember that 1 gramme reduces about 11 grammes of lead from litharge. Starch, which has nearly the same reducing power as flour, is sometimes used. It should be well dried before use. Sugar also possesses similar reducing properties. Its decomposition by heat gives a fair proportion of carbon, but it undergoes a great increase in volume when heated, and on that account is unsuitable for assay purposes.

Tartar or Argol ($2KHC_4H_4O_6$).—This is a crude potassium bitartrate. Cream of tartar, the purified salt, may be used. When heated, tartar is easily decomposed ; it gives off inflammable gases, and agglomerates without fusing or boiling up, leaving a carbonaceous residue. The residue contains about 15 per cent. of carbon when produced from crude tartar, and 7 per cent. when from cream of tartar.

The decomposition may be represented thus :—

$$2KHC_4H_4O_6 = K_2O + 5H_2O + 2C + 6CO.$$

The carbon and carbon monoxide thus set free act as the reducing agents and the K_2O is available as a basic flux. The reducing effect of tartar varies with the purity of the sample, but may be taken as equal to about one-fifth that of charcoal.

Black Flux.—This consists of a mixture of carbon and potassium carbonate obtained by deflagrating crude tartar with potassium nitrate (nitre). For this purpose the mixture of tartar and nitre is poured into a red-hot crucible placed under a chimney with a good draught. A good plan is to stand the hot crucible in an empty wind furnace and add the mixture, in small lots at a time, from a metal scoop. The mixture deflagrates and gives off nitrous vapours, leaving as a residue potassium carbonate and carbon, the proportion of which varies with the quantity of nitre added. When nitre is deflagrated with tartar, the carbon is oxidized by the oxygen of the nitre and potassium carbonate formed.

By varying the proportion of nitre, black flux with different reducing power may be obtained. Black flux with a vigorous reducing action is made from 3 parts of tartar and 1 part of nitre ; for less vigorous reducing power, from

2 to 2½ parts of tartar are used for 1 part of nitre. As black flux is hygroscopic, it must be frequently prepared fresh and kept in a dry place. On this account, a substitute consisting of a mixture of potassium (or sodium) carbonate and flour is used in preference. Usually from 20 to 25 per cent. of wheat flour is added to the alkaline carbonate. Owing to the trouble of preparation and its hygroscopic nature, black flux is only used to a limited extent in assaying.

Potassium Cyanide (KCN).—This salt is easily fusible, and when melted is somewhat volatile. When fused it acts as a powerful reducing and desulphurizing agent, readily removing oxygen and sulphur from metallic compounds. It combines with oxygen, forming potassium cyanate,

$$PbO + KCN = Pb + KCNO;$$

and with sulphur, forming potassium thiocyanate,

$$PbS + KCN = Pb + KCNS.$$

Commercial samples of potassium cyanide vary much in purity, some containing less than 50 per cent. of pure cyanide. The better qualities only should be used in assaying. Potassium cyanide is usually sold in cakes and is powdered for use; but when powdering it, great care must be taken not to inhale it, as it is a powerful poison. During the operation the mortar should be covered with a cloth. The powdered salt should be kept from the air as far as possible, as it absorbs moisture. It is only occasionally used in gold and silver assaying.

Reducing Power.—The actual effect produced by a reducing agent is termed the *reducing power*, and is always given in terms of the lead reduced. It may be defined as follows:—By reducing power is meant the weight of lead yielded (or reduced) by 1 gramme of the reducing agent when fused with litharge. For example, 1 gramme of pure carbon will (theoretically) reduce 34·5 grammes of lead from litharge; the reducing power of carbon is therefore said to be 34·5. But charcoal contains, in addition to carbon, more or less inert ash which has no reducing effect; the actual amount of lead reduced will consequently be materially less. In practice it is usually found to be between 20 and 30 grammes for 1 gramme of charcoal, according to the quality, and these figures express the reducing power of charcoal. The reducing power of the common reducing agents is given in Table VII, but the figures must be taken as approximate only, as the results will vary somewhat, according to the quality and dryness of the sample. For all practical purposes, however, they are sufficiently near to indicate the reducing effect of the different agents most frequently employed in assaying.

TABLE VII.—APPROXIMATE REDUCING EFFECT OF VARIOUS REDUCING AGENTS.

Reducing Agent.	Quantity of lead in grammes reduced from litharge by 1 gramme of reducing agent.
Wood charcoal	22–30 grammes.
Powdered hard coal	25 ,,
,, soft coal	22 ,,
,, coke	24 ,,
Argol (crude tartar)	5–9·5 ,,
Cream of tartar	4·5–6·5 ,,
Wheat flour	10–12 ,,
Starch	11·5–13 ,,
Sugar	12–14·5 ,,
Potassium cyanide	20–30 ,,

It will be seen from these figures that the reducing power of flour on litharge is about one-half, and of argol about one-fourth, that of charcoal. As red lead (Pb_3O_4) is frequently employed in assaying instead of litharge, as a source of metallic lead to collect the precious metals, it is well to remember that it contains an excess of oxygen which will use up part of the reducing agent before any lead separates out. It is obvious, therefore, that 1 gramme of reducing agent will reduce less lead from red lead than from litharge, a fact which has to be taken into consideration when red lead is used in making up assay charges. According to Beringer, the reducing power of 1 part of charcoal, flour, and argol on litharge and on red lead is as follows :—

Reducing Agent.	From Litharge grammes.	From Red Lead grammes.
1 gramme of charcoal reduced ...	22·6 lead.	12·0 lead.
,, flour	11·2 ,,	6·0 ,,
,, argol	5·0 ,,	3·0 ,,

Comparing these figures, it will be seen that the reducing effect on red lead is only about one-half of the effect on litharge.

Reducing Effect of Metallic Sulphides.—The metallic sulphides act as reducing agents and will reduce lead from litharge in the same way as carbonaceous reducing agents.

The following are the sulphides most frequently found in ores :—

Iron sulphides	Pyrite (iron pyrites)	FeS_2.
	Pyrrhotite	Fe_7S_8.
	Arsenopyrite (mispickel) ...	FeAsS.
Copper sulphides	Chalcopyrite (copper pyrites)	$CuFeS_2$.
	Copper glance	Cu_2S.
Lead sulphide	Galena	PbS.
Antimony sulphide	Stibnite	Sb_2S_3.
Copper and antimony sulphide	Tetrahedrite fahlerz (fahl ore)	$4Cu_2S,Sb_2S_3$.
Zinc sulphide	Sphalerite (zinc blende) ...	ZnS.

Iron pyrites usually forms by far the largest proportion of the sulphides found in ores. The weight of lead reduced from litharge by 1 gramme of mineral sulphide varies with the different sulphides, and varies also for the same sulphide, according to a combination of conditions. In discussing the reducing power of sulphides, it is important to remember that sulphur, and in some cases the metals, are capable of two degrees of oxidation, and whether the lower or the higher oxide is formed will depend very largely on the conditions under which reduction takes place. In an assay as ordinarily performed, other substances beside litharge are usually present, and in this connection the character of the slag resulting from the fusion is an important factor in modifying the reducing effect of sulphides. When the slag is acid, owing to the presence of much siliceous material in the charge, the sulphur will in most cases be eliminated as sulphur dioxide (SO_2). On the other hand, if the slag contains a large proportion of a strong base like soda, the tendency will be for the sulphur to be converted into sulphur trioxide (SO_3), with the formation of sulphates. When a sulphide is fused with litharge alone, all the sulphur will be liberated as the lower oxide (SO_2). Taking iron pyrites as an example, the following equation expresses the reaction when it is fused with litharge—

$$FeS_2 + 5PbO = FeO + 5Pb + 2SO_2\,;$$

i.e. 1 gramme of pure pyrites reduces 8·6 grammes of lead. In this case the iron and the sulphur are only oxidized to the lower states, the iron to ferrous oxide, and the sulphur to sulphur dioxide (SO_2), as no alkaline bases are present to induce the formation of SO_3, and consequently of sulphates. If the reaction takes place in the presence of much soda, the sulphur is converted into sulphur trioxide (SO_3), which combines with sodium oxide to form sodium sulphate (Na_2SO_4), the reactions being expressed by the two following equations :—

$$\text{(i)}\ 2FeS_2 + 15PbO = Fe_2O_3 + 15Pb + 4SO_3.$$
$$\text{(ii)}\ 4SO_3 + 4Na_2CO_3 = 4Na_2SO_4 + 4CO_2.$$

In this case, 1 gramme of pyrites exerts its greatest reducing effect and reduces 12·9 grammes of lead. In other words, both the sulphur and the iron have been fully oxidized.

It will be seen from the above equations that the amount of lead reduced from litharge by a sulphide is influenced by the presence of soda, a fact of some importance to assayers which is more fully discussed on page 187. By decreasing the amount of soda or otherwise altering the conditions of the fusion, the sulphur may be converted partly into sulphur dioxide and partly to sulphur trioxide, and it will be obvious that the amount of lead reduced from litharge by 1 gramme of iron pyrites may theoretically vary from 8·6 grammes to 12·9 grammes. In actual practice it may range from about 9 to 11·2 g. according to whether the charge is acid and contains little or no soda, or is basic from the presence of much soda. The approximate reducing power of some of the common sulphides on litharge in the presence of one-fourth of its weight of soda is given in Table VIII. The amount of lead theoretically obtained from 1 gramme of sulphide for the two degrees of oxidation is also given.

TABLE VIII.—Reducing Power of Metallic Sulphides.

Metallic Sulphide.	Weight of lead reduced from litharge by 1 gramme of sulphide.[1]	Theoretical weight of lead reduced from litharge by 1 gramme of sulphide when—	
		(*a*) Sulphur and metal are oxidized to the lowest degree.	(*b*) Sulphur and metal are oxidized to the highest degree.
Antimonite	6 grammes.	5·5 grammes.	8 grammes.
Zinc-blende ...	7–8 ,,	6·4 ,,	8·5 ,,
Copper pyrites ...	7–8 ,,	7·5 ,,	11·6 ,,
Fahlerz	7–8 ,,	...	...
Galena	3 ,,	2·6 ,,	3·4 ,,
Iron pyrites ...	11 ,,	8·6 ,,	12·9 ,,
Mispickel	7–8 ,,	...	...

The figures given in the table represent the reducing power of the different sulphides in a pure state, but it must be pointed out that it is more usual to find several sulphides associated in an ore than to find one sulphide only, and due consideration is given to this fact in dealing later with the assay of sulphide

[1] From experiments by Beringer (*Text-book of Assaying*).

ores. The reducing power of a few metallic sulphides on red lead is given in Table IX from experiments by Beringer. The results obtained with litharge under the same conditions are inserted for comparison.

TABLE IX.—REDUCING POWER OF METALLIC SULPHIDES.

Sulphide.	Lead reduced from litharge.	Lead reduced from red lead.
1 gramme of antimonite reduced ...	6·2 grammes.	3·6 grammes.
,, copper pyrites ...	8·9 ,,	5·9 ,,
,, galena	3·4 ,,	1·7 ,,
,, iron pyrites	11·6 ,,	8·5 ,,

Determination of Reducing Power.—The reducing power is determined from the quantity of lead which is yielded by fusing 1 or 2 grammes of the reducing agent with 60 grammes of litharge and 15 grammes of sodium carbonate. The button of lead obtained is weighed, and the weight divided by the weight of reducing agent used, the result being the reducing power. The presence of soda does not influence the amount of lead reduced from litharge with carbonaceous reducing agents, but it has already been pointed out that soda has considerable influence on the reducing effect of metallic sulphides. In assaying, it is sometimes necessary to determine the reducing power of ores containing more or less pyritic material, and in this case the flux used in determining the reducing power should be the same, and in the same proportion as that to be employed in the charge for the assay of the ore, so that the conditions in both cases may be the same. It is only by working in this way that the true reducing effect of an ore can be ascertained.

Oxidizing Agents.

Oxidation.—When oxygen combines with an element the compound formed is termed an oxide, and the act of combination is known as *oxidation*. In assaying, oxidation usually has reference to the conversion of a metal or element to the state of oxide, or the changing of a compound already in an oxidized state to one of higher oxidation; for example: lead to lead oxide (PbO), sulphur to sulphur dioxide (SO_2), or ferrous oxide (FeO) to ferric oxide (Fe_2O_3). Advantage is taken of oxidation in the operations of roasting, scorifying, cupelling, etc., described subsequently. When the amount of pyritic material (*i.e.* sulphides, arsenides, etc.) present in an ore is in excess of that required to act as a reducing agent, the excess is usually oxidized and passed into the slag. To effect oxidation, special substances termed oxidizing agents are employed. They may be defined as follows:—

An oxidizing agent is a substance that imparts oxygen to an element or compound.

The following oxidizing agents are employed in assaying:—

Nitre (potassium nitrate, KNO_3) is a powerful oxidizing and also desulphurizing agent. Nitre fuses very easily to a watery liquid. It contains nearly half its weight of oxygen, which it readily loses, in two stages, when heated. At a low temperature (a red heat) it is decomposed into potassium nitrite

(KNO_2) and free oxygen, and at a higher temperature more oxygen is set free and potassium monoxide (K_2O) left.

The decomposition is shown thus :—

$$\text{(i) } 2KNO_3 = 2KNO_2 + O_2.$$
$$\text{(ii) } 2KNO_2 = K_2O + 2NO + O.$$

The action of melted nitre is very energetic, sulphides being readily oxidized to sulphates, arsenides to arsenates, and most metals converted into oxides. The oxidizing action of nitre on iron pyrites may be expressed thus :—

$$4FeS_2 + 10KNO_3 = 4FeO + 5K_2SO_4 + 3SO_2 + 5N_2.$$

In the presence of a strong alkaline base such as soda, alkaline sulphates are formed as the result of oxidation, and float as a watery liquid on the surface of the less fluid slag. In this case the oxidizing action of nitre on pyrites is expressed by the following equations :—

$$\text{(i) } 6KNO_3 + 2FeS_2 = Fe_2O_3 + SO_3 + 3K_2SO_4 + 3N_2.$$
$$\text{(ii) } 3SO_3 + 3Na_2CO_3 = 3Na_2SO_4 + 3CO_2.$$

It is for the conversion of metallic sulphides into oxides that nitre is most frequently used in assaying.

Red Lead (Pb_3O_4) is an oxide of lead, frequently used instead of litharge as a source of lead, especially in cases where an oxidizing action is required in the fusion. When heated it parts readily with about one-fourth of its oxygen, which is available for oxidizing purposes without any separation of metallic lead:—

$$2Pb_3O_4 = 6PbO + O_2.$$

The lead oxide (litharge, PbO) that is formed can be used as a flux, or the oxygen it contains can be used for oxidizing purposes, but in this case metallic lead will be liberated to the extent of about 13 parts of lead for each part of oxygen made use of.

Red lead can only be used to a limited extent as an oxidizing agent for assay purposes, as the employment of a comparatively large quantity will give too much lead to be dealt with satisfactorily. The oxidation of iron pyrites by means of litharge is expressed in the following equations :—

$$2FeS_2 + 11PbO = Fe_2O_3 + 4SO_2 + 11Pb,$$
or
$$FeS_2 + 5PbO = FeO + 2SO_2 + 5Pb.$$

Hot Air.—The oxygen of the air is used as an oxidizing agent in all roasting operations. The sulphur of metallic sulphides is oxidized and passes off as sulphur dioxide, and the arsenic of minerals like mispickel, etc., is oxidized to "white arsenic," As_2O_3. The metals present are converted into oxides, mixed with more or less sulphate or arsenate. Thus iron pyrites heated in air is oxidized to ferric oxide and the sulphur to sulphur dioxide :—

$$2FeS_2 + 11O = Fe_2O_3 + 4SO_2.$$

A certain proportion of ferroso-ferric oxide (magnetic oxide), Fe_3O_4, or Fe_2O_3, FeO, is also formed thus :—

$$3FeS_2 + 5O_2 = Fe_3O_4 + 3SO_2 + 3S.$$

Desulphurizing Agents.—When an oxidizing agent is employed to remove sulphur by oxidation, it is also termed a desulphurizing agent. Most oxidizing agents act as desulphurizers, and the substances described above are used for this purpose.

Iron is also a desulphurizing agent frequently employed in assaying. When heated with metallic sulphides the iron combines with the sulphur to form ferrous sulphide, which will be dissolved in the slag or separate out as a regulus according to the conditions of the fusion. Iron is used in the form of hoop-iron, iron rods about ¼-inch in diameter, or of nails.

Examples.—(1) Iron decomposes lead sulphide, forming iron sulphide and metallic lead thus :—

$$PbS + Fe = FeS + Pb.$$

(2) Iron pyrites, when heated with iron, parts with half its sulphur, forming ferrous sulphide :—

$$FeS_2 + Fe = 2FeS \text{ (regulus)}.$$

Alkaline Carbonates.—Sodium and potassium carbonates act not only as basic fluxes but also as desulphurizing agents. The alkaline oxides Na_2O and K_2O react on certain metallic sulphides, separating the metal, with the formation of sodium or potassium sulphate and sulphide. Thus sodium carbonate partially decomposes galena (lead sulphide), setting free some lead and forming sodium sulphide, which combines with the remaining lead sulphide to form a double compound thus :—

$$7PbS + 4Na_2O \, . \, CO_2 = 4Pb + Na_2SO_4 + 3(PbS,Na_2S) + 4CO_2.$$

The double sulphide of lead and sodium is decomposed by iron, and metallic lead set free, thus :—

$$PbS,Na_2S + Fe = FeS + Na_2S + Pb.$$

Sodium oxide (Na_2O) in the presence of carbon (which removes the oxygen) acts as a desulphurizer by combining with sulphur to form sodium sulphide. In this case carbon is employed as an auxiliary to the desulphurizing action of the alkali.

CHAPTER X

PRINCIPLES OF FLUXING: ASSAY SLAGS

To understand the principles of fluxing it is necessary that the student should have a thorough knowledge of the nature of the extraneous earthy or stony matter found associated with precious-metal ores. Such matter is termed *gangue*, *vein-stuff*, or *matrix*. The nature of the gangue is of importance in determining the nature of the flux necessary to convert it into a suitable slag. For convenience, gangue-forming materials are classified according to their chemical character into *acid* and *basic*—silica as a general rule being the acid, and the metallic oxides the basic constituents. The gangue as a whole is classed as acid or basic according to whether acid or basic constituents are in excess. If both acid and basic materials are present in about the same proportion the gangue is said to be *neutral*. In the comparatively rare cases where the proportion of basic material is such that it neutralizes the silica present and combines with it to form a readily fusible compound, the ore is said to be *self-fluxing*.

The principal gangue materials, classified according to their chemical character, are given in Table X.

TABLE X.—GANGUE-FORMING MATERIALS.

Acid Gangues.	Basic Gangues.
Silica, uncombined; as quartz crystals, quartz rock, quartzite, sandstone, etc.	*Metallic oxides*, notably those of iron, manganese, and aluminium.
Siliceous rocks, rocks in which silica, either free or combined, predominates; as granites, porphyry, etc.	*Metallic carbonates*, as those of calcium (limestone), magnesium, iron, etc.
Silicates, or silica, combined with a base; as felspars, mica, hornblende, etc.	*Barytes*, or heavy spar (barium-sulphate).
	Fluor-spar (calcium fluoride).

Quartz gangues will contain from about 80 to 100 per cent. of silica. Siliceous rocks such as granite will contain from 70 to 75 per cent. of silica, and from 13 to 20 per cent. alumina.

The silicates present in the gangues are mainly silicates of alumina, lime, and magnesia, and contain, roughly speaking, from 45 to 65 per cent. of silica, 15 to 20 per cent. of alumina, 5 to 15 per cent. of iron oxide, and 5 to 15 per cent. of lime and magnesia, with a small percentage of alkalis. In the majority of cases the gangue of an ore will consist of rock composed mainly of silica or of siliceous material, and will therefore be acid in character owing to the preponderance of silica. In some cases the gangue may consist of a mixture of acid and basic material and be neutral in character. In comparatively few cases the gangue may consist entirely of basic material and consequently be basic in character. But in addition to the gangue, metalliferous minerals, notably iron pyrites, are found associated with precious-metal ores, and these, when present in any large quantity, play an important part in determining the character of an ore for fluxing purposes, as their oxidation produces basic oxides For example, an ore consisting mainly of iron pyrites with a little quartz would, if roasted, be converted into iron oxides and become strongly basic.

To facilitate description, ores may be classified as follows :—

(1) Ores with siliceous gangue, as ordinary quartz.
(2) Ores with basic gangue, as iron oxide, limestone.
(3) Pyritic ores, as iron pyrites, with or without other metallic sulphides.
(4) Complex ores, as antimonial and telluride ores.

The nature and quantity of the flux to be used depend upon the character and composition of the ore. When smelting ores on a large scale the metallurgist always determines the composition of the ore, and then carefully calculates the quantity of flux to be added to produce a suitable slag. In assay practice, however, it is impracticable to make analyses of ores to determine the composition for fluxing purposes, but with a knowledge of mineralogy the assayer can, by a few simple tests, ascertain sufficiently accurately the nature of the ore which he wishes to flux. Much valuable information may be obtained by a careful examination of a few characteristic lumps of the ore when these are available. If the ore is in the state of powder, a small quantity should be washed or "panned" in a small sheet-iron pan, and the washed material carefully examined with a lens, and, if necessary, with a blowpipe. By this means the gangue is in most cases readily separated from any metalliferous minerals, and the nature of the ore and the approximate proportions of the various constituents are ascertained. For example, it will be comparatively easy to tell at once whether metallic sulphides such as iron pyrites or galena are present, or whether the ore contains limestone, limonite, baryta, or other mineral, and in what general proportions. The information thus gained invariably repays the time spent in obtaining it. The nature of the gangue having been ascertained, the best flux for producing an easily fusible slag is added in accordance with the following simple general rule for fluxing :—

To flux an acid gangue use a basic flux.
To flux a basic gangue use an acid flux.

(1) Ores with Siliceous Gangue.—Taking first the simplest case of a gangue composed almost entirely of quartz or siliceous material, which constitutes the gangue of a large proportion of precious-metal ores. The best and most frequently used flux for silica is soda. It combines with silica to form sodium bi-silicate, which is very fusible and makes a good glassy slag.

With sodium carbonate the reaction is :—

$$SiO_2 + Na_2CO_3 = CO_2 + Na_2SiO_3.$$

With sodium bicarbonate :—

$$SiO_2 + 2NaHCO_3 = 2CO_2 + Na_2SiO_3 + H_2O.$$

Theoretically, 1 part of silica will require 1·77 parts of carbonate or 2·8 of the bicarbonate to form sodium bi-silicate. In practice, from 1½ to 2 parts of sodium carbonate are usually employed to flux 1 part quartz or sand.

An acid gangue may also be composed largely of the natural silicates, which may be considered as combinations of silica and metallic oxides, with usually more or less silica in excess. As previously stated, the most common silicates found associated with ores are the silicates of alumina, lime, and magnesia. The silica in silicates is already partially neutralized by combination with the metallic base present, but as the natural silicates are, in most cases, fusible only with great difficulty, they are, in assaying, generally fluxed in the same way as uncombined silica. Although *soda* is the flux most generally employed for quarts and siliceous material, it is sometimes advantageous to use litharge as a partial or complete substitute for soda. Litharge, when fused with quartz, forms lead silicate, which is a yellow glass, easily fusible, and more fluid than sodium silicate. The reaction is quieter than with soda, as there is no evolution of gas :—

$$PbO + SiO_2 = PbSiO_3.$$

Theoretically, 1 part of silica requires 3·7 parts of litharge. In this reaction the heating must be carefully regulated, otherwise the litharge will fuse and drain away from the unattacked quartz, which is left as a pasty mass on the surface. In using litharge it must be remembered that it is only the lead oxide in excess of that required to give the necessary weight of metallic lead for collecting the precious metals in the ore that is available as a flux.

Borax is also used as a partial substitute for soda to flux silica. Its fluxing action is about equal to that of litharge. With siliceous material, borax forms borate silicates which increase the fluidity of the slag. When borax is used it should be borne in mind that it is an acid flux ; consequently the acid character of the slag (which is a bi-silicate or acid slag) is increased by increasing the proportion of borax. It also tends to make the slag too fluid.

As an example of the use of two fluxes, let us suppose we wish to flux 50 grammes of siliceous ore, which would require theoretically 88·5 grammes of sodium carbonate if this flux alone be used. If we add 71 g. of sodium carbonate (which is only sufficient to flux 40 g. of silica) it leaves 10 g. of silica to be fluxed by other reagents. To flux this we can add 37·2 g. of litharge or the same weight of borax, since the fluxing action of these reagents is about equal. Or, instead of using either litharge or borax separately, we could use both and flux the 10 g. of silica with 18·6 g. of each reagent. A sufficiently fluid slag would still be obtained even if the proportions of the three fluxes were altered by decreasing the quantity of any of them and proportionately increasing either or both the others. In practice it is more usual to rely mainly on soda as the flux for silica and silicates.

(2) Ores with Basic Gangue.—The chief ores which come under this head are those which consist mainly of limonite (hydrated iron oxide), and in most cases are believed to be the result of the decomposition of pyritic material by atmospheric agencies.

Ores containing excess of hæmatite (Fe_2O_3) magnetite (Fe_3O_4), calcite ($CaCO_3$) dolomite ($CaCO_3 + MgCO_3$) or other basic minerals are also included under this head. The bases to be fluxed off in ores of this class are therefore the oxides of iron, lime, alumina, and, usually in smaller proportion, the oxides of magnesium, manganese, barium, copper, zinc, etc. The gangue may consist of one of these oxides only, but it is much more usual to find several oxides present ; and silica, in greater or lesser quantity, is present in practically every ore. The chief flux for basic material is *borax*, which forms fusible compounds with nearly all bases, being especially useful in uniting with metallic oxides. It also forms fusible compounds with silica. When a basic gangue contains a fair proportion of silica it may be partially self-fluxing, as the silica will combine with part of the basic material present to form fusible silicates, leaving only the excess of basic material to be fluxed. In the comparatively rare cases in which an ore is extremely basic and deficient in silica, sand or powdered glass is added. As an exception, the addition of sand is frequently necessary in fluxing the excess of ferric oxide which results from the roasting of very pyritic ores. Sand in suitable proportion forms fusible silicates with alumina, lime, iron oxides, etc., and protects the crucible from corrosion. Ordinary crucibles consist of quartz and clay, and are rapidly attacked by slags that are too basic. Even when silica is added as a flux it is a common practice to add also a little sodium carbonate, as slags consisting of the double silicates of sodium and other bases are much more fusible, and more suitable for the requirements of the assayer, than slags formed of silicates free from sodium. The proportion of soda should be increased, and that of the borax diminished, as the quantity of silica becomes greater. For average basic ores a mixture of equal weights of soda and borax answers very well. An important consideration in the fluxing of basic material consisting mainly of ferric oxide (Fe_2O_3) is the reduction of this higher oxide to the ferrous state (FeO), in which state alone it unites with silica to form a readily fusible silicate. Ferric oxide being infusible, its presence in the slag tends to stiffen it and render it pasty. " When fluxed with as much as six times its weight of borax, ferric oxide yields an unsatisfactory slag, while on the other hand the lower oxide (ferrous oxide) is easy to deal with ; fused borax will dissolve about its own weight of it, and sodium silicate, such as makes up the bulk of most assay slags, will take up at least half as much " (Beringer).

Experience has shown that slags containing much ferric oxide are very liable to retain notable amounts of gold and silver. The reduction of the iron to the state of the lower oxide is effected by the addition of charcoal powder or other reducing agent ($2Fe_2O_3 + C = 4FeO + CO_2$). For ores consisting mainly of ferric oxide, such as pyritic ores after roasting, the quantity of charcoal powder required will be about one-twelfth of the weight of ore taken for assay ; the exact amount needed for any particular ore must, however, be left to the experience of the assayer. In the case of ores oxidized by roasting, experience will ferquently enable the adjustment of the charcoal to be made from the colour of the roasted material. It may be pointed out that brown hæmatites (hydrated ferric oxide) contain in some cases as much as 15 per cent. of water (which is expelled on heating), while red hæmatite (ferric oxide) contains none ; hence more charcoal is needed for the latter than for the former. Manganese peroxide (MnO_2) also acts in the same way as ferric oxide, and when present must be reduced to the lower oxide (MnO). It requires about twice the amount of charcoal for its reduction as compared with ferric oxide. Manganese is not a common constituent of precious-metal ores and when present it is usually in small proportion only. Basic ores containing easily reducible oxides, such as copper and antimony oxides, should be partially fluxed with lead oxide, as it is advantageous in such cases to have an excess of lead oxide in the slag. The lead oxide dissolves the metallic oxides present and prevents any copper, or other metal, from contaminating the lead present for collecting the values. Where there is no such reason for the presence of lead oxide it is generally advisable to use only an amount sufficient to give the desired weight of lead. Ores containing any appreciable amount of copper or antimony belong to the class of complex ores which are more fully dealt with in Chapter XIV, page 210.

(3) **Pyritic Ores.**—Mainly iron pyrites. In assaying ores of this class it is necessary to eliminate the sulphur, otherwise it will pass into the lead, making it more or less brittle, or the lead will be accompanied by a layer of regulus or

"matte," *i.e.* an artificial sulphide of the metals, usually lead and iron. The reducing action of metallic sulphides has already been dealt with on p. 120. When only a moderate amount of pyritic material is present, say under 10 per cent., its effect is simply to act as a reducing agent and increase the weight of the lead reduced, provided there is an excess of litharge in the charge. This effect is easily counteracted by using less charcoal in the assay charge, or by omitting it. When the charge contains only sufficient litharge to yield the desired weight of lead for collecting the values, any excess of pyritic material beyond that necessary to reduce the litharge may render the lead sulphury, or even in some cases give a button of regulus instead of lead. If, on the other hand, an excess of litharge is present, the amount of lead reduced may be considerably in excess of that required to collect the values.

Metallic sulphides are partially decomposed by alkaline carbonates, but carbonaceous matter must also be present. In these decompositions alkaline sulphides are formed, and combine with and retain a certain quantity of the metallic sulphides present (see page 124). The proportion of the sulphides which remains in combination with the alkaline sulphide depends on several circumstances. It is always less when a large proportion of alkaline carbonate has been added to the charge, and when a high temperature has been employed.

According to Beringer, the slag resulting from the fusion of 50 grammes of quartz, with 100 grammes of sodium bicarbonate and a little borax (which represents an ordinary assay charge), will dissolve some 10 or 12 grammes of iron sulphide. In assaying, the decomposition of the metallic sulphides in ores is effected either by means of metallic iron, or some method of oxidation is employed to oxidize the sulphur and thus eliminate it.

Metallic iron acts as a desulphurizer by combining with sulphur to form, in the case of iron pyrites, ferrous sulphide ($FeS_2 + Fe = 2FeS$), which will remain diffused through the slag when it contains sodium sulphide (formed during the fusion) instead of separating as a layer of regulus. This is also true of other metallic sulphides, if not present in too large a quantity and if the charge is not fused at too high a temperature. A moderate amount of sulphur may be successfully converted into ferrous sulphide by metallic iron and dissolved in the slag without a separate layer of regulus forming, but with ores containing excess of pyritic material it is difficult to effect complete reduction by this method, even when large quantities of metallic iron are added to the charge; the lead button obtained is always more or less brittle, owing to the presence of sulphur, and is frequently accompanied by a layer of regulus.

The presence of a large excess of ferrous sulphide in the slag is objectionable, as it is readily oxidized and causes the slag to become pasty, and frequently to retain appreciable quantities of the precious metals, especially silver. Ferrous sulphide in moderate quantities in a slag is not injurious, and hence it is a common and beneficial practice to insert metallic iron in all crucible charges of moderately pyritic ores to decompose the sulphides. It also tends to reduce the metals present in the slag to the lower states of oxidation. It is important to remember that when iron is used as a desulphurizing agent the amount of lead oxide added to the charge should only be sufficient to give a button of lead of the required weight, as any excess will be reduced to metallic lead by the iron. The slag obtained from the fusion should be strongly alkaline, and to ensure this an excess of sodium carbonate should be added to the charge. The desulphurizing action of iron is taken advantage of by assayers as far as possible, as it is a very convenient method of decomposing sulphides, but if the first fusion of a pyritic ore by this method gives a layer of regulus or speise, it is best, in most cases, to repeat the assay by eliminating the sulphur by other methods. Ores containing a large proportion of pyritic material (say over 50 per cent.) are generally submitted to some method of oxidation either before or during fusion; in the first case the ore is roasted to eliminate it, and in the second case the oxidation is effected by the addition of lead oxide or of nitre to the charge.

As the roasting of an ore occupies a considerable amount of time and attention, other methods of oxidation are generally preferred by assayers whenever they are applicable. Roasting can generally be dispensed with in the case of silver ores, but it is sometimes desirable with gold ores where larger quantities have to be used for assay. Fuller details of the methods of oxidation by means of lead oxide and nitre are given on p. 186. When the pyritic material in an ore has been oxidized by a preliminary roasting, it consists of metallic oxides and is consequently dealt with as a basic ore, to which class it now belongs.

(4) **Complex Ores.**—Under this head are included those ores which contain large proportions of metallic sulphides, other than iron pyrites, such as copper pyrites, arsenical iron pyrites, sulphides of antimony, zinc, lead, bismuth, etc.

Those ores are also included in which the precious metals exist in chemical combination with another element, such as tellurium.

Complex ores require special treatment, and they will be dealt with later.

"Universal" Fluxes.—As the selection of the fluxes and the quantity of each to be employed is dependent on the nature of the gangue material and the presence or absence of pyritic material, it is impossible to give definite charges to meet the requirements of all ores: the best fluxes to be used in each particular case must, therefore, necessarily be left to the judgment and experience of the assayer. Typical charges for each class of ores will be found in Chapters XII and XIII. Universal fluxing mixtures suitable for fluxing all classes of ores are frequently recommended, but although these mixtures may be suitable in many cases for certain classes of ore, they sooner or later lead to failure if used for all ores. It must be emphasised that to obtain the best results ores must be assayed on scientific principles, each ore being treated with the fluxes and reagents best suited to its composition. The variation in the chemical character of ores makes it impossible to obtain a perfect mixture that will satisfactorily flux them all. Silver ores especially are often so complex that with a fresh ore it is sometimes necessary to make a trial fusion. In cases where an assayer's work is mainly confined to ores of any particular type or district, experiment should be made with various fluxes to ascertain the best mixture for the ores and this mixture made up as a stock flux. As previously stated, it is usually possible, by varying the proportions of the different fluxes used, to make up several mixtures, all of which are equally suitable for fluxing a particular ore, but, other points being duly considered, preference should be given to the cheapest mixture.

As a general rule, from 2 to 3 parts of flux are quite sufficient to yield a fluid slag with 1 part of ore; in some cases less will suffice, and in other cases more may be required, but usually if it is found that 4 parts of flux are not sufficient, the remedy lies in using a different flux rather than in taking a larger quantity. An excess of flux will sometimes be a distinct disadvantage, as it will tend to stiffen the resulting slag and render it pasty. For example, soda, which is a very strong base, when used to flux silicates of lime and magnesia, may act prejudicially if it is in sufficient excess to set free notable quantities of lime and magnesia, which but for that excess would exist in combination as complex fusible silicates. Also, "there are many minerals which with but little soda form a glass, but with more yield a lumpy, scoriaceous mass. There are many minerals, too, which are already basic (for example, calcite, $CaCO_3$) and, when present, demand either a less basic or an acid flux according to the proportions in which they are present in the ore."[1] When dealing with an unknown ore the first assay may be more or less unsatisfactory, but the result of the fusion will indicate what modification of the charge is necessary to give a good result.

The Assay of Reagents.—Litharge and red lead invariably contain silver in appreciable amounts, and as both reagents are used in the assay of ores, etc., the silver must be determined and taken into account in returning the assay result. If this is not done, it is obvious that the silver will be ascribed to the ore and make the return too high. Litharge is sometimes sold as "silver-free," but it is advisable to test the samples before taking them into use. Gold also is not infrequently present, but the quantity is usually too minute to be taken into consideration in returning the value of the ore.

To determine the silver in litharge or red lead, the following charge is weighed out and intimately mixed :—

Litharge or red lead	100 grammes.
Charcoal	1 gramme for litharge. 2 grammes for red lead.
Sand	7 grammes.
Borax (or soda)	20 ,,

The charge is fused at a red heat in a fireclay crucible for about fifteen to twenty minutes, whereby the lead is reduced. When fusion is completed, the charge is poured into an iron mould and allowed to cool. The slag is afterwards detached from the button of lead, which should weigh about 20 grammes. The lead is then carefully cupelled, and the resulting small button or "prill" of silver weighed. From the weight of this button the quantity of silver in the litharge or red lead used for each assay of an ore, etc., must be calculated, and deducted from the total weight of the silver obtained in the assay. Lead oxides usually contain on an average from 1·0 to 2·0 milligrammes of silver in 100 grammes, which is equivalent to 6·5 dwt. to 13 dwt. per ton of 2240 lbs. The common varieties of litharge usually contain more silver than red lead. To test the button for gold it is flattened out, transferred to a porcelain crucible, and treated with a small quantity of dilute nitric acid. On gently warming, the silver is dissolved, leaving the gold (if any) as a small black speck. The acid is carefully poured off, the gold washed with water, and then heated to redness : when cold, it is, if possible, weighed, but, as stated above, the gold usually amounts to "traces" only and is disregarded.

Granulated lead and sheet-lead (test lead) always contain more or less silver, which must be determined and deducted from the assay result as in the case of litharge, etc. The silver is determined by weighing out 100 grammes and scorifying it down to about one-fourth of its weight and then cupelling. Test lead can be obtained very pure and containing only traces of silver, but it is not safe to assume that a sample is silver free. The silver usually amounts to 1 or 2 mg. in 100 grammes. It is tested for gold in the manner described above.

In cupelling the lead buttons it is very necessary in all cases to remove the cupels from the muffle immediately the cupellation is finished, as the very small silver buttons obtained are very liable to sink into the cupel. When the amount of silver in a new lot of litharge, granulated lead, etc., has been determined, the result should at once be recorded in some convenient place for reference.

[1] Beringer, *Text-book of Assaying.*

The fluxes and other reagents commonly employed in dry assaying do not contain silver or gold, but if they are left exposed in the laboratory they sometimes become accidentally (and perhaps purposely) contaminated or "salted" with material containing gold and silver values. A blank assay (that is, a charge in which the ore is omitted) of the reagents will readily determine whether any precious metal is present. The precaution of assaying fluxes is very seldom, if ever, necessary in well-conducted assay laboratories, but it is very desirable in cases where irresponsible assistants are employed to do the grinding and fluxing of ores and also when these operations are conducted in the same room. It is a wise precaution also in laboratories where rich ores are being dealt with continually.

Assay Slags

The main object of the fluxing of an ore is to convert the infusible gangue materials into a slag, but it is necessary to consider the nature and quality of the slag produced. When fused it must be liquid and homogeneous and not too corrosive on the crucible. If the slag is stiff or not sufficiently liquid when fused, it may retain some of the precious metals in the charge and, if it fuses readily and is very liquid, the particles of reduced lead disseminated in the fused mass may fall to the bottom of the crucible before they come into contact with all the gold and silver present. Generally, the slag resulting from an assay is glassy, and varies much in colour according to the nature of the ore: it should, however, be uniform in colour and structure; not streaked or variegated, as that generally indicates that the mixture has been imperfectly fused. Slags usually consist of artificial silicates formed by the combination, in suitable proportions, of silica with the common bases (*i.e.* metallic oxides, etc.). The silicates are classified according to the ratio of the oxygen in the base to the oxygen in the acid (*i.e.* silica). One molecule of silica consists of one atom of silicon and two atoms of oxygen, so that the various silicates have the composition given below. The metallurgical classification is shown in Table XI, in which the formula RO represents oxides such as PbO, FeO, Na_2O, etc., and the formula R_2O_3 oxides such as Al_2O_3, etc.

TABLE XI.—SERIES OF FORMULÆ FOR SILICATES.

Formula.	Name.	Oxygen ratio in base : in acid.	Example.
$4RO + SiO_2 = R_4SiO_6$	subsilicate	2 to 1	$4PbO, SiO_2$.
$4R_2O_3 + 3SiO_2 = R_8Si_3O_{18}$	,,	2 to 1	
$2RO + SiO_2 = R_2SiO_4$	monosilicate	1 to 1	$2Na_2O, SiO_2$.
$2R_2O_3 + 3SiO_2 = R_4Si_3O_{12}$	,,	1 to 1	$2Al_2O_3, 3SiO_2$.
$RO + SiO_2 = RSiO_3$	bisilicate	1 to 2	CaO, SiO_2.
$R_2O_3 + 3SiO_2 = R_2Si_3O_3$	,,	1 to 2	
$2RO + 3SiO_2 = R_2Si_3O_8$	trisilicate	1 to 3	$2FeO, 3SiO_2$.
$2R_2O_3 + 9SiO_2 = R_4Si_9O_{24}$	,,	1 to 3	
$4RO + 3SiO_2 = R_4Si_3O_{10}$	sesquisilicate	2 to 3	$4MgO, 3SiO_2$.
$4R_2O_3 + 9SiO_2 = R_8Si_9O_{30}$	,,	2 to 3	

In this classification the monosilicate is the neutral silicate: it contains in base and acid equal amounts of oxygen. If a silicate contains but one base it is termed a *monobasic*, or simple, silicate; but if two or more bases are present it is termed a *multibasic*, or double, silicate, and in writing the formula for a double silicate the silicates of the various bases are united by a plus sign. Thus a monosilicate slag containing the three bases Na_2O, FeO, CaO would be written :—

$$2Na_2O,SiO_2 + 2FeO,SiO_2 + 2CaO,SiO_2,$$
$$\text{or} \quad 2(Na_2O,FeO,CaO),SiO_2.$$

The fusibility of silicates is dependent on the proportions of silica and base they contain. The most fusible are the subsilicates or basic silicates, which contain least silica, and, in the silicates classified, the fusibility decreases as the proportion of silica increases. Within certain limits the fusibility of silicates is also greatly dependent on the nature of the particular base present.

The strong bases Na_2O, K_2O, and PbO give easily fusible silicates; FeO, MnO, and CuO give silicates that fuse less readily; MgO, CaO, and Al_2O_3 give silicates that fuse with difficulty. The silicates of lime, alumina, and magnesia are the least fusible, and are almost or entirely infusible at the temperature of the assay furnace, so that in order to obtain a sufficiently fluid slag for assay purposes the proportions of these more infusible bases must be materially smaller than the proportions of the more fusible bases Na_2O, PbO, and FeO. This is effected in practice by the addition of suitable proportions of fluxes containing the more fusible bases. The multibasic silicates are usually more fusible than the simple silicates, and this fact is taken advantage of by the assayer. A slag may be regarded either as a mixture of several silicates or as a solution of one silicate in another. The alkaline bases Na_2O and K_2O and all the common bases FeO, CuO, MgO, Al_2O_3, MnO, ZnO, etc., with which the assayer has to deal form silicates which are soluble in one another.

Borax and the boron trioxide set free by its decomposition act in a similar manner to silica, combining with the bases present to form borates; and as borax is commonly added as a flux, the assay slag in most cases consists of a solution of silicates and borates of various bases.

The following are examples of borate slags :—

(i) $Na_2O,FeO,4SiO_2,2B_2O_3 = FeO,4SiO_2,Na_2B_4O_7.$

(ii) $Na_2O,PbO,Al_2O_3,6SiO_2,2B_2O_3 = PbO,Al_2O_3,6SiO_2,Na_2B_4O_7.$

The assay slag is usually a double silicate, as almost all ores contain one or more of the common bases in greater or lesser quantity. In addition to silicates and borates, the slag may contain dissolved infusible material and also certain metallic sulphides, notably those of sodium and iron. According to Fulton,[1] "the most desirable constitution for an assay slag in general is that of a monosilicate or a sesquisilicate—sometimes, but more rarely, a bisilicate. If the ore is basic a bisilicate may be approached, if acid a monosilicate, or even a subsilicate, in order to ensure complete decomposition." The various assay charges given subsequently will give slags that approximate to the above constitution. The assay slag may be regarded as having a more or less constant constitution, and the assayer, having determined approximately the composition of the ore, adds basic or acid fluxes in correct proportion to produce the desired slag.

It has been pointed out previously that litharge possesses the property of dissolving or holding in suspension certain proportions of other metallic oxides, and when an ore is fluxed with litharge the resulting slag will consist of a mixture of the oxides of lead and other metals with, in some cases, a certain proportion of lead silicate. In the scorification process litharge is produced during the operation, so that the resulting slag is of the nature described.

[1] *Manual of Fire Assaying.*

Calculation of Assay Charge.—The following example will serve to illustrate the method of calculating a charge to produce a desired slag from an ore of known composition.

Let it be assumed that the ore contains the following constituents to be fluxed :—

Silica (SiO_2)	60 per cent.
Ferrous oxide (FeO)	16 ,,
Magnesia (MgO)	3 ,,

In order to simplify the calculations whole numbers are taken, and the percentage of carbonic acid, etc., combined with the bases is disregarded. Let it be assumed also that the ore is to be fluxed so as to produce a monosilicate slag represented by the formula

$$2Na_2O,SiO_2 + 2FeO,SiO_2 + 2MgO,SiO_2.$$

The amount of silica slagged by the bases present is first ascertained by proportion with the aid of the atomic weights, thus—

2FeO	SiO_2	
$2(56 + 16) = 144$	$28 + (16 \times 2) = 60$	144 parts of FeO require 60 parts of silica to form a monosilicate ;

therefore 16 parts of FeO will require $\frac{60 \times 16}{144} = 6{\cdot}66$ parts.

The amount of silica required by the 3 parts MgO is found in the same way :—

$$2MgO,SiO_2 = 80 : 60 \qquad \frac{60 \times 3}{80} = 2{\cdot}25.$$

Thus giving a total ($6{\cdot}66 + 2{\cdot}25$) of 8·91 parts by weight of silica slagged by the bases, so that $60 - 8{\cdot}91 = 51{\cdot}09$ parts of silica remain to be converted into slag as sodium monosilicate. According to the same proportion,

$$2\,Na_2O,SiO_2 = 124 : 60 \qquad \frac{124 \times 51{\cdot}09}{60} = 105{\cdot}6.$$

Hence to flux 51·09 parts of silica 105·6 parts of Na_2O are required, but as this is added in the form of, say, sodium bicarbonate, which contains approximately 40 per cent. of Na_2O, the total weight of this flux required would be $\frac{105{\cdot}6 \times 100}{40} = 264$ parts.

Therefore, if 50 grammes of ore are taken for assay, $\frac{264}{2} = 132$ grammes of sodium bicarbonate must be added to obtain the monosilicate slag desired. If other bases are present or different fluxes are used the quantities required are calculated in the same way after determining what slag is to be produced. When the iron in an ore consists of ferric oxide it is essential to remember that it must be converted into ferrous oxide by the addition of carbon, to form a satisfactory slag with silica.

The amount of carbon required is calculated from the following reaction :—

$$2Fe_2O_3 + C = 4FeO + CO_2.$$

One gramme of Fe_2O_3 requires $\frac{12}{320} = 0{\cdot}037$ gramme of carbon. If charcoal powder, which contains only about 90 per cent. of carbon, is used, the quantity required to reduce a gramme of Fe_2O_3 will be $\frac{{\cdot}037 \times 100}{90} = 0{\cdot}041$ gramme charcoal.

This amount of charcoal would of course be added in addition to that required to reduce the PbO to give the lead necessary for collecting the values.

Magnetite (Fe_3O_4) and manganese dioxide (MnO_2) require reduction in a similar manner.

The accompanying Tables XIIA and B, based on tables computed by Professor Balling, will simplify slag calculations.[1]

TABLE XII.—FOR CALCULATING THE AMOUNTS OF FLUXES REQUIRED TO OBTAIN GIVEN SLAGS.

TABLE A. *For ascertaining the Necessary Amounts of Bases to convert Given Amounts of Silica into Slag.*				TABLE B. *For ascertaining the Necessary Amounts of Silica to convert Given Amounts of Bases into Slag.*			
Base.	Parts by weight of Bases required to convert one Part by weight of Silica into a—			Base.	Parts by weight of Silica required to convert one Part by weight of the Base into a—		
	Mono-silicate.	Bi-silicate.	Sesqui-silicate.		Mono-silicate.	Bi-silicate.	Sesqui-silicate.
Na_2O	2·07	1·02	1·37	Na_2O	0·486	0·972	0·729
PbO	7·36	3·69	4·92	PbO	0·136	0·271	0·203
FeO	2·40	1·20	1·60	FeO	0·416	0·833	0·625
CaO	1·86	0·93	1·24	CaO	0·535	1·070	0·803
MgO	1·33	0·66	0·88	MgO	0·750	1·500	1·125
Al_2O_3	1·14	0·57	0·76	Al_2O_3	0·873	1·747	1·310
MnO	2·36	1·18	1·57	MnO	0·422	0·845	0·633

In order to show the use of this table, the above example may be worked with its aid. For ascertaining the amount of silica converted into slag by the bases in the ore, Table B is used. In the section for the monosilicates, the corresponding figures are found for the bases present, and these figures multiplied together give the following products :—

FeO	$16 \times 0{\cdot}416 =$	6·66
MgO	$3 \times 0{\cdot}750 =$	2·25
	Total =	8·91

Thus 8·91 parts of silica are converted into slag, and of the total amount (60 parts) of silica present there remain $60 - 8{\cdot}91 = 51{\cdot}09$ parts to be converted into slag by the addition of sodium oxide. On referring to Table A, it will be seen that the corresponding amount of sodium oxide is 2·07, the product being $51{\cdot}09 \times 2{\cdot}07 = 105{\cdot}75$.

Several graphic methods for facilitating the calculation of weight of charges are given in *An Introduction to the Study of Metallurgy*, by Sir W. Roberts-Austen and F. W. Harbord.

[1] Quoted by Roberts-Austen & Harbord, *Introduction to Study of Metallurgy*.

CHAPTER XI

ASSAY OPERATIONS

The dry assay of ores involves the operations of roasting, fusion, scorification, and cupellation.

(I) ROASTING

The oxidation of pyritic material in ores is sometimes accomplished by roasting, in which operation advantage is taken of the oxygen of the air as the oxidizing agent. It may be well to point out that the operation is sometimes termed *calcination*, the terms roasting and calcination being used synonymously. But, in a metallurgical sense, the term calcination should be confined to the application of heat for the expulsion of carbon dioxide or volatile matter. In roasting, the sulphur and arsenic in such minerals as pyrites and mispickel are oxidized by the hot air and pass off principally as sulphur dioxide and arsenious oxide (As_2O_3), while the metals generally remain in the final product in the form of oxides. The quantity of ore, 20 to 50 grammes, to be used for the assay must be carefully weighed out before roasting. The roasting is most conveniently effected in a roasting-dish (Fig. 63) sufficiently large to allow of the ore being stirred without loss by spilling. The dish is placed in a muffle heated to a dull-red heat (that is, at a temperature of about 600° C.), and the door removed to allow of free access of air. The ore is frequently stirred, though not violently, to expose fresh surfaces to the air and to prevent clotting. A suitable stirrer for the purpose consists of a piece of stout iron wire flattened at one end, and bent at a right angle about 1 inch from the end. The temperature at which the roasting is started is of importance, as it is during the first few minutes of the operation that the formation of lumps is most to be feared. The employment of too high a temperature at the outset, or neglect in stirring, will cause partial fusion of the more fusible sulphides, such as Sb_2S_3 and PbS, that may be present, the fused matter so formed being troublesome to roast efficiently.

The ore is maintained at dull redness until the blue flame of burning sulphur has ceased, and no incandescent particles are visible on stirring, after which the temperature is raised to redness to decompose sulphates. The operation, which occupies from thirty to sixty minutes for 50 grammes of ore, is complete when the ore remains of a uniform colour on stirring and no further smell of SO_2 is discernible. When finished, the ore is said to be " dead " or " sweet " roasted. In the case of an ordinary sulphide ore, such as pyrite (FeS_2) with quartz, the roasting takes place in two stages. Pyrite, when heated, parts readily with one of the two atoms of sulphur, thus :—

$$FeS_2 + \text{Heat} = FeS + S.$$

The first stage of the roasting consists, therefore, in liberating and oxidizing one-half of the sulphur, which burns with a blue flame, forming sulphur dioxide, which is volatilized. The heat generated by the oxidation of the sulphur causes the ore particles to glow. If a low temperature be employed, a certain proportion of the sulphur is oxidized to sulphur trioxide, which unites with metallic oxides to form sulphates ($FeO + SO_3 = FeSO_4$). The second stage consists in oxidizing the remaining sulphur to sulphur dioxide and the iron to ferrous oxide, thus :—

$$FeS + 3O = FeO + SO_2.$$

Owing to the oxidizing conditions of the roasting, a large proportion of the ferrous oxide is converted into ferric oxide ($2FeO + O = Fe_2O_3$). With prolonged heating ferroso-ferric oxide (magnetic oxide) Fe_3O_4, or Fe_2O_3,FeO, is formed. Any sulphate formed in the first stage will be decomposed as the temperature increases, forming ferrous oxide and sulphur trioxide.

$$FeSO_4 + \text{Heat} = FeO + SO_3.$$

The decomposition of sulphates is essential, otherwise they may be reduced to sulphides in the subsequent fusion of the ore in the crucible, thus :—

$$FeSO_4 + 2C = FeS + 2CO_2.$$
$$2CuSO_4 + 3C + 3O_2 = Cu_2S + SO_2 + 3CO_2.$$

Sulphates are decomposed by a higher temperature towards the end of the operation. Lead and zinc sulphates require a comparatively high temperature to effect their decomposition.

The roasting operation is more difficult and complex when ores contain copper, arsenic, antimony, and lead.

When copper is present, the copper sulphate formed during roasting requires a higher temperature to effect its decomposition than ferrous sulphate. The decomposition is facilitated by the addition of ammonium carbonate, which converts the copper sulphate into ammonium sulphate, which is volatilized, thus :—

$$CuSO_4 + (NH_4)_2CO_3 = CuO + (NH_4)_2SO_4 + CO_2.$$

For this purpose the ore is first roasted until "sweet," then cooled, mixed with about one-fourth of its weight of dry powdered ammonium carbonate, and re-roasted at a low heat until fumes cease to be given off.

Ores containing arsenic and antimony require considerable care and attention during roasting to successfully eliminate these two volatile elements. Both arsenic and antimony, during roasting, are first oxidized to the state of the lower oxides As_2O_3 and Sb_2O_3, which are volatile and largely expelled, but a certain proportion is further oxidized to the higher oxides As_2O_5 and Sb_2O_5, which, in the presence of certain metals, notably silver, copper, and iron, form arsenates and antimonates of these metals. These compounds are stable even at high temperatures, thus causing the arsenic and antimony to be retained in the ore instead of being eliminated. Arsenates and antimonates will carry silver into the slag in the subsequent fusion of the roasted material unless previously decomposed. The decomposition of the arsenates and antimonates is best effected by mixing about an equal volume of powdered charcoal or coal with the ore after it has been roasted "sweet," and roasting again at a red heat until all the carbon is burnt off. During this second roasting the air should be excluded in order to secure a reducing atmosphere, and thus facilitate the reduction of the arsenates and antimonates.

Iron arsenate is readily decomposed when heated with charcoal powder, and practically the whole of the arsenic eliminated, but the reducing action of charcoal on copper arsenate is very slight, only a small elimination of arsenic taking place. The reduction of iron arsenate is shown in the following reaction :—

$$Fe_2O_3,As_2O_5 + 2C = Fe_2O_3 + As_2O_3 + 2CO.$$

Antimonates are also decomposed by charcoal, but less easily than most arsenates, and the antimony eliminated. The presence of sulphur facilitates the elimination of arsenic and antimony by combining with them to form volatile sulphides.

Owing to the readiness with which antimony sulphide fuses, the roasting of ores containing this mineral must be performed at a low temperature to avoid clotting.

Ores containing galena (PbS) are difficult to roast on account of the ready formation of lead sulphate, which is only decomposed at a comparatively high temperature. Under these conditions there is a great tendency for the ore to clot or even fuse, as lead sulphate, and any lead oxide formed may combine to form, in the presence of silica, a readily fusible lead silicate. As a general rule, ores containing large proportions of galena are treated by other methods of desulphurization described subsequently.

The mechanical loss during roasting may be considerable if proper precautions are not taken. The chief causes of loss are spilling through excessive agitation of the ore, and decrepitation of the pyritic material, due in many cases to the coarseness of the sample Certain varieties of pyrites and some metallic sulphides, however, have a great tendency to decrepitate when heated, even when finely powdered, but loss may be prevented by covering the ore with an inverted roasting-dish until all danger of decrepitation is over. It is somewhat difficult to prevent partial fusion or clotting of the ore containing sulphides of antimony or lead, and, in the event of the ore fusing, the only satisfactory procedure is to weigh out a fresh quantity of ore and roast again with greater care. In the absence of fusible minerals the temperature at the commencement of the operation should not be below a dull-red heat, as a low temperature favours the formation of sulphates, which should, as far as possible, be avoided for the reason already stated. The roasting may also be performed in the crucible in which the fusion is afterwards made. In this case the crucible is usually placed in a wind furnace and inclined to the operator, so that the draught of air passing to the furnace flue may impinge as much as possible on the surface of the ore. The top of the furnace must be open, so that air may have free access to the crucible. The ore is stirred with the straight stirrer shown in Fig. 30. The crucible employed should be shallow and wide-mouthed, not deep and narrow, and sufficiently large to allow of the subsequent fusion of the ore being made in it. When the roasting is finished, the ore is turned out into an iron mortar and mixed with the fluxes, taking care to clean the stirrer in the mixture. The charge is then replaced in the crucible used for the roasting and fused in the furnace.

Ores may be conveniently roasted also in an iron pan (Fig. 101) over the crucible furnace. The pan should be coated with a thin layer of chalk or red ochre to protect it from injury through the sulphides attacking the iron.

As previously stated, the roasting of an ore occupies a considerable amount of time and attention, and, when possible, it is better to take advantage of the desulphurizing action of lead oxide or nitre. The roasting of silver ores is seldom attempted, as it is very liable to give low results, especially with high-grade ores. The roasting method of oxidation is frequently used for gold ores consisting mainly of pyritic material, as it permits of a larger quantity of ore being taken than is permissible with other methods of oxidation.

The general tendency of modern assay practice, however, appears to be to obviate the roasting of ores as far as possible.

(II) FUSION

The object of this operation is to concentrate the precious metals in a button of lead, while the gangue, etc., forms (with suitable fluxes) a fusible slag in which the lead sinks. The fusion of an assay charge is effected in a fireclay crucible in either a wind or muffle furnace. A deep and narrow crucible is usually best for the purpose. The crucible is carefully annealed in the ashpit of the furnace before using, otherwise it is very liable to crack when placed in the furnace. The ore and fluxes (excepting a small part of the borax) are well mixed and charged into the crucible, which must not be more than half full. The mixture is then covered with the remainder of the borax, reserved for that purpose, and if iron is required the nails or hoop-iron are inserted in the charge. The crucible is put into the wind furnace at a low red heat (about 600° C.) and packed round with fuel. For this purpose both ordinary and basket tongs are useful (see Figs. 21 and 24). A plan adopted by some assayers is to place an old crucible into the furnace and pile the coke round it; then carefully withdraw it, and lower the charged crucible into the hollow thus formed. If coke is used, all fuel touching the pot should be in the first instance at a black heat. The temperature should not be urged at first, but slowly raised, and after ten or fifteen minutes the charge should have just reached a dull-red heat.

This exposure of the charge to a dull-red heat before fusing is known as "fritting," and is of considerable importance. At this time the lead is all reduced and remains entangled in the charge in the form of minute liquid beads. Chemical action now begins between the fluxes and the ore. Carbonic acid is given off from the sodium carbonate as it unites with the silica, and the charge is continually stirred up by the disengaged gas. The temperature is now gradually raised to a full red heat, and the lead beads unite and sink through the liquefying charge. After a period of 30 or 40 minutes from the time of charging in, the charge is in a state of tranquil fusion, and the lead containing the precious metals is collected at the bottom. The nails (or hoop-iron) are then withdrawn, and all adhering lead shaken off them into the crucible. The crucible is now lifted out of the furnace and tapped on the top of the furnace to assist the lead to settle in the slag. The charge is poured into a warm iron mould and allowed to solidify. It is important that the mould should be clean and quite free from any loose graphite powder, as this would cause the lead to spurt and occasion loss. The lead is found at the bottom of the mould, and, when cold, is detached from the slag by hammering the edges of the mass. It should be soft, malleable, and free from any regulus, and should weigh from 20 to 30 grammes. When freed from the slag, the lead button should be hammered into a disc with rounded edges, or into a cube with rounded corners. If the edges or corners are left sharp, they are very liable to injure the face of the cupel when the button is placed on it. After pouring the charge the crucible may be employed a second time, or even more; but the practice is to be deprecated, unless the ore being assayed is low grade. It should on no account be done with rich ores. The same crucible is used for "cleaning the slag" when this is necessary.

The fusion of the charge too frequently receives but scant attention at the hands of the assayer, but it may be pointed out that the success or failure in collecting all the gold or silver in the charge is to a large extent dependent on the care bestowed on the fusion. If the heat should be too strong at first, the quickly-escaping gases may cause part of the charge to be ejected from the crucible; or the flux might melt and sink to the bottom of the crucible, leaving the quartz, etc. as a pasty mass above. The more slowly the charge is melted the better the chance afforded of bringing the lead into intimate contact with all parts of the pasty mass, and of collecting the precious metals from the ore. With a gentler heat, also, the combination between ore and flux is more complete, and, when the heat is subsequently raised, the fusible compound is simply melted into homogeneous slag. As stated above, the temperature at charging-in should be about 600° C., while " the final temperature " need not, as a rule, be above 1100° C., and may be lower in certain cases. The influence of the rate of heating on the result is very great, as shown by experiments by Dr. Rose.[1]

Four charges, consisting of equal weights of pyritic ore, litharge, and sodium carbonate with a little borax powder and the necessary charcoal, were made.

Two charges were heated very slowly, and remained in the furnace for 50 minutes before fusion was complete. The other two charges were heated as rapidly as possible, and were in the furnace for only 20 minutes. The results were as follows :—

	Gold.	Silver.
Heated slowly ...	oz. per ton (2240 lbs.) (1) 0·130 (2) 0·125	oz. per ton (2240 lbs.) (1) 4·42 (2) 4·28
Heated rapidly ...	(1) 0·075 (2) 0·092	(1) 2·33 (2) 2·71

The mean deficiencies in the charges heated rapidly were thus :—Gold, 34 per cent. ; silver, 42 per cent. ; but the results were also less regular than when slow heating was used. Another example is given on p. 216.

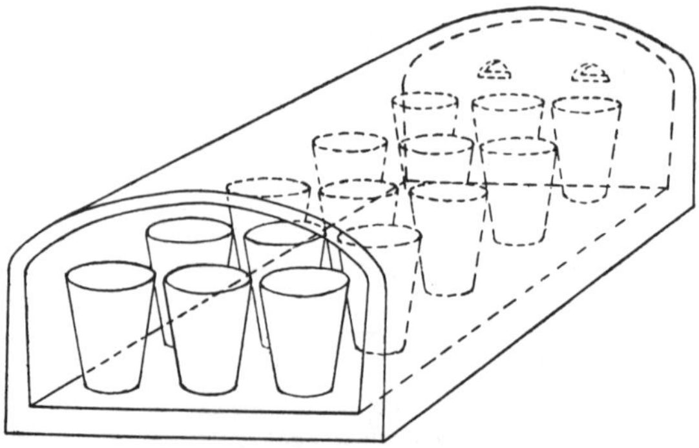

FIG. 103.—MUFFLE CHARGED WITH COLORADO CRUCIBLES.

Fusion in the Muffle (see Fig. 103).—Crucible fusions can be made satisfactorily in a muffle furnace if it is large enough, and in Colorado, and in America generally, muffle furnaces have largely replaced crucible furnaces for general assaying. The temperature required is about the same as that used in scorification, and the conditions of the fusion are similar to those already described. Crucibles of the " Colorado " shape (Fig. 59) are generally used. The advantages claimed for the muffle are that it is cleaner, easier to manipulate, and more uniform in working, the temperature of a muffle being better controlled, and more easily kept constant and uniform than that of an ordinary fusion furnace. When large-size muffles are employed, six or eight or even more fusions can be made at one time, and on that account muffle furnaces are favoured in smelter and mine assay-

[1] *Metallurgy of Gold*, 7th edn., 1937, p. 497.

offices where frequently a large number of assays have to be dealt with daily. The chief disadvantage of the muffle for fusions is that the weight of the ore taken for assay is limited to about 20 grammes, though this objection may be obviated by making several fusions and scorifying together the resulting lead buttons so as to concentrate the precious metals into one button of lead.

Cleaning or Washing the Slag.—The slags of very rich ores may retain enough precious metal to necessitate further treatment. The slag is collected, roughly crushed, and " cleaned " by mixing with the following charge :—

Litharge	30	grammes.
Charcoal powder	1·5	"
Sodium carbonate	10	"

If any regulus or speise forms during the fusion of the ore, and has not been scorified with the lead button should this have been necessary, it must be preserved, carefully crushed, and mixed in with the slag charge and a little more sodium carbonate added. The charge is placed in the crucible previously used for the fusion, and kept melted for ten to fifteen minutes. If the fused mass is thick and pasty, a small quantity of borax may be added with advantage. When tranquil fusion takes place, the contents of the crucible are poured into a mould, and the slag detached from the button of lead and thrown away. The lead button from this second fusion is cupelled either by itself or with the first lead button. It is also a common practice in cleaning slags to fuse the slag alone, and then add a mixture of litharge and charcoal. In this case the crucible is replaced in the furnace as soon as the first charge has been poured, and the slag put back into it, without crushing, after being detached from the lead button. When the slag is fluid, a mixture of 20 grammes of litharge and 1 gramme of charcoal is charged on to the surface with a copper scoop. When the charge is again in a state of tranquil fusion, it is poured and the lead button extracted.

Some assayers consider that it is not necessary to pour the slag resulting from the first fusion before cleaning it, and adopt the following method : after the ordinary fusion of the ore has been made and the crucible contents have become tranquil, a mixture of litharge and charcoal is added, and the charge poured as usual when again tranquil. In this way the " cleaning " or " washing " of the slag forms part of the ordinary fusion and not a separate operation, and is therefore more expeditious (see Balling's method, page 188).

The above methods of cleaning slags are applicable to all ores and products which are assayed for precious metals. The richer the ore or product, the more needful is it to clean the slag ; in the case of poor ores it may generally be omitted. However, in the case of ores with which the assayer is not familiar, it is better not to omit the examination of the slag until experience has shown that this may be safely done. As a general rule, slags from rich ores, slags containing many metallic oxides or metallic sulphides, and slags containing much ferric oxide should be cleaned.

(III) SCORIFICATION

This operation consists of an oxidizing fusion of an ore with metallic lead in the muffle furnace, and has for its object the concentration of the precious metals in a button of metallic lead, and the removal of every other constituent of the ore by the solvent action of molten litharge formed during the operation by the oxidation of part of the lead added.

Scorification is conducted in a fireclay scorifier (Fig. 61), which must be well dried at a gentle heat for ten or fifteen minutes before use. If this precaution is not taken, " spurting " of the lead may occur during the operation. Sometimes the scorifier is coated inside with ferric oxide to protect it as far as possible from the corrosive action of the molten litharge. A thin paste made by mixing powdered hæmatite with water is used for the purpose. The amount of ore and lead used varies according to the nature of the ore, being, on the average, about ten to twenty parts of lead to one part of ore. The charges for the scorification of different classes of argentiferous and auriferous materials are given subsequently in Chapters XII and XIII.

Granulated lead is generally employed for the purpose. The ore is mixed in the scorifier with about half the lead, the mixture smoothed down, the rest of the lead spread evenly over it, and a small quantity of borax glass (0·5 to 1 gramme) placed on the top: the object of the borax being to lessen the corrosion of the scorifier, and to render the slag more liquid. Instead of mixing the ore and the lead as described, it is equally satisfactory, according to Beringer, " to wrap the ore up in thin sheet-lead of the required weight; and if the ore contains much sulphur, the borax may with advantage be added (wrapped in tissue paper) some five or ten minutes after the operation has started."

The scorifier is then put into the muffle at a full red heat with the scorifier tongs (Fig. 26), and the door closed until fusion has taken place. As soon as the lead is melted the door is partly opened to allow a current of air to pass over the scorifier and oxidize the contents. At this stage some of the ore may be seen to be floating on the surface of the molten lead, and any sulphides present are rapidly oxidized, chiefly by the air, but partly by the litharge which immediately begins to form. The sulphur, arsenic, antimony, etc., in sulphides and arsenides are thus soon eliminated as volatile oxides, while the base metals are converted into oxide, and combine with the litharge and borax to form a fusible slag. Silica also forms a fusible compound with the litharge. Thus the litharge, as fast as it is produced, reacts on the various substances contained in the ore, forming with them a clean slag, while the precious metals pass into the unoxidized lead which collects at the bottom of the scorifier.

The slag as it is produced forms a ring, leaving the centre of the lead bath or " eye," as it is termed, exposed to oxidation. As oxidation of the lead continues, the litharge flows to the sides and increases the quantity of slag, until at length the whole of the lead is covered with slag and the eye has disappeared, leaving a flat, uniform surface. The scorification is then considered finished, but it is usual before pouring to clean the slag. This is done by placing a pinch of charcoal or coal powder (about 0·2 gramme), wrapped in tissue paper, on the surface of the slag, with a pair of tongs, and closing the muffle door. A small quantity of litharge is thus reduced, forming a number of globules of lead which sink in the slag, alloying with any silver and gold which the slag may contain, and concentrating them in the molten lead below. The temperature at the finish should be sufficiently high to liquefy the slag completely. A temperature of 1050° to 1100° C. is usually enough. When the slag has again become tranquil, the scorifier is withdrawn, and the contents poured into an iron mould (Figs. 36 and 38). The contents should pour clean and not leave any imperfectly fused material attached to the surface of the scorifier. The assay should be poured as soon as finished, and not left in the muffle, as oxidation of the lead will continue although none is exposed to the air, owing to the interchange of oxygen between litharge and other oxides present. When cold, the slag is detached from the button of lead as already described. The lead button should be comparatively soft and malleable, and weigh from 15 to 20 grammes. In most cases the button may be cupelled direct, but if it weighs much more than 20 g., its weight is reduced by another scorification. If the scorification is prolonged to produce buttons weighing less than 15 g., appreciable amounts of precious metals are liable to be retained in the slags, especially with rich ores. The lead buttons from the scorification of ores containing copper, antimony, etc., are very liable to be hard, owing to contamination with these metals, especially if insufficient lead has been used, and in order to get a pure button for cupellation it is generally necessary to re-scorify them with the addition of fresh lead. When the temperature has been too low, the lead is very liable to retain lead oxide and be more or less brittle. Scorification usually occupies from thirty to forty minutes. The temperature must be well maintained during the operation, and must be high enough to melt litharge when contaminated by silica and oxides of copper, iron, manganese, etc. It ranges usually from 1000° C. to 1100° C., although a higher temperature may be employed for highly siliceous ores.

The slag from scorification should be homogeneous and glassy, while in colour it varies according to the nature of the ore. Scorification slags are essentially oxide slags consisting of metallic oxides dissolved in excess of litharge, and exercise a very corrosive action on the material of the scorifier, especially if they contain copper oxide. If the slag be stony in appearance and white patches of lead sulphate appear on the surface, it generally indicates that too low a temperature has been employed.

The slags from the scorification of average ores may generally be thrown away, as the button of lead will contain practically all the silver; but with rich ores and products it is advisable to save the slags and subsequently "clean" them by fusion in a small crucible, as previously described. In this case the "cleaning" by charcoal powder during scorification is omitted.

The slags from several scorifications are crushed and mixed with about 20 grammes of litharge, 1 g. of charcoal, and 5 g. of borax. The resulting lead button is cupelled.

The slags may also be fused with the powdered cupels used for the cupellation of the lead button obtained from the scorification of the ore. In this way the silver lost in the slag and that absorbed by the cupel is recovered at one operation.

The losses incurred in scorification are chiefly due to the employment of too low a temperature. If the muffle is too cold at first, the scorification is always more or less unsatisfactory, and the silver and gold retained in the slag will be by no means inconsiderable. If the slag appears pasty even when the temperature is sufficiently high, borax may be added with advantage at short intervals in quantities of 0·5 to 1·5 grammes at a time. The borax, wrapped in tissue paper, is placed on the surface of the slag. Only a small quantity of borax should be added at the commencement of the operation, as it would cover the surface of the lead and retard its oxidation. If the ore contains copper, powdered glass or silica may be used with advantage in the place of borax. Scorification is a very convenient and quick method of assay, and needs very little variation for the different classes of ores. It is largely employed for the determination of silver in ores and products. The small quantity of material that can be treated constitutes the chief disadvantage of the method; consequently, gold ores, as a general rule, are better assayed by the fusion method, so that a comparatively large quantity of material may be operated upon.

(IV) CUPELLATION

Cupellation has for its object the removal, as oxides, of base metals such as lead and copper from gold and silver, which are non-oxidizable and are left in the metallic state. The same remarks apply to platinum and metals of the platinum group. The operation is performed in a muffle furnace in a porous vessel, called a *cupel* (Fig. 63), the nature and manufacture of which has previously been described. Cupellation is based on the fact that " when molten lead is exposed to the action of air at a temperature considerably above redness, it combines with the oxygen of the air to form litharge (PbO), which at the temperature of its formation is a *liquid.* Consequently, if the lead is contained in a porous vessel, which allows the fused litharge to drain away as fast as it is formed, a fresh surface of the lead will be continually exposed to the action of the air, and the operation goes on until the whole of the lead has been removed. Silver or gold exposed to similar treatment does not oxidize, but retains its metallic condition ; so that an alloy of lead and silver or gold similarly treated would yield its lead as oxide, which would sink into the porous vessel, while the silver or gold would remain as a button of metal." [1]

There is a considerable difference between the surface tension of molten lead and molten litharge, and while litharge can " wet " the bone-ash surface and hence be obsorbed, molten lead either cannot do so, or does so only to a very limited extent, and hence is not absorbed.

For the same reason the high-surface tension of the molten silver and gold, left on the cupel after the oxidation of the lead, prevents them from being carried into the pores of the cupel, although, as stated subsequently, there is always a small loss of precious metal during cupellation, which is caused by absorption by the cupel.

The oxides of lead and bismuth, in a state of purity, are the only oxides which possess the property of being absorbed by the cupel ; but by the aid of either of these, various oxides which by themselves form infusible scoriæ on the cupel acquire the property of being absorbed. For example, copper, antimony, tin, and most of the base metals form powdery oxides,which are not of themselves easily fusible or absorbed and require the admixture of some solvent—which is invariably lead oxide (litharge)—to render the oxide sufficiently fluid to be absorbed by the cupel.

Advantage is taken of this fact in cupelling material containing a base metal like copper, etc., such as an alloy of silver and copper ; and sufficient lead is added to supply the proper proportion of litharge to dissolve the copper oxide formed in the operation and carry it into the cupel. In the case of antimony and lead, their oxides appear to form compounds with lead oxide, but no compound of copper and lead oxide is formed. The proportion of litharge necessary to cause the absorption of other oxides varies with their nature and other circumstances. The quantity of lead required to supply the correct proportion of litharge in ordinary cases will be given subsequently.

Cupels.—It is extremely important that good cupels should be used for the operation of cupellation ; they must, obviously, have a sufficiently loose texture to allow the fused oxides to be absorbed readily, and at the same time be sufficiently firm to bear handling without breaking, while they must also be made of material that will not be attacked by molten litharge. The method of making cupels has already been described on page 45.

Cupels should be made many months before they are used, and stored in a dry place so that they may be dried very slowly, but completely, as otherwise

[1] Beringer, *Text-book of Assaying.*

cracks appear when they are heated and loss is thereby occasioned. "The square cupels (bone-ash) used at the Royal Mint, London, are made two years before being used, and are dried slowly on shelves at some distance from the furnaces. More rapid drying has been tried on many occasions with less satisfactory results. Nevertheless, fair results, but not the best, can be obtained when using cupels only a few days old" (Rose). It is well to remember that a bone-ash cupel will absorb its own weight of litharge. "About 98·5 per cent. of the litharge formed during the operation is absorbed by the cupel and the remaining 1·5 per cent. volatilized" (Fulton). It is important that the cupel used should be of suitable size for the lead to be cupelled upon it, as it is unsatisfactory to have the cupel too strongly saturated with litharge. As a general rule the cupel (bone-ash) should be at least one-fourth as heavy again as the lead button.

The number of cupellations made at one time is dependent on the size of the muffle, but it is not usual to cupel more than fifty in one batch or "fire," as it is sometimes termed. The cupels must be carefully arranged in a definite order in the muffle when more than one lead button has to be cupelled.

The operation of cupellation is conducted as follows:—The cupels are first cleaned by blowing off the loose dust and by gently rubbing the cavity with the finger; they are then placed in a muffle at a red heat and allowed to remain there for about ten to fifteen minutes until they attain the temperature of the muffle, and in order to expel any moisture or burn off any organic matter that may be present. If this precaution is not taken and the lead is placed direct into a cold cupel, the metal when melted will not clear, and if all the moisture is not expelled steam will escape and bubble violently through the molten lead, causing what is termed "spurting," *i.e.* the ejection of lead particles and consequent loss. A similar result will be produced if the cupels contain particles of carbonaceous matter owing to the escape of bubbles of carbon dioxide resulting from the reduction of the litharge by the carbon.

When the cupels have attained the temperature of the muffle the lead buttons are placed in them, each in its proper cupel, by means of the cupellation tongs (Fig. 28, page 35), and the door of the muffle is closed. Some assayers prolong the heating of the cupels for another ten minutes or more, even after they have attained the temperature of the muffle, before charging in the lead buttons. The "charging in" should be done promptly, so as not to cool the muffle unduly, and with care to avoid damaging the surface of the cupel.

When a large number of assays have to be "charged in" great care must be taken to ensure that the temperature is sufficiently high before charging-in, so that the chilling which necessarily takes place during the operation shall not cool the muffle below the requisite temperature. To start cupellation requires a fairly high temperature, and is a critical part of the operation: the general experience is, however, that beginners are more likely to have the furnace too hot. The exact temperature suitable for cupellation will be discussed later. If the temperature is correct, the lead quickly melts and a grey-black scum of lead oxide forms at once on its surface, but soon fuses (at 875° C.). This is called the "clearing," "uncovering," or "opening up" of the lead button. The molten lead then appears much brighter than the cupel and rapid oxidation commences. The lead buttons should clear as soon as possible, and if this does not take place the buttons are said to be "frozen," in which case the temperature must be raised. The risk of "freezing" at the start is much greater with a cupel which has not been properly heated before introducing the lead.

The melting-point of lead oxide is lowered by admixture of copper oxide, the eutectic mixture freezing[1] at 698° C. If freezing takes place, the lead may be made to clear by placing a piece of charcoal near the cupel so as to raise the temperature locally, or a little charcoal powder may be dropped on to the surface of the lead. Freezing may also take place during the cupellation owing to the temperature falling. In this case the assays should be repeated. The closing of the muffle door at first is simply to increase the heat until the lead buttons have "cleared," and when this has taken place the door is partly opened to admit air and to cool the muffle, as the temperature needed to clear the buttons is higher than that required to keep the cupellation going. When cupellation has started satisfactorily the muffle is cooled to as low a temperature as is compatible with the continuation of the operation. A higher temperature is again required at the finish: the muffle door is therefore closed and the furnace urged towards the end of the operation.

It will be observed that as soon as the lead has cleared, the surface of the metal becomes covered with oily-looking drops of litharge which are rapidly absorbed by the porous cupel and replaced by others. They pass over the surface slowly at first, but as the operation continues they move with greater rapidity. Towards the end of the operation the metal suddenly appears uniformly dull and glowing, and iridescent bands, produced by extremely thin films of fluid litharge, resulting from the oxidation of the last traces of lead, immediately pass over it. On the disappearance of these coloured bands a bright liquid globule or button of gold or silver or an alloy of both is left. If the cupel is withdrawn from the furnace, the button becomes very bright at the moment of solidification, and is said to "flash," "brighten," or "blick." This "flashing" of the buttons of gold or silver is due to the evolution of the latent heat of fusion or "recalescence," which momentarily reheats the cooling globule to its melting-point. The flashing is much more noticeable with gold than with silver buttons. Molten gold has a peculiar green colour which is easy to recognize, and just after solidifying it glows beautifully, even with very small buttons. With very small silver buttons the flashing is rarely noticeable. The flashing is prevented if the button contains metals of the platinum group. Silver buttons require careful attention at the finish of the cupellation.

Silver when molten absorbs oxygen from the air and gives it off suddenly when solidifying, causing a cauliflower-like growth on the surface of the button, and particles of silver may even be ejected out of the cupel and cause serious loss. This "spitting," "spurting," or "vegetation," as it is termed, does not take place readily with silver buttons weighing less than about 5 milligrammes, but all buttons above this weight are very liable to spit. Spitting does not take place if gold is present in the silver button to the extent of 33 per cent. or more (Levol), as the solubility of oxygen in silver is lowered by alloying with gold. If the button is small the cupel may be withdrawn from the muffle as soon as cupellation is finished without risk of "spitting," but with large buttons precautions must be taken.

"Spitting" can be prevented by slow cooling, and in practice the following methods are usually adopted. The cupels are allowed to cool in the muffle itself, the door being closed; or another cupel, previously heated to redness, is inverted over the cupel containing the button, and both carefully withdrawn from the muffle, thus cooling the button slowly between two hot cupels; or the cupel may be withdrawn very gradually towards the door of the muffle. Cooling in the muffle is invariably adopted with a large batch of assays, as, in addition to the danger of upsetting a cupel, it is inconvenient to handle a

[1] J. C. J. Cunningham, *Zeits. für Organ. Chem.*, 1914, **89**, 48.

large number of hot cupels. Buttons that have " spurted " must be rejected and the assays repeated.

The button of precious metal when cold is detached from the cupel with a pair of pliers and the under-surface distorted by squeezing or hammering so as to loosen the adhering bone-ash. It is then cleaned by brushing with a hard brush and weighed. The silver button from cupellation should be bright, round, and slightly crystalline on the upper surface, and adhere slightly to the cupel. When it retains base metals it adheres more firmly to the cupel. Gold may be present in the silver button to the extent of 50 per cent. without showing any yellow colour. Platinum when present destroys the characteristic silver lustre and renders the button dull and grey but more crystalline. Gold, with little or no silver, is easily recognized by its characteristic colour, and the " flash " of gold is so marked that it is a valuable indication of its presence. Gold buttons are more globular than silver buttons of similar size and adhere less to the cupel.

Temperature of Cupellation.—The expression " temperature of cupellation " is usually understood by assayers to mean the temperature of the muffle in which cupellation is conducted, and this meaning of the expression is adhered to in the present volume. It must, however, be pointed out that the true temperature of cupellation is the temperature of the lead being cupelled. Owing to the heat developed by the active oxidation of lead, the actual temperature of the molten lead in the cupel is always higher than the temperature of the muffle where the cupel stands. The muffle temperature always bears a distinct relation to the temperature of the lead being cupelled, and for all practical purposes the muffle temperature affords sufficient indication of the correct temperature at which to start and conduct cupellation. Experienced assayers, however, always closely observe the temperature of the lead on the cupel, and carefully regulate the supply of air through the muffle and the temperature of the muffle so as to keep the lead at the correct temperature. The regulation of the amount of air passing into the muffle during cupellation is of considerable importance, and is a matter which does not always receive sufficient attention. The exact temperature suitable for cupellation can only be ascertained by practice, and the varying light of the day may occasion error in judging of the degree of heat. As previously stated, in general it is well to employ a comparatively high temperature at the commencement so as to well uncover the lead, then to cool down, and increase the heat towards the end of the operation for a few minutes, in order to aid the brightening.

There is a German adage many centuries old which runs :—

> Kühl getrieben, heisser Blick,
> Ist des Probierers Meisterstück.

This may be translated :—

> Cool for the working, hot for the " blick " [*i.e.* brightening],
> That's the assayer's master-trick.

With lead buttons poor in silver or gold, the lowering of the temperature after " clearing " is a matter of importance, especially when the temperature at the start has been excessive ; but when dealing with material such as bullion containing large proportions of the precious metals, it is of more consequence that the muffle should be uniformly hot throughout than that any absolute degree of heat should be attained, as it is usual in cupelling such rich material to use check or proof assays to eliminate uniform errors due to high temperature. The time occupied by the cupellation when small quantities of lead only are used, as in bullion assays, is frequently too short to allow of much variation of temperature being made. The temperature of cupellation has to be varied somewhat according to the nature of the material to be cupelled ; for example, gold, which solidifies at 1064° C., requires a higher temperature than silver, which solidifies at 962° C. A higher temperature is also required for lead containing base metals and a large percentage of the precious metals than for lead containing little gold or silver and free from base metals. Most lead buttons from ore assays will be comparatively pure and contain very much less than 1 per cent. of precious metal, and will melt practically at the melting-point of lead, while in bullion assays the proportion of precious metals is much greater; and base metals such as copper are frequently present and raise the melting-point of the lead alloy on the cupel, so that the temperature of cupellation must be higher in order to prevent freezing. As cupellation proceeds the percentage of lead in the alloy decreases, and that of silver and gold increases.

The lead button from the assay of ores, etc., may be regarded, therefore, as an alloy of lead and silver (or gold) which, during cupellation, undergoes enrichment from practically pure lead melting at 327° C. to pure silver melting at 962° C., or pure gold melting at 1064° C.; and owing to the higher melting-points of the precious metals, the finishing temperature must be above that used during the operation. It will be evident that, as the lead is gradually removed by oxidation, the alloy remaining on the cupel will become gradually enriched in precious metal and will consequently tend to solidify or freeze if the temperature is not raised towards the end, and the solidified metal will contain lead.

This tendency to "freeze" is greater with lead containing base metals and comparatively large amounts of precious metal such as in bullion assays, than it is with lead with smaller silver or gold content, such as is ordinarily obtained from ore assays.

"In the cupellation for silver it would seem at first sight that a final temperature of 962° C. (*i.e.* the melting-point of silver) is necessary in order to prevent freezing and to obtain a silver bead free from lead. However, the phenomenon of the 'surfusion' of the silver, *i.e.* silver in a molten state below its true melting-point, due probably to its separation from its lead alloy by the oxidation of the lead, appears to indicate that this temperature is not necessary. It is true, nevertheless, that the finishing temperature, depending somewhat upon the amount of silver present, may not fall much below 910° C." [1]

The remarks on the cupellation of silver apply also to that of gold, except that, owing to the somewhat higher melting-point of gold (1064° C.), the finishing temperature should be higher. In the case of alloys of silver and gold the finishing temperatures will be intermediate between those for the pure metals, and will be higher as the proportion of gold increases.

During cupellation there is always a slight loss due to absorption by the cupel and by volatilization, but if the temperature employed is much above that actually required for the operation, the loss of silver and gold is very much increased.

Diversity of opinion exists between assayers as to the correct temperature for cupellation, especially for silver, which is subject to greater loss than gold during the operation. As a general rule, the temperature is considered to be properly adjusted when the muffle is at full red heat, the cupel dull-red, and the melted lead much more luminous than the cupel; the fumes should rise slowly to the crown of the muffle, and no crystals of litharge (termed "flakes" or "feathers") should form on the cupel.

"Feathers" are crystals of solid litharge condensed from the vapour and deposited on the rim of the cupel, which is invariably cooler than the molten lead on the cupel. It is stated by Fulton that they will not form above 820° C.

[1] Fulton & Sharwood, *Manual of Fire Assaying.*

When the temperature is too high, the outline of the cupels becomes indistinct, the melted metal is seen with difficulty, and the fumes are almost invisible; when it is too low, the fumes do not reach the top of the muffle, and a ring of indistinct crystals of litharge forms on the top and outer surface of the cupel. Many assayers prefer to conduct the cupellation of silver at a temperature low enough to enable crystals of litharge to form on the cupel. "When cupellation of silver is carried on under these conditions the temperature should not, according to Fulton, be above 820° C., in which case crystals of litharge (feathers) form on the side of the cupel towards the muffle mouth. If the temperature is too low for the cupel successfully to absorb practically all the litharge, these feathers form low down in the cupel. When the temperature is correct, they form near the upper rim of the cupel. It is, however, to be noted that the draught through the muffle influences the formation of feather litharge; *i.e.* if the draught is strong, feathers will form although the temperature is somewhat above 820° C. During cupellation the door of the muffle should never be left wide open, but should be set slightly ajar, so that the cold air will not strike directly upon the cupels."

This description applies more particularly to the cupellation of large lead buttons containing comparatively small amounts of silver, such as are obtained from the assays of ores.

The formation of "feathers" is generally accompanied by a sluggish, heavy movement of the lead fumes, which fall in the muffle. Beyond the fact that it shows that the muffle has not been too hot, it is doubtful whether there is any advantage in cupelling at a temperature sufficiently low to permit of the formation of feathers. It is open to the objection that the silver buttons are very liable to retain lead and other base metals. In any case, the temperature should not be below that required to produce feathers in small quantity only. The author is in agreement with Dr. Rose and others that "it is probably better that the litharge should be completely absorbed."

The practice almost universally adopted by American assayers is to cupel at a temperature sufficiently low to leave a distinct ring of feathery litharge crystals round the edge of the cupel, marking the area covered by the original lead button.

Beads of silver cupelled at this heat always retain traces of lead, but it is assumed that the *plus* error thus introduced is much less than the *minus* error or loss due to volatilization and to absorption by the cupel, so that a rough approximation to accuracy is obtained.[1]

Since the introduction of accurate pyrometers, such as the thermocouple, many attempts have been made to determine the true temperature of cupellation, *i.e.* the temperature of the lead undergoing oxidation in the cupel. In the early experiments many muffle temperatures were taken on the assumption that these would give the temperature of cupellation, but the discrepancies between the different observations clearly indicate that the true temperature of cupellation cannot be ascertained in this way.

The discrepancies between the different observations of muffle temperature are due, among other causes, to the variation of the air-current passing through the muffle, and to the fact that the temperature of the air above the cupel cannot be taken by a thermocouple on account of the radiation of heat from the cupel and from the walls of the muffle. The radiated heat is absorbed by the thermocouple much more than by the air passing through the muffle, with the result that the thermocouple always shows a temperature far above that of the air immediately surrounding it.[2]

[1] Collins, *Metallurgy of Lead*, 2nd edn., 1910, p. 468.
[2] T. K. Rose, *Trans. Inst. Min. and Met.*, 1909, **18**, 463.

As previously stated, the temperature of the lead in the cupel is always greater than that of the muffle, and attempts have been made to ascertain the actual temperature of the molten lead during cupellation. This determination presents many experimental difficulties, as the protective tube of a thermocouple is rapidly destroyed by the corrosive action of the litharge. The difficulty may, however, be overcome by the use of radiation pyrometers, if due precautions are taken.

Another important consideration in determining the temperature of cupellation is the exact freezing-point of litharge, which is most difficult to determine. A cooling curve gives no definite point of arrest. In the determinations by Doeltz and Mostowitsch [1] the freezing-point of litharge was found to be 906° C., and by Mostwitsch [2] alone later as 884° C., this latter figure being in all probability the more accurate. There is no doubt that at 906° C. litharge is completely molten. At 640° C. it is completely solid. Between these temperatures there is a long pasty stage. The exact point above which the pasty litharge is to be called liquid, and below which it is to be called solid, is very difficult to decide. According to Dr. Rose,[3] litharge is not absorbed by a bone-ash cupel maintained at 750° C., but is absorbed by one kept at 825° C.; and Fulton [4] observes that absorption does not take place at 815° C., but is rapid at 883° C. If absorption is to be taken as the criterion of liquefaction, then the melting-point of litharge must apparently be placed at about 825° C. In cupellation, therefore, the cupel itself must be above this, but the temperature of the air in the muffle near the cupel may differ considerably from this for the reasons stated. The results of the most recent attempts that have been made to measure the temperature of muffles during cupellation and also the actual temperature of cupellation are briefly summarized as follows.

The most complete and probably the most reliable experiments to determine the temperature of cupellation were made in 1908 by C. H. Fulton,[5] assisted by O. A. Anderson, I. E. Goodner, and J. D. Ossa. The following is a summary of the results obtained by these authors. The temperatures apply to the cupellation of silver on bone-ash cupels.

(1) In the case of the lead buttons not containing any appreciable amount of copper or iron, etc., a *muffle temperature* of at least 800° C., and, better, one of 850° C., is required to " uncover " or start cupellation.

(2) This temperature may be lowered to about 770° C. during the oxidation of the greater part of the lead.

(3) Toward the end of the cupellation, in the case of silver, it must again be raised to about 830° C. in order to get a pure silver button.

(4) The *actual* temperature of the cupelling lead is always appreciably higher than the *muffle* temperature.

(5) The *actual* finishing temperature of the cupellation cannot be safely carried below about 910° C.

(6) The greatest observed surfusion of silver was 77° C., and this is probably very near the maximum.

(7) Silver beads finishing with surfusion are free from lead.

(8) " Feathers " or crystals of sublimed litharge on the cupel are an indication of the proper cupellation temperature, provided the air-draught is not excessive.

[1] *Métallurgie*, 1907, **4**, 290.

[2] *Métallurgie*, 1907, **4**, 468.

[3] *Trans. Inst. Min. and Met.*, 1909, **18**, 463.

[4] *Western Chemist and Metallurgist*, **4**, 47.

[5] Fulton & Sharwood, *Fire Assaying*, 2nd edn., 1911, p. 97; also West, *Chem. and Met.*, 1908, **4**, 31–54; and *Eng. and Min. Journ.*, 1908, **86**, 326.

(9) It is just as essential to regulate the air-draught of the muffle as its temperature.

Another series of experiments on the temperature of cupellation were made in 1909 by R. H. Bradford,[1] who determined the temperature of the lead button and of the interior of the cupel during cupellation by means of the Le Chatelier pyrometer, the couple at the hot junction being protected by a thin coating of fireclay. It was found that before " driving " (*i.e.* rapid oxidation of the lead) commenced, the temperature of the molten lead rose to 900° C. or above. Temperature determinations at different stages showed that when once the lead commenced to oxidize rapidly, the temperature of the button rose owing to the heat of oxidation of the lead, this heat being sufficient to keep the litharge melted, even when the cupel was drawn forward to a cooler position in the muffle. Cupellation proceeds as long as the heat developed is sufficient to keep the litharge melted, and this can be attained by maintaining the temperature at 650°-750° C., about $\frac{1}{4}$ inch above and near the front of the cupel, and working with a gentle draught through the muffle.

In determining the temperature of cupellation for silver, Lodge obtained a number of results by means of a Le Chatelier pyrometer, the junction being kept about $\frac{1}{4}$-inch above the lead buttons. The results show that at 700° C. crystals of litharge form all about the lead button, at 775° C. crystals of litharge form on the cooler side of the cupel only, while at 850° no more crystals of litharge are formed.

With regard to the temperature for the cupellation of gold, very few experiments have been made. The figures recorded are mainly muffle temperatures. As an example, the temperature of a muffle during the cupellation of standard gold (91·66 per cent. gold and 8·34 per cent. copper) at the Royal Mint, London, was determined by Sir T. K. Rose.[2] The muffle charge consisted of seventy-two assays, arranged in twelve rows of six, the amount of metal taken for each assay being 0·5 gramme wrapped in eight times its weight of lead. The Le Chatelier pyrometer was used, the position of the junction being about 1 inch above the cupel (bone-ash) in all cases.

The muffle was 15 inches long and $6\frac{1}{2}$ inches wide, fired by coke, and it was found that the temperature gradually rose from about 1050° C. to 1080° C. in passing from front to back, whilst along the sides the temperature was 1 or 2 degrees higher than in the middle line. The mean temperature of the muffle was about 1070° C.

It is believed that with gas furnaces and the present arrangements of the muffle the variation in temperature is much less. The temperatures given are very probably too high, for the reasons stated on page 149.

From the various figures quoted it will be apparent that, notwithstanding the amount of work done by different experimenters, the subject of cupellation still offers a field for inquiry, and it is to be hoped that, by the introduction of new methods of investigation, more exact knowledge respecting this most important of assay operations will soon be forthcoming.

As the correct temperature of cupellation can only be ascertained by practice, beginners should cupel weighed quantities of pure silver with lead and note the loss, and if this is excessive the experiment should be repeated until the results show that the correct temperature has been employed. For this purpose six pieces of pure silver, each exactly 0·5 gramme, are carefully weighed and each wrapped in 2 grammes of sheet lead. These are charged into separate cupels that have been heated at least ten minutes previously in the muffle. The cupels

[1] R. H. Bradford, *J. Ind. and Eng. Chem.*, 1909, **1**, 181–184.
[2] *J. Chem. Soc.*, 1893, **63**, 707.

should be arranged in two rows of three in the middle of the muffle. When cupellation is finished, the muffle is cooled slowly until the silver has become solid. The cupels are removed, the buttons cleaned by squeezing sideways and brushing, and then carefully weighed on an assay balance. They should agree closely in weight, and the loss should not exceed 0·5 per cent. The experiment is repeated with fresh quantities of silver until the desired result is obtained.

It is also well to experiment with larger proportions of lead. Thus 0·20 gramme of silver may be cupelled with 20 grammes of lead, in sets of six arranged as before. In this case the loss should not exceed, say, 1·8 per cent. Other series, in which a small addition of copper is made, may also be tried with advantage.

As a guide in cupellation the following scale of "colour temperatures" is given by Fulton[1]:—

	Degrees Centigrade.
Lowest red visible in the dark	470
Dark blood-red or black red	532
Dark red, blood-red, low red	566
Dark cherry red	635
Cherry red, full red	746
Light cherry, light red	843
Orange	900
Light orange	941
Yellow	1000
Light yellow	1080
White	1205

The student may obtain a practical knowledge as to the meaning of these figures by placing in the muffle small pieces of metals or alloys of known melting-points and carefully noting the temperature "colour" of the muffle when they melt.

The metals should be in the form of thin sheet and triangular in shape to give a sharp point. The pieces are placed on clean cupels in the muffle almost entirely closed; the temperature should be very slowly raised, and the appearance of the muffle when each metal begins to melt carefully noted. The following metals and alloys are convenient for the purpose:—

		Melts at
Lead		327° C.
Zinc		419°
Aluminium		657°
Silver eutectic alloy	silver, 72 per cent. copper, 28 per cent.	770°
Pure silver		962°
Silver–gold alloy	gold, 20 per cent. silver, 80 per cent.	1005°
Pure gold		1064°
Copper		1084°

Cupellation on Cupels of Material other than Bone-ash.—Since, as previously stated, a large number of cupels made of materials other than bone-ash (mainly magnesia) are now supplied for assay purposes, it is well to point out that during cupellation there is a considerable difference in the behaviour of

[1] White and Taylor, in *Trans. Amer. Soc. Mech. Eng.*, **21**, 628; H. M. Howe, in *Eng. and Min. Journ.*, **69**, 75; quoted by Fulton & Sharwood, *Manual of Fire Assaying*, p. 80.

these so-called "patent" cupels as compared with bone-ash cupels, due to differences in the thermal properties of the materials used.[1]

The diffusion of heat and the specific heat of magnesia cupels are greater than those of bone-ash cupels. If similar cupellations be conducted in bone-ash and in magnesia cupels side by side, a marked difference will be seen in the behaviour of the lead. The lead on the bone-ash cupel during the cupellation is very bright, whereas the lead on the magnesia cupel is comparatively dull during a considerable part of the operation, and is not so hot, although the muffle temperature is the same for both. "This is due to the fact that the extra heat generated by the oxidation of the lead is diffused as soon as it is generated, owing to the superior diffusivity of the magnesia cupel, and hence cannot serve to raise the temperature of the lead, as is the case in the bone-ash cupel. Hence for the same 'muffle temperature' the actual cupellation temperature of the lead in the magnesia cupels is 50° to 60° C. lower than in the bone-ash cupels."[2]

It has already been pointed out that the heat generated by the oxidation of the lead is sufficient to carry the cupellation to a finish with bone-ash cupels, provided the muffle temperature is not lowered at the end of the operation; but, for the reasons stated above, it is necessary in the case of magnesia cupels to employ a slightly higher temperature during the cupellation, and to raise the temperature towards the end of the operation.

The difference between the temperatures of the lead during cupellation on magnesia and on bone-ash cupels, which can be noticed by observation, is much more marked at the beginning of the cupellation, and, in fact, is hardly discernible at the very end of the operation.

It will be noted also that magnesia cupels retain heat longer than bone-ash cupels, consequently silver beads take longer to solidify and to spit on magnesia than on bone-ash cupels, after being withdrawn from the same muffle temperature. Silver beads are also much less liable to spit on magnesia than on bone-ash cupels, and the nature of the spit is different in the two cases, the spit in the former case generally taking the form of a frosty appearance only instead of the well-known "vegetation" obtained with bone-ash. These important differences in the properties of magnesia and bone-ash cupels are not always recognized by assayers.

An assayer, when asked to test magnesia cupels, usually puts half-a-dozen in the muffle with half-a-dozen bone-ash, and heats up the furnace under conditions suitable for the bone-ash, with the result that he forms an unfavourable opinion of the magnesia cupels. When using magnesia cupels under these conditions the lead is very liable to "freeze," and the results are unsatisfactory; but if the proper conditions for magnesia cupels are employed, the results are quite as satisfactory as those obtained with bone-ash. Although a somewhat higher temperature is required for cupellation on magnesia cupels, the tendency is to have a very much higher temperature than is necessary, in order to make the cupelling lead look like that on bone-ash cupels; and this is a great mistake, whereby one of the most important advantages of the magnesia cupels is lost. The loss of silver due to absorption is usually less with magnesia than with bone-ash cupels.

Portland-cement Cupels.—The behaviour of Portland-cement cupels during cupellation is very similar to that of bone-ash cupels, although, as shown on page 164, the loss by absorption is slightly higher. When using cupels made entirely of Portland cement for gold assays, especial care must be taken to

[1] Bannister and Stanley, "Thermal Properties of Cupels," *Trans. Inst. Min. and Met.*, 1909, **18**, 439–465.

[2] Fulton & Sharwood, *op. cit.*

clean the buttons thoroughly, otherwise when subsequently parting in nitric acid, insoluble silica will remain adhering to the cornet and be weighed as gold. This difficulty may be overcome by "facing" the cement cupel with bone-ash. In this case the cupel mould is filled about two-thirds full with cement, and bone-ash added to fill the mould, the cupel being finished in the ordinary way.

Cement cupels are also made from mixtures of Portland cement and bone-ash, the proportions being usually equal parts of each.

Influence of other Metals on Cupellation.—The materials which are submitted to cupellation may be divided into two main classes, viz. :—

(*a*) Material containing a very small proportion of precious metal and a large excess of lead, such as lead buttons from the assay of ores, etc.

(*b*) Material containing a large proportion of precious metal and requiring the addition of a small quantity only of lead for cupellation, such as bullion.

Various base metals may be present in greater or smaller quantity in any of the materials that can be submitted to cupellation, and may exert considerable influence on the results. When metals such as tin, antimony, and iron are present, the oxides of which dissolve only sparingly in molten litharge and produce a "scoria" or corrode the cupel, they should be previously removed by scorification. The lead buttons from the assays of ores are, as a general rule, practically free from base metals, as it is the usual practice to scorify lead buttons which may unavoidably contain an appreciable quantity of such metals. The materials of the second class may contain comparatively large proportions of base metals, notably copper and zinc, and the satisfactory removal of these during cupellation requires a careful adjustment of the lead. The behaviour of the lead button when hammered usually affords an indication of the presence of base metals and other impurities. Copper, iron, and tin tend to make lead hard, while antimony, zinc, sulphur, and arsenic tend to make it brittle. Lead buttons may also be brittle from the presence of lead oxide, due to the employment of too low a temperature for the fusion of the ore. The appearance of the cupel will also frequently indicate the presence of base metals, as described below, and when the cupel shows signs of these in objectionable quantity it is usually more satisfactory to repeat the assay and scorify, so as to remove them before cupellation. When lead is the only oxidizable metal present, the stained portion of the cupel is yellow when cold except for a small, dark-grey patch where the button has rested. The influence of various base metals, etc., on cupellation is summarized below.[1] (It should be noted that copper exerts its greatest influence towards the finish of the operation, while the influence of the more easily oxidized metals, such as antimony, tin, etc., is exercised at the commencement of cupellation.)

(1) *Antimony* does not interfere if less than 2 per cent. is present. If 4 per cent. is present a slight yellow scoria of lead antimonate is formed, and the cupel is stained dark-brown. The margin of the cupel is sometimes cracked when the antimony exceeds the amount stated.

(2) *Arsenic* at first burns with a blue flame and causes spurting ; after a few minutes the flame turns greenish-white and becomes very brilliant. The cupel is stained pale-brown, and if 4 per cent. of arsenic is present, there is much pale-brown scoria.

(3) *Bismuth* is oxidized and absorbed by the cupel in the same way as lead. Argentiferous bismuth may be cupelled alone, but towards the end of the

[1] See "Notes on Cupellation and Parting," by T. K. Rose, *Chem. Met. and Min. Soc. of S. Africa*, Jan., 1905 ; also "Cupellation of Bismuth-Silver Alloys," E. A. Smith, *Trans. Chem. Soc.*, 1894, **91**, 624.

operation the bismuth appears to act in a similar manner to tellurium, although to less extent, the alloy of bismuth and silver wetting the surface of the cupel and being partially absorbed owing to the breaking-down of the surface tension of the metal. This action does not appear to take place with comparatively large buttons of silver. Traces of bismuth are sometimes present in lead sold for assay purposes, and become concentrated in the silver during cupellation, especially when the cupellation temperature is regulated so as to produce "feathers." Under these conditions the silver beads may retain bismuth to the extent of 2 parts per 1000, varying, however, with the conditions of the cupellation. When bismuth and lead have to be removed from silver by oxidation the former metal is only removed after most of the latter has been oxidized.

(4) *Cadmium* causes a black sooty ring to form inside the cupel near its margin. When cupellation begins, the action is very violent, a yellowish-red flame is seen, followed by much spurting and the appearance of large red-hot scales floating on the lead. Cadmium oxide is not absorbed by the cupel.

(5) *Copper.*—The copper oxide formed during cupellation is dissolved in the molten litharge, and the mixed oxides are absorbed by the cupel, provided the lead is present in proper proportion. If, however, the lead is not present in sufficient quantity, all the copper will not be removed, and the button of silver (or gold) still retaining copper will be found covered with a black coating of copper oxide. The proportion of copper carried into the cupel by litharge varies not only with the temperature, but even for the same temperature, according to the ratio of copper to lead in the alloy. Copper does not oxidize as readily as lead, and tends to concentrate in the button; consequently the lead, which may originally have contained only a small percentage of copper, becomes relatively richer in copper as the operation proceeds. This is an important fact, as experience shows that most of the copper is eliminated during the last stage of cupellation, just before the lead is worked off. It is on this account that the ill-effects of the copper are most marked at the close of the operation, and that copper oxide is found accumulated around the button of precious metal when there is insufficient litharge to remove it. Copper requires at least twenty parts of lead to carry it into the cupel, and in practice a much larger ratio of lead to copper is employed to eliminate the copper in gold and silver alloys (see Chapter XXI.). Since, as shown above, the removal of copper takes place quicker when the ratio of copper to lead is not too great, it is evident that two or three successive cupellations with small quantities of lead will be more effectual than a single cupellation with a larger quantity, and this method of procedure is sometimes adopted by assayers. Copper, if not completely removed, has a very marked effect on the appearance of the silver button: the metal is spread out, is ragged at the edges, and adheres firmly to the cupel. Even when the cupellation has been satisfactory the lower surface of the silver button is flatter, and adheres more firmly to the cupel than buttons from the cupellation of silver free from copper. A small quantity of copper is persistently retained by both silver and gold at the end of the cupellation, especially by gold buttons from which the last traces of copper cannot be entirely eliminated. Copper colours the cupel a dirty green or blackish-green, according to the amount of copper present. Small quantities of copper give a greenish tint. The intensity of the green coloration affords a useful indication of the amount of copper present, and is utilized by experienced assayers to get an idea of the approximate percentage of copper in the material being assayed.

(6) *Iron.*—Lead buttons containing iron melt slowly. Iron oxide is not readily fusible with lead oxide, and a brown scoria is sometimes left on the cupel, which may entangle lead globules and so contain precious metal. The lead may

contain about 4 per cent. of iron without the formation of any scoria. The cupel is stained dark-red, and is moderately corroded.

(7) *Nickel* forms a dark-green scoria and greenish stain. Its effect on cupellation is similar to that of iron, but nickel is more intractable and difficult to eliminate.

(8) *Platinum.*—The surface of the cupel button is dull and crystalline, even when small quantities of platinum are present. The effect of platinum and of the platinum metals on cupellation is more fully discussed in Chapter XXIV.

(9) *Tellurium.*[1]—Tellurium has a great affinity for gold and silver, and exerts considerable influence in cupellation when there is insufficient lead to remove it. The effect of tellurium is greatly to decrease the surface tension of the lead–gold–silver alloy.

Tellurium is removed comparatively slowly during cupellation, and if towards the end there is sufficient left to amount to anything approaching equality of the gold, or gold plus silver, then the surface tension of the globule breaks down completely and the alloy spreads over a wider area, "wets" the cupel, and is completely absorbed. This occurs in the case of bone-ash cupels when the tellurium in the button equals the gold, or gold plus silver, and the lead does not exceed ten times the tellurium.

When the lead exceeds this amount the behaviour is intermediate between the extreme cases of complete absorption and perfect cupellation. Partial subdivision of the lead on the cupel then takes place. This has long been recognized as a possible occurrence when tellurium is present.

It is brought about by portions of the alloy becoming detached at the line of contact between the cupel and the surface of the molten lead during cupellation. The presence of tellurium causes the surface tension of the lead to be weakened, with the result that it tends to adhere to or "wet" the face of the cupel. Portions of the alloy become detached and are either absorbed or leave minute beads on the cupel.

When the gold and tellurium are equal in amount and the lead amounts to 80 times the tellurium, very little loss occurs on a bone-ash cupel, and this loss is not appreciably decreased by additional lead. It is evident, therefore, that in all ordinary cases the button of lead from a crucible fusion of telluride ore weighing 30 grammes or more contains ample lead for cupellation. In the case of silver, more lead is required (from 100 to 120 times instead of 80) to remove the same amount of tellurium as from gold and to leave a rounded bead. Tellurium appears to be held more tenaciously by silver, and the fact that it comes out to a considerable extent in the later stage of cupellation is shown by the white patch of tellurium dioxide which underlies the silver bead.

Silver undoubtedly protects gold from possible losses due to the presence of tellurium, and even when present in amount only equal to that of the gold a considerable saving is effected. With an increase in the proportion of silver the protection is more complete. Tellurium being less easily oxidizable than lead is removed comparatively slowly, and tends to concentrate during cupellation, as shown by the following results by S. W. Smith, obtained by analysing a series of partially cupelled buttons originally consisting of—

Lead	... ...	20 grammes.
Tellurium	... ...	0·05 gramme.
Gold	... ...	0·05 ,,

(i) Cupelled to 11·860 grammes, *i.e.* 59 per cent. of its weight.
Amount of tellurium still remaining, 77 per cent.

[1] S. W. Smith, "Behaviour of Tellurium in Assaying," *Trans. Inst. Min. and Met.*, 1908, **17**, 463.

In this case a very distinct concentration of tellurium has taken place.

(ii) Cupelled to 4·205 grammes, *i.e.* 21 per cent. of its weight.

Amount of tellurium still remaining, 21 per cent.

At this stage the percentage of lead and tellurium which have been removed are equal.

(iii) Cupelled to 1·635 grammes, *i.e.* 8·15 per cent. of its weight.

Amount of tellurium remaining, 5·88 per cent.

At this stage the tellurium is at last leaving the button at a greater relative rate than the lead.

Cupellation of lead buttons containing tellurium should be conducted at a low temperature to obtain the highest results, as it would seem that the tendency towards " wetting " and absorption is greater at higher temperatures, since the effect of tellurium on the cupellation loss is accentuated under these conditions.

Widely different results may be obtained with similar buttons cupelled under different conditions of temperature and draught. Bone-ash cupels show a greater tendency to become " wetted," giving rise to subdivision, than patent cupels. During cupellation, tellurium is oxidized to the dioxide TeO_2.

(10) *Tin* gives a brown scoria if it constitutes more than 3 per cent. of the lead button. In all cases it rises to the surface, oxidizes, and forms a scoria which floats and delays cupellation, but is subsequently carried off by the litharge. The cupel is stained pale brown.

(11) *Zinc* burns with a brilliant green flame at first, and is partly volatilized; it forms a voluminous white or pale-yellow scoria which may or may not be subsequently removed by the litharge, according to the amount of zinc present and the proportion of lead used. The cupel is deeply corroded.

Losses in Cupellation

The weight of the button of silver or gold left on the cupel does not represent the actual amount of precious metal in the material submitted to cupellation, as certain losses, incidental to the operation, always take place. These losses may be divided into (1) losses by volatilization, and (2) losses by absorption, the latter being by far the more important.

(1) **Volatilization.**—Although it has been proved by Makins [1] and others that loss by volatilization does occur during cupellation, the loss is altogether inconsiderable, unless the temperature at which the operation is performed is much too high.

(2) **Losses through Absorption by the Cupel.**—The loss during cupellaton by absorption by the cupel is by no means always inconsiderable. The amount of this loss varies under different conditions and is influenced by :—

(*a*) The temperature of cupellation.
(*b*) The amount of lead employed.
(*c*) The amount of precious metal present.
(*d*) The influence of gold and silver on one another.
(*e*) The nature and amount of the impurities present.
(*f*) The physical condition of the cupel.

Although much experimental data relating to the losses in cupellation, especially of silver, have been published, the results are frequently of little value to the assayer owing to the lack of essential details. Such data are of value only when they give the amounts of lead and gold or silver, etc., used and describe the exact conditions under which the experiments were made. But even when these details are given, the results must be regarded only as affording some indication of the losses that take place during cupellation, and

[1] *J. Chem. Soc.*, 1860, **13**, 77.

not as constants that can be utilized by the assayer for the purpose of correcting the results of an assay. It must be emphasized that the losses during cupellation are dependent on so many conditions that each assayer should determine for himself the exact losses that take place with the cupels he uses and the conditions of working that he adopts.

Very frequently in assay practice the amount of the cupellation loss has to be carefully determined by the methods described subsequently, and the loss duly allowed for in returning the assay results. The figures given in the accompanying tables have been selected from the results of much experimental work and represent the average losses which occur during cupellation under what may be regarded as normal conditions, *e.g.* a temperature of about 750°–775° C.

(*a*) **Effect of Varying Temperature.**—The loss of silver and of gold both by absorption and volatilization is very much increased by the employment of high temperatures, as shown in the case of silver in Table XIII and gold in Table XIV.

TABLE XIII.—EFFECT OF VARYING TEMPERATURE ON CUPELLATION OF PURE SILVER. (*Eager and Welsh.*)[1]

Amount of Silver.	Amount of Lead.	Ratio of Lead to Silver.	Temperature.	Total Losses.
milligrammes.	grammes.	approx.		per cent.
204·62	10	50 to 1	700° C.	1·02
205·00	10	50 to 1	775° C.	1·28
203·00	10	50 to 1	850° C.	1·73
203·00	10	50 to 1	925° C.	3·65
203·00	10	50 to 1	1000° C.	4·87
Results of Similar Experiments by Beringer.[2]				
400	20	50 to 1	About 750° C.	1·75
400	20	50 to 1	„ 1000° C.	4·32

TABLE XIV.—EFFECT OF VARYING TEMPERATURE ON CUPELLATION OF PURE GOLD. (*Eager and Welsh.*)[3]

Amount of Gold.	Amount of Lead.	Ratio of Lead to Gold.	Temperature.	Total Losses.
milligrammes.	grammes.	approx.		per cent.
201	10	50 to 1	775° C.	0·155
201	10	50 to 1	850° C.	0·430
204	10	50 to 1	925° C.	0·460
201	10	50 to 1	1000° C.	1·430
201	10	50 to 1	1075° C.	3·000
Results of Similar Experiments by Beringer.[4]				
1000	20	20 to 1	775° C. ?	0·137[5]
1000	20	20 to 1	775° C.	0·192[5]
1000	20	20 to 1	1000° C. ?	0·604
1000	20	20 to 1	1000° C.	0·620
1000	20	20 to 1	1000° C.	0·645

[1] Lodge, *Notes on Assaying.*
[2] *Text-book of Assaying.*
[3] Lodge, *Notes on Assaying.*
[4] *Text-book of Assaying*, 15th edn., 1921, p. 145.
[5] Highest and lowest results in ten cupellations.

(*b*) **Effect of Varying Lead.**—The loss of silver increases as the quantity of lead employed is increased, as shown in Table XV, but if the quantities of silver and lead employed are increased in the same proportions, the loss slightly decreases, as shown in Table XVI. The experiments by the author were in all cases made on hand-made bone-ash cupels in a gas-fired muffle at a temperature of about 750° C.

TABLE XV.—SHOWING LOSS ON CUPELLATION OF PURE SILVER WITH VARYING AMOUNTS OF LEAD AND SILVER. (*W. H. Kauffmann.*)[1]

(Temperature not stated, but probably about 750° C.)

Amount of Silver, milligrammes.	Total Loss of Silver per cent.			
	With 5 grammes Lead.	With 10 grammes Lead.	With 15 grammes Lead.	With 25 grammes Lead.
25	2·14	2·63	2·69	2·09
50	1·43	2·23	2·14	2·16
100	1·30	1·61	1·68	2·12
200	0·86	1·24	1·40	1·74
Results obtained by E. A. Smith.				
Amount of Silver, milligrammes.	With 3 grammes Lead.	With 6 grammes Lead.	With 9 grammes Lead.	With 12 grammes Lead.
500	0·875	1·140	1·240	1·625
750	0·803	1·000	1·093	1·136

TABLE XVI.—SHOWING GRADUALLY DECREASING LOSS ON CUPELLATION WHEN LEAD AND SILVER ARE INCREASED IN THE SAME RATIO. (*Hambly.*)[2]

Amount of Silver in grammes.	Amount of Lead in grammes.	Ratio of Silver to Lead.	Loss of Silver per cent.
0·065	0·65	1 to 10	1·225
0·650	6·50	1 to 10	1·135
1·625	16·25	1 to 10	1·067
Results obtained by E. A. Smith.			
0·125	1·5	1 to 12	1·40
0·250	3·0	1 to 12	1·32
0·500	6·0	1 to 12	1·20
1·000	12·0	1 to 12	1·14

In the case of gold, Rossler[3] has shown that the loss during cupellation slightly increases with the amount of lead used, but is much less than in the

[1] *Eng. and Min. J.*, 1902, **63**, 829.
[2] *Chem. Gazette*, 1856, **14**, 185–6.
[3] *Dingl. Polytech. J.*, **206**, 185.

case of silver, and the extent of the loss is dependent upon the quantity of gold cupelled. Some results obtained by the author are given in Table XVII.

TABLE XVII.—Showing Effect of Varying Lead on the Cupellation of Pure Gold. (*E. A. Smith.*)

Amount of Gold in grammes.	Total Loss of Gold per cent.			
	With 3 grammes Lead.	With 6 grammes Lead.	With 9 grammes Lead.	With 12 grammes Lead.
0·250	0·113	0·123	0·137	0·145

(*c*) **Effect of Varying Silver.**—The figure in Table XV also shows the cupellation loss with a constant weight of lead and increasing quantity of silver. It will be seen that although the *actual* loss increases with the quantity of silver present, the *percentage* loss is greater on the smaller buttons than on the larger buttons.

Hambly's original figures (Table XVI) are given in grains, and have been converted into grammes by the author.

(*d*) **Effect of Silver on the Cupellation of Gold.**—When gold is alloyed with silver the cupellation loss is diminished, the total loss of gold decreasing as the amount of silver present increases. This protective action of silver is shown by the results of experiments by Rose,[1] given in Table XVIII.

TABLE XVIII.—Loss on the Cupellation of Gold in the Presence of Silver. (*Sir T. K. Rose.*)

Amount of Gold.	Amount of Silver.	Amount of Lead.	Temperature of Cupellation.	Total Loss of Gold.
milligrammes.	milligrammes.	grammes.		per cent.
1	4	25	900° C.	1·20
1	6	25	900° C.	1·05
1	8	10	900° C.	0·90
1	10	25	900° C.	0·80
1	6	25	700° C.	0·45
1	10	25	700° C.	0·39
500	1250	10	900° C.	0·055
Results obtained by Beringer.				
300	300	10	900° C. ?	0·157
300	600	10	900° C.	0·107
300	900	10	900° C.	0·057

(*e*) **Effect of Base Metals on Cupellation Loss:** *Copper.*—The absorption of precious metals by the cupel is always greater when copper is present. The reason of this is not thoroughly understood, but it is usually stated that the copper oxide (Cu_2O) acts as an oxygen-carrier in some special manner towards the end of the cupellation and facilitates the formation of a small quantity of silver oxide (or gold oxide) which is dissolved in the molten mixture of lead and copper oxides and is thus carried into the cupel.

[1] *Eng. and Min. J.*, **79**, 708.

The influence of small amounts of copper on the cupellation of silver and gold is shown in the following results obtained by Lodge, Tables XIX and XX. The temperature of cupellation was 775° C.

TABLE XIX.—INFLUENCE OF COPPER ON THE CUPELLATION OF SILVER. (*Lodge.*) [1]

Silver.	Lead.	Copper.	Percentage of Copper in Lead.	Ratio of Lead to Copper.	Loss of Silver.
milligrammes.	grammes.	grammes.			per cent.
202	10	0·0101	0·1	1000 to 1	1·05
203	10	0·0202	0·2	500 to 1	1·08
202	10	0·0303	0·3	333 to 1	1·29
202	10	0·0404	0·4	250 to 1	1·45
204	10	0·0500	0·5	200 to 1	Retained copper.

Results obtained by the Author. (The cupellation loss with the same quantities in the absence of copper is inserted for comparison.)

Silver.	Lead.	Copper.	Percentage of Copper in Lead.	Ratio of Lead to Copper.	Loss of Silver.
grammes.	grammes.	grammes.			per cent.
0·5	3	None	...	...	0·875
0·5	3	0·04	1·333	75 to 1	1·005
0·5	6	None	...	...	1·140
0·5	6	0·04	0·666	150 to 1	1·380
0·5	9	None	...	...	1·240
0·5	9	0·04	0·444	225 to 1	1·335
0·5	12	None	...	...	1·625
0·5	12	0·04	0·333	300 to 1	1·700

TABLE XX.—INFLUENCE OF COPPER ON THE CUPELLATION OF GOLD. (*Lodge.*) [2]

Gold.	Lead.	Copper.	Percentage of Copper in Lead.	Ratio of Lead to Copper.	Loss of Gold.
milligrammes.	grammes.	grammes.			per cent.
202	10	None	...	...	0·155
202	10	0·0101	0·1	1000 to 1	0·19[3]
201	10	0·0202	0·2	500 to 1	0·20[3]
200	10	0·0303	0·3	333 to 1	0·13[3]
201	10	0·0404	0·4	250 to 1	0·165
202	10	0·0500	0·5	200 to 1	0·250

[1] *Notes on Assaying.*
[2] *Notes on Assaying,* p. 143.
[3] These are actual losses; the buttons retained 0·16 per cent. copper.

Tellurium.—The important influence of tellurium during cupellation has already been discussed. The absorption loss due to the presence of tellurium is a function of the temperature of cupellation. As previously stated, the presence of silver protects gold from possible losses due to the presence of tellurium.

Very few figures appear to have been published showing the loss on the cupellation of pure gold or of pure silver with tellurium. The influence of tellurium on the cupellation of gold in the *presence of silver* is, however, shown in Table XXI.

TABLE XXI.—INFLUENCE OF TELLURIUM ON CUPELLATION OF ALLOYS OF GOLD AND SILVER. (*F. C. Smith.*)

Cupelled with 12 grammes of Lead in every case.

Milligrammes.		Total Alloy, milli-grammes.	Tellurium added, milli-grammes.	Loss by Absorption.		Loss by Volatilization.	
Gold.	Silver.			Gold.	Silver.	Gold.	Silver.
				per cent.	per cent.	per cent.	per cent.
24·76	5·04	29·8	5·0	13·44	27·08	5·65	0·69
23·64	4·81	28·45	15·0	34·22	35·78	5·28	1·75
18·42	3·75	22·17	15·0	29·85	32·01	11·92	17·95

Antimony, Tin, etc.—Antimony and other easily oxidizable metals that tend to form a scoria during cupellation may cause a mechanical loss of silver or gold. When large amounts of scoria form, the gold and silver may be retained as minute beads, which cannot be collected and weighed. The formation of a scoria is, however, not necessarily accompanied by a loss of precious metal. During the cupellation, antimony is oxidized and combines with litharge to form lead antimonate, which is partly absorbed by the cupel and partly remains as a scoria according to the amount of antimony present. The exact losses in cupellation due to the presence of base materials were determined by Rose.[1] In each case 25 grammes of lead were cupelled with 1 mg. of gold, 4 mg. of silver, and 1 g. of the impurity. Bone-ash cupels were used. The results were as follows :—

Impurity added.	Loss of Gold.	Loss of Silver.	Remarks.
	per cent.	per cent.	
No impurity	1·2	12·8	
Antimony	5·3	13·2	No scoria.
Arsenic	3·9	16·3	Much scoria.
Bismuth	21·8	27·9	No scoria.
Cadmium	3·5	13·1	Ring of black feathery scoria.
Copper	10·0	32·6	No scoria.
Iron	4·0	16·6	No scoria.
Nickel	...	...	Total loss due to scoria.
Tin	2·0	13·9	Slight ring of scoria.
Tellurium	55·8	67·9	No scoria.
Selenium	54·1	64·5	No scoria.

In the case of tellurium and selenium practically the whole of the missing gold and silver was recovered by fusing the cupels. The percentage losses observed in these experiments appear very large, but it must be remembered

[1] *Op. cit.*

that the absolute losses were very small. The percentage losses are much smaller when the precious metal beads are larger, as shown by the following results by Beringer :—

Impurity added.	Silver added.	Lead used.	Loss of Silver. per cent.[1]
	grammes.	grammes.	
No impurity added	0·1	20	2·95 } 2·9
,, ,, ,, ...	0·1	20	2·85 }
With 0·5 gramme Antimony ...	0·1	20	3·30 } 3·2
,, 0·5 ,, ,, ...	0·1	20	3·10 }
,, 0·5 ,, Copper ...	0·1	20	5·75 } 4·9
,, 0·5 ,, ,, ...	0·1	20	4·05 }

(*f*) **Porosity of Cupels.**—The amount of precious metal absorbed by a cupel is influenced by the porosity, and this is regulated by the differences in the relative cohesion of the materials employed in making cupels. The porosity of bone-ash cupels varies considerably owing to the differences in the composition and screen analysis of commercial bone-ash and in the amount of pressure used in making them, some assayers preferring to have them well compressed, while others prefer to have them more porous. Cupels made by machine, in which a constant pressure may be obtained, are always more uniform as regards density than cupels made by hand. Magnesia cupels are made by hydraulic pressure and are much denser than those made of bone-ash ; consequently the absorption is, as a general rule, less on magnesia cupels. The nature of the surface of the cupels is also of importance, and no doubt has a considerable influence on the losses due to absorption. At present there is little evidence as to the comparative variations in the absorption of silver and gold by bone-ash and magnesia cupels respectively, but it would appear from the data published that the absorption is less with magnesia cupels than with bone-ash. Whatever cupels are used, the absorption should be tested frequently.

Table XXII gives the results obtained by the cupellation of silver on bone-ash and magnesia (morganite) cupels. Pure silver of the amount given was in each case cupelled with 10 grammes of sheet lead. The temperature employed was sufficiently low to give " feathers."

TABLE XXII.—Comparison of Cupellation Losses with Bone-ash and Magnesia Cupels. (*Anderson and Fulton.*)[2]

Amount of Silver taken.	Bone-ash Cupels. Weight of Silver Bead obtained.	Loss per cent.	Magnesia Cupels (morganite). Weight of Silver Bead obtained.	Loss per cent.
milligrammes.	milligrammes.	per cent.	milligrammes.	per cent.
5	4·85	3·0	4·80	4·0
5	4·94	1·2	4·89	2·2
10	10·00	Nil	10·00	Nil
10	9·68	3·2	9·86	1·4
15	14·36	4·3	14·50	3·3
15	14·70	2·0	14·80	1·3
20	18·92	5·4	19·52	2·4
20	19·98	0·1	19·68	1·6
25	24·60	1·6	24·84	0·64

[1] The highest and lowest results of three experiments in each case.
[2] O. A. Anderson and C. H. Fulton, see Fulton & Sharwood, *Manual of Fire Assaying.*

Benner and Hartmann [1] compared different kinds of cupels with regard to loss of silver by cupelling from 35 to 45 mg. of silver with 10 g. of lead, and found that cupels made of bone-ash of different grades and fineness gave identical results within the limits of experimental error. The average percentage loss of silver with different cupels under similar conditions was as follows:—morganite, 1·99; bone-ash, 2·36; brownite, 2·89; cement and bone-ash (1 : 1), 2·95; casseite, 3·09; cement, 3·38.

The absorption of silver by cement cupels has been ascertained by Holt and Christensen,[2] who tested cement cupels, bone-ash cupels, and cupels made of a mixture of equal parts of cement and bone-ash.

In each case 100 mg. (0·10 gramme) of fine silver were cupelled with about 20 g. of lead. The cupellation was performed at an "average" muffle temperature, and the actual temperature of the cupels measured by inserting a thermocouple into a hole bored beneath the cavity of the cupel.

The results are tabulated in Table XXIII.

TABLE XXIII.—Cupellation Losses with Cement Cupels.
(*Holt and Christensen.*)

Average Temperature of Cupel.	Portland Cement (V.S.). Loss per cent.	Portland Cement (R.D.). Loss per cent.	Equal parts Cement and Bone-ash. Loss per cent.	Bone-ash. Loss per cent.
915° C.	1·30	1·34	1·21	1·26
925° C.	1·81	1·72	1·54	1·70
945° C.	2·53	2·56	2·42	2·42
965° C.	3·37	3·42	3·05	2·96

Table XXIV gives the results of absorption loss with bone-ash and cement cupels obtained by J. W. Merritt [3] on cupelling 10 milligrammes of fine silver with 15 grammes of lead.

TABLE XXIV.—Comparison of Cupellation Loss with Bone-Ash and Cement Cupels. (*Merritt.*)

Temperature.	Absorption per cent.	
	Bone-ash.	Cement.
Light cherry heat (800° C. ?) ...	4·62	4·91
Orange heat (900° C. ?)	6·38	6·64

The following conclusions have been drawn by Mann and Clayton [4] from the results of more than 600 experiments on cupels of different compositions:—

(1) *Losses due to Composition of Cupels.*—A mixture of equal parts of bone-ash and cement gave trouble through spitting, and the beads did not break clean from the cupels. Cupels made of a cement base with bone-ash top were very satisfactory as regards silver losses, but the bond between the base and

[1] *J. Ind. Eng. Chem.*, 1911, **3**, 805–807.

[2] "Experiments with Portland Cement Cupels," *Eng. and Min. J.*, 1910, **90**, 560.

[3] J. W. Merrit, "Cement versus Bone-Ash Cupels," *Min. and Sci. Press*, 1910, **100**, 649.

[4] H. T. Mann and C. Y. Clayton, School of Mines and Metallurgy Bull., University of Missouri, Feb., 1916, 60 pp. Abs. *J. Soc. Chem. Ind.*, 15th Nov., 1916, **21**, **35**, 1113.

the top is very weak. The loss of silver was highest with cement cupels; the beads were held very firmly and had to be scraped. Morganite cupels showed the smallest silver losses; the beads separated very rapidly and clean, but below 850° C. they retained lead. Next to morganite, braunite cupels were best as regards losses, but spitting was very marked. The beads very easily removed, but below 800° C. they retained lead. A few figures showing the actual silver losses are tabulated below:—

TABLE XXV.—SILVER LOSSES WITH VARIOUS TYPES OF CUPELS.
(*Mann and Clayton.*)

Composition of Cupel.	Cupellation Temperature, ° C.	Silver Loss, per cent.
Bone-ash	751	1·5
,, ...	855	4·8
,, ...	951	12·4
,, and Cement (1 : 1) ...	751	2·2
,, ,, ,, ...	849	6·3
,, ,, ,, ...	951	13·2
Cement	751	3·0
,, ...	851	9·2
,, ...	959	19·4
Morganite	775	0·0
,, ...	851	0·6
,, ...	955	2·8
Braunite	755	0·4
,, ...	855	2·5
,, ...	951	5·2

(2) *The effect of moisture, hardness, and surface condition* of the cupels on the loss was very small; with soft cupels slightly less time was required than with hard ones. (3) *Size of bone-ash*: provided the bone-ash was finer than 60-mesh, the size of the particles did not affect the losses. (4) *Composition of assay slag*: the chemical composition of the slag had very little effect upon the silver recovery so long as the charge for the assay was so proportioned that the ore was completely decomposed. A fluid slag was formed; the proper amount of "Collector" was furnished at the proper time; and the lead button was free from objectionable impurities.

Very few determinations of the absorption loss of gold during cupellation have been recorded, but it is always much less than that of silver.

As shown in Table XIV, Beringer found that in three cupellations of 1 gramme of gold with 20 g. of lead made purposely at a very high temperature, the bone-ash cupels absorbed 6·04, 6·20, and 6·45 mg. of gold respectively. Hence at a high temperature there may easily be an absorption loss of more than 0·5 per cent. of gold when cupelled with the proportion of lead given. In ten cupellations with the same quantities of gold and lead, but at an ordinary temperature, the gold recovered from the cupels varied from 1·37 to 1·92 mg., and gave an average of 1·59 mg., or 0·159 per cent.

With ordinary gold-bullion assays the absorption of gold is stated by S. W. Smith [1] to be about 0·5 part in 1000 for bone-ash cupels, and about 0·3 parts

[1] Sydney W. Smith, *Trans. Inst. Min. and Met.*, 1909, **18**, 460.

in 1000 in the case of morganite cupels. It was found that this absorption difference for cupels of these materials was practically constant.

Theory of Cupel Absorption.—Although the absorption of precious metals by the cupel is the predominant cause of loss during cupellation, very little has been published as to the nature of this loss. There seems little doubt, however, that the absorption is due partly to the infiltration into the cupel of minute particles of lead alloy, and partly to the formation of small quantities of the oxides of silver and gold, which are dissolved in the molten litharge and thus absorbed. The loss by absorption occurs mainly towards the end of the cupellation.

The figures in Table XXII, above, showing the absorption with cupels of different materials, clearly indicate that the physical condition of the cupel as regards porosity influences absorption; but that the absorption is not entirely due to the infiltration of the molten alloy is evident from the following experiments by Beringer.[1]

A cupel on which an alloy consisting of 0·80 gramme of silver, 0·47 gramme of gold, and 25 g. of lead had been cupelled, was found to contain 7·5 mg. of silver, and rather less than half a milligramme of gold.

Assuming, for the sake of argument, that the half-milligramme of gold had filtered into the cupel in the form of minute drops of alloy, it would have been accompanied by less than a milligramme of silver, and the presence of the extra 6 or 7 mg. of silver must have been due to a different cause. Similar differences between the ratio of silver to gold in the alloy submitted to cupellation, and the ratio of silver to gold recovered from the cupel, will be noted in the experiments by F. C. Smith, p. 162. There can thus be little doubt that the greater part of the absorption loss must be due to some cause other than the mere filtration of the fused alloy into the cupel, as it is difficult to ascribe the great change in the proportion of the metals to this cause.

This greater part of the loss is ascribed to the formation of oxides of silver and gold, and from the above results it would appear that silver is oxidized to a greater extent than gold.

Both metallic silver and gold are converted into oxides when heated with the higher oxides of lead, copper, and some other metals, and under the conditions existing during cupellation, oxides of the precious metals are undoubtedly formed and carried into the cupel by the molten litharge.

As, however, according to H. Rose and other authorities, silver oxide is decomposed between 250° C. and 300° C. into silver and oxygen,[2] and as the litharge formed during cupellation is subjected to a temperature considerably above this, the existence of silver oxide might be called in question. But although silver oxide is decomposed at a temperature considerably below redness, there is evidence that it is not reduced to metal by heat alone, when mixed with an excess of lead oxide.

In favour of the view that silver is capable of existing as oxide in litharge are the observations of St. Clair-Deville and Debray, and those of Troost and Haute-Feuille,[3] that silver oxide can be formed at high temperatures. In addition, Wait[4] has extracted 18·67 and 19·25 per cent. of the silver from litharge containing 2·94 per cent. of silver by means of acetic acid, and as metallic silver is insoluble in acetic acid it may be assumed that the metal must have been present as oxide. The limiting temperature below which

[1] *Text-book of Assaying*, 15th edn., 1921, p. 101.
[2] Schnabel, *Metallurgy*, vol. I, 1905 edn.
[3] Graham-Otto-Michaelio, *Anorg. Chemie*, 1884, p. 985.
[4] *Trans. Amer. Inst. Min. Eng.*, 1886, **15**, 423.

silver oxide is stable must consequently be regarded as not yet definitely settled.

In the case of gold, Rivot states that it is oxidized to some extent at a red heat in the presence of litharge, cupric oxide, or antimonic oxide, and that it is the oxidized part which is absorbed by the cupel. It is to be noted that the absorption loss of both silver and gold is greater in the presence of copper oxide.

The oxidation of the precious metals during cupellation may be attributed either to the direct action of atmospheric oxygen or to the action of oxides of base metals, which act as " oxygen-carriers " and convey oxygen to the precious metals.

The latter is the explanation usually given to account for the formation of precious-metal oxides during cupellation ; but although this is a very possible explanation, further research is necessary before finally accepting the view that the oxidation is due to the action of base-metal oxides rather than to free oxygen.

This conclusion is hardly borne out by the results of the experiments by Sir T. K. Rose [1] on the oxidation of base metals in argentiferous gold bullion. The question of oxygen-carriers is discussed by Rose, but he remarks that " the whole trend of the results of the work is to show that it is silver that carries oxygen to base metals, not the other way round, and that the oxidation of silver does not take place to any extent until the base metals have been almost entirely removed. This has not been previously recognized, but nevertheless seems to be a perfectly sound generalization, and is of importance both to assayers and refiners."

As already pointed out, the amount of precious metal absorbed by the cupel is greatly influenced by the temperature at which cupellation is conducted, a comparatively slight increase of temperature causing a marked increase in absorption. Whether this increased absorption is due to an increased oxidation of the precious metal, or to a decrease in the surface tension of the lead alloy, is difficult to determine. That the presence of certain base metals in a lead button carrying gold and silver is accompanied during cupellation by a weakening of the surface tension of the molten metal and corresponding increase in absorption is proved by the experiments of S. W. Smith, previously described.

Determination of Cupellation Loss.—In assaying rich ores and alloys of gold and silver it is necessary to make a correction for the loss during cupellation, and this is determined either by fusing the cupel and actually recovering the precious metal absorbed, or by the use of " check " or " proof " assays worked off under the same conditions as the material being assayed. The quantity of silver and gold absorbed by the cupel in the case of ore assays is comparatively small, and is not generally determined except when great exactness is required.

Assay of the Cupel.—All clean bone-ash is broken off the cupel and rejected, and the stained portion crushed so as to pass through an 80-mesh or 100-mesh sieve. Fine crushing is necessary, as bone-ash, owing to its infusible nature, does not combine with fluxes to form a slag, but remains suspended in the fused mass. The charge is made up as follows for bone-ash cupels :—

Cupel (bone-ash)	40 to 60 grammes.
Sodium carbonate	30 ,,
Borax	50 ,,
Litharge	50 ,,
Argol	2·5 ,,

[1] " Refining Gold Bullion with Oxygen Gas, " T. K. Rose, *Trans. Inst. Min. and Met.*, **14**, 1905.

For magnesia cupels the following charge may be used :—

Cupel (magnesite)	40 to 60 grammes.
Sodium carbonate	20 ,,
Borax	20 ,,
Litharge	40 ,,
Silica	15 ,,
Argol	2·5 ,,

Cement cupels may be fluxed with the following charge :—

Cupel (cement)	40 to 60 grammes.
Sodium carbonate	20 ,,
Borax	40 ,,
Silica	15 ,,
Argol	2·5 ,,

If other reducing agents are used instead of argol, the amount added should be sufficient to give a lead button weighing from 30 to 35 grammes.

The addition of fluorspar is frequently recommended as a flux in fusing cupels and is advantageous in producing a fluid slag, but it is not always easily procured. The litharge in the bone-ash acts as a flux. The mixture is fused in a crucible and the resulting button of lead cupelled. The weight of the silver button thus obtained is added to that of the silver obtained from the assay of the ore. To save time and labour, the powdered stained portion of the cupel may be added to the slag from the ore fusion, with any regulus or scoria, and the whole assayed for silver and gold with the addition of fluxes as circumstances may require.

The following method of determining the cuppellation loss, by which a fusion of the cupel is obviated, is given by Beringer.[1] It may be used for determining the cupellation loss in assaying argentiferous lead and rich ores and products. The following example is given.

Suppose we have an alloy of silver and lead in unknown proportions, and that by cupelling two lots of 10 grammes each we get silver weighing (1) 0·1226 gramme, and (2) 0·1229 gramme. We should know from general experience that the actual quantity of silver present was from 2 to 4 mg. more than this. To determine more exactly what the loss is the two silver buttons are wrapped up each in 10 g. of sheet lead and cupelled side by side with two other lots of 10 g. of the original alloy. If now the buttons (1) and (2) weigh 0·1202 g. and 0·1203 g. respectively, they will have suffered in this second cupellation an average loss of 2·5 mg. Suppose the two fresh lots of alloy gave 0·1233 g. and 0·1235 g. of silver, the average loss on these would also be 2·5 mg. Add this loss to each result and take the mean, which is in this case 0·1259 g., thus giving 1·259 per cent. as the correct result for the silver in the alloy.

A similar method is also recommended by S. W. Smith[2] for determining the cupellation loss in bullion assays.

Considering the simplest case of the gold-bullion assay, where the gold is nearly pure and the weight of the gold, silver, copper, and lead taken for the assay is known exactly. A certain number of such assays are cupelled, and

[1] Beringer, *Text-book of Assaying*, 15th edn., 1921, p. 104.

[2] *Trans. Inst. Min. and Met.*, 1909, **18**, 460. See also an interesting paper on "A Rule Governing Cupellation Losses" by W. J. Sharwood. *Trans. Amer. Inst. Min. Eng.*, 1915.

after cupellation the buttons are weighed and the loss made up by the addition of more copper and lead. The assays are then recupelled side by side with a fresh set of similar assays and the resulting buttons parted in nitric acid in the ordinary way. The only difference between the two sets of cornets will be that one set has been cupelled twice and the other set only once. The difference is the cupellation loss.

In the case of bullion assays this is, according to S. W. Smith, the simplest method of determining the cupellation loss.

Assaying by Checks or Proofs.—The method adopted in the assay of gold and silver alloys is to assay " check " or " proof " pieces of pure silver or gold side by side with the metal being assayed, so that they are subjected to exactly the same conditions, and to assume that the loss of weight on the " checks " is the same as that experienced by all the assays worked with them. This method of working is termed " assaying by checks or proofs," and may be illustrated by the following example. Suppose we wish to determine the actual amount of silver in a sample of comparatively pure commercial silver. Two lots of the sample of 1 gramme each are weighed and each wrapped in 5 grammes of sheet-lead. One check consisting of 1 gramme of pure (" proof ") silver is also accurately weighed and wrapped in 5 grammes of sheet-lead. The two samples and check are then cupelled as nearly as possible under the same conditions, the cupels being placed in one row of three across the muffle and the check charged into the middle cupel. Supposing the resulting silver buttons weighed as follows :—

Sample.	Check.	Sample.
0·9968.	0·9992.	0·9964.

The loss on the check is therefore $1 - 0{\cdot}9992 = 0{\cdot}0008$ gramme.

The average weight of silver from the sample is 0·9966, and assuming that the loss on the sample is the same as that on the check, the corrected result will give $0{\cdot}9966 + 0{\cdot}0008 = 0{\cdot}9974$ silver in the sample, or 99·74 per cent. When the alloy submitted to assay contains copper as well as silver, it is necessary to add about the same quantity of copper to the checks as is supposed or known to be present in the samples. When the composition of the sample is quite unknown, a preliminary assay must be made to determine this, as directed in Chapter XVIII, where the subject of checks for gold and silver alloys is more fully discussed.

General Remarks and Further Details of Cupel Manipulation.—*Spurting of Cupels.*—As previously stated, " spurting " of the molten lead may take place at the commencement of cupellation if the cupels are damp or contain carbonaceous matter as an impurity. Little fountains of metal are thrown up, and some part may be ejected from the cupel and occasion serious loss if the lead contains much silver or gold. This " spurting," or " bumping," as it is sometimes termed, is, however, only occasionally due to the escape of carbon dioxide produced by the oxidation of the carbon remaining in the bone-ash. " Spurting " more frequently occurs when, during the heating-up of the muffle into which the cupels have been placed, the draught through the muffle has been shut off by tightly closing the door. Under these conditions the atmosphere of the muffle becomes reducing, and the reducing gases are absorbed by the bone-ash cupels. When the lead is put in and cupellation commences in an oxidizing atmosphere, the absorbed gases escape and bubble through the molten lead and cause the spurting. It may also be caused partly by the escape of carbon dioxide resulting from the reduction of litharge by the reducing gases when cupellation commences. When cupels are allowed to heat up in the muffle and an oxidizing draught is kept going all the time, the spurting does not occur. The author has observed that spurting occurs much more frequently with gas muffles than with coke- or coal-fired muffles, especially if excess of gas is turned on in heating up the muffle. Spurting seldom occurs if the door of the muffle is left slightly open when heating up, so as to allow of sufficient draught of air to keep the atmosphere of the muffle oxidizing as stated. The presence of calcium carbonate ($CaCO_3$) in bone-ash also causes "spurting," as it begins to give off CO_2 at the temperature of cupellation. Firth [1] observed considerable spurting with cupels made from bone-ash containing nearly 8·0 per cent. of calcium carbonate.

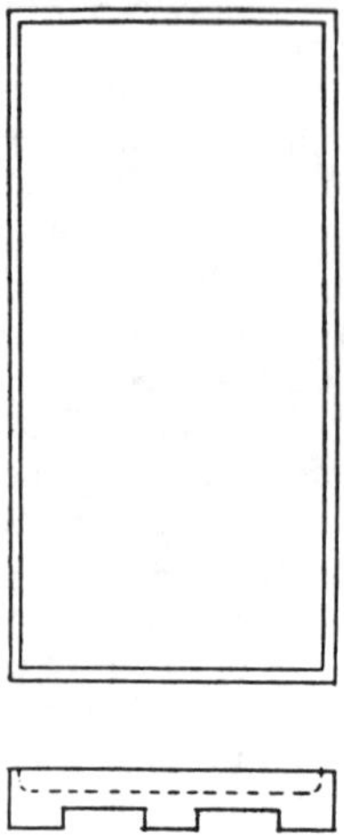

FIG. 104.—GRAPHITE CUPEL TRAY.

At the finish of the cupellation, when the temperature is raised for the brightening of the buttons, the door of the muffle should not be closed too tightly, otherwise the atmosphere becomes reducing and the last traces of lead are not eliminated and the brightening is delayed. This is especially liable to occur with gas muffles if the gas-supply is not properly regulated.

Within recent years special apparatus has been introduced for the quick manipulation of a large number of assays. The following are used in connection with cupellation.

Cupel Tray.—To facilitate the charging-in and withdrawal of a large number of cupels, a cupel tray is sometimes used. A cupel tray used at the Royal Mint, London, is shown in plan and section in Fig. 104. It is 11¾ inches long and 6 inches wide, and holds cupels for 72 assays, the cupels fitting tightly inside the rim. The cupels are square for close packing; the tray is made of "Salamander" graphite, and lasts for some weeks. (Iron trays interfere with cupellation, and are soon destroyed by repeated heating, while fireclay trays soon break.) The tray is sprinkled with bone-ash, and the cupels placed on it before it is charged into the muffle. It is charged in and withdrawn by an iron tool provided with two flat prongs, 1¼ inches wide, and of the same length as the tray, which slide into the grooves underneath the tray. At some assay offices, instead of using a tray, a large number of cupels are made in one block, but the difficulty of maturing the cupels, and the risk of breakage, increases with the size of the block. The advantage of the cupel tray is that single cupels can be rapidly arranged on the tray by hand, before it is put into the muffle. This is an important advantage if the "charging" device described below is used for charging in the assays, as it is necessary in this case to have the cupels carefully placed in position.

Charging Tray.—A device called a "charging tray," for charging a number of assays on to the cupels simultaneously, is used in many assay offices. It is made of cupro-nickel or heat-resisting metal, and consists of a top plate with holes corresponding exactly to the position of the cupels in the muffle. Underneath is a sliding plate with holes corresponding exactly with those of the upper plate. The lead buttons or assay pieces to be cupelled are placed in their proper order in the holes of the upper plate, and rest on the lower plate before introducing the instrument into the muffle. When it is placed in position over the cupels, which have been properly arranged in the muffle, the lower plate is pushed forward to a stop point, thus bringing the apertures of the two

[1] A. T. Firth, *J. Chem. and Met. Soc. of S. Africa*, 1903–4, p. 176.

plates into register, and allowing the lead buttons to drop into the cupels. The action of the charging tray is diagrammatically shown in section in Fig. 103.

The instrument is provided with a long handle and arrangement for working the sliding plate, and is generally fitted with two sidepieces which rest on the floor of the muffle, and thus ensure the plates being at the proper distance above the cupels, and also give some help in placing the instrument into position. When block cupels are used, the lower plate of the charging tray has no holes, and is drawn right out when charging the assays.

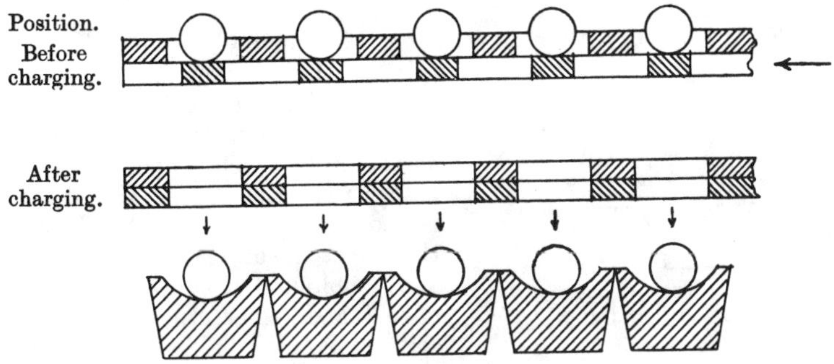

FIG. 105.—ACTION OF CUPEL-CHARGING TRAY.

Cupellation Mirror.—When cupelling a large number of assays, a certain difficulty is experienced in ascertaining when all the cupellations have finished, and to obviate this the simple device shown in Fig. 106 was introduced by H. Westwood, formerly Assay Master of the Birmingham assay office, and is now in use in several assay offices. It consists of a small mirror of plate-glass held in a sheet-iron frame fitted to a long handle at an angle of 30° to the perpendicular, as shown in Fig. 106. The mirror is kept in position by a piece of asbestos card at the back, which also protects it from being blistered by the heat. The mirror is held in the muffle and passed quickly from left to right

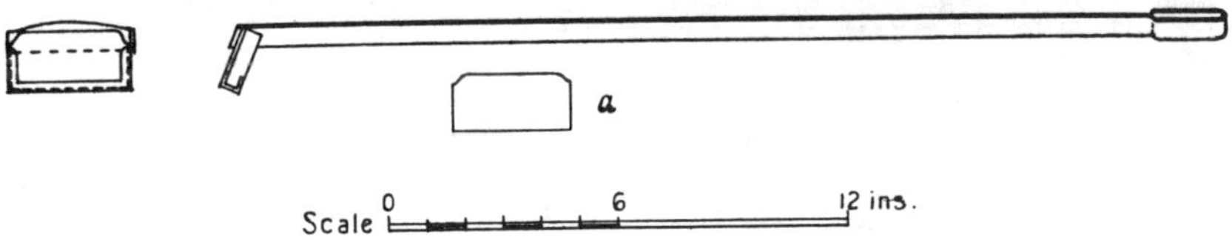

FIG. 106.—MIRROR FOR TESTING PROGRESS OF CUPELLATION.

over each row of cupels in succession, the angle of the mirror allowing a reflection of the buttons to be seen, and the progress of the cupellation ascertained. With a little experience the observation can be made very rapidly, and excessive heating and consequent blistering of the mirror prevented. The mirrors, cut to shape (*a*, Fig. 106), should be kept in stock ready for use, but with care one mirror can be used daily for many weeks before requiring to be replaced. The mirrors should be of plate-glass, as thinner glass is very liable to crack when heated.

Uniformity of Temperature.—To secure uniformity of temperature of the muffle during cupellation as far as possible, the heating of the gas muffles at the Royal Mint,[1] London, was improved by enlarging the combustion chamber

[1] *Fortieth Report of Royal Mint,* 1909, p. 53.

round the muffle, and filling the entire space with clay fire-balls resembling those used in ordinary gas stoves. These act as baffles to the flame, ensuring more complete combustion of the gas before it leaves the furnace. As a result it has been found that, apart from a saving in gas, the furnace is heated more rapidly and is ready for use in less time after being lighted, and that the temperature is more uniform and under better control, so that the quality of the work is improved. When a new muffle has been fitted in a furnace it should be tested for uniformity of temperature by cupelling a batch of proof assays of pure silver and carefully comparing the losses that take place in different parts of the muffle.

Cooling of Assays after Cupellation.—In cooling the muffle to prevent spitting of silver buttons due to dissolved oxygen, the rate of cooling should be such that the cupels cool at the same rate as the muffle. The cupels retain the heat longer than the muffle, and if the cooling is too rapid, the buttons will solidify on the exposed surface while the interior is kept molten by contact with the hot cupel, and when the button solidifies, spitting is very liable to take place. If, however, the cupel is slowly cooled from below, the under surface of the button will solidify first and the dissolved oxygen escape before the silver solidifies as a whole. The rate of cooling is of much more importance than the avoidance of draughts, although, of course, excessive draughts should be avoided. For some years the author cooled down in a gas muffle batches of thirty and forty 1-gramme silver-bullion assays with a gentle draught of air passing through the muffle the whole of the time, an arrangement producing exceptionally few spits. After all the buttons have brightened, the top half of the muffle door is removed, and the small perforated sheet-iron "grid," shown in Fig. 107, inserted in its place and left in position until all the buttons have solidified. The gas is turned down gradually, at about ten-minute intervals, to certain points previously ascertained by experiment. The silver buttons cooled under these conditions are always brighter than those cooled in a muffle from which air-currents are excluded.

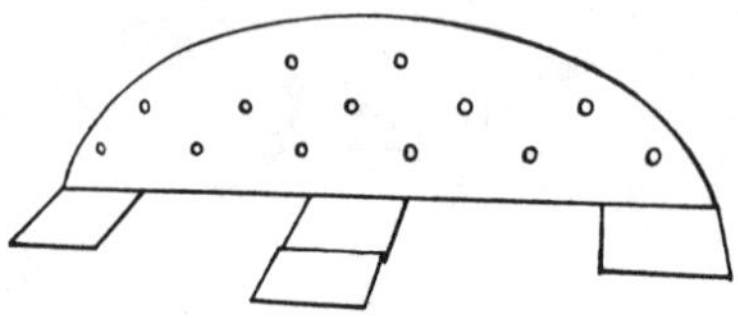

FIG. 107.—GRID FOR CLOSING MUFFLE WHILST COOLING.

Value of System in Assaying—The importance of system in conducting an assay laboratory cannot be too strongly emphasized. If assays are not made in some definite order, and a regular system adopted, confusion is sure to result sooner or later, and when one assay gets displaced it throws doubt on all the rest. It does not follow that a system that is satisfactory for the particular work of one laboratory is suitable for all laboratories. Each assayer must adopt the system best suited to his particular work, but when once the method of procedure is determined it should be strictly adhered to by all who work in the laboratory. With a good system of working, an experienced assayer has little difficulty in following a large number of assays through all the various operations without confusion, but beginners may find it an advantage to mark the scorifiers, etc., so as to identify them. As soon, however, as the system has been learned, the marking should be discontinued. Each day's work should be systematically numbered. The general practice is to work always from left to right in the handling of the assays in the various operations. This systematic order of arrangement of scorifiers, cupels, etc., is kept up, both in and out of the furnace, and the routine of working is never varied, so that a sample can always be identified by its relative position. When a large number of assays of the same kind have to be made, the samples are first placed in order on the balance bench to the left hand of the assayer, ready for weighing. He then takes the first sample, weighs up the quantity required, transfers it to a scorifier or crucible arranged on a tray on his right hand, and duly notes its position. As the weighings are finished, the samples are placed in the same order on his right hand out of the way. The order in which the samples are weighed for assay, and the amount taken, with any other details that may be necessary, should be recorded in the assayer's book as soon as each sample is weighed. These details should never be entrusted to memory alone.

One of the most frequently adopted systems of handling a number of assays from the weighing of the charge to the weighing of the gold or silver buttons is illustrated in the accompanying sketches. The illustrations apply to the handling of fifteen assays, but the system is equally applicable to a larger number. Let it be assumed that seven ore samples, marked *A* to *G*, and some " metallics " belonging to sample *G*, are to be assayed in duplicate by scorification, and that (for convenience of description) the samples are taken in alphabetical order. The charges are weighed and transferred to their respective scorifiers arranged on a tray, the first charge weighed being placed at the left-hand back corner, then the duplicate assay and the rest in succession. Sometimes more than two assays are made of the same sample, but however many assays are made, all the quantities required for that sample should be weighed out before another ore is proceeded with. With duplicate assays the record in the assayer's book would read as follows :—

Assay Number.	Sample.	Assay Number.	Sample.
1	A.	9	E.
2	A. Duplicate of A.	10	E. Duplicate of E.
3	B.	11	F.
4	B. Duplicate of B.	12	F. Duplicate of F.
5	C.	13	G.
6	C. Duplicate of C.	14	G. Duplicate of G.
7	D.	15	G. " Metallics of G."
8	D. Duplicate of D.		

When all the samples are weighed they will occupy the positions on the tray shown in Fig. 108. The tray is carried to the furnace room and the assays put into the muffle in the order shown in Fig. 109, assay No. 13 being put in first and placed at the left-hand back corner of the muffle, then No. 14, and the rest as shown, working always from left to right. When scorification is finished, the assays in the front row are naturally withdrawn first and poured into the ingot mould in the order shown in Fig. 110, thus bringing the assays into the original order again. After cleaning the buttons from slag they are returned to the same position in the tray, and the same method of procedure adopted for cupellation as for scorification. During cupellation the buttons will be in the reversed order, as in Fig. 109, and when withdrawn from the furnace they are placed again in their proper sequence, as in Fig. 110, and are ready for weighing. If the muffle is too small to take all the scorifiers or crucibles, they are divided between two muffles, but the same system of loading and unloading the muffle is adopted.

Another method adopted at a large smelting-works visited by the author is illustrated in Figs. 111 to 114. Large muffles capable of holding twenty-five 3-inch scorifiers are used. The method differs from that described in the fact that the order of the assays is reversed by turning the tray before placing the scorifiers into the muffle, so that in the muffle they occupy the same position as when weighed out. In this case the charges as they are weighed out are placed on a tray in the position shown in Fig. 111, *the handle of the tray being away from the operator as shown.* When carried to the furnace *the tray is turned round,* so that the assays occupy the positions shown in Fig. 112. Assay No. 1 is then put into the muffle first and placed in the left-hand back corner, then No. 2, and so on, until all the assays have been transferred, and occupy the positions shown in Fig. 113. When the assays are withdrawn, No. 21 is poured first into the mould in the left-hand front corner of the mould, and the rest

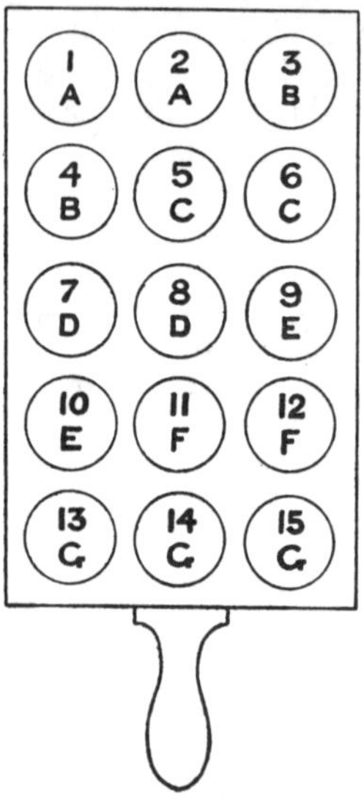

FIG. 108.

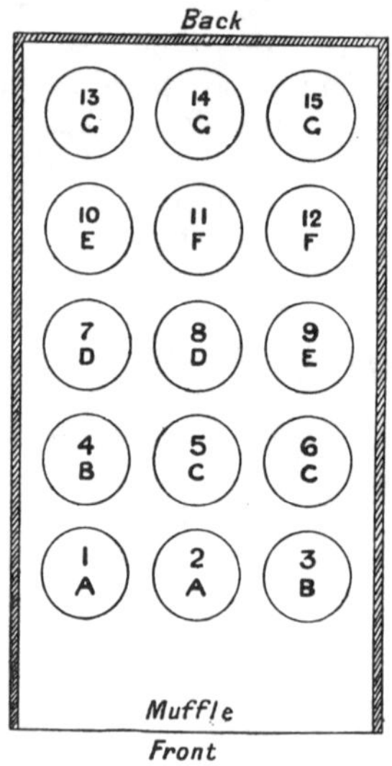

FIG. 109.

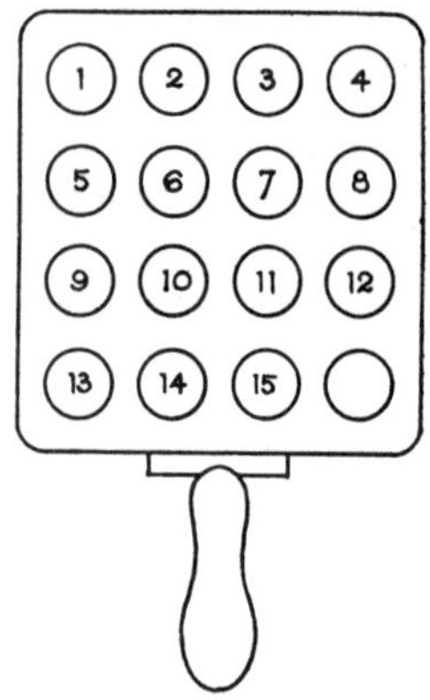

FIG. 110.

POSITIONS OF SAMPLES DURING FUSION, SCORIFICATION, AND CUPELLATION.

in succession, as shown in Fig. 114. By this arrangement the assays occupy the original order of sequence in the muffle, both for scorification and for cupellation. This is a great advantage, since it enables the exact position occupied in the muffle by any particular assay to be easily ascertained, which is very desirable, especially in the case of the cupellations where temperature variations are of such importance.

For cupellation assays only, many assayers charge the assays into the muffle in the same order as that which they occupy in the tray after weighing, as in Fig. 108. In this case assay No. 1 would be put into the muffle first, and placed in the cupel in the left-hand back corner, and then the other assays in succession. When withdrawn from the muffle the assays are returned to their respective places, the cupels first withdrawn being placed in the front row of the tray, and so on. This method obviates the reversing of the order in the muffle, which is necessary in the case of scorifiers, as in Fig. 109, and thus facilitates the calculation for cupellation loss, as the order of the cupels is the same as the silver buttons.

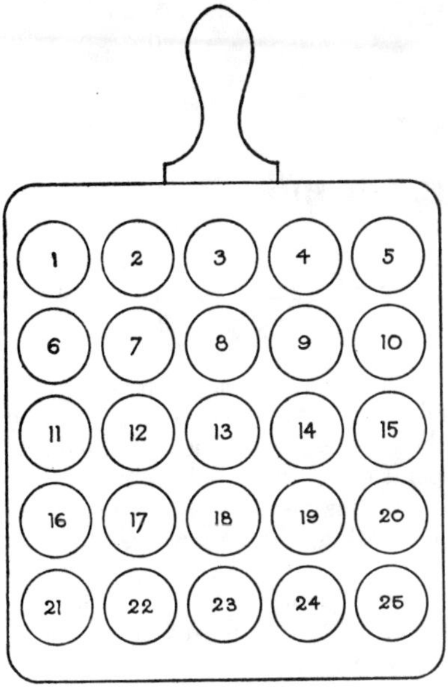

FIG. 111.

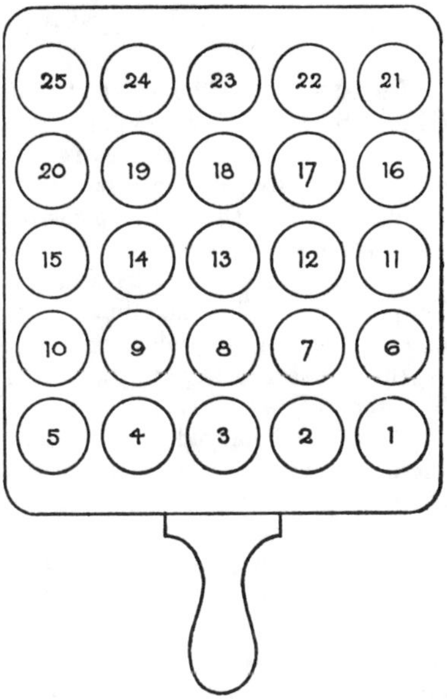

FIG. 112.

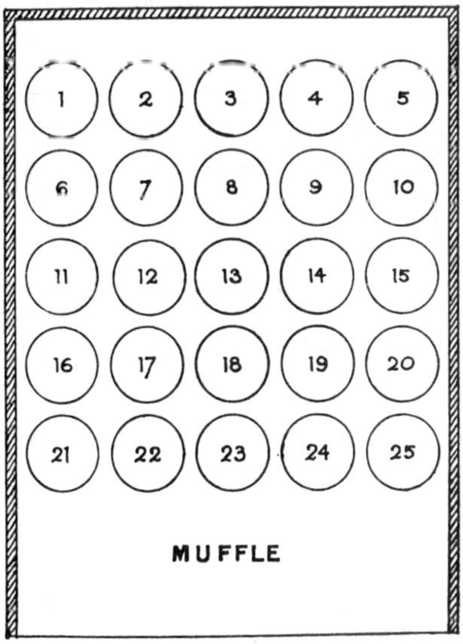

FIG. 113.

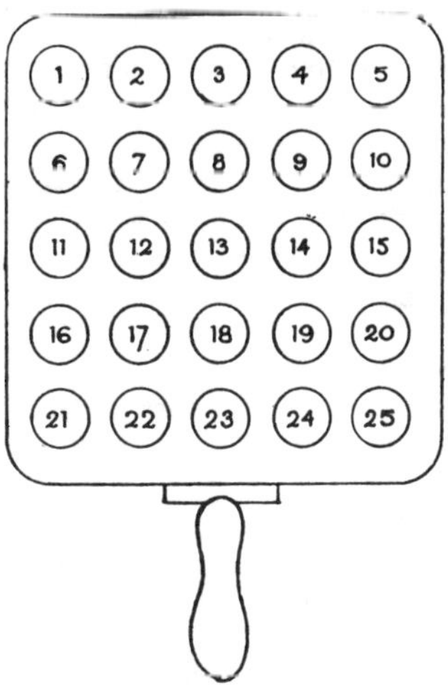

FIG. 114.

ALTERNATIVE METHOD OF ARRANGING ASSAYS.

CHAPTER XII

THE ASSAY OF SILVER ORES

Introduction.—The quantity of silver and gold contained in ores is too minute to allow of its being accurately determined by wet methods, and the assay of precious metal ores is, therefore, almost universally conducted in the dry way, *i.e.* by furnace methods.

The usual method of assay is to fuse the ore with materials yielding metallic lead, which alloys with the gold and silver and sinks to the bottom of the molten charge. Suitable fluxes are added to make the ore readily fusible at a moderate temperature. The lead obtained by this operation is detached from the slag and subjected to cupellation to separate the precious metals. The preparation of the ore for assay by sampling and fine crushing has been described in Chapters VII and VIII.

As already pointed out, opinions differ as to the fineness to which the sample should be crushed for assay purposes. It is usually considered fine enough if all passes through an 80-mesh sieve ; but some ores, such as those containing tellurides, must be crushed through 100-mesh or even a 200-mesh sieve. It is a good rule to crush all ore samples to pass a 100-mesh sieve, unless experience proves that a coarser sample will give satisfactory results with the particular ore undergoing assay.

If metallic particles (*i.e.* metallics) are caught on the sieve, they are collected and assayed separately. The precaution previously enjoined with respect to the metallics is essential.

As the subsequent treatment of an ore is dependent on its nature and richness, a sample should first be carefully examined in order to ascertain as far as possible the nature and proportion of each mineral constituent. Much may be learnt as to the nature of the ore by the examination of unbroken lumps and, after panning, of pulverized ore with a lens. The minerals usually associated with precious metal ores have already been described. The presence of sulphides may be confirmed by the characteristic odour of sulphur dioxide which is evolved when the ore is roasted on a spatula.

Before weighing out the portion of ore required for the assay, the sample must be thoroughly mixed. It is necessary to emphasize the extreme importance of thoroughly mixing the sample, as it is obvious that if the mixing is performed carelessly, the portion selected for assay will not be truly representative of the mass. The value of an assay depends very largely on the care with which the portion of ore used for the assay is taken from the sample submitted. Neglect or deficient skill in this selection renders practically useless all the care that has previously been expended in obtaining a truly representative sample of the material. The amount of ore taken for assay is weighed in grammes or in " assay tons " (A.T.). The " assay ton " system of weights, which is fully explained on page 57, is very frequently adopted in practice where a large number of assays of one kind are made, as it saves calculation.

For general assay-office practice it is more convenient to work on a purely decimal system, and convert when required into ounces per ton, etc. by reference to a set of tables. In this case it is usual first to calculate from the assay result the percentage of precious metal in the ore, and then refer to tables to ascertain the equivalent in ounces per ton, etc. Therefore, to avoid a needlessly troublesome calculation, it is well to take such a quantity of ore for each assay as by a simple multiplication will yield the percentage.

Corrected Assays.—In the assay of ores there are certain losses incidental to the methods used. The chief losses are retention of precious metals in the slag during fusion, and absorption of precious metals during cupellation. As a general rule, the slag losses are considerably higher than the cupel losses. The amount of these losses varies according to the nature of the ore and the conditions under which the assay is conducted ; they are by no means constant, and present no regularity even for the same material. With ordinary gold and silver ores of average grade assayed under normal conditions, both the slag loss and cupel loss are very small and are usually disregarded, unless very great accuracy is required. In the case of rich ores and complex ores it is frequently necessary to determine the losses and correct the assay result accordingly. An assay in which the slag loss and cupellation loss has been determined and taken into account in reporting the assay value of the material is termed a " corrected assay."

The assays made at metallurgical works for ascertaining the value of purchased ores, except those of very high values (and generally for all other purposes), are almost invariably direct uncorrected assays.[1] The presence of copper, bismuth, and especially of tellurium and selenium in the lead button, increases the losses of gold and silver during cupellation. Care must therefore be taken in the scorification or the crucible fusion of the ore to prevent these metals from entering the lead as far as possible.

Although the dry assay of gold ores resembles in its main particulars the dry assay for silver ores, it is convenient to describe the assay of each metal separately.

The systems of working off large numbers of assays have been discussed above.

METHODS OF ASSAY FOR SILVER ORES.

The various ores of silver are described on p. 64.

For assay purposes a silver ore may be regarded as rich when yielding over 200 ounces of silver per ton, and poor if yielding under 50 ounces per ton : ores yielding intermediate amounts may be said to be average ores.

The assaying of a sample may comprise the following operations :—

(1) Concentration of the silver in a button of lead by (*a*) scorification, or by (*b*) fusion in a crucible.

(2) Cupellation of the resulting button of argentiferous lead.

(3) When great accuracy is required, the determination of the silver absorbed by the cupel.

(4) Determination of the silver in the slag produced in the process of scorification or fusion.

(5) Assay of the " metallics."

(6) Weighing the silver.

(7) Examination of the silver for gold.

Two assays at least should be made of each sample.

(1) (*a*) **Scorification Method.**—This is a simple and convenient method of assaying silver ores. It consists in exposing the ore, mixed with granulated lead and placed in a scorifier, to the action of a bright-red heat, in an ordinary assay muffle.

[1] Collins, *op. cit.*, p. 471.

The lead is oxidized by the air, and the silica and other mineral constituents in the ore are fluxed by the litharge, borax, etc. The precious metals are taken up by the lead, which diminishes in amount as oxidation proceeds, until the slag forms a continuous layer over the lead. The charge is then poured into an iron mould, the lead detached, and cupelled to obtain the silver. The operation of scorification is fully described on page 141.

The amount of ore taken for scorification varies from 3 to 5 grammes; but 3 grammes is the amount most frequently taken.

If the ore contains less than 1 per cent. of silver, 5 grammes or about $\frac{1}{6}$ A.T. are taken for a charge; if more than 1 per cent., 2·5 or 3 grammes (from $\frac{1}{10}$ to $\frac{1}{12}$ A.T.); and of very rich ores 1 to 0·5 gramme, or about $\frac{1}{30}$ to $\frac{1}{60}$ A.T.

As a general rule, if more than 5 grammes of ore must be taken, the crucible assay should be adopted. The ore is weighed accurately to 0·001 g. unless very poor, when less accurate weighing will suffice. The proportion of lead added varies from 40 to 100 g. according to the nature of the ore. A very usual charge for ordinary silver ores is 3 g. (0·10 A.T.) of ore and 40 to 50 g. of lead, with a small quantity of borax as a cover.

Table XXVI gives examples of the proportions of lead to be used for different classes of ores :—

TABLE XXVI.—Proportion of Lead for Scorification of Ores.

Character of Ore.	For one part of Ore.	
	Parts of Granulated Lead.	Parts of Borax Glass.
Arsenical	16	0·10 to 0·50
Antimonial	16	0·10 to 1·00
Basic	8 to 10	0·25 to 1·00
Blende	10 to 15	0·10 to 0·20
Cobaltiferous	10 to 20	0·10 to 0·20
Cupriferous	10 to 20	0·10 to 0·15
Fahlerz	12 to 16	0·10 to 0·15
Galena, pure	6	0·15
,, siliceous	8 to 11	0·2
,, zinciferous	8 to 11	0·15 to 0·30
Iron pyrites	10 to 15	0·10 to 0·20
Nickeliferous ores	15 to 20	0·20 to 0·30
Siliceous (quartzose)	8 to 10	0
Tin ores	20 to 30	0·15 to 0·20
Zinciferous ores	15 to 20	0·20 to 0·30

The borax renders the slag more liquid and minimizes corrosion of the scorifier, but its quantity is kept as low as possible to prevent the slag from completely covering the surface of the molten lead too soon. If the ore consists largely of quartz, sodium carbonate should be added instead of borax. In the case of highly cupriferous ores, it is advantageous to add about 0·5 gramme of finely-powdered sand to lessen the corrosive action of copper oxide on the scorifier. Where a large number of assays have to be made, the granulated lead may be measured by the copper ladle or measure, shown in Fig. 115, instead of weighing.

A scorifier, 2·5 inches in diameter, is the size most commonly employed for ordinary scorifications. The scorifiers being charged with the proper proportion of ore, lead, and flux, and the muffle heated to a full red heat, they are removed to the furnace, and as many are introduced as the muffle will accommodate. The introduction of the scorifiers at first greatly reduces the temperature of the muffle; the door of the latter should, therefore, be closed as soon as the charging-in is complete, and remain closed until the lead melts and the assay is thoroughly heated, after which the upper half of the door is removed to allow air to enter.

When scorification is complete, a little powdered anthracite may be added to clean the slag. When the surface is again tranquil the scorifiers are withdrawn from the muffle by means of proper tongs (Fig. 26), and their contents rapidly poured into suitable moulds. When cold, the buttons of lead are readily separated from the adhering slag by a few blows with a hammer. The lead is cleaned from all slag and then cupelled, and from the weight of the resulting silver buttons the assay value of the material is calculated and duly reported. Should the lead exceed 25 grammes in weight, or should it be hard, it must be re-scorified until it is reduced to the required weight and is soft. The slags from very rich ores should always be kept and separately re-treated, as already described.

FIG. 115.

It is essential that the slags should be perfectly and uniformly liquid at the time of pouring from the scorifier; if they are pasty and contain lumps of partially fused material, it usually indicates that part of the ore has been left unacted upon. This may be due to the employment of too low a temperature or to insufficient lead or borax. If the slags should not appear perfectly liquid before pouring, when a sufficiently high temperature is maintained in the muffle, and the other conditions of the operation have been attended to, it will be necessary to add more borax, and, in some instances, even a little nitre. An infusible scoria is very liable to form on the surface of the slag in the case of ores containing large quantities of basic oxides, such as ferric oxide (Fe_2O_3) manganese oxide, copper and nickel oxides. The addition to the charge of borax glass up to 3 grammes greatly facilitates the formation of fluid slags free from infusible scoria. If from any cause the slag is unsatisfactory and contains unfused portions of ore, it is best to make a fresh assay, using less ore or more lead, and if necessary finer ground ore.

Most ores and metallurgical products containing silver may be assayed by scorification; but more experience is required to obtain correct results than by some of the fusion methods. Scorification is, however, not so well adapted for assaying substances poor in silver, as the quantity of material operated upon is comparatively small, though this objection may be obviated by combining the lead buttons obtained from several assays and re-scorifying them so as to concentrate the silver into one button of lead, which is then cupelled.

For the assay of ordinary and rich silver ores, and for products rich in the precious metals, scorification is a very desirable method. It is also a very convenient method for concentrating the precious metals in residues resulting from the treatment of material by combined wet and dry methods. From its convenience and the short time required for the operation, it is very generally employed in establishments where a great number of assays have to be made daily. The small charges used for scorification necessitate the employment of good balances and care in weighing. The silver buttons from duplicate assays of material of moderate richness should agree closely. It should be borne in mind that a difference of 1 mg. in the weight of the silver button represents a difference of 6 to 10 oz. per ton (2240 lbs.), according to the amount of material taken for assay. The assay should be repeated if the difference between the weights of the buttons exceeds 0·5 mg., equivalent to 5·4 oz. per ton when 3 grammes are taken for assay. In the case of "argentiferous lead ores assayed for silver by scorification in triplicate on one-tenth assay ton of ore, the maximum difference between the three silver buttons should not be more than one-tenth milligramme on buttons weighing 1 centigramme each (1 ounce on 100 ounces). When the difference between the heaviest and lightest of the three does not exceed this amount, the mean may be taken, but if the difference is greater, two more assay checks are made."[1]

With rich material, variations in the weights of the silver buttons are not infrequent, even when every care has been exercised, and the only safe plan is to make several assays, and take the mean of the results obtained. The number of assays to be made to control the accuracy of results will vary from 2 to 5 or more, according to the richness and nature of the ore.

(1) (*b*) **Crucible or Fusion Method.**—The object of fusion in a crucible, as of scorification, is to concentrate the silver in a button of lead which is subsequently cupelled: and to retain the gangue material in the slag. The charge for the fusion of any given ore varies somewhat according to its nature, and is made up by the assayer according to his judgment and experience.

The various fluxes employed and the principles of fluxing have been fully dealt with in Chapters IX and X.

The amount and kind of flux required is judged in the first instance from the appearance of the ore and the preliminary tests, and is subsequently modified if necessary, according to the result of the fusion. It is well in this connection to bear in mind the general principle that for a siliceous or acid gangue a basic flux is needed, and for basic material an acid flux. Sodium carbonate being a flux for silica and silicates, and borax for lime, iron oxide, and other bases, the relative quantities of the two must be varied according to the nature of the ore. As a general rule, borax should not be used unless the nature of the ore requires it. As previously stated, from two to three parts of flux are usually quite sufficient to yield a satisfactory slag with one part of ore. Lead oxide in the form of litharge or red lead, and charcoal, or other reducing agent are added to supply the lead necessary for collecting the precious metals.

A mixture of 30 grammes of litharge and 1 to 1·5 grammes of charcoal powder will give a button of lead weighing about 20 g., which is a satisfactory quantity for most silver ores when 20 to 25 g. of ore are used for assay.

Sometimes a comparatively large quantity of lead oxide is added to the charge, and a portion of it allowed to remain unreduced to serve as a flux for the oxides of such metals as copper, antimony, iron, etc. The quantity of litharge or of red lead used must be fairly accurately weighed or measured out, so that the amount of silver it contains may be correctly deducted in accordance with the result of the preliminary assay of the reagent. The charcoal or other reducing agent should also be weighed carefully, and the quantity adjusted so as to produce a lead button of suitable size for cupellation on one cupel. Although it may be desirable, by way of scientific training, for students to weigh out all fluxes, an assayer who possesses the requisite aptitude for manipulation soon acquires the skill which enables him to dispense with the tedious process of weighing out all his reagents. Many assayers either estimate the needed quantity with the eye, or make use of tin or copper measures, which, when filled, contain about the required quantity of flux: this latter is the better of the two methods.

[1] Collins, *Metallurgy of Lead*, 2nd edn., 1910, p. 46.

When the same class of ore is being assayed repeatedly, it is usual to make up a " stock " flux, and use a measured quantity for each assay.

When the charge has been prepared, it is well mixed, transferred to a crucible of suitable size, and fused as previously described. An " E " crucible, Battersea round, is very suitable for silver-ore assays fused in a wind furnace, but if the fusion is to be made in a muffle or furnace of reverberatory type, a crucible of the Colorado shape is generally preferred.

The fusion must be preceded by an exposure to a dull red heat for ten minutes, so as to ensure the complete " fritting " of the charge. The temperature is then raised and fusion effected, and when all action has ceased the charge is poured into a round mould. When cold, the lead is detached from the slag, which on account of its brittleness readily separates from the lead on hammering. If the button of lead weighs less than 20 grammes, the quantity of charcoal must be increased, remembering that charcoal will reduce about twelve times its own weight of lead from red lead and 22·5 from litharge. The lead is cupelled and the assay value of the ore calculated from the weight of the resulting silver button as usual.

In all cases the slag should be kept until it is ascertained from the result of the cupellation whether it is desirable to " clean " it. The slag is generally glassy, but varies with the nature of the ore, and should be uniform in colour and composition. For convenience of description the various ores are divided into four classes. :—

I. Ores with siliceous gangue.
II. Ores with basic gangue.
III. Pyritic ores (mainly iron pyrites).
IV. Complex ores.

The following are typical charges for the several classes of ores :—

I. Ores with Siliceous Gangue.—Ores of this class present little difficulty. For average ores a suitable charge is :—

Ore	20–25 grammes.
Litharge or red lead	30 ,,
Charcoal	1–2 ,,
Sodium carbonate	30 ,,

If the ore contains a small amount of pyritic material, metallic iron in the form of nails or hoop-iron should be added to decompose the sulphides and yield a soft and malleable lead button.

II. Ores with Basic Gangue.—In the silver ores of this class the chief bases to be fluxed are oxides of iron, with alumina, lime, and oxides of manganese. barium, magnesium, and other metals in smaller quantity. The principal fluxes for basic material are borax and lead oxides. More or less quartz or siliceous material is, however, almost invariably present in all basic ores, and a certain proportion of sodium carbonate is usually added to the charge. When the siliceous material in the ore predominates, the charge given above for siliceous ores will be satisfactory, but the proportion of sodium carbonate should be diminished, and that of borax, and if necessary of lead oxide increased as the quantity of metallic oxides in the ore increases. The addition of excessive amounts of borax should be avoided, otherwise the charge may fuse too readily. A mixture of equal weights of sodium carbonate and borax usually answers very well for the majority of basic ores. As previously pointed out, both basic and acid materials are generally present in " basic " ores, and as these combine to form more or less fusible compounds, the quantity of additional flux required is frequently comparatively small.

Ores containing large quantities of iron oxide (or manganese oxide) require more charcoal or other reducing agent to be added to the charge, for the reasons already given on page 128. As an example, 25 grammes of ore, consisting of practically pure ferric oxide, will require 3 grammes of charcoal. The addition of fine sand is necessary to flux basic ores in which silica is deficient.

For average basic ores the following charge may be taken :—

Ore	20–25	grammes.
Litharge or red lead	40	"
Charcoal	2–3	"
Sodium carbonate	20	"
Borax	10	"
Sand or powdered glass	0–10	"

For ores consisting mainly of earthy bases, such as lime, alumina, etc., or of zinc oxide, the quantity of alkaline flux is advantageously increased to 40 grammes.

The quantity of litharge should be increased to 50 to 60 grammes for ores containing oxides of metals such as copper, antimony, zinc, etc., in moderate amount only. The excess of lead oxide passes into the slag and carries the copper oxide, etc., with it, thus preventing the production of hard cupriferous or antimonial lead. Large proportions of lead oxide should, however, be avoided, as slags with a large excess of this oxide are very liable to retain silver and gold. Ores containing large quantities of the oxides of copper, antimony, etc., are more satisfactorily treated by the special methods described subsequently.

III. **Pyritic Ores.**—The important silver ores included in this class consist mainly of iron pyrites with small amounts of other metallic sulphides. The sulphides most commonly present, in addition to the sulphurized minerals of silver, are iron pyrites, copper pyrites, galena (lead sulphide), blende (zinc sulphide), and mispickel (arsenical iron pyrites). Ores containing excess of the sulphides, other than iron pyrites, are dealt with below under the head of "Complex Ores." Ores containing much pyritic material are easily recognized by the ready separation of the heavy sulphides when the powdered ore is washed, or by the characteristic odour of sulphur dioxide which they evolve when a small quantity is roasted on a spatula.

The presence of mispickel is indicated by the odour of arsenious oxide.

As previously stated, metallic sulphides in ores act as reducing agents, and it is of importance that the whole of the sulphur should be either removed or oxidized during the process of assaying, as otherwise, if present in large amounts, it will yield large lead buttons and render the lead sulphury and somewhat brittle, or the lead will be accompanied by a regulus or matte. The following are the methods in general use for the removal of the sulphur :—

(*a*) Desulphurization with metallic iron.
(*b*) Oxidation by roasting.
(*c*) Oxidation with lead oxide.
(*d*) Oxidation with nitre.

(*a*) *Desulphurization with Metallic Iron.*—In this method metallic iron is added to the charge to take up the sulphur forming iron sulphide which dissolves in the slag in moderate quantity and does not form a separate layer of matte. Slags that are strongly alkaline have considerable solvent action on metallic sulphides. The iron added to the charge should be in the form of thick hoop-iron (two or three pieces), in strips about 6 inches long, or small rods of wrought iron. Stout iron nails are frequently employed for the purpose, but they present less surface, and are more troublesome to manipulate. If they are used, from four to six should be added to the charge with the heads downwards, in order to present as large a surface as possible. Sodium carbonate is the chief flux used, as the slags must be strongly alkaline. As a general rule, the amount of sodium carbonate should be twice that of the ore taken. Only as much lead oxide should be added as will give a button of lead of the weight required, since practically the whole of the lead oxide in the charge is reduced to the metallic state by the iron. It is advisable to add sufficient lead oxide to give a comparatively large button of lead, say about 25–30 grammes. The amount of lead oxide may be diminished for ores containing appreciable quantities of galena.

The following is a suitable charge for moderately pyritic ores :—

Ore	20–25 grammes.
Litharge or red lead	40 ,,
Sodium carbonate	30–40 ,,
Borax	10 ,,

with the addition of metallic iron.

The fusion should be performed at a comparatively low temperature, and the crucible should remain in the furnace for about five minutes after the contents are in tranquil fusion. The iron is then taken out and examined while hot, and should any shots of lead be found adhering to it, they must be washed off by immersing it in the molten slag, after which it is withdrawn, and the crucible left in the furnace for a few minutes longer. The crucible is then removed, and its contents poured into a mould in the ordinary way. The fusion should be conducted in a reducing atmosphere as far as possible, as the large amount of ferrous sulphide in the slag is readily oxidized, causing it to become pasty and give trouble in pouring. To preserve the slag from oxidation, many assayers add a layer of salt on top of the charge ; but in the author's opinion it is preferable to use a cover to the crucible, leaving a small opening to permit of the escape of the gases. It is usually advisable to " clean " the slag unless the ore is low grade.

When arsenic is present the iron will decompose the arsenical compound with the formation of a speise, which separates as a hard, greyish-white layer on the surface of the button of lead. The separation of a small quantity of speise may be disregarded, as it seldom retains an appreciable quantity of precious metal ; but if large quantities of arsenic are present, it must be eliminated by other methods. The employment of a low temperature tends to prevent the formation of speise.

Hard and somewhat brittle lead buttons will be obtained from ores containing copper and antimony sulphides, as these compounds are partially reduced to the metallic state by iron, the metals passing into the lead button and causing trouble in the cupellation, unless previously removed by scorification. In this case a cupriferous regulus, which retains appreciable amounts of gold and silver, is also frequently formed. The presence of antimony compounds will produce brittle lead buttons, and give rise to the formation of a scoria on the cupel, which tends to retain part of the precious metals and leads to unsatisfactory results. Ores containing much copper or antimony are best treated by special methods, described subsequently.

This method of assaying pyritic ores is in very general use, as it can be successfully applied to most ordinary sulphide ores. But, as previously stated, only a moderate amount of sulphur can be carried into the slag by the use of iron, and with ores containing a large proportion of metallic sulphide it is very difficult to effect complete decomposition by this method, even when a large quantity of metallic iron is used: the lead button is always more or less brittle owing to the presence of sulphur, and the production of matte is inevitable.

This objection may be overcome by using smaller quantities of ore for assay, but this involves the risk of multiplying errors of assay.

Many assayers use metallic iron for ores rich in sulphides, even when its use necessarily involves the production of matte, but in this case the matte is saved and scorified together with the lead button, prior to cupellation, or the matte is crushed and mixed with the slag so that both are cleaned at the same time. The use of iron is not applicable to ores containing large quantities of arsenic, and metals such as copper and antimony, for the reasons stated above. Ores rich in arsenic, copper, or antimony are best treated by the methods described subsequently.

(b) *Oxidation by Roasting.*—An oxidizing roasting of the ore previous to fusion is a satisfactory method of eliminating the sulphur in pyritic ores, and is sometimes conveniently employed for gold ores; but as smaller quantities of ore are permissible for the assay of silver ores, unless very poor, roasting may generally be dispensed with, and advantage taken of the oxidizing power of red lead or nitre, or the desulphurizing power of iron.

When silver ores are roasted, especially such as consist of argentiferous minerals containing copper, it is generally necessary to employ a very low temperature, as, from their great fusibility, they would otherwise be liable to "clot" or partially fuse, and the further expulsion of sulphur would be rendered difficult.

The methods of roasting ores have already been described. The quantity of ore used for assay must be weighed before roasting. The roasting of silver ores tends to give low results, especially if the ores are high grade. After roasting, the metals exist in the state of oxides, and the ore is therefore basic in character and is fluxed accordingly.

(c) *Oxidation with Lead Oxide.*—In this old method of assay, lead oxide, either litharge or red lead, is employed as the oxidizing agent. Both these oxides readily attack all the sulphides, arsenio-sulphides, etc., and oxidize their constituents, whilst a proportionate quantity of lead is set free. An excess of lead oxide must be used, as it not only acts as an oxidizing agent, but also as a flux for the gangue of the ore and the metallic oxides resulting from the oxidation of the pyritic material. Silica requires nearly 5 parts of litharge, and the metallic oxides from 2 to 10 parts, to form fusible compounds. The proportion of litharge required to form fusible compounds with the principal metallic oxides is as follows [1]:—

Proportion of Litharge to Flux Metallic Oxides.

One part of	As_2O_3.	Cu_2O.	CuO.	Fe_3O_4.	Sb_2O_3.	ZnO.	Fe_2O_3.	MnO.	SnO_2.
Requires parts of PbO	1	1·5	1·8	4	5	8	10	10	13

Charcoal powder is added in cases where the proportion of sulphides in the ore is very small, and not sufficient to give the required weight of lead.

Red lead is used in preference to litharge, on account of the larger amount of oxygen it contains. As a flux, lead oxide has the advantage of being heavy, and consequently occupies very little space in the crucible. It may be remarked,

[1] Hofman, *Metallurgy of Lead.*

however, that care is needed in its employment as a flux by itself, as it has a strong corrosive action on the crucible, and if the fusion of the assay is performed too quickly, it sinks to the bottom of the crucible, owing to the readiness with which it fuses, leaving the lighter earthy matters of the charge as a pasty, semi-fused mass in the upper portion of the crucible, where they escape the fluxing action of the litharge. The fusion should be effected at a comparatively low temperature. It is not desirable to prolong the fusion, on account of the corrosive action litharge has upon the substance of the crucible, which it rapidly destroys.

To lessen this action, it is usual to add fine sand or glass equal to twice the weight of the ore taken, or less according to the proportion of silica in the ore. The proportion of lead oxide required to effect complete oxidation varies according to the nature of the sulphide, but in all cases is large. As a general rule, 1 part of a metallic sulphide requires from 20 to 30 parts of red lead to yield a button free from sulphur. "When less than the requisite quantity of lead oxide is used, only a portion of the sulphide is decomposed, and a corresponding quantity only of lead reduced, whilst the remainder of the sulphide forms, in combination with the litharge and metallic oxide which is produced, a compound belonging to the class of oxysulphides which are generally very fusible" (Mitchell). If oxysulphides are present to a large extent in the slag, they cause the retention of appreciable quantities of the precious metals, notably silver.

An objection to this method of assay is the large quantity of lead produced unless the ore is sufficiently rich to permit of small quantities being taken for assay or contains only a small proportion of metallic sulphides.

The amount of lead reduced by 1 gramme of different sulphides varies from 6 to 11 grammes, as shown by the table on page 122. It is evident, therefore, that if an ore contains a large proportion of these sulphides, and 20 grammes is used for the assay, the quantity of lead reduced will be very much larger than that actually required for an assay. For example:—With an ore containing 75 per cent. of iron pyrites, 20 grammes would yield 165 grammes of lead, and if there were not sufficient lead oxide to yield this weight of metal, the button would be sulphury. This inconvenience may be obviated by oxidizing a part of the pyritic material with nitre, by the judicious employment of which buttons of almost any required weight may be obtained as described below. One gramme of nitre is equivalent in oxidizing effect to 20 grammes of red lead or 26 grammes of litharge. When lead oxide alone is used as the oxidizing agent, the following charge may be used for moderately rich and highly pyritic ores:—

Ore	5–10	grammes.
Red lead (or litharge)	100–200	,,
Sand	0–10	,,

Borax to cover, 10 grammes.

The slags are very fluid and pour readily, and the lead button is generally clean, and separates easily from the slag.

The weight of the button should not be less than 30 grammes, and is usually much above this. If necessary, it should be scorified down to about 15 grammes, and cupelled.

This method can be conveniently applied to highly pyritic ores when sufficiently rich to allow of comparatively small quantities being taken for assay, but is not suitable for low-grade ores, unless the amount of pyritic material present is small.

Sulphide ores rich in copper and antimony are satisfactorily assayed by this method, as these metals are oxidized and dissolved in the slag and not reduced to metal, so that they do not contaminate the lead button as in the case with metallic iron. Except in cases where an excess of lead oxide is required in the slag to flux metallic oxides such as copper, it is the general practice to limit the quantity of lead oxide to that necessary to oxidize the pyritic material, and to add borax and sodium carbonate to flux the gangue material and the metallic oxides formed during the fusion, that are not fluxed by the litharge. It is well to aim at leaving little or no lead in the slag as silicate or as free litharge, as experience indicates that with high-grade ores the presence of an appreciable quantity of lead silicate and oxide in the slag causes low silver results, either by reason of the density of the slag or by dissolving the silver.

The charge must be made up according to the amount and nature of the pyritic material present. In many cases the following charge will be suitable :—

Ore	0·25 to 0·5 assay ton.
Red lead or litharge	50 to 75 grammes.
Sodium carbonate	10 „ 15 „
Borax	10 „ 15 „

When the ores are very pyritic it is preferable to effect the oxidation of a certain proportion of the sulphides by the addition of nitre as already stated.

(*d*) *Oxidation with Nitre.*—Nitre may be employed according to either of the following methods :—

A. To oxidize part of the metallic sulphides.
B. To oxidize the whole of the metallic sulphides.

Method A.—In this method the pyritic material present in excess of that required to act as a reducing agent is oxidized by the addition of nitre. When the proportion of nitre in the charge is insufficient to oxidize the whole of the pyritic material, and when litharge is also present, after the nitre has produced its action, the litharge acts in its turn on the sulphides that still remain unoxidized, producing metallic lead which carries down the silver. Therefore, by employing suitable proportions of nitre and litharge, any desired quantity of lead may be obtained. To ascertain how much nitre is required for a given ore it is necessary first to determine the reducing power of the ore, and from the result obtained to calculate the amount of nitre necessary to oxidize the excess of sulphides.

The approximate quantity of nitre required to oxidize various metallic sulphides is as follows :—

Approximate Oxidizing Effect of Nitre.

	Part of Nitre required to one part of Metallic Sulphide.
Iron pyrites	about 2 to $2\frac{1}{2}$ parts.
Mispickel, copper pyrites, fahlerz, zinc-blende	„ $1\frac{1}{2}$ to 2 „
Antimonite	„ $1\frac{1}{2}$ parts.
Galena	„ $\frac{2}{3}$ part.

Experience has shown that, as a general rule, from 0·25 to 0·30 parts of nitre is required to oxidize the quantity of pyritic material which would act as a reducing agent to produce one part of lead ; or, in other words, the addition of 1 gramme of nitre will decrease the weight of the lead button by 4 to 5 grammes. Conversely, the weight of the lead button may be increased by from 4 to 5 grammes by diminishing the quantity of nitre by 1 gramme. The result of an interesting series of experiments on the oxidation of metallic sulphides with nitre is given in Beringer's well-known book of Assaying, p. 97 (15th edition, 1921). The experiments show that 1 gramme of nitre "kept up" on the average 4 grammes of lead, the range being from 3·2 with acid slags to 5·3 with very basic slags. These facts serve to explain some apparently irregular results obtained in practice.

The reducing power of an ore is determined by a preliminary assay, and to get the best results it is advisable to use a charge of the same composition as the final assay charge.

The following charge may be used :—

Preliminary Assay.

Pyritic ore	5 grammes.
Litharge (not red lead)	100 ,,
Sodium carbonate	10 ,,

Cover of borax, 10 grammes.

If the ore contains little silica, from 1 to 5 grammes of silica should be added.

The lead resulting from the fusion is weighed, and from this weight is calculated how much lead would be obtained from the quantity of ore to be used for the final assay (say 10 grammes). From the weight thus found deduct 25 (the desired weight of lead to collect the values) and divide the difference by 4; the figure obtained gives approximately in grammes the nitre required for 10 grammes of ore.

For example: Suppose the preliminary assay of 5 grammes of ore gave 40 g. of lead, then 10 grammes of ore would yield 80 g. of lead. Subtracting 25 we get $80 - 25 = 55$, and dividing this by 4 we get 13·8 g., the weight of nitre required.

This rule may be used as a general guide for adjusting the nitre, but in practice irregular results are frequently obtained, as the action of the nitre is considerably influenced by the conditions of the fusion.

The difficulty of adjusting the quantity of nitre is largely dependent on the following facts :—

(*a*) The common commercial varieties of nitre, which are those usually employed for assay purposes, frequently contain impurities, consequently, the oxidizing power will vary with the different samples employed.

(*b*) Sulphur, and in some cases the metals, are capable of two degrees of oxidation, and whether the lower or the higher oxide is formed by the action of the nitre will depend very largely on the conditions of the fusion. In this connection the character of the slag resulting from the fusion is an important factor. When the slag is acid owing to the presence of much siliceous material in the charge, the sulphur will, in most cases, be eliminated as sulphur dioxide (SO_2), and the lower oxide of the metal be formed. On the other hand, if the slag contains much soda, the tendency will be for the sulphur to be oxidized to sulphate and the metal converted to the higher oxide.

(*c*) Rapidity or slowness of fusion will also make a great difference in the effect of the nitre; in the former case, the pyritic material sinks before the nitre has time to complete its action, and frequently produces a regulus, and in the latter case this does not occur. The decomposition of potassium nitrate is greatly facilitated by the presence of silica and of silicates.

In making up the charge the litharge is usually added in slight excess of that necessary to provide the required weight of lead. Sodium carbonate equal to twice the weight of the ore, or less, is the chief flux employed.

Part of the sodium carbonate added to the charge will be utilized in the formation of sodium sulphate, which is one of the products of the reaction of the nitre upon the sulphides present, and due allowance must be made for this fact. The sodium sulphate, on account of its extreme fluidity, floats as a watery liquid on the surfaces of the slag, and when present in large quantities causes inconvenience in the pouring of the charge.

If the ore is deficient in silica, some should be added.

The following is a suitable charge for most sulphide ores :—

Ore	10–20 grammes.
Litharge or red lead	40–80 „
Sodium carbonate	10–20 „
Borax	3–5 „
Sand or powdered glass	0–10 „

Nitre sufficient to yield a button of about 25 grammes of lead.

For ores with more than 50 per cent. of pyritic material the weight of nitre required usually varies from an equal weight to twice the weight of the ore.

When fusion is made in a muffle it is not advisable to use more than 20 grammes of nitre in a charge, otherwise the charge is very liable to boil over.

The fusion of charges containing nitre is effected in large crucibles at a comparatively low temperature, care being taken to avoid a sudden increase of temperature, as fusion takes place somewhat quickly, and the slags are in most cases very fluid. When effervescence has ceased, and the contents of the crucible have become thoroughly liquid and tranquil, it is advisable to lift the crucible out of the fire and subject it to a rotary motion in order to mix the liquid contents thoroughly. If this treatment causes effervescence, it indicates that the action of the nitre is not complete, and the crucible must be returned to the fire and the fusion continued.

The button of lead obtained should weigh between 25 and 35 grammes, but if below or above this weight, the charge must be modified to produce the required weight.

The weight of lead obtained will vary somewhat according to the quantity and nature of the fluxes used and the conditions of the fusion. When the charge contains much soda, the tendency will be, as before stated, for sulphates to form ; and as the nitre oxidizes a smaller quantity of pyritic material when the higher oxides of the metals are formed, a larger proportion of the pyritic material is left unoxidized, and acts as a reducing agent, thus producing more lead, if the charge contains sufficient litharge. Until experience has been gained, the first assay of an ore of unknown composition may be more or less unsatisfactory, but the result obtained from the fusion will indicate what modification of the charge is necessary to give a satisfactory result.

With proper precautions the nitre may be so adjusted that a malleable lead button of the desired weight is obtained.

The necessity of making a preliminary assay to determine the reducing power of the ore constitutes one of the chief objections to the nitre method. The assayer at a modern smelting works has no time for preliminary assays. With experience, however, the preliminary assay may frequently be dispensed with.

A good practical assayer will, by a preliminary examination of the ore, or by panning, make an estimate of the amount of reducing agent (pyrites, etc.) present, and from this estimation he will generally, though not invariably, decide correctly the amount of nitre required.

METHOD B (Balling's method).—This method consists in adding sufficient nitre to effect the complete oxidation of the pyritic material present, and when oxidation is complete to add a mixture of litharge and charcoal to supply lead for collecting the precious metals.

By this method a quantity of nitre equal to twice the weight of the pyritic material in the ore is usually sufficient to effect complete oxidation, but in exceptional cases rather more than this amount may be required. Borax and sodium carbonate are added, as in the previous method, to flux the metallic oxides formed, but the red lead is omitted. The fusion is conducted as before, and when the action of the nitre is complete, and the contents of the crucible are in tranquil fusion, it is withdrawn from the fire and allowed to cool until the slag begins to thicken. A mixture of 40 grammes of red lead, 2 grammes of charcoal powder, and 10 grammes of borax is then added, the crucible returned to the furnace, and the charge kept in a molten condition for from ten to fifteen minutes. When all action has ceased, the contents of the crucible are poured into a mould in the ordinary way. By this means the globules of metallic lead obtained by the reduction of the red lead sink through the molten slag and collect the gold and silver.

Granulated lead is sometimes substituted for a mixture of lead oxide and charcoal, but unless the granules of lead are very small there is a great tendency for the result to be low, owing to the rapidity with which the lead sinks through the molten mass. Some assayers prefer to divide the above mixture into two parts, and to add the second portion after a short interval, but from the author's experience there appears to be no advantage in this procedure. Balling's method of complete oxidation requires a little more time than the method of partial oxidation, but the button of lead obtained by the former method is always of the required weight, and is usually comparatively free from impurities, and may generally be cupelled direct. Both methods, however, give equally satisfactory results when due precautions are taken.

IV. **Complex Ores.**—Ores containing large quantities of copper, antimony, zinc, etc., require to be assayed by special methods which are described in Chapter XIV.

(2) **Cupellation of the Lead Buttons.**—The precious metals, after concentration in lead by either scorification or crucible fusion, are separated by cupellation. The operation of cupellation has been described on page 145. Care should be taken to clean the lead buttons from all slag by brushing before cupelling. If the slag is left adhering to the button, it will interfere with the "uncovering" of the lead at the start of the cupellation. The temperature employed should be as low as is compatible with the satisfactory oxidation of the lead. Success in cupellation is also dependent on the quantity of air passing through the muffle, which must be carefully regulated. When the air-current is too rapid, the cupel is cooled, and crystals of litharge (feathers) formed, which may completely cover the metal and protect it from further oxidation, and cause "freezing." When the air-current is too feeble, the completion of the assay is unnecessarily delayed. The temperature should be raised towards the end of the operation to prevent solidification of the silver before the lead has been removed, but excessive heat should be avoided. A plan adopted by many assayers is to place the cupel in the front part of the muffle at the start and keep it there until cupellation is nearly complete, and then to transfer it to the back of the muffle until the "flash" takes place and the operation is finished. The cupel is then gradually moved towards the door in order to cool the silver slowly and prevent "spitting." When the buttons weigh less than 5 mg., spitting does not take place readily. In the case of very small buttons of silver, or silver and gold, it is necessary to watch the end of the cupellation carefully, and promptly to remove the cupels almost immediately after the

button has brightened, as a heavy loss of silver will take place if the buttons are left in the furnace. The buttons are detached from the cupel with a pair of flat-nose pliers and freed from adhering bone-ash by squeezing and brushing and then weighed. Buttons that are too small to be brushed may be cleaned by placing on the palm of the hand and rubbing with the finger. It is advisable to examine the buttons with a lens before weighing.

(3) **Determination of Silver absorbed by the Cupel.**—When great accuracy is required the amount of silver absorbed by the cupel is determined and allowed for in calculating the assay value of the ore. This is desirable for rich ores.

(4) **Determination of Silver in the Slag.**—The amount of silver retained in the slag is not inconsiderable in the case of rich silver ores, and it is frequently necessary to determine it, by the methods already given.

(5) **Assay of Metallics or "Scales."**—The "metallics" may be assayed by cupellation direct, or by scorification and cupellation. Before the metallics are assayed they must be weighed and their weight, and also the weight of the ore from which they have been obtained, carefully recorded. For cupellation direct the "metallics" should be small in quantity and free from siliceous material, etc.: they are rolled up in a weighed quantity of thin sheet lead and then cupelled.

For scorification the metallics are mixed with about 15 to 20 g. of lead, or more if necessary, with the addition of a little sodium carbonate, if silver chloride is present. The scorification is conducted in the usual way and the resulting lead button cupelled. The button of precious metal obtained is weighed, and the results allowed for in calculating the value of the ore.

Metallics vary considerably in their precious-metal content. Some may be very rich in metallic gold or silver, while others may be contaminated with native (metallic) copper, etc., and contain comparatively little precious metal.

As a general rule, if the metallics appear to be rich in native gold or silver, and the total weight is more than about 2 g., they should be divided into two or more smaller lots and each lot scorified and cupelled separately. The resulting precious-metal buttons are then weighed together.

(6) **Weighing the Silver Buttons.**—The methods of weighing are described on p. 58. The balance employed should be sensitive to $\frac{1}{20}$ milligramme or less. Silver buttons from ore assays are generally weighed by direct weighing, the button being placed in the left-hand pan and the weights in the right. Care should be taken to see that the balance is in the proper adjustment before commencing to weigh. The rider should be used for all weights less than 10 milligrammes. The difference between the weights of the silver buttons from duplicate assays of average silver ores should not, as a general rule, exceed 0·5 milligramme. The weight of each button should be recorded as soon as it is ascertained.

(7) **Testing Silver Buttons for Gold.**—Gold is not infrequently present in sensible quantity in argentiferous ores from California, South America, and other localities, and should always be looked for.

For this purpose one or more of the buttons or "prills" are flattened out by hammering and heated with dilute nitric acid in a test-tube or porcelain crucible; the solution is poured off and the gold, which is usually left in the form of a black or brown powder, is washed with distilled water, dried, heated to redness, and, when cold, weighed.

CHAPTER XIII

THE ASSAY OF GOLD ORES AND CALCULATION OF RESULTS (GOLD AND SILVER)

Introduction.—Minerals containing gold are assayed in precisely the same way as ores of silver, but as the former usually contain a very small proportion of the precious metal, it becomes necessary to operate on a larger quantity of material.

Gold ores rarely contain more than a few ounces, often only a few pennyweight of gold to the ton; consequently, the button of gold obtainable from such quantities of ore as may conveniently be employed in assaying is often so small as to require more than ordinary care in its manipulation.

For example: when assaying an ore containing 5 dwt. of gold per ton, if only 20 grammes were taken for the assay, as is usual with silver ores, the button of gold obtained would be so minute that it would not only demand considerable skill and care on the part of the assayer in handling it, but the weight of it would be barely perceptible even on a sensitive balance. In order, therefore, to lessen the strain on the assayer's attention and to bring the button of gold within the scope of the balance, it is usual in the case of gold ores to take from 50 to 100 or even 200 grammes of ore, the larger quantities being used for very poor material. A drawback to large ore charges is the large amount of fluxes required to flux them and the consequent employment of crucibles of large capacity which necessitate the use of some special form of fusion furnace. When dealing with low-grade ore it is preferable to make two or more separate fusions of 50 grammes each, than to attempt one fusion on a larger quantity of ore. If the fusion of gold ores is to be made in a muffle, smaller harges of 20 grammes must be used, and, if necessary, the lead buttons resulting from two or more fusions scorified together so as to concentrate the gold into one button of lead, which is then cupelled. In the author's opinion it is very desirable to employ at least 50 grammes for gold-ore assays, as this quantity is more likely to give results nearer to the average gold content of the parcel of ore from which it is taken, than the results obtained with small charges. It must be remembered also that the errors incident to the assay operations are probably multiplied by taking small quantities for assay.

The actual weight of ore taken for assay varies according to circumstances. If gramme weights are used, a charge of 50 grammes is usually taken for most gold ores; but if the gold content exceeds 5 oz. per ton, less may be taken. When the "assay ton" weights are used a suitable quantity of ore is 1 A.T. if the value in gold is from 0·5 oz. to 10 oz. or more per ton, and the balance for the final weighings is sensitive to 0·02 mg. or less[1]. With poor residues 2, 4, or even 10 A.T. may be taken, and with very rich ores ½ A.T. may suffice. If more than 3 A.T. of ore are taken, the charge is divided between two or more crucibles, and the resulting lead buttons scorified down and cupelled. The same remark applies to charges of more than 100 grammes. It may be

[1] Rose & Newman, *op. cit.*, p. 495.

well here to state that for the purpose of assay the author has considered a gold ore to be rich when assaying over 3 oz. of gold per ton, and poor when yielding less than 10 dwt. per ton. All ores yielding quantities of gold between these limits are considered to be average ores.

As the same methods of assay may generally be applied to both silver and gold ores, it follows that much of what has been said concerning the assay of silver ores applies also to gold.

Silver is invariably present in gold ores and remains alloyed with the gold obtained from the assay, and the main distinction between the methods of assaying ores of the two metals is that in the case of gold ores there is a further operation involved for the separation of the gold from the silver.

The ore is prepared for assay in the manner previously described, and should there be any metallic residue, it must be carefully collected and assayed separately. The ore must in all cases be reduced to a state of fine powder so as to pass through a sieve of at least 80-mesh. Complex ores should be crushed much finer. The remarks on fine crushing of silver ores apply equally to the case of gold ores. The ore is well mixed, and the portion taken for assay weighed accurately on the pulp balance to 0·01 gramme, unless very poor, as in the case of "tailings," when less exact weighing is sufficient.

With regard to the relationship between the pulp balance and weights and the assay balance and its weights, Whitby [1] warns assayers not to place too great reliance on fluxing scales with steel-knife edges and planes, as he has found the sensibility to diminish rapidly from 10 mg. on a 1000-gramme load to 100 mg. on much smaller loads, even when the balance has a case. In regard to the weights, it is always necessary to check them from time to time on a chemical balance. He gives the following table (XXVII) of limitations in reporting assays :—

TABLE XXVII.—Weight Limits in Reporting Assays.
(*Whitby.*)

Charge.	Sensibility of Balance.	
	$\frac{1}{20}$ milligramme.	$\frac{1}{200}$ milligramme.
0·5 assay ton, report to ...	2·0 dwt.	0·20 dwt.
1·0 ,, ,, ...	1·0 ,,	0·10 ,,
2·0 ,, ,, ...	0·5 ,,	0·05 ,,
2·5 ,, ,, ...	0·4 ,,	0·04 ,,
5·0 ,, ,, ...	0·2 ,,	0·02 ,,
10·0 ,, ,, ...	0·1 ,,	0·01 ,,

METHODS OF ASSAY FOR GOLD ORES

The assay of gold ores comprises the following operations :—

(1) Concentration of the gold in a button of lead by fusion in a crucible, or, more rarely, by scorification.
(2) Cupellation of the auriferous lead.
(3) Weighing the gold–silver button.
(4) "Inquartation" and parting to separate the gold.
(5) Weighing the gold.
(6) Assay of metallics.

[1] *J. Chem. Min. and Met. Soc. of S. Africa*, 1906, p. 271.

(1) **Concentration of the Gold in Lead.**—The fusion or crucible method of assay is almost invariably adopted for gold ores. The scorification method is employed occasionally for rich ores, but it is not generally applicable to ordinary gold ores owing to the large quantity of ore required for assay.

The fusion of the ore is usually effected in a wind furnace, a suitable crucible for the purpose being a G Battersea round. The fusion is conducted as previously described. The button of lead should weigh from 25 to 30 g., and should be soft and malleable. If it is hard or brittle it is usually desirable to submit it to scorification before attempting to cupel.

The ore to be assayed may belong to one of the several distinct classes, such as—

I. Ores with siliceous gangue.
II. Ores with basic gangue.
III. Pyritic ores.
IV. Complex ores.

The following are typical charges for the several classes of ore :—

(I) **Ores with Siliceous Gangue.**—Quartz, which constitutes the gangue of most gold ores, is chiefly fluxed with sodium carbonate. About 1½ parts of the dry carbonate will yield a fluid slag with 1 part of quartz, but to make the slag more fluid it is the practice of many assayers to add a small amount of litharge in excess of that required to supply the necessary weight of lead, so as to form lead silicate, which is more fluid when fused than sodium silicate.

A typical charge for quartzose ores would be—

Ore	50	grammes.
Sodium carbonate	70	,,
Litharge or red lead	40–50	,,
Charcoal	1·5	,,

Borax should be substituted for a part of the sodium carbonate for ores containing small quantities of iron oxide or other basic material.

Iron should be added when fusing ores containing pyrites in small amount, but in this case the charge should not contain any excess of litharge, as the iron will reduce any litharge not reduced by the carbon or other reducing agent employed.

(II) **Ores with Basic Gangue.**—The most important basic material to be fluxed in gold ores is iron oxide.

The matrix of surface ores is largely made up of this oxide, and when ores composed mainly of pyrites are roasted, the product will consist of ferric oxide. To produce a fusible slag it is necessary to reduce the iron to the lower oxide (ferrous oxide, FeO), and this is best effected by the addition of a suitable quantity of reducing agent to the charge. Ferrous oxide forms a fusible slag with silica, and when the ore is deficient in silica some must be added.

Sodium carbonate and borax are added, as the addition of alkaline fluxes results in the formation of double silicates, which greatly increases the fluidity of the slag. When suitable proportions of the different fluxes are used no difficulty is experienced in obtaining good slags even when the ores contain large quantities of ferric oxide. If, however, the slag contains any considerable percentage of ferric oxide, owing to insufficient carbon to reduce it to ferrous oxide, it may retain an appreciable amount of gold, although the slag may be perfectly fluid.

To get good results with ores consisting mainly of ferric oxide it is well to add sufficient charcoal to give a larger button of lead than usual, say 40 grammes.

For ores ranging from moderately basic (with about 30 per cent. of ferric oxide) to very basic (practically all ferric oxide), the following proportions may be taken :—

	Moderately Basic.	Very Basic.
Ore	50 grammes.	50 grammes.
Litharge	40 ,,	50 ,,
Charcoal powder	2·5 ,,	4 to 6 ,,
Borax	20 ,,	30 ,,
Sodium carbonate	30 ,,	10 to 15 ,,
Sand or powdered glass ...	Nil.	10 to 15 ,,

For ores containing silicates of alumina, lime, and magnesia the proportion of alkaline flux should be increased, since these silicates are practically infusible. A mixture of 40 grammes of borax and 20 grammes of sodium carbonate usually gives satisfactory results, but the proportion of the two fluxes must be altered to suit requirements. Manganese dioxide (pyrolucite) must be reduced to the lower state of oxidation as in the case of iron. It requires about twice the amount of charcoal for its reduction to the lower oxide as compared with iron oxide, and due allowance must be made for this fact when large proportions of manganese dioxide are present. Should the ores contain small quantities of the oxides of copper, antimony, etc., the quantity of litharge should be slightly increased, the amount required being usually from 60 to 70 grammes for 50 grammes of ore.

Whitby [1] gives the following charge for an ore containing 50 per cent. calcite, siliceous matter, and a little pyrites :—

Ore	1 assay ton or	50 grammes.
Litharge	$1\frac{1}{4}$,,	62·5 ,,
Sodium carbonate	$\frac{1}{3}$,,	17·0 ,,
Charcoal		2 ,,

This charge gives a lead button weighing about 20 grammes.

(III) **Pyritic Ores.**—The sulphides chiefly associated with gold ores are iron pyrites and arsenical pyrites (mispickel).

Ores containing not more than about 25 to 30 per cent. of pyrites may be satisfactorily assayed by the method in which metallic iron is used as a desulphurizing agent, the iron combining with the sulphur to form ferrous sulphide, which passes into the slag. When the percentage of pyrites is much above this, the lead button resulting from the fusion is very liable to be sulphury, and to be accompanied by a matte.

Ores containing a large proportion of metallic sulphides should be roasted. The portion of ore to be used for the assay is weighed, and then roasted in a large roasting-dish or in a crucible, as already described.

Many assayers hesitate to roast gold ores, on the assumption that precious metal is lost during the operation. Although this may be true with certain classes of ore, it may be remarked that the loss of gold sometimes experienced is, in many cases, probably mechanical, due to want of care in conducting the operation. A considerable loss may take place if the ore is stirred too vigorously during the operation. Another very possible source of loss is the retention of gold in the slag, due to the presence of ferric oxide which has not been reduced owing to insufficient reducing agent in the charge when fusing the roasted ore.

It is true that the oxidation of an ore by roasting takes up a considerable amount of time and requires careful attention to get satisfactory results, and

[1] A. Whitby, *J. Chem. Met. and Min. Soc. of S. Africa*, 1906, p. 269.

the tendency of modern assay practice is to make use of other methods of oxidation so far as possible. But although the methods of oxidation described under "Silver Ores" are somewhat more expeditious than oxidation by roasting, their application is limited in the case of gold ores, owing to the large quantity of material that has usually to be taken for assay. As pointed out above, the use of iron is limited to ores containing only moderate amounts of pyritic material, while the large quantities of nitre that would have to be used if this method of oxidation were employed, make its use impracticable in the case of gold ores, except in special cases mentioned later.

However, to save time, many assayers use the minimum quantity of ore permissible for the assay, and make the fusion with metallic iron or with nitre; but with gold ores the results are generally lower than those obtained by roasting, although the risk of precious metal being retained in the slag, owing to the presence of dissolved sulphides, is less with gold ores than with silver ores. Of the two methods, the use of iron is the more satisfactory; but if an ore gives an appreciable amount of matte, or speise, it is best to repeat the assay on a smaller quantity of ore, or to roast before fusion (see "General Remarks on Pyritic Ores," page 199). An objection to the free use of iron is that it tends to alloy with gold and give low results. After roasting, the ore consists mainly of ferric oxide, and is fluxed accordingly on the lines indicated for basic ores, especial attention being given to the amount of reducing agent added to the charge for the fusion. An experienced assayer will frequently adjust the proportion of reducing agent correctly from the colour of the roasted ore.

(IV) **Complex Ores.**—Ores containing large quantities of base metals require to be assayed by special methods which are described in Chapter XIV.

(2) **Cupellation of the Auriferous Lead.**—The cupellation of the lead buttons from gold-ore assays differs very little from that of lead obtained from silver assays. Gold is invariably accompanied by more or less silver, and when the proportion of silver exceeds that of the gold, and both metals have to be determined, the cupellation must be conducted exactly as in the silver assay, the temperature being moderated to lessen the cupellation loss and the button cooled slowly to avoid spitting. When the gold predominates, the temperature should be higher at the finish than in the case of silver cupellation, as the melting-point of gold is 100°C. higher than that of silver. Very small gold buttons should be withdrawn from the muffle as soon as cupellation is finished. The characteristic glow which takes place with gold just after solidifying is a valuable help in finding the position of very minute buttons. When assaying material from which minute gold buttons may be expected, the cupellation should be conducted on a fine-faced cupel. Gold buttons are more globular than silver buttons of the same size, and adhere less to the cupel. The colour of the button left on the cupel affords a useful indication of the proportion of silver present. If the button is yellow, it contains less than half its weight of silver. With more than 50 per cent. of silver the button is white.

The button is removed from the cupel with a pair of pliers, squeezed to loosen the bone-ash adhering to its lower surface, and then cleaned by brushing ready for weighing.

As previously pointed out, the losses of gold and silver during cupellation are mainly due to absorption by the cupel. They amount to about 1 per cent. of the gold and 10 per cent. of the silver in gold ores of moderate value containing small quantities of silver, if the lead button is free from impurities, such as copper and tellurium, when charged into the cupel.[1]

[1] T. K. Rose, *J. Chem. Met. and Min. Soc. of South Africa*, Jan., 1905.

(3) Weighing the Button of Gold and Silver.—The button is weighed on a good assay balance and the weight recorded for the purpose of calculating the amount of silver contained in the ore. The silver in the button is subsequently removed by solution in nitric acid, and the weight of the residual gold taken when the difference between the two weighings represents the silver. When it is not required to determine the amount of silver that accompanies the gold, the weighing of the button of gold and silver may be omitted.

(4) Inquartation and Parting.—The button is treated with nitric acid to separate or part the gold from the silver it contains. Nitric acid will remove silver from alloys of gold and silver without attacking the gold when these metals are present in certain proportions. If the silver is deficient more must be added, otherwise some of the silver will remain undissolved.

This operation is called "inquartation," the word coming from the older assayers, who held the view that the alloy for parting should consist of one-quarter gold to three-quarters silver. As pointed out in Chapter XVIII on bullion assaying, the modern practice is to use the ratio of 2 or 2½ of silver to 1 of gold when parting large buttons of gold such as are obtained in bullion assays, but in parting the small buttons from ore assays it is considered necessary to have at least four times as much silver as gold present. When the amount of gold is very small, it is convenient to use a large proportion of silver in parting. Sir T. K. Rose[1] recommends varying the proportion of silver according to the weight of the gold, and gives the following as the most suitable proportions of silver for different weights of gold :—

TABLE XXVIII.—Proportion of Silver to Gold for Parting Small Gold Buttons (*Rose*).

Weight of Gold.	Weight of Silver.	Ratio of Silver to Gold.
Less than 0·1 milligramme.	2 to 3 milligrammes.	20 or 30 to 1
About 0·2 ,,	2 ,,	10 to 1
,, 1·0 ,,	6 ,,	6 to 1
,, 10·0 ,,	40 ,,	4 to 1
More than 200·0 ,,	450 ,,	2¼ to 1

The ratio of silver to gold in the buttons obtained from ordinary gold ore assays is very variable, so that in practice the proportion of silver is not always under control. If, however, it is less than 4 to 1, silver must be added to bring it up to this ratio at least.

The proportion of silver in the button left on the cupel is estimated from its colour. A pale-yellow button always contains more than 60 per cent. of gold and requires the addition of silver. If the button appears white it is usually assumed that it already contains at least a sufficiency of silver, in the absence of any knowledge to the contrary. But a perfectly white button may not part completely, and if on parting it does not lose at least two-thirds of its weight, a correct proportion of silver must be added and the button again parted. Very frequently, however, the silver in the buttons will be in excess, and on parting, the residual gold will tend to break up and demand very careful treatment to prevent loss. With a little practice the proportion of silver in a gold button may be estimated very closely, and in judging of the amount no ordinary error will very seriously affect the result. When the addition of silver is necessary,

[1] *J. Chem. Met. and Min. Soc. of S. A.*, 1905, **5**, 165–168; and *Precious Metals*, p. 204.

the button with the additional silver is wrapped in as small a piece of lead-foil as possible, and cupelled, and the resulting button parted. Sometimes the addition of silver is effected in the case of small buttons by fusion on a piece of charcoal or on a cupel, by means of the blowpipe; but this method, although expeditious, is not to be recommended, as it frequently occasions loss.

If the addition of silver is known by previous assays of the ore to be necessary, it may be added directly to the charge in the crucible or to the lead during the first cupellation. This course cannot, however, be adopted when it is essential to determine the silver in the ore.

After "inquartation" the button of gold and silver is cleaned, flattened by a hammer on a clean anvil, and "parted" by dropping into nitric acid.

The parting is effected in a test-tube or a glazed porcelain crucible. The strength of the nitric acid used for parting is of importance: authorities differ, however, as to the best strength of acid to employ for small gold buttons. Keller [1] recommends dilute acid consisting of 1 part of acid to 9 parts of water, and states that in this strength of acid the gold almost invariably remains in a coherent mass, even when the silver is 500 times as much as the gold. This strength of acid is recommended also by Fulton [2] for ordinary assay purposes. The buttons should be boiled in the acid for at least ten to fifteen minutes in order to ensure parting.

According to A. M'A. Johnston,[3] dilute nitric acid (1 of acid : 3 of water) is used on the Rand, South Africa, the flattened button being dropped in when the acid is slightly warm. Action starts at once, and unless the silver is in large excess the gold does not break up. The acid is then brought quickly to boiling-point.

Nitric acid of specific gravity 1·25 (4 of acid : 3 of water) is considered by Rose as suitable for use with all buttons, and he expresses the opinion that there is no advantage in using more dilute acids for poorer alloys.

The acid employed should be free from chlorine, otherwise the dissolved silver will be precipitated and contaminate the gold. The freedom of the nitric acid from chlorine is ensured by adding to it a little spent acid containing some silver nitrate.

In parting the buttons, the acid must in all cases be heated previously to boiling, as the gold does not in that case break up into such fine particles as when cold acid is used. The acid attacks the button instantly and violently, turning it black and giving off nitrous fumes. Little buttons containing only a small proportion of gold are dissolved in a few seconds, and decanting of the acid from the residual gold may be at once proceeded with. If the amount of gold is large, the boiling is continued for some minutes. Ten minutes boiling is enough for 10 mg. of gold and 40 mg. of silver. The acid is then poured off, the residue washed at least twice with hot distilled water by decantation, and if the button is very large, fresh acid of specific gravity 1·30 (2 of acid : 1 of water) is added. The boiling is then continued for five or ten minutes longer, when practically all the silver will have been dissolved.

With the buttons from almost all gold ores no second treatment with acid is required.

The gold usually remains as a single piece if it weighs less than 0·1 mg., even if the proportion is only 1 part of gold to 40 or 50 parts of silver.[4] If the ore is rich, the gold sometimes, though rarely, breaks up if hot acid is used,

[1] Keller, *Trans. Amer. Inst. Min. Eng.*, 1908, **37**, 3.
[2] Fulton & Sharwood, *Fire Assaying.*
[3] *Rand Metallurgical Practice*, "Assaying," 1926, vol. I, p. 305.
[4] Rose & Newman, *Metallurgy of Gold*, 7th edn., p. 502.

and invariably breaks up if the parting is begun with cold acid. In the event of the gold breaking up, the washings must be very carefully decanted, otherwise the finer particles may float and be lost. The method of dealing with the residual gold varies slightly according to whether a test-tube or a porcelain crucible has been used for parting.

(*a*) *Parting in Test-Tubes.*—If a test-tube is used for the boiling, 3 or 4 c.c. of acid is usually quite enough. The test-tubes may be heated by supporting them on a piece of asbestos card suitably heated, or by merely resting them in a beaker of boiling water. When the parting is complete and the acid decanted, the parted gold is washed with hot distilled water as stated, and transferred to an unglazed porcelain crucible or "annealing" cup (see Fig. 144). To effect this the test-tube is completely filled with water and the annealing cup inverted over its mouth. The test-tube and cup are then quickly inverted together so that the tube stands mouth downwards in the cup, and the gold falls steadily through the water into the cup. When all the gold has settled, the test-tube is removed in such a way as not to disturb the precious metal. This is best done by lowering the cup, with the test-tube still resting in it, into a porcelain basin filled with water. As soon as the neck of the test-tube is immersed, the cup can safely be drawn away from under it and then lifted out of the water. The test tube should not be taken away first, otherwise the gold may be washed out of the cup by the rapid flow of water which takes place when the test-tube empties.

The water in the cup is poured off and the cup dried at a gentle heat and then gradually raised to dull redness. If the cup is dried too quickly the last drops of water may boil and cause loss of gold by ejection.

The final heating or annealing may be performed in a muffle or over a bunsen or spirit flame.

At a dull-red heat the black gold from parting assumes its ordinary yellow colour, shrinks, and hardens so that it can be transferred to the pan of a balance and weighed. In some smelting works the parting is done in flasks similar to those used for bullion assays, but of smaller size.

(*b*) *Parting in Glazed Crucibles.*—Parting may also be effected in small glazed porcelain crucibles or cups which are used for both boiling and annealing. This method of working, which is largely adopted in mine-assay offices, has the advantage that there is no transference of the gold while in its soft state. On the other hand, the boiling of the acid in a small porcelain crucible requires more care in adjusting the temperature, otherwise there may be loss by ejection. This danger is avoided by using watch-glass covers for the crucibles. Small glazed porcelain crucibles, 1 inch in diameter across the top and $\frac{3}{4}$ inch in depth, are very suitable for the purpose.

The crucible, half filled with dilute nitric acid, is heated on a piece of asbestos sheet, on a hot-plate, until the acid is nearly boiling. The flattened button is dropped into it and the heating continued with, at most, gentle boiling until parting is complete, which even with large buttons takes place in from five to ten minutes. The crucible is then filled with distilled water, which cools it sufficiently for easy handling; and when the gold has settled, the liquor is poured off. The gold is washed in three successive decantations by means of a wash-bottle; and when the last drops of water have been carefully drained off, the crucible is dried and then heated below redness to anneal the gold, precaution being taken that bright metallic gold is obtained. Excessive heating must be avoided, as the glaze on the porcelain crucible softens at a bright-red heat and the gold sticks to it. When cold, the gold is transferred directly to the pan of the balance and weighed.

If the crucible shows a black stain after heating for the annealing of the gold it indicates insufficient washing. The stain is due to metallic silver resulting from the reduction of silver nitrate by the heat.

When a large number of assays have to be dealt with, it is sometimes convenient to number the crucibles with rouge (iron oxide) made into a thin paste with water.

A special apparatus for parting small buttons from ore assay is described on page 273.

The amount of silver retained by the gold when parted in crucibles is stated by J. Phelps to be about 0·5 per cent., an error which is inappreciable in the assay of ores.

(5) **Weighing the Gold.**—The fine gold from the parting should be weighed on a balance turning to $\frac{1}{200}$ milligramme at most, with a half milligramme rider for use on the beam. Direct weighing is usually preferred, but the gold can also be weighed by the method of substitution, by which the bias of the balance is eliminated. The principle of this method, as described on page 59, is to counterpoise the gold, then to remove it from the pan, and compensate for the loss of weight by a rider. Although requiring a little more time, this method is to be recommended when extreme accuracy is desired.

The method of " weighing by swings " is frequently employed to determine the weight of very minute buttons of gold.

It is the practice of some assayers, when dealing with very minute buttons, resulting from the assay of a sample in duplicate, first to weigh approximately the two buttons separately to see that there is no undue discrepancy between the different weights, and then to weigh them together accurately. This, by the law of averages, gives a result of greater accuracy.

(6) **Assay of Metallics.**—This should be made as described for silver ores. For calculation of precious-metal content, see below,

General Remarks on the Assay of Pyritic Ores.—There is considerable diversity of opinion between assayers as to the best method of assaying pyritic gold and silver ores.

With a view to determining the best method to be adopted for ores of this class, experiments have been made from time to time, and the conclusions of some of the more recent investigators may be summarized as follows :—

Experiments were carried out by Keys and Riddell[1] upon a rich silver ore in order to determine the effects of each of the fluxes ordinarily used in the crucible assay upon the silver yield, and to show what type of charge gives the best results on an ore which is high in copper and sulphur and at the same time a typical roasting ore. They concluded :

(1) That a dead roast is not so accurate as a run with a large excess of nitre.

(2) An excess of borax causes low value determinations, especially in the presence of sulphur, copper, and iron.

(3) An excessively acid flux fuses with great difficulty, and it is hard to handle the buttons, and when the ores contain matte-forming materials the value determinations are low.

(4) A very basic flux is open to the same objections. In ores containing copper the loss is not so great, but in zinc ores the results are high. Neither method is practical.

(5) Ores high in copper cannot be run by any crucible method without scorifying the lead buttons.

[1] *Bulletin, Amer. Inst. Min. Eng.*, 1911, 559–568.

(6) Direct scorification of copper-bearing ores does not give as high values as a combination of the crucible and scorification methods.

(7) For ores containing much copper or zinc, a large excess of litharge in the charge greatly improves the buttons, decreasing the time and temperature of fusion with very close actual valuations for the silver content of the ore; and the method is quick, easy, and accurate.

Borrowman[1] gives details of the assay of an ore consisting of almost pure iron sulphide, using the following desulphurizing agents: (1) Nitre alone, (2) preliminary roasting, (3) iron nails, (4) nitre and iron nails. The full charges are given in all cases. It is concluded that with such an ore the maximum amount of nitre can be used with impunity, and that where great accuracy is desired the more troublesome nitre method should be used in preference to the convenient "nail" fusion.

For the assay of sulphides consisting mainly of iron pyrites, with small amounts of copper (up to 3·5 per cent.), zinc sulphide (4 to 8 per cent.), etc., and from 4 to 20 per cent. silica (insoluble), Fulton[2] observes that, "where gold only has to be determined in ores of this character, the roasting method is satisfactory." Roasting, however, proves unreliable for silver, and in many cases (as at Leadville) the silver contents of these sulphides are the most important.

The best method, after many trials, was found to be the nitre fusion on comparatively small lots of ore, the details being as follows:—

Four assays are made on 0·25 assay ton, each with 3 or 4 assay tons of the following stock flux, the amount depending on the reducing power of the ore, *i.e.* on the amount of sulphide present.

Stock flux.		Proportionate amount in 4 A.T. of flux.	
Litharge ...	8 parts.	Ore	0·25 assay ton.
Nitre	1·5 ,,	Litharge ...	62 grammes.
Sodium carbonate	3·0 ,,	Nitre	12 ,,
Borax glass ...	1·5 ,,	Sodium carbonate	24 ,,
Sand	1·5 ,,	Borax glass ...	11 ,,
		Sand	11 ,,

Cover of salt or soda.

The temperature of fusion is brought up gradually to a yellow heat. The buttons are usually clean, and separate well from the slag.

Another method which may be used on this type of ore is the nitre-iron method. This has the advantage that no preliminary assay is necessary to determine the amount of nitre for the proper size lead button, but that only sufficient nitre is added to oxidize the sulphide partially, the iron nails being relied upon to decompose the balance of the ore. On ores of the class described the following charge is successful:—

Ore	0·5 assay ton.
Sodium carbonate	25 grammes.
Litharge	30 ,,
Nitre	15 ,,
Sand	8 ,,
Borax glass	8 ,,
Iron nails	2 or 3 ,,

Thin cover of borax glass.

The nitre and sand are decreased with a decrease in the reducing power of the ores.

[1] *J. Ind. and Eng. Chem.*, 1910, **2**, 251–252.

[2] Fulton & Sharwood, *Manual of Fire Assaying.* See also Hall and Drury, *Eng. and Min. J.*, 1913, **96**, 1125.

Calculation of Assay Results.—The amount of metal contained in ores of the common metals is usually reported in percentage or parts in a hundred, but with gold and silver ores the proportion of precious metal is so small that it is necessary to state the amount present in ounces or pennyweight per ton. Several methods of reporting the assay value of precious-metal ores are in use, and are described below. The calculation of the assay value of the ore from the weight of metal obtained from the assays is a matter of simple arithmetic, but since gold and silver are sold by troy weight, whilst the ton is avoirdupois, it is of importance to remember that the ounces in the two systems are not the same. There are 7000 grains in 1 lb. avoirdupois and 2240 lbs. in an English ton; therefore there are $2240 \times 7000 = 15{,}680{,}000$ grains in a ton.

There are 480 grains in 1 ounce troy, consequently there are $\frac{15{,}680{,}000}{480} = 32{,}666{\cdot}6$ troy ounces in 1 ton avoirdupois. With this data the full calculation can be made if necessary. Thus, suppose 20 grammes of ore gave a button of silver weighing 0·0795 grammes (after deducting silver in litharge, etc.), then :—

Silver per ton (2240 lbs.) $\frac{2240 \times 7000 \times 0{\cdot}0795}{20} = \frac{1{,}246{,}560{\cdot}0000}{20} = 62{,}328$ grains.

$$\frac{62{,}328}{480} = 129{\cdot}85 \text{ ounces.}$$

In practice it would be too fatiguing to calculate each assay result fully, and it is usual to refer to a set of tables which give, either directly or by means of simple addition, the produce per ton.

Table XXIX shows figures for computing from the percentage of gold or silver the troy weight of precious metal either per ton of 2240 lbs. (Long ton) or per ton of 2000 lbs. (Short ton).

The percentage of gold or silver in the ore is calculated from the weight of ore taken for assay and the weight of the resulting button of precious metal, as follows :—

$$\frac{\text{Weight of gold or silver button}}{\text{Weight of ore taken for assay}} \times 100 = \text{Percentage of metal in the ore.}$$

The method of using the table to ascertain the quantity of precious metal per ton of ore is best shown by an example.

Calculation of Silver-Ore Results.—Taking the simplest case of a silver ore assayed by the fusion method.

Suppose a button of silver weighing 0·2135 gramme was obtained from 20 grammes of ore, then,

$$\frac{0{\cdot}2135}{20} \times 100 = 1{\cdot}0675 \text{ per cent. silver.}$$

From the table select the values corresponding to each figure of the weight, thus :

1·0	per cent. =	326·67	ounces per ton (2240 lbs.).	
·06	,, =	19·60	,,	,,
·007	,, =	2·29	,,	,,
·0005	,, =	·16	,,	,,
Total, 1·0675		348·72		

Add these together, and the assay value of the ore is 348·72 oz. per ton (2240 lbs.). If the result is to be returned on the ton of 2000, the figures would of course be selected from the short-ton column accordingly.

The granulated lead, red lead, and litharge used for assay purposes invariably contain more or less silver, and it is consequently necessary to determine beforehand the amount of silver present, and make a corresponding deduction from the weight of the button resulting from the assay of the ore.

TABLE XXIX.—TABLE FOR COMPUTING FROM THE PERCENTAGE OF GOLD OR SILVER THE ASSAY VALUE IN OUNCES AND DECIMALS OF AN OUNCE PER TON.

Per cent.	Per ton of 2240 lbs.		Per ton of 2000 lbs.	
	Ounces.	Equivalent in pennyweight.	Ounces.	Equivalent in pennyweight.
0·0001	0·033	0·65	0·029	0·58
·0002	·065	1·31	·058	1·17
·0003	·098	1·96	·087	1·75
·0004	·131	2·61	·117	2·33
·0005	·163	3·27	·146	2·92
·0006	·196	3·92	·175	3·50
·0007	·229	4·57	·204	4·08
·0008	·261	5·23	·233	4·67
·0009	·294	5·88	·263	5·25
0·001	·327	6·53	·292	5·83
·002	·653	13·07	·583	11·67
·003	·980	19·60	·875	17·50
·004	1·307	26·13	1·167	23·33
·005	1·633	...	1·458	29·17
·006	1·960	...	1·750	...
·007	2·287	...	2·042	...
·008	2·613	...	2·333	...
·009	2·940	...	2·625	...
0·01	3·267	...	2·917	...
·02	6·533	...	5·833	...
·03	9·800	...	8·750	...
·04	13·067	...	11·667	...
·05	16·333	...	14·583	...
·06	19·600	...	17·500	...
·07	22·867	...	20·417	...
·08	26·133	...	23·333	...
·09	29·400	...	26·250	...
0·1	32·667	...	29·167	...
·2	65·333	...	58·333	...
·3	98·000	...	87·500	...
·4	130·667	...	116·667	...
·5	163·333	...	145·833	...
·6	196·000	...	175·000	...
·7	228·667	...	204·167	...
·8	261·333	...	233·333	...
·9	294·000	...	262·500	...
1·0	326·667	...	291·667	...

Also, when the slag and cupels have been separately treated for the recovery of the precious metal they contain, the results of these assays must be allowed for in determining the assay value of the ore.

Thus, using the same example for illustration, we should get the following corrected result after making the above allowances.

"*Corrected*" *Assay Result.*—Ore taken for assay, 20 grammes.

Silver obtained by crucible assay	0·2135	gramme.
Silver recovered from slag	0·0052	,,
Silver recovered from cupel	0·0093	,,
Total	0·2280	,,
Deduct silver in red lead	0·0016	,,
Corrected result for silver in 20 grammes of ore	0·2264 = 1·1320 per cent.	

From the table we find 1·1320 per cent. = 369·79 ounces per ton (2240 lbs.).

Calculation of Silver in Ore with Metallics.—When an ore contains metallics or scales, as is frequently the case with precious-metal ores, the calculation becomes more complicated. The metallics have to be collected and assayed separately, and their precious-metal content calculated independently of the result obtained by the ordinary assay of the ore.

For the purpose of calculation it is necessary to take separately the weight of the metallics and the weight of the ore from which they are obtained. The following example will serve as an illustration :—

Total weight of sample	350·0	grammes.
The metallics weigh	1·25	,,
Hence the sifted portion weighs	348·75	,,

Twenty grammes of the sifted portion, when assayed, gave 0·1315 gramme of silver, equivalent to 0·6575 per cent. The whole of the metallics, when scorified and cupelled, gave 0·831 gramme of silver, equivalent to 66·48 per cent. The total silver content of the ore is calculated as follows :—

Calculate separately the content (best stated in per cents.) of precious metal in each portion of the sample. Multiply each content by the weight of the portion it represents. Add together the resulting products, and divide by the weight of the whole sample.

This may be stated thus :

Let A = weight of the sifted portion.
B = weight of the metallics.
C = assay value of the sifted portion in per cents.
D = assay value of the metallics in per cents.

Then,

$$\frac{A \times C + B \times D}{A + B} = \text{total silver (or gold) content of the ore in per cents.}$$

Taking the above example for illustration we have,

$$\frac{348{\cdot}75 \times 0{\cdot}6575 + 1{\cdot}25 \times 66{\cdot}48}{348{\cdot}75 + 1{\cdot}25} = \frac{229{\cdot}303125 + 83{\cdot}10}{350}$$

$$= \frac{312{\cdot}403125}{350} = 0{\cdot}8923 \text{ per cent.}$$

Referring to the table we find

0·8	per cent.	=	261·33	ounces.
·09	,,	=	29·400	,,
·002	,,	=	·653	,,
·0003	,,	=	·098	,,
0·8923			291·481	

The total silver content of the ore is therefore 291·5 ounces per ton (2240 lbs.).

Calculation of Gold-Ore Results.—In calculating the assay value of a gold ore the gold content and the silver content are calculated separately. The weight of the fine gold from parting is subtracted from the weight of the gold-silver button from the cupellation, and the difference gives the weight of the silver.

Example.—Weight of ore taken, 50 grammes.

Weight of gold-silver button	0·0177	gramme.
Weight of gold from parting	0·0105	,,
Difference equals silver	0·0072	,,
Deduct silver in litharge	0·0015	,,
Silver in ore	0·0057	,,

Gold in 50 grammes ore 0·0105 = 0·0210 per cent. gold.

Silver in 50 ,, ,, 0·0057 = 0·0114 ,, silver.

From the table we ascertain that

0·0210 per cent. = 6·86 ounces gold per ton 2240 lbs.

0·0114 ,, = 3·73 ,, silver ,, ,,

In the case of a gold ore containing metallics, separate assays of the ore and the metallics are made, and the resulting buttons of gold and silver parted in the usual way.

The total yield of fine gold and of fine silver is then calculated independently, as previously described.

Calculation for Assay-Ton System.—The calculation of the assay value of an ore is very much simplified when the quantity taken for assay has been weighed in assay tons (A.T.). As previously explained, the assay-ton system is so devised that if 1 assay-ton of ore yields a button of gold or silver weighing 1 milligramme, then 1 ton of ore yields 1 ounce troy of precious metal per ton of ore. When quantities either larger or smaller than 1 assay ton are taken for assay, the weight of the button of precious metal in milligrammes, divided by the assay-ton weight of the ore taken, will give the number of ounces, per ton, thus :—

$$\frac{\text{Weight of button in milligrammes}}{\text{Weight of ore taken in assay ton}} = \text{Number of ounces per ton.}$$

The following is an example :—

One-tenth (0·1) assay ton of silver ore gave by scorification 0·0413 gramme of silver. Then,

Silver from scorification assay	... ...	0·0413 gramme.
Silver from slag assay ...	... ...	0·0021 ,,
Silver from cupel assay	... ...	0·0020 ,,
		0·0454 ,,
Deduct silver in granulated lead used	...	0·0016 ,,
Corrected weight of silver in $\frac{1}{10}$ A.T. ore	...	0·0438 ,,

·0438 gramme = 43·8 milligrammes.

$$\frac{43{\cdot}8 \text{ mgms.}}{\frac{1}{10} \text{ A.T.}} = 43{\cdot}8 \times 10 = 438 \text{ ounces of silver per ton.}$$

The assay value of a gold ore is calculated in the same way, a separate calculation being made for the fine gold and for the fine silver.

For example :—Ore taken, 2 A.T.

Weight of gold-silver button ...	·0076 gramme	= 7·6 milligrammes.
Weight of fine gold	·0042 ,,	= 4·2 ,,
Weight of fine silver	·0034 ,,	= 3·4 ,,

$$\text{Gold per ton } \frac{4{\cdot}2 \text{ mgms.}}{2 \text{ A.T.}} = 2{\cdot}1 \text{ ounces per ton.}$$

$$\text{Silver per ton } \frac{3{\cdot}4 \text{ mgms.}}{2 \text{ A.T.}} = 1{\cdot}7 \text{ ounces per ton.}$$

The corrections for silver in litharge, etc., have been omitted to simplify the calculations.

Calculation for Ores with Metallics (Assay-Ton System).—The pulverized ore and the metallics are separately assayed, and from the results obtained the total weight (in grammes) of silver (or gold) in the entire sample is calculated. The weight thus obtained, multiplied by 29166·66 and divided by the original weight of ore taken in grammes, gives the value in ounces per ton of 2000 lbs.

The following example will illustrate the method of calculation :—

Let us suppose that a sample of auriferous silver ore, on being pulverized and passed through a 90-mesh sieve, gave the following weights :—

Sifted ore	... ...	142·56 grammes.
Metallics or scales	... ...	1·14 ,,
		143·70 ,,

Each portion when separately assayed gave the following results :—

Sifted Ore, by crucible assay.

One-half assay ton (14·5833 grammes) yielded :—

Silver and gold together	... ...	0·1085 gramme.
Gold by parting	... ...	0·0021 ,,
Silver by difference	... ...	0·1064 ,,
Silver in the litharge	... ...	0·0006 ,,
Silver in the ore	... ...	0·1058 ,,

Metallics, by scorification assay.

The whole of the metallics (1·14 grammes) yielded :—

Silver and gold together	... ...	0·8326 gramme.
Gold by parting	... ...	0·0073 ,,
Silver by difference	... ...	0·8253 ,,
Silver in test lead	... ...	none.

From these results the value of the ore in silver and in gold is calculated separately.

Calculation of the Silver

(*a*) *In Sifted Ore.*—If one-half assay ton yields 0·1058 gramme of silver, then the number of grammes of silver in 142·56 grammes is

$$142{\cdot}56 \times \frac{0{\cdot}1058}{14{\cdot}5833} = 1{\cdot}034 \text{ grammes.}$$

(*b*) *In the Metallics.*—If 1·14 gramme yields 0·8253 gramme, then (as the whole of the metallics were used) the number of grammes of silver will obviously be

$$1{\cdot}14 \times \frac{0{\cdot}8253}{1{\cdot}14} = 0{\cdot}8253 \text{ gramme.}$$

The total weight of silver in the entire sample is therefore :—

Weight of silver in sifted ore	... ...	1·034 grammes.
Weight of silver in metallics	... ...	·8253 ,,
Total weight of silver in entire sample	...	1·8593 ,,

Then the number of ounces of silver per ton of ore is

$$29166{\cdot}66 \times \frac{1{\cdot}8593}{143{\cdot}70} = 377{\cdot}3805 \text{ ounces per ton.}$$

Calculation of the Gold

The number of ounces of gold per ton is calculated in the same way.

(*a*) *In Sifted Ore.*—One-half assay ton yielded 0·0021 gramme, therefore :—

$$142{\cdot}56 \times \frac{0{\cdot}0021}{14{\cdot}5833} = 0{\cdot}0205 \text{ gramme.}$$

(*b*) *In the Metallics.*—The whole of the metallics yielded 0·0073 gramme, therefore :—

$$1{\cdot}14 \times \frac{0{\cdot}0073}{1{\cdot}14} = 0{\cdot}0073 \text{ gramme.}$$

The total weight of gold in the entire sample is

Weight of gold in sifted ore	... ...	0·0205 gramme.
Weight of gold in metallics	... ...	0·0073 ,,
Total weight of gold in entire sample	...	0·0278 ,,

Then the number of ounces of gold per ton of ore is

$$29166{\cdot}66 \times \frac{0{\cdot}0278}{143{\cdot}70} = 5{\cdot}647 \text{ ounces per ton.}$$

Agreement of Results.—The permissible difference between the results of duplicate assays varies according to circumstances. With low-grade ores there should be practically no difference between the results, while with high-grade ores small differences are allowable according to the richness and nature of the ore.

As a general rule, assays of average silver ores should agree very closely, and if properly conducted are of great accuracy. The results of duplicate assays should not differ from each other more than 2 or 3 ounces troy per ton. Should a greater difference be found, an additional assay should be made and if necessary a fresh sample taken.

With regard to gold ores, the allowable differences are much smaller than in the case of silver ores. The results of duplicate assays on ores containing under 5 ounces of gold per ton should not differ more than 1 dwt. per ton or even 0·5 dwt. per ton for low-grade ores. With richer material assaying up to, say, 20 ounces per ton, a difference of 2 dwt. per ton is permissible.

In the purchase of gold and silver ores the difference to be allowed between the assay result of the buyers' representative and the result of the sellers' representative is mutually agreed upon before purchase.

Referee or Umpire Assays.—In the event of the contracting parties failing to agree on the assay results obtained, the sealed referee sample (especially reserved when sampling the ore, etc.) is sent to a mutually-agreed referee or umpire assayer of standing, whose result is accepted as final. The charge for referee assays is usually a fee-and-a-half or twice the ordinary fee and sometimes more, according to the professional standing of the referee. The referee chosen is usually an assayer with a wide experience in the assay of the particular class of ores, etc., in dispute.

Reporting the Assay Results.—Several methods are in use for reporting the assay value of precious-metal ores. As previously stated, in England and Australia the quantity of silver and gold in ores is reported in ounces and decimals of an ounce troy per ton of 2240 lbs. avoirdupois, known as the English or long ton.

In Canada, the United States, and Africa it is reported either in ounces and decimals of an ounce troy per ton of 2000 lbs. avoirdupois, or in the number of dollars and cents worth of precious metal per ton of 2000 lbs. avoirdupois. The ton of 2000 lbs. is known as the American or short ton, and is a sort of compromise between the avoirdupois and metric system. When the quantity of precious metal is less than an ounce it is reported in pennyweight and decimals of a pennyweight.

The old method of reporting results in ounces, pennyweight, and grains troy per ton avoirdupois is now being discontinued. The following is an illustration of the general method of reporting the assay values of gold and silver ores and products. (Sample dried at 100° C.)

Fine gold : 3·24 ounces per ton of 2000 lbs.
Fine silver : 2·40 ,, ,, ,, (or 2240 lbs. as required).

The values are usually reported to the second place of decimals.

The method of reporting "on the ton of 2000 lbs." instead of "on the ton of 2240 lbs." is now being very generally adopted.

In reporting the value of low-grade ores and products, the use of the term trace or traces is to be deprecated, unless when especially designating its application. "Thus in mine samples many mining men prefer that no intermediate stage be adopted between 0·5 dwt. of gold and nil, any distinction between these, except with special work, being generally guesswork on the part of the assayer. In reduction works [*i.e.* cyanide works] samples, reports should be limited to the use of 0·01 dwt. or nil; here again any trace found being merely a speck, unweighable, and sometimes even indeterminable as gold."[1]

The method of reporting the gold and silver content in money value is chiefly confined to America. In calculating the money value of an ore, a troy ounce of fine gold is considered to be worth 20·67 dollars. Fine silver is valued according to the prevailing market price of the metal, which fluctuates considerably. The nominal value of a troy ounce of fine silver may be taken as 60 cents.

As an example, the money value of an ore having an assay value of 5·5 oz. of fine gold per ton and 35·75 oz. of fine silver per ton of 2000 lbs. would be reported thus :—

Gold, 5·5 oz. at 20·67 dollars per ounce	113·68 dollars.
Silver, 35·75 oz. at 0·60 dollar per ounce	21·45 ,,
Total value of ore in gold and silver per ton of 2000 lbs.	135·13 dollars.

This would be termed a 135·13 dollar ore.

American "dollars on the ton of 2000 lbs." may be converted into "ounces in the ton of 2240 lbs." by dividing by 1·1544 in the case of silver, and by 18·457 in the case of gold.

If the value of an ore is to be reported in English currency the value of 1 ounce of fine gold is taken as £4·247 (or £4, 4s. 11·45d.), while the value of fine silver varies but may be taken at 2d. or 2½d. above the market quotation for "standard" or coinage silver, which contains 925 parts of silver per 1000.

As pointed out by Beringer,[2] the practice of reporting assay results in money value is objectionable. "The prices of metals vary with the fluctuations of the market, and if the assayer fixed the price, the date of his report would be all important; if, on the other hand, he takes a fixed price which does not at all times agree with the market one, it leaves a path open for the deception of those unacquainted with the custom."

When it is necessary to state or estimate the money value of an ore, etc. (other than gold), it should be accompanied by the assay value, and the basis on which the former has been calculated from the latter should be stated.[3]

It is the practice of some assayers to report the value of gold ores in terms of "bullion," but this method of reporting is to be strongly deprecated unless the fineness (*i.e.* parts of fine gold in 1000 parts) of the bullion is also stated. The money value of bullion is largely dependent on the proportion of fine gold it contains; thus the value of 1 ounce of bullion 942 fine would be about £4, while bullion 412 fine would only be worth about 35s., so that the statement that an ore contains so many ounces of "bullion" gives little indication of the true money value of the ore. Such reports are very misleading to the uninitiated, and are only favoured by a certain class of company-promoter for reasons that had better not be stated. If the assay value of gold and silver is to be reported as "bullion," it should be accompanied by a statement of the proportion of fine gold and fine silver.

Assay reports should state the exact condition of the sample as to dryness when assayed: it should be stated whether the sample was assayed in the condition in which it was received, or dried at — ° C. As a general rule, the

[1] A. M'A. Johnston, *Rand Metallurgical Practice*, vol. I, 1926, "Assaying," p. 320.

[2] *Text-book of Assaying*, 15th edn., 1921, p. 9.

[3] Recommendation of the Inst. of Min. and Met., *Trans. Inst. Min. and Met.*, 1910, **20**, 530.

assay values of gold and silver ores are reported on the dried ore for the reasons already given.

The amount of moisture is usually stated on the report. Two methods are in use for reporting the amount of moisture present in ores, etc. On the Continent the moisture is invariably reported in percentage, but in England it is still the practice in some cases to report the moisture in grains per pound, thus: if the material contains 1 per cent. of moisture the report will read "moisture $\frac{70}{7000}$" meaning that there are 70 grains of moisture in 7000 grains or 1 lb. avoirdupois.[1]

The common limit for moisture in exported ores is from $\frac{20}{7000}$ to $\frac{50}{7000}$.

How long this old Cornish method of reporting moisture, which has survived in England to the present day, will continue to be used in face of more modern and simpler Continental procedure, it is difficult to say.

The determination of moisture in consignments of ore at British ports or works is carried out by the joint samplers of buyer and seller, who may or may not have anything to do with the subsequent assay of the sample.

Examples of forms used by assayers for reporting the assay value of auriferous and argentiferous ores and products are given in Chapter 11.

[1] H. L. Terry, *Min. Mag*,. 1912, **6**, 62.

CHAPTER XIV

THE ASSAY OF COMPLEX GOLD AND SILVER ORES

THE term complex ores includes all auriferous and argentiferous ores containing excess of substances such as copper, antimony, zinc, arsenic, sulphur, tellurium, etc., which interfere more or less in the ordinary methods of assay and therefore require special treatment.

In many cases complex ores are satisfactorily assayed by combined wet and dry methods. The ore is treated with acid to remove the interfering metal or metals, and the precious metals collected into a button of lead by fusion of the insoluble residue after precipitating the silver in the solution.

Scorification may sometimes be applied to the assay of ores of this class when they are sufficiently rich to permit of small quantities being taken for assay.

The following complex ores are considered :—

(1) Arsenical ores.
(2) Antimonial ores.
(3) Cupriferous ores.
(4) Telluride ores.
(5) Ores containing lead, zinc, nickel and cobalt, tin, bismuth, etc.

(1) **Arsenical Ores.**—Arsenic is most frequently associated with precious-metal ores as mispickel (arsenical pyrites) and occasionally as oxides.

As already mentioned, the presence of arsenic in an ore gives rise to the production of compounds which tend to diminish the accuracy of the assay. It is therefore necessary to eliminate the arsenic, and this is usually effected by subjecting the ore to an oxidizing roasting or to the oxidizing action of nitre. Much diversity of opinion exists as to which is the better of these two methods ; in the author's opinion the former is preferable, but both methods will give good results when care is exercised. For silver ores, scorification may be employed with advantage. Many assayers hesitate to roast arsenical ores on the ground that, owing to the volatile character of arsenic, mechanical loss of the precious metals is caused. It has been pointed out that this may occur when too high a temperature is employed at the outset, thus causing a rapid disengagement of the volatile constituents of the ore, but with due precaution this may be avoided. The formation of arsenates is mostly to be feared, as they are liable to remain undecomposed and cause loss of gold and silver in the slag during fusion. The tendency to form arsenates is greater when nitre is employed as the oxidizing agent, and on that account oxidation by roasting is preferable for the elimination of the arsenic. The ore is roasted, with all the precautions previously mentioned under pyritic ores, until completely oxidized ; then from 5 to 10 per cent. of charcoal or anthracite powder is added, and the roasting continued until all the carbon is burnt out. Owing to the rapidity with which arsenic vaporizes, great care is needed in regulating the temperature. In the early stages of the roasting it is necessary to employ a very low temperature and admit only a limited supply of air.

An attempt to eliminate arsenic by rapid oxidation would be attended with the danger of converting it into arsenates of the metals present in the ore, such as iron, copper, lead, etc. When once formed, arsenates are not readily decomposed, as they resist a high temperature and are only slowly converted into sulphates at a red heat by the sulphuric acid formed during the later stages of the roasting.

As before stated, with the exception of arsenate of copper, the decomposition of the arsenates of the metals most frequently associated with gold ores is effected by the addition of charcoal or anthracite powder, but in many cases prolonged heating is necessary; it is therefore desirable to avoid their formation as far as possible by taking the precautions already mentioned. The expulsion of arsenic, and also of antimony, is favoured by the presence of iron pyrites, as the sulphur distilled from the pyrites in the early stages of the roasting tends to eliminate them as sulphides.

The product obtained by roasting the ore is composed largely of ferric oxide, and is fluxed according to the directions given for the treatment of basic ores, to which class it belongs. By the addition of a small quantity of potassium cyanide to the charge for the fusion, the decomposition of any arsenates remaining in the roasted product is greatly facilitated. In exceptional cases the button of lead may be slightly brittle owing to the presence of arsenic. When an appreciable quantity of arsenic is present in the lead the effect on cupellation is somewhat similar to that produced by antimony. If present in large quantities, it may be removed by scorification. When the oxidation of the arsenical compounds is effected by means of nitre, the ore may be treated by the nitre methods previously described.

If arsenical ores are fused direct with the addition of metallic iron, the lead button will be accompanied by a layer of speise which tends to retain values. Some assayers use the iron-reduction method and scorify the speise with the lead button after crushing it to a coarse powder.

(2) Antimonial Gold and Silver Ores.—Gold and silver ores containing antimony minerals such as stibnite, jamesonite, etc., are usually assayed by the nitre method. Oxidation of the antimonial compounds may be effected by roasting, but owing to the very fusible and volatile nature of stibnite, etc., it is very difficult to roast these ores satisfactorily without caking, especially when the percentage of stibnite is large. The addition of pure siliceous sand to the ore before roasting assists in preventing the formation of lumps.

C. O. Bannister[1] has shown that careful roasting, with the addition of a little charcoal powder towards the end of the operation, followed by a suitable fusion, gives good results. No loss of precious metals takes place in the roasting of antimonial ores when proper precautions are taken, and this method of oxidation is sometimes used by assayers. But in general practice few assayers have the time and patience needed to roast these ores, and the method of oxidation with nitre is most generally employed.

The amount of nitre required is determined if necessary by a preliminary assay, as previously described. The charge must contain excess of sodium carbonate to induce the formation of sodium antimonate. When the ore contains a large proportion of antimony, say over 50 per cent., it is usually more satisfactory to run two charges of 20–25 grammes each, and combine the resulting lead buttons, than to make one fusion on 50 grammes of ore.

The following charge is suitable for ores containing about 75 per cent. of stibnite with gangue mainly silica:—

Ore	20–25	grammes.
Red lead	80	,,
Sodium carbonate	60	,,
Borax	10–20	,,
Nitre	14–20	,,

[1] *Trans. Inst. Min. and Met.*, 1906, **16**, 94.

The fusion should be conducted slowly and at a low temperature in order to keep the button of lead as free as possible from antimony. The button of lead should weigh from 45 to 50 grammes.

With due attention, the nitre may be so adjusted that a malleable button, comparatively free from antimony, will be obtained, and may be cupelled direct. If the button is brittle, comparatively hard, and white, it generally contains antimony, in which case it must be scorified with, if necessary, the addition of more lead before being cupelled.

According to Rivot,[1] antimony does not materially interfere with cupellation if less than 1 per cent. is present. If the lead contains much more than this amount, the antimony in volatilizing takes gold with it, and it also forms antimonate of lead, giving a pale-yellow scoria, and causing the cupel to crack at the edges.

This last effect is very characteristic of antimony, and is more likely to occur with bone-ash cupels than with magnesia cupels. The button of lead is scorified at a moderate red heat with free access of atmospheric air. By this means both the lead and the antimony are quickly oxidized, but the antimony in greater proportion than the lead, a fusible compound of antimony and lead oxide being formed, and the lead practically freed from antimony. The slags from comparatively rich ores should be cleaned in the manner already described.

It is necessary to clean the slags from antimonial silver ores and also in the case of gold ores, if an exact determination of the silver accompanying the gold is required.

Considerable corrosion of the crucible takes place with the highly basic charge employed, but this may be minimized by adding silica to the charge, or, better still, 10 to 20 grammes of fireclay or any common clay. The latter is recommended by Holloway,[2] who points out that it is similar in nature to the pot itself, and its fine state of division renders it more readily acted upon by the fluxes than is silica alone.

The presence of much silica in the charge tends to increase the amount of antimony in the lead button.

Combined Wet and Dry Method.—The following method, involving solution of the ore, may be conveniently employed for antimonial ores as a check upon other methods. It possesses the advantage of producing lead buttons practically free from antimony, and thus avoiding the necessity of scorification previous to cupellation. The weighed quantity of ore (from 25 to 50 grammes) is treated with concentrated hydrochloric acid, and heated until decomposition is complete ; the solution is diluted with a small quantity of water in which a few crystals of tartaric acid have been dissolved. If sufficient tartaric acid has not been introduced, this dilution will cause the precipitation of some of the antimony as oxychloride. The insoluble residue is allowed to settle, and when the solution is clear as much of the liquid as possible is poured off through a filter without disturbing the residue. The residue, if small, is finally transferred to the filter, allowed to drain, and then dried. The filter-paper is burnt, and the ash, with the insoluble residue, is scorified with 10 times its weight of granulated lead and a cover of borax, or it is fused with the following fluxes :—

Red lead	30	grammes.
Charcoal powder	1·5	,,
Sodium carbonate and borax, varying together up to	30	,,

The fusion is effected in the ordinary way, and the resulting button of lead cupelled direct. It may be remarked that the oxidized ores of antimony,

[1] Docimasie, *Traité d'analyse des substances minérales.*

[2] *Trans. Inst. Min. and Met.*, 1906, **16**, 97.

cervantite (Sb_2O_4), and senarmontite (Sb_2O_3) are only partially soluble in hydrochloric acid, the former being practically insoluble. They may be satisfactorily assayed by fusion direct. The oxides, however, are of somewhat rare occurrence, the ore most commonly met with in commerce being the sulphide (stibnite, Sb_2S_3), which is readily soluble in hydrochloric acid.

(3) **Cupriferous Ores.**—Ores of this class are usually very troublesome to assay correctly by the fusion method, as they are generally poor in gold and silver, and frequently contain a large proportion of copper.

Owing to the readiness with which copper compounds are reduced during the fusion of the ore, some difficulty is experienced in passing the copper into the slag, and thus preventing the formation of cupriferous lead buttons. Copper sulphide is decomposed by lead oxide, the sulphur and a part of the copper being oxidized and retained in the slag ; the rest of the copper passes into the reduced lead, thus rendering it hard and difficult to cupel. One part by weight of copper requires about 16 parts of lead for cupellation, and though a smaller quantity of lead will pass the copper into the cupel, yet a sensible amount of gold and silver would in that case accompany the lead and copper, so that an incorrect result would be obtained (Percy). Under these circumstances it becomes necessary to add enough lead oxide to the charge not only to collect the gold and silver and oxidize the sulphur present, but to yield a button of lead sufficiently large to satisfactorily carry into the cupel all the copper present in the lead. With ores containing much copper this would give rise to the production of lead buttons inconveniently large for cupellation direct. In order to obtain a lead button sufficiently free from copper for cupellation, the proportion of copper in an ore should not, according to J. Loevy, exceed 6 per cent.[1] Copper ores (whether oxidized or containing sulphur) which do not contain more than 6 per cent. of copper may be readily reduced by fusion, and yield a lead button which is poor in copper and can be at once subjected to cupellation. With ores containing more than 6 per cent. of copper, Loevy reduces the proportion by the addition of silica. Thus to an ore with 12 per cent. of copper is added an equal weight of sea sand. If the ore contains 24 per cent., three times its own weight of sand is added. The mixtures of ore and sand are then fluxed and fused in the usual way, the larger charges being divided into several smaller charges and the buttons resulting from the fusions concentrated by scorification into one lead button weighing 30 to 35 grammes.

In practice the fusion assay is seldom attempted for ores rich in copper, as they are more satisfactorily assayed by scorification or by a combination of wet and dry methods. For scorification from 2 to 3 grammes ($\frac{1}{10}$ of assay ton) of ore are scorified with from 15 to 20 times its weight of granulated lead, with the addition of 0·25 gramme of borax, and, if necessary, 0·5 to 1·0 gramme of silica.

If the resulting lead button is hard from the presence of copper, it should be re-scorified, with the addition of fresh lead if necessary. In cupelling the lead button the temperature should not be too low at the finish, otherwise the button of silver and gold will retain copper and give too high a result.

In the combination wet and dry method, the ore is first treated with nitric acid to remove the copper by solution. The silver present is also dissolved, and is precipitated by the careful addition of sodium chloride. The insoluble residue containing the precious metals is then scorified or fused in a crucible with fluxes for the recovery of the gold and silver.

To facilitate the settling of the finely-divided precipitate of silver chloride and of the suspended gold it is usual to add a solution of lead acetate, and

[1] *Chem. Zeits.*, 1911, **35**, 278.

precipitate the lead by the addition of sulphuric acid. The precipitate of lead sulphate thus formed settles and carries down the gold and silver, leaving a clear solution.

The method of procedure differs in details with different assayers, but the following method is in very general use :—Stir 25 grammes (or 1 assay ton) of ore into a No. 5 beaker containing 100 c.c. water. Add 50 c.c. commercial nitric acid (sp. g. 1·42). Cover and allow to stand till action apparently ceases. Add another 50 c.c. nitric acid and boil gently until the greater part of the free acid has been expelled. Dilute to 500 c.c., and allow to settle. To precipitate the suspended gold and dissolved silver, add 2 to 4 c.c. normal salt solution, 5 c.c. sulphuric acid, and 10 c.c. lead acetate. Set aside to settle over-night ; filter, wash, and dry. The filtered solution, which should be bright and clear, contains the copper, and may be neglected. The siliceous residue contains the gold and silver.

If the amount of residue is small, it may be scorified. The dried filter is carefully burnt on a scorifier placed in the mouth of the muffle at a low heat, then scorified with about 40 grammes of granulated lead and 1 gramme of borax. The resulting lead button is cupelled, and the gold and silver parted as usual.

In practice, by far the larger number of ores contain so much insoluble material that they are not adapted for scorification. If the ores are pure sulphides or carbonates, as is seldom the case, the residue can be scorified. Usually, however, they are composed chiefly of quartz or insoluble silicates, and consequently it is best to treat them by a crucible method.

The residue may be dried in a roasting-dish, then heated at a gentle heat to burn the filter paper, and mixed with—

Lead oxide	40 grammes	or 1·5 A.T.
Sodium carbonate	30 ,,	or 1·0 A.T.
Borax	10 ,,	
Argol	1·5 ,,	

With cover of borax, 10 grammes.

The charge may be mixed in the crucible with a spatula. Fuse for thirty minutes, then cupel, part, and weigh. Some assayers put about half the lead oxide into the crucible, add the dry residue and filter-paper, then burn the paper, and add the remainder of the lead oxide and other fluxes. The charge is then well mixed and fused as before. The combination method requires careful manipulation to obtain accurate results.

The results for gold are usually a little low : the cause of this has not been fully investigated, but is probably due to the solvent action of copper salts. In precipitating the silver in solution, excess of sodium chloride must be avoided, as silver chloride is appreciably soluble in a strong solution of salt. By the combination method, lead buttons free from copper are obtained, so the operation of cupellation is not liable to be interfered with.

(4) Telluride Ores.—Although the association of gold and silver with tellurium is usually considered to be exceptional, experience has shown that tellurium is more widely distributed and is a more frequent associate of the precious metals than has been generally recognized. The tellurides are very important sources of gold, and occur in veins, impregnations and lenticular deposits associated with pyrites and other sulphides, and with free gold.[1] The chief localities at which gold tellurides are worked are Western Australia (Kalgoorlie), Colorado (Cripple Creek), and Transylvania (Nagyag, etc.). The gold tellurides are

[1] F. Rutley, *Elements of Mineralogy.*

sylvanite and calaverite, both of which are tellurides of gold and silver; and nagyagite, a telluride and sulphide of lead and gold.

Tellurium is also found with selenium in the anode mud or slime obtained in the electrolytic refining of copper and matte.

Selenium has also been found associated with gold and silver ores, and as its behaviour during assay is very similar to that of tellurium, the precautions necessary for the assay of telluride ores apply in general also to selenium ores.

It must be remembered that, although the telluride mineral may be enormously rich in both tellurium and gold, the average telluride ore is one containing only about a pennyweight, or, at the most, a few ounces of gold per ton, the tellurium seldom exceeding the gold in amount, and being often so finely disseminated through the gangue that no portions of it can be picked out. A telluride ore yielding 1 ounce of gold is not likely, for instance, to contain more than 0·005 per cent. of tellurium, and the iron and other minerals which are commonly associated with it interfere seriously with the ordinary methods of detection.

The telluride can to a great extent be separated from the gangue by panning down the finely-ground ore, but the concentrates, although far richer in tellurides, contain usually more of the interfering minerals. It is, however, generally advisable to perform this concentration before applying tests, when sufficient ore is available.

Tests for Tellurium.—Tellurium compounds are very volatile and when heated in a closed glass tube with strong sulphuric acid they give to the acid a brilliant reddish-violet colour. The heat applied must be sufficient to cause the evolution of white acid fumes.

In the case of ores, Holloway [1] suggests concentrating the tellurium in the lead button obtained from the usual crucible assay, then dissolving a portion of the lead in dilute nitric acid, diluting, and precipitating the tellurium with lead-foil. The deposit is then heated with sulphuric acid, to which it imparts a reddish-violet colour.

When gold and tellurium are together in solution, free from chlorine, after treatment of alloys, etc. in aqua regia, both metals may be precipitated with sulphurous acid. The combined precipitate is then treated with nitric acid which dissolves the tellurium as tellurous acid, leaving a residue of gold which may be dried and weighed or inquarted with silver and parted in the usual way.

When tellurium is present in gold alloys it is in great part removed by treatment with nitric acid, but much depends on the amount of tellurium present. The residual gold may then be cupelled and parted with very little loss of gold.[2]

Assay of Telluride Ores.—In assaying telluride ores too much stress cannot be laid on the necessity for careful sampling. Owing to the irregular distribution of the values in these ores fine crushing and thorough mixing are necessary to get a true sample. Experience shows that telluride ores should be crushed to pass a sieve of about 100-mesh.

Tellurium has a great affinity for gold and silver, and also for lead, and when present in the ore in comparatively large quantity some of it will pass into the lead button with the precious metals in the fusion assay and exert its influence in the cupellation, as described on p. 156. With average ores the amount of tellurium left in the gold or silver button is usually insufficient to make any appreciable difference to the assay result, provided proper precautions are taken to prevent it from passing into the lead button during fusion.

[1] *Loc. cit.*, p. 175.

[2] Rose and Newman, *op. cit.*, p. 530.

It is therefore essential to remove as much tellurium as possible before cupellation, and in the assay of telluride ores experience has shown that this is best accomplished by a crucible fusion with an oxidizing charge containing lead oxide as the oxidizing agent.

When ores containing telluride minerals are fused in the presence of a large excess of lead oxide, the tellurium is oxidized, forming probably lead tellurate, and in the presence of soda, sodium tellurate. According to S. W. Smith,[1] tellurium reacts with litharge at moderate temperatures (700° to 900° C.) before the formation of silicates has taken place, in accordance with the equation :—

$$2PbO + Te = Pb_2O + TeO.$$

During the formation of silicates there is a partial reversal of this reaction, although the greater part of the tellurium which has once been oxidized dissolves to a red glass in the slag. When litharge has reacted sufficiently to form silicates there is no longer any oxidizing effect on tellurium.

There appears to be general agreement among assayers in regard to the precautions necessary for successful crucible fusion of telluride ores. The best practice appears to be to make a fusion with a fairly large quantity of litharge or red lead, thus getting a large lead button, and then either to cupel the lead button direct or to partially scorify to reduce the weight and then cupel.

The compositions of the charges recommended by different assayers have the same general character, a large excess of lead oxide (usually litharge) being their chief characteristic.

Holloway and Pearse [2] state that the passage of tellurium into the slag appears to be more complete in the case of a basic than an acid slag. Tellurium has a great affinity for the alkalies-forming alkaline tellurides. Every worker on telluride ores has insisted on the necessity for careful fritting and slow fusion at moderate temperatures. The reasons for this are explained by the conclusions of S. W. Smith, quoted above.

It will be evident that, if the litharge does not exert its full oxidizing effect before the reduction of lead takes place, the tellurium will combine with the lead instead of passing into the slag. It is doubtful, however, whether under the most favourable conditions all the tellurium can be removed from gold and silver by such an oxidizing fusion, owing to the fact that the metallic lead formed during the fusion acts as a reducing agent and reduces some of the tellurium which has been oxidized and passed into the slag.

Quick fusion invariably gives low results and tends to increase the proportion of tellurium passing into the lead. The following results obtained by W. H. Merrett [3] on a telluride ore containing some pyrites, in addition to an ordinary basic gangue, may be given as an example.

In the " ordinary " fusion the fritting and subsequent fusion were conducted slowly, the charge being poured after it had been an hour in the fire.

In the " quick " fusion the fritting and fusion were hastened so that the charge was ready to pour in about twenty minutes.

The results were as follows :—

	Gold. Ounces per ton.	Silver. Ounces per ton.
Ordinary fusion	1·50	1·10
Quick fusion	1·42	0·81

[1] *Trans. Inst. Min. and Met.*, 1908, **17**, 473.
[2] *Ibid.*, p. 189. [3] *Ibid.*, p. 199.

The gangue and the general nature of telluride ores vary so greatly that no charge suitable for general use can be recommended, but it is generally agreed that sufficient lead oxide should be present to yield a large button (not less than 30 grammes) of metallic lead, even when the ore is not rich in precious metal, and even when a very small charge of ore is used.

The charge should contain a comparatively large proportion of sodium carbonate, but borax should be sparingly used: as a general rule, the amount of borax should not exceed one-third the weight of sodium carbonate in the charge.

The following charges may be taken as examples of those used for various telluride ores:—

TABLE XXX.—Charges for Crucible Assay of Telluride Ores. (*Holloway and Pearse.*)[1]

	Nature of Ore.						
	Shaley Ores.			Highly Siliceous Ores.			Basic Ores.
	Rich.	Average (5 ozs. and less).		Rich.	Average.		
	grammes.	grammes.		grammes.	grammes.		
Ore	10	20	50	10	20	50	See note.
Red lead	120	100	120	120	100	120	
Flour	7	6	7	7	6	7	
Sodium carbonate ...	30	40–50	60	40	50	60	
Borax (glass) ...	10	15–20	20	5	5	20	

Note: The charge for basic ores depends entirely on the nature of the ore, and is a matter for experiment. The charge should be such as will produce a fluid and homogeneous slag at a low temperature, and should be as basic as possible without danger to the crucible. In all cases, ordinary borax may replace the borax glass, nearly twice as much being used, and litharge may replace red lead with slight reduction in the amount of flour.

For ordinary siliceous telluride ores of only slight reducing power (*i.e.* containing little pyrites) the following charge is recommended by Fulton and Sharwood[2]:—

Ore	0·5 assay ton (15 grammes).
Litharge	100·0 grammes.
Charcoal	1·1 ,,
Sodium carbonate	30·0 ,,
Borax glass	6·0 ,,
Litharge to cover	10·0 ,,

From 10 to 20 milligrammes of silver foil may be added for parting purposes and to reduce the cupellation loss. The stock flux recommended quite generally by Cripple Creek assayers for the ordinary siliceous ores is given below.[3] With

[1] *Trans. Inst. Min. and Met.*, 1908, **17**, 188.
[2] *Manual of Fire Assaying.*
[3] Fulton & Sharwood, *loc. cit.*

0·5 assay ton of ore about 75 grammes of the stock flux is used, and the charge will contain the different constituents in the following amounts :—

Stock Flux.		Charge for 0·5 assay ton of ore.
Litharge	30·0 parts.	45·5 grammes.
Flour	1·0 ,,	1·5 ,,
Potassium carbonate	7·0 ,,	10·5 ,,
Sodium carbonate	6·0 ,,	9·0 ,,
Borax glass	5·5 ,,	8·5 ,,

It is frequently recommended that the charge for the fusion of telluride ores should be covered with a layer of salt, but there seems little doubt that the use of salt, either in the charge or as a cover, is highly objectionable and invariably yields a lower result.

As previously stated, the fusion of telluride ores must be conducted slowly and at a low temperature, special attention being given to the fritting stage.

The lead button obtained should be large, in order that the proportion of tellurium and gold to the lead may be small for cupellation. This large excess of lead is necessary to assist the removal of the tellurium during cupellation. With sufficient lead the tellurium, although less easily oxidizable than lead, is expelled before the final stage of the operation, and does not exert its injurious effects. An excess of tellurium in the lead button, at the end of the cupellation, causes heavy losses of gold and silver and, as previously stated, may result in the subdivision of the precious metal or even in complete loss by absorption if the amount of tellurium present is equal to one-tenth of the lead. Subdivision and complete absorption cannot, however, occur if sufficient lead is present. In the case of many telluride ores the amount of tellurium in the lead button is comparatively very small when the assay has been properly conducted, and the amount left towards the end of the cupellation is too small to exert any injurious effect.

Opinions differ as to whether the lead buttons should be cupelled direct or submitted to a partial scorification before cupellation.

According to Holloway and Pearse,[1] cupellation after a moderate amount of scorification gives a better result than the direct cupellation of the lead button, but they point out that, if scorification be carried too far, or if the lead be allowed to remain long in contact with the slag, the subsequent loss on cupellation is increased. The cupellation losses are very large after excessive scorification, and are probably caused by a reabsorption of tellurium which is reduced by the lead from the litharge in the scorifier. Holloway and Pearse recommend that the lead button (if not less than 50 grammes) from the crucible assay should be scorified to reduce its weight by about one-third, but not more than one-half, and then cupelled.

Since, however, the scorification of lead buttons tends to concentrate tellurium in the button, most assayers prefer to cupel the large lead button direct.

The relative elimination of tellurium during cupellation and scorification is shown in the following results of experiments by S. W. Smith.[2] A series of lead buttons weighing 20 grammes and containing 0·05 gramme of gold and 0·05 gramme of tellurium were submitted to cupellation and to scorification, and the operation interrupted at intervals for the determination of tellurium.

It will be noted that during cupellation the tellurium tends at first to concentrate in the lead and then begins to be eliminated. In the case of scorification there is a more pronounced concentration of the tellurium, particularly towards the end of the operation.

[1] *Trans. Inst. Min. and Met.*, 1908, **17**, 183. [2] *Ibid.*

The tellurium may be partly removed from the lead button by "soaking" it under an excess of litharge at a temperature between 700° and 900° C., and such treatment of the lead button from the crucible fusion, prior to cupellation, is recommended by S. W. Smith as a desirable substitute for scorification. A low temperature should be employed both for the scorification and cupellation of the lead.

TABLE XXXI.—ELIMINATION OF TELLURIUM DURING CUPELLATION AND SCORIFICATION. (*S. W. Smith.*)

Cupellation.			Scorification.		
Cupelled until Lead weighed grammes.	Per cent. of original Lead remaining.	Per cent. of Tellurium in Lead Button.	Scorified until Lead weighed grammes.	Per cent. of original Lead remaining.	Per cent. of Tellurium in Lead Button.
20·00	100·00	2·48	20·00	100·00	2·48
11·86	59·00	3·25	6·145	30·7	5·94
4·25	21·00	2·49	2·085	10·4	12·90
1·635	8·15	1·80	...	...	...

As previously stated, the presence of silver reduces the loss of gold during the cupellation of buttons containing tellurium, the silver acting as a diluent. It is for this reason that many assayers make an addition of silver, either to the charge for the fusion of the ore, or to the lead button at the beginning of the cupellation. When the assays of telluride ores are carefully conducted, and proper precautions taken, the losses on average ores of, say, 10 oz. and less are very little above those experienced in the assay of average non-tellurous ores.

As, however, the tendency for precious metals, especially silver, to be retained in the slag is somewhat greater in the assay of telluride ores than with ordinary ores, it is very desirable to clean the slags from the fusion of all but low-grade ores. The slags, and also the cupels if necessary, may be cleaned by the methods previously described.

With regard to other methods of assay for telluride ores, experience has shown that generally the scorification method is unsatisfactory and gives low results. Owing to the smallness of the charge which can be taken, only rich ores can be assayed by scorification. The fluxing is less complete in scorification than in the crucible assay, and the tellurium is not so satisfactorily removed; also there is a tendency at the scorification temperature for the lead to be oxidized at the expense of tellurium oxide, which is reduced and enters the lead button and thus becomes concentrated in the lead. Better results are sometimes obtained by the addition of various fluxing materials, but if considerable amounts of these have to be used, the scorification assay approaches the crucible assay in character with none of its advantages.

Combined wet and dry methods of assay have been recommended for the assay of telluride ores, but, as the result of a large number of trials on rich, average, and poor ores, Holloway and Pearse found that even when every precaution was taken the results were lower than by direct fusion, and that duplicate assays seldom agreed within reasonable limits.

The result of their work, which confirms that of other workers, indicates, however, that should a case arise in which preliminary treatment with acid is necessary, the following precautions, which are probably known to most assayers, should be adopted :—

Dilute sulphuric acid (1 to 4), where its employment is permissible, gives less loss than nitric acid. When nitric acid is used, as is generally the case, the weaker the acid the better. One volume of acid to three or four of water is a convenient strength. Care should be taken in adding hydrochloric acid, or salt, to precipitate any dissolved silver, to use the smallest possible quantity, lest gold be dissolved. Ample time should be allowed after such addition for the silver to precipitate, and it is even advisable to test the filtrate to be sure that no gold has passed into solution.

5 (a) Silver Ores containing Lead.—*Galena*, lead sulphide, the chief ore of lead, is found in most mining districts and is frequently associated with zinc-blende and copper pyrites. As galena invariably contains more or less silver, this metal is always tested for and its amount determined. Galena is not usually rich in silver, but it is frequently found in association with rich silver minerals. In the case of lead ores, the silver is generally calculated in ounces per ton upon the lead in the ore as determined by dry assay, as well as upon the ore itself, so that it is usual first to determine the percentage of lead by the fusion method and then to cupel the resulting lead button to determine the silver.

Lead ores are easily concentrated by mechanical operations, so that the samples received for assay are generally comparatively pure. To determine the lead, weigh up 25 grammes of ore, mix with an equal weight of sodium carbonate and 2 grammes of argol; place in a crucible, then cover with a sprinkling of sodium carbonate or borax, and insert a piece of hoop iron bent into a ∩-shape (Fig. 116) to reduce the lead sulphide. Place in a furnace heated to, but not above, redness, and cover the crucible. In about twenty minutes the charge will be fused, and when bubbles of gas are no longer being evolved and tranquil fusion has taken place, the iron is withdrawn and any adhering buttons of lead washed off by dipping it a few times in the slag. Cover the crucible, leave it in the furnace for a minute or two longer, and then pour. Detach the slag, when cold, by hammering, and clean by brushing or washing in hot water and weigh. The weight of the button multiplied by 4 gives the percentage of lead. Duplicate assays for lead should agree within 0·5 per cent., and in practice they frequently agree within 0·2 per cent. To determine the silver, cupel the lead and weigh the resulting silver button. The silver per ton of lead is calculated as follows :—

FIG. 116.

Suppose 25 grammes of ore gave 15·6 g. of lead (equal to 62·4 per cent.) which on cupellation yielded a silver button weighing 0·01 g. Then,

$$\frac{0{\cdot}01 \times 100}{15{\cdot}6} = 0{\cdot}0642 \text{ per cent. of silver.}$$

From the table given on page 202, 0·0642 per cent. is equal to 20·97 oz. of silver per ton (2240 lbs.) of lead.

For the assay of oxidized lead ores, argol or charcoal is employed as the reducing agent instead of iron. Ores comparatively poor in lead may be assayed for silver by the fusion method, but more or less litharge or red lead must be added to the charge to obtain a lead button of suitable weight. The lead button is cupelled, and the slag cleaned in the usual way if the ore is rich. The usual corrections for silver in the litharge used, etc., are made, and the silver calculated in ounces to the ton of ore.

The silver in argentiferous lead ore may also be determined by scorification. In this case take from 5 to 10 grammes of ore, 30 grammes of lead, and 0·5 gramme of borax. Scorify, cupel, and calculate to ounces to the ton. Unless the ore is rich, scorification is liable to give slightly lower results.

(b) Zinciferous Gold and Silver Ores.—Zinc is most frequently met with in ores as the sulphide, zinc-blende. As a general rule, zinc-blende contains very little silver, but occasionally it is comparatively rich in silver. Ores containing a large percentage of zinc are somewhat difficult to assay owing to the fact that zinc oxide is not readily fluxed, and if a good fluid slag is not obtained the results are likely to be low. Hall and Popper [1] have worked out the best conditions for the crucible assay of zinciferous gold and silver ores by direct fusion.

As the result of a large number of experiments, they concluded "that the amount of litharge used should be just sufficient to give a lead button large enough to collect the gold and silver, since lead oxide in the slag seems to interfere with the complete decomposition of the ore, and prevents the formation of a slag which is free from lumps, and which can be readily poured."

The amount of sodium carbonate should be from four to five times that of the ore. Borax glass should be added in amount sufficient to prevent the charge from being entirely basic, and to assist in fluxing the gangue minerals not acted on by sodium carbonate alone. If necessary, the amount of argol sufficient to reduce the whole of the litharge should be added; whilst, if the ore contains more than 15 per cent. of pyrites, the addition of a couple of iron nails will prevent the formation of a brittle button. The following charge was found to be the most suitable for a 20-gramme crucible, the quantities being given in assay tons and in grammes :—

Ore	⅓ A.T. or 10 grammes.
Sodium carbonate	1⅓ A.T. or 40 ,,
Borax glass	½ A.T. or 15 ,,
Litharge	⅘ A.T. or 25 ,,

The most suitable temperature, according to Hall and Popper, is 750° C.–775° C., and the time required for the fusion is 30 to 35 minutes.

The method gives good results with zinc ores containing up to 7·5 per cent. of copper. As zinc does not alloy to an appreciable extent with lead, the lead buttons from the fusion assay of zinciferous materials do not become contaminated with zinc.

A crucible fusion with nitre may be satisfactorily employed for zinciferous ores, the quantity of nitre being varied according to the amount of zinc-blende present. From 1·5 to 2 parts of nitre are required to oxidize 1 part of zinc-blende.

A suitable charge for ores containing about 75 per cent. of zinc-blende is the following :—

Ore	20 grammes.
Litharge	50 ,,
Sodium carbonate	20 ,,
Borax	15 ,,
Nitre	20 ,,
Sand	5 ,,

The sand is reduced in quantity or entirely omitted, according to the amount of silica present in the ore.

[1] *School of Mines Quart.*, 1904, pp. 355–358.

Fusion with nitre is recommended by Lay[1] for the assay of concentrates consisting of zinc-blende and galena such as are obtained at Broken Hill, Australia.

The silver–lead ore at Broken Hill varies widely in its nature, but an average analysis of the crude ore at present being sent to the concentration mill shows it to contain 36 per cent. of silica, 14·4 per cent. of lead, and 19·4 per cent. of zinc, the whole of the silver present being credited to the latter item.[2] For Broken Hill concentrate, containing 60 per cent. of lead, 11 per cent. of zinc, and 30 oz. of silver per ton, Newman[3] states that fusion with excess of lead oxide gives satisfactory results. The charge recommended consists of a quarter assay-ton of concentrate, 30 grammes of red lead, borax, sand, and soda, the latter in slight excess. The resultant lead button weighs about 30 grammes, and separates well from the slag. It is scorified to 10 or 12 grammes, and cupelled at a low heat. Ores containing zinc-blende may be roasted previous to fusion, but it is difficult to roast these ores "sweet," as it requires a full red heat to decompose any sulphate that forms.

When blende is heated in a current of air, decomposition begins at a temperature of about 450° C., with the formation of sulphate and oxide, but the proportion of sulphate is not considerable; its absolute amount increases with the temperature, though but slightly (6 to 8 per cent.), although the ratio of sulphate to oxide is smaller at high than at low temperatures.[4]

Zinc-blende being practically infusible has no tendency to clot, and thus permits of a somewhat high temperature being employed in the earliest stage of the roasting. Zinc-blende is not, however, infrequently accompanied by galena, and if this sulphide is present even in small quantity it is very liable to cause clotting in roasting. There is no loss of gold caused by roasting low-grade gold ores containing zinc-blende, but a loss generally occurs with the increase in the amount of gold. This loss is largely mechanical, and in part due to the volatilization of zinc oxide which causes loss of gold and silver. When the roasted product contains a large proportion of zinc oxide, difficulty may be experienced in satisfactorily melting the charge. Zinc oxide combines with silica in definite proportions with the production of silicates difficult to fuse, but with suitable proportions of borax and alkaline carbonates more fusible compounds may be obtained. With borax, zinc oxide yields a very fluid slag. With litharge it requires at least eight times its weight to form a fusible compound. The following charge is suitable for fluxing an ore, mainly zinc-blende, after roasting :—

Ore (before roasting)	25	grammes.
Litharge	40	,,
Charcoal	1·5	,,
Borax	30	,,
Sodium carbonate	10	,,

The fusion is conducted at a comparatively high temperature and the resulting lead cupelled.

Scorification is frequently employed for zinciferous ores, but if much zinc is present the results are not generally satisfactory, since zinc oxide, as stated above, is only soluble in litharge with difficulty. There is a great tendency for insoluble scoria and crusts to form on the scorifier.

[1] *Mineral Industry*, 1905, **13**, 287.
[2] *Bulletin of Amer. Inst. Min. Eng.*, 1909, pp. 763–793.
[3] J. M. Newman, *Trans. Inst. Min. and Met.*, 1910–11, **20**, 412.
[4] Krutwig, *Congrès de Chim. Pharm.*, Liège, 1906, Sec. XI., p. 419.

For the scorification of ore consisting mainly of zinc-blende the ore is usually mixed with about twenty times its weight of granulated lead and its own weight of borax. Silica equal to half the weight of ore taken is sometimes added to ores consisting largely of zinc-blende.

Some assayers employ a combined wet and dry method for zinciferous ores. The ore is first decomposed by treatment with hydrochloric acid and the zinc removed by filtration after diluting with water. The residue is then dried and scorified or fused in a crucible as directed for copper ores. Accurate results may be obtained by this method if care is exercised in manipulation.

(c) **Silver Ores containing Nickel and Cobalt.**—Arsenical, nickel, and cobalt silver ores are now being mined in the Cobalt district, Ontario. These ores are very rich in silver, often yielding several thousand ounces to the ton. For example, the shipments from one mine average 2500 oz. of silver per ton.[1] The high nickel, cobalt, and arsenic content of the silver ores of the Cobalt district, together with the presence of metallic silver, renders the sampling and assaying especially difficult. D. K. Bullens [2] has described the methods of sampling in present use on two of the mines, the main features being the careful reduction in size of the big sample, the use of mechanical samplers, and the separate assaying of the fine ore and "metallics." The comparative efficiency of scorification and crucible assays is discussed, and it is stated that, with the exception of the very rich ores high in nickel, cobalt, and arsenic, the crucible assay gives the best results. Ores from the oxidized zone present no special difficulty, as the nickel and cobalt are readily prevented from entering the lead button by an excess of litharge. If, however, more than 0·5 per cent. oxide enters the button, the latter becomes covered with a scum of the oxide during cupellation and finally freezes. For sulphide ores a strongly basic slag, high in soda and approaching to a sub-silicate, is employed, since arsenic, in addition to nickel and cobalt sulphides, may be slagged off by an excess of an alkaline flux at a low temperature. The fusion must, however, be conducted rapidly in order to prevent the formation of a speise, so that only small amounts of those ores which have a high nickel, cobalt, and arsenic content can be used, as little as 0·1 or even 0·05 of an assay ton being occasionally employed. For low-grade silver ore rich in nickel and cobalt, a flux containing an excess of litharge has proved satisfactory. The sulphide assay using metallic iron may be employed for comparatively high arsenic-nickel-cobalt ores if a large excess of the alkaline flux is used.

For the scorification of the ores high in silver the following charge is recommended :—

Ore	0·05 to 0·10 assay ton.
Granulated lead	65 to 75 grammes.
Borax glass	3 to 5 ,,
Sand	1 to 3 ,,

Slags and cupels should be treated for the recovery of the silver, and corrections made. It is found to be desirable to check results for silver from time to time by wet analysis. The methods of sampling and assaying silver ores containing the arsenides of cobalt, nickel, and silver, and also metallic silver, have also been described by Handy.[3] Details of the sampling are given on p. 95.

[1] *Trans. Inst. Min. and Met.*, 1909, **18**, 167.

[2] D. K. Bullens, *Eng. and Min. J.*, 1910, **90**, 809–810.

[3] "Sampling and Assaying of Silver Ores containing Cobalt, Nickel, and Arsenic." J. O. Handy, *Eighth Int. Cong. Appl. Chem.*, 1912, Sect. IIIA (abstract, *J. Soc. Chem. Ind.*, 1912, **31**, 880).

The samples are assayed in a gas-fired muffle furnace by the crucible fire-assay method, using the following fluxes :—(i) " Silver " flux containing 40 per cent. each of sodium and potassium carbonates and 20 per cent. of borax glass ; (ii) litharge ; (iii) borax glass for cover. The charge for the crucible contains 0·2 assay ton of pulp, 30 grammes of " silver " flux, 60 grammes of litharge, and 1 gramme of flour. These are mixed by rolling in a rubber cloth, transferred to the crucible, covered with borax glass, and heated to a medium red heat for twenty minutes. The door of the muffle is then shut and the crucible heated to a bright yellow for 30 minutes, the contents poured into a conical iron mould, and when cold hammered free from slag. The metal is then cupelled at as low a temperature as possible, using a freely oxidizing atmosphere without excessive draught ; the temperature and draught control are important factors. The following table gives examples of results obtained by independent assayers :—

Sample.	Silver, oz. per ton.	
Number	A.	B.
37396	142·46	142·64
37691	57·07	53·79
37252	2521·38	2518·95

(d) Auriferous Tin Ore.[1]—Native gold is not infrequently found in tin ores and in the alluvial deposits of tinstone (*cassiterite*, tin oxide), the chief ore of tin. The samples obtained by panning alluvial deposits may be very rich in gold, and contain 25 oz. or more of gold to the ton. The crucible assay gives excellent results, both for silver and gold present ; and as 25 grammes can easily be worked off, it is the most suitable method for most tin ores containing precious metal.

Soda is the chief flux used, as it forms with tin oxide sodium stannate, which is readily fusible. The ore should be crushed to pass a sieve of at least 90 holes to the linear inch. The following charge will give a very good slag with almost all tin ores :—

Ore	25 grammes.
Litharge or red lead	60 ,,
Charcoal	1·5 ,,
Sodium carbonate	40 ,,
Borax	10 ,,

It will be noted that a large amount of sodium carbonate is used in the charge as compared with the amount of ore taken, and that only sufficient charcoal is added to reduce half the lead oxide in the charge. By this means no tin is reduced during the fusion, and the resulting lead button, weighing about 30 grammes, is free from tin and can be cupelled direct without a preliminary scorification. The slags are quite satisfactory, being readily fusible, very fluid, and quite uniform in appearance.

[1] C. O. Bannister, " Assay of Auriferous Tinstone," *Trans. Inst. Min. and Met.*, 1906, **15**, 513.

Experience has shown that it is necessary to clean the slags from all auriferous tin ores carrying over 1 ounce of gold to the ton. For this purpose the slag is collected and fused with—

Litharge or red lead	... ...	30 grammes.
Charcoal	... ...	1·5 ,,
Sodium carbonate	... ...	10 ,,

The resulting lead button is cupelled and parted separately to obtain the values left in the slag. In assaying poor ores containing only a few dwt. of gold per ton it is advisable to work on 50 grammes of ore, and in this case more satisfactory results are obtained by making two separate fusions, each of 25 grammes, scorifying the resulting lead buttons together and then cupelling, than by making one fusion only on 50 g. of ore.

The scorification method invariably gives low results for auriferous tin ores, and this is due to the unsatisfactory nature of the slag obtained. Even when sodium carbonate, equal to the weight of the ore taken, is added to assist in the formation of a fusible slag of sodium stannate, the slags are never quite satisfactory. In the case of rich ores the results may be improved by taking small quantities of ore, such as 2·5 to 3 grammes, and using 5 grammes of sodium carbonate; but in this case any error occurring in the assay is largely magnified in calculating the result, and thus it is not advisable to use this method.

The amalgamation method of assay described on page 230 is used by some assayers for the determination of gold in ordinary alluvial tin ore, because of the necessity for operating on large quantities of such " patchy " ores, first, in order to obtain a fair sample, and second, on account of the extremely low gold values which they commonly carry.

(e) **Auriferous Bismuth Ores.**—Although not of frequent occurrence in nature, bismuth accompanies various ores of silver, and is also found occasionally associated with gold ores. The extent to which gold is found in combination with bismuth ore is referred to on page 365.[1] Bismuth is almost invariably found in nature in the metallic state, but occasionally it is met with as sulphide in bismuthine, and as oxide in bismutite (bismuth ochre). It is also found in combination with tellurium associated with gold.

Assays of auriferous bismuth ores are conducted like those of oxidized gold ores, but on account of the volatility of bismuth it is of importance that a readily fusible slag should be obtained, and for this purpose comparatively large quantities of sodium carbonate or of borax should be employed. When much bismuth is present, the quantity of red lead should be increased in order to effect its oxidation; a little nitre may also be added. Ores containing bismuth should not be roasted, as clotting invariably takes place owing to the low melting-point of bismuth (266° C.).

It is not essential to pass the whole of the bismuth into the slag, but it is desirable to do so, otherwise it alloys with the lead, and although bismuth cupels as satisfactorily as lead, the amount of gold and silver absorbed by the cupel and volatilised is greater when bismuth is present.[2] The time occupied by the cupellation is also lengthened, as bismuth is not quite so readily oxidized as lead.

The fusion of the ore is conducted at a comparatively low temperature. After pouring the charge the mould should be left undisturbed until the slag is quite cold, as the effect of bismuth when combined with lead is to form more readily fusible alloys which remain in a molten condition for a longer period

[1] Consult also *J. Soc. Chem. Ind.*, 1893, **12**, 316.
[2] E. A. Smith, *J. Chem. Soc.*, 1894, **65**, 624.

than lead. If the proportion of bismuth does not exceed that of the lead, the alloys are malleable, but the malleability is diminished with an increase of bismuth. The lead buttons obtained from the fusion, if inconveniently large, may be reduced in weight by scorification, but the temperature employed for this purpose should be low on account of the volatility of bismuth.

The effect of the presence of bismuth during cupellation is to reduce the surface tension of the molten alloy as in the case of tellurium, although the effect is less with bismuth. If the lead button contains a large excess of bismuth and a very small quantity of silver, the alloy of bismuth and silver may "wet" the cupel in the last stage of the cupellation and be partially absorbed. There is less tendency for this to take place with large silver buttons.

CHAPTER XV

SPECIAL METHODS OF ORE-TESTING AND ASSAY

In this chapter brief descriptions are given of the methods that are adopted in testing the suitability of an ore for a particular metallurgical treatment. The tests include (i) Panning or Washing, (ii) Amalgamation test, (iii) Blowpipe assay, (iv) Chlorination test, and (v) Chloridized silver-ore test. The testing of ores for cyanide treatment is dealt with in Chapter XXV.

These preliminary tests must be regarded only as giving some indication of the possibility of treating an ore by the process in question. In the event of these tests proving to be satisfactory, they must be followed by other tests on a much larger scale.

(1) **Panning or Washing Test.**—This method is used for determining the amount of free gold in an ore, and consists in washing the powdered ore in an iron pan, whereby the heavier particles of gold accumulate at the bottom of the pan whilst the lighter particles of sand, etc. are washed away. The accumulated gold, free from sand, is then dried and weighed. The pan is largely used by prospectors for quickly arriving at an approximate estimate of the value and distribution of free gold in an alluvial deposit or vein.

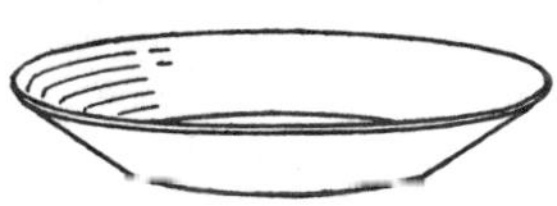

FIG. 117.—GOLD-WASHING PAN.

FIG. 118.—PAN WITH RIFFLE.

The pan (Fig. 117) is usually made of thick sheet-iron, with a flat bottom about 12 inches in diameter, and has sides from 5 to 6 inches in height, sloping outwards at an angle of about 30° to the bottom. Some pans are made with a riffle formed by the thickening or bulging inwards of the side, situated about half-way up the latter and running about half-way round the pan. A pan with a riffle is shown in Fig. 118.

The method of using the pan is as follows :—

For laboratory tests, from 2 to 4 lbs. of the crushed ore or gravel are placed into the pan and submerged in water until the contents are well soaked, in the meantime being stirred with the hand. The pan is then held by the two sides, inclined away from the operator, and raised until the gravel is only just covered with water. Then a half-circular motion is given to the pan, which will cause the finer particles of ore to float off, leaving the coarser material exposed to view. After a second stirring, the coarser material and any pieces of rock are removed, care being taken to wash them properly. This process is repeated until all the gravel has been swept away by the water, and until gold and a certain amount of black sand (magnetic iron oxide) alone remains. From this point onwards great care must be exercised as, the black sand being heavy, the gold is liable to float away with it and so be lost. When the panning has been successfully performed the "colours," *i.e.* yellow specks of gold, are seen distinct from the black sand. The actual motions given to the pan vary with different operators; some use a circular movement until the black sand is reached, whilst others let the water flow on and off the pan without any circular motion whatever. To be of any value a panning must be made with care, that is, with due regard to the laws governing hydraulic separation. Prospectors frequently use a full miner's pan of ore, amounting to about 20 lbs. or 10 kg.

Panning is very frequently performed in a *batea*, shown in Fig. 119, which differs from the ordinary gold-washing pan in not having a flat bottom. It is made of wood turned in a lathe, about 20 inches in diameter, conical, or more rarely basin-shaped, and about 3 inches deep in the centre so that the angle at the apex is about 160°. The gold collects at the lowest point and clings to the wooden surface under conditions such that it would slide over iron. The batea consequently is more rapid and effective in obtaining a "colour" than the pan, especially when the gold is fine. The best material for the batea is mahogany cut with the direction of the grain vertical to the surface of the implement.[1]

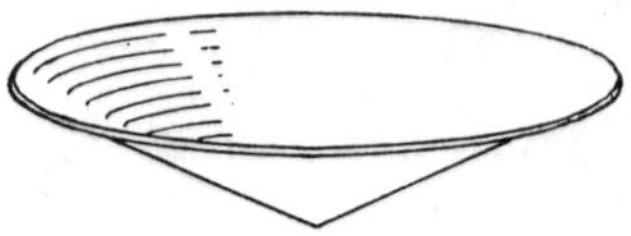

FIG. 119.—BATEA FOR GOLD-WASHING.

The batea is largely used in South America and South Africa, while the pan is in more general use in the United States and Australia.

During panning a loss may result in float-gold, *i.e.* very minute particles of gold, so light that they are often floated away and lost. "The 'colour' or showing of gold obtained on panning an ore carrying iron pyrites (and nearly all unoxidized gold ores carry more or less iron pyrites) is decidedly smaller than that obtained from an oxidized ore of the same assay value. Two reasons probably contribute to this, one being the fact that when mixed with iron pyrites the gold does not separate so well, both on account of the particles being finer, and also because the heavy pyrites tends to cover it up. The other reason is the fact that the pyrites in most cases is, in itself, much richer than the rest of the ore, and thus locks up and conceals an appreciable proportion of the gold."[2]

The amount of gold recovered by panning is determined in several ways. Those who have had considerable experience in panning very frequently judge the value from the apparent amount of the gold particles or "colour" left on the pan. Such guesswork is, however, of little value, as it is almost impossible to value the material obtained by panning by inspection. In some cases the pannings are determined by comparison with weighed standards of gold dust, contained in small glass tubes, but since different samples of gold dust vary considerably in character and physical condition, some being coarse and some very fine, the standards employed should be especially prepared from gold dust obtained from the district in which the panning tests are being made.[3]

The gold may be weighed when sufficient in amount, after drying and separating the black sand by blowing. The residue in the pan is dried and placed in a "blower," consisting of a shallow metal or horn scoop open at one end. This is held with the open end away from the operator and the sand removed by gentle blowing, leaving the particles of gold behind. Unless the blowing is done with extreme care gold is very liable to be lost. The gold when free from sand is weighed.

[1] Rose & Newman, *Metallurgy of Gold*, 7th edn., 1937, p. 106.

[2] W. Frecheville, *Trans. Inst. Min. and Met.*, 1909, **18**, 494.

[3] Consult S. J. Lett, "The Use of Standards in reading Gold Pannings," *Trans. Inst. Min. and Met.*, 1909, **18**, 482–494.

A more general method is to add a little mercury to amalgamate with the gold, and after collecting the liquid amalgam carefully, to volatilize the mercury and then accurately weigh the gold bullion on a portable assay balance. From this weight the value in free gold per ton is calculated. The fineness of the gold can be easily determined by inquartation with the blowpipe and parting in nitric acid. The fineness of the bullion is of importance, as it makes a material difference in the value of the gold-pannings. An objection to the use of mercury is that the gold from many alluvial deposits does not amalgamate readily. The gold is sometimes recovered from the mercury by treating the amalgam with nitric acid. The method yields unsatisfactory results, because the nitric acid dissolves out a portion, but not all, of the silver present in the bullion, especially in the case of coarse gold, which becomes coated and held by the mercury, but is not dissolved in it, so that these particles of gold retain a large and varying proportion of their silver after treatment with the nitric acid.[1]

The most satisfactory method of recovering the gold is to transfer it carefully from the pan to a small piece of lead-foil about 4 inches square which is then wrapped up and cupelled. The bullion button is then inquarted and parted so that the final result is given in fine gold.

When the visible gold obtained by panning is collected and weighed, the results are generally low; when the collected pannings are fused and assayed, the results may be high as regards "free" gold, owing to the inclusion of a certain amount of concentrate containing gold.

The results of the pannings of crushed ores such as auriferous quartz are best returned in ounces or pennyweight of fine gold per ton of 2000 lbs.

Where 2 lbs. of quartz are taken, the following are the weights of gold dust corresponding to the attached values[2]:—

TABLE XXXII.—WEIGHTS OF GOLD-DUST PER SHORT TON CORRESPONDING TO WEIGHTS PER 2 LB. OF QUARTZ. (*Lett.*)

Weight of Gold-Dust.		Equivalent in ounces and pennyweight per short ton.
Grains.	Grammes.	
0·024	0·00156	1 dwt. per ton of 2000 lbs.
0·048	0·00311	2 ,, ,, ,,
0·096	0·00622	4 ,, ,, ,,
0·144	0·00933	6 ,, ,, ,,
0·192	0·01244	8 ,, ,, ,,
0·240	0·01556	10 ,, ,, ,,
0·480	0·03110	1 ounce ,, ,,
0·960	0·06221	2 ,, ,, ,,
1·920	0·12442	4 ,, ,, ,,

Data:
- 1 lb. avoirdupois = 7000 grains troy.
- 1 ton (2000 lbs.) = 907,185 grammes.
- 1 gramme = 15·43235 grains.
- 1 grain = 0·064798 grammes (0·0648 used).

The assay values of alluvials should be reported in grains and decimals of a grain of "fine" gold, or in pence (at 2d. per grain of "fine" gold) or cents per cubic yard.

[1] G. T. Holloway, *Trans. Inst. Min. and Met.*, 1903, **12**, 462.

[2] Lett, *loc. cit.*, p. 485.

It is recommended by the Institute of Mining and Metallurgy[1] that, in the absence of specific information as to boulders, etc., 1 cubic yard of ordinary alluvial be taken as equivalent to 3000 lbs. ($1\frac{1}{2}$ short tons). Samples of alluvial deposits are frequently very poor in gold, yielding only 1 or 2 grains of gold per cubic yard, it is therefore advisable to use 6 lbs. of material for the panning.

Although the results obtained by panning cannot be taken as a correct index of the gold content of an ore, the pan, as previously stated, is undoubtedly, in experienced hands, an efficient method of arriving at an approximate estimate of the value and distribution of gold in a deposit or vein, provided the showing in the pan is checked by some assays of typical samples. When the character of an ore is fairly well known and carries most of its value in "free gold," the relative amount of gold that different samples contain can be ascertained without any great difficulty. Panning can only be learned by experience. Students may practise the use of the pan by weighing out 2 grammes of dressed tin ore, mixing it with 2 lbs. of powdered quartz, and then panning as described. The success of the operation is judged by drying and weighing the tin oxide left on the pan.

(2) **Amalgamation Test.**—The object of this test is to ascertain the amount of gold that can be recovered from an ore by amalgamation. It is not regarded as a method of assay capable of giving the total gold content of the ore.

The test consists in treating a weighed sample of the ore with a known quantity of mercury, and then determining the gold in the resulting amalgam. Preferably three tests are made, namely (i) an assay of the original ore, (ii) the gold content of the amalgam recovered, and (iii) an assay of the tailings after separation of the amalgam.

Amalgamation tests are made either by mixing the ore with mercury in a mortar, or by agitation in a bottle. Bottles are the more convenient and are usually employed when a large number of tests have to be made to control mill practice. The tests are made in strong narrow-mouthed bottles with spring clamps and rubber washers, having a capacity of about 350 c.c. The bottles are shaken by hand, or more frequently in some form of mechanical vertical shaker, such as that described and illustrated on p. 307, and preferably one which allows of the bottles being introduced or removed whilst in motion.

In carrying out the test, 100 grammes of crushed ore or tailings are weighed out, charged into the bottle by means of a metal funnel, 150 c.c. of water added, and then 1·5 to 2 c.c. of mercury run in from a burette.[2] The stopper is then clamped, the bottle wrapped in a thick piece of cloth or flannel, put into the shaker, and agitated for 2 hours. It is then removed and opened, the mouth covered with the thumb, then shaken and inverted over a 3-inch porcelain dish, and as much as possible of the mercury allowed to run out. It is again shaken for a moment, and if necessary a little more water added, and more mercury allowed to run out, the operation being repeated so long as any mercury comes out. If not floured, the mercury is nearly all removed in two operations. The mercury obtained is transferred to a beaker. The bottle is again shaken well and then inverted to let a little sand run out into a dish, and the sand then panned to separate any remaining mercury. Usually only a few small globules are obtained: if much is found the whole charge must be panned. If the mercury has floured, a small globule of liquid sodium amalgam should be added at this stage to quicken amalgamation.

To the mercury, after collecting in a 250-c.c. beaker, is added 0·5 gramme of pure silver-foil. If this has been cleaned by treatment with dilute acid,

[1] *Trans. Inst. Min. and Met.*, 1911, **20**, 530.
[2] W. J. Sharwood, *Amer. Inst. Min. Eng.*, 1915.

or by slightly amalgamating the surface, it may be used to pick up small globules of mercury obtained in panning. The amalgam is treated with about 150 c.c. of warm nitric acid (sp. gr. 1·14 or 1·15) and the residual gold filtered off and washed. The filter-paper is then wrapped in lead foil, placed in a small glazed scorifier, and covered with lead to make up the total to 15 or 20 grammes.

If necessary, additional silver is also added to ensure ready parting. The charge is covered with 2 or 3 grammes of borax glass, the scorifier placed in a muffle, and the charge scorified for a few minutes, after which the lead is poured and cupelled. The button from cupellation is parted in a porcelain cup, and the gold annealed and weighed.

If necessary a correction is made for any gold contained in the silver or mercury used. Using a sample of 100 grammes, each milligramme of gold represents 0·29166 ounces of amalgamable gold per ton of 2000 lbs. If preferred, four assay tons may be used, when each milligramme indicates exactly 0·25 troy ounce; or 120 grammes may be taken, when each milligramme indicates exactly 5 dollars per ton. With these quantities the water must be proportionally increased, and the manipulation in the bottles is not quite so easy as with 100 grammes. In ordinary routine work the tailings may be discarded. Time is saved on the whole by assaying the mercury rather than the tailings, provided sufficient pure mercury can be prepared or obtained, the commercial metal not infrequently containing a little gold. Sufficient mercury must be used to allow of quick action and to avoid serious error in case a few centigrammes are lost.

The addition of chemicals to assist amalgamation is generally to be avoided, and sodium amalgam should be used only when absolutely necessary. The ore to be tested should be ground to a fairly constant degree of fineness of not less than 80-mesh.

Sharwood states that, with the above method of testing, the average percentage of gold amalgamated in some thousands of 100-gramme tests differed very little from the average stamp-mill percentage of recovery for the same period of years.

The chief advantage of carrying out the amalgamation test in a porcelain mortar is that large quantities of ore can be used, from 500 to 1000 grammes frequently being taken.

The ore is placed in a mortar and made into a stiff paste with water, then from 15 to 20 grammes of mercury added, and the whole thoroughly mixed with the pestle for an hour or more. The consistency of the mass after adding water must be such that the globules of mercury do not sink in it but are uniformly distributed through it; and flouring of the mercury is avoided as much as possible. When amalgamation is judged to be complete, the mass is reduced to a thin pulp by the addition of more water, and the whole gently stirred for a few minutes to facilitate the collection of the amalgam at the bottom. The ore is then separated from the mercury by washing it out with a jet of water, the overflow of tailings being collected in a large pan, and further examined for globules of mercury. The procedure for the recovery of the gold is similar to that already described.

Some assayers prefer to heat the amalgam in a scorifier to expel the mercury, then add lead and borax to the residual gold, and cupel the resulting lead button. Others wrap the residual gold in a small piece of lead foil, with sufficient silver for parting, and cupel direct.

The amalgamation assay generally yields low results with ores containing so called "rusty" gold, *i.e.* gold which does not readily amalgamate because the particles are coated with a film of iron oxide.

(3) **Blowpipe Assay.**—This method of assay may be described as a fusion assay in miniature, as it comprises all the operations of an ordinary assay on

a small scale. As a general rule, however, precious-metal ores are too low-grade to permit of a blowpipe assay, and this is especially true of gold ores, with which it is usually necessary to concentrate the values by panning a pound or more of the ore, and to assay the concentrates. Though less exact than furnace methods, it is possible by means of the blowpipe assay not only to detect the presence of gold, etc. in an ore, but also to determine the amount with a fair degree of accuracy.

Since a prospector must be prepared to make an assay, even if only approximate, this method is of importance, as it would obviously be impossible, on prospecting expeditions, to carry the cumbrous apparatus required to make an ordinary assay.

Special blowpipe assay outfits are available.

The method usually adopted is as follows :—A weighed quantity, usually 100 mg. of the powdered ore, or concentrate, is mixed with an equal weight of borax and about 1 g. of granulated lead. The whole is wrapped in tissue paper, placed in a cavity made in a piece of charcoal, and heated in the reducing flame of a blowpipe until fusion is complete. If pyritic material is present, the oxidizing flame may have to be used for part of the time. In some blowpipe outfits small fireclay crucibles and a carbon furnace are provided for the fusion. The lead is then separated from the slag, placed on a small bone-ash cupel, and heated in the oxidizing flame until it is all converted into litharge. The resulting button of silver or gold, which is too minute to permit of weighing, is carefully measured on an ivory scale, which at once gives the percentage amount of precious metal in the ore.

The scale for this purpose consists of a strip of polished ivory on which are drawn two very fine and distinct lines emanating from the zero point at one end and diverging towards the other end. The lines are divided by cross-lines into 100 equal parts, numbered from zero upwards, with numbers which correspond to weights of the buttons in milligrammes.

Thus, if a small button of silver is placed in the space between the two lines, using a magnifying glass to assist the eye in moving it up or down until the diameter of the button is exactly contained within the lines themselves, the diameter of the button can be determined, and from this measurement its weight is found by reference to the figures on the scale. In the case of buttons from gold ores it is usual to separate the gold from the silver by parting in nitric acid.

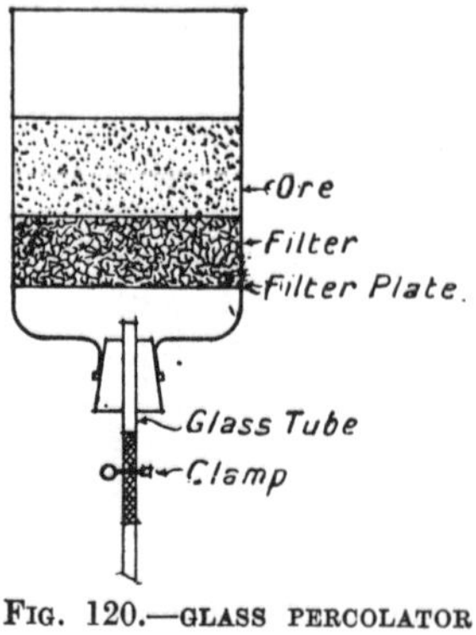

FIG. 120.—GLASS PERCOLATOR FOR CHLORINATION OR CYANIDE ASSAY.

(4) Chlorination Assay.—This method is used in determining the amount of gold present in an ore capable of being extracted by chlorine.

It is applicable to earthy or oxidized ore direct and to ores containing pyrites after roasting "dead." From 200 to 300 grammes of ore (through 60-mesh) are slightly moistened with water, and placed in a glass bell-jar about one litre capacity, with an opening at the bottom for the introduction of a current of chlorine gas (see Fig. 120). A layer of broken quartz or glass, coarse at the bottom and fine at the top, is first put into the jar to act as a filter, and on this is placed the moistened ore, which should not fill the jar more than two-thirds full. Chlorine is then passed into the jar, and when the odour of the gas is noticed above the ore, a cover is put on, the stream of chlorine stopped, and the whole left for one or two hours; after which the reaction is complete if chlorine is still in excess. Hot water is now run through the ore to dissolve the gold chloride formed, and the solution boiled to expel the free chlorine.

A solution of ferrous sulphate and a little hydrochloric acid is added, which precipitates the gold in the metallic state.

The precipitate is collected on a filter, then washed, dried, wrapped in lead-foil, cupelled and weighed. The result obtained is compared with the assay value of the ore before treatment with chlorine. This method fails if more than about 20 per cent. of silver is present, as the silver chloride formed encrusts the gold and protects it from the action of the chlorine. The chlorine may be generated by gently heating a mixture of manganese dioxide and hydrochloric acid; or, instead of employing hydrochloric acid, the materials from which the acid is prepared, namely, sodium chloride and sulphuric acid, may be used. Suitable quantities for one test are as follows:—

Manganese dioxide	3 parts	or	24 grammes.
Sodium chloride	4 ,,		32 ,,
Sulphuric acid (commercial) ...	10·5 ,,		84 c.c.
Water	7 ,,		56 ,,

By adding the water and acid gradually, the gas-supply will run for about $1\frac{1}{4}$ hours.[1] If necessary the tailings left in the jar may be dried and assayed, but care must be taken not to include any of the filter bed when removing the ore.

(5) **Chloridized Silver-ore Assay.**—This test is used to determine the amount of silver present as silver chloride in silver ores that have been chloridized or roasted with salt to convert the silver sulphide present into silver chloride preparatory to the metallurgical treatment of the ore.

Two assays are made, one on the sample of roasted ore as received, and the other on another portion of the same sample after treatment to extract the silver chloride. The difference between the two results gives the amount of silver converted into chloride by roasting the ore. The assay value of the roasted ore is determined by scorification on 3 grammes of ore in the usual manner.

Another 3 grammes are placed in a beaker and digested with a warm 2 per cent. solution of sodium thiosulphate (hyposulphite), which readily dissolves the silver chloride from the ore. The treatment is continued until a small portion of the filtered liquid contained in a test-tube darkens but slightly and does not lose its transparency upon the addition of a few drops of sodium sulphide solution.

When free from silver chloride, the ore is filtered and washed with warm water, then dried, the filter-paper burned, and the residue scorified with 30 grammes of lead and 1 gramme of borax. The lead is cupelled and the silver button weighed. As a confirmatory test, the silver in the clear filtrate may be precipitated as sulphide by the addition of sodium sulphide or by acidulating the solution with hydrochloric acid and warming. The precipitate is filtered off and dried, then gently heated to expel the sulphur and burn the filter-paper, and the residue wrapped in a little lead-foil and cupelled. Any silver sulphate contained in the ore will also be dissolved by the hyposulphite solution and thus be included as chloride. If, as is sometimes the case, it is desired to know the amount of silver sulphate present, a third quantity of 3 grammes of the ore is treated with warm water, which will dissolve the silver sulphate only. The residue

[1] Lodge, *Notes on Assaying*, p. 225.

is then dried and scorified as before. The amount of silver present in the roasted ore as chloride is ascertained as shown in the following example :—

(1) Assay value of roasted ore	195·5	ounces per ton.
(2) Assay value after treatment with hyposulphite	12·3	„ „
Difference equals silver present as silver chloride	183·2	„ „

Therefore, of the total silver in the chloridized ore, the percentage of silver present as chloride will be :—

$\frac{183 \cdot 2}{195 \cdot 5} \times 100 = 93 \cdot 71$ percentage of silver present as chloride in total silver content of ore.

The amount of silver present as sulphate is determined by subtracting the result of the third assay from the result of the first assay and then calculating the percentage. If the ore contains gold, the amount present must be determined by parting the silver buttons and deducting the weight of gold obtained.

The amount of silver present as chloride may also be determined by digesting 5 grammes of the roasted ore in a beaker with dilute ammonia solution, which dissolves the silver chloride. The solution is next filtered, the residue washed with water, and the silver precipitated from the clear solution by acidulating with nitric acid. The precipitate is then washed, dried, and weighed as chloride; or it may be wrapped in a piece of lead-foil and cupelled, and the resulting silver button weighed and the percentage calculated in the usual manner.

This method is more expeditious than that first described, and when carefully performed the results are quite satisfactory.

CHAPTER XVI

GOLD AND SILVER: PHYSICAL AND CHEMICAL PROPERTIES

I. GOLD

Occurrence of Gold.—Originally gold was called by the alchemists a "noble" metal because it was not affected by fire in the same way as copper and other oxidizable metals.

Gold is very widely diffused in Nature and generally occurs in the metallic state, alloyed with a varying proportion of silver and smaller quantities of copper and other metals. Occasionally metals of the platinum group are present. Native gold is never found pure but sometimes contains less than 1 per cent. of impurities. It has been known to contain as much as 99·8 per cent. of gold, but as a rule the gold content ranges from 85 to 95 per cent., the balance being usually silver for the most part.

The British Empire is responsible for about 60 per cent. of the entire world's output of gold, a large proportion of which is produced in South Africa.

Commercial Forms of Gold.—Metallic gold is usually brought into the market as bullion in the form of ingots in a more or less refined state and weighing usually 200 or 400 ounces troy each. The metal sent by the Bank of England to the Royal Mint for coinage is usually in the form of refined ingots, each weighing about 400 ounces and containing over 99 per cent. of gold.

In addition to small ingots of varying weight, various forms of gold are prepared for industrial use. The refined gold used by goldsmiths for the preparation of alloys, etc. is supplied in four different forms, namely, "brown," "grain," "dull," and "sheet" golds.

Brown Gold is in the form of powder, and is the residual gold from refining by the nitric acid parting process, washed and dried. It is principally used for the preparation of gold solutions for electro-gilding, as gold in this form dissolves readily.

Grain or Granulated Gold is prepared by melting "brown" gold and then slowly pouring it into water, whereby it is obtained in the form of small granules or grains. Gold in this form is very convenient for weighing out for alloying purposes, etc.

Dull Gold consists of brown gold which has been heated at a red-heat whereby it changes colour and becomes dull-yellow. On burnishing it assumes the bright lustre of fabricated gold. Dull gold is now very little used, but there are still a few manufacturing jewellers who prefer it to "grain" gold for alloying purposes.

Sheet Gold is prepared by casting the metal into a flat bar or ingot and then rolling it to the desired thickness of sheet. Sheet gold is used for "anodes" in electro-gilding and is cut to the required size.

Gold for Manufacture.—The different carat qualities of gold used in manufacture and in the arts are prepared in the form of sheet metal of all thicknesses, wires both plain and fancy, and seamless tubing etc. Alloyed gold is also manufactured into gold leaf, of various qualities containing from 90 to 98 per cent. of gold, the remainder being silver and copper.

Physical Properties of Gold.—Symbol, Au (aurum); atomic weight, 197·2; specific gravity, 19·32; melting-point, 1063° C.

When pure and in mass, gold is the only metal which has a characteristic yellow colour, the so-called "golden yellow." The colour is greatly modified by impurities, silver and nickel lowering the tint, whilst copper heightens it. Gold leaf is green by transmitted light, and precipitated and other finely-divided gold may assume brown, purple, or black colours.

The melting point of gold is approximately 1063° C. that is, just below that of copper (about 1083° C.), and about 100° higher than that of silver. When molten, pure gold begins to volatilize in ordinary furnaces at a temperature just below 1100° C., but the loss probably seldom exceeds 0·1 per thousand in melting a charge of 1200 ounces when proper precautions are taken in melting (Sir T. K. Rose).

Pure gold is softer than silver, but harder than tin. It is the most malleable and ductile of metals, and can be welded in the cold state. Gold-foil and leaf of extreme thinness can be obtained by beating. The tenacity of gold is comparatively low, being 7 tons per square inch, which is less than that of silver or copper. Unalloyed gold has an elongation of nearly 31 per cent. Its electrical conductivity at ordinary temperatures is 76·7 compared with silver at 100. Its conductivity for heat is also less than that of silver or of copper.

The properties of pure gold are greatly affected by the presence of impurities, even small traces of certain metals, such as lead, bismuth, and antimony having a most marked effect on its malleability, ductility, tenacity and weldability. Thus, if 0·05 per cent. of bismuth is present the metal becomes so brittle that it can almost be crumbled in the fingers (Gowland). A very small quantity of lead also produces extreme brittleness.

Pure gold finds only limited application in industry, and it is in the form of alloys that the metal is chiefly used for coinage and jewellery.

Alloys of Gold.—Although gold will unite with most metals, comparatively few of the alloys are of practical use in the arts or manufactures. The more important alloys of gold are those with copper and silver which are largely used in the jewellery trade and in dentistry. The alloying of gold has in view the hardening of the metal, the lowering of the freezing-point for soldering purposes, and the cheapening of the production; at the same time considerable importance is attached to the colour of the alloys. Advantage is taken of the bleaching action of silver on the yellow colour of gold, to overcome to a great extent the reddening effect of copper. The proportion of gold in these alloys is known as "fineness," and is expressed either in parts per thousand or in carats (twenty-fourth parts), as more fully explained on page 333, where industrial gold alloys are further discussed.

Chemical Properties of Gold.—The most rapid solvent for gold, and the chief solvent used in the laboratory, is hot aqua regia, a mixture of three parts of hydrochloric acid with one part of nitric acid, which evolves chlorine, the active solvent, forming auric chloride, $AuCl_3$. According to Handy,[1] solution is more rapid in a mixture of four parts of hydrochloric acid with one part of nitric acid. Gold is not perceptibly attacked by hydrochloric or nitric acids alone, or by sulphuric acid, or by alkalis. Finely divided gold is slightly soluble in boiling hydrochloric acid, and in boiling nitric acid. Under certain conditions it is also slightly soluble in boiling sulphuric acid.

[1] J. O. Handy, *Chem. News*, 1912, **105**, 186.

It is also soluble in any mixture producing nascent chlorine, as, for example, solutions containing mixtures of (i) nitrates, chlorides, and acid sulphates; (ii) chlorides and some sulphates, *e.g.* ferric sulphate; (iii) hydrochloric acid and nitrates, peroxides, such as permanganates, or chlorates; (iv) bleaching-powder and acids. Gold is soluble in solutions of bromine, or of mixtures producing bromine, and in boiling concentrated solutions of ferric chloride. Iodine dissolves gold only if it is nascent, or if dissolved in an iodide, or in ether or alcohol. The metal is also soluble in solutions of the thiosulphates (hyposulphites) of calcium, sodium, potassium, and magnesium, in the presence of an oxidizing agent (Gowland).

It is slowly dissolved by a dilute solution of potassium or sodium cyanide at the ordinary temperature (this action, however, depends on the presence of oxygen or of an oxidizing agent). A convenient method of dissolving it is to pass a current of electricity through a solution of an alkaline cyanide, using the gold as an anode (Sir T. K. Rose).

Precipitants of Gold.—One of the chief chemical characteristics of gold is the difficulty with which its compounds are formed and the ease with which they are decomposed. Solutions of gold chloride are readily decomposed and the gold precipitated by many reducing agents, of which the following are some of the more important used in assay practice:—

Sulphurous Acid (Sulphur Dioxide).—This is a convenient precipitant of gold and is often used in the laboratory, as it acts almost equally well in cold and hot solutions. It gives a finely granular brown precipitate of metallic gold. Precipitation is retarded by a large excess of hydrochloric acid; and nitric acid and other oxidizing agents must be absent.

Oxalic Acid (*or alkali oxalates*).—This forms a very useful precipitant of gold. It does not act very readily in cold solutions, but in hot solutions of gold chloride metallic gold is precipitated by excess of oxalic acid, or ammonium, or sodium oxalate, either as a scaly brown precipitate, or as a coherent film on the surface of the glass vessel, according to the strength and temperature of the gold solution. Platinum chloride is not reduced by oxalic acid.

Ferrous Sulphate.—In comparatively strong solutions of gold chloride a brown, finely-divided precipitate of metallic gold is formed; and in dilute solutions the fine particles of gold cause the solution to become blue by transmitted light.

Ferrous Chloride.—This precipitant acts in much the same way as ferrous sulphate and is preferable. It should be free from ferric chloride.

Dimethyl-glyoxime.—From feeble hydrochloric solutions of the chloride, gold is completely precipitated in the metallic state on boiling with dimethyl-glyoxime. The bronze-coloured flocculent precipitate settles readily on standing. The washed precipitate may be ignited directly for weighing.

Hydrazine Hydrochloride.—When a solution of chloro-auric acid, $HAuCl_4$, of 0·2 to 0·1 per cent. strength, is treated with a 10 per cent. solution of hydrazine hydrochloride, a spongy form of metallic gold is precipitated which for quantitative determination may be heated to redness and weighed.

Stannous Chloride.—With concentrated solutions of gold chloride, stannous chloride gives a purple-brown precipitate of ill-defined composition, known as "purple of Cassius." The production of the purple tint is facilitated by the presence of a small quantity of stannic chloride, which is always present in a solution of stannous chloride except when freshly prepared. A mixture of stannous and stannic chlorides is prepared by digesting a piece of metallic tin in a solution of stannous chloride with the addition of a little hydrochloric acid.

Hydrogen Sulphide.—From a cold solution of gold chloride, hydrogen sulphide gives a black precipitate of auro-auric sulphide Au_2S_3, and from a boiling solution a brownish precipitate of aurous sulphide Au_2S. These precipitates are insoluble in hydrochloric acid or in nitric acid, but dissolve in aqua regia. Oxidizing agents, such as nitric acid, should be absent before the addition of hydrogen sulphide.

Ammonium sulphide also gives a precipitate of gold sulphide.

Ammonium hydrate gives an orange-red precipitate of ammonium aurate or fulminating gold, $(NH_3)_2\ Au_2O_3$, a compound of auric oxide and ammonia, which detonates when dried and struck. It is decomposed without detonation by hydrogen sulphide, and by stannous chloride. It is insoluble in water, but soluble in potassium cyanide, with the formation of potassium auricyanide.

Hydrogen Peroxide.—This reagent precipitates metallic gold from alkaline solutions of the chloride, the brown precipitate agglomerating readily on heating. Precipitation is facilitated in dilute solutions of gold chloride by the addition of hydrochloric acid and heating.

Caustic Potash.—Neither the hydroxides or the carbonates of potassium or sodium give any precipitate with moderately dilute solutions of gold chloride. From a concentrated solution potassium hydroxide gives a brown precipitate of hydrated auric oxide, soluble in excess and yielding potassium aurate.

Organic Precipitants.—Many organic substances readily precipitate gold from solutions of the chloride. Thus sugar boiled in it yields first a light-red precipitate of metallic gold, which afterwards darkens in colour. Gold is also thrown out of solution by finely-divided charcoal, and by formates, etc.

Metals as Precipitants.—Gold is precipitated from hydrochloric acid solutions of the chloride by most metals, usually in the form of black scales or powder. Zinc and aluminium, in the form of powder, and magnesium in the form of ribbon, are frequently used in the laboratory for precipitating gold.

Preparation of Proof Gold.—The purest gold obtainable is required for use as checks or proofs in assaying gold bullion, etc. It may be purchased, or prepared from a solution of auric chloride by precipitation with sulphurous acid, or oxalic acid.

It is conveniently prepared from cornets from gold assay, which contain about 1 part per 1000 of silver. These are dissolved in 4 parts of hydrochloric acid and 1 part of nitric acid and excess of acid driven off by evaporating nearly to dryness with additional hydrochloric acid on a water-bath. The blackish-red liquid remaining, smelling of chlorine, consists mainly of chloro-auric acid ($HAuCl_4$) which solidifies at 70° C. and is decomposed if strongly heated. When cool it is dissolved in distilled water and diluted to contain about 30 grammes of gold per litre. After vigorous stirring the solution is allowed to stand for some time to enable the finely-divided silver chloride to settle.

The solution of gold chloride is decanted off, further diluted, and allowed to stand until no further settlement of silver chloride takes place. The clear solution is then siphoned off into a saturated solution of sulphurous acid in distilled water, which readily precipitates the gold in a finely-divided state which settles readily. After decanting off the clear liquor the precipitate is thoroughly washed by repeatedly shaking with boiling distilled water. Finally the precipitate is dried and then melted in a clay pot which has previously been washed out with molten borax to close up any pores. The molten metal is cast into an ingot mould, and the bar cleaned by scrubbing with fine sand and water.

The gold is then rolled down to the desired thickness, and again cleaned as before; it is then ready for use.

Oxalic acid is often used as a precipitant instead of sulphurous acid, and is generally regarded as the best precipitant if platinum is present, since platinum is not precipitated from solutions of platinic chloride by this reagent.

In this case the solution of gold chloride should be kept boiling and allowed to stand after adding a large excess of oxalic acid, as precipitation is slower than with sulphurous acid. The precipitated gold is largely in the form of small scales and plates and is very pure. It is washed and prepared for use as before described.

Detection of Gold in Solution.—Ferrous sulphate is frequently used to detect the presence of gold in solution as chloride. Two test-tubes or Nessler tubes are placed side by side on a sheet of white paper, and both nearly filled with the solution to be tested, which should be quite clear. To one tube a few drops of a solution of ferrous sulphate are added. On looking down the tubes a brown coloration, due to finely-divided gold in suspension, will be seen in the tube to which the ferrous sulphate has been added. In this way, with a little experience, the presence of gold in the proportion of only 1 dwt. per ton of water can be detected. The liquid containing the precipitated gold may also be compared with a tube containing water.

A still more delicate test for the presence of minute quantities of gold in solution as chloride is afforded by the reaction of a mixture of stannous chloride and stannic chloride in solution with gold chloride, by which the characteristic precipitate of "purple of Cassius" is produced. The solution containing gold chloride is heated to boiling and poured suddenly into a large beaker containing 5 to 10 c.c. of a saturated solution of the stannous chloride, when a precipitate is at once produced which will be coloured, or tinted purplish-red or blackish-purple, even if the solution contains as little as 1 part of gold in 5,000,000 parts of water, equivalent to 3½ grains per ton of solution.

By this test Sir T. K. Rose[1] has detected the presence of gold in 100,000,000 parts of water (1 grain in 6 tons).

According to Argall[2] the presence of gold in quantities less than 0·1 milligramme may be detected in cyanide solutions by this test.

The cyanide solution to be tested is slightly acidulated and then boiled until most of the HCN is expelled. Potassium chlorate is then added and the solution again boiled until the chlorous gases are driven off. Finally stannous chloride is added, which gives the "purple of Cassius" precipitate as before.

The solution for testing should not be too strongly acid, otherwise the purple colour may not appear. The colour frequently becomes more distinct after standing for some time.

Separation and Determination of Gold.—In using the above precipitants for the quantitative separation and determination of gold, it should be borne in mind that ferrous sulphate and sulphurous acid act well in moderately strong hydrochloric solutions of gold chloride, whilst oxalic acid, hydrogen sulphide, and ammonium sulphide act best in the presence of a slight excess only of hydrochloric acid.

Precipitation with ferrous sulphate is comparatively slow in cold solutions, but more rapid and satisfactory in hot solutions. In strong solutions the precipitate is very finely divided and difficult to wash by decantation without loss. Schoeller and Powell[3] give preference to ferrous chloride for the reduction

[1] *Chem. News*, 1892, **66**, 271.
[2] P. H. Argall, *Mill and Smelter Methods of Analysis*, 3rd edn., 1908, p. 73.
[3] *Analysis of Minerals and Ores of the Rarer Elements*, 2nd edn., 1940, p. 262.

of auric chloride or, if the introduction of iron into the solution is undesirable, to hydrazine hydrochloride.

For ferrous-chloride reduction the hot solution of auric chloride, freed from nitric acid by double evaporation with hydrochloric acid, and containing one-fifth of its volume of hydrochloric acid, is treated drop by drop with a slight excess of a strong, freshly prepared solution of ferrous chloride free from ferric salt. After standing the precipitate is collected and washed with 2 per cent. hydrochloric acid. Oxalic acid or preferably ammonium or sodium oxalate, acts best in boiling solutions, and in the presence of much hydrochloric acid or of alkaline chloride; a large excess of the precipitant should be present. In the cold, and in the presence of much hydrochloric acid or of alkali chloride, the action is very slow and partial. The gold is precipitated in the form of plates or scales which are very pure, and are readily washed. Sodium oxalate is a useful precipitant for solutions containing metals of the platinum group which are not precipitated by it.

Sulphurous acid is an excellent precipitant for most gold solutions. It acts completely in the cold but more rapildy on heating; precipitation is slow until the solution is saturated with the gas. The precipitate is finely granular and is readily collected, and should be washed until the smell of sulphur dioxide has been removed.

Hydrogen sulphide completely precipitates gold as sulphide, both from feebly acid and neutral solutions, but is limited in its application since all other metals whose sulphides are insoluble in hydrochloric acid must be absent. The solution should be boiled, as the precipitate thrown down in cold solutions consists of aurous and auric sulphides together with free sulphur.

In all cases in which gold precipitants are used careful consideration must be given to the nature of any other precious metals, or base metals present, and the precipitant selected which will remove the gold only from solution.

It may be noted here that solutions containing aurocyanides of the alkalis and thiosulphates of gold do not give the ordinary reactions of gold, although hydrogen sulphide precipitates auric sulphide from thiosulphates. Solutions containing cyanides or thiosulphates may be tested for gold by heating with aqua regia to decompose the cyanide, then evaporating in a fume-cupboard with hydrochloric acid to free from nitric acid, and precipitating the gold with ferrous chloride, or other suitable precipitant.

The following separations of gold from other metals in solution are typical of those in use in the assay laboratory:—

Separation of Gold from Silver.—The solution of alloys of gold and silver presents some difficulty. Alloys containing at least 60 per cent. of silver are completely "parted" by boiling in nitric acid, the silver being dissolved and the gold left behind in a brown porous state. The gold, however, obstinately retains about 0·1 per cent. of its weight of silver, and if boiling is continued in strong acid, gold dissolves whilst the amount of silver retained is not appreciably reduced. Hence the need for "proof" assays.

With equal parts of gold and silver, nitric acid leaves half the silver in the gold, and with smaller proportions of silver the acid has less and less effect. Aqua regia attacks these alloys, but its action is soon stopped by the coating of silver chloride which forms on the surface of the metal and protects it from further attack. This impervious film can be removed by treatment with ammonia, and the alternate action of ammonia and aqua regia eventually results in the dissolution of all the gold, but the method is a tedious one.

When the gold contains less than about 15 per cent. of silver, aqua regia dissolves all the gold, whilst the whole of the silver is left as chloride; for this purpose, however, the metal must be in a very thinly laminated state.

More rapid attack on gold–silver alloys results by the addition of ammonium salts to the aqua regia. Pollard [1] recommends a mixture consisting of 5 grammes of ammonium chloride, 5 grammes of ammonium nitrate, and 5 to 10 c.c. of aqua regia. This will dissolve most gold–silver alloys in quantities of 0·5 to 1 gramme, the bulk of liquid being quite small. On diluting the solution with water the silver chloride is completely precipitated. A secondary reaction occurs during solution whereby some of the ammonia is oxidized to nitrogen, and the concentrated solution often deposits purplish-brown crystals of a salt which on analysis proved to have the composition $3AgCl, 4AuCl, 8NH_4Cl$. The crystals are decomposed by water, but are fairly stable in concentrated hydrochloric acid.

In view of the difficulty of getting gold alloyed with silver into solution, it is usual to determine the gold in industrial gold alloys by the parting methods described in Chapter XVIII. If, however, it be desired to dissolve the alloys to effect the solution of the silver and base metals present and to determine them in the wet way, the gold alloy is best mixed with a sufficiency of some other metal, such as silver or cadmium in order to render it amenable to attack by nitric acid. Cadmium is the metal generally recommended for this purpose, and the alloy is made by melting the weighed quantity of gold alloy with five times its weight of cadmium, under a layer of cyanide of potassium or other suitable flux, as described on p. 290. The base metals which pass into solution are determined by precipitation in the usual way.

If additional silver is added to the assay portion to facilitate solution, it must be allowed for in calculating the final result for the silver content of the alloy.

Separation of Gold from Platinum-group Metals.—The increasing use of the platinum-group metals in precious-metal alloys for jewellery and dental purposes has led to the introduction of many methods for separating and determining the gold and silver as well as the individual metals of the platinum group. These methods are more fully discussed in Chapter XXIV.

Gold in solution as chloride with platinum and with other metals of the platinum group is most satisfactorily separated by precipitation with oxalic acid, and warming the solution until precipitation is complete. In some instances, traces of the platinum metals may be present in the precipitate and re-precipitation may be necessary. According to J. Phelps,[2] the best procedure is to evaporate the solution to dryness and heat until copious fumes of oxalic acid are given off. All the platinum metals, except possibly palladium, are thus reduced.

Ferrous chloride may also be used for the reduction of auric chloride, and its separation from the platinum-group metals, but in some cases re-precipitation is desirable.

Separation of Gold from Base Metals.—The base metals most frequently found in industrial gold alloys and products are copper, nickel, zinc, cadmium, and tin.

With the exception of tin, which is converted into metastannic acid, all these metals pass into solution when the alloys are attacked with aqua regia. When gold alloys such as solders containing tin are treated with aqua regia the

[1] W. B. Pollard, *Trans. Chem. Soc.*, 1920, **117**, 99.
[2] Rose and Newman, *Metallurgy of Gold*, 7th edn., p. 529.

colour of "purple of Cassius" is seen. The composition of this compound varies according to the conditions under which it is formed, but it consists essentially of an intimate mixture of tin oxide, and very finely divided metallic gold, or probably of gold oxide. The compound is named after Andreas Cassius, who discovered it and first described it in 1685.

The base metals that pass into solution are separated and determined by the usual gravimetric and other methods, which are beyond the scope of the present work.

Determination of Gold with Hydrazine Hydrochloride.—This useful precipitant for gold has come into more general use in recent years. A solution of auric chloride (0·2 to 0·1 per cent. strength) is treated with a 10 per cent. solution of hydrazine hydrochloride, when a spongy form of gold is precipitated which is heated to redness and weighed. The presence of silver, copper, zinc, cadmium, lead and other base metals in gold alloys gives high results, as they are occluded in the gold sponge unless previously dissolved by appropriate reagents, such as hydrochloric acid or nitric acid. Compared with other existing methods of determining gold by precipitation, the hydrazine hydrochloride was found to be most accurate, the error not exceeding 0·1 per cent.[1]

Schoeller and Powell[2] effect the hydrazine reduction as follows:—

The chloride solution, freed from nitric acid, is diluted with an equal volume of strong hydrochloric acid, heated to boiling, and treated drop by drop with a freshly prepared 2 per cent. solution of hydrazine hydrochloride in slight excess. Re-treatment of the precipitate is desirable in separations of gold from platinum and palladium. The filtrate from the gold is freed from precipitant either by evaporation to dryness, digestion of the residue with strong nitric acid, and renewed evaporation to dryness, followed by double evaporation with hydrochloric acid to destroy nitrates; or by boiling with sodium chlorate.

Determination of Gold in Gold Chloride.—From this salt in the first instance all solutions of gold for electro-gilding are made except those prepared by electrolytic methods. For this purpose it is manufactured on a fairly large scale, and may therefore be readily purchased of a high degree of purity. The formula of pure gold chloride is AuCl, but the commercial salt in its crystallized form as used for making up gilding solutions, whether purchased or made from fine gold in the workshop, contains excess hydrochloric acid and water, and is more correctly described by the formula $AuCl_3$. HCl. H_2O.

According to this formula, the percentage of metallic gold in the salt is 48, but sometimes a slightly higher proportion is found owing to a small loss of hydrochloric acid and water which occurs in drying the crystals.

The percentage of gold is determined by dissolving 0·5 gramme in 25 c.c. of distilled water, and a solution of potassium hydroxide is added until the gold solution is distinctly alkaline to litmus. 5 c.c. of a 10-volume hydrogen peroxide solution are then added and the solution heated at the boiling-point of water for about an hour. The finely-divided precipitate of metallic gold is washed with water slightly acidulated with hydrochloric acid, then dried, carefully ignited and weighed.[3]

The gold may also be determined by precipitation with ferrous chloride or other suitable reagent, as previously described.

[1] Plaksin and Koshuchova, *Zvetnye Metally* (The Non-ferrous Metals), 1931, **6**, (I), 35–41 (in Russian); Abs. *Met. Abs. Inst. Metals*, 1932, (3), **50**, 168.
[2] *Op. cit.*, p. 263.
[3] W. R. Barclay and C. H. Hainsworth, *Electroplating.*

Varieties of gold chloride containing sodium chloride are now largely sold for photographic purposes. These frequently contain only 20 to 30 per cent. of metallic gold, and are, therefore, not suitable for electro-gilding purposes.

Determination of Gold in Gilding Solutions.—The solutions most frequently used for the electro-deposition of gold in industry consist of gold cyanide dissolved in a solution of potassium cyanide. The cyanide solution usually contains from $\frac{1}{4}$ to 1 lb. of cyanide to the gallon, and from $\frac{1}{4}$ to 1 ounce of gold per gallon.

For determining the gold content, 25 to 50 c.c. of the cyanide solution are taken and about twice its bulk of strong hydrochloric acid added. The solution is then boiled with the usual precautions, until there is no further smell of cyanogen gas.

To the clear solution an excess of a solution of ferrous sulphate is added, and the precipitate of metallic gold filtered off, washed, dried, ignited and weighed.

The following procedure has been devised by Kushner as a rapid method for the determination of gold in cyanide solutions.[1] A suitable volume of solution containing 0·05 to 0·2 gramme of gold is titrated with 0·1 N silver nitrate solution to determine the free cyanide. 50 c.c. of concentrated sulphuric acid are then added to decompose the cyanide and precipitate the gold, and the solution is evaporated under a good hood until the precipitated gold turns light-brown in colour and flocculates. The acid is decanted from the gold, which is then heated with a second 50 c.c. of sulphuric acid and the decantation repeated. Finally, the gold is collected, well washed with hot water, dried, heated to redness, cooled, and weighed. The acid filtrates are diluted and treated with hydrochloric acid, and the silver chloride precipitated is collected, washed, dried, heated to incipient melting, cooled, and weighed. By deducting the silver used in the titration, the amount of silver in the solution is obtained.

The methods described for the determination of gold in solutions from cyanide mills are also applicable (see Chapter XXV).

Stripping Electro-deposited Gold.[2]—This problem presents some difficulty when it is necessary to retain intact the article on which the gold has been deposited owing to the fact that any mixture which will dissolve gold will also keenly attack the basis metal on which it has been electro-deposited.

Electrolytic stripping is usually regarded as the most satisfactory method. This is carried out by making the article to be stripped the anode in a solution of potassium cyanide containing about half a pint of cyanide per gallon. A strip of thick gas carbon forms a good cathode, and a voltage of not less than 4 to $4\frac{1}{2}$ volts should be employed. To prevent the basis metal from attack so far as possible, the article should be given a gentle motion and thus ensure uniformity in the solution of the gold deposit. When all the gold has been dissolved further action is prevented by cutting off the current. The gold in the solution can be determined by the method described above.

If the destruction of the article is not of importance, the deposited gold can be stripped by immersing it in dilute nitric acid, thus dissolving the basis metal and leaving the gold as a film which is collected, washed, dried and weighed. In order to facilitate attack by the acid it is usually necessary to cut up the article to expose the base metal to attack. The gold film may also be dissolved in aqua regia and the gold determined by precipitation with ferrous sulphate or other precipitant.

[1] J. B. Kushner, *Ind. and Eng. Chem.*, Anal. Edn., 1938, **10**, (II), 641–642; *Met. Abs. Inst. Metals*, 1939, (2), **6**, 18,250.

[2] Barclay and Hainsworth, *op. cit.*

II. SILVER

Symbol Ag (Argentum); atomic weight, 107·88; Specific gravity, 10·53; melting-point, 961° C.

Commercial Forms of Silver.—As in the case of gold, metallic silver comes into the market mainly as bullion in the form of ingots or bars, which usually weigh from 1000 to 1200 ounces troy. Further details of silver bullion will be found in Chapter XVII.

When silver bullion contains gold it is called "doré silver," and it is then bought by gold and silver refiners. These mixed bars, as cast in the production mills, seldom weigh more than 600 oz. each. For industrial purposes refined silver is frequently cast into smaller ingots, weighing from 50 to 500 oz. more or less, for convenience in handling and alloying.

In addition to bullion bars, fine silver, refined electrolytically, is marketed in the form of sheet metal, approximately 0·03 inch ($\frac{3}{4}$ mm.) in thickness, to be used as anodes in electroplating etc.; and also as "grain" in a granulated form to facilitate weighing out for alloying and other purposes.

The industrial alloys of silver are placed on the market in all the various forms of sheet, wire and seamless tubing, etc., to meet the varied requirements of the arts and manufacture.

Physical Properties of Silver.—Silver is the whitest of metals, and is regarded as the acme of metallic whiteness, all other metals being judged by it. It is extremely malleable and ductile, being exceeded in this respect only by gold. It is harder than gold, and softer than copper, the two metals with which it is most frequently alloyed; in the pure state silver is too soft for many commercial purposes. Silver conducts heat and electricity better than any other metal, and in consequence it is used as one of the standards of comparison, the electrical conductivity of silver being taken as 100.

The melting-point of silver is 961° C., that is, a little below the temperature required to melt gold or copper. When silver is melted in the air, it absorbs about twenty-two times its volume of oxygen, and this is given off with effervescence as the metal solidifies, forming on the surface a number of protuberances, resembling miniature volcanoes, and projecting particles of silver to some distance. This phenomenon is termed "spitting" or "sprouting": it only occurs when the silver is comparatively pure, and is prevented by the presence of small quantities of copper or lead, or by stirring the metal under a cover of charcoal (Gowland).

Under suitable conditions clean surfaces of silver can be satisfactorily welded.

Whilst the chief uses of silver in the past have been for coinage and the production of sterling silver plate, as well as electro-plate, and the manufacture of silver salts for photographic purposes, the considerable reduction in price of the metal in recent years has extended its use to other fields. Its employment, either as pure metal or alloyed, in industrial chemical plants is already reaching appreciable dimensions. In this direction its most extensive application is in pieces of apparatus weighing up to 3 or 4 cwt. which have been and are being made from the pure metal, most of them for handling acetic acid.

Alloys of Silver.—The alloys of silver which find most application in industry are chiefly those legalized for coinage and plate, further particulars of which will be found on p. 336. No alloy of silver which does not conform to the legal standards can be sold as "silver."

Research has shown that the properties of base metals and their alloys are frequently considerably improved by the addition of small percentages of silver, and alloys of this description are being increasingly used in the metal industries. Thus, an important recent use of silver is in the manufacture of commutator copper. Again, investigation showed that the use of a copper–silver alloy, containing a very low content of silver, entirely overcomes the tendency of pure copper to be unduly softened when heated to relatively low temperatures for a short period of time. The silver addition is not sufficient to affect the valuable properties of high-conductivity copper. Another example is the use of the lead–silver eutectic alloy, containing 2·5 per cent. of silver and solidifying at 305° C., as a low-temperature solder.

Chemical Properties of Silver.—Silver is readily soluble in nitric acid, forming argentic nitrate (commonly known as " lunar caustic "), with liberation of oxides of nitrogen. Very strong acid does not act so rapidly as dilute. Its best solvent is dilute acid containing equal parts of acid and water.

Hot concentrated sulphuric acid dissolves silver slowly, converting it into argentic sulphate, with the liberation of sulphur dioxide. Dilute sulphuric acid does not act upon the metal. Silver sulphate is not very soluble in water, even when heated, and it easily crystallizes out of solution.

Hydrochloric acid acts on silver very slowly and only superficially, forming a coating of chloride. The gradual thickening of the film of silver chloride formed by the action of hydrochloric acid, chlorine or aqua regia eventually produces an impervious coating that stops further action. It is insoluble in acid, and creates a difficulty in dissolving gold–silver alloys in aqua regia.

When in fine particles silver is dissolved by potassium cyanide in the presence of air or certain oxidizing agents.

Alkalis have no action on silver, and on that account silver crucibles are used for the fusion of alkaline mixtures.

Silver also dissolves in solutions of sodium hyposulphite to form a double hyposulphite of silver and sodium.

Precipitants of Silver.—Hydrochloric acid and soluble chlorides give a white curdy precipitate of silver chloride, which, on heating and stirring the solution, becomes granulated and settles quickly. On exposure to light the white precipitate assumes a slate colour, which gradually deepens to violet, and finally appears brown or black.

Silver chloride is insoluble in water and dilute acids, but is slightly soluble in concentrated nitric and hydrochloric acids.

It is soluble in a concentrated solution of sodium chloride (brine), and also in ammonium hydroxide, forming the compound 2AgCl, $3NH_3OH$, which is decomposed by the addition of nitric acid with the re-precipitation of silver chloride.

Silver chloride is soluble in potassium cyanide, first being converted into silver cyanide, which dissolves in excess of KCN, forming the double potassium silver cyanide $KAg(CN)_2$. It is also readily soluble in sodium thiosulphate, with the formation of a double thiosulphate.

Silver chloride melts without decomposition at 451° C., to a reddish-yellow liquid. When solidified after fusion it is waxy in appearance, and is translucent, sectile, and tough. After fusion it dissolves with much difficulty.

Potassium (*or Sodium*) *Bromide.*—Soluble bromides precipitate silver bromide from solutions of the nitrate. Its properties resemble those of silver chloride, but it is less soluble in water and other solvents. On that account sodium bromide has long been used instead of sodium chloride at the Utrecht Mint for the precipitation of silver in the volumetric process of assay.[1] Silver

[1] Sir T. K. Rose, *Precious Metals*, p. 149.

bromide is dissolved by potassium bromide. It darkens rapidly when exposed to light.

Potassium iodide gives a yellowish precipitate of silver iodide, insoluble in dilute nitric acid, and only slightly soluble in ammonia, which distinguishes it from silver chloride. Silver iodide is soluble in cyanide and thiosulphate solutions, and in a concentrated solution of potassium iodide.

Hydrogen Sulphide.—Silver sulphide is precipitated from silver solutions by hydrogen sulphide or by soluble sulphides. It is insoluble in dilute acids, except boiling dilute nitric acid, when the formation of silver nitrate occurs with separation of sulphur. It is also insoluble in ammonium sulphide, and ammonium hydroxide, or potassium sulphide.

Silver sulphide is soluble in potassium cyanide, especially in hot strong solutions. This property is useful in removing the tarnish from sterling silver articles when polishing is impossible or undesirable. The so-called " oxidized " silver finish is silver coated with silver sulphide by dipping the articles into a hot solution of potassium sulphide.

Potassium Chromate.—K_2CrO_4 precipitates dark purple-red silver chromate Ag_2CrO_4, soluble in nitric acid and ammonium hydroxide. From weak acid solutions silver dichromate $Ag_2Cr_2O_7$ is precipitated.

Caustic Alkalis.—Potassium or sodium hydroxide yields a greyish-black precipitate of silver oxide, AgO, insoluble in excess but readily soluble in ammonia.

Alkaline Carbonates.—These yield a white precipitate of silver carbonate insoluble in excess; but soluble in ammonium carbonate, ammonium hydroxide, and nitric acid.

Potassium cyanide (or other soluble cyanide other than mercuric cyanide) gives a white curdy precipitate of silver cyanide, insoluble in dilute nitric acid, but readily soluble in excess of the precipitant, forming a double cyanide $KAg(CN)_2$.

Silver cyanide is soluble in ammonium hydroxide, but re-precipitated on addition of nitric acid ; it is soluble also in sodium thiosulphate. Silver cyanide closely resembles silver chloride, but is readily distinguished by the fact that it is decomposed with evolution of hydrocyanic acid when boiled with hydrochloric or nitric acid. When heated alone it gives off cyanogen and leaves a black, non-volatile residue of metallic silver and paracyanogen.

Potassium thiocyanate gives a white precipitate of silver thiocyanate somewhat resembling the chloride in appearance. It is soluble in ammonium hydroxide, but less readily than either silver cyanide or chloride, and is insoluble in dilute nitric acid. It is slowly dissolved by hot concentrated nitric acid. When heated in air it is decomposed and burns with a pale blue flame, leaving a black residue.

Organic Precipitants.—Silver compounds are readily reduced by certain organic substances, such as sugar, tartrates, aldehydes, etc. Thus a solution of sugar when added to a hot solution of silver nitrate gives a greyish-white powder of metallic silver. Many inorganic salts also, which act as reducing agents, *e.g.* ferrous sulphate, precipitate silver from solutions of its salts.

Metals as Precipitants.—Many metals are capable of reducing silver salts in solution. The silver is completely precipitated from solutions of the nitrate or sulphate by copper, iron, zinc, aluminium, magnesium, tin, and lead. The silver is thrown down in a crystalline or spongy state, according to the strength and temperature of the solution. The reducing action of metals on silver solutions is frequently taken advantage of in the separation and determination of silver in alloys, etc., and also for reducing silver chloride to metal. In this case the chloride is covered with dilute hydrochloric acid or sulphuric acid, and thin sheets of iron or other metal inserted.

"Test Acid" for Distinguishing Silver.—The term "test acid" is applied to dilute solutions of silver nitrate, or sulphate, about 2 grammes in 30 c.c. water, which are used to distinguish sterling silverware from white base-metal alloys, such as "nickel silver." Silver nitrate is reduced to the metallic state by copper and other base metals, and produces on the surface to be tested a black stain of finely-divided silver if of base metal, whilst sterling silver is not affected. The test is more fully described on p 360.

For convenience the test acid may be obtained in the form of lunar caustic pencils with which a streak is made on the wetted surface of the article to be tested.

Another delicate test for the same purpose is the Chromic Acid Test. In this case the test acid consists of a strong solution of potassium chromate acidulated with a few drops of strong nitric or sulphuric acid. A drop of this solution placed on the surface of the metal to be tested instantly gives a bright-red stain of silver chromate with fine or standard silver, and only a very faint coloration, or none at all, with other metals and alloys.

This test is useful for distinguishing between sterling silver, silver electroplate, nickel and chromium plate, and stainless steel.

Separation and Determination of Silver.—When volumetric methods are not employed for the determination of silver, it is usually separated from other metals, and determined gravimetrically by adding an excess of hydrochloric acid to a solution of silver nitrate and weighing the silver chloride produced.

Separation of Silver from Gold.—As more fully explained above, alloys of these two metals containing less than about 30 per cent. of gold are attacked by nitric acid, the silver being dissolved and the gold left as a residue which invariably retains a small amount of silver.

The difficulty of effecting the solution of these alloys by attack with aqua regia is described on p. 240.

Separation of Silver from Platinum.—Although alloys of silver and platinum containing excess of silver are attacked by dilute nitric acid, their separation is never complete. The silver is dissolved with a small quantity of platinum, and the residue of platinum persistently retains some silver. The losses and gains vary considerably with the proportions of the constituent metals, and the strength of the nitric acid used. With alloys containing less than about 5 per cent. of platinum, the losses and gains counter-balance one another to some extent, so that the errors in the determinations of the two metals are minimized and the results are sufficiently accurate for many purposes. The separation of these two metals is more fully discussed in Chapter XXIV.

Separation of Silver from Palladium.—Alloys of these metals are attacked by strong nitric acid and both the silver and the palladium dissolved as nitrates. The silver in solution can be determined as chloride by precipitation with hydrochloric acid, and the palladium determined by precipitation with dimethyl-glyoxime or other suitable precipitant as described later.

Separation of Silver from Copper.—Nitric acid is a ready solvent for all alloys of silver and copper. The silver is separated by precipitation as chloride, and the copper in the filtrate determined by any of the usual methods.

Separation of Silver from Lead.—The silver in lead alloys is usually, and most conveniently, determined by cupellation.

If the two metals are in solution as nitrates, the lead is first separated by evaporating a measured quantity of the solution with excess of sulphuric acid, whereby the lead is precipitated as sulphate and is filtered off, washed, dried and weighed. The silver in the filtrate is then precipitated as chloride by the addition of hydrochloric acid.

Separation of Silver from Zinc.—After precipitation of the silver as chloride and filtering, excess of ammonia is added and a rapid current of hydrogen sulphide passed through the heated solution to precipitate the zinc as sulphide.

Separation of Silver from Cadmium.—The silver is precipitated as chloride and filtered off. An excess of ammonia solution is added to the filtrate, and then excess of a solution of potassium cyanide. The cadmium is then precipitated as the bright-yellow sulphide by heating the solution and passing a current of hydrogen sulphide through it.

Separation of Silver from Tin.—When alloys of these two metals are attacked with nitric acid the silver is dissolved, and the tin is converted into metastannic acid. After separation of the tin the silver in the filtrate may be determined volumetrically or as chloride. The tin content is usually taken by difference, but may be determined by weighing the insoluble residue, or it may be fused with a mixture of about equal parts of sodium carbonate and sodium nitrite in a nickel crucible, the melt leached with water and dilute hydrochloric acid, and the tin determined in the solution by the usual iodometric or other method.

Preparation of Proof Silver.—The simplest method of preparing proof silver in quantity is to dissolve commercial silver in dilute nitric acid (1 to 1), allow the liquid to stand, syphon off from any residual gold, etc., dilute with hot water, precipitate the silver with hydrochloric acid, stir well, allow to settle, and wash repeatedly with hot water by decantation. The silver chloride becomes granular by continuous stirring and settles readily. When the decanted liquid no longer shows a trace of impurity, the silver chloride is allowed to settle and the excess of solution decanted off. Sufficient hydrochloric acid is then added to make the remaining solution distinctly acid, and plates of wrought iron (best Swedish bar) are placed in the silver chloride. The white chloride next the iron at once begins to blacken and the whole mass is entirely reduced to metallic silver in the course of a few hours or days, according to the quantity. When reduction is complete, the grey silver powder, which is contaminated with iron chloride, is washed by continuous stirring first with dilute hydrochloric acid, then with pure water, renewed many times.

A portion of the decanted liquid is boiled with a few drops of nitric acid to oxidize any traces of iron present, and is then tested for iron with ammonium thiocyanate, NH_4CNS; when no red coloration is given, indicating that the liquid is free from iron, the silver is dried and melted under a little charcoal in a graphite or clay crucible, and poured into a black-leaded iron mould. Silver thus prepared is usually over 999·9 fine. It is slightly improved in fineness by remelting.

A sheet of zinc or aluminium is sometimes used in place of iron to reduce the chloride. The metallic silver thus produced is washed, heated with hydrochloric acid, washed thoroughly to remove all acid and aluminium chloride, then dried and melted as described. The silver prepared in this way is not quite so pure as that obtained with iron, owing to the difficulty of removing the last traces of metallic salts which contaminate the reduced silver.

The washed silver chloride may also be reduced by transferring it to a porous pot which has been previously steeped in hydrochloric acid for some days and then thoroughly washed by long-continued standing in clean distilled water, frequently renewed. The porous pot is placed in water in an outer vessel, in which a cylinder of wrought iron is dipped as an electrode, and a little hydrochloric acid added to start the action. A cathode of pure silver or platinum is plunged into the silver chloride and connected with the iron anode through an electric battery. The white chloride at once blackens next the cathode, and in the course of a few days the reduction is complete. The chloride of iron which forms diffuses into the inner pot and contaminates the silver, but may be removed by washing. When free from impurity the silver powder is melted as above described. The wrought-iron anode may be replaced by carbon anodes, but the action is slower and the outer liquid becomes strongly impregnated with chlorine gas and should be renewed once or twice a day. The silver in this case is not contaminated with impurity.[1]

Another method frequently used for small quantities is to reduce the silver chloride by fusion with anhydrous sodium carbonate, thus:

$$4AgCl + 2Na_2CO_3 = 4Ag + 4NaCl + 2CO_2 + O_2.$$

When charcoal powder is also added it combines with the liberated oxygen to form CO_2.

The silver chloride after washing is dried and mixed with an equal *bulk* of anhydrous sodium carbonate, and the mixture fused in a clay crucible, first at dull redness, and then gradually raised to bright redness. The reduction is complete in about 30 minutes, and when tranquil fusion has taken place the contents of the crucible are poured into a suitable iron mould and allowed to cool, after which the silver is freed from the adhering slag of sodium chloride and finally cleaned in hot water. As carbonic acid is given off freely during the reduction, care should be taken not to have the crucible more than three parts full, and to avoid a high temperature at the beginning, otherwise the charge may boil over and occasion loss. The silver thus obtained is practically pure, but since pure silver is difficult to cast quite solid, and generally contains minute cavities, the fineness is slightly improved by remelting under charcoal powder.

Determination of Silver in Cyanide Plating Solutions.—For correct control of silver-plating solutions, determinations of the silver and free cyanide at frequent intervals are essential, and also—less frequently—of carbonate. In addition, periodical determinations of sulphate, chloride, iron, copper, nickel, and zinc are desirable. " Strike " solutions are analysed like plating solutions, but in this case it is usually only the silver and free cyanide contents that are important. Usually the silver content of plating solutions is not less than 2 ounces, nor more than 4 ounces (troy) per imperial gallon.

A number of methods have been devised, the silver being separated as silver sulphide either by precipitation with zinc, or by decomposition with acid, after which a determination by any standard method is applicable.

The following procedure, devised by H. Turner and H. C. Wright,[2] is quick and accurate. A measured quantity of the plating solution (usually 10 c.c.) is transferred to a 500 c.c. beaker, warmed on a hot-plate, and then 10 c.c. of a strong solution of ammonium sulphide added. The precipitate of silver sulphide is allowed to settle, the liquid poured off, and the precipitate washed with distilled water by decantation. It is then treated with 10 c.c. of dilute nitric acid (1·2 d.) and boiled until dissolved. The silver solution is diluted to 50–60 c.c., then 5 c.c. of iron alum as indicator added and the solution titrated with standard thiocyanate until a faint permanent pink colour results.

The number of c.c. of N/10 thiocyanate required multiplied by—

(i) 0·1576 = the silver content of the cyanide solution expressed in ounces troy per gallon;

(ii) 0·1729 = the silver content of the cyanide solution in ounces avdp. per gallon.

[1] Sir T. K. Rose, *Precious Metals*, p. 130.

[2] Ref. Paper before Sheffield Soc. of Applied Metallurgy, October, 1913.

The above procedure has been modified by W. J. M. Boyd, who precipitates the silver by the addition of about 2 grammes of zinc dust to the cyanide solution. After treating the precipitated silver with dilute sulphuric acid to remove excess of zinc, it is dissolved in dilute nitric acid and the silver solution titrated as described.

The silver may also be separated by treating the cyanide solution with sodium hydroxide and fine aluminium powder (free from grease). The deposited silver is washed, dissolved in nitric acid, and determined volumetrically with thiocyanate.[1] Another procedure frequently used in works practice is to treat 50-100 c.c. of the plating solution with sufficient excess of hydrochloric acid to convert all the silver into silver chloride.

Owing to the poisonous fumes of hydrocyanic acid liberated, the process should be conducted in a fume cupboard, or where a good draught of air is available.

The silver chloride, after washing and drying, can be weighed as such, and the silver content calculated (AgCl contains 75·27 per cent. of silver), or it can be mixed with about an equal bulk of potassium cyanide and a little charcoal powder, and fused in a small crucible, whereby it is readily reduced to metal. The resulting button of pure silver is then weighed. With a little experience, assays of an accuracy well under 1 per cent. can be obtained by this method.

An alternative procedure is to wrap the dried silver chloride in sheet lead and cupel. The cupel bead of silver is cleaned from adhering bone ash and then weighed.

The free cyanide in plating solutions may be determined by the method described for cyanide mill solutions on p. 461.

Carbonate may be determined by precipitation as barium carbonate, filtering, and titrating with acid, or by titrating a sample with acid after adding silver nitrate to combine with the free cyanide.

Stripping and Determining Silver on Electro-Plate.—As pointed out by Barclay and Hainsworth,[2] "the question of the determination of the weight of silver deposited on electro-plated articles is one which, during recent years, has assumed considerable commercial importance, due to the growing practice on the part of large buyers of such goods to specify the minimum weight of deposit which shall be given to each article. In many cases a guarantee is required from the manufacturer that such a weight actually obtains on the finished article when delivered. It is consequently often necessary to make determinations of the deposit on a sample article taken from the bulk, *e.g.* a spoon or fork, etc."

The quality of electro silver-plate is denoted by special marks, such as "AI," "A," and "B," which represent thicknesses or standards of silver deposit which have long been accepted in the trade for good quality work. In recent years, however, these and similar marks have also been used by makers of very cheap and low-quality electro silver-plate, *e.g.* for a deposit of 3 dwt. of silver per dozen against 20 or 22 dwt. per dozen on similar articles of good quality. Thus the good and the inferior articles carry the same marks, although the quality is vitally different. To remedy this misrepresentation, consideration is being given to the desirability of standardizing, legally or by arrangement, the quality marks on all electro silver-plate. At present the marks carry no guarantee of thickness of silver deposit beyond the guarantee of the maker's name and reputation.

[1] A. Wogrinz, *Z. anal. Chem.*, 1932, **89**, 120–121.

[2] *Op. cit.*, p. 384.

In order to ensure the good quality of their electro-plate, the large shipping and hotel companies have drawn up their own standards and specifications, and samples of all wares manufactured for them are assayed to ascertain whether they comply with the specification. If below the specified quality the consignment is rejected.

The determination of the weight of the silver deposit necessitates its complete removal or stripping from the basis metal, and then the assay of the stripping liquid to determine its resulting silver content. For this purpose it is an advantage to have some idea of the probable amount of silver deposited on the article.

The time taken in stripping affords some indication of the thickness of the deposit, and incidentally of the quality of the goods. The thickness of the silver deposit varies with each class of articles, and is usually recorded by weight in pennyweight. Thus a usual standard deposit for tablespoons and forks is about 35 dwt. per dozen, and is often stamped on them.

Some indication of the minimum strip deposit specified for similar articles is given by the following figures[1]:—For tablespoons and forks, and soup spoons, 35 dwt. per dozen ; for dessert spoons and forks, solid fish eaters, fruit knives and forks, 25 dwt. per dozen; fish eaters and fruit knives and forks having handles other than metal, 20 dwt. per dozen ; teaspoons 15 dwt. per dozen ; coffee spoons 10 dwt. per dozen.

When the weight of the silver deposited is quite unknown it may be approximately determined by carefully weighing the article before and after stripping. This method gives a good idea of the quantity of silver deposited and is frequently used in workshop practice, but it is never quite accurate, even under the most favourable conditions, in view of the difficulty of preventing a slight solution of the basis metal—especially of the cheaper forms—which are more readily attacked by the stripping solution.

For the accurate determination of the silver the stripping liquor, consisting of strong sulphuric acid with the addition of potassium nitrate, and first discovered by Keir[2] in 1790, is still in use and gives satisfactory results.

The article to be assayed is immersed in as little as possible of concentrated sulphuric acid, previously heated to 100°–110° C.

Finely powdered nitre is then sprinkled over the surface of the acid a little at a time, and the liquid stirred. Chemical action results in the liberation of a small amount of nitric acid in the solution which facilitates the solution of the silver deposit and is without marked action on the basis metal, commonly composed of nickel-silver, and similar alloys. Usually the silver deposit is completely dissolved in from 5 to 10 minutes, but much depends on the size of the article and the thickness of the deposit. The action of the acid should be carefully watched, and special attention given to the prominent edges where the plating will be thickest. A little powdered nitre sprinkled just over these points will assist solution. In order to expose the surfaces to fresh liquor, the article should be moved about with a hook of copper wire. Special attention must be given to the stripping, and pitting of the surface should be prevented as much as possible if the article is to be returned for re-plating.

When every trace of silver has been dissolved from the surface the article is removed by means of the copper hook and all adhering liquid washed off into a beaker. The total washings are kept apart from the main liquid and their silver content determined separately. The stripping liquor is allowed to cool, then made up to such a quantity that not more than 50 c.c. need be taken to

[1] Electro-plate Marking Committee, National Assoc. of Goldsmiths (Report in *Goldsmiths' Journ.*, February, 1932, p. 461).

[2] *Phil. Trans. Roy. Soc.*, 1790, **80**, 367.

give 0·5 gramme or less of silver. After thorough mixing, the assay portion is drawn off and diluted to about 400 c.c., iron solution is added as indicator, and the silver determined by titration with standard thiocyanate (Volhard's method) in the usual way.

A proof assay of silver in sulphuric acid should be made at the same time, as experience shows that sulphuric-acid proofs are slightly higher than the usual proofs of silver dissolved in nitric acid, *i.e.* they require slightly more thiocyanate.

The results of the silver content of both the stripping liquid and the wash-water are added together, and multiplied to give the exact weight of the total silver removed from the surface of the article.

The proportions used for the stripping liquid vary with different workers, but usually they are about ½ oz. of powdered potassium nitrate to one pint of strong sulphuric acid.

CHAPTER XVII

BULLION : VALUATION AND SAMPLING

Bullion and Specie Defined.—Bullion may be said strictly to mean gold and silver in the shape of bars or other unmanufactured form, while the term " specie " covers coins of gold, silver, or other metal. Confusion sometimes arises, however, from the fact that the two terms are often used somewhat loosely by business men not directly engaged in the marketing of these commodities. There is, perhaps, a possible explanation of the confusion between these terms in the fact that what is usually known as a " specie " bill of lading is employed for bullion as well as coin. Not infrequently transport vessels carry both specie and bullion.

Bullion varies considerably in composition, and gold and silver may be present in any proportion up to nearly 100 per cent. Any of the base metals may be present, and platinum and its allied metals as well as tellurium and selenium may also be included.

The degree of purity or fineness of bullion is always expressed in parts per thousand, or per " mille " the pure metal being described as 1000 fine. Thus a bar of bullion containing 91·3 per cent. of gold is said to be 913 fine. The weight, fineness, and name or distinguishing mark of the assayers are usually stamped upon each bar of bullion. Although bullion produced by the various methods of ore treatment is often subjected to some preliminary refining operation at the mill or mine and cast into ingot moulds to bring it to a marketable form, the bars are frequently impure and sometimes brittle and require careful handling.

Classification of Bullion

I. Gold Bullion.
- (*a*) Unrefined gold bullion.
- (*b*) Cyanide bullion.
- (*c*) Refined gold bullion.

II. Silver Bullion.
- (*a*) Unrefined silver bullion.
- (*b*) Doré silver.
- (*c*) Refined silver bullion.
- (*d*) Sycee silver (Chinese).

III. Base Bullion.
- (*a*) Lead containing gold and silver.
- (*b*) Copper containing gold and silver.

I. Gold Bullion.—Bars of gold bullion usually weigh 200 or 400 ounces troy.

(*a*) *Unrefined Gold Bullion.*—The gold bullion produced by the various methods of ore treatment is impure and unfit for use in the arts until after it has been refined. The gold content varies according to the process used in its extraction, but unrefined gold bullion always contains some silver and variable quantities of base metals.

(*b*) *Cyanide Bullion.*—The cyanide process always yields impure bullion, the chief impurities being lead, zinc and silver. Cyanide bullion is sometimes called " base bullion," but it is convenient to restrict this term to lead and copper containing precious metals, as described in Section III below. The presence of lead and other impurities in cyanide bullion sometimes gives it a worthless appearance, presenting a fracture somewhat resembling fine-grained cast iron. The gold content of cyanide bullion may vary from about 980 per 1000 down to 500 or 600 fine. It is seldom, however, that cyanide gold is below 800 fine, as the penalty now enforced by the buyer, should this figure not be obtained, is sufficient inducement for the smelter to refine the metal partially before shipment and to guarantee that degree of fineness.

All the gold produced on the Rand is treated at the Rand Refinery, and in 1934, the average assay of the rough gold bars as received from the mills was[1]:—

Gold	889·4 per 1000
Silver	84·5 „
Base metals	26·1 „

Many of the rough bars of cyanide bullion, even if they have the outward appearance of gold, are more or less brittle, and require careful handling to prevent loss. Even so little as 0·05 per cent. of lead is sufficient to make gold very brittle. Bismuth or tellurium has a similar effect.

(*c*) *Fine or Refined Gold Bullion* is practically free from silver, and contains from 990 to 1000 parts of gold per 1000.

The purity of the gold produced by the refining processes in general use is as follows :—By sulphuric-acid parting, 996 to 998 fine ; by chlorine gas, 996 to 997 fine ; and by electrolytic processes, 999·8 to 999·9 fine.

II. Silver Bullion.—Silver bullion is usually cast into bars which weigh from 1000 to 1200 ounces troy.

(*a*) *Unrefined Silver* varies in silver content from 500 to 995 per 1000, and may or may not be free from gold, but invariably contains some base metals, notably copper.

(*b*) *Doré Silver.*—Silver bullion containing gold is termed " doré bullion " or " doré silver," from the French *doré*, literally " gold-bearing." It usually contains only small amounts of gold, 1·0 to 50 parts, or somewhat more, per 1000, and is bought by gold and silver refiners. Doré bars are seldom cast of greater weight than 600 oz. at the mills. The anodes of doré silver for electrolytic parting by the Moebius process are 18 inches long, 10 inches wide, and about ½-inch thick.

(*c*) *Refined Silver Bullion.*—This term is applied to silver after being refined to remove the gold and other impurities. It is nearly pure metal and is usually 997·5 to 998·5 fine, or occasionally over 999·0 fine. Fine silver is chemically pure silver such as that refined electrolytically. Silver refined by the cupellation process is sometimes known as " test silver " or " fire refined " silver, and may be 997 or 998 fine or sometimes higher.

(*d*) *Sycee Silver* (*sī cē′*), Chinese bullion.—This term is applied to a special form of bullion used by the Chinese as silver currency, and is frequently mentioned in London bullion-market reports. The term is the foreign rendering of a Chinese word meaning " fine silk," and evidently relates to the silky fracture of a bar of fine silver. Although the Chinese make free use of silver coin, the actual basis of accounting is sycee.[2]

[1] Rose and Newman, *Metallurgy of Gold,* 7th edn., p. 476.
[2] B. White, *Silver,* 1919, p. 112.

The term denotes silver melted into small hemispherical ingots known in native parlance as " shoes," from their resemblance to the bound foot of a Chinese woman. These peculiar shaped ingots vary considerably in weight; the principle range is about 56 oz., although they are sometimes made as light as one ounce. Private enterprise alone provides this form of money. The quality varies but is usually about 982 fine, and is determined by cupellation by an expert called a " Kungkoo," who assays it and then inscribes upon the surface of the ingot the premium or otherwise to which it is entitled with regard to the quality.

The bankers of the different districts appoint their own Kungkoo, and the fineness of the sycee made is only good for that locality—hence the variations in fineness. On the whole the melting of the silver and its alloying-down to the required quality is reliably performed, and sycee therefore enjoys a reputation denied to coin from native official mints. Nevertheless, B. White remarks that " when sycee has been shipped and refined in Europe, the melters have occasionally discovered pieces of iron that certainly had no right to be there."

III. Base Bullion.—This name is applied to bars or pigs of lead and copper containing gold and silver, and in some cases the platinum metals, resulting from the smelting of argentiferous and auriferous lead and copper ores. The precious metal content of these bars varies considerably.

(*a*) *Base Lead Bullion.*—The argentiferous pig lead from smelting processes usually contains less than 650 oz. of silver per ton, as loss of silver in the slag is likely to result if the pig lead is richer in silver. The lead frequently contains more or less gold. Pig lead from the Pattinson refining process contains from 500 to 650 oz. of silver per ton, whilst the Parkes refining process produces a richer lead for cupellation containing from 2000 to 5000 oz. of silver per ton.

(*b*) *Base Copper Bullion.*—Owing to the great affinity of gold and silver for metallic copper, this metal is also used for the concentration of the precious metals in smelting processes. The copper containing the gold and silver is cast into bars or pigs constituting base copper bullion. Base bullion is more fully discussed in Chapter XX.

Marketing Bullion [1]

An assayer should be familiar with the manner and terms of selling gold and silver bullion. The following notes are therefore given.

Shipments of bullion to London by the gold-mining companies are received by their bankers. On the day of arrival, bids are asked from the Bank of England and from foreign banks which may want the gold for export. The highest bidder gets it. The market for gold is intimately associated with the banking transactions whereby international balances are settled.

Prices of gold and silver are given regularly in the daily press. At one time, London was supreme as the market for gold. Owing however, to the developments consequent on the war of 1914 it now shares that distinction with New York, but whilst New York is essentially a free market in which gold can always be dealt in, transactions in London are subject to licence. Yet, in spite of this fact, most of the orders come to the London market. London still remains the chief market for silver.

The following sentence from a press report of the gold bullion market may be regarded as typical:—" Gold fell 4d. to 91s. 9d. per oz. ex London and 91s. 4d. ex Durban." The difference in the price represents, to some extent,

[1] See *The Times Trade Supplement*, **15**, p. 172.

the difference between a c.i.f. (cost, insurance, freight) and an f.o.b. (free on board) quotation, because the price in London includes the cost of the freight and insurance of the gold from South Africa, which is the principal source of supply. Gold is now sent direct from South Africa to India.

The price of gold is always quoted per ounce "fine," and the greater part of the metal produced in South Africa is marketed in bars of 400 oz. each. Transactions take place, therefore, on the basis of a unit of 400 oz. Bars, however, of half that size are specially manufactured for India. Gold is also made up into bars of only 10 oz. each and stamped with the name of the bank which imports them into that country.

Each time gold is bought and sold a calculation has to be made, because the price is for "fine" gold, and each bar which is assayed contains a certain proportion of alloy. This proportion varies from 2 to 5 per mille, and an allowance has to be made to the buyer for this small proportion of metal other than gold. As a rule, each bar contains 996 parts of "fine" gold.

The price at which the Bank of England must buy gold is fixed by Act of Parliament, and it is unable to buy gold when the price is abnormal. After the Continent and India have fulfilled their requirements the bulk of the world's gold is now taken by the United States. India is also a buyer of sovereigns, in which dealings take place in the London market at a premium approximately equivalent to the current premium on bar gold over the normal price in London.

The Silver Market.—Turning now to the basis of silver prices, those firms which deal in a large way in gold also specialize in silver. Silver is sold in the state in which it arrives in this country, usually as bars of various sizes, unless it be low quality, in which case it is generally remelted here by one of the Bank of England melters and recast after partial refining, when necessary, to raise it to marketable fineness.

A representative report of the London silver market would be to the effect that "Silver remained unaltered at $34\frac{5}{8}$d. per ounce for cash and $34\frac{7}{8}$d. for forward delivery." Silver is usually marketed in bars of brick shape, each weighing 1100 oz., or about 91 lb. (troy weight). As in the case of gold, special calculations have to be made each time silver is bought and sold, but of a reverse character to those necessary in the case of gold. As previously indicated, the basis of the silver quotations is a standard of 925 parts per 1000 parts of silver and alloy together. This standard assumes, consequently, that there are 75 parts of alloy in each 1000 parts, whereas the actual bars delivered contain at least 996 parts of fine silver in each 1000 parts, and often as many as 999 parts. Thus a buyer of 100,000 standard ounces at say 24d. per standard ounce would obtain an amount of fine silver in the form of bars, each weighing about 1100 ounces, and assaying from 996 to 999 fine, which will contain as nearly as possible 100,000 standard ounces of silver at a cost of £10,000 plus brokerage charges. In New York the price of silver is quoted for fine ounce troy.

The silver market is intimately connected with banking transactions, large exports being made from London to the Far East in settlement of balances. The silver market of the world, more than any other metal market, is determined by London. The silver brokers of London meet daily and fix a price for that day. They are bound to sell at the price thus fixed any quantity required, but are not bound to take any quantity that may be offered for sale. If the quantity offered is more than can be disposed of it necessarily goes over to the day following, and then has an effect in fixing the price for that day.

The quotation fixed by the brokers is frequently called the "official price," and is the price commonly adopted in business and by the press. Occasionally there are quotations for "commercial bars," which are a little higher than the "official price." These represent the retail business that is done with jewellers, manufacturers of silverware, etc.—a business into which the large refiners do not enter.

There is a large demand for silver in industry, and the wholesale prices of articles manufactured with the metal vary, naturally, with the prices of the raw material. In the retail trade other factors come into play, such as the prices ruling when the stock was bought.

In the case of "forward delivery" transactions, the term "forward delivery" means that the silver is to be delivered two months from the date on which the contract is made.

The chief sources of silver supply are North, Central and South America, Mexico being the largest producer. Large numbers of silver bars are brought each week to London from these sources, and are dispatched thence to all parts of the world, the East and Far East especially being large consumers. To suit the convenience of the Indian market, the silver refined in London is usually cast into bars that are 996 fine.

The term "seigniorage," used in connection with bullion transactions, is applied to a charge levied on bullion brought by private persons to the Mint to be coined.

Melting, Refining, and Casting of Bullion

The assayer is frequently called upon to melt and cast bullion. It may be well, therefore, to describe briefly the melting and refining of small lots of crude bullion and the casting and sampling of the bars for assay purposes. For full details of these operations the description given by Rose and Newman in *Metallurgy of Gold* should be consulted.[1]

Gold bullion is generally melted in graphite crucibles in a crucible furnace such as that described in Chapter III. A fireclay crucible (fluxing pot) of fine texture is sometimes used for melting small lots of bullion.

All crucibles must be thoroughly annealed before being used, as they are very liable to crack when placed in a hot fire unless this precaution is taken. The crucible should be stored in a warm place near the furnaces for several days before it is wanted. When required for use it is first placed on the top of the furnace or in the ashpit for a few hours, and then held over the open furnace by means of the crucible tongs for a few minutes until it is thoroughly warm throughout. The crucible is then lowered, rim downwards, upon the burning fuel (coke). As soon as the rim becomes red-hot, all risk of cracking is past, and the crucible may be turned over and placed in position for the reception of the bullion. Salamander crucibles are well annealed before being sold and do not therefore require such careful annealing before use. The crucible stands on a firebrick about 3 inches thick, placed on the fire-bars. The fuel is built up round the pot until it reaches to its rim, and the fire urged until the pot is uniformly heated to a full red-heat. Borax is then charged into the crucible by means of a scoop to act as a flux for the base metal oxides and to assist the bullion to melt. As soon as the borax has melted, the introduction of the bullion is commenced. Before commencing to melt, the ashpit of the furnace should be cleared in order to avoid unnecessary trouble in case of accident.

When the quantity of bullion is small it may be introduced by means of a copper scoop or wrapped in paper. With large quantities it is safer to use the

[1] The account given here is mainly based on Rose and Newman's description.

shoot shown in Fig. 121, which is held in the left hand so that the tube is inside the crucible, while the bullion is transferred to it in a scoop by the right hand. Large pieces of metal are added by the crucible tongs. Bars of metal to be remelted should be placed on top of the furnace to warm before being introduced. When the first supply has melted down more is added in the same manner, and the operation repeated until the crucible contains sufficient metal to form a bar. The crucible is not allowed to become more than two-thirds full at any time. During melting a cover is kept on the crucible as much as possible to prevent coke-dust contaminating the metal.

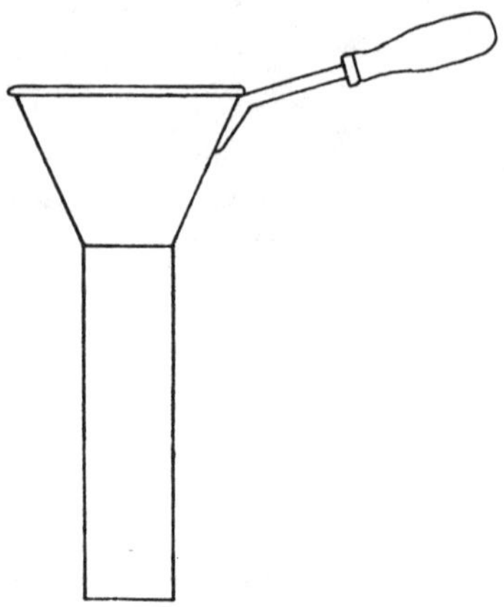

FIG. 121.—SHOOT FOR INTRODUCING BULLION INTO THE CRUCIBLE.

When the metal is thoroughly molten it is ready for casting, unless the bullion is impure, in which case it is usually subjected to partial refining before casting. If the bullion is of a high degree of purity containing little dirt or base metals, very little flux is added, a spoonful or two of sodium carbonate and nitre being enough. In this case the slag is not skimmed off but poured with the metal.

When gold bullion containing a large percentage of lead, zinc and other metals (*e.g.* some bullion produced by the cyanide process) is to be refined, it is subjected to a preliminary operation called "toughening."[1] The gold is melted, and nitre is thrown upon its surface and stirred in. Much oxygen is evolved and vigorous bubbling ensues. The base metals are oxidized and skimmed off the surface of the molten metal, and more nitre is added until the gold is sufficiently pure, according to the judgment of the melter, who is guided partly by the increasing pastiness of the metal. Some more nitre is then stirred in and the pot covered over and its temperature raised for pouring. The gold after toughening still contains copper as well as silver, but most of the other base metals will have been removed. Toughening is a tedious operation, and may involve considerable losses of precious metal by spirting and volatilization. The losses are mainly temporary, as most of the gold can be recovered by grinding and washing the ashes, and by the use of condensing-chambers between the furnace flue and the stack.

To protect the crucible from corrosion by the molten oxides, etc. bone-ash is sometimes thrown into it and pushed back from the centre of the molten metal towards the periphery, thus forming a ring round the inside of the pot. The bone-ash does not melt but absorbs the oxides, thus clearing the surface of the metal. The nitre additions are projected on to the clear surface of metal within the ring of bone-ash. In casting gold ingots the mould should be warm, well oiled or smoked with pine-wood, and the crucible moved backwards and forwards along the length of the mould during pouring.

A little oil should be poured upon the gold, as soon as it is cast, and the mould covered with an iron plate. Immediately the gold has solidified, the ingot should be turned out, brushed well all over with a cocoa-fibre brush, and then plunged into cold water.[2]

[1] Sir T. K. Rose, *The Precious Metals*, p. 185.

[2] Gowland, *Metallurgy of the Non-Ferrous Metals*, 4th edn., 1943.

Silver Bullion.—When silver bullion has been treated for a long time by nitre and raised to a high degree of fineness, it is affected by a peculiar bubbling which is probably due to the evolution of oxygen previously absorbed from the nitre. If this continues it is necessary to stir continuously with a graphite rod, keeping the surface covered with charcoal powder, until the bubbling ceases. Iron and zinc can be removed from bullion by the use of nitre ; but the oxidation of lead is more tedious, and bismuth, tellurium, and copper are very troublesome. When lead is present, alternate additions of sal-ammoniac and nitre are recommended. Iron is used to remove arsenic, antimony, sulphur, etc., the molten metal being stirred briskly with an iron rod for a few minutes. Antimonides, arsenides, or sulphides are formed and separate from the molten silver.

The ingot mould for receiving the metal after refining is thoroughly cleaned and smoothed inside by rubbing with emery paper and oil, or with a dry piece of pumice stone. It is then wiped with an oily rag, or blackleaded inside, and placed on the top of the furnace until it is too hot to touch with the hand but not sufficiently hot to ignite the oil. When the bullion is ready to pour, the mould is placed on a level surface and a little oil poured into it. Any cheap, non-volatile oil is suitable for the purpose, and sufficient should be used to cover the bottom of the mould to a depth of about $\frac{1}{16}$-inch. The charge must be stirred thoroughly before pouring, so as to obtain a bar as homogeneous as possible to ensure a correct assay. The stirring is usually done with a graphite rod made expressly for the purpose ; an iron stirrer should not be used, otherwise alloying may take place. The graphite rod is annealed and raised to a full red-heat before being used, and is held firmly by a pair of tongs. With very small meltings it is sufficient to lift the crucible out of the furnace with the tongs and to give it a rotary motion just previous to pouring. Since segregation may occur on cooling, a sample for assay is often dipped out immediately after stirring, as described later.

To cast the metal, the crucible is lifted from the furnace, usually with basket tongs, and the contents poured rapidly but steadily into the mould, the crucible being held for a little time in an inverted position to allow the last portion of the metal to flow from it. The oil is ignited and burns on the top of the cast metal, thus keeping it from tarnishing.

In small castings the slag is allowed to flow out and remain on the top of the metal in the mould ; in large castings the slag is usually skimmed off with a ladle before pouring. The bar, while still too hot to be handled, is turned out by inverting the mould, and the slag is separated by a few light taps with a hammer. The bar is then momentarily dipped into water to assist in the complete removal of the last fragments of the slag, or it may be dipped first into dilute sulphuric acid and then into clean water, the bar retaining sufficient heat after removal to expel all moisture. This treatment gives a bar free from tarnish, which is then stamped. In casting the bar, a mould of suitable size for the weight of metal to be cast should be selected. It may be noted that 1 cubic inch of fine gold weighs 10·168 troy ounces, and 1 cubic inch of fine silver weighs 5·568 troy ounces. Table XXXIII, of sizes and capacities of bullion moulds of the shape shown in Fig. 122, will be found useful.[1]

Segregation or Liquation in Bullion.—It is well known that when two or more metals are melted together and allowed to cool, it seldom happens that the resulting alloy solidifies or " freezes " as a whole and at a definite temperature. Usually one portion freezes first, rejecting another portion of different

[1] These moulds are made by Messrs. Fraser & Chalmers.

composition, which then solidifies at a lower temperature. This segregation of the constituents is frequently termed *liquation*.[1]

The researches of Péligot, Roberts-Austen, Rose, and others have shown that practically no segregation occurs in gold alloyed with either silver or copper

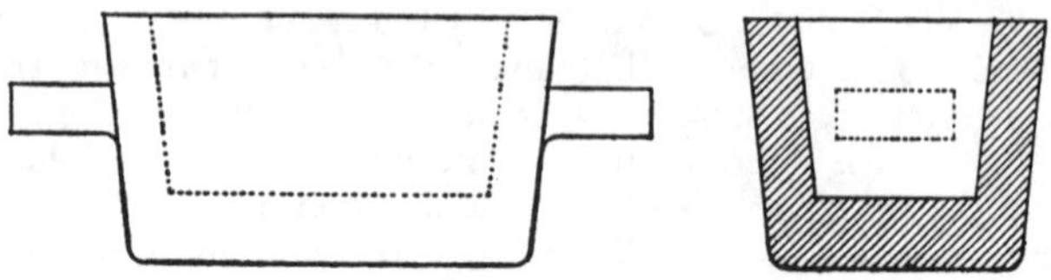

FIG. 122.—BULLION MOULD.

TABLE XXXIII.—SIZES AND CAPACITIES OF INGOT MOULDS.

Inside Measure.			Capacity in ounces.		Weight.
Length.	Width.	Depth.	Gold.	Silver.	
inches.	inches.	inches.			lbs.
1	5/8	1/2	4	2	1
1 1/2	1	3/4	10	5	1
2 1/4	1 1/16	1	25	12	1
3 3/8	1 3/8	1 1/8	50	25	3
3 1/2	2	2	95	50	6
4	2	1 3/4	100	56	7
4 1/4	2 1/4	2	136	76	9
4 1/2	2 1/2	2 1/4	180	100	10
5	2 1/2	2 1/4	244	134	10
5 1/8	2 3/4	2 1/4	250	140	10
5 1/2	2 5/8	2 3/4	295	166	11
5 1/2	3	2 3/4	365	200	12
5 3/4	3	2 3/4	375	208	13
6 1/2	3 1/4	3 1/4	550	300	15
6 3/4	3 1/2	3 1/4	620	340	19
7 1/2	3 1/2	3 1/4	730	400	28
8	3 3/4	3 1/2	910	500	35
9	3 3/4	3 1/2	1015	600	36
9 1/2	4	3 1/2	1285	700	40
9 1/2	4 1/2	3 1/2	1448	800	41
10	4	4	1470	800	42
10 1/2	4	4	1650	900	55
11	4 1/2	4	1830	1000	65
11	4 1/2	4 1/2	2200	1200	72
11 1/2	5	5	2750	1500	76

or with both when rich in gold and free from all impurities; but when base metals other than copper are present, segregation occurs and the solidified metal is not uniform in composition. The result is that the exact value of the bullion cannot be determined, and the buyer must make some allowance to guard against loss. In the case of bullion containing a large proportion of base metals, such as cyanide gold, a considerable amount of segregation generally takes place, the precious metals being driven outward or inward in the cooling mass according to

[1] See E. F. Law, *Alloys*, p. 35.

the nature and amount of impurities present. The question of segregation in connection with the alloys of gold, silver, lead, and zinc produced in cyanide mills has been investigated by Matthey.[1] He found that one such ingot weighing about 120 ounces contained 662 parts per 1000 of gold at the bottom corner, and only 439 parts at the top. In another case, when 16·4 per cent. of lead and 9·5 per cent. of zinc were present, the standard fineness of an ingot weighing 400 ounces as shown by actually separating the whole of the precious metals, was, gold 614·0, silver 75·8, and its true value £1028, while the value as deducted from the average of fourteen assays made on it (gold 576·0, silver 90) would have been only £965. Seven dip assays made on this ingot varied from 562 to 622 fine. Other cases of irregular distribution were even more remarkable. Some results obtained by Stockhausen with cyanide bars are given on p. 264.

In the case of silver bullion the bars are rarely perfectly homogeneous, even when over 950 fine, but the amount of segregation is in many cases too small to have any serious effect.

More or less segregation always occurs in silver bars containing copper, the silver being concentrated towards the centre, or to the outer surface of the cooling mass, according to whether silver or copper is in excess. The presence of other base metals in silver produces similar effects. For example, Gowland and Koga[2] have shown that when silver bullion containing 984·4 parts of silver and 15 parts of bismuth is melted and cast into an open ingot mould so as to give an ingot weighing about 1000 oz., the portions of the ingot which remain longest fluid are richer in silver than the others.

Silver and gold present in bars of lead and of copper are also subject to the same irregularity of distribution, the segregation in some cases being considerable and causing great difficulty in determining the value of the bars.

The liquation of base bullion is discussed later.

Sampling of Bullion

Since segregation may occur in either gold or silver bullion on solidification, the method of sampling for assay purposes is of considerable importance. When the bar is clean, small portions should be taken from different parts for assay; and if no segregation has taken place, the difference between the assay results of the several portions should be very small. Before sampling a bar, it should be stamped with a distinguishing number, and the same number placed with each sample taken from the bar. This number should represent the bar through every stage of the assay by which its value is determined.

Some assayers stamp the initials of their name on the cut faces so that no further portion can be removed after it leaves their hands without detection.

When the bar has been pickled by dipping in acid after casting, the surfaces must be well scraped before sampling, as the outside is usually finer than the interior, in consequence of the pickling. For this reason some assayers always reject the first portions of metal removed in taking the sample.

Four methods are in use for sampling gold and silver bars, viz. (1) cutting or chipping, (2) drilling, (3) dipping, (4) granulation. The sampling of base bullion is dealt with later.

(1) Sampling by Cutting.—This method is in very general use, and consists in cutting pieces from diagonally opposite corners of the bar, as shown in

[1] *Proc. Roy. Soc.*, 1896, **60**, 21, quoted by Rose & Newman, *Metallurgy of Gold*, 7th edn., p. 516.

[2] *J. Chem. Soc.*, 1887, **51**, 410.

Fig. 123. Experience proves that with bullion of high grade, these samples are usually representative of the whole bar, but they are not reliable when the bars are not uniform in composition.

In the case of fine gold ingots weighing 400 ounces which are received from the Bank of England for coinage, an ordinary sample is cut from the middle of one of the lower edges of the ingot.

Fine silver bars are usually sampled in the same way. Cutting samples are taken with a narrow steel chisel, which must be used with care, otherwise the chip when cut off is liable to " fly away " and be lost. From 1 to 1·5 gramme

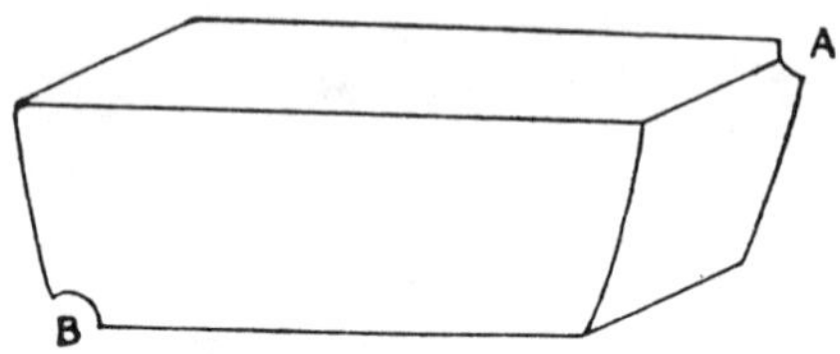

FIG. 123.—METHOD OF SAMPLING BULLION BAR BY CUTTING.

is taken for each cut. A sampling-machine is usually employed if a large number of bars have to be dealt with. Sampling by taking cuts is not always satisfactory if the bar is brittle, as a much larger piece may break off with the chip than is required. The sample is prepared for assay by flattening on a bright anvil with a hammer or by rolling.

(2) Sampling by Drilling or Boring.—In this method the bar is drilled, top and bottom, with a small drill. This may be done in a lathe or with a ratchet drill. An $\frac{1}{8}$- or a $\frac{3}{16}$-inch bit is generally employed for the purpose. A twist-drill as shown in Fig. 124 is frequently used. The drill is preferably made of high-speed tool steel as this avoids contamination of the sample with fine particles of steel. No lubricant should be used in drilling. The bar to be sampled should be placed in a clean metal tray in order to retain the drillings so that no loss may occur. The surface borings resulting from the first few revolutions of the drill should be rejected. Those that follow, to the extent of a little more than 1 gramme, are collected and reserved for assay, each lot being kept separate. Metal templates are sometimes used to locate the right place to drill.

FIG. 124.—TWIST DRILL FOR SAMPLING.

One boring is usually taken from a top corner of the bar and one from an opposite bottom corner, as in the case of a cut sample. Some assayers mix the two lots of drillings to form one sample. When it is known that the bar is not uniform in composition, four borings are sometimes taken, two from the top and two from the bottom, as shown in Fig. 125. The four lots of drillings may be assayed separately, or the two lots from the top may be mixed to form one sample, and the two lots from the bottom mixed for a second sample, according to circumstances.

The character of the drillings varies according to the nature of the bullion. With tough bars the drillings are soft, long and spiral in form, and can be rolled and cut up for mixing. When there is a tendency for the bars to be brittle the drillings are hard and short. In this case they are well mixed and the portion for assay selected at random.

The mixing of drillings from different parts of the bar is to be deprecated, unless it is known from the source of the bullion that the bar is practically uniform in composition.

(3) Dip Sampling.—The sample is dipped from the molten metal just before casting into a mould. The metal is melted in a graphite crucible under a cover of borax, and one or two dip samples taken by means of a graphite rod with a cavity at the end, capable of holding about 5 grammes of metal.

Sometimes a small graphite crucible, fastened to an iron rod with iron wire, is used, but is not desirable as iron alloys with gold. At the Rand Refinery, however, iron dipping-rods, heavily coated with a graphite mixture and having two small cups, 6 to 8 inches apart, are used. By this device two samples are obtained simultaneously from two different parts of the molten metal.

In the usual procedure, one sample is dipped from about the middle of the pot before pouring, or two samples are taken, one from the top just before pouring, and a second sample towards the end of pouring. The sample is either allowed to solidify in the dipper, or is poured into a small ingot mould under borax, or may be granulated by pouring into water. The borax is then detached, and the cleaned metal flattened and rolled for assay. Dip sampling is usually resorted to when the bullion bars are not uniform in composition.

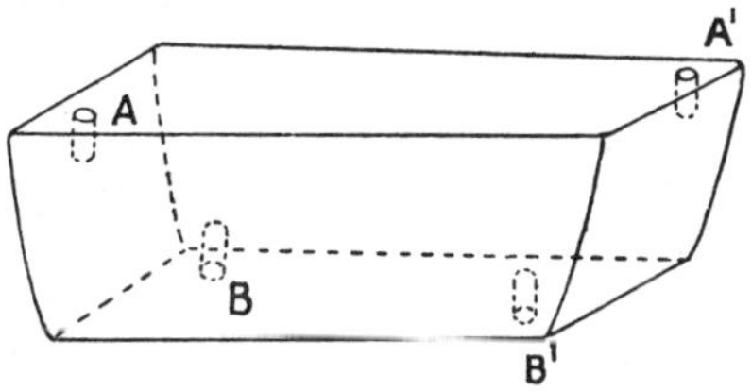

FIG. 125.—METHOD OF DRILLING LOW-GRADE BAR FOR SAMPLING.

A comparison of results of assays on drill and dip samples of cyanide bars is given by Hance[1] as follows;—

(i) Drillings near the edge of the bar are very variable but generally higher than dips; they may be 1 to 2 per 1000 higher.

(ii) Drills near the centre are generally lower than dips.

(iii) In general, top drillings and top dips are higher than the bottom ones (gravity, however, may also act in enriching the bottom).

(iv) Variations in the drillings (often 2 or 3 per 1000, occasionally 5 per 1000) are greater than in dips (sometimes 1 or 2 per 1000).

(v) The difficulties in drill determinations vanish if some silver is present, as lead, zinc, etc. are soluble in the gold–silver alloy, but not in gold free from silver. The presence of from 90 to 100 parts of silver per 1000 is ample and prevents discrepancy in drill assays due to segregation.

Dip sampling is very rarely necessary in silver bullion when the fineness exceeds 975 per 1000, and assay pieces can be cut from one of the bottom edges of an ingot, preferably at a point one-third distant from one of the ends (Gowland).

(4) Sampling by Granulation.—The metal is granulated by pouring into water. When the bullion has been melted, and if necessary refined, it is well stirred, and two samples are taken with a scoop or small ladle, one from the

[1] Quoted by Rose & Newman, *op. cit.*, p. 576.

bottom of the molten mass, and the other from the top. Each ladleful is poured, steadily and continuously in a narrow stream, into a clean copper bowl or wooden bucket containing warm water, which is slowly rotated with a paddle. The granulated metal resulting from each ladleful is kept separate. The water is drained off and the granules carefully dried. After drying, the granules should be hammered flat or rolled, in order to remove any enclosed water. This method of sampling is sometimes adopted for bullion of high grade, but it is not satisfactory for impure bullion, as the pouring into water may result in partial oxidation of some of the base metals present and thus alter the composition of the sample so that it is not representative of the mass.

General Remarks on Methods of Sampling.—The importance of employing a suitable method of sampling, especially in the case of bullion that is subject to liquation when cast into bars, is shown by the irregularity of the assay results obtained by Stockhausen,[1] when drill and cutting samples were taken from cyanide bars. It was pointed out that the variation is sometimes due to imperfect mixing and sometimes to liquation, and experience showed that a dip assay (from molten metal) alone was to be relied upon with this material. In the case of two bars exhaustively examined by means of carefully checked assays of the dipped sample, and of drillings and cuttings taken from different parts of the bars, the assays gave the following results :—

TABLE XXXIV.—Assays of Cyanide Gold Bars sampled by Different Methods. (*Stockhausen.*)

	Average Assay.	Extremes.	Maximum Difference.
	Bar 1.		
Dip sample	809·0	808·9–809·6	0·7
Drillings	810·3	810·0–811·0	1·0
Cuttings	807·2	804·0–810·9	6·9
	Bar 2.		
Dip sample	679·5	679·1–680·3	1·2
Drillings	680·0	679·5–680·2	0·7
Cuttings	677·5	676·3–678·9	2·6

From an extensive series of tests made by F. P. Dewey[2] in 1909 at the San Francisco Mint, it was found that, as a rule, in the cyanide bars from several California plants the chip samples taken from the outside of the bars would be about 2·5 per 1000 fine less in gold than the borings when taken away from the edges of the bar, and that the borings gave satisfactory samples of the bars. Dewey also gave the results obtained by sampling, in three different ways, three cyanide bars high in gold (about 835 fine) and very low in silver (8·0 fine) and each weighing about 1500 ounces. "The assays show a wide variation on the chip samples. While the drill-sample assays are fairly concordant for this class of material, the dip-sample assays agree much better, and are to be preferred." These conclusions confirm those of Stockhausen, quoted above.

[1] "Liquation in Cyanide Gold Bars," *Chem. Met. and Min. Soc. S.A.*, 1897.

[2] F. P. Dewey, "The Assay and Valuation of Gold Bullion," Spokane meeting of *Amer. Inst. Min. Eng.*, September, 1909.

In many instances the composition of the bullion is such that a preliminary refining becomes necessary before the gold content can be determined with accuracy. "The difficulties in valuing base brittle gold and silver bars may be insuperable. Mints and Refining Offices are often compelled to refine the bars partially before sampling them."[1]

Sampling Gold Bullion containing Platinum-group Metals.—It has been found in practice that the results of assays of samples taken in the ordinary way by cutting a small portion from one end of a gold bar or ingot containing platinum and its allied metals do not indicate the actual percentage of gold and of platinum, etc. present in the entire mass. It is therefore evident that the platinum metals have been redistributed during the cooling and solidification of the mass.

Experiments by E. Mattey[2] have shown that ingots of gold containing platinum metals are not uniform in composition owing to segregation of the constituents—the platinum metals, which are the less fusible constituents, moving towards the centre of the ingot on solidification. Thus the assay of a sample cut from one end or corner gives a result too high in gold, sometimes by several per cent., and correspondingly low for platinum and its allied metals. Osmium and iridium, in the form of osmiridium, when present, do not alloy with the gold, but remain as hard grains in the solidified mass, and being of high density are found chiefly at the bottom of gold ingots. To separate the osmiridium, the greater portion of the molten gold is carefully but rapidly poured into a mould; the remainder, which contains almost all the osmiridium, is allowed to cool in the crucible until it has solidified and then assayed for osmiridium. An alternative plan is to allow the whole charge to solidify in the crucible, and then cut off the lowest portion, which is set aside.

As osmiridium settles better from an alloy chiefly consisting of silver than from pure gold, the rich "bottoms" are melted several times with silver, the lowest part being cut off each time. The gold is thus gradually replaced by silver, which eventually forms by far the greater part of the mass. The alloy is then granulated and the granules sampled for assay.

The osmiridium is extracted commercially by parting the granulated metal in sulphuric acid to dissolve the silver and then treating the residue of gold and osmiridium with aqua regia, by which the gold is dissolved, and the osmiridium separated as a black powder.[3]

Iridium is not invariably separated, however, from the gold bars in which it is contained, and traces can be observed in some of the refined commercial bars met with in London.

Preparation of the Bullion Sample for Assay.—The sample obtained by any of the methods described above is prepared for assay by flattening on a polished steel anvil with a special hammer weighing about 11 lbs., until thin enough for cutting conveniently with shears, or for stamping out small discs of approximately equal weight in a fly-press. The anvil and hammer should be kept perfectly clean and used for this purpose only.

In the case of samples of more or less brittle bullion, a heavy iron ring about 8 inches in diameter and $1\frac{1}{4}$ inches high is placed on the anvil to enclose the assay piece and retain any metal that tends to fly away. Some assayers use the anvil only to flatten the assay piece to a convenient thickness for rolling, and then pass the metal through the rolls until of the desired thickness.

[1] Rose & Newman, *Metallurgy of Gold*, 7th edn., p. 516.
[2] *Phil. Trans. Roy. Soc.*, 1892, **183**A, 629–652.
[3] Rose & Newman, *op. cit.*, p. 463.

Bullion samples are usually wrapped in "assay papers" or "dockets" about $6\frac{1}{2}$ inches by 4 inches, on which the number and any other details relating to the sample are recorded. In many cases the papers bear the name of the assayer. The general method of folding the assay paper so that the sample shall not fall out is shown in the accompanying illustration, which gives the details of the folding and is self-explanatory (see Fig. 126). Tough paper of

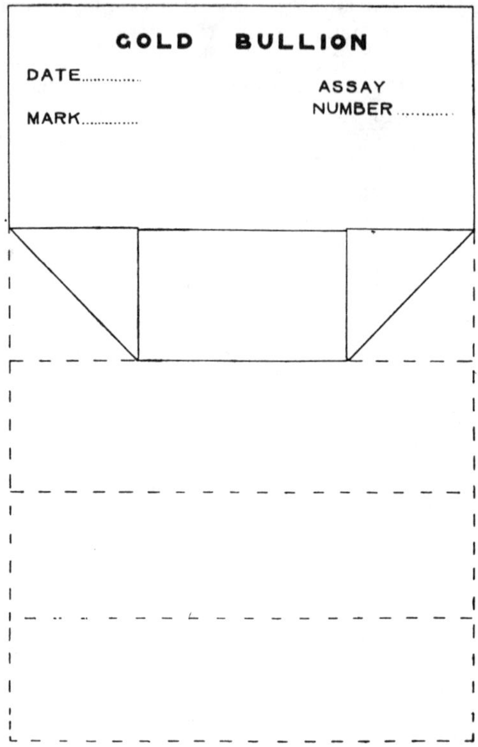

FIG. 126.—METHOD OF WRAPPING BULLION SAMPLE.

good quality should be used, otherwise the metal is liable to cut through and occasion loss.

Owing to the high value of the precious metals, the unused portion of the sample is returned with the gold cornet from the assay when the value of the bullion is reported, or the value of the metal taken for sample is allowed for in the settlement for purchase of the bullion.

CHAPTER XVIII

THE ASSAY OF GOLD BULLION

Apparatus used in Bullion Assaying

The following apparatus is used in connection with bullion assaying.

Hammers and Anvil.—These are used for flattening bullion " assay-pieces " and gold and silver buttons. The face of both the hammer and the anvil should be kept quite bright and clean, and they should be used for the above purpose only. Suitable anvils and hammers are illustrated and described on p. 39.

Rolling-Mill.—The ordinary jewellers' rolling-mill (Fig. 127) is the form usually adopted for assay purposes ; but considerable care should be devoted to accuracy of manufacture, as it is of great importance to be able to roll large numbers of slips of metal to the same thickness. In one form of mill the rolls are accurately adjusted by one motion, as shown in the illustration. The size of the rolls used is dependent on the nature of the work to be done. A pair of 2-inch rolls is often used, but 3-inch rolls will be found a more generally convenient size. The rolls should be fixed on an iron or wooden stand bolted firmly to the floor. The handle should have a radius of not less than 15 inches, and time may be saved by introducing an inclined plane of sheet brass between the two uprights of the mill, by means of which the strips, after passing through, are immediately returned to the front. A section of an inclined plane or shoot is shown in Fig. 128.

The rolls should be kept quite clean and bright and should be used for bullion work only. In assay offices where a large amount of general assaying is done it is usual to have an additional and cheaper pair of rolls for rolling out metals such as lead for cupellation, etc., and for " breaking down " small flat ingots of fine gold or silver.

Steel Forceps.—The ivory-tipped forceps used for the manipulation of weights are not convenient for handling bullion, etc., when weighing out the quantity for assay, steel forceps are therefore generally employed. The forceps illustrated in Fig. 129 are very suitable for the purpose : they are about 5 inches long and are serrated to prevent slipping.

Shears.—Metal shears with blades about $1\frac{3}{4}$ inches in length are employed in cutting off the portion of metal for assay. The shears should be strong, of good quality, and have a keen cutting edge. The cutting pliers illustrated in Figs. 50 and 51, p. 40, are also useful for cutting bullion samples.

File.—In some assay offices the adjustment of the portion of metal for assay to the correct weight is completed by means of a fine file, about $\frac{3}{4}$ inch wide, which may conveniently be fixed on a shallow wood tray in front of the assayer's balance. The file should be flat and smooth ; if too rough, the adjustment of the weight is rendered more difficult.

Sample Rack.—This consists of a block of hardwood, such as oak or walnut, provided with a number of slits $\frac{3}{16}$-inch wide cut at an angle of about 80°, as shown in Fig. 130. It is used to stack the assay papers (Fig. 126) after the sample of bullion has been weighed, and is placed either to the left or right of the balance when weighing out the samples. The rack is about $2\frac{1}{2}$ inches wide, $1\frac{1}{8}$ inches thick, and the length is varied according to the number of slits required, usually twenty or less. Other forms of sample racks are in use, but that described will be found very convenient.

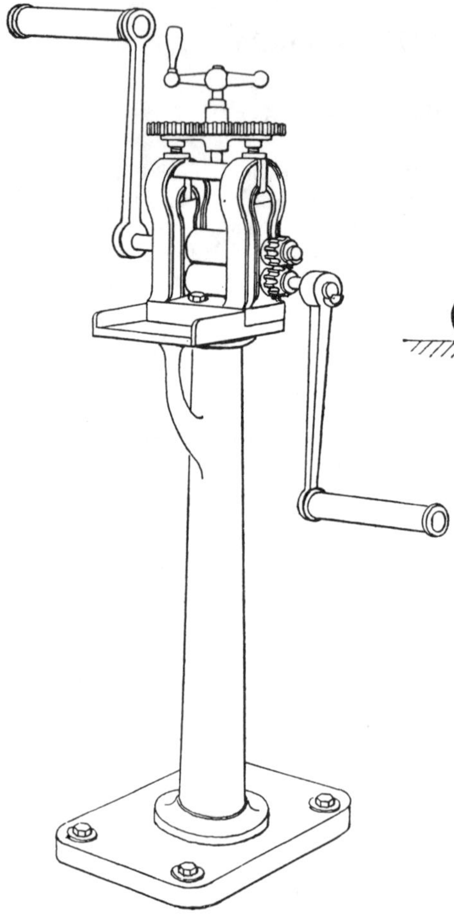

FIG. 127.—JEWELLERS' ROLLING-MILL.

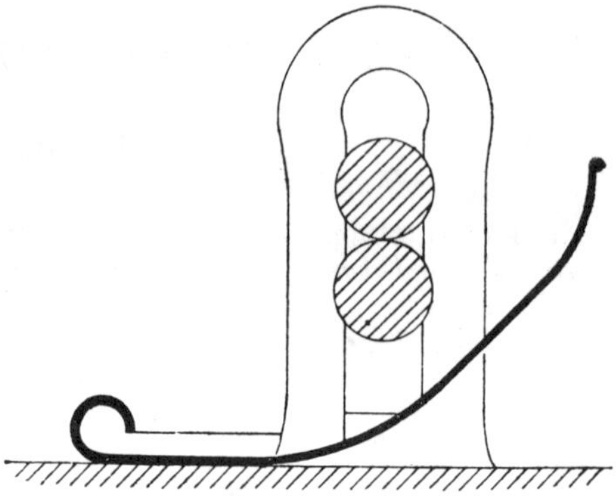

FIG. 128.—SHOOT FOR ROLLING-MILL.

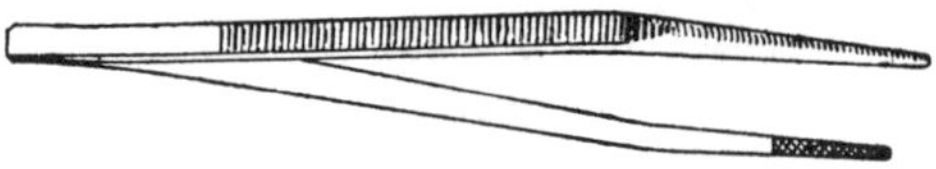

FIG. 129.—STEEL FORCEPS.

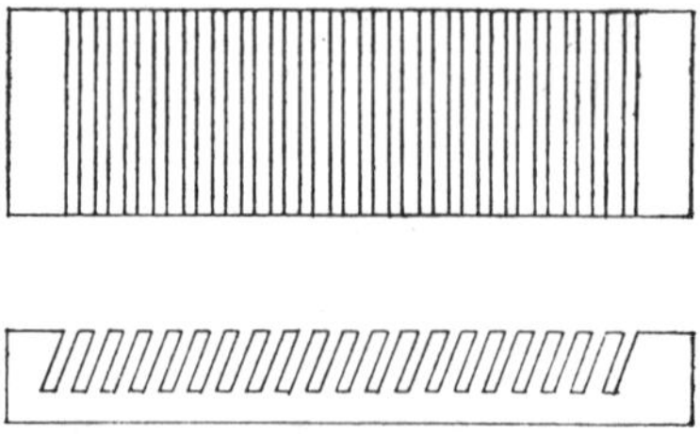

FIG. 130.—SAMPLE RACK FOR ASSAY DOCKETS.

Trays.—Several forms of trays are employed to keep the assay-pieces and the buttons produced in cupellation, etc., in order through the various operations. The tray shown in Fig. 131 is of iron, and those in Figs. 132 and 133 are of wood. A piece of sheet zinc is inserted in the bottom of the latter to protect the wood from burning through the tongs becoming heated when charging in a tray full of lead buttons for cupellation. The compartments are $1\frac{1}{4}$ inches square and about $\frac{1}{2}$-inch deep. The sloping sides of the compartment enable larger numbers to be pasted on than is possible in the case of the tray in Fig. 132 so that the number of any particular assay is more readily seen.

Lead-Foil (test lead).—This should be free from silver, and should be tested for silver before use.

Lead-foil in strips about 1½ to 2 inches wide, and rolled to a thickness of from No. 2 to No. 5 Birmingham metal gauge (0·009 inch to 0·014 inch), is

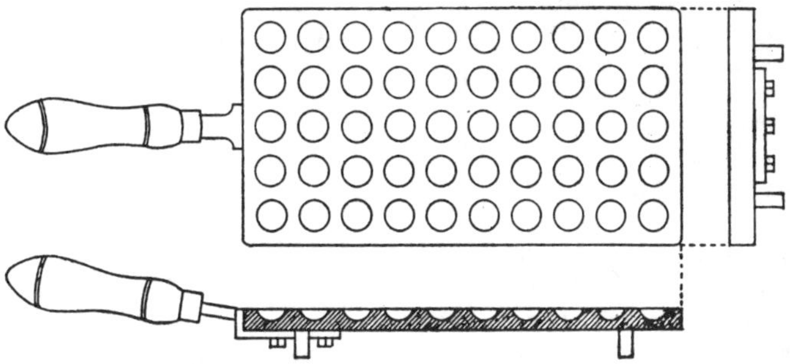

Fig. 131.—Iron tray for assay-pieces.

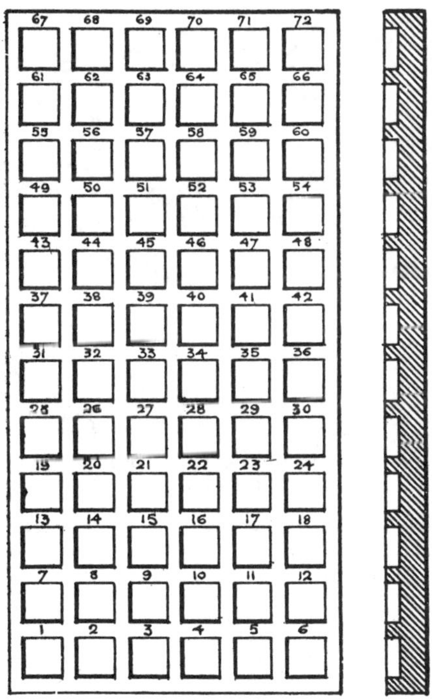

Fig. 132.

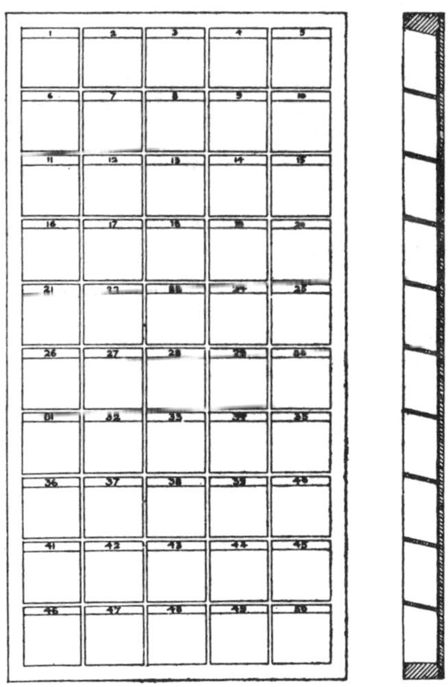

Fig. 133.

Wood trays for assay buttons.

usually employed. For bullion assays the required quantity is cut off, weighed, and folded into a packet for the reception of the portion of bullion weighed for assay.

It is important that the packet should be securely folded after the bullion has been introduced, otherwise part of the sample may be lost by cutting through the foil.

A convenient and very usual method of folding the lead-foil is shown in the accompanying illustrations, Fig. 134. *A* shows the lead ready for folding. *B*, the lead with left-hand corner folded over to the right, and then the right-hand corner folded over to the left, thus forming a packet for receiving the weighed sample. *C* shows the packet, after inserting sample, pressed flat and folded under the line *a-b* ; *D* the packet with the corner c_1 folded on the top, to make it convenient for rolling ; *E* the packet rolled between the fingers as indicated by the arrow.

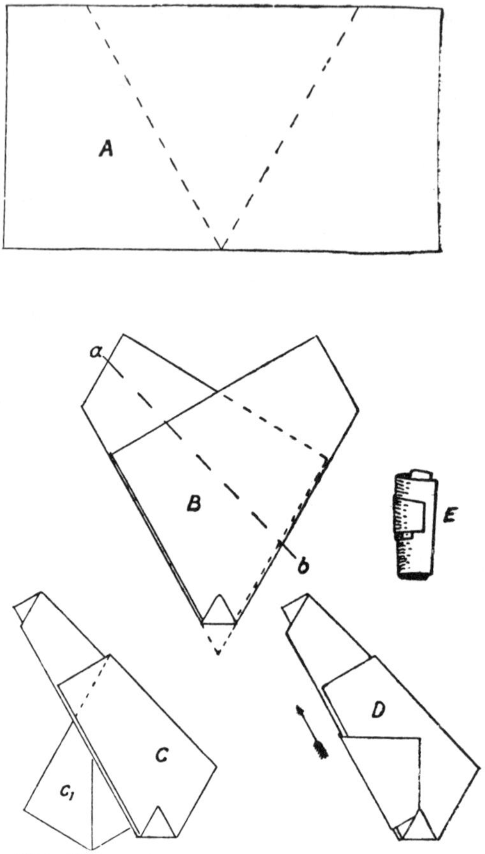

FIG. 134.—METHOD OF FOLDING LEAD-FOIL FOR ASSAY SAMPLE.

By folding in this way the sample is always enclosed in several thicknesses of lead and does not therefore tend to cut through.

When a large number of assays of the same kind have to be made the lead-foil is conveniently purchased cut ready to size and weight.

If varying weights of lead are required, much time is saved by measuring off a strip of foil of uniform width instead of weighing.

A convenient device[1] for this purpose is shown in Fig. 135. It consists of a sheet-iron plate with an oblong slit cut in the middle, and graduated into a number of divisions corresponding to different weights of lead in grammes. The graduations

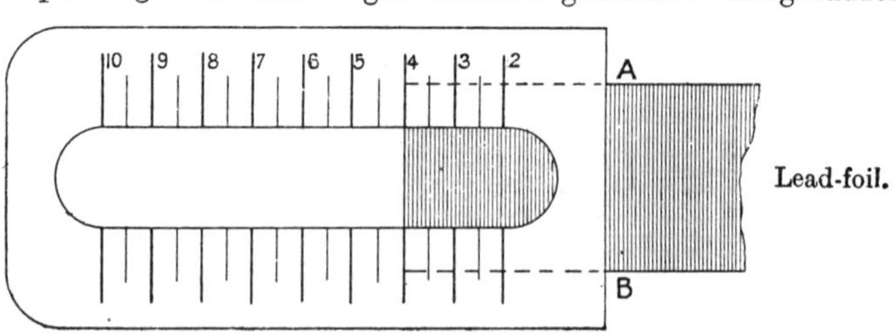

FIG. 135.—DEVICE FOR MEASURING OFF LEAD-FOIL.

are made by noting the positions occupied by different weights of lead-foil of uniform width. To use the instrument, the end of the strip of lead-foil is placed

[1] First suggested by W. F. Lowe, A.R.S.M., of the Chester Assay Office. The instrument described s a modification of that used by Lowe.

under it in position against the lines corresponding to the weight of lead desired, the plate held down firmly, and the lead then torn off along the line A-B. The illustration shows the position of the strip for obtaining 4 grammes of lead.

A plate is very readily made for use with any desired thickness and width of lead. The lead packets should always be prepared by the assayer himself, and a quantity kept on hand ready for use.

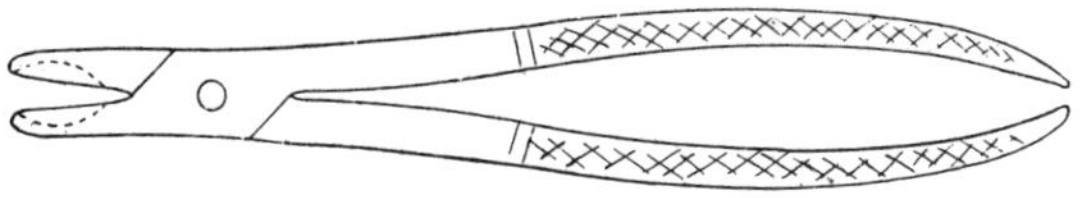

FIG. 136.—BALLING NIPPERS.

Balling Pliers or Nippers.—These are used for squeezing the lead-foil packets into the form of a bullet for convenience of manipulation. They consist of a pair of strong steel pliers with smooth concave chops as shown in Fig. 136. The diameter of the cavity in the chops corresponds to the size of lead bullet desired: from $\frac{3}{8}$-inch to $\frac{1}{2}$-inch is a very usual size for squeezing bullion assays.

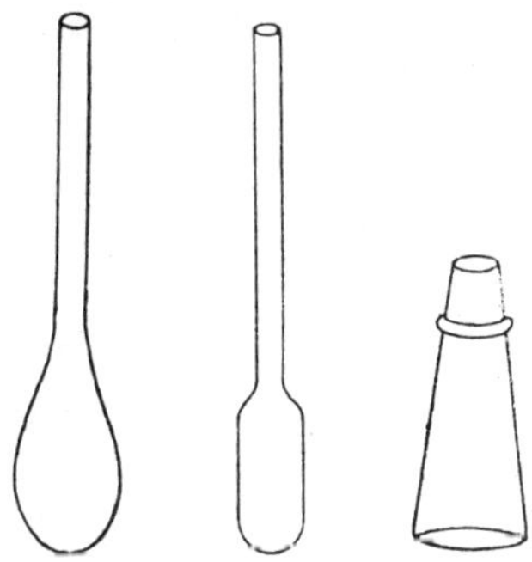

A B C

FIG. 137.—TYPES OF GLASS FLASKS FOR PARTING.

The lead to be "balled," after being folded as described above, is placed between the chops fully opened, and then squeezed repeatedly by alternately opening and closing the pliers held in the right hand. The chops of the pliers are lightly supported between the thumb and second or third finger of the left hand and the lead is gently rotated each time the pressure is released, by a slight upward movement of the first finger, thus gradually causing it to be converted into a ball. With a little practice the pliers can be readily manipulated and a batch of forty or fifty assays "balled up" in a very few minutes. When the lead is balled in this way the assays are much more convenient for charging into the furnace than when the lead packets are irregular in shape.

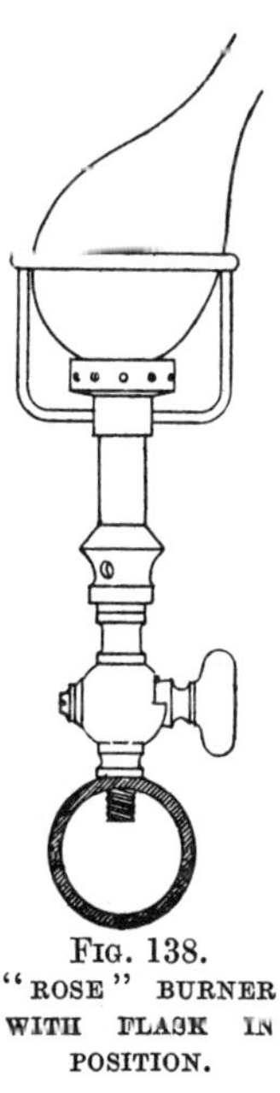

FIG. 138. "ROSE" BURNER WITH FLASK IN POSITION.

Parting Apparatus.—The parting operation, to be subsequently described, in which the "cornets" are treated with boiling nitric acid, may be conducted either in glass flasks or in platinum cups.

Flasks.—Several forms of glass flasks are employed, as shown in Figs. 137 A, B, and C. When flasks are used, a series is so arranged over a number of bunsen "rose" burners, that several can be heated simultaneously. A section of a burner with a flask in position is shown in Fig. 138. The burners are suitably arranged in a fume cupboard with flues connected with a chimney for exhausting the fumes.

An arrangement for heating ten parting-flasks is shown in front elevation and section in Fig. 139. The cupboard is glazed at the sides, and has a window in front which slides up and down. The flasks are kept in position on the burners by means of a wire support as shown in Fig. 138, or more generally by means of a perforated strip of sheet iron as shown in Fig. 139. The top of the burner is made slightly concave (see Fig. 138) so as to keep the bottom of the flask in the centre of the gas jets.

Another form of flask sometimes used for parting is illustrated in Fig. 137 c. In this case the flask is heated on a hot-plate covered with asbestos board.

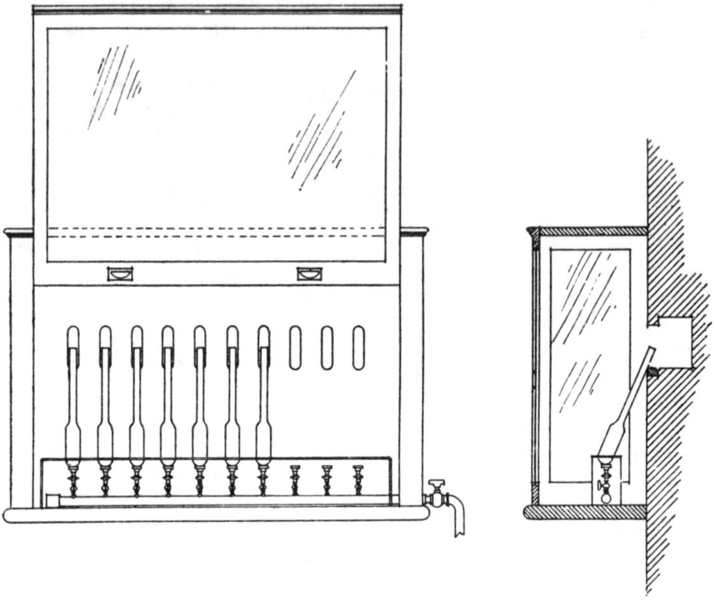

FIG. 139.—ELEVATION AND SECTION OF GLAZED CUPBOARD FOR HEATING PARTING-FLASKS.

Platinum Cups.—When it is desired to treat several cornets simultaneously, short thimble-like platinum cups are used. A cup is shown at *A*, Fig. 140; at the bottom there are five slits, which are fine enough to prevent any gold from escaping, and large enough to allow of the free passage of nitric acid or water. A modern design of frame made of stout platinum wire for holding twenty cups is illustrated in Fig. 140. It is supported on four legs, and is provided with a handle. The cups are slightly enlarged at the top, as shown, to support them in the tray. One of the cups is provided with a thin wire handle and is placed in the left-hand corner at the back of the frame so as to identify No. 1 in the series. A platinum hook fitted in a wooden handle is used for lifting the frame in and out of the hot acid (Fig. 140, *B*), The frame, with its cups, is immersed in boiling nitric acid in a large glass beaker or porcelain boiler. Platinum boilers were formerly used, but are now being discontinued owing to cost..

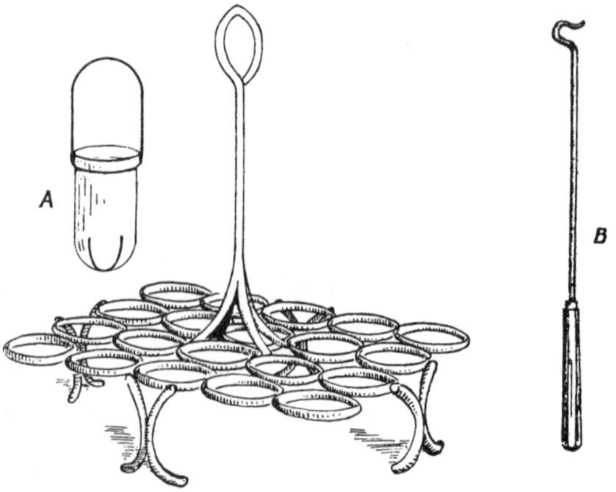

FIG. 140.—PLATINUM FRAME FOR HOLDING PARTING-CUPS (*A*); AND (*B*) PLATINUM HOOK.

Platinum parting-frames to hold any desired number of cups are obtainable from Messrs Johnson, Matthey & Co., Hatton Garden, London, the well-known platinum refiners. Frames to hold fifty or sixty cups are in general use in bullion assay offices.

A special and somewhat novel form of parting apparatus has been devised by Monkton for the treatment of small cornets resulting from the assay of gold cyanide solutions carrying about 0·2 to 2 dwt. of gold per ton, or any other routine gold assays giving cornets that are not liable to break up on treatment with acid. The apparatus is shown to scale in Fig. 141 which is self-explanatory. It is made of white glazed porcelain, and consists of seven oblong cavities, about ⅜-inch deep, arranged in a row, the two end cavities being made slightly deeper so as to act as legs or supports to prevent rocking. A lip is provided

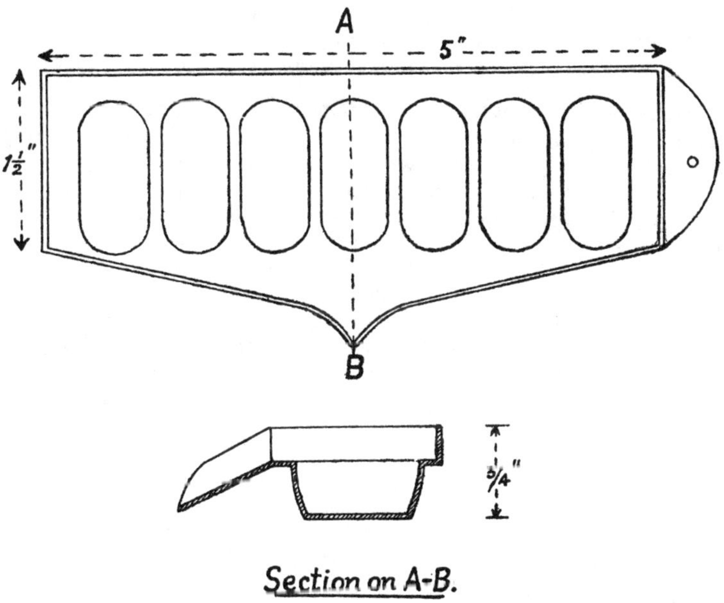

Fig. 141.

as shown for pouring off the spent parting acid. The apparatus is manipulated by means of a bent wire "holder" which is inserted in the ⅛-inch hole in the lug or thumb-piece at the right-hand end. The cornets are placed in position in the cavities, and parting acid added from a tap fitted to a small reservoir of acid, or by means of a burette held in the hand and passed over the cavities in succession. The apparatus is then heated on an electric or other form of heater, which serves both for parting and for subsequent annealing of the cornets. The apparatus, which is very convenient and handy, is in use in some of the South African gold mills.

Method of Assay of Gold Bullion.

Introduction—Gold is always determined by the *parting assay*, which consists in cupelling the bullion with from two to three parts of silver, followed by the removal of silver by boiling in nitric acid, and the weighing of the gold residue. Cupellation removes the greater part of the base metals present in the bullion as oxides dissolved in litharge, and an alloy of gold and silver is left on the cupel. This is " parted " by nitric acid, which dissolves the silver and leaves the gold unattacked. In the assay of refined gold bullion the gold alone is determined—silver, if any, being disregarded. Unrefined gold bullion contains some silver and variable quantities of base metals, and it is usually necessary to determine the proportion of both gold and silver.

Determination of Gold only.—The method of conducting the " parting assay " varies slightly as to details of manipulation in different assay offices, but the following method of procedure may be regarded as descriptive of the general practice. The parting assay of gold in bullion comprises the following five distinct operations :—

(1) Preparation of the assay-piece for cupellation.
(2) Cupellation of the assay-piece.
(3) Preparation of the assay-piece for parting.
(4) Parting and annealing the cornets.
(5) Weighing the cornets and reporting.

(1) Preparation of the Assay-Piece for Cupellation.—The sample or assay-piece is selected and prepared for assay by flattening or rolling in the manner previously described. Each assay-piece is placed in an assay paper, and its number and any other details relating to it duly recorded on the paper. The assay-piece is then prepared for cupellation by weighing about 0·5 gramme and wrapping it in lead-foil with the silver necessary for parting.

Weighing.—In the assay of gold bullion the " millième " system of assay weights is used, a millième being 0·5 milligramme. In this system, fully described on p. 57, the half-gramme weight is stamped 1000, and the decimal subsidiary weights stamped 900, 800, etc., down to 0·5. These stamped numbers denote the number of half-milligrammes (*i.e.* millièmes) contained in the weight. Ordinary weights in the gramme system may of course be used, each milligramme corresponding to 2 millièmes in the assay system. The special millième set of " gold assay weights " is, however, very generally used in all bullion assay offices and in many mines where bullion assays are being made regularly. The report finally made gives the number of parts (in millièmes and tenths) of pure gold in 1000 parts of the bullion. In the Colonies it is usual for assayers on gold mines to use for weighing their parted gold the set of weights in which the weight marked 1000 weighs 1 gramme, and therefore the weight marked 1 equals 1 milligramme.[1] This set of assay weights is described on p. 58, and is generally used for the assay of silver bullion only.

The assay-piece is in many cases first rough-weighed (*i.e.* adjusted approximately) by an assistant upon a " preparing balance " capable of turning to 1 or 2 millièmes. About 1001 millièmes are weighed from each piece, in duplicate or in triplicate, and the weighed pieces folded in a small square of paper and tucked inside the paper containing the rest of the sample or placed in wooden partitioned trays until they can be checked by the assayer. The method of weighing is conducted as follows :—The assayer seats himself directly before the balance, having the assay-piece in a convenient position in front of him. The 1000-millième weight is placed on the right-hand pan of the balance and portions of the assay-piece added to the left-hand pan until in slight excess. The portions are adjusted approximately to the exact weight by cutting with shears and filing. In using the shears the metal to be cut must be held firmly between the forefinger and thumb of the left hand, and care taken to keep the plane of the assay-piece perpendicular to the cutting faces of the shears. The adjustment of the assay-piece is completed by carefully filing on a very fine-

[1] H. T. Durant, *Mining Mag.*, 1911, **5**, 435.

grained file. With practice, weighing becomes a very quick operation, a skilful assayer being able to perform forty to fifty weighings in an hour. The "fine" weighing or final adjustment of the assay-pieces is effected on a "fine" balance. "When the assay is reported to $\frac{1}{10,000}$ part, it is evident that the fine balance used must clearly indicate a difference in weight of 0·1 per 1000 or 0·05 milligramme. With such a very sensitive balance it is extremely tedious to adjust the assay-piece to exactly half a gramme, and in practice it is usual to adjust it until almost correct, and then to note the slight excess or deficiency in weight and record it on the assay paper as a plus or minus 'weighing in' correction. Thus an excess weight of three-tenths of a milliéme would be recorded on the corner of the assay paper as +3. This correction is allowed for in weighing the final gold cornet. It is convenient to have the balance adjusted so that one subdivision of the ivory scale traversed by the pointer corresponds to 0·05 milligramme (*i.e.* one-tenth of a millième)."[1] The weight of the assay-piece should not differ from 1000 by more than four subdivisions on the scale (*i.e.* four-tenths), and if the deviation is greater than this the assay-piece is adjusted until it comes within this limit. In some bullion assay offices "substitution weighing," as previously described is adopted.

The weighed assay-piece is wrapped in lead-foil (folded as previously described) together with the fine or proof silver necessary for parting, and some copper, unless this metal is already present in the bullion being assayed. As each lead packet is prepared it is placed in order in a numbered compartment of a wooden tray and its position duly recorded on the assay paper. The lead packets are next squeezed in the balling nippers to form "bullets," and the assays are then ready to be charged into the furnace.

The amount of silver, copper, and lead used in the assay varies according to circumstances.

Experience has shown that the alloy for parting should not contain less than two nor more than three parts of silver to one part of gold. If the silver is in large excess, the residual gold, instead of holding together in a form easy to manipulate, falls to a powder which requires great care in its treatment to avoid loss. If the silver is insufficient, the gold protects it from the action of the acid and some will remain undissolved, so that the alloy will not be properly "parted."

The modern practice is to aim at getting an alloy with $2\frac{1}{2}$ parts of silver and 1 part of gold. After a long series of experiments at the Royal Mint, however, it has been found that the most exact results are obtained by using a ratio of silver to gold of about $2\frac{1}{6}$ to 1.[2] In the American Mints a ratio of 2 of silver to 1 of gold is used.

If silver is present in the gold bullion, it is allowed for in calculating the quantity of silver to be added. In order to form an alloy of gold and silver in the correct proportions for parting it is obvious that the amount of silver already present in the bullion must be known, and if this is not previously known with a fair degree of accuracy a preliminary assay must be made.

Preliminary Assay.—Experienced assayers can frequently judge the fineness of the gold bullion with sufficient accuracy by the colour and hardness of the metal, and add the correct amount of silver to ensure parting. In some cases the source of the bullion is sufficient indication of the amount of silver required. With cyanide gold it is generally necessary to make a preliminary assay. When the composition of the bullion is quite unknown, it may be determined by means of the touchstone already described, or more satisfactorily by direct

[1] Sir T. K. Rose, "Practice at the Royal Mint," vide *Metallurgy of Gold.*

[2] Rose, *Precious Metals*, 1909, p. 207.

cupellation and parting. For this purpose weigh up two lots of 100 mg. (0·1 gramme) of the bullion, add 300 mg. of silver to one lot, and then wrap each lot separately in 3 grammes of lead. Cupel both side by side. Flatten the button containing the added silver and boil in 15 c.c. of dilute nitric acid (1 to 1) for fifteen minutes; the resulting gold is washed, dried, heated to redness, and weighed.

The weight in milligrammes gives directly the percentage of gold. The weight (in milligrammes) of the other button gives the percentage of gold and silver together; the difference between the two gives the percentage of silver. The remainder will be base metal. Thus:—

Weight of gold and silver	= 98·4	mg. (or per cent.).
Weight of gold	= 94·0	,, ,,
Difference equals silver	4·4	,, ,,

and 100–98·4 equals 1·6 base metal.

If the base metal consists of copper, its presence will be indicated by the appearance of the cupel. In cases where it is known that the alloying metal is copper only, or that silver is not likely to be present in large quantity, a simple cupellation with lead is alone necessary to determine the fineness of the bullion. When the approximate composition of the alloy has been ascertained, the amount of silver to be added for parting is readily determined. For example, suppose the bullion has the composition given above and that 0·5 gramme of it is to be taken for an assay. Then the 0·5 gramme will contain:—

	Composition per cent.		Proportion in 0·5 gramme.
Gold	94·0	Gold	0·470 gramme.
Silver	4·4	Silver	0·022 ,,
Base	1·6	Base	0·008 ,,
	100·0		0·500 gramme.

The total silver required to part the gold is $0{\cdot}47 \times 2{\cdot}5 = 1{\cdot}175$ grammes. Allowing for the 0·022 g. of silver already present, the amount of silver to be added is $1{\cdot}175 - 0{\cdot}022 = 1{\cdot}153$ g. As previously stated, the silver is incorporated with the gold, and at the same time the base metals, such as copper, eliminated by wrapping in sheet lead and cupelling.

The quantity of lead to be used is dependent on the weight of bullion taken for assay and on the proportion and nature of the base metals present. Table XXXV shows the proportions recommended for ordinary gold bullion of different degrees of fineness when using 0·5 gramme for assay.[1]

TABLE XXXV.—Weight of Lead to be taken for Cupellation of 0·5 gramme Gold Bullion.

Fineness of Bullion (Gold and Silver together).	Weight of Lead (for 0·5 gramme Bullion).
950 to 1000 fine.	2 grammes.
900 to 950 ,,	4 ,,
800 to 900 ,,	6 ,,
700 to 800 ,,	8 ,,
600 to 700 ,,	10 ,,
500 to 600 ,,	12 ,,
100 to 500 ,,	12 ,,

[1] Recommended by the Committee appointed by the Chemical, Metallurgical, and Mining Soc. of S. Africa to investigate the causes of the differences in gold bullion assays, see *Proc. of the Society*, 1897–99, **2**, Appendix I, p. 791.

When more than about 6 grammes of lead are required, some assayers wrap the assay-piece in part of the lead and then for cupelling charge in the excess lead either before or after introducing the lead containing the assay-piece.

M'Arthur Johnston states[1] that when the bullion is below 700 fine it is best to scorify 0·5 gramme with from 10 to 12 grammes of lead before cupelling and to use checks and submit them to exactly the same treatment.

If the base metal present in the gold consists mainly of copper, as in the case of industrial gold alloys, other quantities of lead than the above are used, as described in Chapter XXI. If the bullion is known to be free from copper it is usual to add a piece of copper-foil or wire weighing from 10 to 20 mg. The copper is added as the small quantity that remains in the button tends to prevent " spitting " or spurting after cupellation and also to increase the malleability of the button for rolling.

(2) Cupellation of the Assay-Piece.—The cupellation may be conducted in any of the muffle furnaces described in Chapter III, but gas-muffle furnaces are preferable for bullion assays when a good supply of coal gas is available, and are now very generally used. Opinions differ as to the number of assays that should be cupelled at the same time when a large amount of bullion work has to be dealt with daily. Some assayers cupel only twenty assays at the same time, in order to avoid extreme variations of temperature,[2] while at the Royal Mint seventy-two assays constitute a " fire," the cupels used being in sets of four in the middle of the muffle (see p. 170). In most assay offices a " fire " will consist of twenty assays or less. The cupels should be arranged carefully in rows of four abreast in the central part of the muffle and surrounded by blank cupels, so that the temperature may be kept as uniform as possible for those cupels in which the assays are placed. The cupels made by the French firm of Deleuil are very suitable for bullion assays. The cupels are heated in the furnace for about half-an-hour before use, and the muffle brought to a good orange-red heat before charging the assay into the cupels. When the cupels are ready, the assays are charged into them singly with tongs or by means of some charging device such as that described on p. 170. The charging-in is performed carefully, but as rapidly as possible, so as not to cool the muffle unduly, and the muffle door is then partly closed. The exact temperature employed for cupellation, provided it is not too low, is of less importance than uniformity of temperature, as in bullion assaying check or proof assays are invariably used to correct uniform errors due to high temperatures. At the same time excessive heating should be avoided. The conditions for gold cupellation are discussed in Chapter XI.

The time occupied by the cupellation will necessarily vary with the amount of lead used; with 2 grammes of lead it is complete in from eight to twelve minutes. The completion of cupellation takes place in the front rows of cupels and proceeds regularly backwards. The cupels are sometimes removed from the furnace whilst the assay is still molten, and if any detached globules of metal are observed, they are washed down very carefully by a circular movement of the cupel when removed from the furnace, but this requires considerable experience in manipulation. The method of removing the assays whilst still molten is adopted at the Royal Mint and other offices, where all the cupels are removed at the same time on a slab, as described below. With single cupels, each cupel as it is withdrawn is placed in its proper position in the cupel tray. The characteristic " flashing " of gold ensues in a few seconds, and it is most marked

[1] *Rand Metallurgical Practice*, 1926, vol. I, 306.
[2] *Vide* A. C. Claudet, *Trans. Inst. Min. and Met.*, 1906–07, **16**, 137.

in the purer buttons, in which little copper or lead remains. Slight effervescence may occur in these cases, but if the bullion being assayed contains more than 50 parts of copper per 1000, the buttons from the cupellation are never sufficiently freed from base metals (notably copper) for "spitting" to take place (Sir T. K. Rose). To avoid losses by "spitting," etc., a small quantity of copper is frequently added to all assays of fine gold, and most assayers allow the assays to remain in the furnace until the buttons have solidified. To effect this the muffle-door is closed and the temperature lowered. The buttons when "set" (solidified) should show a depression on the top, and no "spurting" (Fig. 142, *a*). The detection and the effect of the presence of metals of the platinum group during cupellation and the subsequent parting are discussed in Chapter XXIV.

(3) **Preparing the Assay-Piece for Parting.**—The buttons are removed from the cupels by means of a pair of sharp-nosed pliers and squeezed to free them

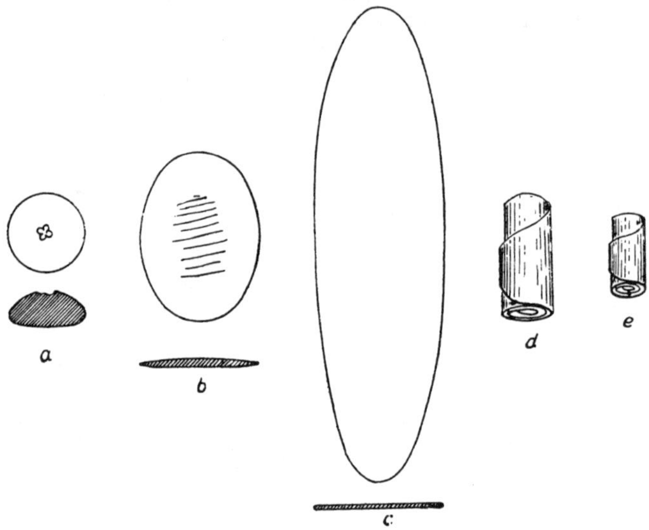

FIG. 142.—PREPARATION OF CORNETS FROM ASSAY-PIECES.

a, button (plan and sectional elevation); *b*, button after flattening (plan and sectional elevation); *c*, fillet (plan and sectional elevation); *d*, cornet before annealing; *e*, cornet after annealing.

from adherent bone-ash. They are then cleaned on the under side with a stiff brush and placed in the compartments corresponding to their cupels in a metal or wooden tray. The complete removal of the bone-ash is not of extreme importance, since bone-ash is readily dissolved by nitric acid on parting. The buttons are flattened on an anvil with a polished hammer weighing about 7 lbs. (Figs. 44 and 45), both perfectly clean and bright and used for this purpose only. A heavy blow is first delivered on the centre of the button, the diameter of which is thereby increased to about half-an-inch. A blow is then given on the edge, so directed as to elongate the metal in one direction, and a similar blow is next given on the opposite edge, reducing it to the form shown at *b*, Fig. 142. The flattened buttons are then annealed at a low red-heat on a clean "fireclay" tile in the muffle or by means of a blowpipe. In some assay offices the buttons are annealed on a piece of iron gauze which is heated by means of a "grid" burner.

At the Royal Mint the annealing is effected in an iron tray especially made for the purpose, which is placed in the muffle.[1] The object of the annealing is to soften the buttons to facilitate rolling, the hammering having hardened them. After annealing, the buttons are passed in succession through the "flatting" rolls to form elongated "fillets" all of the same length (about 2¼ inches) and thickness, the rolls being adjusted so that one passage through them reduces the buttons to the required thickness, which is about 0·01 inch (No. 3 Birm. metal gauge), or about the thickness of an ordinary visiting-card (see *c*, Fig. 142). The edges of the fillets should be quite smooth and not ragged, otherwise loss may take place during the boiling.

After being rolled they are again annealed to soften them, and then separately rolled up between the finger and thumb, or round a glass rod, into a "cornet" or spiral (*d*, Fig. 142). It is the practice in some of the American assay offices to stamp a distinguishing letter or number at one end of the fillet before it is rolled into a cornet and then to commence rolling at the unstamped end so that the number remains in sight when the cornet is completely rolled. All the cornets should have the same number of convolutions, and care is taken to leave that which was formerly the lower side of the button, outside, for the reason stated on p. 419. This face is always dull and is easily recognized.

(4) Parting and Annealing the Cornets.—The "cornets" prepared as described are now ready for "parting" the silver from the gold by treatment with acid. This is effected either in platinum cups contained in a frame, or in glass "parting-flasks" such as those described on p. 271. Parting-frames and cups made of fused silica are now coming into use as a substitute for platinum, and appear to give satisfaction.

(*a*) *Parting in Platinum Cups.*—These are now used whenever possible, as they effect a considerable saving of time. The cornets are placed in the frame in their respective cups, and the frame immersed in boiling dilute nitric acid and boiled for ten to fifteen minutes, or until no more nitrous fumes are observed. The frame is taken out and drained from acid liquor, then washed by dipping in and out of a vessel of hot distilled water, drained again, and immersed in a second lot of stronger boiling nitric acid. The cornets are boiled in this for from fifteen to twenty minutes, when they are drained and washed as before, and are then ready for annealing.

It is usual to keep a stock of the acid suitably diluted to the two strengths required for the parting. These are known as the "parting acids" The first or No. 1 parting acid is the weaker and is used in the first attack on the cornets. After treatment in this acid nearly all the silver has gone into solution as silver nitrate. The second or No. 2 parting acid is stronger, and is used to remove the small quantity of silver still remaining in the cornet. The proper strength of parting acid is of importance. When the ratio of gold to silver in the cornet is 1 to 2½, the specific gravity generally recommended for No. 1 acid is about 1·2 (equal parts of acid and water), and for No. 2 acid a specific gravity of about 1·3 (2 parts of acid to 1 of water). In practice, the No. 1 acid is usually prepared by diluting No. 2 acid after use, the small proportion of silver nitrate this contains not being harmful for this purpose, but an advantage, since it serves to remove any chlorine that may be present. The nitric acid used should be free from hydrochloric acid or chlorine in any form, and a small quantity of silver is kept in solution in the stock acid, so that chlorine, if present, would be instantly detected. The nitric acid should also be free from sulphuric acid. The washing of the cornets after treatment in No. 1 acid is frequently omitted,

[1] Consult Rose, *Metallurgy of Gold*, and Percy, *Metallurgy of Silver*.

but experience proves that the "final" cornets retain less silver when they are well washed with distilled water at this stage. Platinum boilers are still in use in many offices, but owing to the high price of platinum, glass and porcelain boilers are being introduced. The platinum boilers are fitted with a platinum cover, the lower edge of which drops into a channel extending round the upper edge of the boiler and kept filled with water to form a "water-joint." The top of the cover is connected by means of a platinum tube with any suitable arrangement for removing the fumes and condensing nitric acid.[1]

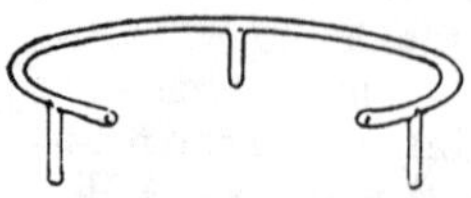

FIG. 143.—ANTI-BUMPING DEVICE.

With frames for parting twenty cornets or less, the boiling is usually effected in glass beakers (preferably of heat-resisting glass) with a clock-glass cover, a separate beaker being used for each parting acid. The beakers are placed on a gauze-covered quadripod stand and heated by means of two bunsen burners with "rose" tops.

Bumping is very liable to occur during the boiling, especially with the stronger acid, and to prevent this the addition of a small ball of well-burnt clay of about the size of a pea is generally recommended, as it lessens the tendency to irregular and vigorous boiling. Bumping is also very effectively prevented by the use of the simple device shown in Fig. 143, which the author has used for many years with entire satisfaction.[2]

It consists of a piece of glass rod, bent to form a semicircle, and to which are fused three short legs, about 1 inch long, of glass tubing of about $\frac{1}{8}$-inch bore, so that the top end is sealed while the lower end is left open. The apparatus is simply placed in the beaker containing the cold acid, so that the open ends of the tubes rest on the bottom of the beaker, and is allowed to remain there during use. The diameter of the apparatus is such that it fits close to the sides of the beaker and allows ample room in the middle for introducing the platinum frame.

After the final washing of the cornets, the frame is placed on a pad of felting or a folded duster to absorb the last drops of water. It is then introduced into the muffle at a bright red heat, kept especially for the purpose, and allowed to remain there until the cornets are thoroughly red-hot. The cornets, which before annealing are of a dull-brown colour and friable, now assume a bright gold-yellow colour, diminish considerably in bulk, and harden. The annealing must not be conducted at a temperature sufficiently high (orange-red) to cause adhesion between the platinum and the gold. The cornets when cold are ready for weighing.

When gold accidentally adheres to the platinum cups it may be removed by brief immersion in dilute aqua regia or by several hours immersion in dilute potassium cyanide solution (0·5 to 1·0 per cent.)

(*b*) *Parting in Flasks.*—The flasks most frequently used are either glass bulbs with long necks (Fig. 137 A) heated on rose burners of special construction, as described on p. 271, or they are small, flat-bottomed conical flasks which are

[1] Platinum boilers such as are used at the Royal Mint are fully described and illustrated in Percy's *Metallurgy of Silver*, p. 263.

[2] This device was suggested to the author by a paper on "Prevention of Bumping during Boiling," by Heyward Scudder, *J. Amer. Chem. Soc.*, 1903, **125,** 163–165.

Theory.—In a general way the theory of the action of such a tube is that, when heated, the air in the capillary expands and passes through the liquid in bubbles. The vapour of the liquid gradually replaces the air, and the stream of bubbles is continuous as long as the temperature around the capillary is at the boiling-point of the liquid. This constant bubbling prevents superheating and consequent explosive boiling.

conveniently heated on a hot-plate (Fig. 137 c). The bulb flask is generally used in England, while in the U.S.A. conical flasks are more frequently employed.

About 50 c.c. of the first parting acid are placed in each flask and raised to boiling-point. The flask is then withdrawn and tilted a little to one side, whilst the cornet is cautiously dropped into it. After introducing the cornet, the hand should be withdrawn promptly, to avoid scalding by the sudden issue of hot acid vapours. The flask is replaced on the burners or hot-plate, and the acid kept boiling for ten or fifteen minutes (*i.e.* for about five minutes after nitrous fumes cease to be given off). The flask is then withdrawn with a piece of stout paper folded into a strip, or with a pair of wooden tongs, and the acid liquor is carefully decanted off into a clean beaker, which is best placed upon a white glazed tile, so that any detached particles of gold accidentally carried over during decanting are readily detected. The decanted silver-nitrate solution is transferred to a bottle marked " Waste Silver," and the silver subsequently recovered by precipitation as chloride.

The residual gold in the flask is washed by adding hot distilled water and again decanting. About 40 c.c. of the second parting-acid, previously heated, are then poured into the flask, which is now replaced on the burner or hot-plate and the boiling continued for fifteen or twenty minutes. As it is at this stage that bumping is most liable to take place, the heating of the flasks should be carefully regulated. The addition of clay peas or a piece of capillary glass tube lessens the tendency to bump, and in the case of flat-bottomed conical flasks it may entirely be prevented by adopting the device previously described.

After boiling with the second acid, the flask is removed and the acid carefully decanted off. The liquor is preserved in an " Acid Waste " bottle and may be used for making up the first parting acid, as previously indicated. The flask is then washed twice with hot distilled water by decantation. If any small particles of gold have become detached from the cornet, time must be allowed for them to settle before each decantation. After the last decantation the flask is filled to the brim with water, a small unglazed porous crucible (annealing-cup Fig. 144) is inverted over the mouth, and then the flask and crucible are quickly but carefully inverted together. The pure gold, which is of a dark-brown colour and exceedingly fragile, falls steadily through the water and collects in the crucible.

FIG. 144.

When time has been allowed for even the finest of any gold particles to settle into the crucible, the flask is gently raised until the mouth is on a level with the edge of the crucible, when with a quick side-motion the flask is removed, the water from it of course running to waste. If the flask is raised too quickly, the inrush of air will cause a rapid displacement of the water, with consequent risk of the gold being washed out of the crucible. The manipulation of the flask which is illustrated in Figs. 145 and 146 requires experience, and beginners will do well to do the transfer over a basin as a safeguard. The water in the crucible is poured off, the cornet dried, and then annealed at a red heat over a gas flame or in the muffle, when the gold assumes its characteristic bright-yellow colour and hardens, so that it is in a state fit to be transferred to the balance and weighed. The spongy cornet from parting is probably an unstable form, or allotrope, of gold ; it is considerably reduced in volume on heating.

A convenient and simple form of electrical annealing apparatus devised by P. J. Perry[1] consists of a square chamber made by placing four small fireclay tiles on end, and suitably suspending on the inside walls short coils of resistance

[1] Communicated to the Author.

wires, arranged vertically, and about $\frac{1}{4}$-inch apart as in the ordinary domestic electric fire. By passing a current through the wires the chamber is quickly heated to the desired temperature for annealing. The size of the chamber should be such that ample clearance is given for the platinum tray when it is lowered into it. During annealing the chamber may be closed by a tile or a sheet of asbestos card.

In many assay offices the cornets are annealed in the same muffle as that used for cupellation, a practice that is to be deprecated owing to possible contamination with lead fumes.

Comparative Advantage of Platinum Cups and Parting-Flasks.—The use of platinum cups effects a great saving of time in decanting and washing, the manipulation is less tedious than with parting-flasks, and the treatment to which the cornets are subjected is more uniform, so that the correction afforded by the use of check assays is more trustworthy. There is also a considerable saving in acid when a large number of partings have to be made. Platinum cups can, however, only be employed where the amount of silver present bears such a relation to the gold that no cornet is in danger of breaking-up—as it will be obvious that, if one cornet in a series breaks up, fragments may adhere to a number of others. An objection to the use of platinum cups is the difficulty

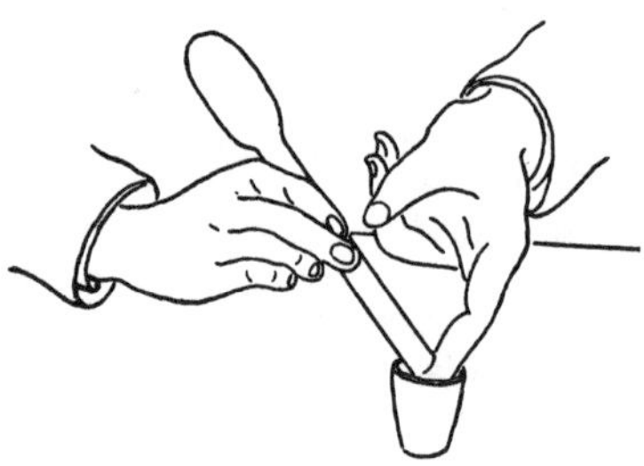

FIG. 145.

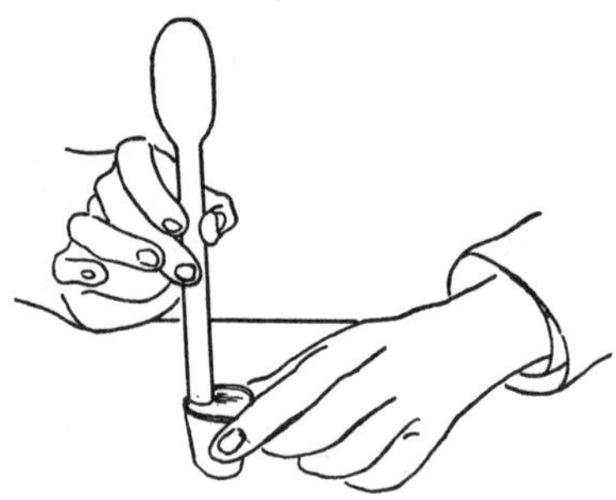

FIG. 146.

of detecting in any individual cornet the presence of such metals as platinum and palladium, which impart a straw-yellow or orange colour respectively to the acid, so that less information on the samples of bullion being assayed is obtained.

If the approximate proportion of gold in the bullion or gold alloy is unknown, so that it is not certain that the cornet will remain entire in the acid, it must be boiled separately, and the use of flasks becomes necessary. When the parting is effected in flasks, the results always differ more or less from those obtained in the platinum apparatus, especially with high-grade bullion, even when check assays are used, as it is impossible to ensure that all the cornets are subjected to exactly the same treatment, however carefully the heating of the flasks may be effected. As these irregularities are eliminated when all the cornets are parted under exactly the same conditions, it is very desirable that platinum cups or the recently-introduced fused quartz cups should be employed in all cases where a large number of assays of bullion or industrial gold alloys of approximately known composition have to be made. Flasks are used in all metallurgical works where a large number of bullion samples of unknown and irregular composition have to be assayed, and are of course used for all preliminary assays of bullion. In some gold mills the parting is done in test-tubes with slits or small holes in the bottom, the glass tubes being used in place of the platinum cups.

(5) **Weighing the Cornets.**—The platinum frame is taken to the balance room, and the cornets, when cold, are usually weighed on the same balance and pan in which the assay pieces were weighed out originally. The balance must readily indicate differences of 0·05 millième per 1000, or $\frac{1}{40}$ milligramme, when loaded with a weight of $\frac{1}{2}$ gramme in each pan. In order to ensure this high degree of sensitiveness, it is very necessary that assay balances should be used with the greatest care, and periodically examined by some experienced person. The weights and riders used should also be compared with a standard set from time to time. A full description of balances and weights and methods of weighing has been given in Chapter V.

In weighing the cornets, the "check" or "proof" assay is weighed first, and the excess or deficiency in weight (usually excess) is applied as a correction to all the cornets from assays worked with it. The correction or "surcharge" to be applied varies according to the nature of the bullion being assayed, and is fully discussed subsequently. An excess or plus surcharge is subtracted from the weight of each cornet, and the minus surcharge added. The surcharge correction is conveniently applied by means of a light rider. The "weighing-in" correction is also allowed for. Thus, if the original weight taken was, say 1000·4 (recorded as + 4), it is sufficient to deduct 0·4 from the final weighing. It must be pointed out that in this correction 0·4 of bullion is reckoned as fine gold, but the error is inappreciable when bullion differing but little from pure gold is under examination, and it will remain inappreciable if the difference from the correct weight is kept at less than 0·5. In the case of bullion 900 fine, nine-tenths of the weighing-in correction must be added to or deducted from the weight of the cornet. If the bullion is 500 fine, one-half of the weighing-in correction must be applied; and in general, if the weighing-in correction is x and the weight of the cornet is y, the correction to be applied in weighing out is expressed by the formula (due to Sir T. K. Rose)—

$$\frac{xy}{1000}.$$

With the millième system of weights the weight of the cornet (after correction) at once indicates, without further calculation, the fineness of the bullion expressed in parts (in millièmes and tenths) of pure gold in 1000 parts of the bullion.

A. O. Watkins[1] has described two methods (in use at the Perth Mint, Western Australia) of weighing the cornets, by which the balance readings give the fineness directly, *i.e.* the weights of the cornets corrected for surcharge.

Reporting.—The degree of accuracy now attained in most assay offices reduces the probable error in the report of a gold bullion assay to about 0·1 per 1000; but, to prevent the error from rising above this amount, all weighings must be correct to 0·05 per 1000, which is not always the case in ordinary bullion assays (Rose).

A. C. Claudet,[2] one of the Bank of England assayers has stated that with fine gold bullion (*i.e.* over 999·0 fine) it is difficult, in spite of all precautions, for the reports to be nearer, on individual bars, than 0·1 to 0·2 per millième, and he considers this to be the limit of accuracy that can be obtained, outside research work.

Gold bullion purchased by the United States Mints is reported to the nearest quarter millième.

[1] Paper before Nat. Hist. and Science Soc. of Western Australia, 1912; also *Chem. News*, 1912, **106**, 248.

[2] *Trans. Inst. Min. and Met.*, 1906–7, **16**, 139.

With mine gold bullion the results are very generally reported to the nearest ½ millième for the gold (*i.e.* 0·5 per 1000); thus, 899·2 is reported 899·0 fine, 899·3 as 899·5 fine, 899·7 as 899·5 fine, and 899·8 as 900·0 fine. It is, however, perhaps more satisfactory for the assayer to report the exact result obtained for the assay.

Liability to Error of the Parting Assay.—The accuracy of the result of the parting assay is liable to be diminished by certain errors incidental to the process; the weight of gold, therefore, as indicated by the balance, does not represent the amount originally present in the assay piece.

The errors are both positive and negative, and are due to certain losses and gains, viz.:—

(*a*) A loss of gold on the cupel, and to a less extent by solution in the acid.

(*b*) An apparent gain of gold due to the retention of silver in the parted gold.

"Both errors are small, and as they are of an opposite character they tend to neutralize each other. Hence they are altogether without effect on the accuracy of the assays of ores when the total gold is reckoned in milligrammes. And even with the larger amounts present in bullion assays their influence is so small that an uncorrected result is still fairly accurate; the resultant error would not be more than one part in two or three thousand" (Beringer).

The errors may be summarized as follows:—

(*a*) *Losses of Gold.*—As fully discussed in Chapter XI, gold is lost during cupellation, chiefly by being carried into the cupel. This constitutes by far the largest proportion of the loss in the gold assay. The amount varies with the conditions of the cupellation, and it is considerably affected by the presence of base metals, especially copper. An increase in the percentage of copper in the assay piece is accompanied by an increase in the loss by absorption. The amount varies with the proportion of copper, being about 0·3 or 0·4 per 1000 when the copper in the gold does not exceed 10 per cent., and more if larger quantities of copper are present.

The *loss of gold by solution in nitric acid* during parting is very small, but experiments have proved that gold is actually dissolved. Beringer states that on a 500-milligramme assay-piece of bullion it may amount to from 0·05 to 0·15 milligramme, *i.e.* from 0·01 to 0·03 per cent.[1] According to Rose, 8 per cent. of the loss in the parting assay is due to the gold going into solution in the nitric acid. F. P. Dewey[2] has shown that, on boiling about 25 grammes of gold in concentrated nitric acid for two hours, the gold passed into solution to the extent of 660 milligrammes per litre, and he observes that the solution of the gold is influenced as much by the temperature (120° C.) required to boil the concentrated acid as by the strength of the acid.

(*b*) *Silver retained by the Gold after Parting.*—The gold obtained by parting always retains more or less silver, the amount varying in a marked degree with the proportion of gold to silver in the alloy before parting, and with the thickness of the cornets. Experiments by Sir T. K. Rose at the Royal Mint show that the amount of silver retained by the cornet is least when the ratio of 2 parts of silver to 1 part of gold is used, and a larger or smaller ratio than this results in a less pure gold after parting. If the ratio of silver to gold is lower than 1·75 to 1, the cornets are not properly parted by boiling in two acids in the ordinary way. In the assay of bullion, the parted gold resulting from treatment with the first acid retains on an average 0·25 per cent. of silver.

[1] *Text-book of Assaying*, p. 154.
[2] *J. Amer. Chem. Soc.*, 1910, **32**, 318.

After treatment in the second acid, the gold cornet retains from 0·06 to 0·09 per cent. of silver, according to the length of time the boiling is continued. The amount can be reduced to about 0·05 per cent. by a third boiling. W. F. Lowe, working on half an ounce of gold cornets obtained in assaying gold bullion in the ordinary way, found it to contain 0·123 per cent. of silver.[1]

Occluded Gases.—Cornets retain about twice their volume of gases (mainly carbon monoxide) in occlusion after annealing.[2] This amounts to 0·02 per cent. by weight, which is reckoned as silver retained. The quantity of gas retained appears to vary slightly with the temperature at which the annealing of the cornets is conducted. If they are not heated sufficiently strongly, they remain porous and retain gas.

Surcharge.—*Net Losses and Gains.*—The net sum of the above losses and gains incurred in the various operations is called the " surcharge," since, in the assay of high-grade bullion, the silver retained is usually in excess of the gold lost, so that the cornet will weigh more than the gold originally present in the assay piece ; if the reverse is the case, the result is considered by some assayers to be less accurate. When there is a gain in weight it is referred to as a "plus surcharge," while a loss in weight is designated a " minus surcharge." With material rich in gold, the silver retained more than compensates for the loss, and, as just stated, the surcharge is positive ; but with a decrease in the proportion of gold the loss is greater and the surcharge is negative. Thus there is usually no surcharge for bullion of about 700 to 800 fine: above this fineness there will be a " plus surcharge," and below a " minus surcharge." The surcharge is reported in parts per 1000 in the same way as the fineness of bullion. Thus a surcharge of + 0·25 means that the gold as weighed was 0·25 part per 1000 more than the gold actually present, while a surcharge of – 0·25 means that on the whole there was a loss of 0·25 part per 1000 in the assay. As previously stated, a " plus surcharge " is subtracted from the weight of the cornet, and a " minus surcharge " added.

The actual amount of the surcharge not only varies with the fineness of the bullion, but is also influenced by the conditions under which the assay is conducted, such as the temperature of the furnace, strength and purity of acid, etc.

The loss of gold by absorption during cupellation is increased by employing too high a temperature, while by prolonged boiling in acid, and especially by annealing the fillets at a high temperature, the amount of gold dissolved in the acids is increased.

With fine gold bars (say over 990 fine) assayed under normal conditions there is always a " plus surcharge " varying from about + 0·7 to + 1·0 part per 1000. Claudet[3] states that during the operations the assay approximately *loses* 0·3 millième of gold and retains 1 to 1·3 millième of silver, not dissolved out in the parting process ; the difference between these is the surcharge.

With bullion from about 800 to 900 fine, cupelled in batches of thirty-six assays, Gowland[4] found that the surcharge varied from 0·5 to 0·7 per 1000, the average being about 0·55.

The surcharge on mill and cyanide bullion is stated by M'Arthur Johnston to vary from 0 to 0·5 per 1000.

In assaying English standard gold (916·6 fine), Sir T. K. Rose[5] has found the surcharge, under normal conditions of working, to be 0·4 to 0·8 part per 1000.

[1] *J. Soc. Chem. Ind.*, 1889, **8**, 687.
[2] Graham, *Phil. Trans. Roy. Soc.*, 1886, p. 433.
[3] *Trans. Inst. Min. and Met.*, 1906–7, **16**, 138.
[4] *Trans. Inst. Min. and Met.*, 1906–7, **16**, 142.
[5] *J. Chem. Soc.*, 1893, **63**, 710.

With this surcharge the total loss of gold is about 0·4 per 1000, of which about 82 per cent. is absorbed by the cupel, 8 per cent. is dissolved in the acid, and the remaining 10 per cent., which is unaccounted for, is probably volatilized. These ratios, however, vary considerably, according to the conditions of working.

The surcharges obtained in the assay of industrial gold alloys containing gold, silver, and copper are given in Table LI on p. 349.

"The more closely a routine is adhered to—the same from time to time in all its minor details of temperature, using cupels of the same make, acid from the same bulk, etc.—the more uniform will the surcharge remain from day to day. But even when this approximate uniformity is secured, there still remains an influencing cause, the neglect of which will lead to reports of a comparatively inaccurate character: *the loss on the cupel depends upon the quantity of gold in the assay*, so that the surcharge will be greater in samples of gold of high fineness than in those of low fineness passed in the same fire and parted in the same acids" (Percy).[1]

Assaying with Checks or Proofs.—From the statements in the preceding paragraphs it will be seen that the errors in the parting assay, although comparatively small, are of sufficient magnitude to invalidate more or less the accuracy of gold assays; and in mints and places where bullion and gold alloy assays must be made with the highest attainable accuracy, the surcharge is always determined by the use of "checks" or "proofs" of similar composition to the bullion or alloys being assayed.

For this purpose pure gold and silver are necessary, and the preparation of these is described on pages 238 and 248 respectively.

The "proof" metals are usually rolled to sheet conveniently thick for cutting up with the shears, or stamping out into discs of convenient size and weight for final adjustment. The metals may also be drawn into flat wire from which pieces of the required weight are cut off with a gauged cutter. Some assayers prefer to granulate the proof metals in water and use the finer granules that pass through a sieve of about 13-mesh (0·04 inch).

Others use the metals in a powdered form obtained in a pure state by precipitation, and heating at a red heat.

The proofs are subjected to the same operations side by side and under identical conditions with the assay-pieces. The composition of the check is, as mentioned above, the same as the composition of the bullion, etc. being assayed, in order to make the assays absolutely comparable. Being of the same composition and weight, and undergoing exactly the same treatment, the check may reasonably be expected to have the same surcharge as the assay-piece. The checks are made up from the data obtained by a preliminary assay.

Suppose the preliminary assay of a gold alloy gave the following composition:—

Gold:	0·4600	gramme, equal to	920·0	parts per 1000	
Silver:	0·0174	,, ,,	34·8	,,	,,
Copper:	0·0226	,, ,,	45·2	,,	,,
	0·5000		1000·0		

For the ordinary assay using 0·5 gramme (1000 millièmes) the total weight of silver required would be 0·460 gramme $\times 2\frac{1}{2} = 1{\cdot}150$ grammes. Deducting the silver already in the alloy, we get $1{\cdot}150 - 0{\cdot}0174 = 1{\cdot}1326$ grammes, the actual weight of silver to be added to the assay-piece.

[1] *Metallurgy of Silver and Gold*, p. 276.

The check assay is then made up from the following weights of pure metals, viz. :—

Pure gold	0·4600 gramme.
Pure silver	1·1500 ,,
Pure copper	0·0226 ,,

These quantities are wrapped in the same weight of lead as that used in the assays, and then the check and assays are cupelled, parted, etc., as nearly as possible under the same conditions. Suppose the parted gold from the check weighed 0·4604 gramme, the gain in weight or plus surcharge is 0·0004 gramme, and this weight would be deducted from the results of all the assays worked with it. Since the amount of surcharge varies with bullion of different fineness, a surcharge correction is never applied except to bullion of the same fineness as that represented by the "check assay" it was calculated from.

It will be evident that, if the gold used in making up the checks is not of the highest degree of purity, an error will be introduced, and the amount of gold found in the bullion, after correction, will be in excess of the true content.

If a sample of proof gold is not pure an allowance must be made. Thus, if it is only 999·7 fine, a deduction of 0·3 is made from all results of assays checked by it. The fineness of the gold, when carefully prepared by the method given is usually taken as 1000 fine, but for very accurate work it may be regarded as 999·9 fine. In this case a deduction of 0·1 must be made, and this has proved to be a very close approximation to the correct one (Rose).

It is the practice of some assayers to use "cornets" for "pure" gold in making up checks ; but unless the gold is first melted into a bar and carefully assayed to determine its purity, the practice is to be deprecated, since as previously stated, parted gold invariably contains more or less silver.

The number of checks to be used for a given number of assays varies according to the nature of the bullion to be assayed. Thus, for the assay of bullion over 999·0 fine, one check is usually used with each nineteen assays, and its relative position is always the same—the eleventh cupel from the front—the cupels being arranged in five rows of four.[1]

For bullion of lower fineness, such as cyanide gold bars, a larger number of checks are used. Assayers on the Rand, South Africa, use one check for every four bullion assays, that is, three checks in a batch of twenty cupellations.[2] At the Royal Mint, London, a batch of seventy-two assays of standard gold (916·6 fine) includes six checks.[3]

In order to get correct results in the parting assay of gold and to utilize the checks to the greatest advantage, it is essential that all the operations should be carried on under exactly the same conditions. A systematic and uniform method of working must be adopted, and although the details of practice may differ from those described, equally accurate results may be obtained so long as all the assay pieces and checks are treated alike.

A series of experiments was made by Gritton[4], at the Sydney Mint, with synthetic assay-pieces composed of gold–silver–copper, to determine the corrections to be made when dealing with rough coppery gold bullion. The object aimed at was to draw up a table of corrections, based on carefully conducted assays, that could be applied to the results of coppery bullion of

[1] A. C. Claudet, "The Assay of Gold Bars," *Trans. Inst. Min. and Met.*, 1906–7, **16**, 137.

[2] *Rand. Met. Prac.*, 1926, vol. I, 306.

[3] Rose, *Precious Metals*, p. 210.

[4] H. B. Gritton, *Forty-Fifth Report of Royal Mint*, 1914, p. 99.

varying composition in order to arrive at the true assay value without the necessity of using a special check for each particular sample.

Each experimental assay-piece consisted of 0·5 gramme, inquarted with silver in the ratio of 2·66 to 1 of gold, and cupellation lead 4·0 grammes. When more than 4 grammes of lead were required, 1 or 2 grammes were added after the first portion had been cupelled off.

As the result of the experiments the following table was drawn up :—

TABLE XXXVI.—CUPELLATION LOSSES WITH COPPERY GOLD BULLION (*Gritton*)

Gold, fineness of Bullion (millièmes).	Corrections for Gold Assays of Rough Gold Bullion containing :—					
	8·33 per cent. Copper.	9·5 per cent. Copper.	14·5 per cent. Copper.	19·5 per cent. Copper.	29·5 per cent. Copper.	39·5 per cent. Copper.
905·0	0·1	0·3	...	...	...	...
805·0	0·3	0·4	0·5	0·6	...	...
705·0	0·4	0·5	0·6	0·7	1·0	...
605·0	0·4	0·6	0·7 a	0·8	1·1	...
505·0	0·6	0·8	0·8	1·0	1·4	1·5
405·0	0·7	0·9	1·0	1·1 a	1·7 b	1·8
305·0	0·9	1·1	1·2	1·4	2·1	1·9
205·0	0·9	1·2	1·2	1·6	2·3	2·3 b
105·0	1·0	1·2	1·3	1·8	2·5	2·5
						2·7

(*a*) Extra lead addition of 1 gramme. (*b*) Extra lead addition of 2 grammes.

The corrections given in the table are to be added to the weights of the cornets after the surcharge, determined by the usual standard gold proof, has been deducted. It will be observed that the corrections gradually increase as the copper content of the bullion increases. The increased quantity of copper in the assays causes an increased absorption of gold by the cupels, and also, in those assays which have required extra lead to complete the cupellation, an increased loss of gold by volatilization.

Limits of Accuracy of the Gold Bullion Assay.—The degree of accuracy of the gold bullion assay as conducted in most assay offices is probably about 0·1 per 1000, but in 1893 T. K. Rose[1] showed that by the use of suitable precautions the probable errors could be reduced to $\pm$ 0·02 per 1000 in the case of high-grade refined gold.

The probable errors in the results of gold assays made in the course of routine work in the assay office of the Royal Mint, London, and by Bank of England assayers, were investigated by J. Phelps[2], of the Royal Mint, in 1910.

The degree of accuracy obtained in gold bullion assays in the United States mints and assay offices, and at the Mint Bureau at Washington, has been discussed by F. P. Dewey,[3] of the Mint Bureau, Washington. Table XXXVII gives particulars of an attempt made by Phelps to compare the probable errors of a single gold-bullion assay made in these different assay offices, as deduced from the figures given in the papers quoted.

[1] *J. Chem. Soc.*, **63**, 700–714.
[2] *Ibid.*, **97** (1910), 1272–1277.
[3] *Trans. Inst. Min. Eng.*, 1909.

The figures assigned to the Bank of England assayer's results and to those of the United States Mint are deduced from the comparison of the results with subsequent assays of the same material at the Royal Mint; the others are calculated by means of the usual formulæ from the differences of individual results from their mean.

TABLE XXXVII.—COMPARISON OF RESULTS OF GOLD BULLION ASSAYS MADE IN DIFFERENT ASSAY OFFICES (*Phelps*).

Authority or assayer.		Probable error per 1000.
Royal Mint	Ordinary assays—Single assay	±0·043
	Trial plate comparison—Single assay ...	±0·023
Trade assayers (London)	Single assay	±0·081
	Triple assay	±0·057
United States Mint—Ten ingots by two assays ...		±0·062
United States Mints (F. P. Dewey)—Single assay ...		±0·150

Stamping the Gold Bars.—In mine and other assay offices where the gold bars after being assayed are sold, they are weighed and then stamped with steel punches giving the following data: Name or mark of the assayer; number of the bar; total weight of the bar in troy ounces and decimals; fineness of the gold; fineness of the silver; and the date of assay. In America, the total money value of the bar in dollars is also added.

Determination of Silver in Gold Bars.—In the case of mine gold bullion, the amount of silver present in association with the gold is usually required. The silver is generally determined by difference. Two or more assays for gold are made by parting in the usual way, and duplicate assay pieces of 0·5 gramme are at the same time cupelled without the addition of silver, but with sufficient lead to remove, as far as practicable, all the base metals present. The difference in weight between the gold-silver button so obtained and the amount of gold determined by the ordinary parting assays gives the amount of silver present in the bullion. The weight of lead to be used is generally guessed at from experience: if the button left on the cupel is not quite bright, it is re-cupelled with the addition of a very small quantity of lead and weighed again. In practice, this assay for the determination of silver is utilized as a preliminary assay to ascertain the approximate fineness of the bar. The colour of the gold-silver button indicates, to an experienced assayer, the right proportion of silver to be added for the gold assay proper. The two chief sources of error in this method are the absorption of gold and silver by the cupel, and the retention of base metal by the precious-metal bead. These errors may or may not balance each other, but they may be controlled to some extent by the use of checks, which should always be used where extreme accuracy in the results is desired.

Another method of determining the silver in gold bars now being increasingly used is described below.

Parting Assay of Gold by Inquartation with Cadmium (Balling's method)[1].—Cadmium may be substituted for silver in the operation of parting; it is very fusible (melting-point 322° C.), readily unites with the gold, and renders it possible to effect the union of the metals in a short time in a porcelain crucible over a spirit lamp.

[1] *Oestr. Zeits. für Berg. und Hüttenwesen*, 1879, p. 597; and 1880, p. 182.

The resulting gold–cadmium alloy is brittle, and cannot be flattened, but some assayers crush it to powder to facilitate parting. This method can be used for the determination of both the gold and the silver in bullion.

(*a*) *Determination of the Gold.*—A weighed quantity of the bullion, usually 0·5 gramme, is melted in a porcelain crucible with cadmium in the proportion of $2\frac{1}{2}$ to 1 of gold, and sufficient potassium cyanide to cover the metals completely during the fusion to protect them from oxidation. If an appreciable amount of silver is present in the bullion, less cadmium is used, so that the combined weight of cadmium and silver will be equal to $2\frac{1}{2}$ times that of the gold. The fusion should be complete in about five minutes, and, after cooling, the cyanide is dissolved out with hot water and the clean button of alloy placed in a flask with 20 c.c. of water. The water is heated to boiling and 40 c.c. of nitric acid added in four successive additions of 10 c.c., and the boiling continued for an hour. The acid solution is then carefully decanted off, and the gold residue washed by boiling for a few minutes in distilled water. The gold is then dried, annealed and weighed.

When the proportion of cadmium is $2\frac{1}{2}$ times the gold, the gold after parting is obtained in a coherent form. In the case of gold bullion practically free from silver, an increase in the proportion of cadmium up to 5 times the weight of the gold, is recommended by some assayers. After parting, the gold retains its original form, but shows upon its surface numerous cracks from which the cadmium has been dissolved. Parting is sometimes effected in two acids, the button being boiled once with dilute nitric acid of 1·2 specific gravity and twice with acid of 1·3 specific gravity. The boiling in the first acid is continued for not less than half-an-hour, or longer according to the amount of gold present. Each boiling in the second acid is continued for ten to fifteen minutes. By this method the losses of gold and silver incidental to cupellation are entirely avoided. A "check" assay is run at the same time, a preliminary assay being made to ascertain the approximate composition of the bullion.

A similar method employing zinc in place of cadmium, had been recommended previously by von Juptner.[1] Donau[2] recommends inquartation of the alloy with the eutectic zinc–cadmium alloy (83 per cent. cadmium, melting-point 266° C.) at 400° C., in a glass tube through which H_2 is passed. Two inquartations are necessary when the gold alloy contains palladium. The beads are parted first in 1 : 3 then in 1 : 1 nitric acid. If tin is present it is removed by heating the residual gold with ammonium chloride prior to weighing.

(*b*) *Determination of the Silver.*—Balling's method of parting is recommended by some assayers for high-grade bullion in which it is desired to determine the silver, as difficulties are encountered when attempts are made to determine the very small amounts of silver present in this class of bullion by the usual parting methods previously described.

The content of silver in the bullion can be determined in the parting solution either volumetrically after it has been carefully poured off from the gold, or by precipitation as chloride.

The following method for the determination of the silver has been in use at the Great Boulder Perseverance Mine, Western Australia.[3] The acid solution from the parting, including the washings, is diluted to 150 c.c., 10 c.c. of ferric alum indicator added, and the silver titrated with a standard solution of

[1] *Zeits. für anal. Chem.*, 1879, p. 104, quoted by Rose, *Met. of Gold.*

[2] Julius Donau, *Z. anal. Chem.*, 1936, **104**, (7/8), 257–270; *Met. Abs.*, 1936, **3**, 212.

[3] E. H. Taylor, *J. Chamber of Mines of West Australia*, July 31, 1908; also *Eng. and Min. Journ.*, 1909, **87**, 543. Consult also J. E. Clennell, *Eng. and Min. Journ.*, 1907, **83**, 1909.

ammonium thiocyanate containing 1·6 grammes of the pure salt to the litre (see Volhard's method, p. 313). The thiocyanate solution is standardized against pure silver, each c.c. being equivalent to about 4·483 parts of silver per 1000. A "check" is run at the same time, the composition of the bullion having been ascertained by a preliminary assay in the usual way.

Whitehead[1] recommends the following procedure. The 0·5 gramme of bullion is heated in a small crucible with 10 grammes of potassium cyanide, and when the latter is fused 1 gramme of cadmium is added. The molten alloy is poured on to a slab, and when cool it is washed to free it from cyanide. The alloy is powdered, transferred to a flask along with 1·004 grammes of pure silver foil, and then treated with nitric acid of specific gravity 1·278. When solution is complete, the total silver is determined either as chloride or by means of standard thiocyanate. The result is compared with a check assay made with 1·004 grammes of pure silver dissolved in the same amount of nitric acid. The difference between the two results, gives the amount of silver in the bullion.

According to Dewey[2] the procedure recommended by Whitehead gives results that are fully as concordant as those obtained by cupellation, and in his opinion it may possibly supplant cupellation in large works laboratories where close accuracy is desired.

Assay by the Touchstone or Streak Testing.—This old method of determining the fineness of a gold alloy consists in rubbing the alloy to be tested on a small block of hard, smooth, dark stone (usually lydian-stone), called a *touchstone*, and comparing the appearance and colour of the streak thus produced with those made by a series of small bars of carefully prepared alloys of definite composition, known as "touch-needles" (Figs. 147 and 148). The effect of the action of a drop of nitric acid and of dilute aqua regia on these streaks is also noted; the streak from the less pure alloy will be more readily acted upon, with the production of a more or less green colour, according to the proportion of copper present. Several series of touch-needles are usually employed, consisting of alloys of gold and copper, gold and silver, and gold, silver, and copper, the alloys being made either to correspond to legal standards or in series in which the proportion of gold gradually decreases, say, from 1000 fine to 500 fine with a difference of 20 parts between the fineness of each needle. For convenience the touch "needles" are frequently soldered to the points of a star-shaped piece of metal, as shown in Fig. 148. Each "needle" is stamped with a number denoting the fineness.

The valuation of an alloy is made by determining to which of the touch-needles the streak it produces most nearly corresponds. In order to get a correct streak of the alloy to be tested, the surface of the metal should first be slightly filed away, as this may have been made somewhat richer than the bulk of the alloy by boiling with acid to remove the base or inferior metal from the surface, as in the "colouring" process used by goldsmiths.

The touchstone is generally used for the approximate assay of small articles of jewellery which it would be necessary to destroy in order to obtain samples for a correct assay; it is also of use to the assayer in determining the approximate fineness of gold bullion and of industrial gold alloys; but it cannot be relied upon for very accurate assay, although it yields sufficiently good results for a preliminary test, and for some purposes is sufficiently exact. It requires, however, a sharp and very practised eye. "The trial is more sensitive for

[1] *Journ. Franklin Inst.*, 1891, **132**, 365–369.

[2] F. P. Dewey, *Ind. and Eng. Chem.*, 1914, **6**, 728.

alloys below 750 fine than for higher standards. The amount of gold in alloys between 700 and 800 fine can be determined correct to five parts per 1000."[1]

In some cases where the same class of bullion is frequently assayed, the touch-needles are made from pieces cut from gold bars. Thus, in assaying large quantities of gold-dust from Korea, Gowland[2] put aside pieces cut from the ingots of the gold-dust, which varied from about 790 to 990 per 1000 of gold, and made touch-needles from them. By the use of these he was able to determine the proportion of gold present to within 5 or 10 millièmes (*i.e.* five to ten parts per 1000), and thus to decide on the amount of silver to be added for parting.

Although regarded in this country merely as a useful preliminary test to ascertain the approximate carat fineness of an alloy, on the continent, on the other hand, the touch-needle has long been in use as one of the recognized methods of determining the quality of jewellery before receiving the official mark of guarantee.

In recent years endeavours have been made to render the streak testing of precious metals applicable to the assay of all the complex gold, silver,

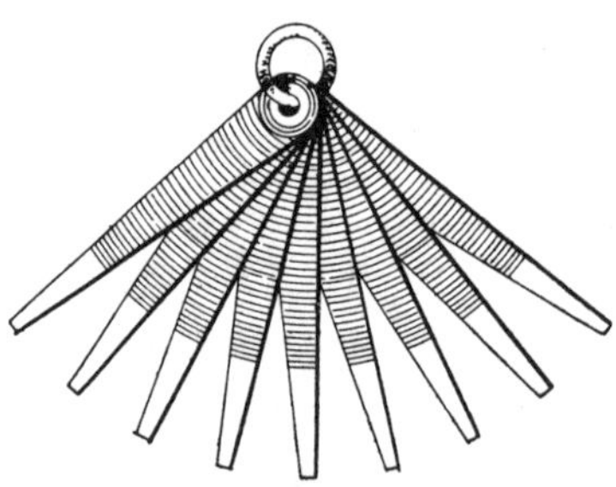

FIG. 147.

FIG. 148.

TWO TYPES OF TOUCH "NEEDLE."

and platinum alloys which are now used in the jewellery trade. According to Hradecky,[3] it is possible with experience to obtain results which are within 0·5–1·0 per cent. of the true gold content by the use of specially-prepared standard needles; and this applies to the examination of white golds as well as to the usual copper, silver, and copper–silver carat alloys. Naturally, however, it is not only the colour of the streak that is compared, but also its behaviour under the action of various reagents, a list of which is given in the work referred to with details of their preparation and use.

The great advantage of the streak test is that only a minute amount of the alloy is used and no damage is done to the article; this test is therefore of considerable value in many cases where it is impossible to take a large enough sample for the ordinary assay, but as previously stated the results obtained are not sufficiently accurate for hall-marking purposes as required in this country.

In an effort to increase the accuracy of the test, recent work[4] has shown that it is possible to identify the constituent metals of the streak by suitable microchemical tests.

[1] Rose & Newman, *Metallurgy of Gold*, 7th edn., 1937, p. 531.

[2] *Trans. Inst. Min. and Met.*, 1906–07, **16**, 142.

[3] *Die Strichprobe der Edelmetalle* ("The Steak Test for Precious Metals"), by Karl Hradecky, Vienna, 1930. Review *J. Inst. Metals*, 1931, **47**, 126.

[4] *Mikrochemie*, 1930–31.

Volumetric Determination of Gold in Electrolytes, etc.

Volumetric methods of gold assay find application in mints and refineries, where bullion is refined by electrolytic processes, for determining the gold content of the electrolyte of auric chloride and of wash waters. Several volumetric methods have been proposed, but few of these appear to have been applied in practice. Those in use include the following:—

Juptner's (Ferrous Sulphate) Method.—In this method, devised by Juptner,[1] the solution of gold chloride is treated with a measured excess of a standard solution of ferrous sulphate to precipitate the gold, and the free ferrous sulphate then determined by titration with standard permanganate.

Solutions of ferrous ammonium sulphate rapidly precipitate definite quantities of gold from solutions of auric chloride with a marked degree of concordance, provided that no free nitric acid or nitrates are present which would affect the re-solution of the finely-divided precipitate. The strength of the ferrous solution is important: if it is too weak the precipitate will remain partially suspended, and if too strong the burette readings are less exact, as each drop will obviously throw down a larger amount of gold. The strength of solution usually employed in practice is such that 1 c.c. will precipitate 25 mg. of gold.

The permanganate solution is prepared so that 1 c.c. will exactly correspond to 1 c.c. of the ferrous sulphate solution. It is then only necessary to deduct the number of c.c. of permanganate used in titrating from the amount of ferrous ammonium sulphate solution added to the gold solution, in order to determine the amount of gold in milligrammes.

The procedure adopted at the Mint Refinery, San Francisco, for determining the gold in the electrolyte, etc., using ferrous sulphate solution, has been described by H. French.[2] The electrolyte is a solution of gold chloride, containing, in the first set of cells, 70 g. of gold per litre and from 10 to 12 per cent. of hydrochloric acid, and in the second set only 60 g. of gold per litre, but with the same amount of acid.[3] The ferrous sulphate solution used is of such strength that 1 c.c. will precipitate 25 mg. of gold. It is prepared by dissolving 153·5 grammes of pure ferrous ammonium sulphate in warm distilled water and diluting to one litre. About 1 c.c. of concentrated sulphuric acid is added to maintain the solution in the ferrous state. Such a solution will remain unaltered for months.

A permanganate solution is prepared by dissolving 12·3 grammes of potassium permanganate and diluting to one litre. This solution has the power of oxidizing an equal volume of the standard ferrous solution to the ferric state.

Usually, after weighing and dissolving the permanganate, 5 c.c. will oxidize from 4·9 to 5·1 c.c. of the ferrous sulphate solution. The permanganate solution and the ferrous sulphate solution are brought into a perfect balance of chemical-conversion power by the addition of from 10 to 20 c.c. of water to the stronger solution. For all practical purposes these two solutions give satisfactory results, so long as they are within one-fiftieth of their theoretical proportion of strength. Occasionally it may be desirable to be more exact, and then the amount of ferrous iron in solution is determined by titration with a standard solution of potassium permanganate in the manner adopted for iron assays, care being

[1] *Oestr. Zeits. f. Berg. und Hüttenwesen*, 1880, p. 182.

[2] *Mining and Engineering World*, 1912, **37**, 853–855. See also "Volumetric Determination of Gold," V. Lenher, *Min. and Scientific Press*, S. Francisco, June 21, 1913.

[3] Gowland, *Metallurgy of the Non-Ferrous Metals.*

taken to prevent any oxidation of the ferrous sulphate solution before titrating. As a rule the crystals of ferrous ammonium sulphate average between 14·2 and 14·3 per cent. of iron.

The actual precipitating power of the ferrous ammonium sulphate solution is usually checked by the gravimetric method, a carefully measured quantity, say 5 to 10 c.c., of the ferrous sulphate solution being added to a solution of gold chloride, and the resulting precipitate of metallic gold weighed. After precipitation the gold is allowed to settle, the clear solution being decanted through a filter, and the residue of gold is then boiled in dilute hydrochloric acid, and washed on to the filter with distilled water.

Particles or films of gold adhering to the sides of the beaker are removed by means of a small piece of filter-paper. After carefully washing the gold until all trace of iron is removed, it is transferred to a porcelain crucible and dried, then heated to redness, cooled, and weighed. The results are a little higher than when the dried precipitate is cupelled. To be exact the gold should be carefully wrapped in a small piece of lead-foil and cupelled. In this way a careful check is made upon the titrations.

Unless the approximate percentage of gold in solution is known, a preliminary test is made to determine the probable amount of ferrous sulphate solution required.

Having determined this, a measured quantity of the gold solution, usually 5 c.c., is run from a pipette into a beaker, 10 c.c. of hot water added, and about one-fifth c.c. of sulphuric acid shaken in from a " dropping " bottle. Then, according to the judgment of the operator, the standard ferrous sulphate solution is run in rapidly from a burette and in sufficient quantity to turn the faint yellow tint of the gold chloride solution to a pale green in which a turbid mass of finely-divided brownish-black gold is suspended. The contents of the beaker are vigorously stirred with a glass rod and allowed to settle.

If the orange-yellow colour prevails over the green, another c.c. of the precipitant is added until an excess is obvious by the permanent pale-green hue of the solution in the beaker. After allowing the precipitate to settle, water is added to the solution, increasing its volume to nearly 100 c.c. Then the permanganate solution is run in carefully from a burette until all of the excess of ferrous sulphate solution has been oxidized, as indicated by the last drop of the permanganate turning the solution a permanent pink. To calculate the result :—If 5 c.c. of the gold solution were taken and 18 c.c. of the standard ferrous sulphate solution had been run in, effecting a complete precipitation as above described, plus a slight excess, and the latter had been determined by adding 3 c.c. of permanganate, then 15 c.c. was the exact volume of the ferrous sulphate solution required to precipitate the gold contained in 5 c.c. of the sampled solution. Therefore $15 \times 0{\cdot}025$ gramme $= 0{\cdot}375$ gramme of gold in 5 c.c. (*i.e.* $\frac{1}{200}$ part of a litre) of the gold solution tested, and one litre of this solution would contain ($0{\cdot}375 \times 200$) 75 grammes of gold.

For the determination of gold in solutions containing less than 1 gramme per litre, the iodide method is more exact.

Potassium Iodide (Peterson) Method.—This volumetric method of determining gold was originally devised by Peterson,[1] and developed by later workers. Under certain working conditions it is of sufficient accuracy for routine work. It is based on the formation of the complex compound $Au(AuI_4)$ according to the reaction :—

$$2AuCl_3 + 6KI = 6KCl + Au(AuI_4) + I_2.$$

[1] *Zeits. anorg. Chem.*, 1899, **19**, 59.

The solution of auric chloride to be tested is treated with excess of potassium iodide to precipitate the gold as aurous iodide, and the free iodine liberated determined by titration with standard thiosulphate, using starch as indicator. 1 c.c. of N/100 sodium thiosulphate, corresponds to 1·97 mg. gold.

From 10 to 20 c.c. of the gold solution are diluted to 200 c.c., 0·02 gramme of potassium iodide added, and the solution titrated with standard thiosulphate in the usual way. It is advisable to add a very slight excess of thiosulphate and then to titrate back with N/100 iodine solution until a faint pink tint is detected. Provided the gold solution contains no free chlorine, and not more than 1 gramme of gold per litre, the method yields accurate results. The accuracy is said to be increased by using a mixture of potassium iodide and iodate for precipitating the gold.[1]

Under certain conditions the method is applicable to the determination of gold in the presence of copper.[2]

The corresponding reaction for copper is :—

$$2\ CuCl_2 + 4KI = 4KCl + Cu_2I_2 + I_2.$$

This second reaction occurs only with a large excess of concentrated potassium iodide solution, and by using very dilute reagent (0·1 per cent.), formation of cuprous iodide can be supressed entirely. With 3·078 mg. of gold present and 20, 30, 40, 50, 60, 80 mls. of 0·05 per cent. copper sulphate solution added, the quantities of gold found were 3·19, 3·15, 3·15, 3·23, 3·19, and 3·21 mg. respectively. The consistent positive error is attributed to difficulty in recognizing the blue starch-iodide end-point in the presence of the blue copper sulphate solution.

Pollard Volumetric Method for Gold Wares.—A volumetric method of assaying gold wares was devised by W. B. Pollard in 1916, and was introduced at the Cairo and Alexandria Assay Offices in Egypt where the quality of such wares is tested, and where it displaced the usual furnace assay owing to the greater speed and cooler working conditions and lower costs. The original procedure, however, has been improved upon, and the method, as now worked, is described by Pollard[3] as follows.

For making the assays the following four solutions are prepared :—

(*a*) *Sodium arsenite reagent.* 400 grammes of sodium arsenite are dissolved in 2 litres of water, filtered, diluted to 30 litres, and made acid with 350 c.c. of hydrochloric acid.

To this arsenite solution are added 60 g. of mercuric chloride dissolved in 1 litre of water, and 0·3 g. of potassium iodide dissolved in 25 c.c. of water. The reagent thus prepared is further diluted as required, so that 50 c.c. will be equivalent to 0·5 g. of gold.

(*b*) *Urea–potassium bromide solution,* containing 10 g. urea, 5 gramme potassium bromide, and 2 c.c. strong hydrochloric acid dissolved in water and made up to a litre.

(*c*) *Decimal hydroquinone solution,* containing 0·8372 g. of hydroquinone and 10 c.c. strong hydrochloric acid per litre.

(*d*) *Indicator solution.* A saturated solution of *o*-toluidine in saturated sodium fluoride, acidified with acetic acid.

To determine the gold in bullion or in an alloy rich in gold, the sample containing about 0·5 g. of gold is dissolved by heating in 3 c.c. of hydrochloric acid and 1 c.c. of nitric acid and the solution cooled, whilst a current of air is

[1] Vanino and Hattwagner, *Zeits. anal. Chem.*, 1916, **55**, 377.

[2] Brull and Griffi, *Ricerca scientifica*, 1938 (iii), **9**, (9/10), 622–623. *Cf. Met. Abs.*, 1940, **66**, 174.

[3] W. B. Pollard, *Trans. Inst. Min. Met.*, 1932, **41**, 434.

passed through it to remove gaseous by-products. 50 c.c. of the urea–potassium bromide solution are then added and the temperature raised to 40° C. The arsenite solution is then run in from a burette until no more brown colour, due to the conversion of chloro-auric acid to bromo-auric acid, is observed. The gold is precipitated and settles well, and re-solution is prevented by the urea, which destroys all traces of nitrous acid. After addition of 20 c.c. of *o*-toluidine indicator the last traces of gold are titrated with decimal hydroquinone until the yellow colour disappears.

The number of c.c. of decimal hydroquinone used gives the number of mg. of gold remaining in solution at the end of the arsenite titration, so that the weight of gold precipitated by the volume of arsenite used is obtained.

The arsenite and hydroquinone solutions are standardized against pure gold and adjusted before use so that 1 c.c. of the arsenite solution is equal to 0·01 gramme of gold, and 1 c.c. of the hydroquinone solution is equal to 0·001 gramme of gold. Greater accuracy may be obtained by the use of a 50-c.c. Stas pipette for the titration, as used in the wet assay of silver.

When the alloy to be tested contains much silver it may be inquarted with copper to assist dissolution in aqua regia. Of all the metals usually found in gold alloys only palladium and a large excess of cadmium interfere. Full details, including the effects of interfering metals, are given in Pollard's papers.

Electrolytic Assay of Gold.—Gold may be accurately deposited from cyanide solution. Potassium cyanide is added to the gold solution in the proportion of 2·5 g. per 100 c.c. of solution, and a current is passed through the solution between platinum electrodes. According to Classen,[1] the current density should be about 0·3 ampère per square decimetre, the fall of potential 2·7 to 4 volts, and the temperature 50° to 60° C. The gold is deposited in about 1½ hours: it is washed with water and alcohol, dried, and weighed together with the cathode. The increase in weight of the cathode equals the amount of pure gold in the quantity of alloy taken.

S. P. Miller[2] states that from 150 c.c. of a solution containing 1 gramme of potassium cyanide and 0·1291 gramme of gold, the latter is deposited completely in 2½ hours with a potential difference of 1·8 volts and $N.D._{100} = 0{\cdot}2 - 0{\cdot}04$ ampère during the first 1½ hours and then increased so as to maintain a voltage of 1·8. The temperature employed is 55° to 65° C.

Messrs. Perkin and Prebble[3] recommend the use of ammonium thiocyanate in place of the potassium cyanide which is generally recommended. With a current density of 0·4 to 0·5 ampère per square decimetre the deposition of 0·05 to 0·08 gramme of gold is complete in 1½ to 2 hours at ordinary temperatures.

The solutions employed contained 5 grammes of ammonium thiocyanate in 120 to 150 c.c., the gold chloride solution which was used for the tests being added in measured volume. The gold is deposited on a platinum electrode in the usual way. The completion of deposition is judged by testing a sample of the electrolyte with stannous chloride after the thiocyanate has been decomposed with sulphuric acid.

Removal of Deposited Gold.—" The best way of removing deposited gold from platinum electrodes is by means of a solution of potassium cyanide containing an oxidizing agent such as hydrogen peroxide, sodium peroxide, or an alkali persulphate."[4]

[1] *Quantitative Analysis by Electrolysis.*

[2] *J. Amer. Chem. Soc.*, 1904, **26**, 1255–1269.

[3] F. M. Perkin and W. C. Prebble, " Electrolytic Analysis of Gold " (paper before the Faraday Society), *Elec. Chem. and Met.*, 1904, **3**, 490–494.

[4] Sir T. K. Rose, *Metallurgy of Gold.*

The electrolytic method of assay is used in some mills for the determination of the gold in cyanide solutions resulting from the cyanide process. In this case the cathode consists of a cylinder of lead-foil notched at the base to facilitate circulation of the liquid. The solutions, which vary in gold content, are electrolysed for four hours with a current varying from 0·06 to 1·2 ampère.

The gold separates as a bright-yellow deposit. When precipitation is complete, the cathode is washed in water, dried, rolled up, then scorified with test lead, and cupelled. Details of this method, as recommended by Crichton, are given on p. 471.

Electrolysis of Platiniferous Gold.—Balling[1] has proposed the separation of gold from platinum and metals of the platinum group by electrolysis. A weighed quantity of the platiniferous gold, in the form of thin sheet, is used as the anode, while an accurately weighed sheet of pure gold or of platinum is used as the cathode. Both sheets are placed in a solution of gold chloride and subjected to the action of an electric current. The gold anode is dissolved and is deposited on the cathode, while the platinum metals are liberated and fall to the bottom as a grey-black powder, which is carefully collected and assayed for platinum, etc.

Spectroscopic Analysis of Precious-metal Alloys.—In recent years the possibility of applying spectrographic methods of quantitative analysis to metals and alloys has been closely investigated, both in this country and abroad, with promising results. So far as the precious metals and their alloys are concerned, spectrum analysis appears at present to be limited, in practice, to the detection of impurities and of the constituent metals, with perhaps semi-quantitative analysis in some few cases.

The spectroscopic analytical control of precious metals has recently been studied by Pastore and Occhialini[2] and their conclusions are of interest in drawing attention to the best form in which to use the metals to be tested. To avoid distortion of the incandescent thin metal wire in Occhialini's method (*Atti R. Accad. Lincei*, 1929, **9**, 537), they wind it into a spiral 3 mm. in diameter and insert it at right-angles to the arc between two carbon poles. The length of lines of the main constituent gives an idea of the quantity of alloying elements present without the need for the determination of each element individually, and this result can be obtained very rapidly. For platinum, gold, and silver the diameter of the wire used was 0·05 – 0·025 cm. and the spectral lines used were 5301·05, 5230·3, and 5465·43 λ, respectively. Examples of determinations are given in the original paper. An attempt at the quantitative analysis of silver–gold alloys by optical spectroscopy was made by Mankin.[3] No definite relations could be found between relative intensities or corresponding lines and composition of the alloy. The pairs of lines selected for comparison were 4669–4241: 5465–4793: 5472–4811. The first number in each pair is the wave-length of the silver line and the second of the gold line.

The chief value of the spectrograph in the laboratory of platinum refiners is to provide a rapid method of identification of alloys; to investigate the cause of local failures in platinum ware due chiefly to contamination with

[1] Balling, *Fortschritte im Probirwesen*, Berlin, 1887, p. 116.

[2] Salvatore Pastore and Emilio Occhialini, *Ricerca scientifica*, 1938, [ii], (9/10), 619–622: abs.: *Met. Abs.*, 1940, **7**, 175. See also F. Twyman, "Industrial Applications of Spectrography in the Non-Ferrous Metallurgical Industry," *J. Inst. Metals*, 1939, **64**, 379–388.

[3] Winifred Mankin, *J. Proc. Roy. Soc. N.S. Wales*, 1936, **70**, 95–99: *Met. Abs.* (*Inst. Metals*), 1937, **4**, 105.

arsenic, silicon, or phosphorus; and to control the purity of platinum and its alloys for special purposes, *e.g.* gauze for ammonia oxidation catalysts.[1]

Since it is frequently difficult to dissolve alloys of the platinum metals, the spectrographic test is conveniently applied to small beads of the metal which are inserted into the graphite arc; usually 50 mg. are sufficient with a 20-minutes' exposure.

Determination of Tellurium and Selenium in Gold Bullion.—Tellurium is found in small quantity principally in gold bullion from certain localities in America and Western Australia. Its action in the assay of gold bullion is discussed on p. 162.

To determine the tellurium, a weighed quantity of the bullion is dissolved in aqua regia and then evaporated with a large excess of hydrochloric acid until no more chlorine is given off.

The solution is diluted and the gold and tellurium precipitated with a current of sulphurous acid. The joint precipitate is attacked with nitric acid which dissolves the tellurium as tellurous acid, leaving the gold as a residue which may be washed, dried and weighed, or inquarted with silver and parted.

Selenium may also be present in small quantity as an impurity in commercial gold bullion.

For the separation of gold from tellurium and selenium,[2] the chloride solution is adjusted to 1·2 N hydrochloric acid, and the gold precipitated by boiling with hydroquinone. In the filtrate, tellurium is separated with hydrazine and SO_2 in 9N hydrochloric acid. Both precipitates are dried at 105° C. in CO_2 for weighing.

Schoeller[3] gives stannous chloride as a quantitative reagent for selenium and tellurium. The method depends on the collection of the selenium and tellurium (as dioxides) with ferric hydrate, dissolution of the washed precipitate in 1 : 2 hydrochloric acid and precipitation of the selenium and tellurium by addition of 20 per cent. stannous chloride solution to the boiling solution.

The precipitate is collected, washed, and dissolved in hydrochloric acid saturated with bromine, and the selenium precipitated by treating the solution with a slow current of SO_2. The filtrate from the selenium is diluted with an equal volume of water, and the tellurium precipitated with hydrazine, 2 HCl, and SO_2.

According to Ljung,[4] as little as 0·05 parts per million of selenium in hydrochloric acid solution gives a distinct reaction when boiled with ammonium thiocyanate, which reduces selenium dioxide (SeO_2) to selenium; the colour varies from pinkish-white through bluish-red to deep red, according to the selenium concentration.

Gold may be separated from tellurium by precipitation with sodium nitrite from a boiling solution of the chlorides buffered with sodium potassium tartrate to a *p*H of 1–3. For the determination of both elements the sum is first found by precipitation with $N_2H_4.2HCl$ and SO_2, the weighed precipitate is dissolved in aqua regia, the nitric acid expelled by evaporation with hydrochloric acid, and the gold precipitated as above.[5]

[1] H. E. Stauss, *Proc. Amer. Soc. Test. Mat.*, 1935, **35**, (II), 57–60; *cf. Met. Abs.* (*Inst. Met.*), 1935, **2**, 707.

[2] J. Seath and F. E. Beamish, *Ind. and Eng. Chem.* (Anal. edn.), 1937, **9**, (8), 373–374.

[3] W. R. Schoeller, *Analyst*, 1939, **64**, (758), 318–323.

[4] H. A. Ljung, *Ind. and Eng. Chem.* (Anal. edn.), 1937, **9**, (7), 328–330; *cf. Met. Abs.*, 1937, **4**, 533.

[5] Lenher, Smith and Knowles, *Ind. and Eng. Chem.*, Anal. edn., 1934, **6**, 43–45; *Met. Abs. Inst. Metals*, 1934, **1**, 251.

CHAPTER XIX

THE ASSAY OF SILVER BULLION

THE impurities generally present in silver bullion are copper, lead, and less frequently zinc, bismuth and selenium. Usually they are not in sufficient quantity seriously to affect the assay result. The silver in silver bullion and alloys is determined by one of the four principal methods, namely :—

(I) By cupellation.
(II) By titration with sodium chloride (Gay-Lussac method).
(III) By titration with thiocyanate (Volhard method).
(IV) By conversion into chloride (gravimetric or Indian Mint method).

The quality of silver is also tested by means of the touchstone ; the amount of the metal contained in rich silver–copper alloys (over 850 fine) may be determined by the oxidation colour test ; and electrolytic methods of analysis are in use for nitrate or cyanide solutions. These processes are described at the end of the present chapter.

(I) Cupellation Assay of Silver.

This method of assay has largely been displaced by wet methods for bullion assays, but is still in use for the assay of ores, etc. It is accurate to about 0·5 part per 1000 when nearly pure silver is being assayed. As the operation of cupellation has already been described in Chapter XI, it will only be necessary here to point out the precautions which must be observed in conducting the operation for silver assays.

The cupellation assay of silver may be divided into three distinct operations, namely :—

(1) The preparation of the assay-piece.
(2) Elimination of impurities by cupellation.
(3) Weighing the buttons and reporting the results.

(1) The Preparation of the Assay-Piece.—The sample is selected by any of the methods described in Chapter XVII, but additional precautions are necessary to ensure that the assay-piece represents the mass from which it is taken ; for, as previously stated, alloys of silver and copper are peculiarly liable to undergo liquation. The assay-piece is flattened and a portion adjusted to an exact weight, usually 0·5 gramme, in the manner already described under the assay of gold bullion. The assay-piece is wrapped in a packet of sheet lead, the weight of which bears a definite ratio to that of the copper present. The lead should be at least five times the weight of the assay-piece, and should be increased in proportion to the amount of copper present : the exact proportions usually recommended are given in Table XXXIII. The corners of the lead packets are squeezed down so as to fit the cupels by pliers, or are " balled " in the balling nippers (Fig. 136) especially designed for the purpose.

The assay-pieces thus prepared are placed in their proper order in a wooden tray with numbered compartments, from which they are transferred direct to the muffle. The amount of lead to be employed in the cupellation depends on the composition of the alloy under examination; it is obviously necessary, therefore, that this should previously be approximately determined.

An experienced assayer will frequently be able to decide this with sufficient accuracy from the colour and hardness of the metal; or the composition may be roughly ascertained by means of the touchstone streak, or by noting the amount of oxidation produced on a clean piece of the bullion when gently heated in an oxidizing atmosphere.

Preliminary Assay.—When necessary a preliminary assay may be made by cupelling 0·5 gramme of the bullion with 7 grammes of sheet lead, and working the assay in the usual way. The silver button is cleaned and weighed. The loss will give the amount of base metal present: this is usually copper, and its presence may be detected by the colour stain of the cupel.

The composition of the bullion having been approximately determined, the following Table XXXVIII gives the proportions of lead to be used with one part of bullion of different fineness.

TABLE XXXVIII.—PROPORTIONS OF LEAD REQUIRED FOR CUPELLATION OF SILVER BULLION.

Fineness of Bullion. Silver in 1000 parts.	Parts of Lead for one part of Bullion.	Ratio of Lead to Base Metal.
1000 to 950	5	100 to 1
950 ,, 900	7	70 to 1
900 ,, 800	8	40 to 1
800 ,, 700	10	33·3 to 1
700 ,, 600	12	30 to 1
600 ,, 500	14	28 to 1
500 ,, 250	16 to 18	33·3 or 24 to 1
250 ,, 0	18 ,, 20	18 or 20 to 1

For the cupellation of English standard silver (925 per 1000), six times its weight of lead is usually employed.

(2) Elimination of Impurities by Cupellation.—The assay-pieces are introduced into the cupels, arranged in rows corresponding to those on the tray, the muffle being at a bright red heat. The cupellation is conducted in the manner previously described on page 145. The proper adjustment of the heat and gradual cooling of the assays are of the first importance. It is advisable to keep the temperature as low as possible, so as to reduce the volatilization of the silver and its absorption by the cupel to a minimum; but, on the other hand, care must be taken to avoid all risk of the buttons setting before the oxidizable metals have been removed. The heat should be so adjusted that only slight fumes are visible over the cupels, and the fused metal should be very luminous and clearly distinguishable from the cupel. With too low a temperature crystals of litharge collect round the button, while with an excessive heat the fumes rise rapidly to the crown of the muffle. With "patent" magnesia cupels the conditions of cupellation differ somewhat from those just described for bone-ash cupels (see page 47).

In cooling down the muffle prior to withdrawing the cupels, special care must be taken to prevent "spurting." The methods generally adopted for preventing this have already been discussed.

When the cupellation has been satisfactorily performed the silver buttons are bright, crystalline, and slightly depressed in the centre, and dull white on the under surface. The presence even of a minute quantity of copper renders a silver button less globular. After all the buttons are set, the cupels are withdrawn and placed in their proper positions in the cupel tray. The buttons are then removed by means of a pair of pliers, squeezed to loosen the adhering bone-ash, and cleaned with a stiff brush. It is the practice in some assay offices to remove the buttons from the cupels, and give each button a blow on its side with a small hammer to loosen the adherent bone-ash before it is brushed.

The buttons are now ready for the final operation of weighing.

(3) **Weighing the Buttons and Reporting.**—In ascertaining the absolute amount of pure silver present in the assay-piece, the method of weighing and the system of "weighing-in" correction, described in Chapter V, is employed.

Checks for Silver Assays.—The results of silver-bullion assays are controlled by check assays of pure silver or of silver with the addition of copper or other base metal to make them of the same composition as the assay pieces. The checks are made up in the same manner as gold checks, previously described. The losses during silver cupellation are discussed on page 157.

The absorption of silver by the cupel is large, and may amount to as much as 10 parts per 1000: it varies considerably, being increased by a higher temperature of the furnace and by the presence of a large amount of copper or of lead. It is therefore necessary to use several checks in different parts of the muffle; and since the loss of silver is much greater than in the case of gold, and varies in different parts of the muffle, it is necessary that the checks should be more numerous. It is a common practice to place one check in each row of assays in the muffle, and to vary the position of each in the rows. Thus in a batch of forty assays of fine silver bullion including eight checks, arranged in the muffle in eight rows of five abreast, the positions of the checks would be 4, 7, 15, 18, 21, 29, 32, 40—cupel No. 1 being at the extreme left of the back row.

The correction to be applied to an individual assay-piece is deduced from the checks in its immediate neighbourhood. Considerable experience and knowledge are necessary in applying such corrections to the weights obtained, as well as a thorough knowledge of the variations of temperature in the muffle used. The variations of temperature are determined in the first instance by working off a complete set of fine silver cupellations. This must be done each time a new muffle is taken into service.

The amount of the correction due to cupellation loss need not exceed 10 parts per 1000. When carefully conducted the cupellation assay is, as before stated, accurate to about 0·5 per 1000 when nearly pure silver is being assayed.

Reporting the Results.—The results of silver bullion assays are reported in parts per 1000 to the nearest whole number, *i.e.* to the nearest 0·1 per cent. Thus a result 995·6 is reported 996 fine and 994·4 as 994 fine. The results of assay-pieces from the top and bottom of the silver bars should agree closely; in the case of fine silver bars, within 0·5 per 1000. In making the report, either the average of the results or the lowest result is given: it is customary to give the lowest result of assays of fine silver. If the cupellation has been carefully performed and proper checks used, there is no difference between the results with silver bars over 980 fine. Bars between 980 and 725 fine show a difference of 0·5 to 3 parts per 1000; from 720 to 710 fine, no difference, or only an infinitely small one.[1] This is due to the retention of small quantities of base metals, which counteract the loss of the silver absorbed by the cupel. The greatest differences occur in low-grade bullion below 400 fine. Very considerable differences may occur between the assay results if the bars have not been properly melted and mixed before casting to ensure uniformity.

[1] Kerl, *Assaying*, 1889, p. 146.

Apparatus Employed in Wet Assay of Silver—(Methods II, III and IV).

The following apparatus is used in connection with the wet methods of silver assaying :—

Weights.—The special set of "silver assay weights," described in Chapter V, are very frequently used in the assay of silver bullion. In this system the numbers stamped on the weights denote their weight in milligrammes. Thus the 1-gramme (*i.e.* 1000 milligrammes) weight is stamped 1000, the 0·5 gramme (500 milligrammes) weight 500, etc. It will be noted that in this system the value of the weights is double that in the "millième" system used in gold assaying, in which, as already stated, the 0·5-gramme (500 milligrammes) weight is stamped 1000. In offices where the two systems are in use it is essential to keep the weights distinct, as the confusing of the weights used may lead to serious error in reporting the value of bullion.

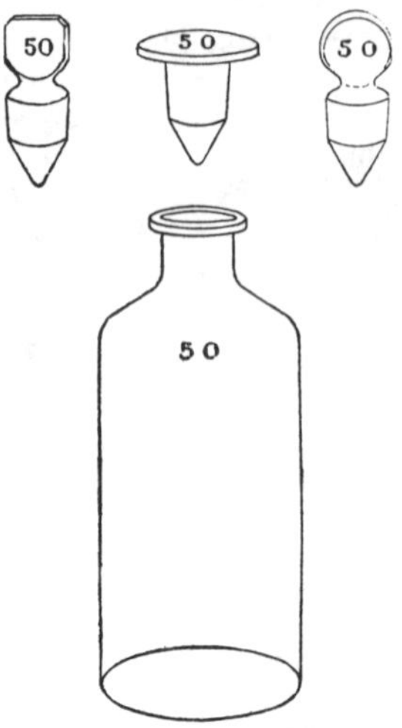

FIG. 149.—ASSAY-BOTTLE FOR SILVER BULLION.

Assay-Bottles.—Round, narrow, stoppered bottles, capable of holding 8 to 10 ounces (225 c.c. to 300 c.c.), both bottles and stoppers being numbered, are usually employed. The bottles should be of good clear white glass and capable of being heated in a water-bath without cracking. The usual shape of bottle is shown in Fig. 149 with three kinds of stoppers in common use. Some unnumbered bottles, provided with a "sand-blasted" patch, should be kept in stock to replace any numbered bottles in the series that are accidently broken. The required number is best painted on the ground patch with black enamel paint. The number on the bottle into which each assay-piece is introduced is noted on the corner of the assay paper in which the sample is wrapped.

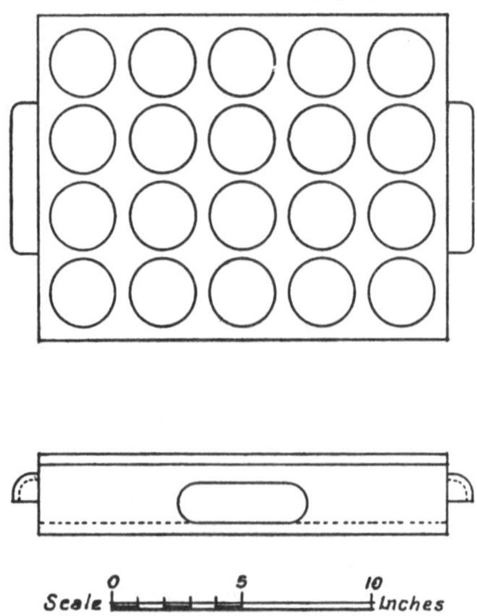

FIG. 150.—WOODEN TRAY FOR ASSAY-BOTTLES.

FIG. 151.—STOPPER TRAY.

Water-Bath.—A water-bath for heating the bottles is best made of copper with a loose plate with holes for keeping the bottles in position. A water-bath to hold twenty bottles is similar in size and shape to the wooden tray shown in

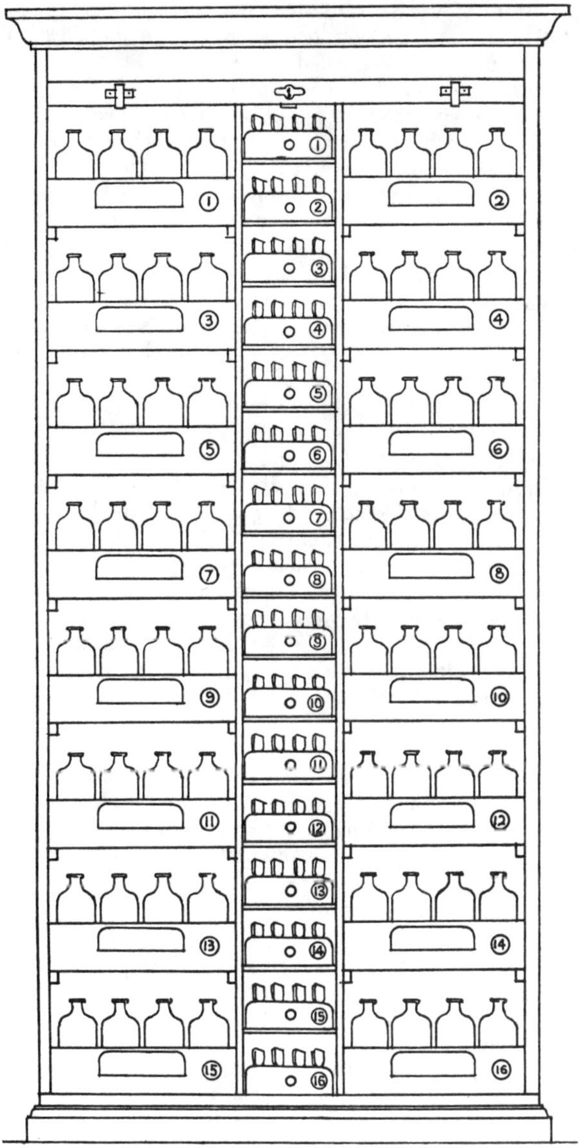

Fig. 152.—Assay-bottle cabinet.

Fig. 150. The bath is supported on an iron quadripod stand and heated by three or four rows of air-gas burners.

Bottle Trays.—When a large number of assays have to be dealt with it is convenient to work them in batches of twenty, and to facilitate the handling of the bottles they are placed in wooden trays (see Fig. 150). The tray has a thin wooden top, with holes to admit the bottles, so that they are always kept in correct position: the top is made of three-ply wood to prevent warping. A hole is cut in each side of the tray, as shown, to permit of a cloth being inserted to wipe the bottom of the tray in the event of a bottle breaking. The tray should be well made and be provided with wooden handles. The trays are distinguished by numbered enamelled plates.

Stopper Trays.—The numbered stoppers for the bottles are kept in their proper order in the tray shown in Fig. 151. This is made of a block of hardwood with holes drilled in a diagonal direction so that when the stoppers are placed in position the numbers are readily seen. The tray is placed in front of the bottle tray to the right of the operator when titrating the assays, as described later. The trays are distinguished by numbers corresponding to those on the trays containing the bottles to which the stoppers belong.

Bottle Cabinet.—In offices where several hundred assays are made daily, some convenient method of storing the assay-bottles becomes a necessity. A very convenient form of cabinet for the purpose is shown in Fig. 152. It holds 16 trays, each containing 20 bottles—a total of 320 bottles—and their stoppers, which are placed on a series of shelves between the bottle trays, as shown. The bottle trays are slipped into position on runners. The cabinet is provided with a roll front to keep out the dust.

Burettes.—A burette of large bore and with large opening for quick delivery is used for adding the requisite quantity of acid to the assays. A 200-c.c. burette, 24 inches long and $1\frac{1}{8}$-inch bore, graduated into 10 c.c., is very suitable for adding the acid to a batch of twenty assays. The burette is held in the left hand and passed over the mouth of each bottle in succession, the time occupied in adding acid to twenty assays being about forty seconds. A hand pipette, similar to that shown in Fig. 156, is frequently used for the purpose, but is not so suitable.

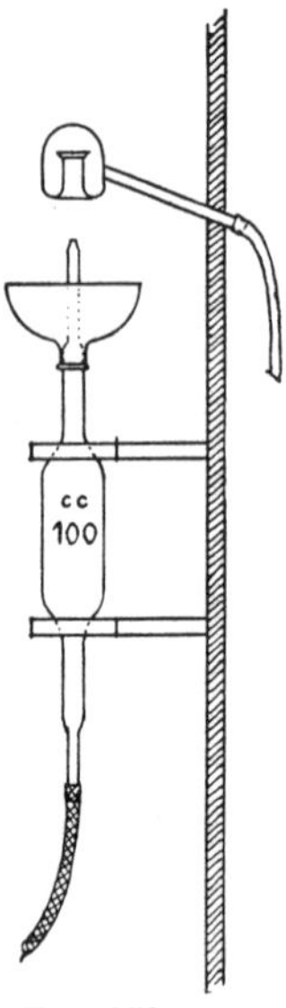

FIG. 153.—STAS PIPETTE.

Pipettes.—A large pipette for adding 100 c.c. of a standard solution to each assay is required. The form of pipette devised by Stas, shown in Fig. 153, is used for the purpose. The solution is introduced from below by means of an india-rubber tube, and the pipette when full is closed at the top by the finger with the left hand. The tube is then removed with the right hand, the bottom of the pipette gently "wiped" between the thumb and first finger, and the solution allowed to run into the bottle placed ready to receive it. The bottom end of the pipette is sometimes sand-blasted for a short distance to prevent the slipping of the rubber tube. The pipette is fitted with a cup, and also provided with a glass hood or cap to catch any overflow when filling. The hood is connected by rubber tubing to a receptacle to receive the overflow of solution, which can be used again.

The Stas pipette is one of the most important pieces of apparatus used in silver assaying and it must be made with great care. It is not so important that the pipette shall deliver exactly 100 c.c. as that in two consecutive deliveries the volume shall not differ by more than 0·005 c.c. In well-made pipettes the error in the deliveries is often not more than 0·002 c.c.

Where large numbers of assays have to be made, a set of four pipettes is mounted in a suitable case side by side. Experience proves that there is no advantage in having more than four pipettes in a series. A convenient

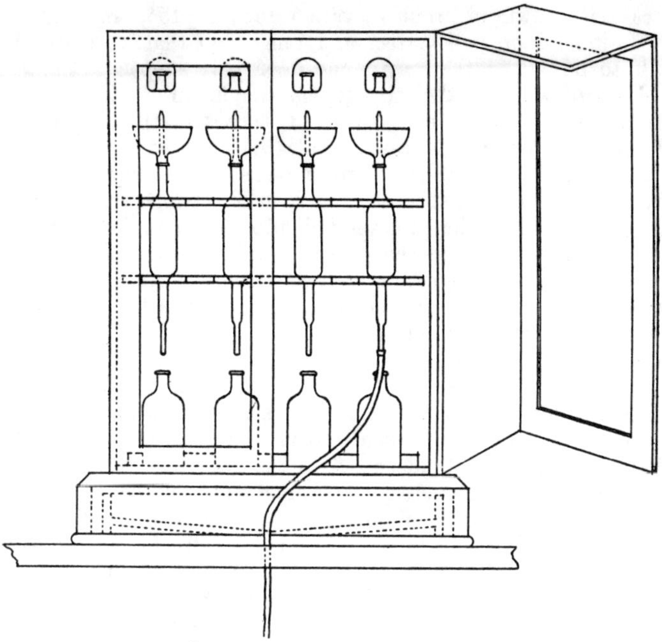

FIG. 154.

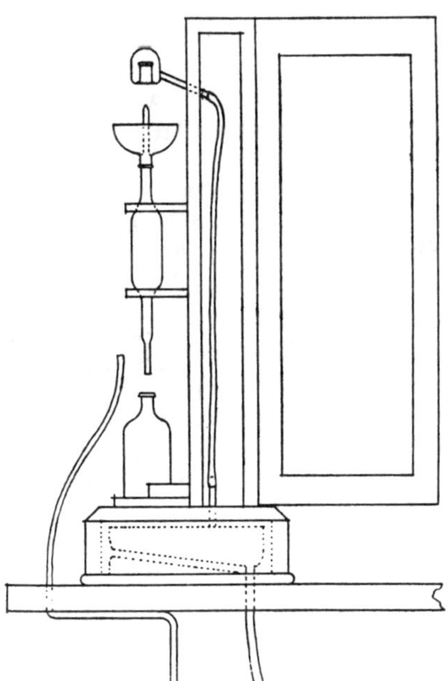

FIG. 155.

FRONT AND SIDE ELEVATIONS OF CASE TO HOLD FOUR STAS PIPETTES AND ASSAY-BOTTLES (ONE GLAZED DOOR SHOWN PARTLY OPEN).

form of case is shown in front elevation in Fig. 154, and side elevation in Fig. 155. The doors are glazed, and hinged so that they can be swung right back to be out of the way: one door partly open is shown in Fig. 154. The overflow from the pipettes is caught in a glazed earthenware trough from which it passes into a jar beneath the bench on which the case stands. When the assay-bottles are placed on the stand they fit into a "guide," so that they are at once in the correct position under the pipettes. Holes are bored in the stand, in direct line with the pipettes, so that any "drops" from the pipettes fall into the earthenware trough. The tube for supplying solution to the pipettes passes through a clip fixed to the floor and operated by means of the foot, so that both hands are left quite free for manipulating the pipettes and bottles. A standard solution of common salt may be conveyed from the supply tank to the pipettes through a tin-lined lead pipe, but a glass tube should be used for thiocyanate solutions.

In working with a series of pipettes it is of the utmost importance that each pipette should deliver the same volume of solution. All pipettes should be accurately tested before being taken into use. For accurate assay work it is well to have the pipettes standardized at the National Physical Laboratory. Before working off a batch of assays, the pipettes should be filled and emptied several times until the volume of solution delivered is constant. The pipettes should be cleaned periodically with a warm solution of caustic soda.

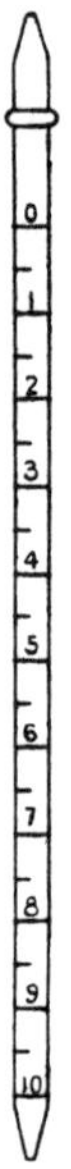

FIG. 156.—HAND PIPETTE.

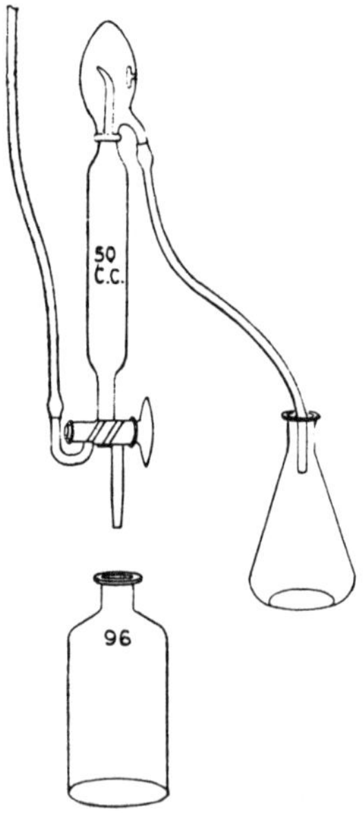

FIG. 157.—AUTOMATIC PIPETTE.

Small Pipettes.—For the addition of dilute solutions small hand pipettes are sometimes used (Fig. 156). They have a capacity of 10 c.c., graduated in $\frac{1}{2}$ c.cs. Most assayers use ordinary 50-c.c. burettes for the addition of dilute solutions.

Automatic Pipette.—For diluting a large number of assays with the same volume of water or dilute indicator solution, as in the Volhard method described later, the automatic pipette shown in Fig. 157 is convenient. It is fitted with a two-way cock, and has a large outlet tube to give quick delivery. The pipette is connected with a large supply of solution by means of a rubber tube, and is filled from below in the same way as the Stas pipette. The overflow solution is caught in the hood and delivered into a separate vessel placed to receive it.

Testing Shelf.—The end of the reaction in the Gay-Lussac method, described subsequently, is best seen against a black surface. A shelf painted dead-black should be fixed in front of a window, which may be glazed with yellow glass, to receive the bottles during the addition of the decimal solution; it is provided with a blackened upright back 3·5 inches high, extending along its entire length, and the bottles are so arranged on it that the light passes through the upper part of the liquid they contain.

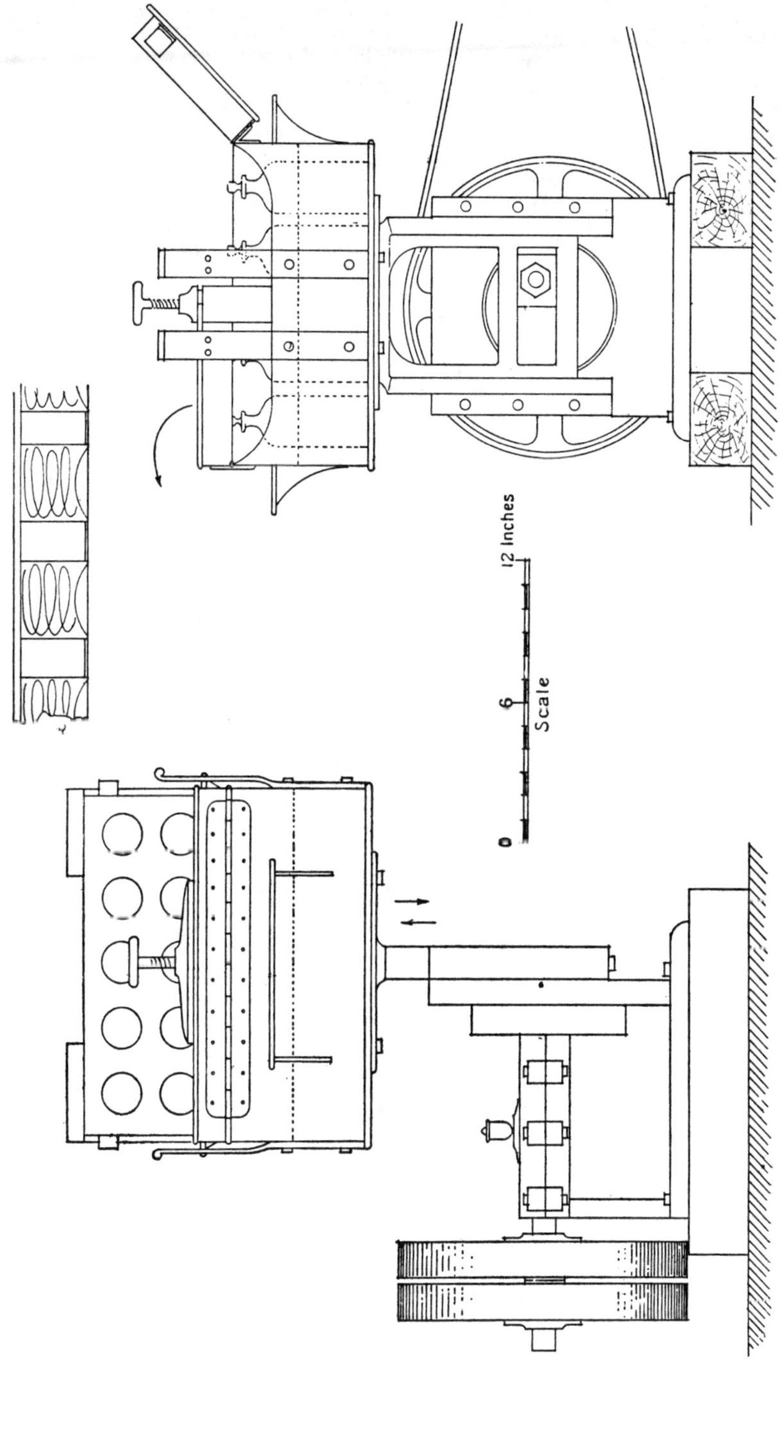

FIG. 158.—FRONT AND SIDE ELEVATIONS OF SHAKING-MACHINE TO HOLD 20 ASSAY-BOTTLES.

"When artificial light has to be used, it has been found that by placing screens so that the light from an electric arc-lamp passes through the upper portion of the silver solution, while the eyes of the observer are protected from the direct rays, accurate readings of the 'clouds' or fine layers of floating silver chloride can be made without any undue straining of the sight." The facilities thus secured are especially valuable during the fogs which are more or less frequent in large cities in winter.[1]

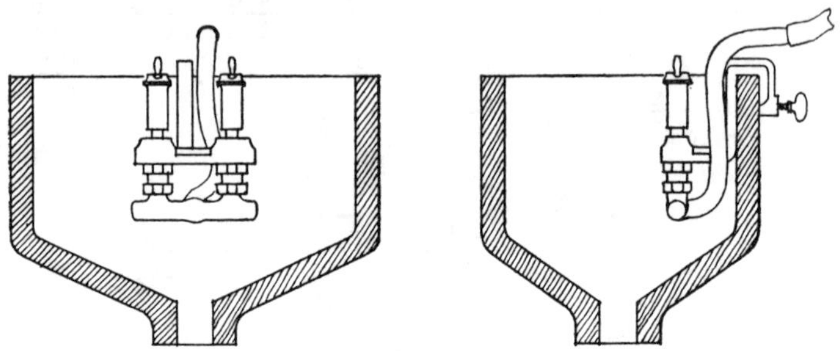

FIG. 159.—TWIN JET BOTTLE-WASHING APPARATUS.

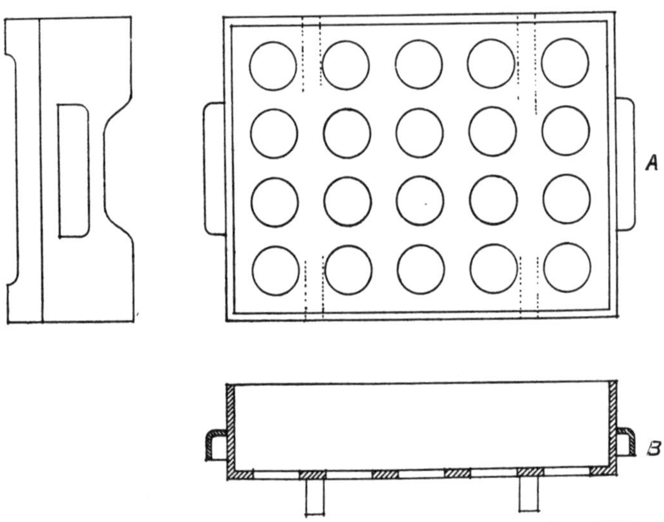

FIG. 160.—DRAINER FOR USE WITH ASSAY-BOTTLE TRAY (FIG. 151).

Shaking-Machine.—Shaking is always resorted to in the wet method of silver assay to induce the clearing of the solution. It may be conducted by hand, two bottles being shaken at a time; but if many assays have to be dealt with, it is more convenient to employ some form of shaking-machine or agitator.

One of the shaking-machines in use in large assay offices is shown in Fig. 158. It is made of copper, but may be made of wood, and holds twenty bottles. The general arrangement of the appliance is shown in two elevations. The shaker has a reciprocating movement with a 2-inch throw, the speed of the fly-wheel being 300 revolutions per minute. The power is supplied by a $\frac{1}{2}$ H.P. continuous-current motor. The box in which the bottles are placed is fitted with a diaphragm to keep the bottles in position, and springs are fitted in the lid,

[1] *Eighteenth Mint Report*, 1887, p. 53.

as shown in the enlarged section, to prevent the stoppers from coming out during the shaking. The lid, which is in two halves, is secured by a clamp. Springs are fitted to the side of the box to break any sudden fall of the lid, which might result in breakage of the bottles. When the assays are worked off in batches of twenty in the trays described (Fig. 150), a considerable saving of time is effected by having the shaker so arranged that the tray is placed directly into it without the necessity of transferring the bottles separately.

Bottle-washing Apparatus.—An important consideration in assay offices dealing with a large number of wet silver assays is the washing of the bottles. This may be accomplished very rapidly by means of the patent[1] automatic washing jets shown in Fig. 159. These jets, which are connected up with an ordinary water-supply by means of a hose-pipe, are so constructed that when a bottle is inverted over them and gently pressed down, a jet of water rises in the bottle and washes it. It is convenient to use two jets, clamped to the side of a sink, so that two bottles can be washed simultaneously. A jar is placed below the sink to receive the precipitate from the bottles.[2] The operator places a tray of bottles on one side of the sink, then takes a bottle in each hand, empties the contents, holds the bottles over the jets for a few seconds, then returns the bottles to their places. With this apparatus a batch of twenty bottles and stoppers is readily washed in 1½ minutes. After being washed, the "drainer" shown in plan in Fig. 160 A, and elevation in Fig. 160 B, is placed over the tray of bottles, the drainer and tray then inverted together, and the bottles allowed to drain until quite dry (see Fig. 161). The bottles are then transferred to the bottle cabinet ready for use.

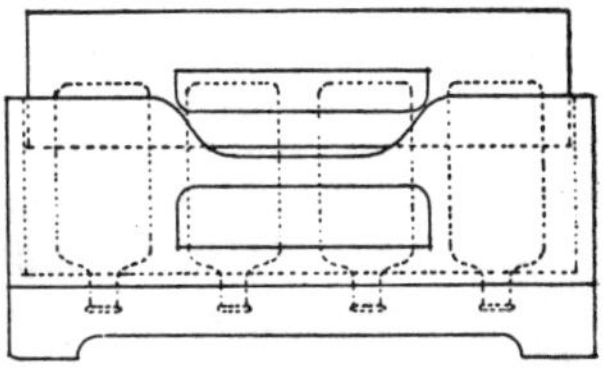

FIG. 161.—METHOD OF DRAINING ASSAY-BOTTLES.

(II) Assay by Titration with Common Salt (Gay-Lussac Method).

This method was invented by Gay Lussac in 1832, and has displaced the cupellation process in almost all mints and in many private assay offices. It is the most accurate method of silver-bullion assaying at present known. It is a volumetric method, in which the volume of a standard solution of sodium chloride required for the precipitation of a known amount of silver in solution as silver nitrate is measured. The end of the operation is judged from the appearance of an exceedingly faint cloud produced by sodium chloride in a solution from which almost all the silver has been precipitated.

Three solutions are required, namely :—

(1) Standard or normal solution of sodium chloride (or of sodium bromide).
(2) Decimal solution of sodium chloride.
(3) Decimal solution of silver nitrate.

When large numbers of assays are made, it is usual to prepare periodically a considerable bulk of these solutions, which should be preserved in well-closed vessels of stoneware or glass.

(1) "*Normal Salt*" *Solution.*—This is made up of such a strength that 100 c.c. will precipitate about 1 gramme of silver: this requires

[1] Proprietors of patent, Messrs. D. G. Binnington, Kingston Machine Works, Hull.

[2] The silver chloride from the assays is allowed to settle, the clear liquor siphoned off, a little hydrochloric acid added, and the chloride then reduced by plates of wrought iron, as described on p. 248.

5·416 grammes of dry sodium chloride per litre. This is usually termed the "normal" solution, although not correctly of "normal" strength. The strength of the solution is standardized by means of check assays of pure silver.

(2) "*Decimal Salt*" *Solution.*—This, as its name implies, is one-tenth the strength of the normal solution: it is frequently called the decinormal solution. It is prepared by taking 1 part by volume of the normal solution and adding to it 9 parts of distilled water. A convenient method is to introduce the contents of the 100-c.c. Stas pipette into a litre flask and fill up with water. As 1 c.c. of this solution is equal to one-tenth c.c. of the normal solution, it will obviously precipitate 0·001 gramme of silver.

(3) "*Decimal Silver*" *Solution.*—One gramme of pure silver is dissolved in a small quantity of pure dilute nitric acid, and the solution diluted to one litre. One c.c. of this solution will contain 0·001 gramme of silver: it follows therefore that 1 c.c. of the decinormal salt solution will just precipitate the silver in 1 c.c. of the decimal silver solution.

Standardizing the Solutions.—A weighed quantity, in duplicate, of proof silver of exactly 1·003 grammes is transferred to an 8-ounce (205-c.c.) glass bottle and dissolved in 15 c.c. of dilute nitric acid (sp. gr. 1·26) and heated until the nitrous fumes are expelled. When cold, 100 c.c. of the normal solution is run into the bottle from the Stas pipette (see Fig. 153), and the contents are shaken for three or four minutes either by hand or a mechanical shaker, until the silver chloride has curdled and will settle to the bottom, leaving the supernatant liquid clear. If the normal solution is made up correctly, it will have precipitated just 1·0 gramme (1000 mg.) of silver, leaving ·003 gramme (3 mg.) unprecipitated. When the precipitate has settled, the solution is tested by the addition from a pipette or burette of 1 c.c. of the decimal salt solution, which is allowed to run down the side of the bottle so that it forms a layer resting on the assay solution. If any silver remains in solution, a cloudy layer will be formed near the surface of the liquid. This is best observed against a black background, and the shelf described above is used for the purpose. If a cloudiness is seen, the bottle is again shaken to clear the liquid, and then another c.c. of decimal salt added. If this second addition of decimal salt solution gives a precipitate, the shaking and settling are repeated and further additions of decimal salt made of 1 c.c. at a time, shaking between each addition, until no further cloud appears.

An experienced assayer soon learns to judge by the density of the cloud whether only part of the c.c. has been used up, and readily decides when all the silver has been precipitated. The end reaction is marked by the gradual formation in the course of about ten minutes of a very faint cloud, the value of which to the fourth of a c.c. (or 0·25 mg.) is judged by the assayer. In very exact work experienced assayers judge the clouds even more closely. In this way the exact strength of the normal solution is determined in duplicate. As an example, suppose the total additions of decimal salt were 4 c.c. (equivalent to 0·004 gramme of silver), and that the cloud produced by the last addition was declared to be equivalent to only one-half of a c.c. (equal to 0·0005 gramme), then $4 - 0{\cdot}5 = 3{\cdot}5$ c.c., the amount of solution actually required to complete the precipitation of the silver.

Since 100·35 c.c. of normal solution was required to precipitate 1·003 grammes of silver, the strength of the solution is calculated in terms of 100 c.c. as follows. Thus:

$$100{\cdot}35 : 1{\cdot}003 :: 100 : x$$
$$x = 0{\cdot}9995 \text{ gramme, or } 999{\cdot}5 \text{ mg.}$$

The strength of the solution is finally recorded on the bottle as follows :—

$$100 \text{ c.c.} = 999{\cdot}5 \text{ mg. Ag.}$$

Some assayers prefer to weigh out exactly 1 gramme of silver for the check assays, and in this case the strength of the " normal " solution is adjusted so that after the addition of 100 c.c. to the assay solution 1 or 2 mg. of silver will still be left in solution and thus ensure the production of a cloud when decinormal salt solution is added.

If the strength of the solution is incorrect to the extent of more than 2 points (*i.e.* 2 mg.), it should be corrected by the addition of either water or salt, and then re-standardized. A solution that is too strong is dealt with as follows :—

If the first addition of decimal salt solution causes no cloudiness, the normal solution contains an excess of salt, and 5 c.c. of decimal silver solution (equivalent to 5 mg. of silver) are added to give an excess of silver in solution, and then the additions of decinormal salt are proceeded with in the usual way. The 5 mg. of silver must be allowed for in calculating the strength of the solution (see note on page 313).

Method.—It is evident from the preceding paragraph that the method requires a practically constant quantity of silver, that is, one which varies by a few milligrammes only in each determination, and thus avoids undue additions of the decinormal solutions. If many additions have to be made, the operation not only becomes tedious, but the solution also ceases to clear after shaking, so that it becomes impossible to determine the finishing-point.

Under these conditions the composition of the bullion must be known approximately before the weight to be taken for assay can be determined. If this is not known, a preliminary assay must be made by cupellation or by the Volhard method, described later. If the composition is, say, approximately 900 fine, then the weight to be taken is

$$\frac{1{\cdot}003}{900} \text{ or } 1{\cdot}115 \text{ grammes (1115 mg.).}$$

This amount of bullion is weighed out in duplicate, placed in bottles, dissolved in acid, and then titrated as described above.

The fineness of the bullion is calculated as follows :—

Suppose the strength of the normal solution is 100 c.c. = 1001 mg. of silver, and that 100·2 c.c. of normal solution was used in the titration (*i.e.* 100 c.c. normal salt + 2 c.c. of decimal salt), then :

$$100 : 1001 :: 100{\cdot}2 : x$$
$$x = 1003 \text{ mg.,}$$

which is the amount of silver in the 1·115 grammes of bullion taken. The fineness of the bullion is calculated thus :—

$$1115 : 1003 :: 1000 : y$$
$$y = 898{\cdot}65 \text{ parts per 1000,}$$

which is the fineness of the bullion.

If the assay contains less than 1 gramme of silver in solution, the first addition of the decinormal salt solution of course produces no precipitate. Five milligrammes of silver in solution (*i.e.* 5 c.c. of decimal silver solution) is then added, and the assay proceeded with in the usual way, 5 mg. of silver being deducted from the amount found. Thus, employing the sample given above, suppose the total additions of normal and decimal salt solution required to precipitate all the silver in solution (including the 5 mg. added) amounted to

100·2 c.c., then the quantity of normal solution used to precipitate the silver actually present in the bullion will be 100·2 — 0·5 = 99·7 c.c., and

$$100 : 1001 :: 99{\cdot}7 : x$$
$$x = 997{\cdot}99 \text{ mg.},$$

the amount of silver in the bullion taken, and

$$1115 : 997{\cdot}99 :: 1000 : y$$
$$y = 895{\cdot}05,$$

which is the fineness of the bullion.

Since the volume of normal solution measured varies with the temperature, a difference of 5° C. making a difference of 0·1 c.c. in measuring the 100 c.c., it is always desirable to run a check assay with each batch of assays.

In the case of low-grade bullion where it is necessary to take a large quantity in order to obtain 1 gramme of silver in solution, it is usually more satisfactory to weigh 1 gramme of the bullion and add a carefully weighed quantity of fine silver to make up the total silver to about 1 gramme. By adopting this method, excess of copper in the solution is avoided and the end-point decided without difficulty, which is not always possible when the solution is intensely coloured with copper nitrate.

Limit of Accuracy of the Gay-Lussac Method.—As ordinarily carried out by experienced assayers the method is accurate to 0·1 part per 1000, but with proper precautions and the expenditure of sufficient time it is capable of giving results accurate to 0·01 per 1000.[1]

Interfering Metals.—Mercury interferes with the method, as it is precipitated as mercurous chloride (Hg_2Cl_2) with the silver, but its presence is detected by its prevention of the darkening of the silver chloride when exposed to light. The addition of 20 c.c. of sodium acetate and a little free acetic acid to the assay before titration will prevent the precipitation of the mercury. Mercury is sometimes present in mill bullion which has been retorted at too low a temperature when expelling it.[2]

Tin also interferes, as it is converted into metastannic acid by the nitric acid, and remains as a white amorphous powder suspended in the solution, thus producing a cloudiness which interferes with the finish of the assay. The presence of as little as 0·05 per cent. of tin was found by Salas [3] to be sufficient to cause the obscuration of the end-point, while 0·5 per cent. rendered the assay impossible. In order to prevent the formation of a precipitate of metastannic acid, the following procedure is recommended by Salas.

A quantity of bullion containing about 1 gramme of silver is accurately weighed and placed in the assay-bottle. Two grammes of tartaric acid crystals and 3 or 4 c.c. of water are added, and the bottle is heated till the acid is dissolved. After cooling, 10 c.c. of dilute nitric acid (1 : 1) are added, and when the silver has dissolved, the solution is titrated as usual. Under these conditions the liquid remains clear, but it must be kept cool during and after the solution of the silver. If the bullion contains more than 5 per cent. of tin, a smaller quantity than 1 gramme should be taken for the assay and some pure silver added.

Silver bullion from Mexico and Bolivia frequently contains tin, and it is also met with (sometimes in large quantities) in silver bars resulting from the treatment of jewellers' sweeps. Tin is sometimes present to the extent of over 5 per

[1] Rose, *Precious Metals.*
[2] Fulton & Sharwood, *Fire Assaying.*
[3] L. E. Salas, *Trans. Amer. Inst. Min. Eng.*, March, 1912, 267–278.

cent. in English standard silver (925 fine) as a substitute for copper. Silver bullion containing tin may be satisfactorily assayed by cupellation or by the Volhard method.

Note on the End-point in Gay Lussac's Method.—In 1857, Mulder [1] showed that when the most exact chemical proportions of silver and salt are made to react on each other, the addition of either salt or silver to the clear supernatant liquid will produce a precipitate, indicating the presence of free silver nitrate and sodium chloride in a state of equilibrium, either of which is thrown down on the addition of the other. This is known as Mulder's "neutral point." No trouble is experienced in determining the end-point in the ordinary assay if care be taken that rather more than 1 gramme of silver is present. When less silver is present, so that the solution contains excess of sodium chloride after the addition of 100 c.c. of normal salt solution, an excess of silver should at once be added, rather than an attempt made to finish the assay by the addition of decinormal silver. The existence of the "neutral point" was explained by Hoitsema [2] by the solubility of silver chloride. As this compound is also present in the solid state, the product of the concentrations of its ions must remain constant, and hence the additions of either silver or chlorine ions must cause a precipitation of the salt.

A similar neutral point is not observed in titrating with a bromide or iodide, because the solubility is less than the amount necessary to produce the apparent precipitate. It is for this reason that some assayers use a normal solution of sodium bromide instead of the chloride.

Titration with Chromate Indicator.—A modification by Mohr in which potassium chromate is used as an indicator, is especially applicable to the determination of the strength of photographic and other silver solutions. A measured quantity of the silver solution to be tested is mixed with an excess of a standard solution of sodium chloride, a few drops of potassium chromate solution (the neutral or yellow chromate K_2CrO_4) are added, and the excess of the chloride in solution is determined by titration with a standard solution of silver nitrate until the orange colour of the silver chromate is persistent.

Silver chromate being soluble in acids, the solution to be titrated, if acid, must first be carefully neutralized with sodium or calcium carbonate.

For certain purposes this method is very serviceable as it is simple and quick and the end-point distinct.

(III) Assay by Titration with Thiocyanate (Volhard Method).

This method was first described by M. Carpentier[3] in 1871, and improved in 1874 by J. Volhard,[4] by whose name it is generally known. It is based on the fact that ammonium thiocyanate, when added to a solution of silver nitrate containing a ferric salt, does not produce the distinctive deep-red coloration with the ferric salt until the whole of the silver has been precipitated as thiocyanate. The presence of a ferric salt (obviously not the chloride) therefore acts as a very delicate indicator and clearly defines the completion of the precipitation of the silver.

In conducting the assay, a ferric salt is added to the silver solution and a standard solution of thiocyanate then run in. The white silver thiocyanate

[1] *De Essayeer-methode van het zilver scheikundig onderzocht*, Utrecht, 1857. Quoted by Percy, *Metallurgy of Silver*, p. 291.
[2] *Zeits. physikal. Chem.*, 1896, **20**, 272–282.
[3] *Compt. Rend.*, 1871.
[4] *J. für. prakt. Chem.*, 1874, (2), **9**, 217–222.

precipitate settles readily on shaking, leaving the liquid clear; and a persistent red-brown coloration in the liquid when the thiocyanate is in excess indicates the finish. The silver solution must be cold, and all the water used must be free from chlorine, otherwise silver will be precipitated and errors introduced.

The following solutions are prepared :—

Standard Thiocyanate.—This is prepared from either ammonium thiocyanate or potassium thiocyanate, both of which are too deliquescent to allow of very exact weighing.

From 3·5 to 4 grammes of crystallized ammonium thiocyanate (or 4·5 to 5 grammes of potassium thiocyanate) are dissolved in a litre of distilled water with pure salt. 100 c.c. of this solution are equivalent to 0·5 gramme of silver.

Standard Silver Nitrate.—Dissolve 5 grammes of fine silver in 50 c.c. of dilute nitric acid (1 : 1), boil to expel nitrous fumes, and dilute to 1 litre. 100 c.c. of this solution will contain 0·5 gramme of silver.

Indicator.—Several ferric salts are used as indicators. That most frequently employed is a saturated solution of iron alum, 2 c.c. being used for each assay. Another convenient ferric salt is the sulphate, a solution of which may be prepared by dissolving a few crystals of ferrous sulphate in water, adding about half the volume of strong nitric acid in order to oxidize the iron, and then briskly boiling the mixture for a few minutes to expel all the nitrous fumes. This solution is slightly diluted with water, and 3 or 4 c.c. used for each titration.

A saturated solution of iron nitrate is also very suitable for the purpose. It is prepared by dissolving iron wire in dilute nitric acid (1 : 1). About 2 c.c. are used.

When a number of titrations are to be made, the ferric solution should be made up to some definite strength, and the same measured quantity used for each assay. The addition of a short piece of fine iron wire to each assay bottle has been suggested; this dissolves at the same time as the silver, and supplies the ferric indicator.

Standardizing the Solution.—The thiocyanate solution is standardized by transferring 50 c.c. of the silver nitrate solution to a flask, adding 10 c.c. of dilute nitric acid (1 : 1), 2 c.c. of ferric indicator, and diluting with 100 c.c. of water.

The solution of thiocyanate is then run in from a burette. As each drop falls into the flask, the red colour of ferric thiocyanate appears for a moment, but at once disappears on gently shaking. The solution is cautiously added until the first indications of a permanent red coloration are seen. The colour thus developed should remain permanent after continuous shaking. After shaking, the precipitated silver settles readily, enabling the colour of the clear supernatant liquid to be distinctly seen. From the volume used, the exact strength of the thiocyanate solution is ascertained, and the amount of dilution, or the addition of crystals, which it requires in order to bring it to the desired strength, is readily calculated. After the solution has been thus adjusted, it should be once more titrated to verify the strength.

To obtain reliable results, the nitrous fumes must be expelled, and the silver solution must be cold. The effect of variations of temperature may be seen from the following experiments by Beringer,[1] in which 20 c.c. standard silver nitrate were used.

Effect of Varying Temperature.

Temperature ...	10° C.	30° C.	70° C.	100° C.
Thiocyanate required	19·6 c.c.	19·3 c.c.	19·0 c.c.	18·6 c.c.

[1] *Text-book of Assaying*, p. 122.

The effect of varying amounts of silver in the assay solution is as follows :—

Silver solution added	1 c.c.	10 c.c.	20 c.c.	50 c.c.	100 c.c.
Thiocyanate required	1·0 c.c.	9·7 c.c.	19·6 c.c.	49·4 c.c.	99·0 c.c.

From these results it will be seen that the error in the determination of the silver increases with an increase in the amount of silver present. This error is eliminated in the ordinary assay by the use of checks of the same composition as the silver being assayed.

The Volhard method, worked as described, is valuable for determining silver in bullion and alloys of unknown composition, where no more than an ordinary degree of accuracy is demanded. It is quick, easy in manipulation, and applicable under most of the usual conditions, and on that account is very frequently employed in determining the approximate composition of bullion which is to be more accurately assayed by the Gay-Lussac method. The presence of other metals, as lead, copper, nickel, zinc or cadmium, exerts no disturbing effect so long as the solution contains free nitric acid, and the metals are not present in large excess. The effects of palladium, tin, and large quantities of copper or nickel are discussed below.

The assay is conducted exactly as in the titration for standardizing the thiocyanate solution, the silver being dissolved in 15 c.c. of dilute nitric acid (1 : 1), and diluted to a convenient volume. The greatest disadvantage of the method when worked in this way is the brown coloration produced by the thiocyanate when the assay is nearly, but not quite, finished, and the slowness with which this is removed on shaking up with the precipitate. This is worse with large quantities of precipitate, and if about 1 gramme of silver is present, it gives an indefiniteness to the finish, which lowers the precision of the method to about 2 parts per 1000, a degree of accuracy which is useless for the assays of bullion[1].

In determining silver in silver bullion by Volhard's method, therefore, the general practice is to work on the principle of the Gay-Lussac method, by adding a measured quantity of normal thiocyanate solution to precipitate the bulk of the silver, and then to make successive additions of decinormal solution until the whole of the silver is precipitated. Checks of similar composition to the bullion are run with each batch of assays to eliminate errors. Working in this way upon samples of pure silver, Van Riemsdijk[2] found that by successive additions of decinormal ammonium thiocyanate solution, each sufficient to precipitate 0·25 milligramme of silver, results are obtained which are accurate to 0·25 part per 1000.

According to Van Riemsdijk, this marks the limit of accuracy of Volhard's method when the above method of working is adopted, as he points out that an appreciable colour cannot be obtained without an excess of about this quantity of the thiocyanate solution.

For some years it has been the practice at several assay offices to modify slightly the ordinary method of finishing the assay by using a " normal " solution sufficiently strong to give a permanent red colour to the check assay, and using this colour as a standard of comparison. By this means the finishing-point of the assay is perceived more sharply and with greater certainty than is possible when the assay is finished in the usual way by cautiously adding the decinormal solution until the first indications of a permanent red coloration are seen.

Method.—In conducting the assay, a quantity of bullion is taken such that about 1 gramme of silver is present. It is dissolved in 10 c.c. of dilute nitric

[1] Beringer, *loc. cit.*, p. 123.

[2] *Algemeen. Verslag. Munt Collegie*, 1878, Appendix 38.

acid (sp. gr. 1·2) in a bottle of about 250 c.c. capacity, and heated in a water-bath until all traces of nitrous acid are expelled. The solution is then diluted with 50 c.c. of water containing 2 c.c. of a saturated solution of iron alum decolorized by nitric acid, and 100 c.c. of "normal" ammonium thiocyanate is added from a Stas pipette. This solution contains about 7·04 grammes of ammonium thiocyanate per litre, and 100 c.c. is sufficient to precipitate about 1 gramme of silver and leave a very small amount of thiocyanate in excess, which in the presence of the ferric salt gives a pale red colour to the check assays. For the assay of fine silver bars a convenient strength for the normal solution is such that 100 c.c. is equivalent to about 1·0003 to 1·0005 gramme of silver.

The bottle is then well shaken for about two minutes, when the white precipitate of silver thiocyanate settles to the bottom and leaves the liquid clear. If some silver is still in solution, the liquid is colourless or retains the pale-green colour due to the dissolved copper originally contained in the bullion. Decinormal thiocyanate is now run in from a burette until the colour *approaches* that of the check; the liquid is then well shaken and allowed to clear; after which the decinormal solution is cautiously run in, drop by drop, until the colour corresponds exactly to that of the check. The addition of 1 drop (0·05 c.c., equivalent to 0·05 mg. of silver) of decinormal thiocyanate makes a perceptible difference to the colour, which is readily recognized.[1] The assay should be well shaken before the final comparison is made.

Experience has shown that equally accurate results are obtained whether the amount of silver left in solution after the addition of normal solution is 1 mg. or 20 mg., or more.

The comparison of the assays with the check is best made by placing the bottles so that they catch the reflected light of a white surface. A shelf with a sheet of white opal glass, arranged so that it faces the direct light from a window, is very suitable for the purpose. A side light is liable to give rise to error of judgment in comparing the tints. If after adding 100 c.c. of normal solution the ammonium thiocyanate is in excess, the liquid is coloured red from the action on it of the ferric salt, the intensity of the colour varying with the amount of thiocyanate in excess.

In this case the assay may be finished either by adding decinormal silver solution until the colour of the assay solution corresponds with that of the check, or an excess of silver may be added and the assay completed in the ordinary way, the added silver being of course allowed for in calculating the results. The latter method is to be preferred. The method of calculating the results is the same as that described for the Gay-Lussac method.

By finishing the assay as described, the exact composition of the bullion can readily be ascertained correctly to 0·2 per 1000, or even more closely.[2] When the bullion contains copper or nickel, the blue colour of the solution somewhat masks the colour of the ferric thiocyanate, and to obtain a satisfactory tint for comparison it is desirable to use a slightly stronger "normal" solution.

For example, the strength of the solution for assays of standard silver (925 silver and 75 copper) should be such that 100 c.c. will precipitate 1·001 gramme of silver. It is very important in assaying silver–copper alloys to

[1] J. W. Leather has drawn attention to the extreme delicacy of this reaction in an interesting paper on "The Determination of Small Quantities of Iron" (see *J. Soc. Chem. Ind.*, 1905, **24**, 385).

[2] "Assay of Silver Bullion by Volhard's Method," E. A. Smith, *Trans. Inst. Min. and Met.*, 1907, **16**. 154.

use checks of pure silver, with the addition of copper approximately equal to that in the bullion. With silver–copper alloys below 900 fine, more satisfactory results are obtained by taking about 0·5 gramme only for assay, and adding fine silver to bring up the total silver to about 1 gramme, thus reducing the amount of copper in the assay solution.

When tin is present, the metastannic acid formed tends to remain in suspension and clouds the solution, but does not seriously interfere in comparing the final tints; to get accurate results, however, an equal amount of tin must be added to the check assays. By allowing the solutions to stand for five or ten minutes before making the final comparison, the bulk of the tin will settle with the silver precipitate, leaving the solution almost clear. The author has obtained very satisfactory results with this method with bullion containing as much as 10 per cent. of tin.

Antimony, when present, is converted into antimonic oxide by the action of nitric acid, which produces a cloudiness in the solution as in the case of tin. By the addition of a few c.c. of a solution of tartaric acid the antimony is retained in solution and the assay then finished satisfactorily.

Palladium interferes as its salts give a deep red solution with excess of potassium thiocyanate, and mask the finishing-point. Platinum also gives a yellow coloration that interferes with the finishing-point. Alloys of silver containing palladium or platinum may be accurately assayed by the Gay-Lussac method.

The Volhard method is particularly suitable for the assay of silver plate at the hall-marking offices. The modification of the method described was proposed by Arthur Westwood, the assay master of the Birmingham Assay Office, and has been used there since 1892 for doing over 1000 silver assays a day. Later it was introduced at the Sheffield Assay Office and elsewhere.

Combined Gay-Lussac and Volhard Method.—To obviate the tediousness of determining the residual silver by adding successive small quantities of salt in the last stage of the Gay-Lussac method, Beringer[1] proposed a combination of the Gay-Lussac and Volhard methods which answers admirably when applied to the assay of silver bullion.

The combined method is conducted as follows:—

Prepare the assay as described for the Gay-Lussac method and precipitate the bulk of the silver by adding 100 c.c. normal salt solution: shake till the liquor clears, and filter into a flask, washing with a little distilled water. Filtration from silver chloride is necessary, and the wash-water must be free from chlorine. Add 2 c.c. of "ferric indicator" to the filtrate and titrate with a dilute thiocyanate solution made by diluting the ordinary standard solution to such an extent that 100 c.c. of the diluted solution shall be equivalent to 0·1 gramme of silver or 1 c.c. = 1 mg. To prepare this, multiply the *standard* by 1000 and dilute 100 c.c. of the standard solution to the resulting number of c.c. Thus, with a solution of a standard 0·495, dilute 100 c.c. to 495 c.c. using, of course, distilled water.

The titration is continued until the characteristic colour of the ferric thiocyanate is permanent. The absence of the silver precipitate from the solution enables the titration to be completed much more readily. An advantage of this modification is that an excess of 15 mg. of silver may be determined as easily and exactly as 5 mg.

The solution is standardized by means of fine silver, about 1 gramme of which is accurately weighed and treated as described above. The standard

[1] *Text-book of Assaying*, 15th edn., p. 124. Consult also A. E. Knorr, *J. Amer. Chem. Soc.*, 1897, **19**, 814–816.

of the salt solution is then found by deducting the number of c.c. of the thiocyanate used for the check from the number of mg. of the fine silver initially weighed out. Thus: Suppose 1015·0 mg. of silver required the addition of 13·5 c.c. of thiocyanate solution, equivalent to 13·5 mg. of silver, then the silver precipitated by the salt is 1015·0 − 13·5 = 1001·5 mg. (1·0015 gramme), which is the standard. The strength of the salt solution may therefore be expressed as 100 c.c. = 1001·5 mg. Ag.

The total amount of silver in the bullion taken for assay is calculated by adding to the standard of the salt solution the milligrammes of silver equivalent to the number of c.c. of thiocyanate used.

The fineness of the bullion is calculated in the ordinary way from the silver in the assay piece and the weight of bullion taken.

According to Knorr,[1] the limit of accuracy of this combined method of assay is 0·2 to 0·3 per thousand.

Attempts have been made to increase the accuracy of the Volhard method as ordinarily worked by filtering off the precipitate obtained on the addition of 100 c.c. of normal thiocyanate and then finishing the assay by adding decinormal thiocyanate to the clear filtrate.

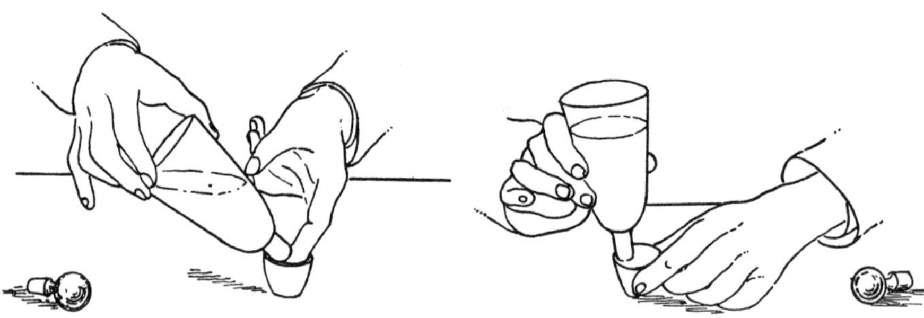

FIG. 162.—TRANSFERRING SILVER CHLORIDE TO WEDGWOOD CRUCIBLE (INDIAN MINT METHOD).

By finishing the titration in a clear solution, free from precipitate, the end of the reaction is much more readily distinguished, but experiments by Hoitsema [2] proved that the filtration of the solution impairs the accuracy of the results. He concluded, as the result of experiments, that the filtration method is not reliable when more than about 1 mg. of silver remains in the clear solution, the error increasing with the increase in the amount of silver remaining in solution. Thus in six tests with 1000, 1001, 1002, 1003, 1004, and 1005 mg. respectively of pure silver, after adding 100 c.c. of standard thiocyanate (100 c.c. = 1000 mg. of silver) and filtering, the amounts of dilute thiocyanate required to produce a pale red colour in the filtrate were 0, 1, 3, 7, 14, and 20 drops instead of there being a regular increase of 20 drops (equal to 1 mg. of silver) in each case. In other tests also variable results were obtained, which were attributed by Hoitsema to absorption by the silver thiocyanate precipitate of soluble silver salts, and also of the soluble thiocyanate.

Tests made by Beringer [3] and also by the author confirm the conclusions of Hoitsema that for accurate work the precipitate must be allowed to remain in the solution when finishing the assays by Volhard's method.

[1] *J. Amer. Chem. Soc.*, 1897, **19**, 814–816.
[2] *Zeits. angew. Chem.*, 1904, **17**, 647–650.
[3] *Text-book of Assaying.*

(IV) Gravimetric Assay of Silver (Indian Mint Method).[1]

This method was introduced by Dr. Dodd, of the Calcutta Mint, in 1851, and is of a special character. It consists in adding an excess of hydrochloric acid to a solution of silver nitrate and weighing the silver chloride produced. The weight of silver bullion taken is 18·817 grains, which is equivaleut to 25 grains of silver chloride if the bullion consists of pure silver. After precipitation, the bottle is well shaken, and the chloride is allowed to settle and is washed by decantation. The silver chloride is transferred to a Wedgwood cup by placing the latter over the mouth of the bottle and inverting it (see Fig. 162). Afterwards the excess water is poured off, the chloride broken up by a glass rod, and dried. The drying is finished on a hot-plate, and the cake of chloride is weighed while still warm. A set of weights is used in which the weight marked "1000" weighs 25 grains, and the smaller weights are in proportion, so that the composition of the alloy is indicated without calculation.

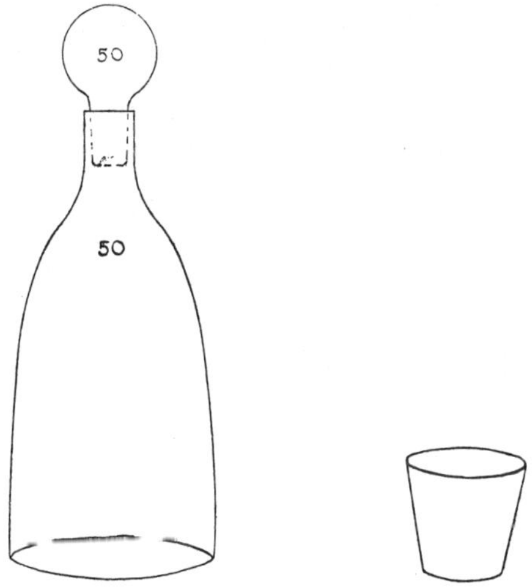

FIG. 163.—ASSAY-BOTTLE AND CRUCIBLE USED IN INDIAN MINT METHOD.

In this method any gold present in the bullion remains unchanged and is weighed with the silver chloride.

The special form of bottle and the Wedgwood crucible employed in this method are shown in Fig. 163. The method is practically confined to the Indian Mints, where the conditions of the country with regard to climate and cheap native labour render it suitable.

Other Methods of Bullion Assaying.

Touchstone Test for Silver.—The quality of silver is sometimes tested by means of the touchstone, as previously described for gold. If the streak is silver, it will dissolve immediately on wetting with a drop of dilute nitric acid; to confirm the presence of silver, a drop of dilute hydrochloric acid may be added, when a white precipitate of silver chloride will be produced. If there is no white precipitate the metal is probably "nickel silver," a white alloy of nickel, zinc, and copper. If a white residue remains after wetting the streak with nitric acid the metal is tin or an alloy containing tin.

The streak may also be tested by moistening it with a solution of 3 g. of potassium chromate and 2 c.c. of sulphuric acid in 32 c.c. of water.

[1] *J. Asiatic Soc.*, Bengal, 1870, **39**, pt. ii, No. IV; also J. Percy, *Metallurgy of Silver*.

If the article is silver or silver-plated the streak becomes bright red owing to the formation of a film of silver chromate. The "test acid" described on p. 247 is also used for the recognition of silver alloys and silver-plated wares.

Oxidation Test for Silver–Copper Alloys.[1]—When alloys of silver and copper are very slightly heated, superficial oxidation and consequent discoloration occurs, increasing in intensity with the increase in the copper content. By this means some indication may be obtained as to the proportion of copper in certain of these alloys. A classification of the colour changes obtained in various alloys is given in the following table :—

Silver in 1000 parts of the Alloy.	Oxidation Colour.
1000 fine (*i.e.* pure silver)	Dull but quite white.
950 ,,	Uniform grey-white.
925 ,,	Dull grey-white, reddish fringe at edges.
900 ,,	Dull grey-white, black fringe at edges.
880 ,,	Dark grey, almost black.
860 ,,	Almost black.
840 ,,	Quite black.

In carrying out the test, the surface of the alloy must be clean and free from grease, and too high a temperature must be avoided since that would give different results owing to more rapid oxidation.

The surface of the alloy should be cleaned by scraping to remove any silver that may have been electro-deposited in finishing standard silver wares.

The alloy can be heated over a Bunsen flame or in a muffle. This is a very convenient and approximately reliable method of distinguishing between fine and standard silver, and is very useful in locating the soldered seams on silver ware.

The test is quite unmistakable in distinguishing the richer silver alloys but it ceases to be applicable in the case of alloys containing more than 150 parts of copper per 1000 of the alloy, that is, below 850 fine.

Electrolytic Assay of Silver.—This is sometimes effected in a solution of silver nitrate, but more frequently in a cyanide solution.

The conditions of silver deposition from solutions of the nitrate were investigated by Kuster and von Steinivehr in 1898.[2] It was found that silver up to 2 grammes in weight could be deposited in compact adherent form, suitable for quantitative determination, from solutions of the nitrate containing from 1 or 2 c.c. of free nitric acid (1·40 sp. gr.) and 5 c.c. of alcohol, using an E.M.F. of between 1·35 and 1·38 volts. The temperature of the electrolyte should be from 55° to 60° C. With an E.M.F. above or below the limits given, unsatisfactory results are obtained.

The electrolytic assay of silver is, however, best conducted in a cyanide solution prepared from the silver nitrate solution by adding sufficient potassium cyanide solution to dissolve the precipitate of silver cyanide first formed.

The usual conditions for stationary gauze are current from 0·8 to 1 ampère; current density, 0·2 to 0·5; E.M.F. 3·7 to 4·8; temperature 20° to 60° C. If allowed to run overnight the current density should be reduced to 0·08 to 0·15.

To determine whether deposition is complete, the spent solution should be boiled to free from hydrocyanic acid, and tested with hydrogen sulphide.

According to Benner and Ross,[3] silver may be rapidly and accurately deposited from cyanide solutions with stationary gauze and using higher

[1] Chaudet, *L'Art de l'Essayeur*, quoted by J. Percy, *Metallurgy of Silver*, p. 157.
[2] *Zeits. f. Elektrochem.*, 1898, **4**, 451–455.
[3] *J. Amer. Chem. Soc.*, 1911, **33**, 493, 1106.

currents than those stated. If the quantity of silver to be deposited does not exceed 0·15 gramme, high currents do not cause sponginess, and as much as 5 ampères may be used, reducing the time required for deposition to one-half of that required when 1 ampère is used.

Price and Humphreys [1] state that for the rapid deposition of silver, stationary electrodes are not so reliable or satisfactory as rotating cathodes. Experiments showed, however, that the electrolyte which gave satisfactory results with the rotating electrode could not be used with the fixed electrodes, since the deposit was spongy and quite unweighable, even when a current of 1 ampère only was used. For stationary electrodes the electrolyte used was silver nitrate with slight excess of potassium cyanide, and for rotating electrodes, silver nitrate with the addition of 10 c.c. of concentrated nitric acid and 15 c.c. of concentrated ammonia.

Electro-analysis of Silver–Copper Alloys.[2]—The alloy (1 gramme) is dissolved in 30 c.c. of 1 : 2 nitric acid, the solution made alkaline with ammonium hydroxide, cooled, treated with 10 c.c. of ammonium hydroxide (d., 0·88), and electrolysed with a rotating platinum-gauze cathode and a platinum spiral anode to deposit the silver in about ten minutes. Then 10 c.c. of 1 per cent. hydrogen peroxide are added and electrolysis continued for 20 minutes at 0·2 ampère. The cathode is washed first with water, next with alcohol, dried at 110° C., and then weighed to determine the silver.

The spent electrolyte is acidified with nitric acid, an excess of 10 c.c. of acid added, and the copper deposited with a current of 2·5 ampères, using a rotating gauze cathode which is weighed to determine the copper.

Determination of Gold in Doré Bullion.—The gold content of doré bullion is very variable, and it is frequently desirable to make a preliminary assay in order to ascertain the best procedure to adopt in determining it. When small quantities of gold are present, from 1 to 2 grammes of the bullion are treated with dilute nitric acid, not stronger than 1·2 specific gravity, and the residual gold collected and weighed. In some cases it may be necessary to take from 5 to 10 grammes of bullion in order to obtain a suitable quantity of gold to weigh. Since gold, when in a very finely divided state, is soluble in hot nitric acid, the strength of the acid used becomes a matter of importance.

Dewey [3] recommends the following method for collecting very small quantities of finely-divided gold from solutions rich in silver. The solution is heated to free from excess of nitric acid, and if necessary diluted, and then a small quantity of a saturated solution of hydrogen sulphide added. By this means any gold that may be in suspension is collected, and any in solution is precipitated as gold sulphide conjointly with a small amount of silver sulphide.

The solution is well stirred, allowed to stand for two or three hours, and the sulphides then filtered off. The precipitate is washed, dried, carefully burnt, and the metallic residue wrapped in a small piece of sheet lead, then cupelled, and the gold recovered by parting in nitric acid in the usual way. The collection of the gold by precipitating a small quantity of the silver as chloride was found to be less satisfactory.

When the amount of gold in the bullion is comparatively large it may be determined by the usual parting assay as described for gold bullion.

In cases where there is more silver than is desirable for parting, some assayers add a weighed quantity of fine gold to the assay portion in order to

[1] *J. Soc. Chem. Ind.*, 1910, **29**, 307–309.

[2] W. N. Miller, *Ind. and Eng. Chem.* (Anal. Edn.), 1936, **8**, (6), 431–432 : Abs. *Inst. Met. Metallurg. Abs.*, 1937, **4**, 55.

[3] *J. Amer. Chem. Soc.*, 1910, **31**, 323.

make the proportion of gold to silver 1 to $2\frac{1}{2}$, and then part as usual. The weight of gold added must obviously be allowed for in calculating the result.

If the gold is to be separated by one boiling in nitric acid, at least eight parts of silver to one part of gold must be present.

The amount of gold in mixed bars is frequently reported in grains per pound when the amount is small.

Determination of Platinum in Doré Bullion.—Traces of platinum are occasionally present in doré bullion. Small amounts of platinum in doré bullion containing gold up to about 50 parts per thousand discolour the gold residue in the first nitric acid used in parting, so that as little as 2 per mille of platinum can be detected.[1] When gold is absent, the presence of platinum may readily be detected in the nitric acid if only 0·1 per mille is present. Traces of platinum in doré bullion may be collected and determined as described above for gold.

Determination of Copper in Silver Bullion.—It is sometimes desirable to determine the small amounts of copper in silver bullion, and this is frequently done by the colorimetric method, in which advantage is taken of the deep blue colour imparted to copper solutions by the addition of ammonia, the depth of colour depending on the quantity of copper present. By comparing the assay solution with a similar solution of equal intensity of colour, containing a known weight of copper, and by taking the volume into consideration, it is possible to determine with a fair degree of accuracy the percentage of copper in the bullion. According to Beringer,[2] the quantity of copper in 100 c.c. of the assay solution should not be more than 15 mg., or less than 0·5 mg.

In the case of low-grade silver bullion containing appreciable amounts of copper, it is necessary to determine it by chemical methods.

Determination of Selenium in Silver Bullion.—Selenium is not an infrequent impurity in commercial silver bars from certain sources, and may be present up to about 0·15 per cent. Its presence is usually detected in assaying the bullion by the Gay-Lussac method, as it passes into solution and produces a peculiar shimmering effect which is continuous and makes the end-point more difficult to detect.

The selenium is usually derived from the use of selenious sulphuric acid and, in the process of refining, is precipitated with the silver by the metallic copper used for precipitating the silver.

Silver is freed from selenium by exposing the molten metal to oxidation, or by small additions of nitre.

Selenium [3] may be determined by dissolving 5 to 10 g. of the bullion in nitric acid, precipitating the silver as chloride, and filtering. The filtrate is then evaporated with excess of hydrochloric acid, and a little sodium or potassium chlorate to destroy any nitric acid that may be present, and to reduce selenic acid (H_2SeO_4) to selenious acid (H_2SeO_3). The solution is diluted with water and treated with a solution of sodium sulphite, then warmed, and finally boiled. The selenium separates as a red precipitate which, on boiling, becomes denser and black. It is collected on a weighed filter, washed with hot water, dried at 100° C. and weighed as pure selenium.

Selenium may also be precipitated with hydrogen sulphide, and the precipitate oxidized with hydrochloric acid and potassium chlorate, and the selenium separated in the manner described.

Schoeller recommends stannous chloride as a quantitative reagent for selenium.

[1] H. J. Rawlins, *Trans. Inst. Min. and Met.*, 1914, **23**, 177.

[2] *Text-book of Assaying.*

[3] Beringer, *op. cit.*

CHAPTER XX

THE ASSAY OF BASE BULLION (LEAD AND COPPER)

As previously stated, the term "base bullion" is conveniently restricted to argentiferous and auriferous lead and copper bars or pigs. The sampling and assay of these two types of base bullion are described separately below.

(I) Base Lead Bullion (Crude Lead or Work Lead).

The lead bullion obtained from different smelting firms naturally varies considerably in composition, since this depends on the nature of the ores and methods used in the extraction of the metal. Silver ores are frequently mixed with lead ores in order to extract the silver and give a rich base bullion.

Segregation in Lead Bullion.—Silver and lead mix together in all proportions while they are molten, but on solidification they separate so that the alloys are not homogeneous. Consequently, the argentiferous lead produced in smelting operations, and cast into cast-iron moulds, undergoes segregation on solidification, and the bars are not uniform in composition. A result of this is that the valuation of argentiferous and auriferous lead is a matter of difficulty. Moreover, in many smelting establishments the surface of the melted bullion is skimmed, and the clear lead ladled into the mould, until the latter is filled to within an inch of the top, and when it has solidified the mould is filled completely. This results in a bar of good appearance but composed of good lead above and below with more or less dross in the middle, which has separated from the lead first cast, while practically no dross separates from the thin layer of clean metal cast on the top. Bars that are cast in this way are very difficult to sample accurately.

Large quantities of lead carrying gold and silver are sold to refiners in bars weighing about 80 to 110 lbs. each. The bars are approximately 24 by 5 inches at the top, 20 by 3 inches at the bottom, and $3\frac{1}{2}$ to 4 inches in depth. Some bars are cleaner from dross than others. Lead bars are received from South Australia, Queensland, Mexico, Greece, South America, and other places. It has been found that different parts of the same bar of lead carrying silver or gold vary very much in their content of these metals, and this circumstance affects the sampling of argentiferous and auriferous lead bullion on a large scale. As a rule, the top of a bar is poorer in silver and gold than the bottom, but when the gold content becomes high, the reverse appears to be the case.[1]

Sampling Lead Bullion.[2]—The ordinary methods of sampling are as follows :—

(*a*) Cutting chips out of the top and bottom of the bars.
(*b*) Punching half-way through the bars.
(*c*) Sawing, by a circular saw, through the bars.
(*d*) Dip sampling after melting the bars in mass into clean lead and dross.

[1] Consult Percy's *Metallurgy of Silver*, p. 172 ; J. and H. S. Pattinson, *J. Soc. Chem. Ind.*, 1892, **11**, 321 ; G. M. Roberts, *Trans. Amer. Inst. Min. Eng.*, 1898, **28**, 413 ; A. C. Claudet, *Trans. Inst. of Min. and Met.*, 1898, **6**, 29.

[2] A. C. Claudet, *loc. cit.*

(*a*) *Chip Sample.*—Chips are cut by a gouge or chisel out of the top and bottom of every bar, or alternately the top of one bar and the bottom of the next, the chips being taken from opposite sides and ends of the bars.

(*b*) *Punching Sample.*—A punch resembling a leather punch, such as those shown in Fig. 164, is very frequently used. It is driven about half-way through the bars by the use of a sledge hammer, and gives cores about 2 inches long and about ¼-inch in diameter. When sampling a large lot of bullion it is usual to sample five bars at a time. In this case it is customary to place the five bars side by side on the sampling platform, and to drive in the punch at one end of the first bar and at the opposite end of the last bar, and on the others in intermediate positions in such a manner that all the holes will be along a diagonal of the rectangle enclosing the bars. The bars are then turned over and similar portions punched out in the bottoms of the bars and along the other diagonal (see Fig. 165 A). Or one set of five may be sampled along the top and the next set along the bottom of the bars.[1] The chips or cores are then carefully melted into one bar, or those from the tops and bottoms melted separately into two bars.

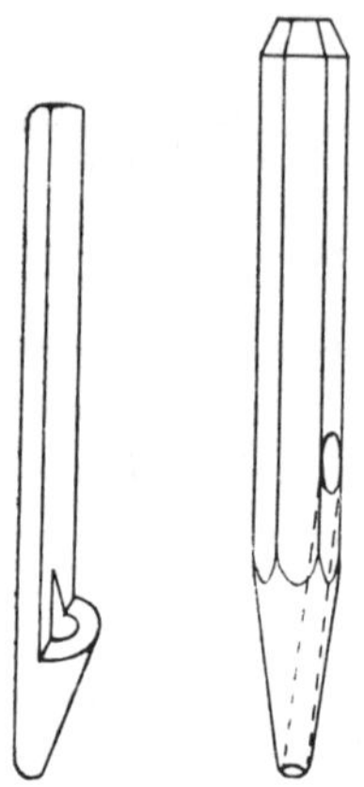

FIG. 164.—PUNCHES FOR SAMPLING LEAD BULLION BARS.

The molten metal is well stirred and then poured into a mould to give a thin flat bar about 10 to 18 inches long, about 3 inches wide, and ½ to ¾ inch thick. This is sampled by cutting out portions from the ends and sides, as shown in Fig. 165 B, or cutting strips right through the bar, or by punching out pieces at regular intervals, diagonally across the bar. The four samples thus obtained are, as a general rule, each approximately of the weight required for an assay.

On assaying the bars, the results are compensated for enrichment on melting, if necessary. This method should only be employed with lead containing 10 to 20 ounces of silver per ton. Punching methods have now been replaced in many cases by dip sampling.

(*c*) *Sawing.*—The bars are sawn transversely, either half-way through or right through, and usually alternately in the middle of one bar and at one end for the next bar, and so on. The sawings are either melted into small bars and assayed, or they are assayed without melting; in the latter case, the result must be corrected for impoverishment, due to the oil which has to be used as a lubricator when sawing the bars. Sawing is done by a circular saw, operated by steam power, and the teeth should be rather small, to prevent too frequent breakages, and to produce small-sized sawings.

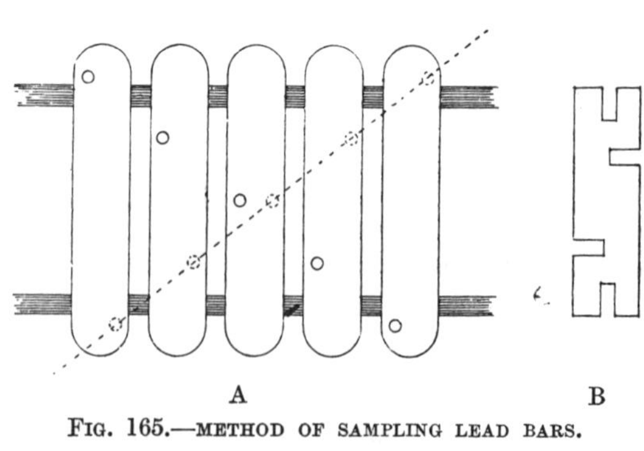

FIG. 165.—METHOD OF SAMPLING LEAD BARS.

[1] Beringer, *Assaying*, p. 160.

The "sawings," as the sample obtained in this way is termed, which may amount to two or three hundredweight, are thoroughly mixed and reduced in bulk by quartering in the usual way or by means of a mechanical sampler, such as Clarkson's. The final sample, weighing about 2 to 3 lbs., is sent to the assayer. It is contaminated with oil, and requires special treatment, as described later. This method may be satisfactorily used with lead containing silver up to 100 or 200 oz. per ton.

(*d*) *Dip Sampling.*—The lead is melted in large iron pots, holding from 10 to 30 tons. When the charge is thoroughly liquid it is well stirred to ensure uniformity, then carefully skimmed to remove dross, and a dip sample taken with a small iron ladle. The dross, amounting to from 1 to 5 per cent. according to the impurities in the lead, is broken down when cold and a sample taken. The two samples are then assayed, and the results corrected and calculated on the weight of the original charge. This method takes the longest time, but is stated to yield the most reliable results; it should be employed in all cases where the silver is above 200 ounces per ton, and where the gold is above 1 or 2 ounces per ton.

In American refineries the usual practice is to take the dip sample with a ladle giving a cone-shaped sample weighing from 12 to 16 grammes. The samples are taken either from the drossing kettle just previous to tapping, or from the stream of molten bullion as each tenth bar is cast into the moulds from the drossing kettle. The cones are sometimes termed "gum drops."

A comparison between the different methods of sampling lead bullion is made in the following tables, XXXIX, XL and XLI, which give the results of assays obtained by A. Claudet[1] for gold and silver in commercial lead bars from various sources.

TABLE XXXIX.—Comparison between Chip and Sawing Sampling. Argentiferous Lead.

Lot.	Total Weight.		Silver per ton.		
			Chip Sample (London).	Chip Sample (foreign).	Sawing Sample.
	tons	cwts.	ounces.	ounces.	ounces.
1	13	5	114·20	120·20	116·20
2	12	4	104·80	110·65	107·75
3	12	14	132·90	140·65	136·25
4	12	17	135·85[2]	137·75	135·50

TABLE XL.—Comparison between Dip and Sawing Sampling. Argentiferous Lead.

Lot.	Total Weight.		Silver per ton.	
			Dip Sample.	Sawing Sample.
	tons	cwts.	ounces.	ounces.
1	19	4	119·80	120·20
2	22	14	131·10	132·50
3	25	10	125·40	127·40
4	18	16	126·10	127·55
5	68	7	70·80	72·20

[1] *Loc. cit.*

[2] Average result. Top chips, 130·5 oz. Bottom chips, 140·8 oz.

TABLE XLI.—COMPARISON BETWEEN DIP AND SAWING SAMPLING AND THE PRECIOUS METALS (GOLD AND SILVER) ACTUALLY RECOVERED BY REFINING THE LEAD.

Lot.	Total Weight.			Sawing Sample.	Dip Sample (Colonial assay).	Recovered by Refining.
	tons	cwts.		ounces per ton.	ounces per ton.	ounces per ton.
1	2	13	gold .	1224·0	1305·0	1261·0
			silver .	305·0	296·0	296·0
2	6	0	gold .	...	342·0	340·0
			silver .	...	102·0	99·0

In lot No. 2 two quantities of 1 ton each were sawn separately and the sawings melted and assayed. Dip samples were also taken and assayed. The results for the gold were as follows :—

	Gold per ton.	
	Sawing Sample. ounces.	Dip Sample. ounces.
Lot A, about 1 ton	303	312
Lot B, „	332	329

Preparation of Lead Bullion Sample for Assay.—The sample of chips, punchings, or sawings sent to the assayer usually weighs about 2 or 3 lbs. For assay purposes it is carefully melted, and cast into an ingot mould to obtain a small " sample bar " or " assay bar " from which portions are cut for assay.

Great care has to be exercised during the melting to prevent oxidation, as this causes an enrichment of the precious metals and consequent error in the assay value of the lead unless the results are compensated for enrichment, as described below. The error due to this cause may be considerable when the lead contains impurities, such as antimony, copper, sulphur, etc., which produce much dross or scum, *i.e.* a mixture of oxides, sulphides, etc., contaminated with metallic lead.

In the case of samples taken by means of a saw, the " sawings " are always contaminated with the oil used for lubricating the saw. The sawings are first freed from oil by washing with ether, benzene, or carbon bisulphide and carefully dried. The removal of the oil may have decreased the weight of the sample by about 0·2 to 0·5 per cent.[1] The sawings thus cleaned are thoroughly mixed, and from 100 to 200 grammes carefully weighed and melted under 10 or 20 grammes of potassium cyanide to prevent the formation of dross. A sample consisting of chips or cores is melted in the same way.

The molten metal is well stirred and then poured into a mould to give a thin, flat bar which necessarily varies in size according to the weight of the sample, but is usually about 10 to 15 inches long, about 3 inches wide, and ½ to ¾ inch thick. The mould should be warm, and in casting the metal the crucible should be moved backwards and forwards along the length of the mould so as to distribute the lead and ensure a uniform bar. Any slag adhering to the lead is readily removed by washing in hot water. When the bar is quite clean the portions required for assay are cut from the ends and sides, as shown in Fig. 165 B, or by punching out pieces at regular intervals, diagonally across the bar, or cutting strips right through the bar. The four portions thus obtained are, as a general rule, each approximately of the weight required for an assay, viz. 15 to 20 grammes.

[1] J. and H. S. Pattinson, *loc. cit.*

The addition of potassium cyanide or other flux during melting is necessary, as with some classes of lead the loss of weight, due to oxidation of the impurities when the sample is melted alone, is very considerable. The percentage loss on melting is reduced to less than one-half by the use of potassium cyanide. With samples of lead of average quality the loss when melted under potassium cyanide does not, as a general rule, exceed 0·25 per cent., and is frequently about half this amount.

This loss may at first sight seem to be of little importance, but, actually, the correction of the assay corresponding to this small loss makes a difference on lead containing, say, 500 ounces of silver per ton, of about 1 ounce per ton; and this amount, when deducted from the assays of the sample, gives the true content of silver in the original lead. With richer leads the correction is of course proportionately greater. With auriferous lead the correction for loss on melting is of considerable importance, owing to the greater value of gold compared with silver.

It is obvious that, if the sample has not been freed from oil before melting, the melting loss will include also the loss due to the oil, and an error would be made which would lead to an *excessive deduction* amounting probably to more than 2 ounces per ton in the case of lead containing 500 ounces of silver per ton.

Determination of Silver in Lead Bullion.—The four or more portions selected for assay are adjusted to exactly 15 or 20 grammes (or to one-half assay ton), according to the percentage of silver in the lead, and then carefully cupelled at a low temperature and the resulting buttons of silver weighed. If the lead is very impure, it should first be scorified with about 1 gramme of borax, and if necessary with the addition of about 30 to 50 grammes of sheet lead. The resultant buttons, which should weigh about 15 grammes, are then cupelled. If the lead is very poor in silver, from 40 to 50 grammes may be scorified down to about 15 grammes and then cupelled. In the case of lead rich in silver (say over 500 ounces per ton) the cupels are crushed, fused and assayed, and the correction for cupel absorption made. Or the cupellation loss may be determined by wrapping each silver button obtained from the cupellation separately in 20 grammes of sheet lead, and recupelling side by side with two or more fresh lots of 20 grammes each of the sample of lead bar. Then weigh the resulting silver buttons, calculate the loss incurred, and add this to the weight of the silver buttons from the two fresh portions cupelled.

It is usually preferable, however, in the case of rich argentiferous lead, to run check assays after first determining the approximate composition of the sample by a preliminary assay.

The silver in lead bullion may also be determined by a volumetric method.[1] The procedure is as follows:—The lead is dissolved in nitric acid (1 c.c. for each gramme of lead) and 10 grammes of tartaric acid (to keep any antimony in solution). The solution is diluted to 250 c.c. and the silver precipitated with a slight excess of ammonium thiocyanate. The precipitate is collected, washed, and dissolved in ammonia, and the silver re-precipitated by the addition of hydrochloric acid until slightly acid. The filtrate from the hydrochloric acid is diluted to 400 c.c. and titrated with potassium permanganate.

Determination of Gold in Lead Bullion.—To determine the gold the silver buttons, after carefully weighing, are flattened and parted in a small parting-flask by heating with 15 to 20 c.c. of dilute nitric acid of about 1·2 sp. gr. until the silver is dissolved and all nitrous fumes expelled. The residual gold

[1] W. R. Wagstaff, *Ind. and Eng. Chem.*, Anal. edn., 1932, **4**, 51; Abs. *Met. Abs. Inst. Met.*, 1932, **50**, 310.

is washed, and heated for a few minutes with stronger nitric acid of about 1·3 sp. gr. and then washed, dried, annealed, and weighed.

Calculating and reporting the Results.—The silver and gold in lead bullion is always reported in ounces per ton, in the same way as for ores. The amount of silver may vary from 10 ounces to about 1000 ounces per ton, while gold may be present to the extent of 500 ounces per ton or even more, but is usually much less. As previously stated, the smelters generally keep the proportion of silver in the lead below 2 per cent. (650 ounces per ton), in order to avoid undue losses in the slag.

The results of the several assays should agree very closely, say within half an ounce per ton, an average bullion assaying from 200 to 250 ounces of silver per ton. With richer argentiferous lead assaying over, say, 500 ounces, the variation between the results will be greater than this, and it is advisable in assaying such material to make a larger number of assays, from five to ten, and take the average of all the results. There should be practically no difference in the weights of the gold results, and if they are not almost identical the assays should be repeated. The assay value of the lead both in silver and gold is reported in ounces and decimals of an ounce to the second place of decimals on the ton of either 2240 lbs. or 2000 lbs. The former is used in England, but the latter is becoming more universal.

Correction for Loss during Melting.—In calculating the assay value of the lead bars it is necessary, as previously stated, to make a correction for the loss during melting, otherwise the value reported will be too high. The following is an example :—

Weight of sample before melting	200·0 grammes.	
Weight of sample bar after melting	199·4 ,,	
Loss on melting ...	0·6 ,,	= 0·3 per cent.

Therefore 100 grammes of the original sample before melting is represented by 99·7 grammes after melting. If 20 grammes of the lead after melting gave 0·1671 gramme of silver, equal to 0·8355 per cent., then the percentage of silver in the original lead will be :—

$$\frac{0{\cdot}8355}{100} \times 99{\cdot}7 = 0{\cdot}8329 \text{ per cent. of silver.}$$

And 0·8329 per cent. equals 272·08 ounces of silver per ton of lead of 2240 lbs. If gold is also present it is calculated in the same way.

With lead bars which yield a large amount of dross on melting, the dross is removed and sampled separately. It is then assayed, and the results obtained for the silver and gold present allowed for in calculating the total assay value of the lead bars, as shown below.

Assay of "Dross" from Lead Bullion.—The dross skimmed off from the surface of the molten lead contains, as a rule, from 80 to 95 per cent. of lead present as metal, lead oxide, sulphate, etc. and usually contains copper and antimony in fair proportion. The skimmings vary considerably in content of silver according to the richness of the bullion melted. They are assayed by a crucible fusion, the following being a suitable charge :—

Dross	20 to 30 grammes.
Sodium carbonate	50 ,,
Borax	10 ,,
Argol	1·5 ,,
Litharge	20 ,,
A piece of hoop-iron or nails.	

The charge is fused in the ordinary way as for silver-ore assays, the resulting lead button cupelled, and the silver button cleaned and weighed. If gold is present the buttons are flattened and parted. The slags from the fusion assay of dross from rich bullion should be collected and assayed to recover the silver, as previously described.

Calculation of Assay Value of Lead Bars yielding Dross.—The value in silver and gold of impure lead bullion with dross is calculated from the following data :—

Melting.		Assays of Bar.		Assays of Dross.	
Bar weighs	4·23 lbs.	Silver ...	374·22 oz.	Silver ...	776·15 oz.
Dross	1·02 „	Gold ...	1·14 „	Gold ...	2·95 „
Total (after melting) ...	5·25 lbs.	Per ton of 2000 lbs.		Per ton of 2000 lbs.	

Then

The Silver in the Bar is :

$$\frac{4{\cdot}23}{2000} \times 374{\cdot}22 = 0{\cdot}791475 \text{ ounces.}$$

The Silver in the Dross is :

$$\frac{1{\cdot}02}{2000} \times 776{\cdot}15 = 0{\cdot}395836 \text{ ounces.}$$

The total silver in the bar and dross is $0{\cdot}791475 + 0{\cdot}395836 = 1{\cdot}187311$ ounces.

Hence the assay value of the silver in the sample is :

$$1{\cdot}187311 \times \frac{2000}{5{\cdot}25} = 452{\cdot}308 \text{ ounces per ton of 2000 lbs.}$$

The gold is calculated in the same way.

The Gold in the Bar is :

$$\frac{4{\cdot}23}{2000} \times 1{\cdot}14 = 0{\cdot}002411 \text{ ounces.}$$

The Gold in the Dross is :

$$\frac{1{\cdot}02}{2000} \times 2{\cdot}95 = 0{\cdot}001505 \text{ ounces.}$$

The total gold in the bar and dross is $0{\cdot}002411 + 0{\cdot}001505 = 0{\cdot}003916$ ounces.

Hence the assay value of the gold in the sample is :

$$0{\cdot}003916 \times \frac{2000}{5{\cdot}25} = 1{\cdot}49 \text{ ounces per ton of 2000 lbs.}$$

The assay value of the sample of impure lead bullion will therefore be :

Silver ... ... 452·31 ounces per ton of 2000 lbs.
Gold ... ... 1·49 „ „ „

(II) Base Copper Bullion.

Copper ores frequently contain notable amounts of silver and gold which render their treatment profitable. During smelting operations these metals become concentrated in the final impure copper, which is cast into bullion bars, or more frequently into anode plates for electrolytic refining.

Segregation in Copper Bullion.—During the solidification of argentiferous and auriferous copper bars, a rearrangement of the constituents takes place so that the silver and gold are not evenly distributed. Ledoux [1] has shown that while there is no absolute rule, yet highly refined material free from arsenic generally has a concentration of the precious metals along the centre-line, the maximum being at the middle of the bar. The distribution of silver in copper bars is shown in Table XLII, which gives the results of assays of samples taken from different parts of the bars.

TABLE XLII.—ASSAYS OF ARGENTIFEROUS COPPER BARS (*Hixon*).[2]

Assay No.	Part of Bar sampled.	Bar No. 1. Silver in ounces per ton.	Bar No. 2. Silver in ounces per ton.
1	End	358·1	231·2
2	End	403·4	247·4
3	Top	608	451·7
4	Bottom	366·2	203·6
5	Side	393·9	252·9
6	Side	372·5	235·9
7		423·6	351·4

Sampling Copper Bullion.—The sampling of gold- and silver-bearing copper bars is governed by the same principles as the sampling of lead bars.

Argentiferous copper bars are usually sampled by boring, by taking dip samples, or by sawing. Ledoux [3] recommends sampling by sawing, the sample being taken from five bars laid side by side. A diagonal line is drawn across the bars, and each bar sawn through at the point where the diagonal intersects the centre-line of the bar, as illustrated for lead bars in Fig. 165 A.

Buyers and sellers are inclined to accept the "dip" method of sampling if done at their own works. In view of the want of uniformity in these bars, Ledoux urges that both buyer and seller should agree in their contracts as to the method of sampling and assay to be employed.

A comparison between three methods of sampling copper melted and cast into the form of anodes at the Washoe Smelter, Anaconda, Montana, has been made by Wraith [4] with special reference to the silver content. The smelter's method of sampling consists in "shotting" a small portion of the molten stream of copper, during casting, by "batting" the stream into water with a wooden paddle and selecting the portion passing a 4-mesh and remaining on a 10-mesh screen. The refinery method consists in drilling a $\frac{1}{2}$-inch hole through each

[1] A. R. Ledoux, *J. Canadian Min. Inst.*, 1899, **2**, 108–118.

[2] Hixon, *Lead and Copper Smelting*, p. 83.

[3] *Loc. cit.*

[4] W. Wraith, *Trans. Amer. Inst. Min. Eng.*, 1910, 209–214 (abstract from *J. Soc. Chem. Ind.*, 1910, **29**, 493). See also "The Principles and Practice of Sampling Metallic Metallurgical Materials with special reference to the Sampling of Copper Bullion," by E. Keller, 1916, Bureau of Mines, Washington, Bulletin No. 122.

fourth anode, a template the size of the anode being used having ninety-nine holes arranged at equal distances over the surface, each hole being used in turn. The total drillings are mixed and ground to pass a 16-mesh screen. A ladleful of copper was also taken from the molten stream and "shotted." It was found that the smelter and refinery methods of sampling gave the true silver content of the copper and agreed within practical limits, but ladle samples tend to give high results on account of the segregation of silver towards the middle of the cast sample bar.

Determination of Silver and Gold in Copper Bullion, etc.—Silver is found in copper bullion and in most samples of commercial copper. The bars resulting from the smelting of argentiferous copper ores may contain from 100 to 1000 ounces or even more of silver per ton. Low-grade bullion contains from 10 to 50 ounces of silver per ton.

To determine the silver, duplicate samples of the borings of about 25 to 30 grammes (or one assay ton) are weighed, transferred to a beaker, and 100 c.c. of water and 50 c.c. strong nitric acid (sp. gr. 1·42) added. When the violent action has ceased, another 50 c.c. of nitric acid is added, and the solution heated, stirring from time to time, until everything soluble is dissolved, and until the oxides of nitrogen have been expelled. The removal of these is facilitated by blowing a gentle current of air through the solution. A dilute solution of sodium chloride is then added to precipitate the silver as chloride, and the solution boiled for about one minute to facilitate the collection of the precipitate. The usual strength of salt solution used contains 0·542 gramme of sodium chloride per 1000 c.c., and 1 c.c. of this solution will precipitate 1 mg. of silver. An excess of about 5 c.c. above that required to precipitate the silver present should be added. If the amount of silver in the copper is small (say under 50 ounces), 10 c.c. of a concentrated solution of lead acetate (or nitrate) and 2 c.c. of strong sulphuric acid are added in order to obtain a precipitate of lead sulphate to aid the settling of the finely-divided silver chloride. The precipitate is allowed to settle for twelve hours, the solution is then filtered through a double Swedish paper, and the precipitate washed into the point of the filter-paper. The filter is carefully dried, and when dry transferred to a scorifier containing 2 grammes of granulated lead. This is placed in the muffle, heated just to dull redness, and the filter-papers burnt off, but only until the flame disappears, and not to a white ash. Then add another 8 grammes of lead and 0·5 gramme of borax glass and raise the muffle temperature sufficiently to render the charge molten, and then pour, as no scorification is necessary. The lead button is cleaned and cupelled at a low temperature, and the resulting silver button weighed. Some assayers carefully burn the filter-paper over a small piece of lead-foil and allow the ash to fall on to the foil, which is then wrapped up and cupelled direct. The results should agree within about 0·25 ounce of silver per ton.

With a view to avoiding loss of gold by solution, sulphuric acid is sometimes recommended in place of nitric acid for dissolving the copper, but the action in this case is much slower, and a longer time is required to effect solution. The sulphuric acid method is stated to give results equal to those obtained by the "all-fire" method described below.

Argentiferous copper is also assayed by scorification or the "all-fire" method, as it is sometimes termed, but this method is not so suitable, as small quantities only can be taken for assay. When this method is used, from 5 to 10 separate quantities, each of about 3 grammes (or one-tenth assay ton), are separately scorified with 15 to 20 times its weight of granulated lead and a cover of 1 gramme each of borax and sand. The lead buttons obtained are cupelled separately, and the beads of precious metal are weighed, first separately, and then together according to the richness of the bullion. With high-grade material the cupels are ground up and fused in several lots, and the silver obtained from the cupellation of the resulting lead buttons is added to that of the first assay.

Gold.—When gold is present in the bullion it is left as an insoluble residue after treatment with nitric acid, as described for the determination of silver. Since, however, copper salts in the presence of nitric acid exert a solvent action on finely-divided gold, it is advisable not to heat the solution too strongly when dissolving the copper. The assay is conducted as previously described for silver, and if enough silver is present to allow the button from the cupellation to be parted, it is flattened and parted in dilute nitric acid in the manner described for gold ores. When the bullion contains gold and silver, both metals are usually determined in the same quantity taken for assay. If there is insufficient silver, a small quantity of silver nitrate may be added to the solution or a little silver-foil added during the cupellation. The scorification method is used as a check upon other methods, especially for gold. The results for gold should agree very closely.

Reporting the Results.—The gold and silver content of copper bars are generally reported in ounces and decimals of an ounce per ton either of 2240 lbs. or 2000 lbs., as in the case with lead bullion.

In the case of gold, the return when below an ounce is made in pennyweight and decimals of a pennyweight per ton.

Owing to the small amount of precious metals present in copper bullion it is very necessary, in order to obtain reliable results, to use not less than one assay ton of the copper borings and to make all the final weighings on a good assay balance.

CHAPTER XXI

THE ASSAY OF INDUSTRIAL GOLD AND SILVER ALLOYS

(I) ASSAY OF GOLD ALLOYS.

Introduction.—For the purpose of this chapter the term " industrial alloys " includes the gold alloys of definite composition especially prepared for industrial purposes, as distinct from those included in the term " bullion " Also included in this chapter are the so-called " trade bars " which consist of ingots of varying size and composition obtained from the treatment of scrap metal, lemel, sweep, etc., resulting from the working of the precious metals. In some cases these bars have a high precious-metal content, whilst in others it is very low. The residual metal from coinage operations is termed " scissel " and is remelted.

Legalized Gold Alloys.—The industrial gold alloys consist mainly of the numerous alloys used for coinage and jewellery, and in dentistry. The more important are those made to conform to the legalized standards. The gold content of jewellery alloys is expressed in the old carat system in which pure gold is arbitrarily divided into 24 parts or " carats," but the carat bears no relation to that used for the valuation of diamonds.

Formerly five standards were legalized in this country for gold work, viz. 22-carat (91·7 per cent. gold), 18-carat (75 per cent. gold),, 15-carat (62·5 per cent. gold), 12-carat (50 per cent. gold) and 9-carat (37·5 per cent. gold). In 1932, however, the two standards 15-carat and 12-carat were abolished, and a 14-carat (58·5 per cent. gold) standard legalized in their place, mainly to encourage export trade, since 14-carat gold has long been used for jewellery purposes on the Continent and in America. The gold content legally adopted closely approximates to the true 14-carat gold which contains 58·33 per cent. of pure gold.

The 22-carat alloy is the legal standard for the British coinage and is known as " Standard Gold." It finds only limited application in the jewellery trade. Best-quality jewellery is made of 18-carat gold, but 9-carat gold is the quality most frequently used in the trade, more than fifty per cent. of the gold wares submitted for hall-marking being of this standard.

Gold alloys made to standards other than the legal standards are also used by jewellers for special purposes, and some of these are as low as 6-carat or 7-carat quality.

All the legalized gold alloys vary considerably in composition, for whilst the gold content must be kept constant, the proportions of the added metals may be varied at will, so that it is possible to produce for each carat quality a considerable range of alloys of different colour and varied mechanical properties to suit all requirements. Thus, by suitable alloying a 9-carat alloy is produced which closely resembles 18-carat gold in colour, and is being largely used.

In the industrial alloys the gold is usually alloyed with copper or silver, or with a combination of both of these metals, whilst in the lower carat alloys small quantities of zinc are added, but the proportion seldom exceeds 10 per cent. as it induces brittleness.

In special cases small proportions of cadmium or nickel may be present. Cadmium may be a constituent of low-melting gold solders. Alloys in which the silver content predominates are usually pale-yellow in colour, whilst those in which copper predominates have a decided red tint and constitute the red golds of the trade. Nickel has a bleaching effect when added to gold and is used in the production of the white gold alloys introduced as substitutes for platinum. These are usually made to 18-carat, 14-carat, and 9-carat standards.

The 9-carat white golds may be regarded as good-quality nickel silver alloyed with the legal quantity of gold. The two chief groups of white gold alloys are the gold–nickel–zinc and the gold–palladium alloys. Of these, the latter are softer and have the more satisfactory working properties, but are more expensive.

The compositions of some typical industrial gold alloys are given in Table XLIII.

TABLE XLIII.—Compositions of Typical Industrial Gold Alloys of Legal Standards.

	22-Carat.		18-Carat.			
Gold, per cent.	91·7	91·7	75·0	75·0	75·0	75·0
Silver, „	2·0	5·0	25·0	16·0	12·5	...
Copper, „	6·3	3·3	...	9·0	12·5	25·0
	14-Carat.					
Gold, per cent.	58·5	58·5	58·5	58·5	58·5	58·5
Silver „	37·3	17·5	15·0	12·5	10·0	6·5
Copper, „	4·2	24·0	26·5	29·0	31·5	35·0
	9-Carat.					
Gold, per cent.	37·5	37·5	37·5	37·5	37·5	37·5
Silver, „	20·0	17·5	11·0	6·0	5·0	9·5
Copper, „	37·5	42·0	51·5	55·0	57·5	50·0
Zinc, „	5·0	3·0	...	1·5	...	3·0
	White Gold Alloy (Nickel).					
	18-Carat.		14-Carat.		9-Carat.	
Gold, per cent.	75·0	75·0	58·5	58·5	37·5	37·5
Nickel, „	16·5	12·0	17·0	12·5	17·5	18·5
Copper, „	3·5	7·0	16·0	24·0	27·5	33·0
Zinc, „	5·0	6·0	8·5	5·0	17·5	11·0

Gold alloys similar in composition to those used in the jewellery trade are used for dental purposes. The gold sheet for crown work is usually 22-carat, but apart from this the standards used for dental work seldom exceed 18-carat or 20-carat (83·3 per cent. gold) the former being the most common for general purposes. In some cases the dental golds are alloyed with a certain percentage of platinum, which imparts greater hardness and elasticity. Such alloys are known as "platinized gold" and are more fully dealt with in the chapter on the Platinum Metals.

Descriptions of Imitation Gold Jewellery.—Many terms have come into use in the imitation gold jewellery trade to denote the method of manufacture. These are not generally understood by the public and their use has frequently led to confusion, if not in some cases to misrepresentation. It therefore became desirable, in the interests of both manufacturers and the public, to attempt some clear definition of the terms. The following definitions have been proposed by the various jewellers' associations, and have been generally accepted:—

Gold Front.—A sheet of gold of a standard not lower than nine carats which is made in a separate portion, and afterwards attached to a silver or base-metal back in such a way that the separate sheet of gold can be removed from the back of silver or base-metal.

Rolled Gold.—A sheet of gold of a standard not lower than nine carats sweated or soldered to a sheet or bar of silver or base metal, the whole being rolled down together; or in the case of wire, a core of silver or base metal completely surrounded with gold of a standard not lower than nine carats, sweated or soldered on, the whole being drawn down in such a way that on annealing and pickling a surface of gold will be shown.

Gold Filled.—A sheet of gold of a standard not lower than nine carats sweated or soldered to each side of a bar or ingot of silver or base metal, the whole being rolled down together; or, in the case of wire, a core of silver or base metal completely surrounded with gold of a standard not lower than nine carats, sweated or soldered on, the whole being drawn down together in such a way that on annealing and pickling a surface of gold will be shown.

Gold Shell.—A casing of gold of a standard not lower than nine carats on silver or base metal, the casing being of such thickness that when the base is dissolved in acid the gold shell is left intact.

Gold Cased.—Silver or base metal having thereon an electro-deposit of gold of such density that neither nitric acid of 1·40 specific gravity nor lunar caustic applied for two minutes will discolour or stain it.

Gilt.—Silver or base metal having gold deposited thereon by a chemical or electro-deposition process.

Fire Gilt or *Mercurial Gilt.*—A covering of gold dissolved in mercury, *i.e.* as an amalgam, from which the mercury has been expelled by heating.

Gold Solders.—These consist of alloys of gold–silver–copper in suitable proportions to give alloys of low melting-point, and are similar in composition to the gold alloys given in Table XLIII. Small quantities of zinc, and less frequently of cadmium and tin, are sometimes added to increase the fusibility still further, but the additions of these metals combined seldom exceed ten per cent.

In the jewellery trade, suitable solders for use on work of 14-carat and above are frequently made by adding about 5 parts of silver to every 24 parts of the same carat gold as that to be soldered.

The colour of gold solders is made to correspond as closely as possible to the carat gold on which they are used, but they are, of course, of lower melting-point.

The following are the compositions of typical gold solders:—

TABLE XLIV.—PERCENTAGE COMPOSITIONS OF GOLD SOLDERS.

Use.	Gold.	Silver.	Copper.	Zinc.	Cadmium.	Tin.
For 18-carat gold ...	65	15	14	3·5	...	2·5
Hard or Best	62·5	22·5	15	...	...	...
For 14-carat gold ...	58·5	27	14·5	...	...	...
Medium	50	30	20	...	...	...
Soft, Easy-flowing ...	42·5	32·5	25	...	...	...
For 9-carat gold ...	37·5	40	14·5	8·0	...	...
" " ...	37·5	40	12·5	5·0	5·0	...

To comply with the Assay Office Regulations, the solders used in the manufacture of 9-carat gold wares not infrequently contain cadmium in addition to zinc to lower the melting-point, as they must "assay at not less than the standard declared by the sender" of the wares to be hall-marked.

Gold and Silver Standards for Coin and Plate.

The standards now in force for gold and silver coin and plate in various countries of the world are as follows:—

Gold Standards for Coin and Plate.[1]—For coin the gold standard in most countries is 900, among exceptions being Great Britain, which has adopted 916·6. This higher standard is also in use in Brazil, Chile, India, Newfoundland, Peru, Portugal and Turkey, whilst that of Egypt is 875.

For gold plate and jewellery the most universal standards are 750 and 585 (or 583).

Lower standards of 375 are or were used by Great Britain, Estonia and Tunisia, and 333 by Germany, Italy, Bulgaria and Portugal.

Silver Standards for Coin and Plate.—The standard of 900 is very general for silver coin. In Great Britain the standard is 925 (temporarily lowered to 500). In some countries lower standards are in use, especially for subsidiary coins.

For silver wares various standards are in use. The 925 standard is the one current in the British Empire and the United States of America, the so-called *sterling silver*. The silver ware exported by England, too, is generally of this standard. There is also another British alloy for plate, containing 958·3 parts of silver, the so-called *Britannia silver*, but it is seldom used as it is less durable than sterling silver.

France manufactures larger objects in 950 standard for the home market, and frequently 800 in smaller articles.[2]

The Scandinavian countries, the Netherlands, Czecho-Slovakia, Austria, Portugal, Egypt and Bulgaria normally use standards between 850 and 830.

Germany manufactures 800 for the home market as well as for export; and in many countries, owing to German influence, the 800 standard is current. Before the Second World war, however, manufacturers in Germany were discussing a change from 800 to 835, as the lower-quality alloy tarnishes somewhat readily. The 835 standard, on the contrary, is regarded as satisfactory.

[1] From Report of Committee for International Unification of Precious Metal Standards, 1928.

[2] Consult *Monnaie, Médailles et Bijoux*, A. Riche, Paris, 1889, p. 356.

Assay Offices for Hall-marking.[1]

British law requires that the quality of all gold and silver and "plate" articles, with certain exceptions including most jewellery, shall be determined by assay at offices or "halls" in various parts of the kingdom duly authorized for that purpose; and that if the article be found equal to the required standard, it shall be stamped at those offices with a series of marks known as the "hall-mark" —so-called because originally the testing and marking was done at the Goldsmiths' Hall, London.

The earliest of the hall-marking laws was enacted over six hundred years ago in the reign of Edward I.

As previously stated, the present legal British standards for gold wares are 22-carat, 18-carat, 14-carat, and 9-carat. For silver wares the most important legal standard is 925, known as Standard or Sterling silver; and the seldom-used standard 958·4 commonly known as Britannia silver because it is marked with the figure of Britannia to distinguish it from sterling silver which is marked with a lion rampant. The wares sent for hall-marking are carefully sampled by scraping or cutting so as to obtain a representative portion of the metal of which they are made, and the samples are subsequently assayed. If the ware, after testing, is found to comply with one of the legalized standards, the marks prescribed by statute for that respective standard are then struck upon it. In the event of any ware failing to pass the official test, a fine is imposed, and the article is broken up and returned to the manufacturer. A description of the methods of assay for gold and silver alloys used for the manufacture of wares has already been given.

Before it can be accepted for hall-marking, every ware must bear the mark of the manufacturer. This mark consists of the maker's initials or, in the case of a company, the initials of the name of the firm. The maker's mark must be approved by the assay authorities and duly registered before being used.

The hall-marks struck upon the wares are usually placed quite close to the maker's mark and are as follows:—

(1) The city mark showing the hall or assay office at which the ware was tested;
(2) The standard or quality of the gold or silver, as the case may be; and
(3) The year in which the ware was hall-marked.

Formerly another mark, known as the "duty mark," consisting of the sovereign's head, was struck upon the wares to denote the payment of duty, but this was discontinued in 1890.

Offices for hall-marking are at London (Goldsmiths' Hall), Birmingham, Chester, Sheffield, Dublin, Edinburgh, and Glasgow. Offices formerly existing at Exeter, York, Newcastle, Norwich, and other places have long been closed.

The City Marks to denote the place of assay are as follows:—

London ...	A leopard's head (the arms of the Goldsmiths' Company).
,, ...	A lion's head erased (*i.e.* without the body), for silver wares of 958·4 standard.
Birmingham	An anchor.
Chester ...	A sword between three garbs.
Sheffield ...	A crown for silver wares.
,, ...	A York rose for gold wares.
Dublin ...	A harp crowned.
Edinburgh	A castle.
Glasgow ...	A tree with a fish across the trunk, a bell hanging from one of the branches, and a bird on the top branch.

[1] For information on hall-marks, consult C. J. Jackson, *English Goldsmiths and their Marks*, 1905.

The Standard or Quality is denoted by the following marks :—

Standard Marks for Silver.

Standards.	Marks.
925·0 (Sterling or standard silver) ...	A lion passant.
958·4 (Britannia silver)	A figure of Britannia.

Standard Marks for Gold.

Standards.	Marks.
916·6 or 22 carat	A crown and figures 22.
750·0 or 18 ,, ...	A crown and figures 18.
585·0 or 14 ,, ...	Figures 14 and decimal figures ·585.
375·0 or 9 ,, ...	Figure 9 ,, ,, ·375.

The Date-Mark to denote the year of assay consists of a single letter of special design, and different for each assay office, which is used throughout the year, and is changed every year, usually in June.

A complete hall-mark on any single piece of plate will therefore consist of four separate marks, viz. : (1) maker's mark ; (2) city mark ; (3) standard mark ; and (4) date letter.

Foreign silver or gold wares imported for sale must first be submitted for assay, and if of the required standard, they are stamped with distinctive marks, to denote that they are of foreign manufacture.

Hall-marks for Foreign Imported Gold and Silver Plate.[1]

City Marks (place of assay).

London	Sign of constellation Leo.
Birmingham	Equilateral triangle.
Chester	Acorn and two leaves.
Sheffield	Libra.
Dublin	Boujet.
Edinburgh	St. Andrew's Cross.
Glasgow	Double block-letter F inverted.

Standard or quality marks (silver).

For standard or sterling silver	Numerals 925, or decimal figures ·925
For Britannia standard ...	Decimal figures ·9584

Standard or quality marks (gold).

For 22-carat gold wares ...	Figures 22 and decimals ·916
,, 18-carat ,, ,, ...	,, 18 ,, ,, ·75
,, 14-carat ,, ,, ...	,, 14 ,, ,, ·585
,, 9-carat ,, ,, ...	Figure 9 ,, ,, ·375

The date-mark stamped on foreign gold and silver wares is the same as that used for wares manufactured in the United Kingdom.

So fully is the protection afforded by the Hall-mark recognized and demanded by the public that manufacturers send in for hall-marking articles, including most jewellery, which in the ordinary course are exempt. It is very necessary,

[1] See "Hall-marking of Foreign Plate," Order in Council (illustrated), 1906 (*London Gazette*, May 15, 1906) ; and the Order of 1932 legalizing 14-carat gold.

however, to distinguish between the genuine hall-mark and certain unauthorized marks, such as "18-Ct." and "9-Ct.," which some makers are accustomed to stamp on goods which they have not had assayed and marked at one of the authorized Assay Offices. Such marks, which purport to give the standard of the article, represent merely the personal guarantee of the maker; and from the author's experience many of the goods so marked are below the standard stated on them.

Most other countries have some form of hall-marking system, but none have the reputation of the British marks. In America, silver wares are marked "sterling," whilst goldsmiths stamp their names and the carat quality under their own guarantee only. France has an elaborate system of "control" marks, whilst in Germany the marking of jewellery and plate is left to the manufacturer, who is responsible for the correctness of standard and its conformity with the marking.

Jewellery marked with any standard must also bear the initials or trade-mark of the maker.

Silver plate is marked with the half-moon and the imperial crown, and gold plate with a sun, orb and the crown. The gold and silver standards for jewellery, plate and coinage in most countries have been given above.

Trial of the Pyx.—The term "pyx" is applied to the box at the Royal Mint in which samples of gold and silver coin are deposited to be tested at the annual "trial of the pyx" by a jury of the Worshipful Company of Goldsmiths at Goldsmiths' Hall, London. Formerly sample coins pending the assay or trial appear to have been deposited in the pyx (literally, a box) in the Pyx chamber at Westminster Abbey, which in Norman times (11th century) probably housed the Royal Treasury. It is from this circumstance that the trial no doubt took its name. This ancient ceremony of testing the standard of the coinage was practised in the country probably as early as Saxon times.[1]

Sampling of Gold Alloys and Wares.—The method of sampling gold alloys varies according to their condition. When in the form of ingots, slabs, or bars they are sampled in the ordinary way by cutting a single sample from a corner or the middle of one side. Experience proves that in the case of alloys of high standard these samples are usually representative of the whole mass. In dealing, however, with bars of low-standard alloys, it is very advisable to take two samples by cutting diagonally opposite corners from the upper and lower sides, as alloys containing a large proportion of copper are always subject to more or less segregation during solidification, even when every precaution has been taken in melting, and they are consequently not perfectly uniform in composition. As previously stated, researches have shown that practically no segregation occurs in the gold–silver and gold–copper alloys rich in gold and free from all impurities; but in the case of many of the ordinary trade bars of gold alloy, more especially those of low standard, the discrepancies frequently met with between the assays of samples taken from different parts of the bars prove that they are seldom uniform in composition, although in many cases the differences in the assay results are comparatively small.

The errors due to the non-homogeneous character of the alloys are to some extent modified by rolling the bars. Yet the general experience is that assays of samples from the middle of the strips especially of low-standard alloys, invariably differ slightly from samples taken at the edges. These differences in the gold-content may be considerable when proper care has not been exercised in the preparation of the alloy. The question of segregation in ternary alloys of

[1] H. J. Chaney, *Our Weights and Measures*, 1897.

gold–silver–copper is of considerable interest and importance, and one upon which more experimental work remains to be done.

When the alloys are in the form of sheet or wire, or of manufactured articles, considerable care has to be exercised in sampling, as the surface is invariably richer in gold than the interior owing to the " pickling " or " blanching " to which the metal is subjected to remove the coating of oxide formed during the operations of annealing, soldering, etc. Pickling dissolves the base metal on the surface leaving behind a thin film of gold. The removal of this enriched or " coloured " surface before sampling is very necessary, as it is obvious that if it is included in the sample the result obtained for the gold assay will be too high and will not represent the true fineness of the alloy. The errors introduced in this way are largely dependent on the thickness of the sheet[1] or wire and the standard of the alloy. As a general rule, the errors are smaller with the alloys of higher standard, as there is less base metal to be dissolved out. Assays made by the author[2] show that the errors on sheet metal may vary from 0·5 to 2 or 3 parts per 1000 or more above the correct standard of the alloy.

Sampling Gold Wares.—In the case of gold wares where it is usually necessary to scrape the surface with special steel scrapers in order to obtain a sample, the removal of the " colour " becomes a matter of extreme importance : if it is not removed before sampling, the errors on the gold assay will be much greater than those given above for cutting samples of sheet metal, but the extent of the errors will necessarily be dependent on the depths to which the scraping or " drawing " (as it is sometimes termed) is carried, and also the area covered in taking the sample. In sampling gold-plate and articles of jewellery the assayer is confronted with difficulties which do not arise in connection with the sampling of sheet metal and gold bullion, as the area from which a sample can be taken in the former case is frequently very limited. When a number of small articles of the same kind have to be tested, one (or more) is selected as representative of the lot and is cut up for assay. The errors arising from the inclusion of the coloured surface in the case of scraping samples may vary from a few parts per 1000 to anything up to 100 parts per 1000, or even more in exceptional cases.

At the assay offices for hall-marking it is the usual practice to remove the " colour " from gold wares by a preliminary scraping, or by " buffing " before scraping to obtain the sample proper.

Assay of Lemel or Trade Bars.—The term *lemel* is derived from the French *limaille* (filings of metal), and is applied to the fine particles of gold, etc. produced in the manufacture of jewellery by the processes of filing, cutting, turning, etc. This rich material is " dressed " to free it from organic matter by burning, and the metallic residue, after removing iron particles with a magnet, is mixed with fluxes and fused in a crucible. The molten metal is cast into a mould, producing bars of all sizes and shapes, known as " lemel bars," " trade bars," or " parting-assay bars," which are sent to the refiners for valuation and sale.

If the lemel is received in the condition as collected in the workshop it is " dressed," as described on p. 376, before being assayed. It is to the manufacturer's advantage to keep the lemel as clean as possible and free from iron filings, etc., which is a common impurity derived from the tools used.

Owing to the segregation of the constituent metals, which almost invariably takes place, lemel bars are frequently very irregular in composition, especially

[1] The minimum thickness allowed by the assay offices for gold wares is 0·0076 inch (No. 36 Imperial Standard Wire Gauge).

[2] *Chem. News*, 1906, **93**, 225.

if they contain platinum metals or have not been melted and cast at a sufficiently high temperature to thoroughly melt the metal, or have been allowed to solidify very slowly after casting. It is frequently advisable to remelt them under a layer of borax to prevent oxidation, and then, after stirring well, to recast at a suitable temperature, in order to obtain a more uniform bar for sampling. But even when every precaution has been taken to produce a sound bar it is very difficult to prevent some segregation, and in consequence the sampling of trade bars requires careful consideration.

A method of sampling frequently adopted is to draw a diagonal line on the top surface of the bar, and then to reverse the bar and draw a diagonal line in the opposite direction on the lower surface. Three or more holes, equidistant along the diagonals, are then drilled half-way through from the top and the bottom of the bar respectively. A fine drill should be used, without oil of any kind, and the drillings well mixed before taking the portion to be used for assay. For bars rich in platinum metals it is advisable to take ten or more drillings, or preferably to take a dip sample from the molten metal before casting (when this is possible) and after stirring well to ensure uniformity of composition. After thorough stirring the molten metal may also be granulated by pouring it in a steady stream into water, and sampling the granules.

Very small lemel bars may be sampled by drilling in at least two or three different places ; or by hammering or rolling them into moderately thin sheet which is then cut up into small pieces and a portion taken for assay after well mixing. Before drilling these bars it is essential to remove all surface dirt.

Samples of lemel bars are usually assayed by the refiners by double cupellation and parting, whereby the gold and silver contents are determined in the same assay-piece. From three to five separate assays may be made, especially if the bars are very rich in precious metals. For each assay from 0·25 to 0·5 gramme is taken and cupelled separately with about sixteen times its weight of lead, thus removing the base metals. The resulting cupel buttons are weighed accurately, and if they have practically the same weight, the bar is considered uniform. The buttons are again cupelled in sheet lead, with the addition of two-and-a-half times their weight of silver to obtain parting buttons. These are then rolled into a cornet, and parted in nitric acid, and the residual gold weighed. The comparison of the weights is a further check upon the homogeneity of the bar.

When platinum metals are present, the procedure must be modified, as described in Chapter XXIV. The weight of the gold deducted from the weight of the first cupellation button gives the silver content of the lemel bar.

Methods of Assaying Industrial Gold Alloys.—The methods employed for the assay of gold alloys are similar in principle to those universally adopted for the assay of ordinary gold bullion, and involve cupellation to remove the oxidizable metals and inquartation with subsequent parting in nitric acid to separate the gold.

The proportion of copper, however, in industrial gold alloys is greater than in ordinary gold bullion, and on this account more lead has to be employed for cupellation. The amount of lead required is dependent on the following facts :—Copper has a greater affinity for gold than for silver, and it is necessary to use a larger proportion of lead to ensure its oxidation when combined with gold than when in combination with silver. This proportion varies according to the composition of the alloy and the temperature used for cupellation, but experience has shown that for any given standard the amount of lead necessary for gold–copper alloys is nearly twice that required for silver–copper alloys of the same standard.

As the result of repeated experiment it has been found that the maximum quantity of lead required to eliminate copper in the case of gold--copper alloys is about thirty-two times the weight of alloy taken, while for silver–copper alloys about twenty times will suffice. It must, however, be pointed out that although the greater part of the copper present in a gold–copper alloy will be eliminated during cupellation by the addition of a suitable quantity of lead, the resulting gold button in all cases obstinately retains a small quantity of copper which cannot be entirely eliminated even by a second cupellation with a fresh quantity of lead.

The separation of copper from gold is also greatly facilitated by the presence of silver, and as this metal is usually present in industrial alloys or is added for parting purposes, it has to be taken into account when making up the cupellation charge.

It appears from the work of Sir T. K. Rose[1] that silver acts as a carrier of oxygen to base metals, and that the oxidation of silver does not take place to any extent until the base metals have been almost entirely removed.

From these remarks it will be evident that the quantity of cupellation lead is largely regulated by the amount of copper and of silver alloyed with the gold. If insufficient lead is used the copper will not be completely removed, and if too much lead is employed the cupellation losses will be excessive.

It is the coexistence of these two facts that makes it somewhat difficult to decide upon the exact quantity of lead to use, especially when the approximate composition of the alloy is unknown, and it is not surprising to find a want of agreement on this point amongst different authorities, as Table XLV shows.

For alloys in which the alloying metal is copper only, the proportions recommended by D'Arcet, Kandelhardt, and Cumenge and Fuchs respectively, are as follows :—

TABLE XLV.—PROPORTIONS OF LEAD FOR CUPELLATION OF GOLD–COPPER ALLOYS.

Gold in 1000 parts.	Amount of Lead employed for one part of Alloy.		
	D'Arcet.[2]	Cumenge and Fuchs.[3]	Kandelhardt.[4]
1000	1	1	8
900	10	14	16
800	16	20	20
700	22	24	24
600	24	28	24
500	26	32–34	28
400	34	32–34	28
300	34	32–34	32
200	34	32–34	32
100	34	30	32
50	34	28	32
0	...	11	32

Kandelhardt's table is modified to make it uniform with the others. As Rose has remarked, " It is difficult to understand the small quantity of lead assigned by Cumenge and Fuchs to the alloys containing less than 200 parts

[1] *Trans. Inst. M. M.*, 1905, **14**, 441.
[2] Pelouze and Frémy, *Traité de Chimie Générale.*
[3] *Encyclopædie Chimique*, vol. III, *L'Or*, p. 154.
[4] *Gold-Probirverfahren*, Berlin, p. 3.

of gold per 1000." The majority of the industrial gold alloys are ternary alloys of gold–silver–copper, and for the cupellation of these less lead than the quantities just given is required.

As the result of experiment on the cupellation of industrial gold alloys of different compositions, Riche suggested that the most suitable proportions of lead would be the means of the quantities recommended by D'Arcet for the cupellation of gold–copper and of silver–copper alloys respectively. The quantities obtained in this way are given in Table XLVI, which is compiled from D'Arcet's well-known tables.

TABLE XLVI.—Lead for Cupellation of Gold–Silver–Copper Alloys. (*Suggested by Riche.*)

Standard of Alloy.	Lead employed for 1 part of Alloy.		
Gold in 1000 parts.	D'Arcet's quantities for gold–copper alloys.	D'Arcet's quantities for silver–copper alloys.	Mean quantities for gold–silver–copper alloys.
1000	1	0·3	0·65
950	...	3	3
900	10	7	8·5
800	16	10	13
700	22	12	17
600	24	14	19
500	26	16–17	21·5
400	34	16–17	25·5
300	34	16–17	25·5
200	34	16–17	25·5
100	34	16–17	25·5

The quantities recommended by Riche agree fairly closely with those in general use, as will be seen from Table XLVII, which gives the quantities of lead for the cupellation of gold alloys of the legal standards recommended by different assayers experienced in assaying these alloys. The table is compiled from notes kindly communicated to the author at various times. Riche's quantities are inserted for comparison, and are modified to make them uniform with the others.

TABLE XLVII.—Quantities of Lead used by different Assayers for the Cupellation of Legal Standard Gold Alloys.

Standard.		Amount of Lead employed for 1 part of Alloy.							
Gold in parts per 1000.	Carats.	Riche.	Assayer A.	Assayer B.	Assayer C.	Assayer D.	Assayer E.	Assayer F.	Assayer G.
916·6	22	8	8·3	14·4	13·0	12	8	6	16
750·0	18	15	16·6	14·4	15·6	16	15	8	16
625·0	15	18	16·6	16·8	22·4	22	18	22	24
500·0	12	21	16·6	16·8	22·4	22	20	28	30
375·0	9	25	25·0	16·8	24·0	24	24	38	30

The author prefers the quantities given in column E, which are very similar to those recommended by Riche.

As an interesting exception to the above quantities, it may be mentioned that W. F. Lowe, of the Chester Assay Office, established the practice of using the same weight of lead for alloys of all standards, viz. about twenty times the weight of the sample taken. The method of adding the lead varied with individual assayers, but when using the larger quantities necessary for low-standard alloys, it was not usual to add all the lead in one charge. The general practice would be to wrap up the assay-piece and the necessary parting silver in a piece of lead-foil of convenient size, and to " ball up " separately the remainder of the lead required. In charging the assays into the furnace, the lead containing the assay-piece was charged either before or after the ball of extra lead, the second lot being charged as soon as the first lot had melted.

When a furnace load consists of assays of alloys of different standards, the charging in of the different quantities of lead should be so arranged that all the cupellations will finish at the same time. It was with this object in view that the same weight of lead was used at the Chester Assay Office for all standards as above-mentioned.

Well-formed, clean, and bright buttons may be obtained with less lead when the latter is added in two portions. In this case the assay-piece and parting silver is usually wrapped in two-thirds of the total lead and charged in first, and when the cupellation is nearly finished, the remainder is added in the form of a bullet. This practice was adopted for the three lower standards by the assayer B (Table XLVII), hence the smaller quantities of lead given in the table for these alloys.

The practice has also been adopted at the Royal Mint for the assay of industrial gold alloys of low fineness. In this connection, Rose[1] has pointed out that, in making an assay of a gold alloy for the determination of the gold only, the elimination of the whole of the copper is not necessary. The buttons for parting must be malleable, and should be clean and bright in order to make sure that no insoluble foreign material is left adhering to them. It is, therefore, necessary to remove most of the copper by cupellation, as otherwise the buttons blacken on cooling. The exact proportion of copper which can be left in the button without disadvantage has not been determined, and is not of practical importance, since if the button is clean, the amount of copper it contains is immaterial.

The removal of copper entails an increase in the loss of gold by cupel absorption, and therefore in making an assay for gold only it is advantageous to eliminate as little as possible of the copper.

In considering the removal of copper it is easy to exaggerate the importance of the precise amount of lead to be used. In practice any quantity less than about eight times the weight of the gold alloy is inconvenient, because of the risk of small globules being " hung up " on the cupel and separated from the remainder of the assay-piece. The upper limit of the quantity of lead is also largely a matter of individual choice and convenience, influenced by the length of time in working off, the size of the cupel, etc. An increase in the amount of lead, within moderate limits, has very little effect on the loss of gold ; thus, Rose compared the effects of 4 parts, 8 parts, and 12 parts of lead on 22-carat gold, and found the results after parting to be identical.

According to Sir T. K. Rose, there is little advantage in increasing the lead beyond about twenty times the weight of the alloy (Lowe's choice), because in a diluted mixture of lead and copper very little copper is oxidized. In fact, most of the copper, as previously pointed out on p. 342, is eliminated during the last stage of cupellation, just before the lead is worked off.

[1] *J. Inst. of Metals*, 1910, **3**, 130.

It was due to these considerations that the method of using successive doses of lead was developed at the Royal Mint. The quantities of lead used there are as follows :—

TABLE XLVIII.—QUANTITIES OF LEAD USED AT THE ROYAL MINT FOR CUPELLATION OF GOLD ALLOYS.

Standard.	Amount of Lead used for 1 part of Alloy.		
	Lead for 1st Charge.	Lead for 2nd Charge.	Total Lead.
916·6	8·0	...	8·0
750·0	8·0	...	8·0
625·0	8·0	4·0	12·0
500·0	8·0	8·0	16·0
375·0	8·0	8·0	16·0

The approximate surcharges by this method are those given in Table L. The quantities of lead given are said to be convenient, and to give concordant results. The absorption of gold by the cupel is greater by this method, but it is not excessive when precautions are taken to avoid too hot a fire : the buttons are always bright, well-formed, and clean.

The exact method of procedure adopted in assaying industrial gold alloys varies according to requirements. For many commercial purposes an exact determination of the gold is all that is required ; in other cases a determination of the silver is also required, or the complete composition of the alloy may be desired.

Assay for Gold only.—For the determination of gold only, from 0·25 to 0·50 gramme (500 to 1000 millièmes gold assay weights) of the alloy is weighed and cupelled, with the addition of silver equal to two-and-a-half times the weight of the gold assumed to be present, and sufficient lead to remove the copper and zinc if present. The resulting gold–silver button is flattened and parted, and the gold cornet weighed in the ordinary way. For the cupellation the quantities of lead given in Table XLVII are used.

As described on p. 285, certain losses and gains (the sum of which is called the *surcharge*) take place during the cupellation and parting of gold alloys, the losses being due to (1) absorption by the cupel, (2) volatilization, and (3) solution in the acid, and the gains to the retention of silver by the gold cornet. The amount of the surcharge varies with the conditions of working and the composition of the alloy, the general experience being that the losses under all three heads are greater as the percentage of copper in the alloy increases.

With regard to the losses in cupellation, it has been proved experimentally that the gold lost by volatilization is inconsiderable compared with the absorption by the cupel, although it would appear that the loss by volatilization, though small, is proportionally greater with an increase in the temperature of cupellation.

Much of the experimental work done in connection with the loss of gold in cupellation relates to gold alloyed with either silver or copper, and the results are not therefore altogether applicable to ternary alloys of gold, silver, copper, of which most of the industrial alloys consist.

A comprehensive series of experiments made by A. Riche, formerly Director of the Mint Assay-office in Paris, and recorded in his well-known book *L'Art de l'Essayeur* (Paris, 1888), appears to be the first record of experimental work relating to the assaying of industrial gold alloys. The results obtained by Riche were subsequently confirmed by the author.[1]

The loss of gold in cupellation is considerably influenced by the presence of copper, experience showing that an increase in the percentage of copper results in an increase in the loss of gold. According to Rose, the loss of gold is about 0·3 or 0·4 per 1000 when the copper in the gold does not exceed 10 per cent., and more if larger quantities of copper are present. The loss on the cupel is also dependent on the quantity of gold in the assay piece. In ordinary assays of gold alloys it is usual to take a constant weight of the alloy for all standards; hence, as pointed out by Beringer,[2] it will be obvious that as the weight of copper in the cupel charge increases, the weight of gold decreases. The silver, on the other hand, is always approximately two-and-a-half times as much as the gold, whatever its quantity may be. But the cupellation loss, as previously pointed out, is smaller with less gold and greater with more copper, and it so happens that in assaying these alloys these two opposites may, under suitable conditions, nearly neutralize one another.

In this connection some interesting results were obtained by W. F. Lowe[3] of the Chester Assay Office. He found that the gold recoverable from the cupels on which 20 grains of gold alloy had been assayed was almost identical, although the standards of the alloys varied from 916·6 to 375·0. The results are given in Table XLIX :—

TABLE XLIX.—Gold absorbed by Cupel during Cupellation of Gold Assays. (*W. F. Lowe.*)

Standard.	Actual weight of Gold in 20 grains of Alloy cupelled.	Loss of Gold per 1000 of actual Gold present.	Weight of Gold obtained from 1 cupel (= 4 Assays of 5 grains each).	Absorption in part per 1000.
	grains.		grains.	
916·6	18·3	0·765	0·014	0·70
750·0	15·0	0·966	{ 0·014 0·015	0·70 0·75
625·0	12·5	1·160	{ 0·014 0·015	0·70 0·75
375·0	7·5	1·866	{ 0·014 0·014	0·70 0·70

Lowe remarked " that the above were all cupelled on different days, so that the temperature of the furnace would not account for this regularity; but as the amount of lead used is the same (about twenty times the weight of the sample taken) for all standards, this most probably is the cause of the regularity." The heavier losses shown in the third column, added by the author, are mainly due to the increase in the proportion of copper in the separate alloys. By altering the weight of lead, or in any other manner modifying the working conditions, other figures would of course be obtained.

In practice it is more usual, as already explained, to vary the lead according to the fineness of the samples, and under these conditions the general experience appears to be that the loss due to absorption into the cupel is greater with alloys of low standard than with those of high standard.

[1] " The Assay of Industrial Gold Alloys," *J. Inst. of Metals*, 1910, **3**, 98–137.

[2] *A Text-book of Assaying*, 15th edn., p. 145.

[3] " Assaying and Hall-marking at the Chester Assay Office," *J. Soc. Chem. Ind.*, Sept., 1889.

In addition to the cupel losses there is the loss of gold by solution in acid during parting, but experience proves that this loss is comparatively small.

The total losses may or may not be counterbalanced by the silver retained by the gold after parting, which amounts to about 1 part per thousand under normal conditions of working. In practice it is found that with alloys rich in gold the silver retained by the cornet more than compensates for the various losses, and the surcharge is positive; but with alloys of low standard the losses are greater and the surcharge is negative, the cupellation loss being usually greater on account of the larger proportion of copper present, as previously pointed out.

Examples of the surcharge obtained in the assay of gold alloys of various standards by different assayers are given in the following Table L. The figures represent only the relative surcharges; the absolute amounts vary with the treatment, but when a very systematic and uniform method of working is adopted the variations are very slight.

TABLE L.—Surcharge per 1000 on Assays of Gold Alloys obtained by Different Assayers.

Standard.	Pelouze and Frémy.[1]	Rose.[2]	Lowe.[3]	Smith.[4]
916·6	...	+ 0·6	+ 0·26	+ 0·1 to + 0·35
900·0	+ 0·25			
800·0	+ 0·50			
750·0	...	+ 0·5	0·0 to − 0·13	0·0 to + 0·4
700·0	0·00			
666·6				
625·0	...	+ 0·4	0·0 to − 0·13	0·0 to + 0·2
600·0	0·00			
546·0				
500·0	− 0·50	0·0	...	− 0·1 to − 0·5
400·0	− 0·50			
375·0	...	− 0·6	− 0·52 to − 1·05	− 0·2 to − 0·6
333·3				
300·0	− 0·50			
200·0	− 0·50			
100·0	− 0·50			

The necessity of using checks and of working all assays under exactly the same conditions in all cases where results of extreme accuracy are desired, is well known to all bullion assayers, and cannot be too strongly emphasized in dealing with the assay of gold alloys. The more closely a routine is adhered to, the more uniform will the surcharge remain from day to day.

To make the alloys absolutely comparable, it is essential, especially when the alloys contain much copper, that the composition of the check should approximately correspond to the composition of the alloy being assayed.

When the approximate composition of an alloy is not known, a preliminary assay must be made to determine this, as already described. There is, however, considerable variation in the composition of the alloys of low standard; and when a large number of different alloys of any particular standard have to be assayed it would be extremely laborious to have to make up checks to correspond in composition to each alloy. It is the usual practice, therefore, in such cases to make up the checks to conform to an alloy of average composition. Typical

[1] "Experiments on Synthetic Alloys of Gold and Copper," *Traité de Chimie*, 3rd edn., **3**, 1230.

[2] Compiled from results obtained at Royal Mint, *J. Inst. of Metals*, 1910, **3**, 131.

[3] Results obtained at Chester Assay Office, *J. Inst. of Metals*, 1910, **3**, 128.

[4] Compiled from a number of results obtained by the author.

compositions of industrial gold alloys are given later in Table XLIII, and will serve as a guide in preparing suitable checks. In each case silver is added to make the total weight of silver equal to two and a half times that of the gold.

In practice it is usual (except in the case of alloys containing a large proportion of silver) to disregard any silver that may be present in the alloy being assayed, and to add the same total weight of silver to both checks and samples.

In cases where copper only is present as an alloying metal, and extreme precision is not required, cupellation alone is necessary to determine the gold, and the quantities of lead given in Table XLV are used for the purpose. The principal difficulty which the direct cupellation of such an alloy presents, consists in the fact that some gold is absorbed by the cupel when the temperature is high, and that it is impossible, as stated above, to completely remove the copper and lead when the temperature is low.

Inquartation with Copper.[1]—In this method, which was introduced at the Birmingham Assay Office for the assay of gold wares, cupellation is omitted, and copper substituted for silver for inquartation and parting. A weight of about 0·3 gramme is a convenient quantity to take for assay. Sufficient copper is added to the sample to give, together with the alloying metals in the sample, a total amount equal to 2–2¼ times the weight of gold present, and the whole is melted in a small clay cup placed in a silica tube through which a current of steam is passed to prevent oxidation. Heating is continued for five minutes after the copper has melted. After cooling, the button is hammered and rolled into a fillet and parted as usual.

For carrying out a large number of assays simultaneously by this method, the separate clay cups are replaced by ten cup-like cavities formed in a clay rod 9 inches long, and semicircular in section for convenience in sliding it into the silica tube for melting.

The silica tubes are 1⅛-inch bore and 3 ft. long, and five or six are arranged in an ordinary gas muffle furnace from which the muffle has been removed. Half of the length of each tube projects in front of the furnace, and the molten assays are drawn into the unheated half to set and cool. Proof assays are necessary in each batch of assays for accurate work.

Experience shows that there is no appreciable loss of gold by this method: and the surcharge is always "plus" with any carat alloy assayed.

The gold cornets after annealing are duller in appearance than those resulting from the usual silver inquartation.

Determination of Gold and Silver in Industrial Alloys (see also separate assays above).—Silver in gold alloys is usually determined by difference, the gold and silver being weighed together after cupellation, and the gold determined separately in the same or another portion of alloy.

When one assay-piece only is used for the determination of both gold and silver, from 0·2 to 0·5 gramme of the alloy is cupelled with enough lead to remove all the copper, but *without* the addition of silver for parting purposes. In this respect it differs from the method described above, in which the copper is eliminated and the silver at the same time incorporated with the gold by a single cupellation with lead.

The gold–silver button resulting from the cupellation is carefully weighed and then subjected to inquartation and parting in the ordinary way. By subtracting the weight of the gold cornet from the weight of the gold–silver button, the proportion of silver is at once ascertained.

To obtain reliable results by this method considerable practice is required, and special attention must be given to the proportion of lead employed and to

[1] A. Westwood, "The Assay of Gold Bullion," *J. Inst. Metals*, 1922, (1), **27**.

the temperature of cupellation. These must necessarily to a large extent be left to the judgement and experience of the assayer, based on a preliminary examination of the alloy.

The quantities of lead given in Table XLVII are representative of those in general use for the cupellation designed for the removal of copper, but some assayers prefer to use the quantities given for gold–copper alloys in Table XLV, or even larger quantities. Experience has shown, however, that the copper can be more completely separated from gold alloys and less silver lost by cupelling at a slightly higher temperature with a small quantity of lead than by employing more lead and working at a lower temperature. As the loss in cupellation is largely dependent on the amount of gold and copper present, the temperature must be varied in accordance with the composition of the alloy in order that the loss may be minimized as much as possible.

In cases where gold predominates, finished cupellation should be at a higher temperature than in the case where the quantity of gold is small. For alloys of low standard, the temperature at the beginning should be very little above that ordinarily employed for silver assays, and should be increased gradually until at the end it approaches more to that employed for the cupellation of gold. Assayers who use the larger quantities of lead invariably cupel at the lowest possible temperature, *i.e.* with formation of feathers. Gas muffles are undoubtedly the best form of furnace to use for these assays, as the temperature is far more readily controlled in these than in coke-fired furnaces.

The effect of varying quantities of lead in cupellation has been ascertained by Riche in the series of experiments already referred to. The following table gives the results of his experiments on 0·5-gramme synthetic alloys of gold–silver–copper cupelled in the quantities of lead recommended by D'Arcet for gold–copper and gold–silver alloys respectively (see Table XLVI). The results have been tabulated by the author for convenience of reference :—

TABLE LI.—SURCHARGE ON CUPELLATION OF GOLD–SILVER–COPPER ALLOYS IN VARYING LEAD. (*Riche.*)

Composition in parts per 1000.			Amount of Lead used for 1 part of Alloy.	Cupellation Surcharge. Parts per 1000.	Amount of Lead used for 1 part of Alloy.	Cupellation Surcharge. Parts per 1000.
Gold.	Silver.	Copper.				
900	50	50	10	+ 5·0 to + 8·0	7	+ 12·0 to + 15·0
900	25	75	10		7	
800	150	50	16	No surcharge	10	+ 8·0 to + 10·0
800	100	100	16	Slight surcharge	10	
800	50	150	16		10	
700	200	100	22	Very slight surcharge	12	+ 5·0 to 8·0
700	100	200	22		12	
700	50	250	22		12	
600	300	100	24	− 2·0 to − 5·0	14	No surcharge
600	200	200	24		14	+ 4·0 to + 9·0
600	100	300	24		14	
600	50	350	24		14	
500	400	100	26	− 3·0 to − 10·0	16–17	Assays approach exactness perhaps a little high
500	300	200	26		16–17	
500	200	300	26		16–17	
500	100	400	26		16–17	
500	50	450	26		16–17	
400	600		34	...	16–17	Exact
300 and less	700		34	Exceeds − 10·0	16–17	Exceeds − 10·0

The author has repeated many of Riche's experiments, and confirmed his conclusions that the surcharge resulting from the cupellation may vary considerably according to the amount of lead employed and the temperature of cupellation: the latter is the more important factor. The "surcharge" in this case refers to cupellation losses or gains only, and does not include silver retained in "parting."

The method previously mentioned of using less lead for the cupellation and of adding it in two separate quantities (viz. adding one-third after the other two-thirds have worked off) may also be adopted with advantage for direct cupellation without the addition of parting silver. The resulting gold–silver buttons are always much brighter and more satisfactory, and, provided the quantity of lead and the cupellation temperature are carefully regulated, the loss of gold and silver on the cupel, although generally a little greater than when the lead is added in one charge, is not excessive.

It would appear from tests made by the author that when a suitable quantity of lead is employed the copper is more satisfactorily eliminated with two successive charges than with one charge. With a little experience the best quantity of lead for two charges for any particular alloy is readily ascertained and undue cupellation loss avoided. The temperature of cupellation should be kept low until the first charge has nearly but not completely worked off, and then raised during the working off of the second charge.

To ascertain the percentage of silver in the alloy it is necessary to determine the gold, and, if this has not been done in a separate assay-piece, the gold–silver button from the cupellation is weighed, inquarted with silver, and parted. The amount of silver to be added for inquartation can be ascertained from the preliminary assay, or by touchstone, but an experienced assayer can frequently judge the quantity required with sufficient accuracy from the colour of the gold–silver button. Buttons containing from 20 to 40 per cent. of silver have a greenish-yellow tint, but, when the silver amounts to 50 per cent., the yellow tint is scarcely perceptible. With more than 50 per cent. of silver the buttons are silver-white.

The inquartation is best effected by cupellation with the smallest convenient quantity of lead. The practice of alloying the silver by means of the blowpipe is to be deprecated.

The results obtained for gold by this method are approximately accurate when due care is exercised, but they are invariably lower than those obtained for the gold when determined in a separate assay-piece by direct cupellation with parting silver, as previously described. The difference is, no doubt, partly accounted for by the fact that, in the latter case, the presence of the parting silver facilitates the elimination of the copper, and also helps to minimize the loss of gold.

The actual differences between the two methods obtained with gold alloys of different standards is largely dependent on the amount of copper present. The author found that the losses by the double cupellation method exceed the losses by direct cupellation by about -2 parts per 1000 for alloys 375 fine, and about $-0{\cdot}5$ per 1000 for alloys 917 fine.

In the hands of an inexperienced assayer, the results obtained for the determination of the gold and silver by the double cupellation method may be far from accurate, and, since the losses and gains previously detailed are dependent on so many conditions, it is always necessary, in cases where extreme accuracy is desired, to use checks. This method of double cupellation, or "parting assay," in which the gold and silver are determined in the same assay-piece, is largely used by trade assayers for the valuation of lemel bars, etc., and in the hands of experienced assayers who have ascertained the best working conditions for the particular class of alloys with which they have to deal, the results are sufficiently accurate for commercial purposes.

It is not the usual practice of trade assayers to use checks except in special cases, but it may be remarked that the increasing demand for greater accuracy in trade assays is causing checks to be more frequently used than was formerly the case. For this purpose a rolled sheet of metal of the desired composition is usually assayed and reserved for check or proof purposes. The results recorded above show to what extent the method is reliable when checks are not used. In making assays by this method, it is well to remember that it is the gold content of the material which is the more important, because of its value, and the working conditions should be such that no undue loss of gold is occasioned.

As uniformity of temperature is very necessary with this method, a few assays only should be cupelled at the same time in order to avoid extreme variations of temperature, and the batch of cupels should be surrounded by empty cupels.

When the exact determination of the gold and silver in an alloy is required, the author prefers to determine the gold by direct cupellation with parting silver, and to determine the silver in a separate assay-piece by direct cupellation with lead only, checks being used in both cases.

Another method sometimes recommended for the determination of silver in gold alloys, consists in adding parting silver and cupelling as already described, and then deducting the weight of the parting silver from the weight of the gold-silver button resulting from the cupellation. In this case the silver added for parting must be accurately weighed, but, unless checks are worked off at the same time as the alloys, the results are of little value, except, perhaps, as a guide to the approximate composition of the alloy. If checks are not used, the whole of the cupellation loss will be thrown on to the silver in the alloy.

This will be evident from the following figures, which give the results of an assay on a 0·5-gramme sample of gold bar from clean lemel :—

Weight of gold and silver button from cupellation	0·7660 gramme.
Less parting silver added	0·4750
Gold and silver in alloy	0·2910
Gold after parting	0·1896
Difference = silver in alloy	0·1014 = 202·8 per 1000.

The total loss on the check cupelled at the same time was ·0065 gramme, and if this loss is added to the above result, it will make the silver in the alloy 0·1079 gramme or 215·8 part per 1000, instead of 0·1014 gramme or 202·8 parts per 1000.

Experiments conducted by the author show the average loss on alloys of different standards cupelled with parting silver direct to be as follows :—

Standard of Alloy.	Loss in parts per 1000.
916·6	26·1
750·0	23·0
625·0	22·5
500·0	20·3
375·0	19·5

These figures represent the relative losses only ; the actual losses will vary with the composition of the alloys, the proportion of lead, and temperature of cupellation.

Assay to determine Composition of Gold–Silver-Copper Alloy.—When the composition of an alloy is desired, the gold and silver are determined by the methods described, and the copper determined by difference. The following is an example.

An ingot resulting from the melting scrap gave—

	Parts per 1000.
Gold and silver ...	683·8
Gold	503·7
Difference = silver	180·1

1000 — 638·8 = 316·2 copper.

Composition.	
Gold	503·7
Silver	180·1
Copper (by difference)	316·2
	1000·0

If zinc enters into the composition of an alloy, it will be detected on the cupel as a scoria during the first stages of the cupellation. When it is required to know the amount present, a weighed quantity of the alloy is treated with aqua regia and the zinc determined by any of the well-known analytical methods.

Preliminary Assay.—When the composition of an alloy is quite unknown, a preliminary assay must be made to ascertain this approximately in order to determine the correct quantities of silver, copper and lead to be added.

It very frequently happens, however, that the standard of the alloy is stated when the sample is sent for assay, and even when this is not the case a knowledge of the source of the metal will often be a sufficient guide for adjusting the lead, etc.

An experienced assayer will frequently be able to decide the approximate composition from the hardness and appearance of the metal and the effect of the action of nitric acid or other "test" acid. The colour alone is not a sufficient indication of the standard, as alloys of low standard are made to resemble alloys of higher standard by the careful regulation of the proportion of silver and of copper added to the gold. The "test" acid used by the author is that recommended by Wigley,[1] and consists of nitric acid, 8 parts; hydrochloric acid, 1 part; and water, 6 parts. The ordinary nitric acid test only distinguishes gold which is below 9-carat quality: but the mixture given will differentiate gold having a standard below 18-carat. For instance, nitric acid has no effect on 15- or 18-carat gold, but with the mixed acids the 15-carat will assume a dark or nearly black colour, whilst the 18-carat gold will be unaltered; and 9-carat gold will become green like brass under the influence of the acid. Each quality of gold below 18-carat will show a dark colour increasing in intensity according to the standard.

The testing is made by moistening the stopper with the acid and applying it to a polished surface free from grease: the result is evident in the course of a few seconds. If there is a doubt whether the article is plated or well gilt, the surface is scraped or filed, and the acid again applied.

After considerable experience it is possible to determine, with a fair degree of accuracy, the standard of any given alloy, but it may be remarked that the action of the acid is not quite the same in every case with alloys of the same standard but with different ratios of copper and silver. For example, an alloy of 375 standard containing a large percentage of copper is more readily acted upon than an alloy of the same standard in which the alloying metal is mainly silver. If these tests are not sufficient to indicate the quality of the gold, it may be determined by the touchstone or by cupellation. In the latter case about 0·25 gramme of the alloy is cupelled with three times its weight of silver

[1] Wigley, *Art of Goldsmith and Jeweller.*

and from 15 to 30 times its weight of lead, and the resulting button thus inquarted with silver is parted. Should the cupellation button be unsatisfactory, it must be recupelled with more lead.

It is the practice of some assayers to make the preliminary assay serve the purpose of an assay for the determination of the silver, and then to make a separate assay for gold based on the information obtained.

Reporting.—The results of assays of gold alloys are reported either decimally in the ordinary way or by the " carat " system, according to requirements. The former method is now being more generally adopted, but the older " carat " system is still in use, as the generality of workers and dealers in gold and silver alloys are more conversant with this system of computation than with the decimal equivalent; it is, therefore, the usual practice of trade assayers in reporting the proportion of gold in jewellery alloys to refer always to the legal gold standard of 22-carat (916·6). These reports are readily understood when it is remembered that " standard gold " (916·6) contains in 24 parts, 22 of gold and 2 of other metal; these parts, as is well known, are called carats, the gold carat containing 4 grains and the grain being subdivided into 8 parts. In reporting how much gold in carats, grains, and " eighths " is present in an alloy above or below the legal " standard," the conventional terms " betterness " and " worseness " are used, the metal being reported as B^r (abbreviation for better) or W° (abbreviation for worse).

For example, a sample of 18-carat gold would be reported as 4 carats worse, the report reading thus :—

Gold Report.

	Carats.	Grains.
W°	4	...
Standard ... ... 22 carats.		

" Parting " assays, in which both the gold and silver are determined, are reported in ounces of fine gold and fine silver in one pound troy, the abbreviations F.G (fine gold) and F.S (fine silver) being very frequently employed, thus :—

Parting Assay Report.

	Ozs.	Dwt.	Grs.
F.G	8	10	6
F.S	2	7	0

In pound troy.

The remaining 1 oz. 2 dwt. 18 grs. would be base metal, probably copper. The result is also sometimes reported in pennyweight per ounce.

The less cumbrous method of showing how many parts of gold or silver are contained in 1000 parts of the alloy is now universally adopted for all bullion assays, and will no doubt in time replace the above methods for jewellers' and goldsmiths' reports.

Remedy or Tolerance of Fineness.—It is not possible to make all standard alloys of exactly the legal standard, and the maximum difference or variation in fineness above or below the legal standard allowed by law is called the " remedy."

In England the remedy allowed in the case of the gold coinage (916·6 fine) is 2 parts per mille[1] (*i.e.* 2 parts per 1000), and for silver coinage (925 fine) 4 per mille.

The remedy of fineness allowed in the countries which have adopted the standard 900 fine for both the gold and the silver coinage, is in most cases 1 part

[1] *Annual Rep. Roy. Mint* (London), 1907, **38**, 49.

per mille, which is only one-half of that allowed on the alloys of higher fineness used in England.

"Test Acid" for distinguishing Gold Alloys.—Gold and its alloys may be distinguished from base metals of a golden colour, such as brass, etc., by testing the surface with nitric acid, or a mixture of acids, to which the term "test acid" is applied. The action of the test acid produces a green stain on base-metal alloys containing copper, whilst gold alloys are not acted upon, or only show a discoloration which varies in intensity according to the carat quality of the alloy tested. Further details of this test, which is largely used by goldsmiths and jewellers, are given above.

Distinguishing White-Gold Alloys from Platinum.—The white-gold alloys introduced as substitutes for platinum may be distinguished by application of the test acid given in Chapter XXIII.

Distinguishing Test for Electro-gilded and Rolled-gold Articles.[1]—As a preliminary test the touchstone may be used to give some indication of the carat fineness of the gold surface.

The most usual test applied to distinguish between electro-gilt and rolled-gold articles consists in immersing the article in dilute nitric acid (about one part pure acid to three or four parts of water), either cold or only slightly heated to 40° C., the test being best carried out in a small porcelain basin. The acid attack is continued until all the basis metal has been dissolved, leaving a residual film of gold which will vary in thickness and appearance according to the method of manufacture of the imitation gold article. As a general rule an electro-gilt article will leave a film of gold which is bright on both the upper and lower surfaces. At the same time the thickness of the film indicates the quality of the gilt article. In the case of good-quality articles the gold coating is usually coherent and pliable, and retains perfectly the shape of the original article. On the other hand, a very thin gold deposit will simply leave flakes of gold, resembling gold leaf, floating on the surface of the acid.

The same test is applied to rolled-gold articles, but there is a decided difference in the appearance of the gold film that remains from acid treatment. Instead of being bright on both upper and lower surfaces, the upper surface is usually bright, but the lower surface is dull and brown or black in colour, due to the diffusion of a certain amount of base metal during the process of manufacture.

As in the case of gilt articles, the quality of the rolled-gold goods is indicated by the thickness of the residual gold film. (See also "Assay of Rolled Gold" below.)

Microscopic Distinction between Rolled Gold and Electro-plated Gold.[2]—Edges of the articles which have been cut, sawn, or punched when examined under the microscope show the absence of gold on the cut side when the article is made of rolled gold, whereas in plated articles all portions have the same thickness of gold. Soldered places in rolled-gold articles are also free from gold on the surface, but are covered in plated articles. The thickness of gold in pressed parts is the same on the prominences and in the recesses in the case of rolled gold, but is thinner in the recesses in the case of plated gold. In the links of chains, gold fails to plate at the places where two links touch, but in rolled-gold chains there is an equal thickness of gold all round the link.

[1] For further particulars on these tests see *Co-ordinated Report on the Rolled Gold and Gilt Jewellery Investigation*, by E. A. Smith, published by the Birmingham Jewellers' and Silversmiths' Association, 1927.

[2] E. Raub, *Mitt. Forschungsinst. Edelmetalle*, 1932, **6**, 57–63 : Abs. *Met. Abs. Inst. Metals*, 1933, **53**, 188.

To prepare the specimens for examination they should first be plated with copper from a cyanide bath, then from an acid bath until a heavy coating is obtained: the article is then sectioned, polished, and etched in concentrated nitric acid for 1 to 2 seconds.

Assay of Gold Jewellery Alloys, etc.—The composition of these alloys has been discussed on page 333. They are conveniently assayed by the parting methods previously described.

A procedure devised by Beamish [1] and others for the determination of gold in precious-metal alloys by a wet method is as follows:—The alloy is dissolved in aqua regia and the solution evaporated to dryness with about 1 gramme of sodium chloride, the residue being moistened with hydrochloric acid and dried three times to expel nitric acid. The final residue is extracted with 1 : 9 hydrochloric acid, the solution filtered, and any insoluble matter washed with 10 per cent. hydrochloric acid. The gold is precipitated from the boiling solution with 5 per cent. hydroquinone solution added drop by drop, 3 c.c. being required for every 25 mg. of gold. The coagulated precipitate is collected on a close filter-paper, washed with boiling water, and ignited for weighing. Platinum and palladium do not interfere unless the amount present is more than double that of the gold, in which case re-precipitation is necessary. When much platinum or palladium is present it is better to precipitate the gold from a cold solution, allowing two hours for completion of the reaction.

Assay of Nickel White-Gold Alloys.—In these substitutes for platinum the yellow colour of gold is bleached by the addition of nickel. (For palladium white-gold alloys, see page 385.) The alloys are made in 18-carat (75 per cent. gold), 14-carat (58·5 per cent. gold), and 9-carat (37·5 per cent. gold) qualities. The 18-carat alloys contain from 18 to 20 per cent. of nickel, 5 to 7 per cent. of zinc, and the remainder copper. The 9-carat alloys consist essentially of a good white nickel–silver alloy to which the requisite quantity of gold has been added.

The gold in these alloys is determined by the methods described for the assay of gold alloys and bullion, but the intractable nature of nickel in the cupellation makes it necessary to use a larger quantity of lead, the usual quantity being from 20 to 30 times the weight of the assay piece. About one-third of the lead should be added first and when the bulk of this has worked off the remainder is added. This procedure ensures a much more satisfactory oxidation and removal of the nickel. The temperature of cupellation should be well maintained towards the finish, and magnesite cupels should be used. The requisite quantity of silver necessary for parting purposes is added during cupellation. The resulting gold–silver cupel bead is parted in nitric acid, and the residual gold washed, dried and weighed.

So long as the cupellation conditions are arranged to give a satisfactory bead for subsequent treatment, the removal of the whole of the nickel, and other base metals, is more or less immaterial, since these are eventually dissolved out in the nitric acid parting.

An alternative and more satisfactory procedure is to omit cupellation and to alloy the assay-piece direct with excess of silver, copper, cadmium or zinc, in an annealing cup under a cover of borax. The resulting button is then parted in the usual way. Inquartation with copper instead of silver is particularly effective, and may be resorted to when silver is present in the white-gold alloy and has to be determined. The silver passes into solution with the copper during parting and can be separated by precipitation as chloride.

[1] F. E. Beamish, J. J. Russell, and J. Seath, *Ind. and Eng. Chem.*, Anal. Edn., 1937, **9,** (4), 174–176: Abs. *Met. Abs. Inst. Metals*, 1937, **4,** 254.

White-gold alloys of 9-carat quality, containing 37·5 per cent. of gold, are attacked by nitric acid, the nickel, copper and other base metals going partly into solution. The impure gold residue is washed, dried, and wrapped in sheet lead, with the requisite silver for inquartation, and cupelled. The cupel button is then parted in nitric acid, and the residue of pure gold washed, ignited, and weighed.

The solders used in white-gold manufacture consist of alloys of the same metals but in different proportions, to render them more fusible, the gold content being usually about 50 to 60 per cent. in the high-grade solders.

It is advisable to use checks in assaying white-gold alloys by cupellation, especially those of 9-carat quality.

The nickel in these alloys may be determined by precipitation with dimethyl-glyoxime, the characteristic red compound being purified, if necessary, by dissolving the precipitate in hydrochloric acid, and re-precipitating under the usual conditions. The results may be inexact if zinc has been separated previously as zinc sulphide in the presence of $CH_2Cl \cdot CO_2H$. Exact results are obtained if the solution is treated with SO_2, and the precipitation of the nickel complex with dimethyl-glyoxime carried out in aqueous ammonia solution.[1]

Assay of Gilt Articles.—This term is applied to articles of silver or base metal plated with gold by electro-deposition or by mercurial gilding. The latter process is now seldom used. To determine the gold, the whole article, or a weighed portion, is either cupelled direct and the resulting gold bead weighed, or the assay portion is treated with dilute nitric acid as described above and the residual gold washed, dried and weighed, or preferably wrapped in lead and cupelled. From the weight of the gold bead the total amount of gold deposited on the article can be calculated.

Assay of Rolled Gold.—This material,[2] in the form of sheet, wire, and seamless tubing, finds wide application in the production of imitation jewellery and in the optical trade. It consists of a thin layer or veneer of some standard gold alloy which has been soldered or sweated on to a thicker sheet of brass or bronze, or in some cases of nickel silver or sterling silver, as a basis metal, thus making a composite plate, wire or tube for manufacturing purposes.

The carat quality of the gold surface depends on requirements, but usually rolled gold is of 18-carat, 14-carat, or 10-carat quality. Rolled gold with a 22-carat gold surface is made to a limited extent for best-quality umbrella mounts, etc., and 12-carat and 9-carat are used for certain classes of cheaper imitation jewellery. The wire used for spectacle frames is usually of 14-carat, 12-carat or 10-carat quality.

For many purposes the gold plating is on one side of the sheet only, but for watch-cases and other requirements the base metal is plated with gold on both sides. The French term *doublé* better conveys the nature of the material. The Swiss term is *plaque*. Rolled gold is also known under various trade names, such as, gold plate, rolled-gold plate, gold filled, gold shell, etc.

In rolling down the composite bar, the gold follows the reduction of the base-metal backing, so that the original ratio in thickness between the two metals is retained in the finished sheet. This ratio will depend on the nature of the articles to be manufactured from the sheet, and is usually 1 of gold-alloy to 10 or 20 or more of base metal. The composition of the gold alloys

[1] Azzarello *et al.*, *Brit. Chem. Abs.*, 1937, (B), 353.

[2] *Rolled Gold: Its Origin and Development*, Ernest A. Smith, *J. Inst. Metals*, 1931, 175–205.

used for its preparation is essentially the same as that used for jewellers' alloys.

The carat quality of the gold surface, and its thickness in relation to that of the basis metal are frequently stamped on the article. Thus the rolled-gold or gold-filled spectacle frame manufacturers and wholesale distributors place on the market frames bearing a label, stamp or some other mark which indicates to the purchaser the actual amount of gold the frame should contain, *e.g.* $\frac{1}{10}$th 10-ct., $\frac{1}{10}$th 12-ct., $\frac{1}{7}$th 14-ct., etc. These descriptions mean that the gold coating is of the carat quality stated, and forms one-tenth or one-seventh of the material as a whole, the remaining nine-tenths or six-sevenths being base metal or silver. The actual thickness of the alloyed gold surface, however, varies considerably, being extremely thin for cheap, low-grade material, and going up to about 0·003 inch or more in thickness for the better grades.

The method of distinguishing rolled gold articles from electro-gilded wares is described above.

The gold content may be determined by cupelling a weighed portion of the rolled gold direct with a suitable quantity of lead, as described for gold alloys ; or by treating it with dilute nitric acid and then weighing the residual gold, or cupelling it after wrapping in lead-foil. The weight of the resulting gold bead is then compared with the quantity of gold which by calculation should be present in the material as described. Thus for $\frac{1}{10}$th 14-ct., the result should be 58·5 parts of fine gold per 1000, and for $\frac{1}{10}$th 10-ct., 41·7 parts per 1000.

Experience shows that rolled gold frequently assays below the quality marked on it, partly because the solder used in manufacture is usually included in the assay portion.

From an enquiry into the manufacture of gold-filled spectacles it appears that in most cases the quality was much below that specified. The best frame specified as $\frac{1}{10}$th 10-ct. (41·7 thousandths) assayed at 46·0 thousandths (probably actually a $\frac{1}{10}$th 12-ct. frame), and the worst at 18·2 thousandths ; some frames marked $\frac{1}{10}$th 10-ct. were no higher than $\frac{1}{20}$th 10-ct. The average quality was found to be between $\frac{1}{12}$th 10-ct. and $\frac{1}{14}$th 10-ct.[1] As the result of this enquiry it was suggested that the assays of the whole frame should conform to the specified standard of fine gold per 1000 *excluding solder*, and allowing 5 per cent. tolerance. It would appear that American manufacturers of gold-filled frames are permitted by law a tolerance of 10 per cent. on the assay result. The tolerance is allowed to cover any slight loss of gold during manufacture.

Assay of Gold Solders.—The gold alloys used for soldering purposes consist mainly of gold, silver and copper in varying proportions, according to the fusibility required, with small additions of zinc, and in some cases cadmium or tin. The composition of some typical gold solders is given on page 336. Gold solders are assayed by the methods given for gold alloys used in jewellery manufacture, etc.

Assay of Gold Alloys containing Platinum Metals.—Gold is an important constituent of many industrial platinum alloys, the assay of which is dealt with in Chapter XXIV.

(II) ASSAY OF INDUSTRIAL SILVER ALLOYS.

The silver alloy known as " standard " or " sterling " silver is by far the most important of the industrial alloys of silver. It consists of 92·5 per cent.

[1] *Enquiry into the Manufacture of Gold-filled Spectacles*, The Joint Council of Qualified Opticians, 1932.

of silver and 7·5 of some other metal, usually copper. It is the legal standard for the manufacture of silver ware and, until 1920, was the standard for the silver coinage. The silver coinage alloy in use until 1947, when the silver coinage was temporarily withdrawn, contained silver 50, copper 40, nickel 5, and zinc 5 per cent. For silver wares another legal standard exists, containing 95·83 per cent. of silver, but it is seldom used as it is softer and less durable than sterling silver. It is commonly known as "Britannia Silver" since silver wares of this standard are hall-marked with the "figure of a woman commonly called Britannia"[1] instead of the lion passant as in the case of sterling silver.

Although, speaking generally, sterling silver consists of silver alloyed with copper, it must be pointed out that, in the quest in recent years for an alloy of standard fineness that will be more resistant to tarnishing, the copper may be partially or wholly replaced by other metals such as zinc, cadmium, antimony, manganese, and beryllium. This fact should not be overlooked in assaying samples of standard silver.

Silver alloys 900 fine, and also of lower fineness down to 600 fine, are in use to a limited extent for the manufacture of jewellery and articles which are exempt from hall-marking. The silver standards employed for coinage and for plate in other countries are also fixed by law and are given on p. 336, from which it will be seen that the alloy 900 fine is more universally adopted, especially for coinage, than any other silver alloy.

Next in industrial importance are the numerous silver solders used in the manufacture of plate and for soldering and brazing purposes in the engineering and other industries.

Silver Solders.—The silver alloys used as solders for silver wares and for nickel silver in electro-plate manufacture usually consist of alloys of silver, copper, and zinc; but tin and cadmium, and occasionally arsenic and phosphorus, are also used with some solders. The following are typical compositions of solders used by silversmiths[2]:—

TABLE LII.—COMPOSITION OF SILVERSMITHS' SOLDERS.

Description.	Silver.	Copper.	Zinc.	Tin.
	per cent.	per cent.	per cent.	per cent.
Hard Solder	80	13·2	6·8	...
Medium Solder	66·7	23·3	10·0	...
Ordinary Solder for Plate ...	64·5	22·5	13·0	...
Quick-running Solder	62·5	20·87	10·38	6·25
Common quick Solder ...	57·0	27·7	11·53	3·85

The solders used for nickel–silver work that is to be electro-plated usually contain less silver than the above, and where cheapness is an important consideration, alloys containing as little as 10 per cent. of silver may be used.

Alloys of lower silver content constitute the silver brazing alloys which are being increasingly employed in various branches of the engineering industry as they are superior to the ordinary brazing brasses.

[1] Britannia silver must not be confused with the alloy known as "Britannia metal,' consisting of tin, antimony, and copper.

[2] E. A. Smith, "Silver Alloys of Industrial Importance," *Proc. Sheff. Soc. of Eng. and Metallurgists*, Nov., 1901.

The following are the specifications of the British Standards Institution for silver solders for brazing purposes :—

Grade.	Composition per cent.		
	Silver.	Copper.	Zinc.
A	60 to 62	27·5 to 29·5	9 to 11
B	42 to 44	36 to 38	18·5 to 20·5

Grade B is frequently termed " turbine solder " as it is used for brazing turbine blades.

A low-melting-point brazing alloy containing a small percentage of silver, known as " Sil Fos," contains also phosphorous which has a deoxidizing effect when the alloy is melted.

When cadmium is a constituent of silver solders it frequently replaces about one-half of the zinc content.

Sampling Silver Alloys and Wares—*Standard Silver.*—The remarks in the sampling of industrial gold alloys apply equally to industrial silver alloys. As previously stated, more or less segregation always takes place during the solidification of silver–copper alloys, so that it is not possible to obtain ingots absolutely uniform in composition, the middle portion being slightly richer in silver. In preparing standard silver which is subsequently to be manufactured into plate and hall-marked, it is the usual practice of manufacturers to add a small quantity of pure or virgin silver (usually about half a pennyweight to the troy pound) above that required by the legal standard, in order to counteract any irregularities in composition due to segregation, and also because the commercial fine, or virgin silver used for melting is never perfectly pure. The variations in the silver content due to the non-homogeneous character of the alloy are to some extent modified by rolling the bars, but experience shows that the middle of commercial sheets or strips of standard silver (925 fine) often contains at least 0·25 parts per 1000 more silver than the edges. The actual differences are dependent on the width of the strip, but from the results of assays made by the author from time to time, it would appear that in the case of the fillets about 18 inches wide, most generally sold for silversmiths' work, the middle differs from the edges, on the average, by about 0·8 to 1·0 part per 1000. Standard silver ingots or flat bars cast for rolling purposes, etc., are usually sampled by cutting one or two portions from diagonally opposite corners, and flattening them for assay.

Silver ingots of the Continental standard, 835 fine, are also irregular in silver content, owing to segregation. Inverse segregation occurs in the middle porous portion of the upper part of the bar ; the copper-rich eutectic is there sucked outwards by the contraction which occurs in solidification. The bearing of this fact on the sampling and assay of the ingots is discussed by Bauer and Deiss.[1]

The sampling of silver coins also requires careful consideration. The relatively wide variations in fineness which occur in different parts of the same silver coin of " 500 " standard calls for accurate sampling.[2] The British coins

[1] O. Bauer and E. Deiss, *Z. anal. Chemie*, 1933, **92,** 161–164 : *Met. Abs. Inst. Metals*, 1934, (2), **1,** 250.

[2] S. W. Smith, *Annual Rep. Roy. Mint* (London), 1934, **65,** 61–62.

of this standard now in use contain, as mentioned above, silver 50, copper 40, nickel 5, and zinc 5 per cent. respectively. Coins of moderate size which have been cut in double rows from the strip are sampled by taking equal portions of sectors from opposite sides, the larger coins being first rolled to about half their thickness.

Large coins cut in single rows are sampled by taking equal portions of three small discs at regular intervals at a distance from the edge of the coin equal to one-seventh of the diameter.

Silver Wares.—In sampling silver wares by scraping, the first scrapings should be rejected, as a certain amount of enrichment may have taken place on the surface, due to the " pickling " in dilute sulphuric acid (or " boil ") to which the articles are subjected in the process of manufacture.

In the case of finished wares, the surface may consist of pure silver, as it is a common practice to " finish " silver wares by coating them with a thin layer of silver by electro-deposition, as in electro-plating.

Preliminary Tests for Silver Alloys: Test Acids.—To determine whether an alloy having the appearance of silver, consists of silver or of some white base-metal alloy, a dilute solution of silver nitrate (or sulphate) is invariably employed. The " test acid " is prepared by dissolving 2 grammes of silver nitrate in 50 c.c. of distilled water, or by dissolving 1·27 grammes of pure silver in a small quantity of dilute nitric acid, expelling the excess of acid by evaporation, and diluting to the required volume. A drop of this dilute silver nitrate has no effect on silver alloys above 925 fine, but with alloys below this fineness a brown stain is produced, which increases in intensity as the proportion of base metal in the alloy becomes greater. With base metals, the silver nitrate instantly produces a black stain, due to the precipitation of finely-divided metallic silver by the base metal. The surface of the alloy should be cleaned by scraping before applying the test, so as to remove any silver that may have been electro-deposited upon it.

Another test acid consists of dilute nitric acid (1·2 sp. gr.) in which some potassium chromate is dissolved. This solution gives a red deposit on the surface of silver, which thus distinguishes it from other white metals such as tin.

Methods of testing silver alloys by means of the touchstone and by oxidation are described on pp. 319 and 320.

The oxidation test is very useful in locating the " seams " or joints in silver wares, since the solder used is almost always of lower fineness than the article. For this purpose the article is gently warmed over a bunsen flame in the place where the seam is suspected.

Methods of Assay.—Silver alloys are universally assayed by the method previously described for bullion in Chapter XIX.

The dry or cupellation method is still largely employed by trade assayers, but is gradually being replaced by the wet methods. For the cupellation assay 0·5 gramme of alloy is usually taken and cupelled with the quantities of lead given in Table XXXVIII, page 300. The loss of silver in cupellation varies with the amount of copper present in the alloy and with the temperature employed.

Table LIII gives the average loss of silver which occurs in the cupellation of alloys of silver and copper in different proportions. In all cases where great accuracy is desired, check assays must be used to determine the actual losses, and the assays corrected accordingly.

Standard silver in which the silver is alloyed with zinc, cadmium, or tin may also be assayed by cupellation without increasing the quantity of lead; but to obtain satisfactory results a good supply of air must be allowed to pass through the muffle, and the temperature must not be allowed to get too low. With these precautions the whole of the scoria first formed is entirely removed before the completion of the cupellation and the buttons obtained are quite satisfactory.

TABLE LIII.—AVERAGE CUPELLATION LOSS ON SILVER–COPPER ALLOYS.

Fineness of Alloy.	Silver obtained.	Loss of Silver.
parts per 1000.	parts per 1000.	parts per 1000.
1000	998·97	1·03
975	973·24	1·76
950	947·50	2·50
925	921·75	3·25
900	896·00	4·00
875	870·93	4·07
850	845·85	4·15
825	820·78	4·22
800	795·70	4·30

The wet assay of industrial silver alloys is conducted as described for silver bullion. When the composition of the alloy is entirely unknown, a preliminary cupellation assay is made to determine the approximate fineness.

Reporting.—The results of assays of silver alloys are usually reported in parts per 1000 to the nearest whole number, as in the case of silver bullion; but the old method of reporting trade assays of silver alloys as better or worse than the legal silver standard of 11 ounces 2 dwt. of fine silver per pound troy (*i.e.* 925 fine) is still in use. Thus an alloy 926 fine, which is $\frac{1}{4}$ dwt. above standard, would be reported $\frac{1}{4}$ dwt. better; and an alloy 900 fine, which is 6 dwt. below standard, would be reported 6 dwt. worse.

Assay of Silver Solders.—Silver solders differ considerably in composition, the silver content varying from 10 to 75 per cent. The compositions of typical silver solders are given in Table LII, page 358. Low-grade silver solders may be regarded as ordinary brazing brasses to which silver is added, sometimes less than 10 per cent.

In assaying silver solders the usual determinations are of silver, copper, zinc, and cadmium. In routine analysis, however, it is customary to take zinc by difference. Impurities such as lead, tin, iron and nickel are, if necessary, determined in a separate 5-gramme sample.

The following procedure for the assay of silver solders is recommended by the American Society for Testing Materials :—

After solution of the alloy in nitric acid the silver is precipitated as chloride and either weighed as such or cupelled with lead, and the resulting silver button weighed. The filtrate is evaporated with sulphuric acid, and the copper determined by electrolysis. The cadmium and zinc in the spent electrolyte are separated by three precipitations with hydrogen sulphide, and the dissolution of the cadmium sulphide is effected in nitric acid. The cadmium is then determined by electro-deposition after addition of caustic potash and potassium cyanide, or by evaporation of the solution with sulphuric acid to obtain cadmium sulphate for weighing. The zinc is determined volumetrically by the ferrocyanide method or gravimetrically as zinc oxide.

For the determination of the impurities that may be present the following methods are recommended. Tin is determined as oxide obtained by the treatment of 5 grammes of the alloy with 1 : 4 nitric acid. Lead is determined as lead sulphate by the "lead acid" method. Iron is volumetrically determined with standard potassium permanganate after reduction in the Jones' reductor. Nickel is determined by precipitation with dimethyl-glyoxime.

The precipitation of cadmium as sulphide and its separation from zinc can be effected if the solution is free from chlorine and nitrous oxide and contains 15 c.c. of sulphuric acid per 100 c.c.[1] Passage of hydrogen sulphide through the boiling liquid until it is cold precipitates a granular, readily filtrable cadmium-sulphide free from zinc. The precipitate is collected in a glass filter-crucible, washed, and dissolved in hot HCl. Evaporation of this solution with sulphuric acid affords cadmium sulphate for weighing.

The separation of cadmium from zinc by hydrogen sulphide[2] is best effected in a sulphuric acid solution rather than in a hydrochloric acid solution. For the separation of little cadmium from much zinc the solution should contain 4 c.c. of sulphuric acid (1·84 d.) per 100 c.c.: double precipitation is generally advisable. Determination of cadmium by electrolysis is more convenient than weighing as cadmium sulphate or as cadmium phosphide, $Cd_2P_2O_7$. For 1 to 10 mg. of cadmium, electrolysis is carried out in a feebly alkaline cyanide solution, but for more than 10 mg. of cadmium a slightly acid sulphate solution is better. Copper-plated platinum cathodes are used; after weighing the deposit it can be removed without appreciably affecting the copper by treatment with 2 per cent. nitric acid.

Assay of British Silver Coinage Alloys.—The following procedure is given by S. W. Smith[3] for the assay of the silver coinage alloy now in use, containing silver 50, copper 40, nickel 5, and zinc 5 per cent.

The silver is determined by the Gay-Lussac method. The liquid from this is filtered and boiled and the copper therein determined by the iodide method. The precipitate is converted into cuprous oxide which is filtered through pulp and the nickel determined in an ammoniacal solution with standard potassium cyanide, using silver nitrate as an indicator. The zinc in the residual liquid is not easily determined by the usual methods of precipitation or titration, but is readily precipitated by 8-hydroxy-quinoline in an alkaline solution, with re-solution and re-precipitation if calcium or magnesium be present. The zinc compound is dried at 140° C. and weighed, and contains 18·49 per cent. of zinc.

Rapid determinations are made by volatilizing the zinc from samples diluted with an equal weight of tin, side by side with synthetic check samples.

Assay of Standard and Britannia Silver.—These alloys usually consist of silver and copper, containing 92·5 and 95·83 per cent. of silver respectively. When a standard alloy of special ductility is required the copper is partially or wholly replaced by cadmium. Other metals may (as noted above) be present in special sterling alloys made for tarnish-resisting purposes.

The silver content of these alloys is usually determined by the Gay-Lussac or the Volhard method (Chapter XIX). It is customary to take the copper content by difference, but it may be determined in the solution from the Gay-Lussac method by a coloration test, or determined gravimetrically. If zinc or cadmium are present they may be determined by the procedure described under silver solder assay.

The presence of cadmium in sterling silver may be detected by removing the silver by precipitation as chloride, and adding excess of ammonia to the

[1] C. Zollner, *Z. anal. Chem.*, 1938, **114**, (1/2), 8–15: *Met. Abs. Inst. Met.*, 1939, **6**, 64.

[2] L. Baumeister, *Chemist and Analyst*, 1932, **21**, (3), 7–8: *Met. Abs. Inst. Metals*, 1933, **53**, 35.

[3] *Annual Rep. Roy. Mint* (London), 1928, **59**, 51–52.

filtrate ; a deep-blue coloration denotes the presence of copper. Excess of potassium cyanide solution is then added until the blue colour is discharged, and a current of hydrogen sulphide passed through the solution. A bright-yellow precipitate denotes the presence of cadmium.

Assay of Silver-Lead Alloys.—Alloys of these two metals are now finding industrial application. Low lead–silver alloys containing 3·5 to 5 per cent. of lead are being advocated for bearings ; while the lead–silver eutectic, containing 2·5 per cent. of silver, is used as a low melting-point solder : in some cases 0·25 per cent. of copper is added.

These alloys are best assayed for silver by cupellation as described for the determination of silver in lead bullion.

Alloys of high silver content may be dissolved in nitric acid and the silver determined by a volumetric method after separating the lead as sulphate.

Assay of Silver-Zinc Alloys.—Silver–zinc alloys have long been used by native silversmiths in India for the production of silver wares. This explains the whiter appearance of such wares when compared with the silver–copper alloy in use in this and other countries. The Indian alloy usually contains from 5 to 10 per cent. of zinc.

Alloys of these two metals may be assayed by the usual volumetric methods, or by the gravimetric determination of the silver as chloride, after solution in nitric acid.

Assay of Silver Alloys containing Platinum Metals.—The industrial silver alloys in which the platinum metals are constituents are comparatively few, the more important being the binary alloys of silver–platinum, and silver–palladium, the assay of which is dealt with in Chapter XXIV.

Assay of Silver-Tin Alloys for Dental Amalgams.—These alloys consist essentially of silver and tin to which small amounts of other metals are added. They are usually in the form of filings in order to facilitate mixing with mercury to produce the amalgams for dental purposes. The composition varies from about 50 to 70 per cent. of silver, and 28 to 52 per cent. of tin, with from 3 to 5 per cent. of copper, from 1 to 5 per cent. of gold, and 0·5 to 3 per cent. of zinc. Special alloys may contain up to 0·5 per cent. of platinum.

The best alloys are usually considered to be those which do not contain less than 65 per cent. of silver. The composition of a modern alloy was found to be silver 71·6, tin 25·5, zinc 2·7, copper 0·2 per cent.[1]

Usually it is only a determination of the silver and gold that is required. For this purpose from 0·5 to 1 gramme of the filings are mixed with granulated lead and scorified to remove the tin and other base metals. The resulting lead button is cupelled, the cupel bead parted in nitric acid, and the silver precipitated as chloride, and weighed as such. The residual gold is washed, ignited, and weighed.

The alloy may also be attacked by boiling in sulphuric acid which dissolves the silver and tin, leaving a residue of gold (and platinum). The silver is separated by the addition of hydrochloric acid and the silver chloride precipitate washed, dried, and weighed. The residual gold is dried and weighed.

The copper and zinc in solution are determined by any of the usual methods.

[1] G. C. Paffenbarger, *J. Amer. Dental Assoc.*, 1939, **26,** (6), 990–994.

CHAPTER XXII

THE ASSAY OF PRODUCTS CONTAINING GOLD AND SILVER

INCLUDED in this chapter are descriptions of the methods employed for the determination of gold and silver in some of the more important products and by-products obtained in the metallurgical treatment and manufacture of the precious metals. They are conveniently arranged in alphabetical order under three heads, viz. :—Gold and silver in (i) base metals, (ii) metallurgical products, and (iii) manufacturing products.

(I) Determination of Gold and Silver in Commercial Base Metals.

The chief base metals assayed for gold and silver are lead and copper, constituting base bullion, which has been fully dealt with in Chapter XX. Other commercial base metals, however, frequently contain more or less gold and silver, and the assayer is sometimes called upon to determine these.

In Antimony.—Gold and silver have long been known to accompany antimony ores from Borneo, Australia, Portugal, and other places. Owing to the difficulty of separating the precious metals from antimony ores, the ingots of "star antimony" obtained by smelting the ores not infrequently contain appreciable amounts of gold and silver. Thus the following amounts of gold and silver were found in different samples of commercial metallic antimony assayed by the author[1] :—

TABLE LIV.—GOLD AND SILVER IN COMMERCIAL ANTIMONY.

Details of Sample.	Ounces per ton (2240 lbs.).			
	Gold.		Silver.	
	oz.	dwt.	oz.	dwt.
(1) Commercial "star antimony"	0	19·5	1	15
(2) ,, ,, ,, ...	1	12	0	5
(3) ,, ,, ,, ...	0	4·5	6	4
(4) ,, ,, ,, ...	0	13·5	0	8
(5) Smelted from antimony ore from Yugoslavia ...	0	0·75	0	15
(6) ,, ,, ,, Oporto, Portugal	1	15	1	12

Metallic antimony may be assayed for gold and silver by the following crucible fusion method, which consists in oxidizing part of the antimony with litharge, using the lead thus reduced as a collecting agent, and oxidizing any excess of antimony by means of nitre. The sample should be powdered in

[1] *J. Soc. Chem. Ind.*, 1893, **12**, 316.

an iron mortar, to pass an 80-mesh sieve, and then well mixed. Satisfactory results are obtained with the following charge :—

Powdered antimony	25 grammes.
Litharge or red lead	50 "
Sodium carbonate	25 "
Nitre	10 "

The charge is fused for fifteen or twenty minutes at a dull-red heat, and poured when tranquil fusion has taken place. The lead button, which should be malleable and weigh about 25 to 30 grammes, is cupelled direct, and the resultant gold–silver button parted in the usual way.

In Bismuth.—A considerable portion of the bismuth of commerce is derived from the native metal, which frequently contains gold and silver. The precious metals are also found in most bismuth ores.

Commercial bismuth comes into the market in the form of circular " cakes," and may contain appreciable quantities of gold and silver. Samples of commercial bismuth from different sources, assayed by the author, gave the following precious metal contents [1] :—

TABLE LV.—GOLD AND SILVER IN COMMERCIAL BISMUTH.

Details of Sample.	Ounces per ton (2240 lbs.).	
	Gold.	Silver.
	ozs.	ozs.
(1) " Australian " bismuth	3·59	108·42
(2) " American " bismuth	0·16	23·03
(3) " German " bismuth	0·09	23·82
(4) Commercial bismuth (source unknown) ...	0·59	72·84
(5) Crude bismuth, Bolivia[2]	...	26·13
(6) Refined bismuth, Saxony	...	22·87

Samples of bismuth are prepared for assay by crushing to pass an 80-mesh sieve, and mixing. As bismuth is capable of being cupelled as readily as lead, the assay may be made by cupellation direct ; but experience shows that the cupellation loss is greater than in the case of lead, and the buttons left on the cupel are liable to retain bismuth (see page 155). If cupellation is adopted, the sample need not be finely crushed ; a weighted quantity (say 25 to 30 grammes) is wrapped in a piece of sheet lead, and cupelled direct at a low temperature. To obtain silver buttons free from bismuth, it is advisable to add 1 or 2 grammes of lead when the cupellation of the bismuth is nearly complete. Checks should be used if great accuracy is desired.

It is more satisfactory to oxidize the bismuth by fusion with litharge or nitre, then flux it and collect the precious metals in a button of lead. For this purpose the powdered sample is fused with a charge similar to that given above for metallic antimony. A small quantity of bismuth may be retained by the lead, but the button from the fusion may be cupelled direct without difficulty.

In Copper.—Commercial copper may be assayed for gold and silver by scorification, or by a combined wet and dry method, as previously described under " Base Copper Bullion," p. 330.

[1] *J. Soc. of Chem. Ind.*, 1893, **12**, 316.
[2] Nos. 5 and 6, *Bull. Imp. Inst.*, 1912, **10**, 628–641.

In Commercial Lead.—Commercial, or soft, lead is generally of a high degree of purity, containing over 99·9 per cent. of lead, the chief impurities being copper, antimony, bismuth, iron, silver and zinc. The silver content is usually very low, owing to desilverization during refining, and varies with the source of the metal, and the method of refining. The desilverized lead from Parke's process will contain about 4 to 6 dwt. silver, and 0·6 to 0·7 per cent. zinc per ton.

Analyses of typical brands of refined lead gave the following results for silver content :—

TABLE LVI.—SILVER CONTENT OF COMMERCIAL LEAD.[1]

Source.	Silver per cent.	Ounces per ton of 2240 lbs.
English (ordinary)	0·0003 to 0·0015	0·098 to 0·490
,, (chemical)	0·0002	0·065
Australian (ordinary)	0·0009 to 0·0012	0·294 to 0·392
,, (special)	0·0002	0·065
American (San Francisco) ...	0·001	0·327
Electrolytic lead	0·001 to 0·0025	0·327 to 0·816

The silver in commercial lead is best determined by cupelling from 40 to 50 grammes or more, either direct, or first reducing the weight by scorification and subsequent cupellation. The weight of the lead after scorification should be about 15 grammes.

Whether the silver determination be made by direct cupellation or by scorification and subsequent cupellation, the result in both cases is the same, provided proper attention is given to the temperature of both operations. When the silver content is very small and large quantities of the sample have to be taken, scorification before cupellation is preferable. In view of the small quantity of silver usually present in commercial lead it is necessary to watch the last stage of the cupellation, and to remove the cupel as soon as the last traces of lead have disappeared.

Cupellation is also a good test for the presence of traces of gold. In this case the bead remaining on the cupel must be dissolved in very dilute nitric acid, and any gold residue weighed and the weight deducted from the weight of the cupel bead.

In Commercial Tin.—As previously stated, gold and silver are sometimes found in tin ore, and pass into the bars of tin obtained by smelting the ore. The bars may be sampled either by drilling, and the drillings well mixed and a weighed portion assayed direct, or the whole sample melted under a layer of potassium cyanide. The resultant button is freed from slag by treatment with hot water, and then rolled into a thin strip from which a weighed quantity is then cut into small fragments and fused with the following charge:—

Metallic tin	25	grammes.
Litharge or red lead	50	,,
Sodium carbonate	50	,,
Nitre	10	,,

[1] W. Gowland, *Metallurgy of Non-Ferrous Metals*: J. A. Smyth, *Lead* (Common Commodities and Industries Series).

The fusion of the charge is conducted at a dull-red heat for fifteen to twenty minutes, and poured when tranquil. The lead button obtained should be sufficiently free from tin to permit of direct cupellation; but if it is hard and white, owing to the presence of an appreciable amount of tin, it must be scorified with the addition of its own weight or more of lead prior to cupellation.

The assay may be conducted by scorification, but in this case a smaller quantity must be taken. From 2 to 5 grammes of tin are scorified with sixteen to twenty times its weight of lead and 5 grammes of sodium carbonate to form sodium stannate. The lead button is very liable to retain tin, and may require rescorifying with the addition of lead. As the amount of precious metals present is usually small, it is frequently necessary to scorify five or more separate lots of the sample and to re-scorify the resultant lead buttons together so as to concentrate the gold and silver into one lead button, which is cupelled and parted in the ordinary way.

The determination of silver in argentiferous tin by treatment with hydrochloric acid or nitric acid and solution of the silver is sometimes recommended, but, from the author's experience, the method is not reliable.[1]

In Commercial Zinc.—The following combined wet and dry method is recommended by K. Friedrich[2] for the determination of silver in commercial zinc. It is a combination of Pufahl's and Kerl's methods, the former consisting of solution in hydrochloric acid and fusion of the residue with potassium cyanide, followed by cupellation, and the latter consisting of fusion with excess of lead and borax and subsequent cupellation.

The method is as follows:—The sample is granulated, and a suitable quantity (100 to 1000 grammes, according to the expected silver content) weighed off into a beaker (or beakers in the case of large amounts). A small quantity of hydrochloric acid is added, and when nearly saturated it is poured off through a filter, and a second quantity poured on to the sample; by adopting this plan solution is greatly hastened. As soon as, or just before, solution is complete, the whole residue is rinsed on to the filter and washed free from chlorides. The filter is placed in a crucible, dried and burnt; to the residue, from 7·5 to 15 grammes of assay lead and some borax are added, and the whole heated to fusion. The button of argentiferous lead is then cupelled, and the silver button weighed in the usual way. For accurate results a lead–silver button of about the same silver content should be cupelled at the same time as a check to determine the cupellation loss. The loss is allowed for in calculating the silver content of the zinc.

As the result of experiments on alloys of known composition, Friedrich concluded that—

(1) The cupellation method gives accurate results even when zinc as well as lead is present in the alloy.

(2) If the excess of acid, after dissolving the zinc, be allowed to stand upon the residue, the silver gradually goes into solution: hence the direction to filter off before complete solution of the zinc.

(3) In the presence of hydrochloric acid or zinc chloride, very considerable quantities of silver chloride may be formed and volatilized during the burning of the filter: hence the need for thorough washing of the residue.

In the paper referred to, a table is given showing the amount of silver in grammes per metric ton in various samples of commercial zinc.

This method may be used for the determination of silver in the rich argentiferous zinc resulting from Parke's process for the desilverization of lead;

[1] Consult also Bannister, *Trans. Inst. Min. and Met.*, 1905–6, **15**, 518.

[2] *Zeits. angew. Chem.*, 1904, **17**, 1636–1644.

but in this case a smaller quantity, from 5 to 10 grammes, of material is taken for assay. The amount of silver in the rich argentiferous zinc varies considerably; it may contain from 5 to 10 per cent. of silver or even more, with about 80 per cent. lead and the remainder zinc. In some works this rich material is assayed by scorification, from 3 to 5 grammes being scorified with fifteen times its weight of lead and 1 or 2 grammes of borax. The temperature should be well maintained to produce a fluid slag.

(II) Determination of Gold and Silver in Metallurgical Products.

Under this head are included a number of products and by-products containing gold and silver, resulting from the metallurgical treatment of ores by smelting, refining, etc. The precious-metal content of these materials varies very considerably.

In "Burnt Ore."—"Burnt ore" is the term given to roasted cupriferous iron pyrites from sulphuric-acid manufacture. It usually contains small quantities of gold and silver: burnt ores from Spanish pyrites carry about 0·005 per cent. of silver and a very small quantity of gold. Burnt ores, which consist mainly of ferric oxide, are assayed by crucible fusion.

Beringer recommends the following charge which gives a fluid and satisfactory slag[1]:—

Ore	100 grammes.
Litharge	100 ,,
Charcoal	7 ,,
Borax	50 ,,
Fine sand	50 ,,

Mix; place in a large crucible; cover with salt; and melt down under cover. When fused, insert an iron rod into the molten charge for two minutes, then withdraw it, and pour the charge quickly into a large conical mould. The button of lead should weigh about 50 grammes. Cupel and weigh the silver. The litharge may be replaced by red lead, in which case another gramme of charcoal powder must be added.

The following combined wet and dry method may also be employed:—

"Take 100 grammes of the ore, place in a large beaker of 2½ litres capacity, and cover with 375 c.c. of hydrochloric acid. Boil for half-an-hour, until the oxides are dissolved and the residue looks like sand and pyrites; then add 20 c.c. of nitric acid, and boil till free from nitrous fumes. Dilute to 2 litres with water, and pass a current of sulphuretted hydrogen till the iron is reduced, the copper and silver precipitated, and the liquor smells of the gas. This takes about one hour and a half.

"Filter off the precipitate (rejecting the solution), and wash with warm water. Dry and transfer to an evaporating dish, adding the ashes of the filter-paper. Heat gently with a bunsen burner until the sulphur burns, and then roast until no more sulphurous oxide comes off. When cold, add 30 c.c. of nitric acid, boil, and dilute to 100 c.c. Add 1 c.c. of very dilute hydrochloric acid (1 to 100), stir well, and allow to stand overnight. Decant on a Swedish filter-paper, dry, and burn.

"Mix the ashes with 100 grammes of litharge and 1 gramme of charcoal, and fuse in a small crucible. Detach the button of lead and cupel. Weigh, and make the usual corrections for silver in the litharge used."

In "Concentrates."—The methods of assaying pyritic and other concentrates resulting from the mechanical dressing of ores are the same as those

[1] *Text-book of Assaying*, 15th edn., p. 116.

given for ores containing sulphides (see Chapter XII). They may be assayed for gold and silver, etc., by crucible fusion direct or roasted before fusion.

Charges used for the direct fusion of concentrates obtained on the Rand are given on page 479.

Beringer recommends the following charge for the assay of pyritic concentrates after roasting:[1] take one assay ton (30 grammes) of the concentrates, roast, and fuse the oxidized product with :—

Litharge	45	grammes.
Charcoal	2	,,
Sodium carbonate	30	,,
Borax	30	,,

Fuse, detach the slag, cupel and part in the ordinary way.

In Copper Matte.—Copper regulus or matte consists of copper sulphide, with varying proportions of iron sulphide, obtained in smelting copper ores. The amount of gold and silver in copper mattes varies considerably: a matte of average grade will contain about 20 per cent. of copper and 5 ounces of gold and 40 ounces of silver per ton. High-grade mattes will contain from 50 to 60 per cent. of copper, and mattes high in gold and silver may contain 15 ounces of gold or more and as much as 200 ounces of silver per ton. The gold in copper mattes is generally determined by the scorification or "all-fire" method, as the results are higher than those obtained by combined methods. Silver is sometimes determined by scorification, but more generally by a combined wet and dry method. Crucible fusion methods are also employed.

(*a*) *Scorification Method.*—Ten portions of the sample, each of 0·10 assay ton, are placed in 3-inch scorifiers with 50 grammes of test lead, one-half of which is mixed with the matte and the other half used as a cover. The charge is then covered with 2 grammes of a mixture of equal parts of borax glass, and fine sand. The scorification is conducted at a moderate temperature, and usually occupies from twenty-five to thirty minutes. The resultant buttons, which weigh about 15 grammes, and are hard from the presence of copper, are cleaned from slag and rescorified with the addition of 25 grammes of test lead. The lead buttons obtained weigh from 10 to 12 grammes: they are cupelled separately, and the gold–silver buttons weighed separately, and then together. To determine the gold, the buttons are grouped into two lots of five each, and parted in dilute nitric acid in the ordinary way.

As a correction for the cupellation loss is frequently allowed in practice, the cupels (after removing the unstained portion) are crushed to pass a 100-mesh sieve, and then re-fused in five lots of two each with the following charge[2] :—

Litharge	90	grammes.
Sodium carbonate	50	,,
Borax	50	,,
Argol	3	,,

The lead buttons are cupelled and the gold-silver buttons weighed and parted. The weight of silver and gold thus obtained is added to the weights obtained in the original assay. Sometimes no correction is allowed. It is not usual to clean the slags. In calculating the results, allowance must of course be made for the silver in the test lead used.

(*b*) *Combined Wet and Dry Method.*—Weigh out, in duplicate, 1 assay ton of matte, place in large beakers of a litre capacity, provided with glass covers, and treat with 100 c.c. of distilled water and 50 c.c. concentrated nitric acid.

[1] *Text-book of Assaying*, p. 140.
[2] A. H. Ledoux, *Amer. Inst. Min. Eng.*, Oct., 1894.

Allow the violent action to subside, add 50 c.c. more of concentrated nitric acid; heat until red fumes cease to come off, dilute to 500 c.c., and add 10 c.c. of a concentrated solution of lead acetate, 5 c.c. of concentrated sulphuric acid, and 10 c.c. or more of salt solution, stir briskly, and let it stand overnight to allow the lead sulphate to settle. The strength of salt solution frequently recommended contains 0·542 gramme of NaCl per 1000 c.c. (1 c.c. = 1 milligramme Ag), and an excess of 4 to 8 c.c. beyond that required to precipitate the silver assumed to be present should be added. The precipitate is filtered off and washed: the filtrate should be perfectly clear. The filter-paper and precipitate are dried, then wrapped in about 8 grammes of sheet lead and scorified with 40 grammes of test lead, with a cover of 1 gramme of borax glass. The button is cupelled and weighed, and then the gold separated by parting.

The cupels and slags are in many cases re-assayed and the resultant gold and silver added as a correction to the original assay. Some assayers use sodium bromide instead of sodium chloride, on account of the greater insolubility of silver bromide. The lead acetate is added to cause the precipitate of silver-chloride and the finely-divided gold to settle quickly, and to enable them to be filtered effectively.

(*c*) *Crucible Fusion.*—Crucible methods are sometimes used for the determination of gold and silver in copper mattes, as they permit of a larger quantity of material being used.

The following method is used by the Standard Smelting Company at Rapid City, South Dakota, for copper mattes containing up to 20 per cent. of copper and high in gold and silver[1]:—

The sample of matte is put through a 120-mesh sieve and well mixed. Then 0·25 assay ton of matte is fused with 3·5 assay tons of the following stock flux and a thin cover of borax glass:—

Stock flux.		Proportion of fluxes for one assay.	
Litharge	70 parts.	Matte	0·25 assay ton.
Silica	11 ,,	Litharge ...	67·0 grammes.
Sodium carbonate	25 ,,	Silica	10·5 ,,
Nitre	5 ,,	Sodium carbonate	24·0 ,,
		Nitre	5·0 ,,

Fusion is effected quickly at a comparatively high temperature, giving a clean fluid slag and a bright lead button weighing approximately 20 grammes. The button is cupelled directly for gold and silver. The lead buttons are always more or less cupriferous, but no scorification is made before cupellation. The assay should be made in duplicate: for control assays four fusions, each of 0·25 assay ton of matte, are made. In this case the slag and cupel losses are determined. The slag and cupel from one assay are crushed and fused together in the same crucible as was used for the original fusion. The slag and cupel from each assay is treated in this way, and the sum of the four corrections thus obtained added to the sum of the original buttons. According to Fulton, the average correction by this method, on copper matte of ordinary grade (*i.e.* about 20 per cent. copper, 5 ounces gold, and 40 ounces silver), is 2·5 per cent. for the gold and 5·5 per cent. for the silver. The results obtained by this crucible method are usually slightly higher than those obtained by the scorification method.

Another procedure is to boil a weighed quantity of the matte with strong hydrochloric acid which dissolves the sulphides and thus avoids roasting to eliminate the sulphur. The resulting solution is evaporated almost to dryness

[1] Quoted by Fulton and Sherwood, *Manual of Fire Assaying.*

with the residue of silica, etc., and the whole then mixed with lead and scorified, or mixed with the usual fluxes, etc. and subjected to crucible fusion.

The resulting lead buttons are then cupelled, etc. and the gold and silver determined in the usual way.

In Electrolytic Mud, Slimes, or Sludge.—Anode sludge, resulting from the electrolytic refining of copper and lead, contains considerable quantities of gold and silver. After accompanying the copper in the smelting processes the gold, silver, and platinum metals become concentrated in the cast anodes, and during electrolytic refining settle at the bottom of the vats in the form of mud. Selenium and tellurium when present pass into the mud along with the precious metals. Lead also passes chiefly into the mud. After screening to remove nodules of metal, and filtering through muslin to remove excess of the electrolyte, the sludge is washed, dried, and treated for the extraction of the precious metals.[1]

The gold and silver content varies greatly according to the nature of the material treated. Slimes from the electrolytic refining of copper usually contain from 40 to 55 per cent. of silver, from 0·2 to 1·3 per cent. of gold, and from 10 to 20 per cent. of copper. Lead anode sludge frequently contains only a small amount of copper. Small quantities of platinum and metals of the platinum group are not infrequently present in electrolytic slimes and should be tested for, since they become concentrated, and can be detected in the sludge, when the quantity is too small to permit of detection in the copper anodes and in the ore from which the copper was obtained.

The following are examples of the precious metal content of anode muds :—

Gold and Silver Content of Anode Mud from Copper Refining.[2]

	A.	B.	C.	D.
	per cent.	per cent.	per cent.	per cent.
Silver	15·725	25·0	40·0	46·58
Gold	1·450	0·05 to 1·0	2·0	0·15

(A) Raw "anode mud" from electrolytic copper refinery at Great Cobar Syndicate at Lithgow, New South Wales.[3]
(B) Raw "anode mud" from refining copper from ores of Butte, Montana.
(C) Typical "anode mud."[4]
(D) Source not given.[5]

The gold and silver in anode mud is determined by the scorification or the crucible fusion method.

The moisture in the sample is first determined by carefully drying 1 lb. (454 grammes) in a stove ; not in an open pan. Assay results are returned on the dry material. The sample is prepared for assay by grinding to pass a sieve of 80- or 90-mesh, and the metallics which are invariably left in the sieve are collected and assayed separately. The sample must be thoroughly mixed after grinding.

The assay of slimes by scorification is conducted as described above for copper matte. When great accuracy is desired ten scorifications should be made, and the slags and cupels should be assayed and corrections made for the gold and silver they contain.

[1] Consult *Met. and Chem. Eng.*, 1911, **9**, 417–420.
[2] Quoted by W. J. Sharwood, *Economic Geology*, 1911, **6**, 28.
[3] G. H. Blakemore, *Proc. Aust. Inst. M.E.*, 1910.
[4] L. Addicks, *J. Franklin Inst.*, 1905.
[5] A. Holland, *Ann. Chim. anal. appl.*, 1899, **4**, 123–125.

The following mode of procedure is given by T. Ulke[1] as a commercial method of determining gold and silver in electrolytic slimes.

Two samples are weighed out and scorified in the usual manner. One of the lead buttons is cupelled, the resulting silver bead weighed, then wrapped up in lead-foil until it approximately equals the weight of the other lead button, and then both are cupelled, and the resulting beads weighed. The loss in weight of the first bead gives the loss in cupellation, which must be added to the weight of the other bead as a correction. The slag from the scorification is re-scorified with borax and 16 grammes of test lead, and the resulting lead button cupelled. The weight of the bead thus obtained is added to the corrected weight of the second of the beads previously obtained, and the combined beads are parted, etc., in the usual way.

For a crucible fusion assay the charge given on page 373 for silver precipitate will be found to be satisfactory.

The Volhard method may also be applied to the determination of silver in electrolytic copper refinery slimes.[2] Heat slowly 0·5 grammes of the slime powder with 5 g. of potassium hydrogen sulphate and 0·5 g. of sodium nitrate in a porcelain dish until the mass is well melted and a quiet fusion is obtained. Cool, leach with 50 c.c. of hot water, filter, add 30 c.c. of concentrated nitric acid (free from chlorine), 5 c.c. of ferrous ammonium sulphate indicator solution, and titrate with standard ammonium thiocyanate solution.

In Lead Dross.—Lead dross results from the melting of base lead bullion, and consists of oxides of lead and other base metals contaminated with metallic lead. The method of assay for lead dross is described on page 328.

In Lead Fume.—Lead fume consists of the solid particles which escape from lead-smelting furnaces and are condensed. The fume is deposited in the state of fine powder or dust, varying in colour from grey to white. It contains considerable proportions of lead sulphide, oxide, sulphate, and carbonate, together with more or less silver, and also gold in minute proportions.

Lead fume requires care in handling, owing to its fine state of division and tendency to loss by dusting. The fume is assayed by crucible fusion, the following charge being suitable :—

Lead fume	25	grammes.
Sodium carbonate	25	,,
Argol	5	,,
Borax	5	,,

A piece of hoop-iron (or iron nails) is added to the charge to decompose the sulphur compounds. The charge is fused at a red heat, the time usually occupied being about twenty minutes. The resultant lead is freed from slag and cupelled.

In Litharge and Red Lead.—As previously stated, the oxides of lead invariably contain silver, which must be determined when litharge or red lead is used in an assay charge. The silver in litharge is determined by fusing 100 grammes with 1 gramme of charcoal and 20 grammes of borax. In the case of red lead the same charge is used with 2 grammes of charcoal. The lead is cupelled, and the button of silver weighed. As the silver buttons are very small, the cupellation must be carefully watched towards the end of the operation.

In "Cement Silver" (Silver Precipitate).—This product consists of metallic silver precipitated from solution by metallic copper in the lixiviation processes of ore treatment, and in the sulphuric acid refining of silver.

[1] *Eng. and Min. Journ.*, 1899, **68**, 727–729.
[2] F. de Chaves, *Anales Soc. Espan. Fis. Quim.*, 1931, **29**, 651–652 : *Chem. Abs.*, 1932, **26**, 937.

The "cement silver" from sulphuric acid refining contains a high percentage of silver. Usually the washed precipitate is strongly compressed into brick-like cakes which retain not more than 8 to 10 per cent. of water. After drying, the cakes are melted and cast into ingots which usually assay from 996 to 998 fine.

The cement silver from ore lixiviation processes contains, in addition to metallic silver and gold, sulphates of lead and lime, oxides of zinc, copper, and iron, and more or less organic matter. The amount of silver in these precipitates is very variable; an average precipitate contains about 5 to 10 per cent. silver, 0·02 per cent. gold, 20 per cent. lead, and 2·5 per cent. zinc.

Samples of this silver precipitate are generally received in a dried condition, *i.e.* free from "water at 100° C." Since, however, it rapidly absorbs water, it is advisable to determine the moisture on about 250 grammes before making the assay. The sample should be assayed as soon as it is dried, and care should be taken in weighing it to prevent absorption of water. The determination of the silver and gold is made by crucible fusion. Scorification cannot suitably be adopted in some cases owing to the combined water contained in the precipitate. If this method is used, from 5 to 10 scorifications of about 3 grammes each are made and the resultant buttons re-scorified together or treated separately and the mean of all the results taken.

Beringer[1] recommends the following crucible assay:—

Weigh up 3 to 5 grammes of the precipitate, mix with 100 grammes of litharge and 1 gramme of charcoal. Melt in a crucible at a moderate heat, and pour. Detach the slag, replace it in the crucible, and when fused "clean" it by adding a mixture of 20 grammes of litharge and 1 gramme of charcoal. When the fusion is again tranquil, pour, and cupel the two buttons of lead together. The cupellation loss should be determined by the use of checks or other method and added to the result of the assay.

It is also advisable to fuse the cupels and make allowance for the silver they contain. The charges for the fusion of cupels have been given previously. In calculating the percentage of silver, allowance must be made for the silver contained in the litharge used.

The gold is determined in the ordinary way by treating the cupellation button with dilute nitric acid.

Owing to the richness of these precipitates it is difficult to get close agreement between the assay results, and from five to ten fusions should be made and the average of the results taken. The gold may also be satisfactorily determined by a combined method. For this purpose take 25 grammes of precipitate, treat with 100 c.c. dilute nitric acid (1 acid : 1 water), and, when all action has ceased, dilute and filter. Wash and dry the residue, burn the filter-paper, then mix the residue and filter ash with 50 grammes of litharge, 1 gramme of charcoal, and 50 grammes of sodium carbonate. Transfer to crucible, fuse, and pour; detach the slag, cupel the lead button, and part in the ordinary way.

The insoluble residue usually contains sufficient silver for parting the button from cupellation. An exact determination of the silver in the precipitate cannot be made by treatment with nitric acid, since not all the silver usually passes into solution.

In Silver-Sulphide Precipitate.—This consists of sulphide precipitated in the lixiviation process of silver extraction. In general these precipitates carry from 25 to 40 per cent. (8000 to 14,000 ozs. per ton) of silver, 20 to 33 per cent. of copper, a little lead, some free sulphur, and the balance various impurities in small amounts. In some cases a small quantity of gold is also present. These sulphide precipitates are very disagreeable to handle; when

[1] *Text-book of Assaying*, 15th edn., p. 116.

dry they are finely divided and dusty; they soil everything they touch, and when caked they are sometimes semi-malleable.

Considerable difficulty is experienced in sampling this very rich material. According to Furman,[1] the following method answers very well for sampling comparatively large lots. Spread the material evenly on a clean iron floor, and divide it into "squares" about 1 foot square. From each square dig out five samples, and put these together into one pile. Well mix the pile, spread out, and reduce as before. The final bulk sample, weighing about 5 lbs., is pulverized in a coffee mill and reduced still further. The portion taken for the assay sample is ground on a bucking-plate until it passes a sieve of 80- to 100-mesh and then very thoroughly mixed. On account of their richness, only small quantities can be taken for assay, and even then large buttons of silver are obtained which are liable to spit, and the presence of so much copper requires the use of large amounts of lead. They may be assayed by either scorification or crucible fusion, but the former method is usually advocated and is used in settling contracts for handling the sulphides.

For scorification, from 6 to 20 separate charges are run on each lot of sulphides.

One-twentieth or one-tenth of an assay ton of the dried samples of sulphides is scorified with 55 grammes of lead and a cover of 3 to 5 grammes of borax. One half of the lead is put in the bottom of the scorifier and hollowed out; the sulphides are put into the hollow, and the remainder of the lead poured over them; the borax is then placed on top. The resultant lead buttons are cupelled separately at a low temperature.

The slags from all the scorifications are carefully collected, crushed to 20-mesh and mixed with the following charge:—

Litharge	20 grammes.
Sodium carbonate	15 ,,
Argol	2 ,,
Cover of salt.	

Fuse, pour, and cupel the lead button.

The cupels are crushed (after removing unstained portions) to pass 60-mesh, then fused with the following or other suitable charge:—

Litharge	30 grammes.
Sodium carbonate	30 ,,
Borax glass	30 ,,
Argol	2 ,,

The average silver content of the slags and cupels, which may amount to as much as 2 per cent., is added as a correction to the average of the original assays.

In the crucible methods for the assay of silver sulphides, no metallic iron is employed to decompose the sulphides, but a sufficient amount of litharge or red lead is added to oxidize completely the sulphur in the charge, and leave an excess of lead oxide to supply the metallic lead to collect the gold and silver. The result of the crucible assay, when carefully conducted, is as accurate as that obtained by scorification. The following stock flux may be used:—

Stock Flux.		Quantities for one Assay using one-tenth Assay Ton of Sulphide Precipitate.	
Litharge	600 grammes.	Litharge	40 grammes.
Sodium carbonate	350 ..	Sodium carbonate	25 ,,
Borax glass ...	100 ,,	Borax glass ...	7 ,,
Silica	100 ,,	Silica	7 ,,
Charcoal	8 ,,	Charcoal	0·75 ,,

[1] *Assaying*, 6th edn., 1908, p. 20. See also F. P. Dewey, *Trans. Amer. Inst. Min. Eng.*, 1896, Colorado Meeting, and *J. Soc. Chem. Ind.*, 1896, **15**, 434–436.

With one-tenth assay ton of sulphides use 80 grammes of stock flux, and twice this amount for one-fifth assay ton.

The precipitate is mixed with two-thirds of the charge and the remainder of the charge used as a cover. Fuse the charge at a good red heat for twenty-five or thirty minutes, then pour, detach the slag from the lead button, and cupel. The cupel absorption is determined and added. It is also advisable to collect the slag and clean it: this may be done by adding the slag to the charge for the fusion of the cupels, thus cleaning the slag and cupels in one operation. Smelters do not, as a rule, allow the slag absorption, and if the fusion is properly conducted the values retained by the slag are very small with the above flux. The charges given for the assay of cyanide precipitate may also be used for silver sulphide precipitate (see page 477). At least five fusions should be made, and the results averaged.

Wet methods of assay may also be adopted for these sulphide precipitates. From 25 to 50 grammes of the dried precipitate are dissolved in nitric acid, and all nitrous fumes expelled. Any gold present will be left as an insoluble residue. The solution, containing mainly silver and copper in neutral or slightly acid solution, is treated with a solution of ammonium oxalate, whereby the silver is precipitated as silver oxalate, insoluble in excess of the precipitant. Ammonium oxalate is added until the precipitate appears of a pure white colour.

The copper remains in solution as a soluble ammonium oxalate. The precipitate is filtered off and washed first with ammonium oxalate, and finally with cold water. It is then dissolved in dilute nitric acid, the solution transferred to a litre flask (1000 c.c.) and diluted to the mark.

Several portions of from 25 to 50 c.c. are then accurately measured off, and each separately transferred to suitable assay bottles, and the silver determined by the Volhard or Gay-Lussac methods in the usual way. The mean result of a number of determinations should be taken.

The silver may also be separated from the copper by adding a slight excess of ammonia and passing a current of sulphurous-acid gas through the warm ammoniacal solution, thus precipitating the silver in the metallic state, which may be treated as described.

Another method consists in adding to a slightly acid solution of the two metals, sufficient potassium cyanide solution to dissolve the precipitate first formed, then warming the solution and passing a current of hydrogen sulphide, which precipitates silver sulphide but does not precipitate the copper in the presence of potassium cyanide, in which copper sulphide is readily soluble.

The silver precipitate is filtered off, washed by decantation, then dissolved in hot nitric acid, and the silver determined as already directed.

Wet methods have the advantage that larger quantities of the precipitate can be taken for assay without inconvenience: and by diluting the solution to large measured bulk and taking portions for determining the silver, the errors of manipulation are reduced.

In "Tailings."—"Tailings" or "tails" consist of the finely pulverized ore which has been treated for the extraction of the values, but still contains small quantities of gold and silver. The assay of tailings differs only from that of ores in the fact that a larger quantity must be taken, frequently from 3 A.T. to 5 A.T., for the assay, since they are always poor in gold, containing usually less than 1 dwt. per ton. "Tail Assays" check extraction results.

It is not unusual to take the portion for assay without any further grinding, as tailings have already been crushed and passed through a sieve of at least

30-mesh. It is advisable, however, to crush all tailings samples to pass a 90- or 100-mesh sieve, as is done for the assay of the original ore or " heads." From 2·5 to 5 assay tons of material should be taken for the assay. The charges given for residues on p. 479 may be used.

III. Determination of Gold and Silver in Manufacturing Products.

In the various processes of manufacturing precious metals there are produced a number of so-called " waste products," which frequently contain considerable amounts of gold and silver (and platinum metals) and are received for assay, valuation, and sale to the refiners.

The assay of products containing metals of the platinum group is dealt with in Chapter XXIV.

In Silversmith's Chloride Precipitate.—This consists of silver chloride precipitated from the pickling solutions, etc. used in silversmiths' work. Although termed " chlorides," these precipitates frequently consist of metallic silver precipitate due to the use of copper or iron to throw the silver out of solution instead of common salt. The precipitate is usually very rich in silver, of which it may carry over 60 or 70 per cent. ; pure silver chloride contains 75·27 per cent. of silver. The precipitate is frequently moist when received, and must be carefully dried and the result of the assay reported on the dry material. It is important to mix the dried precipitate thoroughly before sampling. The assay may be made by scorification or by crucible fusion, as described above for silver sulphide precipitate, corrections being made for silver lost in the slag and by cupel absorption.

A more satisfactory procedure is to fuse a comparatively large lot of the precipitate with about an equal bulk of sodium carbonate and a little charcoal, and to cast the reduced silver in an ingot, which is sampled and assayed by any of the usual methods for bullion, etc.

Another procedure consists in mixing from 2 to 5 grammes of the dry precipitate with about an equal bulk of potassium cyanide and a little charcoal, and fusing to reduce the silver.

The resulting button of silver is freed from adhering slag and weighed, and the result accepted as the silver content ; or the button may be flattened, dissolved in nitric acid, and assayed volumetrically, or precipitated as chloride and weighed.

In Goldsmith's Lemel (from Fr. *limaille*, metal filings).—Lemel from goldsmith's work is usually mixed with fluxes and fused at the works to produce a lemel bar which is sent to the refiners for valuation and sale.

Frequently, however, small lots of lemel are received for assay in the condition as collected without being subjected to any preliminary treatment or dressing.

The undressed lemel contains impurities, such as organic matter, iron filings, etc. from the files and other tools, bits of binding wire, etc., which are removed by passing the lemel through a sieve of about 30-mesh. This also separates any pieces of scrap metal contained in the lemel. If the metallic portion left on the sieve is large in quantity it is treated separately by melting down with fluxes to obtain a small lemel bar which is assayed, and the precious-metal content is added to the result obtained from the fine sifted portion. If, on the other hand, the metallic portion is small, it may be added to the finer portion of the lemel and the whole parcel fused with fluxes to obtain a small lemel bar, which is sampled and assayed, and the value of the lemel reported.

Before assaying, the finer portion of the lemel is first burnt in an iron pan to remove the organic matter, and the iron particles removed by means of a magnet. When the lemel contains particles of nickel white gold, special precautions have to be taken, since particles of these alloys are frequently magnetic and in consequence will be removed with the iron filings.

Tests have shown that the loss of white gold increases with the strength of the magnet used and the amount of white gold filings in the lemel. In order to prevent losses of white gold particles it is advisable to use as weak a magnet as is compatible with the separation of the iron filings, and then to treat the latter with hot hydrochloric acid to dissolve out the iron and thus recover any particles of white gold that may be entangled in it.

After treatment with a magnet the sifted portion is ground to pass a sieve of 60- to 80-mesh, then well mixed, and a sample for assay taken by quartering down.

The lemel may then be treated as a gold ore and assayed by scorification, or by crucible fusion as described for gold ores.

If the whole parcel of lemel is mixed with the fluxes and fused to obtain a small lemel bar, the following charge may be used :—

Lemel	20	parts.
Sodium carbonate	3	,,
Borax (dried)	1	,,
Sodium chloride	2	,,

The charge is melted at a bright red heat and, when fused, a few small pieces of nitre about the size of a pea are added at short intervals. When tranquil fusion results, the contents are poured into an ingot mould and the metal cleaned from slag. The bar thus obtained should be re-melted with a little borax or charcoal as flux and then poured into an ingot mould and weighed. If necessary it may be partially refined by the addition of nitre.

Lemel bars vary considerably in composition, and their valuation is attended with difficulty, as they are rarely uniform in composition. The sampling and the assay of lemel bars is described in Chapter XXI.

In Goldsmiths' and Silversmiths' " Sweep."—Sweep consists of the sweepings of the benches and floors of gold- and silversmiths' workshops. This material is very heterogeneous in character, and generally contains considerable amounts of foreign matter. The gold and silver are usually present in the metallic state, but the amount varies very considerably. The sampling of sweep presents considerable difficulty. The material is first " dressed " by burning to destroy organic matter. The mass is then ground sufficiently fine to pass a sieve of 20- to 40-mesh. The metallic portion left on the sieve is freed from iron by means of a magnet, and then melted with fluxes to obtain a bar, which is weighed, sampled, and assayed as described under "Lemel Bars " (p. 340). The sifted portion is freed from iron filings etc., with a magnet, then quartered down and sampled. The precautions to be taken when the sweep contains particles of nickel white gold, which are magnetic, are mentioned above. The final portion of sweep taken for " assay sample " is ground to pass a sieve of 80- to 90-mesh and thoroughly mixed. It is then assayed, and the precious metal content, including that from the " metallics," reported upon.

Samples rich in metal may be assayed by scorification, as follows :— Scorify 5 grammes of the material with 10 to 15 times its weight of lead and 1 gramme of borax. Cupel, weigh, and part as usual. Several scorifications should be made according to the richness and the average of the results taken.

Poor samples are best assayed by crucible fusion. No charge can be given to suit all samples of this material, but the following will be found generally satisfactory :

Dressed sweep	25 to 50 grammes.
Litharge	50 ,,
Charcoal	2 ,,
Sodium carbonate	30 ,,
Borax	20 ,,

If the sample contains much ferric oxide it may be necessary to increase the amount of charcoal in the charge.

The lead button from the fusion is cupelled and the resulting gold–silver button parted. It is advisable to make duplicate assays, or more, depending on the precious-metal content of the sweep, which is frequently indicated from its source.

The assay of sweep containing metals of the platinum group is dealt with in Chapter XXIV.

When the sweep contains appreciable quantities of nickel and copper from the working of nickel white gold, etc., these metals pass into the lead button resulting from the crucible fusion and cannot be completely removed by scorification. Such buttons, upon cupellation, yield a greenish-black scoria surrounding the silver bead which contains minute prills of silver and retaining rhodium and ruthenium which would normally form a black incrustation on the bead.

Materials containing appreciable quantities of nickel and copper should therefore be treated by a combination method, involving acid treatment to remove the base metals, and crucible fusion or scorification of the residue. A weighed quantity of the sweep is treated with dilute nitric acid to dissolve the base metals so far as possible, a little hydrochloric acid added to precipitate the silver in solution, and the residue then subjected to crucible fusion or scorification to collect the precious metals in lead, as previously described.

A partial precipitation of the base metals in the filtrate with hydrogen sulphide is desirable in order to collect any traces of precious metals that may be present. The precipitate so obtained is added to the main residue.

The precious metals in the lead button are separated and determined by cupellation and parting. The valuation of sweeps is described below.

In Crucible and Melting Sweep.—This sweep includes the pounded crucibles used for melting gold and silver, and ash and refuse from the furnace; when the amount of this sweep is relatively small it is added to the workshop floor sweepings. Graphite crucibles, stirrers, etc. that have been used in melting bullion always contain particles of the precious metals mechanically held in the pores of the material. They are crushed and washed and sometimes amalgamated with mercury to separate shots of metal. The sampling of the crushed material (frequently termed " blacks ") is difficult, and must be carried out with care. The portion selected for the assay sample should be crushed to pass a sieve of 100-mesh and the metallics carefully collected and assayed separately. The assay of this material is also troublesome owing to the large percentage of graphite it contains and the difficulty of effecting its oxidation readily. The graphite is usually oxidized by roasting or by the use of nitre. If the sample is roasted, this is best done in the muffle in a roasting-dish sufficiently large to permit of the material being spread out into a thin layer. To facilitate the roasting, the sample must be very finely powdered and must be stirred constantly: even under these conditions the oxidation is very slow.

Samples of "blacks" may be assayed either by scorification or by crucible fusion, but in both methods it is usual to take only a small amount, generally one-fourth assay ton of the material, as the graphite may cause trouble if larger amounts are used. Scorification is usually adopted for rich blacks, a charge of granulated lead equal to about ten times the weight of roasted blacks being used, with a sprinkling of borax as cover (about 0·5 gramme). The slag from several scorifications is re-melted in a crucible and a mixture of litharge and charcoal thrown on to the surface of the fused material. In this way from 0·5 to 2·5 per cent. of the total values is recovered from the slag. The cupellation loss is usually from 0·5 to 1 per cent.

According to Loewy,[1] correct results may be obtained by scorifying without previous roasting, the charge being 5 grammes of material, 60 grammes of lead, and a little borax or glass powder as a cover.

Fulton recommends fusion in a scorifier with oxidizing agents.[2] From 0·05 to 0·10 assay ton is taken and mixed with a little more than one-half of its weight of nitre and 30 grammes of litharge, placed in a 2·5-inch scorifier, covered with 30 grammes of litharge and afterwards with a thin cover of borax glass, then placed in a muffle, and fused finally at a yellow heat. The lead buttons are cupelled, the gold–silver button weighed and parted as usual.

This material may also be assayed by the crucible method by using nitre and litharge to effect the oxidation of the graphite.

The following charges are given by M'A. Johnston as typical of those in use on the Rand for the direct fusion of samples of graphite crucibles[3] :—

	A.	B.
Ground product ...	0·25 assay ton.	0·25 assay ton.
Sodium carbonate ...	0·5 ,,	0·5 ,,
Litharge	4·0 ,,	2·0 ,,
Borax	0·25 ,,	0·25 ,,
Sand	0·5 ,,	0·5 ,,
Nitre	nil.	about 0·3 ,,
Weight of lead button obtained	55 to 60 grammes.	Approx. 40 grammes.

The weight of lead obtained is dependent on the amount of graphite present in the material. Charge A is stated to give the higher and more correct result.

In Goldsmiths' "Polishings" and "Lappings."[4]—These materials result from the finishing processes in goldsmiths' work, and are usually treated separately and not mixed with the ordinary shop sweep as they are invariably richer in the precious metals—which is especially true of polishings from the working of the best-quality gold standards. "Polishings" consist of the materials, used in the polishing processes, such as waste emery, pumice-stone, rotten-stone, rouge (iron oxide), etc. ; and in addition worn-out threads ; cotton-waste used for cleaning the laps and absorbing the polishing refuse from the lathe ; worn-out brushes, emery, and sandpaper, all of which contain gold and silver. "Lappings" consist of fine emery resulting from the lapping process, a special mode of finishing with the lap certain kinds of work such as gold chains, flat surfaces, and articles that require faceting.

[1] *Proc. Chem. and Met. Soc. of S.A.*, 1898, **11**, 205.
[2] *Manual of Fire Assaying.*
[3] *Rand Metallurgical Practice.*, vol. I, p. 317.
[4] See T. B. Wigley, *Art of the Goldsmith and Jeweller.*

This class of workshop refuse is burnt in an iron pan to remove organic matter and allowed to cool, then passed through a fine sieve, of from 60- to 80-mesh, and the coarse parts pulverized until all the refuse has passed through the sieve. It is then thoroughly mixed, and sampled and sold to the refiners.

The assay of these materials is conducted either by scorification or crucible fusion in the same way as sweeps and ores, the fluxes being selected in accordance with the nature of the material. If the polishings are known to be very rich they can be mixed with fluxes and fused into a button which is sampled and assayed. If relatively poor they can be subjected to sampling and assay in the same way as sweep and sold to the refiners. The valuation of polishings is described below.

In Goldsmith's " Colour-water."—This consists of the acid mixtures whereby a gold alloy of low standard has imparted to it the rich yellow colour of fine gold, owing to the superficial removal of the other metals alloyed with the gold. As articles which have been coloured lose from 12 grains to 1 dwt. per ounce of their weight, according to the carat quality and the surface area of the articles, great care should be taken of the colour-water, which contains appreciable quantities of both gold and silver.

The gold and silver are precipitated with ferrous sulphate and filtered by decantation through a sawdust filter-bed.

The sawdust is dried and burned, and if sufficiently rich in precious metals can be mixed with fluxes and fused to obtain a button, which is assayed and sold. If comparatively poor in precious metals it is mixed with the polishings and competitive offers invited from the refiners.

In Silversmith's Pumice.—Pumice is a siliceous igneous rock which has been cooled very rapidly from a molten state. It is generally used as an abrasive and also for polishing in the form of powder, but pumice blocks are also used.

Pumice powder of various grades has to a large extent superseded the use of sand for sand-blasting purposes, particularly in the preparation of sterling silver and electro-silver plated surfaces for an " oxidizing " finish or for gilding.

A polishing composition consisting of fine powdered pumice mixed with a little oil to the consistency of paste is placed upon the article when polishing goldsmith's work. During these finishing processes the pumice becomes charged with more or less of precious metal, and when sufficiently rich is sold to the refiners.

This material is assayed in the same way as siliceous gold and silver ores by crucible fusion and cupellation of the resulting lead button, followed by parting of the cupel bead in the case of pumice containing gold.

In Silversmiths' " Buffing " Sand and Lime.—Fine siliceous sand and lime are used in the polishing process usually known as " buffing," in the silver and electro-plate industries. (These materials are not usually employed in finishing goldsmiths' work.) Both materials become enriched in silver during the process and are subsequently sold to refiners for the recovery of the precious metal. They are sampled and sold on much the same terms as sweep as described below.

Buffing sand which is only used for the first smoothing-down is usually very poor in silver, so that about 50 grammes should be taken for assay. The silver is best determined by crucible fusion, the material being treated as a siliceous ore in the usual way. Buffing lime is usually comparatively rich in silver and may contain up to 3 or 4 per cent. or more. From 20 to 50 grammes are taken for assay, according to its richness in silver. The determination may be made either by scorification or by crucible fusion, but the latter is preferable. The material is fluxed and treated in the same way as a lime-bearing ore. In this connection fluor-spar is a useful flux to obtain fluidity.

Valuation and Sale of Goldsmith's Sweep and Polishings.[1]—In view of increased competition, considerably more attention has been given in recent years to the collection of the various so-called waste products from the working of precious metals, and to their preparation for sale by competitive offers invited from several firms of refiners.

When the sweep and polishings (including lappings) have been prepared by dressing, *i.e.* burning off all combustible matter, and crushing to remove lumps and facilitate mixing, two or three refiners are invited to take samples, determine their value by assay, and then make an offer for the "lot."

The general rule is for each competitor to take two pounds weight of sweep and two ounces of polishings; if the latter material is rich, they agree to pay for one ounce, whether they purchase or not.

If the refiners are not satisfied with the mixing, they remove the sweep from the bin, mixing it while doing so, and then throw it back again; in some cases they reserve a small portion for sample each time the operation is performed, and as this may be repeated on an average eight or ten times a thorough mixing is ensured. The several portions thus taken are well mixed and transferred to sample bottles and sealed.

The polishings are also submitted to careful mixing with a trowel, but as the bulk is much smaller and the material much richer in value than sweep, the process is performed with more care and the weighing performed with more accuracy.

After the samples have been assayed by each refiner, each separate lot is sold to the highest bidder.

[1] See T. B. Wigley, *The Art of the Goldsmith and Jeweller.*

CHAPTER XXIII

METALS OF THE PLATINUM GROUP: SOURCES, PROPERTIES AND USES

The six metals which comprise the platinum group are platinum, palladium, iridium, rhodium, ruthenium, and osmium. Platinum was the first discovered, is the best known, and is the most extensively used in industry. The other five metals of the group are given above in the order of their importance to the jewellery, chemical, electrical, and other industries in which they are finding increasing application.

All the metals of the group are white in colour, but each possesses a faint characteristic tint which distinguishes it from the others. Under normal atmospheric conditions, and when in compact metallic form, they are remarkably permanent in air, and show extraordinary powers of withstanding chemical reagents; properties that chiefly give them their value and desirability for use in industry.

The physical and chemical properties of the individual metals of the group are described below.

Sources of the Platinum Metals.—The chief sources of platinum are the grains of native metal found principally in alluvial deposits derived from the disintegration of primary rocks, and frequently in the same alluvial deposits as gold. The crude platinum contains about 3 to 8 per cent. of the other metals of the group, which are recovered in the refining of the crude platinum.

The first commercially important discovery of native platinum in a primary deposit was made at the Onverwacht Mine, Lydenburg, South Africa, which has yielded considerable quantities of the metal in recent years.[1] The principal alluvial deposits of crude platinum at the present time in order of importance are found in the Ural Mountains, Colombia, and Abyssinia. Typical analyses of Ural, Colombian, and Onverwacht platinum are given in Table LVII.

TABLE LVII.—COMPOSITION OF NATIVE OR CRUDE PLATINUM.

Source.	Per cent.			
	Platinum.	Palladium.	Iridium. Rhodium. Ruthenium. Osmium.	Iron.
Ural platinum	73 to 86	0·3 to 1·8	2·5 to 7·7	8 to 17
Onverwacht platinum ...	80 to 84	0·4	0·5 to 1·0	11
Colombian platinum ...	84 to 86	0·5 to 1·0	3 to 6	5 to 8

[1] Atkinson and Raper, "Metals of the Platinum Group," *J. Inst. Metals*, 1936, (2), **59**, 180.

Crude platinum is separated from associated metals and refined by a somewhat complicated series of operations during which the other metals of the platinum group are recovered. These metals also occur in small quantities in other ores, notably those of copper and nickel, and are recovered in the electrolytic refining of these metals. Thus the chief source of supply of palladium at the present time is the copper–nickel ore of Sudbury, Ontario, the palladium becoming concentrated during smelting operations, and finally recovered as a slime or mud in the electrolytic refining of the copper–nickel matte.

Iridium is recovered during the refining of crude platinum, and also from the natural alloys of iridium and osmium in variable proportions and known respectively as osmiridium or iridosmine, according to which of the two metals preponderates.

Rhodium occurs in crude platinum, which formerly was the chief source of the metal, the amount present being usually under 2 per cent. Appreciable quantities of rhodium, however, are now being recovered from the Canadian mines of the International Nickel Company, and new uses for the metal are being sought; hence its increased use in industrial alloys.

Ruthenium occurs in crude platinum, which furnishes the chief supply of the metal.

Osmium is recovered in small quantity from the refining of crude platinum, but its chief source is from the natural alloy, osmiridium.

From the analyses of crude platinum given in Table LVII it will be seen that the total quantity of all the metals associated with platinum is comparatively small and seldom exceeds 8 per cent.

Platinum-metal Alloys and their Industrial Uses.[1]—In the unalloyed state platinum finds only limited application in industry as it is too soft for many purposes, and in consequence it is generally alloyed with other metals, including those of the platinum group. Platinum of 99·9 per cent. purity and over is used for electrodes, thermocouples and resistance pyrometers. Platinum crucibles and chemical wares are not often made of chemically pure metal, and there may be present small amounts of iridium and rhodium and traces of iron to stiffen them.

The use of platinum for the resistance windings of high-temperature laboratory furnaces is well known. Pure platinum for this purpose has some disadvantages, but it has been found that the alloy of platinum with 10 to 13 per cent. rhodium is more satisfactory.

In Jewellery.—The most important platinum alloys are those used for the manufacture of fine jewellery, for which a standard of 950 is most generally used. In this standard iridium counts as platinum, and base metals may be present in amounts up to 5 per cent. Two commonly-used alloys for jewellery purposes contain 3 per cent, and 4·5 per cent. of copper respectively, the addition of which is said to improve the colour of pure platinum. Instead of copper, the hardening of platinum is also effected with the other platinum metals such as iridium, rhodium, and ruthenium. These alloys are superior in several respects to the copper-containing alloys. The alloys of platinum and iridium are very hard and elastic, unalterable in the air, and take a high polish.

The highest-quality platinum jewellery is made from 5 per cent. and 10 per cent. iridio-platinum. Other very high-grade alloys can be made with the addition of 5 per cent. rhodium plus ruthenium, the ruthenium content being varied according to the hardness required.

[1] Atkinson and Raper, *loc. cit.*, 204.

A similar range of palladium alloys may be prepared with similar strength characteristics. Iridium, rhodium, and ruthenium harden palladium to a rather less extent than they harden platinum for equivalent additions. Palladium alloys were largely used for jewellery purposes during the Great War (1914) when platinum was scarce. In normal times this use largely disappears, though on account of appearance and working properties the palladium alloys are considered to be superior to the "white gold" alloys introduced as substitutes for platinum, some of which have a high palladium content.

In Dentistry.—After jewellery the next most important group of alloys are those used in dentistry, and here platinum and palladium are alloyed chiefly with gold, silver and copper.

A subdivision of the alloys into wrought and casting alloys is convenient. The wrought alloys are used in the form of circular blanks of sheet metal for swaged plates and in the form of wire and thin bands for correcting malformations of the mouth. Two of the most useful alloys for swaged plates are "dental alloy" of the composition platinum 30 and silver 70 per cent. and the alloy palladium 50 and silver 50 per cent. The high-strength wires and bands used for correcting malformations of the mouth are of complex and varied composition and may contain up to 30 per cent. of platinum and/or palladium, the remainder being gold, silver, and copper. The so-called platinized gold alloys used in dentistry usually contain from 2 to 8 per cent. of platinum for hardening purposes, according to the carat quality of the alloy.

Dental casting-alloys should not have a melting-point greater than 1050° C., hence the amount of platinum and/or palladium which may be added is restricted. Alloys in the 18-carat range may contain up to 12–15 per cent. platinum and/or palladium, but cheaper alloys of approximately 9-carat quality may contain up to 25 per cent. of palladium.

Industrial Uses.—Various other alloys containing metals of the platinum group are fabricated to meet the special needs of industry. The alloy platinum 95 and nickel 5 per cent. is particularly suitable for the filaments used in amplifying tubes in communication systems. A large variety of platinum-metal contact alloys are used where conditions are severe or where great reliability is required. Aeroplane and automobile magneto contacts, for instance, are made of 15 to 25 per cent. iridio–platinum.

The alloy palladium 60 and copper 40 per cent. is finding many uses among which may be mentioned telegraph relays. Very hard alloys containing varying proportions of iridium, osmium, and ruthenium are used for fountain pen points.

In the chemical industry use is made of gauzes of the alloy rhodium 10 and platinum 90 per cent. for the oxidation of ammonia in the production of nitric acid. Alloys of gold, platinum, and rhodium are in use as rayon-spinning nozzles. An alloy composed of 96 per cent. of palladium and 4 per cent. of ruthenium is said to be suitable for high-class jewellery.

Typical compositions of industrial alloys containing metals of the platinum group are given in Table LVIII.

The electro-deposition of rhodium is finding increased application as a tarnish-resisting coating on silverware etc. (see p. 398).

Osmium, the remaining metal of the group, finds only very limited industrial application. The natural alloys of osmium and iridium, are very hard and, as mentioned above, are used for tipping the nibs of gold pens.

The above details will give the assayer some indication of the class of materials containing metals of the platinum group that may be presented for assay. In addition there are the numerous products such as scrap metal, sweep, etc., resulting from the fabrication of the metals in works practice.

TABLE LVIII.—COMPOSITION OF VARIOUS INDUSTRIAL GOLD ALLOYS CONTAINING PALLADIUM.

	18-ct. White Gold.		Dental Pins.		Dental Casting Golds.	
	per cent.	per cent.	per cent.	per cent.	per cent.	per cent.
Gold	75·0	75·0	43·5	25·0	10 to 20	10 to 15
Palladium ...	17·5	16·0	15·0	70·0	20 to 30	22 to 30
Platinum ...	...	2·0	38·0	...	...	...
Iridium ...	...	...	...	...	...	0·5 to 5
Silver	7·0	7·0	...	4·0	40 to 50	27 to 50
Copper	0·5	...	3·5	...	10 to 18	10 to 20
Zinc	...	...	...	...	1 to 4	...
Nickel	...	...	...	1·0	...	...

Hall-marking of Platinum.—The increasing popularity of platinum jewellery has directed attention to the desirability of including it under the hall-marking regulations so that it can receive an official quality mark in the same way as gold and silver wares. The existing hall-marking laws recognize only gold and silver as "precious metals," all other metals, including those of the platinum group, being regarded as base metals.

Owing to the time required for assaying platinum alloys, a more complicated process than in the case of gold or silver alloys, it has not yet been found possible to evolve a suitable scheme for hall-marking platinum jewellery. At present most manufacturers adhere to a 950 platinum standard which has been agreed to by all the important European countries and by American manufacturers. Iridium counts as platinum in this 950 standard. In America however, legislation requires that platinum jewellery shall consist of "not less than 95 per cent. of *metals of the platinum group.*"

In the absence of a hall-mark for platinum there is no guarantee that the 950 standard is being maintained beyond the standing and reputation of the manufacturer, and in consequence, when platinum is scarce, the cheaper metal palladium is not infrequently substituted for platinum in the alloys made for jewellery purposes.

Physical and Chemical Properties of the Platinum-group Metals.

The characteristic physical and chemical properties of the individual metals of the platinum group are described below, but it is well in the first place to briefly consider them collectively.

Although, as stated above, each metal of the group possesses a faint but distinctive tint, it must be admitted that when the metals are in the usual commercial forms of sheet or wire it is extremely difficult to distinguish between them by appearance only, and frequently a confirmatory test has to be applied. The physical properties of the metals are very dependent on purity, and on their condition, that is, whether in the cast or in the wrought state.

The platinum metals are included amongst the heaviest substances known, the specific gravities falling into two distinct groups, namely, the lighter metals, forming the ruthenium group, and the heavier metals, viz. the osmium group, as shown in Table LIX, in which it will be observed that the specific gravities fall with increasing atomic weights.

All members of the group are very difficult to melt. The most easily fusible is palladium, which melts at 1553° C., *i.e.* at about 500° C. higher than the

TABLE LIX.—Specific Gravity of Platinum-group Metals.[1]

Group I.			Group II.		
Metal.	Specific Gravity.	Atomic Weight (0 = 16).	Metal.	Specific Gravity.	Atomic Weight (0 = 16).
Ruthenium ...	12·20	101·7	Osmium ...	22·48	191·5
Rhodium ...	12·41	102·9	Iridium ...	22·42	193·1
Palladium ...	12·16	106·7	Platinum ...	21·40	195·2

melting-point of gold. The mechanical properties of the two very ductile metals of the group, namely platinum and palladium (which can be worked as easily as platinum), are now known with accuracy and fairly completely. Information about the properties of the other four metals is at present incomplete, although reliable data is gradually forthcoming.

In discussing the chemical properties of the platinum-group metals it is necessary to bear in mind that the extent to which they are attacked by acids, etc. is very much dependent on their purity and physical condition. When they are very finely divided, as in the case of sponge, or when separated from alloy with silver or some other metal, they are more readily attacked than when in mass. For instance, platinum, when alloyed with a certain proportion of silver, is attacked by nitric acid and partly dissolved, whilst osmium-black dissolves in nitric acid, and rhodium-black is attacked by hot concentrated sulphuric acid.

The behaviour of the metals when attacked by acids in compact metallic form is as follows: Iridium, rhodium, and ruthenium are not attacked by aqua regia. Platinum is dissolved by aqua regia, but is not attacked by any single acid. Palladium is easily soluble in aqua regia, and it is also attacked by hot concentrated sulphuric acid and to a less degree by nitric acid, but with these exceptions it is not acted upon by the common acids.

The metals of the group that are not attacked by acids are rendered soluble by fusion with certain chemical reagents, or by means of chlorine gas. Thus, finely-divided rhodium, when heated with potassium bisulphate together with a small proportion of ammonium bisulphate, is converted into bisulphate of rhodium and potash, which is soluble in boiling water.

(1) **Platinum.**

Physical Properties.—Symbol, Pt; atomic weight, 195·2; specific gravity (annealed metal), 21·43; melting-point, 1773° C.

Platinum is not so white as silver, having a distinctive greyish tinge, but when polished its lustre is very little less brilliant than that of silver. The colour of finely-divided platinum is grey or black according to the method of preparation. The grey powder, known as "platinum sponge," which is a common commercial form of the metal, usually contains about 99·5 per cent. of platinum.

Best-quality commercial platinum is usually 99·8 per cent. or purer. It is melted in the electric furnace, or by the oxy-hydrogen or oxy-coalgas flame, in a furnace constructed of well-dried lime blocks. When molten it absorbs oxygen which is given off on solidification, causing spitting in some cases.

[1] Atkinson & Raper, *loc. cit.*, 199.

When pure, platinum is harder than copper, silver or gold. Its tenacity is between those of silver and copper. It is very malleable and ductile, and can be rolled into thin sheets or foil, and drawn into fine wire. These properties are impaired by the presence of impurities. Small quantities of the other metals of the platinum group harden it and lower its ductility; whilst as little as 0·03 per cent. of silicon makes it hard and somewhat brittle. This explains the brittleness sometimes met with in platinum pins used in false teeth manufacture. Platinum can be welded at a red heat, and commercial bars are frequently made by consolidating platinum sponge by forging when hot.

In the form of a finely-divided powder, known as "platinum black," it is largely used as a catalytic agent, as it absorbs large quantities of gases, especially of oxygen, with great rapidity. This substance is produced by precipitation from alkaline platinum solutions by means of certain reducing agents such as formic acid. Both platinum black and platinum sponge become lustrous and greyish-white in colour when burnished.

Chemical Properties.—When in mass and in a state of purity, platinum resists the action of most chemical agents. It is not acted on by either pure hydrochloric, nitric, or sulphuric acids. It is dissolved, with the formation of platinum tetrachloride $PtCl_4$, by aqua regia and other mixtures which evolve chlorine, but far less readily than gold. Advantage may be taken of this fact to remove gold that has accidentally adhered to the platinum cups in the parting assay of gold, when the gold cornets have been annealed at too high a temperature. By carefully immersing the cups in dilute aqua regia, moderately heated, the gold may be dissolved off without injury to the platinum cups.

Platinic chloride, like gold chloride, will unite with hydrochloric acid to form the complex chlorplatinic acid H_2PtCl_6. The potassium and ammonium salts of this acid, K_2PtCl_6 and $(NH_4)_2PtCl_6$, are yellow crystalline precipitates which are often used in the assay of platiniferous materials. When decomposed by heating they yield "platinum sponge."

Platinum sponge is slowly dissolved by boiling potassium cyanide with the evolution of hydrogen and formation of a double cyanide.

Platinum is oxidized by fusion with alkaline hydroxides or with potassium nitrate, and is also attacked by fused alkaline cyanides, forming platinocyanides.

Precipitants of Platinum.—*Hydrogen sulphide* slowly produces, in solutions acidulated with hydrochloric or sulphuric acid, a black precipitate of platinic sulphide PtS_2. The precipitate forms more quickly and becomes more granular on boiling the solution and passing a rapid current of hydrogen sulphide. The precipitate is insoluble in nitric and in hydrochloric acids separately, but is soluble in aqua regia. It is soluble with difficulty in normal ammonium sulphide, but more readily in the yellow sulphide, yielding the thio salt $(NH_4)_2PtS_3$. Heated out of contact with the air it is decomposed into the monosulphide PtS and sulphur.

Ammonium Sulphide. This reagent also precipitates platinum as sulphide, soluble with difficulty in excess of the reagent.

Ammonium Chloride. This is an important precipitant of platinum and is used to separate it from solutions containing platinum and other metals. It produces a yellow crystalline precipitate of the double ammonium platinichloride $(NH_4)PtCl_6$. Dilute solutions of platinum must be concentrated by evaporation with a small excess of the reagent to ensure precipitation in the form of coarse, orange-yellow crystals. The precipitate is slightly soluble in cold water, and completely soluble in boiling water. It is insoluble in alcohol, and in excess of the reagent. Upon heating to redness, ammonium platini-chloride is decomposed, leaving spongy platinum, which distinguishes it from the corresponding potassium platini-chloride which, on heating, leaves spongy platinum mixed with potassium chloride.

Potassium chloride also produces a yellow precipitate of potassium platini-chloride, analogous in its appearance and properties to the precipitate just described.

Dimethyl-glyoxime[1] precipitates platinum in feebly acid solutions of platino-chlorhydric acid, H_2PtCl_6 as bronze-coloured crystals of $PtC_8H_{14}N_4O_4$, but about 1 milligramme of metal per 100 c.c. remains in solution; this may be recovered by evaporation. The precipitate does not ignite well and should therefore be heated with concentrated sulphuric acid to convert it into sponge, any platinum dissolved by the acid being precipitated by dilution.

Hydrazine Hydrochloride. In a chloride solution, neutralized with sodium carbonate solution and treated with a moderate excess of 10 per cent. hydrazine hydrochloride solution, platinum is quantitatively precipitated in the metallic state. This reagent is useful in the determination of platinum.

Titanous Salts. Titanous chloride, or preferably titanous sulphate, reduces solutions of platinum with precipitation of the metal (see p. 400).

The following reactions may be compared with those given for gold on p. 237.

Ferrous sulphate gives in neutral solutions a brown precipitate of metallic platinum, but only after prolonged boiling. No reduction takes place in acid solutions.

Oxalic Acid. Platinic chloride is not reduced by oxalic acid. As oxalic acid reduces gold solutions it affords a very convenient precipitant for separating the two metals. (Platinum is slightly reduced by ammonium oxalate but not by sodium oxalate.)

Sulphurous acid does not precipitate platinum from solutions of its salts, but it may be remarked that traces of platinum are usually carried down when sulphurous acid is used to separate gold and platinum, and re-precipitation is necessary for accurate work.

Stannous chloride $SnCl_2$ produces a blood-red coloration, due to the reduction of platinic chloride $PtCl_4$ to platinous chloride $PtCl_2$; but no precipitate is formed. The colour is extracted by adding ether and shaking. This affords a useful reaction for platinum, since it is not given by iridium; but organic matter such as filter-paper must be absent.[2] (See also Gold Reactions, pp. 237 *et seq.*)

Potassium nitrite gives no precipitate at first, but on standing, yellow crystals of a double nitrite of potassium and platinum are deposited.

Metallic zinc (magnesium and other metals) precipitates metallic platinum in a finely-divided state.

Detection of Platinum.—A simple test for distinguishing platinum from palladium, which closely resembles platinum in colour, consists in placing a drop of iodine solution on the surface; this, on evaporation by warming, produces a black stain on palladium but not on platinum.

A similar test reagent is suggested for use in the case of streaks made on a touch-stone, as in the case of gold.

Distinguishing Platinum from White Gold.—In view of the use of white-gold alloys as substitutes for platinum, the following test is useful in distinguishing between them (gold is decolorised with palladium or nickel):—

The test acid consists of 3 parts of hydrochloric acid and 1 part of nitric acid. As this mixture deteriorates after some lapse of time it should be prepared

[1] Thompson, Beamish and Scott, *Ind. and Eng. Chem.*, Anal. Edn., 1937, **9**, (9), 420–422.
[2] Wohler, *Chem. Zeits.*, 1907, **31**, 938.

in small quantities only. A drop of the mixture is applied to a perfectly clean surface of the metal to be tested. It is best to use the acid slightly warm, and to allow the action to go on for about half a minute, or one minute if the test acid is used cold. The drop is then soaked up with a piece of white blotting-paper, and the colour noted. Platinum shows only slight action but no colour stain; palladium white-gold alloy shows a light-brown stain; and nickel white-gold alloy shows a bright golden-yellow stain.

Gravimetric Determination of Platinum. (*a*) *Precipitation as Ammonium Chlorplatinate* $(NH_4)_2PtCl_6$.—This is a very convenient and frequently-adopted method of determining platinum, but the platinum solution must be concentrated and freed from nitric acid by repeated evaporations with hydrochloric acid. The platinum solution is then nearly neutralized with ammonia and evaporated until crystallization begins. Just sufficient hydrochloric acid is then added to dissolve the crystals, and excess of a saturated solution of ammonium chloride added. After thoroughly mixing, an excess of alcohol is added and the solution allowed to stand for twenty-four hours, after which the precipitate of ammonium chlorplatinate is filtered off, and washed well with alcohol or ammonium chloride. It is then dried, ignited, and weighed as metallic platinum.

The ignition should be performed gently at first, to avoid volatilizing some of the platinum chloride, and then afterwards more intensely. With comparatively large quantities of platinum, the ignition should be performed in an atmosphere of hydrogen, and it is advisable to break up the mass to ensure complete decomposition.

Potassium chloride is sometimes recommended for precipitation instead of ammonium chloride, but is not suitable.

Both chlorides are practically insoluble in alcohol and but slightly soluble in cold water, the ammonium salt being the least soluble. When, however, the ammonium salt is heated to redness, it is reduced and pure spongy platinum is left as a residue; whilst in the case of the potassium salt the residue consists of spongy platinum mixed with potassium chloride which is not completely removed by water when washing the precipitate.

(*b*) *Precipitation as Platinum Disulphide* PtS_2.—This is best effected by passing hydrogen sulphide rapidly through the slightly acid platinum solution, which is maintained at the boiling-point. It is necessary to effect the precipitation at boiling-point, otherwise coloured precipitates are obtained, consisting most probably of thio platinic acids. Platinum disulphide precipitate is usually regarded as somewhat difficult to filter, but when formed under the above conditions it will filter readily and satisfactorily.

Single acids attack the precipitate with difficulty; and it is practically insoluble in alkali sulphides. It readily dissolves, however, in aqua regia. When heated to a red heat in air it is decomposed, leaving a residue of metallic platinum which is weighed.

Heating should be gentle at first and under cover, as decomposition of the sulphide is somewhat sudden and violent.

In order to facilitate precipitation of the sulphide, Ivanov[1] recommends the addition of magnesium chloride to the platinum solution. It is removed from the precipitate by washing with acidulated water, leaving the platinum disulphide in a pure state.

Gaze[2] recommends the addition of from 5 to 10 per cent. of mercuric chloride to the platinum solution, whereby the platinum disulphide comes down with

[1] Ivanov, *J. Russ. Phys. Chem. Soc.*, 1916, **48,** 527.
[2] R. Gaze, *Apoth. Zeits.*, 1912, **27,** 959; Abs. *J. Soc. Chem. Ind.*, 1913, **32,** 48.

mercuric sulphide in a form that filters readily. The latter is expelled by heating the precipitate.

Gilchrist and Wichers[1] dilute the platinum solution to 400 mls., add 5 mls. of hydrochloric acid per 100 mls., and treat the hot solution with a rapid stream of hydrogen sulphide, the treatment being continued whilst the liquid cools. The precipitate is collected, washed with 1 per cent. hydrochloric acid, dried, and ignited in a porcelain crucible. The resultant metal is leached with dilute hydrochloric acid, again collected, strongly ignited, and weighed. The metal thus obtained contains on an average 0·0005 gramme of sulphur per gramme, which cannot be expelled by ignition.

(*c*) *Precipitation as Metal.*—Precipitation with zinc (or magnesium) is a convenient means of recovering platinum, and of separating it from base metals, etc. which are not precipitated by zinc, and it is frequently used in the course of platinum metal analysis.

The acid chloride solution is heated to about 80° C., and excess of pure zinc added. When precipitation is complete, the excess of zinc is dissolved by the addition of strong hydrochloric acid. The precipitate of finely-divided platinum is collected and washed twice with N hydrochloric acid, and finally with 0·1/N acid.

Precipitation of platinum as metal is also effected with hydrazine hydrochloride. In the usual procedure the chloride solution, freed from nitric acid if necessary by double evaporation with a slight excess of sodium chloride (0·6 gramme per gramme of platinum), is diluted to 100 to 200 mls., neutralized with sodium carbonate solution, and boiled with a moderate excess of 10 per cent. hydrazine hydrochloride solution until the precipitated platinum has coagulated. This is left to settle, filtered by decantation, washed several times with hot water, and digested with hot 1 per cent. hydrochloric acid, which removes any absorbed alkali. The precipitate is collected, washed with acidulated water, ignited, and reduced in hydrogen if bulky.

Volumetric Determination of Platinum.—Peterson[2] recommends pouring a measured quantity of the platinum solution into a concentrated and cold solution of potassium iodide, and determining the free iodine liberated by titration with thiosulphate and starch as indicator, as previously described for the volumetric determination of gold. The reactions are :—

$$PtCl_4 + 4KI = PtI_2 + I_2 + 4KCl$$
$$I_2 + 2Na_2S_2O_3 = 2NaI + Na_2S_4O_6.$$

From these equations it will be seen that two molecules of thiosulphate (316) correspond to one atom of platinum (194). The accuracy of the method has been proved by titrating known quantities of platinum.

Colorimetric Determination of Platinum.—The red coloration produced in platinum solutions with stannous chloride, due to the reduction of chlorplatinic to chlorplatinous acid, is utilized by C.W. Davis[3] for the determination of platinum in dilute solutions. The aqua-regia solution, freed from nitrates by evaporation, is taken up with 2 c.c. of dilute hydrochloric acid (1 : 1). Two drops of stannous chloride solution (5 grammes stannous chloride and 2 grammes water in 10 c.c. of the above hydrochloric acid) are added and the red colour produced compared with that of solutions containing known quantities of platinum prepared in the same way. The intensity of the colour is inversely

[1] R. Gilchrist and E. Wichers, *J. Amer. Chem. Soc.*, 1935, **57**, 2565 ; quoted by Schoeller and Powell, *loc. cit.*, 280.

[2] H. Peterson, *Zeits. anorg. Chem.*, 1899, **19**, 59.

[3] C. W. Davis, U.S. Bureau of Mines, Tech. Paper No. 270, 1921.

proportional to the acidity, hence the solution to be matched must be prepared with standardized hydrochloric acid (33 mls. of concentrated acid per litre). The method is not recommended for quantities of platinum exceeding 0·0002 gramme.

Palladium and gold interfere, as stannous chloride gives a brown coloration with both these metals.

(2) **Palladium.**

Physical Properties.—Symbol, Pd; atomic weight, 106·7; specific gravity, 11·3; melting-point, 1553° C.

Next to platinum itself, palladium is the most plentiful and most important industrially of the platinum-group metals. It is a white metal, closely resembling platinum in colour, and, being less expensive, is frequently used in place of platinum for jewellery purposes, etc. Its density is about one-half that of platinum, being 11·3 compared with 21·5 for platinum. This is an important consideration when palladium is used as a substitute for platinum, since it means that, weight for weight, palladium has about twice the bulk of platinum and can therefore be manufactured into practically twice the number of similar articles.

Palladium is the most fusible of the platinum-group metals, its melting-point being 1553° C., that is, nearly 220° C. below that of platinum. When molten it absorbs oxygen, and tends to spit like silver when the gas is liberated on solidification. It has approximately the same hardness as platinum, the Brinell hardness being 49, compared with 47 for platinum. It has good malleability and can be rolled into thin sheets, and its extraordinary ductility is demonstrated by the fact that it can be beaten out into leaves in the same way as gold. Its physical properties being closely akin to those of platinum, it can be mechanically worked by the methods that are applicable to the industrial working of platinum.

Chemical Properties.—Palladium oxidizes more readily than platinum, and when heated to about 700° C. undergoes superficial oxidation, with the production of a faint blue film which disappears on further heating.

Palladium is readily attacked by concentrated nitric acid, forming palladous nitrate $Pd(NO_3)_2 \cdot 2H_2O$, but the dilute acid has little action, except when the metal is finely divided or when it is alloyed with silver and some other metals, in which case it is either partially or completely dissolved, according to the proportion of palladium in the alloy. Concentrated hydrochloric acid slowly attacks palladium, and boiling concentrated sulphuric acid converts it into palladous sulphate. Aqua regia rapidly effects its solution.

Precipitants of Palladium.—Dilution with water causes partial precipitation of palladium as a brown basic salt from slightly acid solutions of the nitrate.

Hydrogen sulphide or *Ammonium sulphide* throw down from acid or neutral solutions black palladous sulphide PdS, insoluble in alkaline sulphides but soluble in boiling hydrochloric acid, and readily soluble in aqua regia. The precipitation is best made in a boiling solution.

Alkaline hydroxides precipitate a yellowish-brown basic salt of palladium, soluble in excess of the reagent.

Soluble carbonates precipitate brown palladous hydroxide $Pd\,(HO)_2$, soluble in excess and re-precipitated on boiling.

Ammonium hydroxide or *Ammonium carbonate* produce no precipitate from the nitrate solution, but discolour the dark-brown solution, with formation of double palladammonium salts. With palladous chloride, ammonium hydroxide gives a flesh-coloured precipitate soluble in excess, forming the diammoniate $PdCl_2 2NH_3$. This compound is precipitated by diluting with a large volume of water, and acidulating with hydrochloric acid. It is dried at 110° C. to 120° C.

Potassium iodide (and other soluble iodides) give in very dilute solutions a black flocculent precipitate of palladous iodide PdI_2, somewhat soluble in excess of KI to a red solution. This is a very characteristic reaction of palladium.

Mercuric cyanide gives a yellowish-white, gelatinous precipitate of palladous cyanide $Pd(CN)_2$ readily soluble in potassium cyanide, and in ammonium hydroxide. It is slightly soluble in hydrochloric acid. On heating it leaves a spongy mass of metallic platinum. This also is a most characteristic reaction of palladium, and is unique amongst the platinum metals.

Ammonium chloride does not readily precipitate palladium from solution unless in the palladic form, and unless oxidizing agents are present.

Potassium chloride precipitates red chlorpalladate K_2PdCl_6, insoluble in alcohol and in water. The palladium must be in the palladic state and oxidizing agents must be present.

Stannous chloride produces a brownish-black precipitate, soluble in hydrochloric acid to an intense green solution.

Titanous Salts. Solutions of palladium are reduced by titanous salts, either titanous chloride or sulphate (preferably the latter), with precipitation of the metal.

Ammonium thiocyanate gives no precipitate, even after the addition of SO_2. This reaction distinguishes palladium from copper.

Potassium thiocyanate. In solutions of potassium chlorpalladite, KCNS produces a brick-red precipitate, soluble in excess of thiocyanate to a rich ruby solution of potassium pallado-thiocyanate, $K_2Pd(CNS)_4$, which crystallizes out in ruby needles upon evaporation.

Sulphurous acid gives no precipitate with solutions of palladium salts.

Dimethyl-glyoxime. A one-per-cent. solution of this reagent in 2 per cent. hydrochloric acid, when added to a moderately acid chloride solution of palladium, gives a bulky orange-yellow precipitate of the compound $C_8H_{14}N_4O_4Pd$, which on ignition yields the metal. The precipitate is insoluble in water, in dilute hydrochloric and dilute acetic acids, and in 50 per cent. ethyl-alcohol. The corresponding nickel compound is soluble in dilute mineral acids.

This is a characteristic reaction of palladium and is extremely useful in the quantitative determination of the metal (as described below).

Persulphates. When a solution of potassium palladochloride or pallado-bromide is heated with a persulphate, the corresponding palladichloride or palladibromide is precipitated. This reaction may be used for the separation of palladium from other metals.

Metallic magnesium or *zinc* reduce palladous chloride solutions, with deposition of finely-divided metallic palladium. These metals are useful precipitants for palladium in analysis.

Gravimetric Determination of Palladium.—Of the few reagents used in practice for the separation and determination of palladium, dimethyl-glyoxime is perhaps the most important and the one most generally used, as it is highly sensitive and is unaffected by the presence of other platinum metals. (For co-precipitation of platinum, see below.) The precipitate, however, is very bulky and inconvenient for the determination of large quantities of palladium, consequently, it is usually more convenient in such cases to precipitate the palladium with ammonium chloride. In some cases precipitation as sulphide may be applied, and in others displacement of palladium from solution with metallic zinc or magnesium.

(*a*) *Precipitation with Dimethyl-glyoxime.*—This is a convenient and accurate method for the determination of palladium, and for its separation from other metals of the group, none of which is precipitated by dimethyl-glyoxime with the exception of platinum itself, which, as shown below is partially precipitated under certain conditions.

Any gold in the solution as chloride is completely precipitated by dimethyl-glyoxime ; but the determination of palladium is not hindered by the presence of copper or iron, provided a sufficient excess of dimethyl-glyoxime is present to effect the reduction of any ferric salt present.

This reagent is very sensitive, traces only of palladium being readily detected and determined ; but when the amount of metal in solution is comparatively large there is a tendency for the precipitate to creep up the sides of the beaker, and due precaution is necessary in collecting it.

Aqua regia solutions must be evaporated with hydrochloric acid to free from nitric acid, and to convert the palladium into chloropalladous acid, before precipitation with dimethyl-glyoxime.

The palladium precipitate being very voluminous, the use of dimethyl-glyoxime becomes impracticable with large quantities of palladium, as in the analysis of the commercial metal. In such cases the bulk of the metal may be precipitated as iodide with potassium iodide, and the balance with dimethyl-glyoxime in the filtrate ; or the palladium may be determined with ammonium chloride.[1]

The precipitation of palladium from sulphate solutions is not quite quantitative.

In the usual procedure,[2] the moderately acid chloride solution of palladium is diluted to 200 mls., treated with a slight excess of a 1 per cent. solution of dimethyl-glyoxime in 2 per cent. hydrochloric acid (0·1 gramme of palladium requires 0·22 gramme of the precipitant), and set aside for half-an-hour. The bulky orange-yellow precipitate is collected and washed with 1 per cent. hydrochloric acid, then with hot water. Suction filtration is employed with advantage for the voluminous precipitates from metal exceeding about 0·05 gramme. It may be advisable to dissolve larger precipitates in aqua regia and repeat the precipitation.

The washed precipitate, wrapped in the filter, is transferred to a weighed porcelain crucible, and carefully dried on an asbestos mat ; the heat is then gradually increased until the paper is charred, and finally the metal is ignited more strongly, reduced in hydrogen, cooled under carbon dioxide, and weighed as pure palladium.

Alternatively, the precipitate produced by dimethyl-glyoxime may be collected on a weighed filter or porous porcelain crucible, washed as before, and gently heated to constant weight at 110° C. (palladium factor, 0·3167).

During ignition, palladium usually assumes a deep-purple tint due to a surface film of oxide, which may be reduced by treating the warm ignited metal with a few drops of formic acid, instead of by reduction in hydrogen. After washing and drying the metallic palladium is weighed.

When other metals are present and have to be determined the filtrate from the palladium precipitate is freed from excess of dimethyl-glyoxime by evaporation almost to dryness and digestion with strong hydrochloric acid and small additions of sodium chlorate.

As stated above, gold is quantitatively precipitated with the palladium by dimethyl-glyoxime but can be separated by the following procedure :—

[1] Schoeller and Powell, *op. cit.*, 272.

[2] M. Wunder and V. Thuringer, *Zeits. anal. Chem.*, 1913, **52**, 101.

The washed combined precipitate is rinsed back into the beaker, and boiled with dilute aqua regia until it has dissolved. The solution is cooled, diluted until the acidity is reduced to 2 to 3 per cent., and the palladium re-precipitated with dimethyl-glyoxime; the nitric acid present in the solution prevents the co-precipitation of the gold.[1] For the same reason, palladium alone will be precipitated if nitric acid is present in, or is added to, the original solution.

Platinum is also co-precipitated to a certain extent with the palladium if the solution is hot and an alcoholic solution of the reagent is employed, part of the platinum being reduced to the platinous condition, in which condition it reacts like palladium.

The co-precipitation of platinum is detected by the greenish tint of the palladium precipitate, and is prevented by working in cold dilute solution as prescribed above, or by addition of nitric acid.

Palladium precipitates containing platinum or gold may be purified by dissolving in aqua regia and re-precipitating.

Precipitation with dimethyl-glyoxime is a more attractive procedure for the determination of palladium than hydrogen-sulphide precipitation, but the latter has its uses in some cases.

(*b*) *Precipitation as Ammonium Chlorpalladate.*—The precipitation of palladium as ammonium chlorpalladate with ammonium chloride is convenient in some cases. The chloride solution of palladium, containing a liberal amount of ammonium chloride and excess of nitric acid, is slowly evaporated until the ammonium chlorpalladate separates out as a bright-red crystalline precipitate and the solution becomes colourless.

After cooling, the precipitate is collected, washed with strong ammonium chloride, ignited gently until the ammonium salt is expelled, and then, with the heat raised to redness, reduced in hydrogen and weighed as metallic palladium.

Ammonium chloride will also precipitate any platinum present. In this case the palladium is first separated by precipitation with dimethyl-glyoxime, and the platinum determined in the filtrate after evaporation with hydrochloric acid and the addition of small amounts of sodium chlorate for the destruction of the excess of dimethyl-glyoxime in the solution.

(*c*) *Precipitation as Sulphide.*—Palladium is the most readily precipitated as sulphide (palladous sulphide, PdS) of all the metals of the platinum group of metals. It is best precipitated in this form from slightly acidulated solutions of its salts by boiling the solution and passing a current of hydrogen sulphide gas. Under these conditions the precipitate becomes granular and is filtered off, washed with hot water, then dried, ignited and weighed as metallic palladium.

If the palladium solution is not perfectly clear after passing the gas, a little more hydrochloric acid should be added and the solution again boiled.

The precipitation may also be effected with ammonium sulphide but this reagent is not quite so satisfactory as hydrogen sulphide gas. When ignited, the palladium-sulphide precipitate does not part with the whole of its sulphur, and the impure metal tends to run into one or several globules when much precipitate is involved. In this case the precipitate should be dissolved in aqua regia, and re-precipitated with dimethyl-glyoxime or ammonium chloride, after evaporation of the solution with hydrochloric acid to expel nitric acid, etc. The ignited precipitate is reduced in hydrogen, cooled under carbon dioxide, and weighed.

[1] *Ausgew. Methoden,* 137, per Schoeller and Powell, *op. cit.,* 272.

(*d*) *Precipitation as Pallados-ammine Chloride.*[1]—The palladium present as chloride is converted into soluble tetrammino-palladous chloride by the addition of excess ammonia solution, according to the equation

$$PdCl_2 + 4NH_3 = [Pd(NH_3)_4]Cl_2.$$

When excess of hydrochloric acid is added, the sparingly soluble yellow compound known as pallados-ammine chloride is precipitated according to the equation

$$[Pd(NH_3)_4]Cl_2 + 2HCl = [Pd(NH_3)_2Cl_2] + 2NH_4Cl.$$

If other metals are carried down in small quantity the impure pallados-ammine chloride is purified by dissolving in ammonia and re-precipitating with hydrochloric acid; the pure salt is converted into sponge by igniting and weighed as pure palladium.

Colorimetric Determination of Palladium.[2]—This method was devised by F. C. Robinson, of the Sheffield Smelting Co., Ltd., in 1913, for determining small quantities of palladium in parting acids, etc., and consists in matching the ruby colour of the assay solution of palladium with a standard palladium solution of the same volume. In common with other colorimetric methods, it is based on the assumption that equal volumes of different solutions of the same colour will contain equal quantities of the substance that produced the colour. As the tint imparted by palladium salts to sulphuric acid is not quite the same as that imparted to nitric acid, it is necessary to make up standards of both acids.

The standards are prepared from a weighed quantity of pure palladium sponge, finely powdered and dissolved in half-strength nitric acid; a little powdered silver added to the palladium is found to assist solution.

In making the assay, the whole or a carefully measured portion of the assay solution, such as the sulphuric or nitric acid solutions from parting, is placed in a white glass cylinder, standing on a white surface to facilitate comparison. Into an exactly similar cylinder is placed nearly the same quantity of concentrated acid (sulphuric or nitric), and then standard palladium solution added from a burette until the colour approaches that of the assay solution, the liquid being well mixed after each addition.

More concentrated acid is then added until the bulk of the two solutions is equalized, and then more standard palladium solution carefully added until the colour of both the assay solution and the standard are the same. The burette reading gives the quantity of standard solution added, from which can be calculated the palladium content of the solution tested.

The acids used must be free from chlorides, since any silver present will be converted into chloride, which gives a slight opalescence that makes comparison and the accurate matching of colours very difficult or impossible.

Metals other than palladium which give a coloured solution—such as platinum, ruthenium, iron, copper, and nickel—must obviously be absent. The base metals should be eliminated during the cupellation of the precious metals on magnesite cupels.

In sulphuric-acid parting, traces of ruthenium give a pink tint; and excessive boiling may dissolve sufficient platinum to produce a brown tint, thus making comparison not only difficult but in some cases unreliable.

In nitric-acid parting solutions, any platinum present should not exceed 5 mg. for every 50 mg. of palladium, and above this quantity platinum imparts a visible colour of its own and makes comparison results high.

[1] See "Metals of the Platinum Group," Atkinson and Raper, *J. Inst. Met.*, 1936, (2), **59**, 185.

[2] *Trans. Inst. Min. Met.*, 1926, 35.

Solutions containing very little palladium may, if necessary, be concentrated by evaporation in order to intensify the colour.

With experience the method gives accurate results under the conditions named, but it is necessary when testing solutions from double parting that the sample assayed should contain gold in amount equal to ten times the platinum and palladium together, and silver about three times the combined weight of the other three metals. If a preliminary assay shows a deficiency of these metals, weighed quantities must be added to make the correct ratio.

For a more accurate determination of the total palladium a special cupellation assay should be made, and the bead parted in nitric acid only. This reduces the solution to be tested to one kind only, and minimizes inaccuracy. But to correct the silver and platinum results obtained in the usual double parting assay, it is necessary to know the amount of palladium in the sulphuric and nitric acids, respectively; and the colorimetric method is quick and convenient for determining this in serial works practice.

(3) Iridium.

Physical Properties.[1]—Symbol, Ir; atomic weight, 193·1; specific gravity, 22·4; melting-point, 2454° C.

Iridium is distinguished physically from platinum by its greater hardness, its brittleness, and its slight grey tint. It can be melted in the high-frequency induction furnace. Experience shows that cold-working of iridium is impossible owing to its brittleness, the metal crumbling to small fragments after a very small reduction in the rolls. It can best be worked at a white heat. As a starting-point either sintered sponge or fused metal may be used. Iridium is largely used as a hardener for alloying with other metals of the platinum group, and is usually supplied industrially in the form of sponge.

Chemical Properties.—In compact form, iridium is not attacked by any single acid or by mixtures of acids. It is soluble, more or less, in aqua regia if alloyed with a large excess of platinum. When alloyed with a large excess of silver, lead or zinc, and the alloying metal subsequently removed with nitric acid, the iridium is left in a state of fine sub-division and is more susceptible to attack by aqua regia. When iridium powder is mixed with 2 parts of sodium chloride and strongly heated in a current of moist chlorine, sodium chlor-iridate is formed (Na_2IrCl_6). This salt is very soluble in water and in alcohol. It may be obtained also by solution of the hydroxide or sulphide in aqua regia and evaporation with sodium chloride.

Iridium, in the state of very fine powder, when fused with sodium peroxide, yields a residue containing the sesquioxide Ir_2O_3 with alkali, which is practically insoluble in water. On treating the residue with aqua regia a dark brownish-green solution of sodium chlor-iridate is obtained. This compound is isomorphous with potassium chlorplatinate.

Precipitants of Iridium.—*Hydrogen Sulphide.*—In a slightly acid and nearly boiling solution of chlor-iridate, hydrogen sulphide first decolorizes the dark solution, owing to the reduction of the tetrachloride to trichloride, and subsequently precipitates the sesquisulphide Ir_2S_3.

When iridium sulphide is dissolved in aqua regia and the solution evaporated with sodium chloride, a dark-brown mass of sodium chlor-iridate is obtained, which is decomposed on heating, yielding metallic iridium.

Ammonium sulphide yields the same precipitate, soluble in excess.

[1] R. H. Atkinson & R. A. Raper, "Metals of the Platinum Group," *J. Inst. Met.*, 1936, **59**, 179–210.

Alkali Hydroxide. By the action of alkali hydroxide or carbonate upon a boiling solution of sodium chlor-iridate, or by bromate hydrolysis, a deep-blue precipitate of hydrated iridium dioxide is formed.

Ammonium Chloride. When a cold chlor-iridate solution is treated with excess of solid ammonium chloride, the ammonium salt $(NH_4)_2IrCl_6$ is precipitated as a more or less amorphous, dark red powder.

A solution of sodium chlor-iridate, when evaporated with a small excess of ammonium chloride, yields a lustrous, black, crystalline precipitate of ammonium chlor-iridate.

Potassium chloride yields the corresponding potassium chlor-iridate.

Reducing agents such as ferrous salts, oxalic acid, stannous chloride, sulphur dioxide, hydrazine, etc. readily reduce alkali chlor-iridates, especially in hot solutions, to double salts of iridium trichloride known as chlor-iridites (M_3IrCl_6). Sodium, potassium, and ammonium chlor-iridites form olive-green crystals soluble in water, but insoluble in alcohol. Chlor-iridites are oxidized to chlor-iridates by the action of aqua regia.

Metallic zinc or *magnesium* reduce iridium salts to metal, which is deposited as a fine black powder. In this condition metallic iridium is less resistant to attack by aqua regia than when in mass.

Determination of Iridium.—In the determination of iridium advantage is usually taken of its insolubility in the residue obtained from aqua regia treatment. The finely-divided iridium is collected washed, ignited, reduced, and weighed. It is also obtained as metal when thrown out of solution with zinc or magnesium.

For the separation and gravimetric determination of iridium several precipitants are available. The finely-divided metal is first converted into a soluble compound by fusion with sodium peroxide.[1] The powder is thoroughly mixed with from 15 to 20 grammes of sodium peroxide, 2 or 3 grammes of sodium hydroxide added, and the whole fused in a nickel crucible. Iridium is precipitated as metal from the alkaline sodium hydroxide solution by means of hydrazine hydrochloride. Before ignition the precipitated iridium is leached with 1 per cent. hydrochloric acid.

Ammonium chloride is frequently employed as a precipitant for iridium. The procedure for its precipitation by this reagent from solutions of ammonium chlor iridate is the same as that given for the corresponding platinum salt. The precipitation is quantitative within the limits of experimental error. The ignited iridium must always be reduced before weighing.

Iridium may also be determined by precipitation as hydroxide (bromate hydrolysis) as described above. The flocculent precipitate of iridium hydroxide is washed with 1 per cent. ammonium chloride solution, dried by gently warming in a porcelain crucible, moistened with a few drops of saturated ammonium chloride solution (to prevent decrepitation), ignited very gently at first, then strongly, and reduced in hydrogen. The ignited metal is leached with dilute hydrochloric acid, again collected, ignited, reduced, and weighed.

When thrown out of solution with metallic zinc or magnesium, the finely-divided deposited metal before ignition is treated with dilute sulphuric acid to extract any oxide of zinc or magnesium with which it may be contaminated ; it is then washed with hot water, dried, ignited in hydrogen and weighed. Should platinum be deposited with the iridium, it may be dissolved out by treating the ignited residue of the combined metals with aqua regia diluted with four to five times its volume of water, and heating to a temperature of 40° to 50° C. Iridium is not dissolved in this strength of aqua regia, unless it

[1] Schoeller and Powell, *op. cit.*, 281.

is too hot. The residue of combined metals must not be ignited at too high a temperature, otherwise the platinum and iridium will alloy and the final residue will contain platinum.

Gold cornets from the nitric-acid parting assay which contain iridium are usually digested with dilute aqua regia to dissolve the gold and leave the iridium as a finely divided residue, which is washed, dried, ignited and weighed.

(4) **Rhodium.**

Physical Properties.[1]—Symbol, Rh ; atomic weight, 102·9 ; specific gravity, 12·4 ; melting-point, 1966° C.

Rhodium is a hard white metal, somewhat resembling aluminium in appearance. At various times rhodium has been melted, cast, and mechanically worked, but high-purity metal is essential for ease of working. It may be worked starting either with sintered sponge or a cast ingot.

Rhodium can be melted readily in the high-frequency induction furnace. During melting, oxygen is absorbed in considerable quantity, and the metal tends to spit and sprout during solidification. When molten it has a considerable affinity for carbon, a property shared by the other metals of the platinum group. The forging or swaging of rhodium ingots does not present any unusual difficulties. They cannot be forged at room temperature, but become quite malleable above a red heat. For the production of wire the ingots may be hand-forged at about 1100° C. to bars of a suitable size for reduction by swaging.

Apart from its use for alloying purposes, rhodium is being electro-deposited on sterling silver and other wares to give an untarnishable finish. The colour of electro-deposited rhodium is similar to that of platinum but slightly whiter, and much of the finest jewellery to-day is finished with a thin rhodium surface, in order further to enhance the brilliance of the gems, and render soldered joints untarnishable and not conspicuous. Rhodium plating is also being widely used for reflectors, and by manufacturers of imitation jewellery to give platinum-like finish and protection against tarnishing.

Chemical Properties.—When heated in the air the surface of rhodium becomes slightly tarnished or oxidized.

The compact pure metal is practically insoluble in all acids, including aqua regia, although when alloyed with certain other metals it is attacked by aqua regia, and a portion of the rhodium passes into solution. Thus, alloys of platinum and rhodium containing less than 5 per cent. of rhodium are soluble in aqua regia. By the action of hot strong sulphuric acid, spongy rhodium is converted into rhodium sulphate $Rh_2(SO_4)_3$.

Finely-divided rhodium is attacked by fused bisulphate, giving an orange-red melt which contains the double sulphate $M_2SO_4Rh_2(SO_4)_3$.

As in the case of iridium, pure finely-divided rhodium—such as that precipitated from solution by zinc—when intimately mixed with two parts of sodium chloride and strongly heated in a current of moist chlorine, yields a garnet-red melt, soluble in water to a rose-red solution.

When finely-divided rhodium is fused with excess of lead and the button dissolved in dilute nitric acid, the rhodium remains as a residue, which dissolves on being vigorously boiled in strong sulphuric acid. Treatment of the lead button with nitric acid may result, in certain circumstances, in part of the rhodium passing into solution, as nitrate of rhodium may exist in solution with lead nitrate, as in the case of silver nitrate.

[1] Atkinson & Raper, *loc. cit.*

Precipitants of Rhodium.—The reactions of rhodium are less pronounced than those of the other platinum metals, and its presence in solution is easily masked by preponderating amounts of iridium, platinum, or palladium. Rhodium is distinguished from the other platinum metals by the fine rose-red colour of its "double" chlorides, and this characteristic colour affords a means of identifying rhodium.

Hydrazine hydrochloride. Rhodium is quantitatively precipitated as metal when a solution of the chloride is rendered alkaline with sodium carbonate and boiled with hydrazine hydrochloride solution in excess until the precipitated metal is coagulated.

Ammonium sulphide precipitates rhodium sesquisulphide, RhS, from hot solutions of rhodium salts. The precipitate is insoluble in excess of this reagent, in which respect it differs from platinum and iridium sulphides, both of which are soluble in excess of ammonium sulphide.

Hydrogen Sulphide. The precipitation of rhodium by hydrogen sulphide is somewhat erratic. In some cases quantitative precipitation is achieved, whilst in others the passing of a rapid current of hydrogen sulphide through a chloride solution for a long period fails to completely precipitate the rhodium as sulphide. Consequently this precipitant cannot conveniently be used for the quantitative separation of the metal.

Rhodium sulphide is soluble in aqua regia, and evaporation of the resulting solution with sodium chloride yields a rose-red solution of the double chloride.

Alkalis. In concentrated solutions, alkalis effect the precipitation of yellowish-brown rhodium hydroxide, Rh_2O_3, $3H_2O$, on boiling the solution. Precipitation is also facilitated by the addition of alcohol. The precipitate is soluble in excess of the reagent, in contradistinction to the corresponding ruthenium salt which is insoluble. Potassium hydroxide gives no precipitate from cold solutions of rhodium tri-chloride, but rhodium hydroxide is precipitated on the addition of alcohol. This reaction is characteristic of rhodium.

Ammonium hydroxide gives a pale-yellow flocculent precipitate which forms only in concentrated solutions and after standing. The precipitate is soluble in hydrochloric acid, forming a rhodammine salt.

Potassium Iodide. On adding a warm concentrated solution of potassium iodide to one of rhodium bromide a black precipitate of rhodium tri-iodide, RhI_3, is obtained. No precipitate is given in cold dilute solutions. Potassium iodide gives no precipitate in solutions of the double sulphate or chloride.

Potassium nitrate yields, in solutions of the trichloride, an orange-yellow precipitate which is slightly soluble in water, and is but slightly decomposed by concentrated hydrochloric acid. This reaction also is characteristic of rhodium.

Potassium chloride effects the precipitation of the red crystalline double salt K_2RhCl_5, but only on adding excess of potassium chloride to a solution of rhodium trichloride and allowing it to crystallize. The precipitate is slowly soluble in water, and insoluble in alcohol.

Metallic zinc or *magnesium* precipitate finely-divided metallic rhodium from solutions of its salts, insoluble in aqua regia.

Determination of Rhodium.—Rhodium is usually determined as metal, and for this several reagents are available.

The chloride solution is rendered alkaline with sodium carbonate and boiled with excess of hydrazine hydrochloride until reduction is complete and the precipitated metal coagulates. The precipitate is leached with 1 per cent. hydrochloric acid, then ignited in a weighed crucible, reduced in hydrogen, and cooled in carbon dioxide.

Another procedure is to precipitate the rhodium as hydroxide in neutral chloride solution by bromate hydrolysis, and reduce to metal.

Rhodium may also be quantitatively precipitated from chloride or sulphate solutions by displacement with magnesium or zinc, but the results tend to be high, especially with zinc. When, however, small amounts of rhodium only are involved, the results are usually sufficiently accurate for technical work. The precipitated metal is washed with dilute hydrochloric acid, ignited, reduced, and then digested with N hydrochloric acid for half-an hour to remove traces of the reducing metal ; after which it is again collected, washed, ignited, reduced, and weighed.

The determination of rhodium in alloys is dealt with later in this chapter.

The titanous sulphate method [1] is also applied to the determination of rhodium and its separation from iridium.

The chloride solution of the metals is evaporated with 10 c.c. of sulphuric acid and 2 to 3 c.c. of nitric acid until heavy fumes are evolved ; after diluting to 200 c.c. the boiling solution is treated with a 20 per cent. titanous chloride solution until the rhodium is precipitated as metal and the solution remains purple.

The rhodium is washed with dilute sulphuric acid, redissolved by strongly heating with 10 c.c. of sulphuric acid and fuming nitric acid, and then reprecipitated as before.

The second rhodium precipitate is dissolved in sulphuric acid and the solution boiled with 20 c.c. of water and 10 c.c. of hydrochloric acid and diluted to 400 c.c.: the rhodium is then precipitated with hydrogen sulphide and the rhodium sulphide, Rh_2S_3, converted into metallic rhodium for weighing.

The filtrates from the titanous chloride treatment contain the iridium and titanic sulphate, $Ti(SO_4)_2$; the latter is removed by precipitation with cupferron at 0° to 5° C., the excess of the reagent is destroyed by evaporation of the solution with nitric acid, and the iridium precipitated by neutralizing to pH 6 and boiling with sodium bromate.[2] The black IrO_2, aq. is washed with 1 per cent. ammonium chloride, dried, ignited, and reduced in hydrogen to iridium for weighing.

Stripping Electro-deposited Rhodium.—The stripping of defective rhodium deposits, and the determination of the amount of rhodium electro-deposited on silverware, etc. as a protective coating against tarnish, present considerable difficulty owing to the resistance of the metal to chemical attack. The deposit is very thin and normally only of about 7 to 10 millionths of an inch, which has been found sufficient to prevent tarnishing, but even this extremely thin film is very difficult to remove.

One method,[3] based on experience, is to immerse the article in a mixture of molten potassium and sodium cyanides ; if the article is made the anode and a current passed, the removal of the rhodium film is accelerated. In this way rhodium plating can be removed from sterling silver articles without any significant attack of the silver. (Further details are given in British Patent No. 419,819.) It is obvious, however, that the process is only suitable for operation by experts, and there is still need for a simple and effective process which can be carried out in any plating shop.

[1] R. Gilchrist (*U.S. Bur. Stand. J. Research*, 1932, **9**, 547–556): Research Paper No. 489 : Abs., *Met. Abs. Inst. Metals*, 1932, (3), **50**, 749.

[2] Abbreviation $pH = \log \{1/(\text{H-ion concentration})\}$.

[3] Atkinson & Raper, "Metals of the Platinum Group," *J. Inst. Met.*, 1936, (2), **59**, Discussion, p. 210.

In regard to the rhodium content of the plating solutions, the following electrolyte recommended by Atkinson and Raper[1] may be taken as typical of those in use. It is a sulphate solution containing 2 grammes of rhodium and 33 c.c. sulphuric acid per litre, and is used with platinum anodes at 40° C. and 5 amperes per square foot.

(5) Ruthenium.

Physical Properties.[2]—Symbol, Ru; atomic weight, 101·7; specific gravity, 12·2; melting-point not known, but higher than that of iridium.

Ruthenium is a white metal, resembling iridium in appearance, is very hard and more or less brittle, and inherently difficult to fabricate.

During melting there is a considerable loss of metal due to the formation of the volatile ruthenium dioxide and tetroxide. When the molten metal solidifies it tends to spit like rhodium. On melting ruthenium in a graphite crucible in a high-frequency induction furnace it dissolves about 1·2 per cent. of carbon, thus conforming to the general behaviour of the platinum-group metals.

So far little is known of the mechanical properties of pure ruthenium, but there are indications that sintered ruthenium powder is forgeable into a compact mass at very high temperatures.

Chemical Properties.—In compact metallic form, pure ruthenium is not attacked by any single acid, or by aqua regia. When alloyed with a large excess of certain other metals, such as silver and lead, which are subsequently removed by acid treatment, the ruthenium remains in a state of fine subdivision which is more or less acted upon by aqua regia, depending upon the degree of subdivision. This attack is less pronounced than in the case of rhodium.

Finely-divided metallic ruthenium dissolves in alkaline hypochlorite solution, to which it imparts a deep yellow to orange colour.

Ruthenium is best obtained in solution by fusion of the finely-divided metal with sodium hydroxide and nitrate, chlorate, or peroxide. The melt when treated with hot water yields a deep orange-red solution of sodium ruthenate.

The characteristic compound of ruthenium is the volatile *tetroxide* RuO_4. It is less readily formed and more unstable than the corresponding oxide of osmium, and advantage is taken of this fact in separating the metals by distillation. Ruthenium tetroxide is a yellow solid which is decomposed at 107° C. with a smoky flame, forming the dioxide, which is very volatile and yields a vapour that is explosive, poisonous, and of an offensive odour. By the action of hot hydrochloric acid, the tetride is converted into non-volatile ruthenium trichloride, the yellow solution turning a deep brownish-red, which is characteristic and distinguishes it from osmium.

The tetroxide is as important for its detection of the metal as it is for its quantitative separation by distillation.

The reddish-brown distillate of ruthenium trichloride obtained under the conditions of the chlorine distillation is a very sensitive reaction, and absolutely specific for ruthenium.[3]

Precipitants of Ruthenium.—*Hydrogen Sulphide.*—When hydrogen sulphide is passed through a solution of ruthenium trichloride partial reduction is liable to occur, as revealed by the blue colour of the solution.

Ammonium Sulphide. When a feebly acid solution of ruthenium trichloride is boiled with excess of ammonia and ammonium polysulphide for about ten minutes, then acidified with dilute hydrochloric acid and again boiled for some time, ruthenium sulphide is precipitated, and under these conditions coagulates completely.

[1] Precious Metal Supplement, *Watchmaker and Jeweller*, 1934.
[2] Atkinson & Raper, *loc. cit.*, 196.
[3] Schoeller and Powell, *op. cit.*, 286.

Zinc or *Magnesium.* Ruthenium is quantitatively precipitated from solutions of the chloride by displacement with zinc or magnesium.

Sodium Thiosulphate. A dilute solution of ruthenium trichloride, when made ammoniacal and boiled after the addition of a crystal of sodium thiosulphate, assumes an intense violet colour, which is characteristic of ruthenium. Other platinum metals should be absent.

Ammonium Chloride. Ruthenium may be precipitated from the concentrated distillate of the chloride by the addition of solid ammonium chloride until the liquid is saturated. The brown crystalline precipitate of ammonium chloro-ruthenite is compact, but comparatively soluble.

Alkalis. The addition of alkali hydroxide, carbonate or bicarbonate to solutions of ruthenium trichloride results in a black precipitate of hydrated ruthenium sesquioxide, $Ru_2O_3 . 3H_2O$.

Determination of Ruthenium.—The separation and determination of ruthenium is always preceded by fusion with sodium peroxide, followed by distillation and the production of ruthenium trichloride. From this solution the ruthenium is usually precipitated as sulphide, which is ignited and the ruthenium weighed as metal. For this purpose, the ruthenium chloride solution is freed from excess of acid by evaporation on the water-bath almost to dryness, and the moist residue dissolved in 100 to 200 mls. of 2 per cent. hydrochloric acid. The solution is boiled with excess of ammonia and ammonium polysulphide for 10 minutes, acidified with dilute hydrochloric acid, and again boiled for some time until the mixed precipitate of sulphur and ruthenium sulphide coagulates completely. The precipitate is collected, washed with hot water, ignited wet in a weighed crucible, reduced as usual in hydrogen, cooled in carbon dioxide, and weighed as ruthenium.

This procedure becomes inconvenient on account of the bulkiness of the precipitate when more than about 0·2 gramme of ruthenium is in the solution.[1] When large quantities of ruthenium have to be determined the bulk of the ruthenium may be precipitated from the concentrated distillate by addition of solid ammonium chloride until the liquid is saturated. The brown crystalline precipitate of ammonium chloro-ruthenite is collected, washed twice with ammonium chloride solution, and ignited to metal with the usual precautions. The pale brown filtrate and washings are then precipitated with polysulphide as described, the balance of the ruthenium thus being recovered as sulphide.

The double-chloride precipitate is crystalline and compact but comparatively soluble, hence the necessity for a combination process.

Zinc precipitation is also a convenient means of recovering ruthenium from solutions, and of separating it from base metals, etc., not precipitated by zinc. The acid chloride solution free from nitrates is heated to about 80° C., and treated with excess of pure powdered zinc, or thin strip zinc. When action is complete, the excess of zinc is dissolved by small successive additions of strong hydrochloric acid to the hot solution. The precipitated ruthenium is collected on a loose filter and washed twice without delay with N hydrochloric acid, the washing being completed with 0·1 N acid. It is then dried, ignited, reduced, and weighed as metallic ruthenium.

Gilchrist Method.—The following method for the determination of ruthenium and its separation from platinum, palladium, rhodium and iridium has been devised by Gilchrist.[2] The method of separation differs essentially from existing methods in that the ruthenium is distilled from a solution of its sulphate in

[1] Schoeller and Powell, *op. cit.*, 280.

[2] Raleigh Gilchrist (*U.S. Bur. Stand. J. Research,* 1934, **12,** 283–290 : *Met. Abs. Inst. Metals,* 1934, **1,** 251.

diluted sulphuric acid containing sodium bromate. A very suitable reagent for the quantitative absorption of the distilled tetroxide was found to be diluted hydrochloric acid saturated with SO_2. The recovery of ruthenium is accomplished by precipitating the hydrated oxide from a boiling solution at *p*H 6, after sulphite compounds have been decomposed by treatment with hydrochloric acid. The hydrated oxide is ignited in the air to an anhydrous oxide, which is then reduced to metal under hydrogen.

The separation of ruthenium from osmium and other metals by distillation, and its determination in the hydrochloric acid distillate are described below.

(6) **Osmium.**

Physical Properties.—Symbol, Os; atomic weight, 191·5; specific gravity 22·48; melting-point, not determined, but higher than ruthenium.

Very little indeed is known about the working properties of both osmium and ruthenium. Osmium is the most infusible metal of the platinum group. Osmium sintered on heating to a temperature estimated to be equivalent to the melting-point of rhodium (1966° C.), and the metal had a pronounced metallic lustre showing a bluish tinge.[2] It is hard and brittle, and the heaviest substance known.

Chemical Properties.—When strongly heated in air, osmium is oxidized to the volatile tetroxide OsO_4, which has a very penetrating odour somewhat resembling that of chlorine, is intensely poisonous, and may produce permanent injury to the eyesight. The tetroxide is the most important and most characteristic compound of osmium. It is soluble in water to a colourless liquid devoid of acid reaction, and emitting the characteristic smell of the compound, especially when warmed. The same poisonous vapour is evolved when osmium compounds are heated with concentrated nitric acid.

The compact metal is insoluble in all acids, but when very finely divided, osmium is more or less soluble in fuming nitric acid, and is slightly attacked by aqua regia.

The finely-divided metal may be rendered soluble by fusion with sodium peroxide, or by a mixture of sodium peroxide and nitrate or chloride, or with barium peroxide and nitrate.

When solid or gaseous osmium tetroxide acts on sodium hydroxide in the presence of a reducing agent (*e.g.* alcohol), sodium osmate Na_2OsO_4 is obtained, the solution assuming a deep purple colour.

Osmium is attacked by chlorine at high temperatures, the tetrachloride $OsCl_4$ forming at 650° to 700° C. Finely-divided osmium, obtained by ignition of the sulphide or hydroxide in hydrogen, dissolves slowly in strong, alkaline hypochlorite solution, giving an orange-yellow solution.

The finely-divided metal, when ignited in air, volatilizes as tetroxide and emits a very penetrating smell resembling ozone.

The outstanding property of osmium, shared by no other element, is the conversion of its soluble compounds into a volatile oxide (OsO_4) by the action of nitric acid, a property which is utilized in the quantitative separation of the metal by distillation. The lower oxide OsO_2 is not volatile.

Precipitants of Osmium.—*As hydroxide.*—When the alkaline osmate solution is approximately neutralized with sulphuric acid, the colour of the solution changes from mauve to brown. Upon heating the solution to about 60° C., then accurately neutralizing with sulphuric acid and gently boiling for some hours, a deep-black pulverulent precipitate of osmium hydroxide is formed and settles readily.

[2] Atkinson & Raper, *loc. cit.*, 196.

As Sulphide.—Osmium is precipitated as sulphide when the alkaline osmate solution is heated with ammonium polysulphide, boiled for 10 to 15 minutes, and acidified with hydrochloric acid. After slight cooling, and treating the solution with a moderate excess of ammonia and more polysulphide, again boiling for 10 minutes and then re-acidifying, the osmium is quantitatively precipitated as sulphide. The final boiling should be continued until the sulphur in the precipitate coagulates.

Determination of Osmium.—Processes for the determination of osmium are based upon the fact that the soluble compounds of osmium, normally derived from a fusion with oxidizing alkaline fluxes, are converted into the volatile tetroxide when their solutions are boiled with excess of nitric acid; this property allows osmium to be quantitatively separated from all other elements by distillation.

The osmium in the osmate solution from distillation is usually determined either by precipitation as hydroxide, or as sulphide.

If the amount of osmium to be determined is substantial, as indicated by the more or less intense mauve colour of the osmate solution, it should be precipitated as hydroxide; if, on the other hand, the solution is barely or faintly coloured, precipitation as sulphide is to be preferred.[1]

For precipitation as hydroxide,[2] the alkaline osmate solution is transferred from the distillation receiver to an 800-ml. beaker, and approximately neutralized with sulphuric acid (1:1), about 20 mls. of acid being required for an initial charge of 25 grammes of sodium hydroxide used as precipitant. The solution is heated to about 60° C., then accurately neutralized with 2 N sulphuric acid (spot test on litmus paper), and gently boiled for some hours or until the deep-black pulverulent precipitate of osmium hydroxide settles readily. After allowing to settle, the precipitate is collected in a porous porcelain filtration crucible, and thoroughly washed with saturated ammonium chloride solution; the crucible is allowed to dry over sulphuric acid, placed inside a Vitreosil Rose crucible, and warmed with a micro-burner in a stream of dry hydrogen, reduction to metal taking place below 200° C. Once started, the reduction proceeds readily, the ammonium chloride from the wash-liquor tending to lessen decrepitation. The heat is gradually increased, the Rose crucible being finally heated with the full flame of a bunsen burner for half-an-hour. The reduced osmium is allowed to cool under hydrogen, which is then displaced by dry carbon dioxide to prevent oxidation. The porous crucible and contents is accurately weighed, and then heated in an efficient draught-cupboard until the osmium has been volatilized. The crucible is ignited to constant weight, the fixed residue, consisting of silica derived from the action of the alkali on the beaker, is weighed and the weight subtracted from the first weight, the difference being the weight of the osmium.

For precipitation as sulphide the alkaline osmate solution is heated with ammonium polysulphide, boiled for 10 to 15 minutes, and acidified with hydrochloric acid. After slight cooling, it is treated with a moderate excess of ammonia and more polysulphide added, again boiled for 10 minutes, re-acidified, and finally boiled until the sulphur coagulates. The precipitate is collected in a porous porcelain crucible, washed, and ignited under hydrogen, etc., exactly as described above for the hydroxide precipitate.

The results obtained by this method have a tendency to be slightly high, since osmium obtained by ignition of the sulphide in hydrogen generally retains

[1] Schoeller and Powell, *op. cit.*, 278.

[2] Procedure of Schoeller and Powell, *op. cit.*, 278, after O. Ruff and F. Bornemann, *Zeits. anorg. Chem.*, 1910, **65**, 429.

a little sulphur, which is volatilized with the metal during the final ignition. This procedure should be applied only to small amounts of osmium.

Quantitative Separation of Platinum-group Metals from one another and from other Metals.

The quantitative separation and accurate determination of the metals of the platinum group is difficult and tedious, requiring expert chemical knowledge and experience, and is complicated by the presence of gold and other metals.

In solution the platinum-group metals form very stable complex salts, many of which are isomorphous. For this reason clear separations of the metals are rarely accomplished. Experience shows that their separation from one another is seldom complete with one precipitation, and in consequence traces of metal have almost invariably to be recovered from the solution after filtration of the precipitate when accurate results are necessary.

The following separations are representative of those in use in assaying materials containing metals of the platinum-group. Other methods of separation and the procedures adopted for the assay of materials containing all the metals of the platinum group, together with gold and silver and base metals are given in Chapter XXIV.

Platinum from Gold.—The hot solution of platinum and gold chlorides, freed from nitric acid by double evaporation with hydrochloric acid, and containing about one-fifth of its volume of hydrochloric acid, is treated drop by drop with a slight excess of a strong, freshly-prepared solution of ferrous chloride free from ferric salt. The precipitate of gold is allowed to settle, then collected on a small close-textured filter and thoroughly washed with 2 per cent. hydrochloric acid, and finally with water; then dried, ignited and weighed as metallic gold. If much platinum is present the ignited precipitate should be dissolved in aqua regia, the solution freed from nitric acid as before, and the precipitation with ferrous chloride repeated.

The platinum in the clear filtrate may be determined by precipitation with ammonium chloride, and subsequent ignition to metal, in which condition it is weighed.

Instead of ferrous chloride the gold may be precipitated in the chloride solution, after dilution with an equal volume of strong hydrochloric acid, by heating to boiling and adding a slight excess of a freshly-prepared 2 per cent. solution of hydrazine hydrochloride. The procedure is the same as described above, including re-treatment of the precipitate.

The separation of gold from platinum in alloys may be effected by a parting process, or by direct action of aqua regia. The aqua regia solution is freed from nitric acid by double evaporation with hydrochloric acid; the gold is precipitated with a slight excess of ferrous chloride, or other reducing agent. The filtrate containing the platinum is heated with a slight excess of sodium chlorate for the oxidation of iron and platinum, and the latter precipitated with ammonium chloride. Unless the platinum content is quite small, the precipitate of chlorplatinate should be purified by the chlorine process.

Platinum from Silver.—When these two metals are in solution together, the silver is precipitated as chloride, and the platinum determined in the filtrate by precipitation with ammonium chloride. The silver chloride precipitate tends to retain small amounts of platinum, according to the amount present in solution, and if it is tinted pink or brown it is dissolved in ammonia and re-precipitated as described under the separation of palladium and silver.

The separation and determination of platinum and silver in alloys is described on p. 411.

Platinum from Palladium.—Platinum is first separated in a chloride solution of the two metals by precipitation with ammonium chloride, and the precipitate washed with a saturated solution of ammonium chloride and alcohol. It is then dried, ignited, reduced, and weighed as metallic platinum.

The filtrate from the ammonium-chlorplatinate precipitation, containing a liberal amount of ammonium chloride, is slowly evaporated on the water-bath with excess of nitric acid. The palladium separates as a bright red crystalline precipitate, the solution becoming colourless. After cooling, the precipitate is collected, washed with strong ammonium chloride solution, ignited gently, and reduced in hydrogen. The filtrate usually contains a small but significant quantity of palladium, which is recovered by precipitation with dimethyl-glyoxime.

Palladium can also be separated from platinum by precipitation with dimethyl-glyoxime, but platinum is co-precipitated to a certain extent with the palladium if the solution is hot and an alcoholic solution of the reagent is employed, part of the platinum being reduced to the platinous condition, in which it reacts like palladium. The co-precipitation of platinum is detected by the greenish tint of the palladium precipitate, and is prevented by working in cold dilute solution, or by addition of nitric acid.

Platiniferous palladium precipitates may be purified by re-precipitation from dilute aqua regia solution.

Platinum from Iridium.—When alloys of these two metals are attacked with aqua regia, the platinum is dissolved and the iridium left as a residue which is washed, ignited, reduced, and weighed. The platinum in solution as nitrate may be precipitated with ammonium chloride, or it may be taken by difference. Other methods for the separation of these two metals are described on page 397.

Platinum from Rhodium.—The methods of separation for these two metals are somewhat similar to those employed for the separation of platinum and iridium, the rhodium remaining as an insoluble residue after acid treatment. When both metals are in solution the platinum may be separated by precipitation with ammonium chloride, provided the rhodium content is small. Procedures for separating rhodium from platinum by bromide-bromate hydrolysis are described on page 437.

Palladium from Gold.—The separation of gold from palladium may be effected by precipitation of the gold by ferrous chloride, or other reducing agent, from a chloride solution of the two metals. The precipitated gold usually retains some palladium and should be dissolved in aqua regia, the solution freed from nitric acid by evaporation, and then re-evaporation with hydrochloric acid, and the precipitation of the gold repeated.

Palladium from Silver.—The silver is usually separated from solution of the two metals with hydrochloric acid, and the palladium precipitated in the filtrate by dimethyl-glyoxime. When, however, a nitrate or sulphate solution of palladium and silver is treated with hydrochloric acid, the silver chloride carries down palladium, and is thereby coloured pink to brown, according to the amount of palladium present. For complete separation, therefore, dissolution of the precipitate and re-precipitation is necessary.

It has, however, been shown by Schoeller and Powell[1] that silver chloride free from palladium can be obtained in one operation if the palladium is converted into a tetrammine palladous salt prior to the precipitation. The

[1] Schoeller and Powell, *op. cit.*, 261.

procedure adopted is to treat the hot solution of the nitrates with ammonia until the precipitate first formed redissolves and the solution becomes colourless. A small piece of litmus-paper is dropped into the solution, which is slightly acidified with acetic acid, and treated whilst hot with dilute hydrochloric acid drop by drop, with constant stirring, until no further precipitate forms, after which it is left in the dark till clear. The precipitate is filtered by decantation and washed with very dilute nitric acid. The filtrate is treated with a little more hydrochloric acid and the palladium precipitated with dimethyl-glyoxime.

Palladium from Platinum.—See p. 406.

Palladium from Iridium.—The palladium is separated by precipitation with dimethyl-glyoxime, and the iridium in the filtrate with ammonium chloride, after destroying the excess of dimethyl-glyoxime in the solution. See p. 393.

Palladium from Rhodium.—The same procedure may be adopted for the separation of these two metals, as described for palladium and iridium above.

Rhodium from Gold.—Gold may be quantitatively separated from rhodium in one operation by precipitation with ferrous chloride or other reducing agent, as described for the separation of platinum and gold.

Rhodium from Silver.—Rhodium is one of the only two metals of the platinum group that form normal nitrates, and may therefore co-exist in solution with silver nitrate. To effect their separation the boiling nitrate (or sulphate) solution is precipitated drop by drop with very dilute hydrochloric acid, until no further precipitate forms. The silver chloride is allowed to settle, filtered off, and washed as usual. The filtrate is evaporated with sulphuric acid for the removal of nitric acid (sulphate solutions free from nitric acid need not be treated in this manner), and the rhodium is precipitated as metal with hydrazine or with magnesium.[1]

Rhodium from Platinum.—See page 406.

Rhodium from Palladium.—Same procedure as for palladium and iridium, above.

Rhodium from Iridium.—These two metals are satisfactorily separated after fusion with excess of lead, and subsequent treatment with nitric acid, thus giving a rhodium–iridium residue from which the finely-divided rhodium is separated by boiling in strong sulphuric acid. If the rhodium content is high, it is advisable to submit the combined rhodium–iridium residue from the acid treatment of the lead button to a second fusion with lead, and to repeat the whole process.

Gilchrist and Wichers apply the titanous sulphate method to the separation of iridium from rhodium, but the presence of titanium tends to interfere with the iridium determination.

Platinum Metals from Copper.[2]—Platinum and iridium may be separated from copper by precipitation as " double chlorides " with ammonium chloride, from which the two metals are subsequently separated. The procedure for the separation of palladium from copper varies according to the amount of palladium present. For substantial quantities of palladium, precipitation as ammonium chlorpalladate is convenient, but for small amounts of palladium dimethyl-glyoxime may be used with advantage.

For the separation of rhodium from copper the nitrite method may be used. This method is also applicable for separating all the platinum metals from copper, as well as from lead and bismuth.

[1] Schoeller and Powell, *op. cit.*, 261.
[2] Ibid, 266.

CHAPTER XXIV

ASSAY OF THE PLATINUM-GROUP METALS

Introduction.—The extensive use of the platinum metals and their alloys in jewellery manufacture and other industries, and in dentistry, makes it necessary for assays of the increasing number of such metals and alloys to be made frequently. In addition the assayer has to deal with trade and lemel bars, and with scrap metal, sweep, etc. resulting from the working of the precious metals.

Platinum ores are rarely received for assay, but gold, silver and base-metal ores and concentrates are not infrequently tested for the presence of metals of the platinum group. Slimes or mud from electrolytic refining processes are also assayed for the platinum metals.

The separation and determination of platinum and its allied metals in these various products may conveniently be considered under three main heads, namely: (i) the assay of industrial alloys and bullion; (ii) the assay of manufacturing products, etc.; and (iii) the assay of ores and concentrates.

Before dealing in detail with the assay of these three classes of material, reference must be made to the behaviour of platinum-metal alloys when parted in acids.

Parting Assay of Platinum.—Platinum alloys and bullion may be assayed by parting in acids, as described for gold alloys, but it is necessary to determine first of all the approximate composition of the alloy and then to make an exact assay. The parting assay of platinum is based on the fact that platinum, when alloyed with a suitable proportion of silver and some other metals, is attacked and dissolved by nitric acid with the formation of platinum nitrate. On the other hand, when alloys of platinum and silver containing excess of silver are attacked by strong sulphuric acid, the silver is dissolved as silver sulphate, leaving the platinum as a residue.

The determination of platinum in alloys of gold, silver and platinum, therefore, involves double parting, the alloy being first attacked by boiling in concentrated sulphuric acid, which dissolves the silver, and leaves the gold and platinum as a residue. The residue is then alloyed with silver and attacked by boiling in nitric acid, thus effecting the separation of the platinum, which is dissolved with the silver, leaving the gold as a residue. The difference between the weights of the residues from the two partings gives the platinum content of the alloy.

For satisfactory separation the three metals must be present in certain well-established ratios, and in some cases the parting operations must be repeated, as described later.

The parting assay of platinum involves cupellation as the first stage, the object of the operation being less to remove the whole of the base metals that may be present than to provide a clean, globular, and perfectly-mixed button or prill of the precious metals for subsequent treatment. The presence of platinum metals, however, not only necessitates a very high cupellation temperature, but also has a marked effect on the button left on the cupel;

and it will be well to consider briefly the effects produced by the individual metals of the group.

Owing to the high density of the platinum metals they tend to settle in the bottom of the cupellation bead. This is particularly noticeable with platinum, and with iridium and osmium, the specific gravities of which are nearly twice as great as those of the other metals of the group.

Platinum and its associated metals, especially palladium, absorb more or less oxygen when molten and tend to spit like silver when the gas is expelled on solidification.

The most fusible metal of the platinum group is palladium which melts at 1553° C., that is, nearly 500° C., higher than the melting-point of gold.

Cupellation of Platinum metals.—Experience and research[1] have shown that the appearance of the surface of cupelled gold and silver beads containing platinum metals is greatly influenced by the nature of the metals present, by the cupellation temperature, and by the rate of cooling. The presence of more than one platinum metal often complicates the effect produced. The various platinum metals all interfere with each other's surface effects, so that it is impossible, without considerable experience in this class of work, to say definitely from the appearance of the bead which metals are present, or even if present at all. In all cases chemical tests are essential for a reliable detection.

Although characteristic surface effects are observed which serve to indicate the presence of platinum metals in a cupellation bead, even these may be influenced by the presence of lead in the bead when the cupellation temperature at the finish has not exceeded 900° to 950° C.

The surface effects of the individual platinum metals on silver cupellation beads have been studied by Bannister and Patchin,[2] and their observations may be summarized as follows :—

Platinum. The surface of the bead is duller than that of pure silver or gold, and more or less crystalline according to the amount of platinum present. 1·6 per cent. of platinum in silver produces changes visible to the naked eye, and 0·3 per cent. can be detected with certainty under the microscope. The boundaries of the crystals become distorted, and a banded structure develops as the platinum content increases. The effect of platinum on gold is similar but less marked. Beads containing equal quantities of silver and gold have a frosted surface appearance when over 2 per cent. of platinum is present.

Palladium. The effect is similar to that of platinum. The presence of palladium may be detected by a yellow coloration of the parting acid, even 0·0002 gramme of palladium producing a faint yellow tint.

Iridium. The presence of iridium makes the bead more spherical. Under the microscope the crystal faces appear strongly marked with lines, crossing one another after the manner of slip bands, which are ascribed to internal stresses resulting from occluded oxygen. Black spots appear on the surface of the bead, and the cupel is stained black. (As shown later, the presence of iridium frequently results in the precious metals being left on the cupel in a subdivided state.)

Rhodium. Traces of rhodium in silver beads cause a distinct crystallization visible to the naked eye. With 0·03 per cent. rhodium and upwards, the beads spit persistently during solidification and assume a blue-grey colour. Gold with 0·2 per cent of rhodium has a ruby tint, and with 0·8 per cent. is covered with a black film.

[1] Forbes and Beamish, *Ind. and Eng. Chem.* (Anal. Edn.), 1937, **9,** (8), 397–400.
[2] *Trans. Inst. Min. Met.,* 1914, **23,** 163–190.

Ruthenium. This metal produces a black crystalline deposit at the edge of the bead near the bottom, and gives a " single-sided herringbone structure " to the surface of the bead. The black deposit is small but distinctly visible to the naked eye with 0·004 per cent. of ruthenium.

Osmium. Up to 2 per cent. osmium gives no characteristic appearance to the silver bead. Osmiridium gives results similar to those obtained with iridium alone.

The presence of platinum and its allied metals raises the melting-point of the gold or silver button left on the cupel, so that cupellation must be conducted at a higher temperature in order to prevent freezing of the precious metal bead and to remove lead so far as possible. The cupellation requires a higher temperature in proportion as the percentage of platinum is greater, and alloys containing over 50 per cent. of platinum cannot be freed from lead without the help of the oxygen-gas blowpipe.

The appearance of any of the effects noted above is an indication of the probable presence of platinum metals in the cupellation bead, but not necessarily of their nature or amount.

According to A. M. Smoot,[1] ruthenium produces a blue incrustation and rhodium an iridescent film on gold beads ; osmium and—to a smaller extent—ruthenium and iridium are lost by volatilization during cupellation ; and ruthenium, rhodium, and iridium separate completely or almost completely from solution in gold when the gold solidifies, iridium sinking to the bottom and ruthenium and rhodium rising to the top. The presence of more than one platinum metal often complicates the effects produced.

With regard to the cupellation of platinum itself, it is evident from experiments by Sharwood[2] on platinum–silver alloys that the lead retained after cupellation decreases with an increase in the ratio of silver to platinum, as shown by the results given in Table LX. The alloys were cupelled at the temperatures ordinarily employed for gold cupellation.

TABLE LX.—LEAD RETAINED IN THE CUPELLATION OF PLATINUM–SILVER ALLOYS AND OF PLATINUM–SILVER–GOLD ALLOYS. (*Sharwood.*)

No.	Composition of Alloy.			Lead retained.	Character of Button left on Cupel.
	Platinum taken.	Silver taken.	Gold taken.		
	mgs.	mgs.	mgs.	mgs.	
1	100	...	...	37·5	Hard silvery.
2	100	25	...	31·0	,,
3	100	50	...	26·2	Dull grey.
4	100	100	...	25·0	,,
5	100	101	48	24·0	,,
6	100	206	48	22·0	Smooth, brittle.
7	100	206	6	10·0	,, ,,
8	100	310	...	10·0	Slightly crystallized.
9	100	427	...	5·0	Smooth and silvery.
10	100	470	19·4	2·0	,, ,,

The lead is almost entirely eliminated with a ratio of 10 parts of silver to 1 part of platinum. The retention of lead appears to depend mainly upon the

[1] *Min. and Met.*, 1932, **13**, 364–366.
[2] *J. Soc. Chem. Ind.*, 1904, **23**, 412.

fusibility of the button, oxidation ceasing for practical purposes when the button solidifies.

Action of Acids in Parting Platinum.—Much experimental work has been done in connection with the action of sulphuric and nitric acids on platinum when alloyed with other metals, but in view of the many factors involved it is perhaps not surprising that the results in some cases are conflicting. Some few facts, however, have been definitely established and may be summarized as follows :—

(a) **Action of Sulphuric Acid on Platinum Metals.**—When alloys of platinum and silver are attacked by boiling concentrated sulphuric acid, the silver dissolves as silver sulphate and the platinum remains undissolved as a black residue.

To ensure the complete removal of the silver, however, the proportion of the two metals must be at least 10 parts of silver to 1 part of platinum. According to results obtained by Thompson and Miller,[1] silver remains with the platinum residue when the alloy contains 20 per cent. or more of the platinum, the amount of silver remaining undissolved increasing with the increase in the percentage of platinum present. The following results (Table LXI) were obtained with platinum–silver alloys of different composition, 300 mg. of the alloy being taken in each case and heated for fifteen minutes in 10 c.c. of concentrated sulphuric acid. The solution was then decanted off and the residue re-treated for fifteen minutes with 5 c.c. more of concentrated sulphuric acid. The residues were finally tested for silver and the amount determined. In some cases, traces of platinum were found in the acid filtrate, showing that platinum is slightly soluble in sulphuric acid.

TABLE LXI.—Action of Concentrated Sulphuric Acid on Platinum–Silver Alloys. (*Thompson and Miller.*)

Approximate Composition of Alloys used.		Actual Composition of Alloys used.		Silver retained in Platinum Residue.
Platinum.	Silver.	Platinum.	Silver.	
per cent.	per cent.	per cent.	per cent.	per cent.
10·0	90·0	10·39	89·61	Trace
20·0	80·0	20·59	79·41	0·59
30·0	70·0	31·46	68·54	0·98
40·0	60·0	37·89	62·11	2·24
50·0	50·0	57·05	42·95	2·70

The state in which the platinum is left after parting is dependent on the proportion of silver in the alloy. With more than five parts of silver the platinum is left in a state of fine division (but not as colloid), and loss may take place in decanting unless great care is exercised.

Any lead that may be present in the cornets as the result of cupellation will be converted into lead sulphate, which forms a white deposit on the residual platinum. It must be removed by treatment with hot ammonium acetate before annealing and weighing.

When ternary alloys of platinum, gold, and silver are attacked by boiling in concentrated sulphuric acid, the silver dissolves and leaves the gold in the

[1] Thompson and Miller, "Platinum–Silver Alloys," *J. Amer. Chem. Soc.*, 1906, **28** 1115-1132.

platinum residue ; but experience shows that the silver is only satisfactorily removed when the platinum, gold, and silver are present in definite ratios. Very little work has been done, however, to ascertain the best ratios to be employed.

A number of experiments were made by H. Carmichael[1] with a view to separating the silver from various ternary alloys of gold, platinum, and silver in which the metals were present in different ratios. The results were as follows (Table LXII.) :—

TABLE LXII.—Separation of Silver from Alloys of Platinum, Gold, and Silver by means of Sulphuric Acid Parting.

No.	Weight of Metal used, milligrammes.			Ratio of Metals.			Weight of Residue. Platinum + Gold.	Weight of Silver retained in Residue.
	Platinum.	Gold.	Silver.	Platinum.	Gold to Platinum.	Silver to Platinum.		
							mgs.	mgs.
I.	100	100	500	1	1	5	204·7 204·7	4·7 4·7
II.	100	100	500	1	1	5	204·0 200·6	4·0 0·6
III.	5	100	300	1	20	60	105·5 105·4	0·5 0·4
IV.	5	100	300	1	20	60	105·3 105·2	0·3 0·2
V.	5	50	300	1	10	60	55·3 55·2	0·3 0·2
VI.	5	25	300	1	5	60	30·3 30·3	0·3 0·3

Note.—All the alloys were cupelled and then treated as follows :—

No. I alloy. Parted in strong sulphuric acid.
No. II ,, Parted in strong sulphuric acid and the residue treated with strong nitric acid.
No. III ,, Parted in dilute sulphuric acid.
No. IV ,, Parted in dilute sulphuric acid.
No. V ,, Parted in dilute sulphuric acid, washed and treated with strong nitric acid.
No. VI ,, Parted in dilute sulphuric acid, washed and treated with strong nitric acid.

The residual platinum was treated with nitric acid in the cases stated, with the object of dissolving out, if possible, the small amount of silver still remaining in the cornet.

It may be remarked that in the above table the ratio of silver to platinum alone is given, but in adjusting the proportion of silver for parting ternary alloys it is usual to consider the ratio of the silver to one part of platinum and gold together.

The above results show that the proportion of silver may be decreased if the proportion of gold to platinum is increased.

In practice it is usual to employ about 3 parts of silver to 1 part of platinum and gold together in order to obtain a satisfactory cornet, as the cornet is very liable to break up if this proportion is much exceeded. The silver may therefore

[1] *J. Soc. Chem. Ind.*, 1903, **22**, 1325.

be regarded as a constant quantity, and the ratio of platinum to gold is regulated accordingly, the ratio generally employed being 1 part of platinum and its allied metals to 10 parts of gold.

As previously stated, 1 part of platinum requires about 10 parts of silver for sulphuric-acid parting, and gold requires about $2\frac{1}{2}$ parts of silver. But since the proportion of silver is fixed at three times the weight of the platinum and gold together, it becomes necessary to dilute the platinum with gold to such an extent that the ratio of silver to platinum is such that practically no silver remains in the cornet. Experience shows that this is accomplished by diluting the platinum with ten times its weight of gold as stated.

This proportion is arrived at thus :—

	Parts of silver.
1 part of platinum requires 10 parts of silver	10
1 part of gold requires $2\frac{1}{2}$ parts of silver, therefore 10 parts require $10 \times 2\frac{1}{2} = 25$ parts	25
Total ...	35

Thus 11 parts of alloy (1 part platinum + 10 parts gold) would require 35 parts of silver, which is approximately three times the proportion of platinum and gold together, *i.e.* $11 \times 3 = 33$, the actual quantity that would be used.

Satisfactory results may be obtained with a ratio of about 8 of gold to 1 of platinum, but if the ratio is less than this the cornets retain more silver. Some assayers prefer to work with a larger ratio of silver and smaller ratio of gold to platinum, such as those given in Table LVII.

The figures given in the table show that equally good results may be obtained by varying the ratios. Thus the surcharge in Experiments V and VI is the same, viz. 0·3 mg., for the same weight of platinum (5 mg.) ; but in Experiment V the platinum was alloyed with ten times its weight of gold, and the silver added was equal to about five times the gold + platinum, while in Experiment VI the platinum was alloyed with only five times its weight of gold, and with silver equal to ten times the gold + platinum.

It is doubtful whether any ratios of metals can be selected that will give cornets which are entirely free from silver.

The parting should be effected in slightly diluted sulphuric acid, since it is proved by the researches of Conroy[1] and others that hot concentrated sulphuric acid, even when pure, exerts a marked solvent action on platinum, especially at a temperature of about 250° C. to 280° C. The temperature of boiling pure sulphuric acid is 338° C.

According to Steinmann,[2] boiling concentrated sulphuric acid dissolves considerable quantities of platinum from platinum–silver alloys, the loss being usually from 20 to 30 parts per 1000 parts of platinum, but may be as much as 50 parts. By using slightly diluted acid the silver can be extracted without loss of platinum. Steinmann recommends rolling the alloy to a thickness of 0·2 millimetre, and heating for fifteen minutes with dilute sulphuric acid, (82 vols. of concentrated acid to 18 vols. of water) at a temperature not exceeding 240° C. The treatment with acid is repeated twice, and the residual

[1] "The Action of Sulphuric Acid on Platinum," J. T. Conroy, *J. Soc. Chem. Ind.*, 1903, **22**, 465 ; "The Dissolution of Platinum by Sulphuric Acid," M. Delépine, *Bull. Soc. Chem.*, ser. 3, **35**, No. 1, 1906 (abstract *Chem. News*, 1906, **193**, 108–109) ; and also "Action of Boiling Sulphuric Acid on Platinum," Le R. W. M'Cay, *Eighth Int. Cong. Appl. Chem.*, 1912, Sect. I, Orig. Comm., **1**, 351–359.

[2] A. Steinmann, "Platinum Assay," *Schweiz. Woch. Chem. Pharm.*, 1911, **49**, 441–444, 453–457 (abstract *J. Soc. Chem. Ind.*, 1911, **30**, 1216).

platinum is washed, dried, and weighed. It is stated that by this treatment hardly any platinum is dissolved; also, if some copper is added before cupellation lead and silver are not retained by the bead left on the cupel.

According to Frohlich,[1] the errors due to dissolved platinum in sulphuric acid parting are minimized by using acid containing 0·5 per cent. arsenic, prepared by dissolving 10 grammes of As_2O_3 in 1500 grammes of sulphuric acid at 200° C. After adding the requisite quantity of fine gold to the sample to give a ratio of 1 platinum to 10 gold, and silver equal to 3 or 4 times the combined weight of platinum and gold, it is cupelled with 1 gramme of lead, finishing at 1150° C. The resulting bead is flattened, annealed, rolled into a cornet, and heated with the arsenical sulphuric acid at 280° to 300° C., until evolution of SO_2 ceases. By this procedure it is stated the errors amount to about 0·3 per thousand for platinum, and about the same for gold.

Many assayers use slightly diluted acid consisting of 90 volumes of concentrated sulphuric acid to 10 volumes of water, and boil the cornets for fifteen minutes, or until all visible action has ceased.

The residual platinum is then treated once or twice with strong sulphuric acid. The use of dilute acid lessens the tendency of the cornet to break up, especially if from 4 to 5 parts of silver are used for parting.

The sulphuric acid used for parting must be free from nitric and hydrochloric acid.

The action of concentrated sulphuric acid on other metals of the platinum group present in the platinum cornet is very much dependent on the quantity of such metals present. Palladium is attacked during the parting and goes into solution as palladous sulphate, giving a cherry-red liquid. In the author's experience, solution of the palladium is largely dependent on the amount present, and on the ratio of palladium to the gold, silver and platinum present, more especially the latter metal. An increase in platinum content appears to retard solution of the palladium. It is never safe to assume that when the palladium content is small, say under 2 mg., it has all passed into solution.

Experience indicates that as a general rule about one-third of the palladium is removed by sulphuric-acid parting, and the remaining two-thirds in the subsequent nitric-acid parting; but, as already stated, much depends on the amount of palladium and other metals in the cornet and their ratio to one another. The extent to which the parting acids have been coloured by the presence of palladium and other platinum metals is best seen by decanting the liquid into a white porcelain vessel.

Ruthenium appears to be acted upon to a small extent by sulphuric acid under certain conditions, giving a rose-coloured solution of ruthenium sulphate. The tetroxide is soluble in sulphuric acid, and it seems very probable that this compound of ruthenium is formed to a small extent under the oxidizing influences of cupellation, and remains dissolved in the button left on the cupel.

According to experiments by Beamish and Russell,[2] when a silver bead containing 12 times as much silver as total platinum metals is parted in boiling sulphuric acid, most of the palladium and a small part of the platinum, rhodium, and ruthenium dissolve, but the insoluble portion always retains some palladium.

The sources of error in sulphuric-acid parting are (i) the conversion of any lead remaining in the cupel bead into lead sulphate, which contaminates the cornet; (ii) the solution of some platinum if the cornet becomes finely divided,

[1] K. W. Frolich, *Zeits. Elektrochem.*, 1935, **41**, 207–211.

[2] F. E. Beamish and J. J. Russell, *Ind. and Eng. Chem.* (Anal. Edn.), 1936, **8**, (2), 141–144.

and to avoid which the ratio of gold to platinum should be at least 10 to 1; (iii) the retention of silver by the platinum cornet.

With regard to this last source of error, it should be noted that when an alloy of 75 per cent. silver and 25 per cent. platinum begins to solidify, the first crystals to separate contain 45 per cent. of platinum and sink to the bottom of the solidified mass. It is these crystals, rich in platinum, which cannot be parted with sulphuric acid and tend to retain the silver in the platinum–gold cornet. By annealing the cornet, however, before treatment with sulphuric acid, sufficient diffusion of the metals can easily be induced to ensure that the crystals richest in platinum shall contain not less than 65 per cent. of silver and 35 per cent. platinum.

This diffusion and redistribution of the platinum is effected by annealing the cornet at a temperature somewhat higher than that ordinarily employed for gold assays, and prolonging the annealing for some minutes.

(b) Action of Nitric Acid on Platinum Metals.—As previously stated, platinum, when alloyed with silver, is soluble to some extent in nitric acid, so that it can under certain conditions be separated from gold by parting in nitric acid.

The platinum goes into solution in the nitric acid in colloidal form, giving to the solution a brown to blackish colour according to the amount of platinum

TABLE LXIII.—SOLUBILITY OF PLATINUM-SILVER ALLOYS IN NITRIC ACID OF 1·10 SP. GR. AND OF 1·40 SP. GR. (*Thompson and Miller.*)

Composition of Alloy.		Parted in Nitric Acid, 1·10 sp. gr.		Parted in Nitric Acid. 1·40 sp. gr.	
Platinum.	Silver.	Platiniferous Residue.[1]	Platinum dissolved.[2]	Platiniferous Residue.[1]	Platinum dissolved.[2]
per cent.	per cent.	per cent.	per cent.	per cent.	per cent.
0·5	95·5	0·42	0·08	0·22	0·28
1·0	99·0	0·85	0·15	0·42	0·58
2·0	98·0	1·74	0·26	1·09	0·91
3·0	97·0	2·19	0·81	1·81	1·19
4·0	96·0	2·98	1·02	2·42	1·58
5·0	95·0	3·56	1·44	2·62	2·38
10·0	90·0	...	...	4·53	5·47
13·0	87·0	3·33[3]	9·67	5·79	7·21
14·0	86·0	4·26	9·74	4·97	9·03
15·0	85·0	4·32	10·68	7·93	7·07
16·0	84·0	4·55[3]	11·45	11·54	4·46
18·0	82·0	4·54[3]	13·46	11·65	6·35
20·0	80·0	...	...	13·94	6·06
25·0	75·0	16·62	8·38	20·66	4·34
30·0	70·0	...	...	29·29	0·71
31·5	68·5	38·58	See Table LIX.	...	...

present. The formation of colloidal platinum is very marked, especially with alloys containing about 20 per cent. of platinum[4]. the solution in this case being opaque and almost black. With different percentages of platinum the intensity of the colour decreases, while with 10 per cent. the solution is almost

[1] This includes silver retained (see also Table LVI for silver in residues).
[2] Difference between original percentage of platinum and percentage of residue.
[3] Average of two results.
[4] Thompson and Miller, *loc. cit.*

colourless. On allowing the platinum solution to stand for two or three days, the colloidal platinum separates as a very fine black powder, and the solution becomes almost colourless. The colloidal platinum may be collected by the addition of ammonium or sodium nitrate.

In practice, however, it is usual to add a small quantity of some salt, such as lead acetate or barium nitrate, and sufficient sulphuric acid to give a heavy precipitate that will mechanically carry down the finely-divided platinum, which is subsequently recovered.

The results of experiments made by different workers to determine the extent of the solubility of platinum in nitric acid are conflicting. The most complete investigation on the action of nitric acid on platinum–silver alloys appears to be that conducted by Thompson and Miller,[1] who worked on alloys containing from 0·5 to 57 per cent. of platinum. The tests were made with nitric acid of two different strengths, viz. 1·1 sp. gr. and 1·4 sp. gr., but it was found that in no case would the alloy dissolve completely in the acid; some platinum was always dissolved with the silver, and some silver always remained with the platinum residue.

The proportion of platinum passing into solution varies with the increase in the ratio of silver to platinum and also with the strength of the acid, as shown by the results given in Table LXIII.

The platinum residue from nitric acid parting invariably retains silver, the amount of which increases with increase in the percentage of platinum in the alloys treated. Thus Thompson and Miller obtained the following results with alloys containing more than 10 per cent. of platinum (Table LXIV) :—

TABLE LXIV.—Showing Amount of Silver retained in Platinum Residue from Treatment of Platinum-Silver Alloys with Nitric Acid of 1·10 sp. gr. (*Thompson and Miller.*)

Composition of Alloy.		Total Residue.	Silver in Residue.	Platinum in Residue.	Platinum dissolved.
Platinum.	Silver.				
per cent.	per cent.	per cent.	per cent.	per cent.	per cent.
10·39	89·61	3·86	0·27	3·59	6·80
20·59	79·41	8·58	1·81	6·77	13·82
31·46	68·54	36·59	12·09	24·50	6·96
37·89	62·11	49·13	13·64	35·49	2·40
57·05	42·95	65·16	12·19	52·97	4·08

From a study of the cooling curves and micro-structure of the silver–platinum series of alloys, Thompson and Miller conclude that the alloys consist of a number of platinum–silver compounds of different solubilities, the relative proportions of these compounds in the alloys being dependent on the rate of cooling. These conclusions help to explain many of the irregular results obtained in parting with nitric acid. Thus an alloy containing 31.46 per cent. of platinum when *rapidly cooled* and treated with nitric acid 1·10 sp. gr. gave 36·59 per cent. of residue (including silver retained), while the same alloy *very slowly cooled* gave 33·58 per cent. of residue (including silver retained), a difference of 2 per cent. These results emphasize the importance of using a correct annealing temperature for cornets containing platinum.

[1] Thompson and Miller, "Platinum–Silver Alloys," *J. Amer. Chem. Soc.*, 1906, **28** 1115–1132.

The solubility of platinum in nitric acid is influenced by the presence of gold in the platinum–silver alloys, the actual amount of platinum dissolved being dependent on the ratio that exists between the three metals present.

As gold very frequently accompanies platinum in the materials submitted for assay, the action of nitric acid on ternary alloys of gold–platinum–silver is of considerable importance to assayers; but at present very few data are available to show what ratio of metals will give the best results. The figures obtained by the few investigators who have worked on these ternary alloys show divergences which make it difficult to decide what individual metal is responsible for any given set of results.

From a few tests made by Sharwood[1] it would appear that an increase of gold, with the platinum and silver remaining constant, retards solution of the platinum, but more extended work is necessary on these lines.

As an example of the action of nitric acid on ternary alloys of gold–platinum–silver, the following figures (Table LXV) may be given. They are the results of experiments by H. Carmichael[2] to determine the amount of platinum dissolved by nitric acid from gold–platinum–silver alloys of different composition.

The alloys were cupelled, rolled, and parted first in dilute nitric acid 1·16 sp. gr. (26 per cent. HNO_3) and then in stronger nitric acid 1·26 sp. gr. (41 per cent. HNO_3). The resulting cornets were washed, dried, and weighed.

TABLE LXV.—Parting of Platinum–Gold–Silver Alloys in Nitric Acid. (*Carmichael.*)

Weight of Metals used, milligrammes.			Ratio of Metals.			Weight of Cornet (Gold + Platinum).	Weight of Platinum retained by Cornet.[3]
Platinum.	Gold.	Silver.	Platinum.	Gold.	Silver.	mgs.	mgs.
20	100	300	1	5	15	102·7	2·7
15	100	400	1	6·6	26·6	101·2	1·2
						100·2	0·2
10	100	300	1	10	30	100·8	0·8
						100·4	0·4
10	100	500	1	10	50	100·2	0·2
						100·2	0·2
10	200	600	1	20	60	200·0	0
14	200	800	1	14·3	57·1	200·3	0·3
						200·3	0·3
14	300	900	1	21·4	64·3	300·0	0
7	100	400	1	14·3	57·1	100·2	0·2
5	100	500	1	20	100	100·0	0

In these experiments it was found that 400 milligrammes of added silver parted as succesfully as 500 milligrammes, and at the same time gave a more compact cornet less liable to break up.

The result of the last experiment appears to indicate that it is possible to use ratios of the metals such that the alloy may be successfully parted by one treatment in nitric acid, but since this conclusion is contrary to experience it needs confirmation. Thompson and Miller concluded, as the result of the experiments already referred to, that the separation of platinum from gold, in

[1] *Loc. cit.*

[2] *Loc. cit.*

[3] This may also include a little silver retained.

one operation, by means of alloying with silver and parting with nitric acid, is impossible.

In practice, as a general rule, no attempt is made to remove all the platinum from gold by one parting, but authorities differ as to the best ratios of metals to be employed to effect the solution of the platinum.

From the result of the last experiment given in the above Table LXV, it seems that the platinum is most satisfactorily removed when there is 7 per cent. of platinum to gold, or a ratio of 1 to 14, and the silver amounts to about four times the platinum and gold together. Although more platinum may be removed in one operation by using a large proportion of gold, this will result in producing very large cornets if much platinum is present; and if the total weight of the metals is reduced so as to obtain a cornet of convenient size, the original weight of sample that will have to be taken for assay may be too small to give satisfactory results. Also the addition of more than 3 parts of silver renders the cornets very liable to break up, although 4 or even 5 parts may be satisfactorily employed if the strips of alloy are not rolled too thin. In this case, however, the cornets are very porous and tender, and there is a tendency for more silver to be retained.

In considering the ratios of metals to be used, it is well to draw attention to the results of some experiments by Carmichael, from which it appears that the action of mass plays a part in the separation of the platinum. Thus, an alloy of 7 mg. of platinum, 100 mg. of gold, and 400 mg. of silver parted successfully, but when double the quantity of each metal was taken, the cornet retained 0·3 mg. of platinum; by increasing the gold, however, to 300 mg. the alloy parted successfully, practically all the platinum being removed. This would show that as the weight of platinum to be parted increases, it is desirable to increase the ratio of gold to platinum. This fact has been recognized in practice, as there is more or less agreement amongst assayers that the total weight of platinum present in an assay should not exceed 200 mg., and that the quantity of material taken for assay should be such that about this weight or less of platinum is present. Some assayers use a smaller ratio of gold to platinum, say 5 to 1, for the parting when a small quantity of platinum only is present, and increase the ratio to 10 to 1 or even more when much platinum is present.

The ratios of gold, platinum and silver most generally employed for nitric-acid parting are the same as those given for sulphuric-acid parting, namely, 1 part of platinum to 8 or 10 parts of gold, and silver equal to 2½ or 3 times the weight of the platinum + gold. These ratios ensure a satisfactory cornet, and, as shown in the above table, permit of a considerable proportion of the platinum being dissolved in the first parting. To separate the remaining platinum it is necessary, after weighing the residue of platinum and gold, to alloy it with more silver by cupellation and then to part again in nitric acid. The inquartation and parting is repeated until all the platinum has been removed.

The extent to which the platinum has been separated in each parting is judged by the colour of the residue or cornet obtained. When much platinum is present the cornet is black, but as the platinum is gradually removed in successive partings the colour of the cornet changes, passing through a series of steel-grey tints until finally it assumes the characteristic yellow colour of pure gold.

The action of nitric acid during parting on other metals of the platinum group that may be present is as follows:—

Palladium is dissolved in parting if the weight of silver is at least three times that of the palladium, yielding an orange-coloured solution; but when much palladium is present double parting appears to be necessary to ensure complete solution.

Iridium, if present, always sinks to the bottom of the cupelled button, as it is very dense and is not usually fused at the temperature of the muffle, but occurs in the state of fine black crystalline particles. Hence when the button is rolled into a cornet with the lower face outwards, as directed on p. 278, iridium shows itself as black sooty spots or streaks which are seen by a lens to fill up depressions in the surface of the gold. Iridium, being unacted upon by nitric acid, remains with the gold, consequently cornets containing iridium require special treatment, as described subsequently. Under certain conditions, however, small quantities of iridium are carried mechanically into the acid solution, and may be deposited as a very finely divided black residue. When osmiridium is present, it is not dissolved in nitric acid, and therefore remains with the gold on parting.

Rhodium is slightly dissolved during parting in nitric acid, but the greater part remains with the gold. Ruthenium also may be dissolved to a slight extent.

The nitric-acid parting of silver assay beads containing platinum metals has recently been the subject of investigation by Seath and Beamish,[1] who concluded that when the silver : platinum ratio is 15 to 1, all the platinum cannot be dissolved by three successive treatments with nitric acid, although with similar beads containing palladium instead of platinum all the palladium is dissolved in one parting. The presence of gold in the beads accelerates the solution, whilst iridium, rhodium, and ruthenium considerably retard dissolution of the platinum and palladium.

Apparatus for Sulphuric-Acid Parting.—The parting of cornets containing platinum by boiling in concentrated sulphuric acid is a delicate operation, as the boiling-point of sulphuric acid (338° C.) is so high that glass vessels are liable to be cracked by cold draughts. Moreover, the acid boils explosively with heavy bumping, which may occasion loss by ejection and even lead to the breaking of the flask. Boiling sulphuric acid is also terribly corrosive and inflicts dangerous wounds. On this account it is frequently advised that the platinum frame described and illustrated on page 272 should be used with a platinum boiler for sulphuric-acid parting, but experience proves that the platinum is attacked during the parting, and in some cases in sufficient amount to give a brown tint to the liquid. The solution of the platinum is probably accelerated by galvanic action between the gold cornets and the platinum cups.

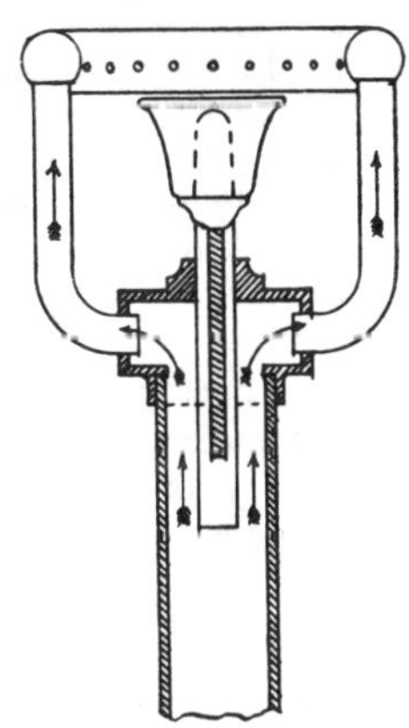

FIG. 166.—QUENNESSEN BURNER FOR ANNULAR HEATING.

Under these conditions it is advisable to use a parting frame and cups made of fused silica, or to use glass.

If the ordinary glass-bulb parting-flask (Figs. 137A and B, p. 271) is used, it must be protected from the air and heated at the sides more than at the bottom. Small clay peas or pieces of carbon are usually added to assist the acid to boil quietly.

Quennessen,[2] of the Paris Mint, devised an improved form of burner for heating the flasks, which obviates the drawbacks mentioned above and permits of sulphuric acid being boiled readily without bumping. The burner is shown

[1] J. Seath and F. E. Beamish, *Ind. and Eng. Chem.* (Anal. Edn.), 1938, **10,** (11), 639–641; Abs., *Met. Abs. Inst. Metals*, 1939, **6,** 17.

[2] "New Gas Burner for Heating Parting-flasks," L. Quennessen, *Chem. News*, 1903, **88,** 66 (illustrated).

in section in the accompanying illustration (Fig. 166). The gas comes out through the holes arranged round the interior of a crown, so that the heating is annular and the boiling is effected, not at the bottom as in the burner illustrated in Fig. 138, page 271, but round the sides of the flask. Thus ejections during parting are prevented. To enable the heat to be applied at various heights the bottom of the flask is held by a copper support in the form of a cup, which is brazed to the upper part of a screw so that it can be raised or lowered as desired.

When a glass parting-flask is used it should be allowed to cool after boiling ; otherwise it will tend to crack during decantation of the acid, owing to sudden cooling.

The acid should be decanted from the cornets as completely as possible, and the washing with water done with care, in view of the fact that sulphuric acid has a great affinity for water, the addition of which causes hissing and spurting from the energy of combination, and considerable evolution of heat.

Double Parting Assay.—In practice, double parting in sulphuric and nitric acids is mainly restricted to alloys containing platinum, gold, silver, and copper, with relatively small amounts of the other platinum metals.

Preliminary Assay of Platinum-metal Alloys (and Bullion).—It is necessary, frequently, to make a preliminary assay of the alloy in order to ascertain how much gold and silver must be added to the assay portion to bring these metals up to the required ratios.

In the approximate assay the amount taken is usually 0·25 to 0·5 gramme, which is accurately weighed, and then cupelled with about 5 grammes of lead, preferably on a magnesite cupel, at a high temperature to remove base metals. If the cupel button is flat and irregular in shape it is again cupelled with more lead. Frequently the appearance of the cupel bead will give some indication of the nature of the platinum metals present. When a rounded bead is obtained it is weighed and the weight accepted as that of the gold, platinum, and silver combined. The difference between this weight and that of the metal taken for assay gives the base metal present. Allowance must, however, be made for considerable loss of silver as a result of the high temperature employed for cupellation.

The button is next wrapped in sheet lead and re-cupelled with three times its weight of gold (accurately weighed) and silver equal to three times the weight of the button plus the gold added.

In some cases, in order to get a good cupel bead in the first instance, it is well to add the requisite gold and silver during the first cupellation, and allow for these additions when weighing the final cupel bead. The alloy thus obtained is flattened, annealed, and parted by boiling in concentrated sulphuric acid for fifteen minutes, which dissolves the silver and leaves the gold and platinum as a residue. The residue is washed, dried, annealed, and weighed. The difference between this weight and that of the original cupellation button is silver. The residue is inquarted by cupellation with three times its weight of silver, and the resulting button rolled and parted in dilute nitric acid as in the ordinary gold-bullion assay.

If the colour of the cornet inclines to steel-grey instead of the characteristic colour of pure gold, it is desirable to inquart, re-cupel, and part a second time, in order to obtain a closer approximation to the amount of platinum present. In most cases one parting in nitric acid will give a result sufficiently near for a preliminary assay.

The cornet represents the original gold content (after allowing for the amount added), and the platinum is taken by difference.

Thus, to take an example :—

Weight of alloy taken for assay 0·25 gramme.

Weight after cupellation (silver + platinum + gold) = 0·215 gramme. The difference between this and the weight of alloy taken gives the *base metal* 0·250 – 0·215 = 0·035 g.

Weight of residue after parting in sulphuric acid (platinum + gold) = 0·833, of which 0·645 was added as pure gold ; therefore 0·833 – 0·645 = 0·188, and this deducted from the cupellation button, gives the *silver* 0·215 – 0·188 = 0·027 g.

Weight of residue after parting in nitric acid (gold only) = 0·770, of which 0·645 was added as pure gold ; therefore 0·770 – 0·645 = 0·125 equals the *gold* in the alloy, and this weight, subtracted from that of the above platinum + gold residue, gives the *platinum*, thus 0·188 – 0·125 = 0·063 g.

The approximate composition of this alloy is thus determined to be as follows :—

Gold	0·125 × 4 =	50·0	per cent.
Platinum	0·063 × 4 =	25·2	,,
Silver	0·027 × 4 =	10·8	,,
Base metal (probably copper) ...	0·035 × 4 =	14·0	,,
		100·0	

Final Assay.—The amount of alloy taken for assay is usually 0·25 gramme, but as it is advisable not to have more than about 50 milligrammes (0·05 gramme) of platinum present, the weight taken should be proportional to the platinum present.

The alloy is cupelled with lead, preferably on a magnesite cupel, to remove base metals, the amount of lead required for cupellation being about twice that required for a gold-bullion assay, as given in Table XXXV, p. 276. The cupellation is finished at a very high temperature, and if the button is not satisfactory it is re-cupelled with 1 or 2 grammes of lead.

A check assay is run at the same time as the ordinary assay. If there is a surcharge on the check of more than 1 or 2 parts per 1000 in the first cupellation, showing retention of base metal (copper or lead), the buttons are re-cupelled with fresh lead as stated. The cupellation button is cleaned and weighed, and the difference between this weight, after correction for loss and the weight of the alloy taken gives the base metal present.

The button is then re-cupelled with the addition of gold if necessary, and with silver equal to three times the weight of platinum and gold in the alloy plus any gold that is added. Thus, using the example given above :—suppose the corrected weight of the gold–platinum–silver cupellation button from 0·25 gramme of alloy was 0·220, then 0·25 – 0·220 = 0·030 g. base metal.

The amount of gold and silver required to be added for parting purposes would be determined thus :—

Platinum present	= 0·063	gramme	approximately.
Gold already present	= 0·125	,,	,,
Silver ,, ,,	= 0·027	,,	,,

One part of platinum requires 10 parts of gold, therefore 0·063 requires 0·063 × 10 = 0·63 gramme gold, and this weight minus the gold already present gives 0·63 – 0·125 = 0·505 gramme, the weight of gold to be added. This must be accurately weighed.

One part of platinum + gold requires $2\frac{1}{2}$ parts of silver, therefore 0·063 + 0·125 + 0·505 (added gold) = 0·693 parts require 0·693 × $2\frac{1}{2}$ = 1·732 grammes silver, and this weight minus the silver already present gives 1·732 – 0·027 = 1·705 grammes, the weight of silver to be added.

The amount of sheet lead required for this second cupellation is about 5 grammes.

The cupellation is conducted as described above and the button cleaned, hammered, and rolled into a cornet, as described in the assay of gold bullion. It is very important to anneal the fillet or strip thoroughly before rolling into the form of a cornet. The cornet is boiled in slightly diluted sulphuric acid (9 volumes of concentrated acid to 1 volume of water) for twenty to twenty-five minutes, after which the liquid is decanted as completely as possible and the cornet again boiled for ten to fifteen minutes in concentrated sulphuric acid and then washed, transferred to an annealing-cup, dried, annealed, and weighed.

If, after washing, a white deposit of lead sulphate is seen upon the cornet, it must be treated with a warm, strong solution of ammonium acetate, again washed with distilled water, and then annealed. Lead may be present in consequence of the employment of too low a temperature for cupellation.

Some assayers boil the cornets a third time in sulphuric acid, and this is desirable when the cornets are large.

The weight of the cornet gives the amount of platinum and gold in the alloy plus the added gold, in this case, say, 0·696. Then 0·696 minus 0·505 added gold equals 0·191 gold + platinum in the alloy, and the difference between this weight and the weight of the gold–platinum–silver cupellation button gives the silver in the alloy.

For the determination of the gold, the cornet is re-cupelled with 2½ times its weight of silver and 3 grammes of sheet lead.

The resulting button is hammered, rolled, well annealed, and then parted in nitric acid of two strengths, as described in the assay of gold bullion on page 278, and the gold cornet weighed. If, as frequently happens, there is a surcharge in the check assays of more than 1 or 2 parts per 1000 showing the retention of platinum, the inquartation and parting is repeated. When the cornet retains an appreciable amount of platinum it will be steel-grey in colour and the inquartation and parting must be repeated until the cornet obtained is of the characteristic pure gold–yellow. The final weight is then, say, 0·6306, and after deducting 0·505, the result 0·1256 is obtained for the amount of gold present in 0·25 g. of the original alloy. The platinum is taken by difference: thus the gold plus platinum from the sulphuric-acid parting was 0·191, and subtracting the weight of gold from the nitric-acid parting 0·1256, we get 0·0654 g. as the amount of platinum present in 0·25 g. of the original alloy.

It is assumed that the necessary corrections as shown by the check have been made in all the operations.

When iridium is present in the alloy it remains with the gold, and the cornet requires special treatment, as described on p. 419.

The final result of the assay will be reported in parts per 1000 as follows :—

Gold	502·4	per 1000
Platinum	261·6	,,
Silver	116·0	,,
Copper (by difference)	120·0	,,
Total	1000·0	

According to Trenkner,[1] the error in assaying alloys of gold and platinum, which is considered by the Paris Mint to amount to 1 per cent., may be greatly

[1] *Metallurgie*, 1912, **9**, 103–105.

lessened by the following method, which gives the results within 0·5 per 1000, or closer in experienced hands. The alloy is subjected to a preliminary assay to determine the approximate composition. Then 0·5 gramme of bullion if finer than 250 per 1000, or 1 gramme if baser, is mixed with pure silver in the proportion of about 10 of silver to 1 of gold and platinum. From 3 to 15 grammes of lead are added, and the whole cupelled. Allowing for added silver, the weight of the button gives silver, gold, and platinum together. The silver loss during cupellation is best determined by control assays. The addition of so much silver results in the entire removal of lead from the obtained button. No appreciable amount of gold or platinum is lost. The button is parted, without flattening, in 25 c.c. of concentrated sulphuric acid, nearly boiling, for thirty minutes. The residue is coherent, and is washed and dissolved in aqua regia, and the solution evaporated to small bulk, diluted, and the silver chloride filtered off. To the filtrate is added 15 c.c. of hydrochloric acid, and 1 gramme of hydrazine hydrochloride. After one hour, with occasional stirring, the gold is filtered off, ignited, and weighed. As it often contains some platinum, it is cupelled with 5 to 8 parts of silver, and the button parted in nitric acid. Silver and platinum dissolve. The pure gold is weighed. Platinum is precipitated from solution by an excess of ammonium or potassium hydrate, ignited, and weighed.

Hollard and Bertiaux[1] find that by the usual methods for the analysis of platinum–gold–silver alloys the results are too low for the gold and too high for the platinum and silver, but are very accurate for the gold and platinum together. The following method is recommended by them for the determination of the silver :—The alloy, cut into small pieces or rolled out to thin foil, is dissolved in aqua regia (1 vol. of nitric acid + 5 vols. of hydrochloric acid), the solution is evaporated to the consistence of a syrup, and then evaporated to dryness three times with the addition of nitric acid, the residue boiled with a few c.c. of water, 2 c.c. of nitric acid, and 2 drops of hydrochloric acid, the liquid diluted to 100 c.c., the silver chloride filtered off, washed well, dissolved in 30 c.c. of a 20 per cent. solution of potassium cyanide, the solution diluted to 150 c.c. and electrolysed. The deposited silver is dissolved in nitric acid, and determined by titration with standard thiocyanate solution.

Effect of Platinum Metals on Parting Assay.—When other metals of the platinum group are also present some modification of the usual double parting method of assay, described above, becomes necessary. The exact procedure adopted will depend on the amount of the platinum metals present.

As previously stated, palladium when present is partially separated in the sulphuric-acid parting and the remainder dissolved out in the nitric-acid parting ; hence it becomes necessary to determine the amount of palladium in both acid solutions, and to correct accordingly the platinum determination made by difference. In these circumstances it is usually more satisfactory to part direct in nitric acid to ensure complete solution of all the palladium in one acid solution. The coloration method described on p. 395 affords a convenient and ready means of determining the palladium both in nitric and sulphuric-acid solutions.

Iridium will remain with the gold in the cornet from nitric-acid parting in the form of black particles, and in order to determine the gold it is necessary, after weighing the cornet, to digest it in dilute aqua regia (1 to 5 water) whereby the gold is dissolved whilst the iridium remains as an insoluble residue, which is carefully collected, washed with dilute aqua regia, and with ammonia, then dried, ignited in hydrogen, and weighed.

[1] *Ann. Chim. Anal. Appl.*, 1904, **9**, 287–292 (abstract *J. Soc. Chem. Ind.*, 1904, **23**, 952).

The weight of the iridium deducted from the weight of the gold–iridium cornet gives the gold present, after making allowance for the gold added to secure the correct ratio for parting purposes.

The residue will also contain traces of any rhodium or ruthenium that may be present, but the amount is usually very small, and in many cases in commercial work it is usual to include these in the insoluble residue and to make the return as " Iridium, etc." If the amount of rhodium or of ruthenium is comparatively large, special attention must be given to their separation and determination by the methods subsequently described.

Osmium, if present in the original material, is practically all eliminated by volatilization during cupellation.

Whilst the determination of platinum by the double-parting method is still largely employed in assaying industrial alloys, the increase in the production of alloys containing notable quantities of other platinum metals makes it desirable to adopt other procedures, as described later.

In some cases the cornet from sulphuric-acid parting is dissolved in aqua regia, and the metals separately determined in the solution, whilst in other cases sulphuric-acid parting is omitted, and the cupel button treated direct with nitric acid, and the cornet dissolved in aqua regia.

I. ASSAY OF INDUSTRIAL PLATINUM-METAL ALLOYS.

This section deals more particularly with the numerous alloys containing platinum metals used for jewellery, dental work, and in the electrical, chemical, and other industries. Also included are the lemel or trade bars containing platinum metals. (The assay of loose lemel and workshop sweepings is dealt with in Section II, p. 429, and also on p. 376.) As previously stated (Chapter XXIII), many of these alloys are complex in composition, following the increased use of metals of the platinum group, other than platinum itself, in their make-up.

For assay purposes the alloy samples are usually in the form of sheet, wire, or drillings made as previously described.

Experience shows that it is frequently more satisfactory to take two or more separate assay portions and to determine one or two of the main constituents in each, rather than to attempt the determination of all the constituents in one assay-piece.

As a general rule the platinum metals are determined gravimetrically, being precipitated in the metallic state and weighed as such, or precipitated in the form of some compound—usually the hydroxide, sulphide, or double chloride—for subsequent separation.

Some useful notes on the usual precipitants of the platinum metals have been given in the preceding chapter. The preliminary preparation of platinum-metal alloys for assay by fusion with silver, zinc or lead is discussed below.

General Procedure for Assaying Platinum-metal Alloys.—Any of the platinum-group metals may be present in the industrial alloys, some in larger proportion than others, together with gold and silver, and more or less base metal, especially copper. Osmium and ruthenium are comparatively rare constituents, except as impurities. The procedure adopted for the assay of these alloys is entirely dependent on their composition, and in order to determine the best method to adopt in any particular case it is generally necessary to make a preliminary assay as described above.

In some cases the complete solution of the alloy can be effected by attack with nitric acid or with aqua regia, but this procedure is somewhat restricted by the insolubility of some of the platinum metals when direct attack is attempted. More generally the separation of the platinum metals is effected by first alloying the assay-piece with a suitable excess of silver (with or without gold), then wrapping it in sheet-lead, cupelling, and treating the cupel button with acids. In some cases where cupellation is undesirable, the alloying of the assay-piece with the requisite excess of silver (or in some cases zinc) is effected by melting in a small carbon crucible or annealing-cup, as described below.

Although cupellation, as a first step, is very advantageous in many cases, especially for the elimination of copper, nickel, and other base metals, its applicability is restricted since it may result in appreciable losses of the platinum metals present. An excess of silver equal to from 30 to 50 times that of the combined platinum metals plus gold lessens the risk of cupellation loss, particularly volatilization loss of palladium, and ensures the more complete removal of lead by cupellation, and also the solution of the silver present in small quantity in the original alloy when the cupel button is subsequently parted in acids.

The concentration of platinum and palladium (with or without gold) into a silver bead by cupellation is very satisfactory, provided there is sufficient silver (at least 30 parts to 1 part of platinum and palladium combined) to prevent solidification of the bead before all the lead has been oxidized.

In the case of iridium, rhodium, and ruthenium, cupellation is limited to materials containing very small amounts of these three metals because they oxidize more or less readily, thus giving rise to low results. A large excess of silver tends to counteract oxidation by keeping the metals in solution in the molten alloy.

Osmium is partly or wholly volatilized as the tetroxide (depending on the amount present) during the later stages of cupellation; hence this procedure is not admissible for materials containing osmium, when this metal has to be determined. The precious metal bead, containing excess of silver, whether obtained by cupellation or by direct fusion in an annealing-cup, is parted either in sulphuric or in nitric acid.

Parting of the silver bead in sulphuric acid is largely restricted to alloys of platinum and silver in which silver preponderates, and containing relatively small amounts of platinum, palladium, and gold. For alloys richer in platinum, attack by nitric acid should be applied.

Treatment of the silver bead with nitric acid is applicable to alloys containing high or low percentages of platinum and palladium, free from or containing only subordinate amounts of iridium, rhodium, and ruthenium; osmium, if present, is not recovered. This treatment results in the solution of the platinum, palladium and silver and a residue containing the iridium, rhodium and gold. The residue is treated with dilute aqua regia which dissolves the whole of the gold, leaving the iridium, rhodium and ruthenium, which may be weighed together or further treated for their separation and determination.

The assay of complex materials liable to contain all the metals of the platinum group as well as gold and silver, together with some base metals, is considered at the end of this chapter.

Many of the methods introduced in recent years for the assay of platinum-metal alloys are the outcome of research especially devised to separate and determine as quickly as possible the constituents in the complex industrial alloys now in use. The descriptions of these procedures not infrequently begin with the assumption that all the metals involved are already in solution, and seldom give details as to how solution can best be effected. Some of these procedures are briefly described in this Chapter.

For convenience of description the assaying of the industrial alloys will be dealt with under four sub-headings, viz. (*a*) Jewellery alloys, (*b*) Dental Alloys, (*c*) Miscellaneous Alloys, and (*d*) Commercial Metals.

Before, however, dealing with these sub-headings, brief reference will be made to the action of aqua regia on platinum-metal alloys, and their preliminary preparation for assay.

Action of Nitric acid and of Aqua Regia on Platinum-metal Alloys.—Of the six platinum metals only rhodium and palladium form normal nitrates, and may therefore co-exist in solution with silver and other nitrates.

When a complete analysis of platinum-metal alloys has to be made it is frequently necessary to effect their solution, either completely or partially, by attack with nitric acid or with aqua regia (3 volumes of hydrochloric acid to 1 of nitric acid). The extent of the attack of these acids is entirely dependent on the noble and base metals present, and their proportions in the alloy.

The action of acids on the platinum metals is modified when they are alloyed with other metals, as they are frequently liberated in a finely-divided state and are more susceptible to attack. Thus, palladium when alloyed with silver is soluble in nitric acid ; and platinum, although insoluble in nitric acid, dissolves to a brownish colloidal solution when platinum–silver alloys containing excess of silver are boiled with nitric acid.

Aqua regia, however, is the solvent most frequently used in the attack on alloys containing metals of the platinum group, and the preparation of the solution for analysis calls for special care, as it is essential for the subsequent separation of the metals that the excess of acid should be removed, so far as possible, by evaporation, and that the solution should be free from nitrous compounds.

Attention has already been drawn to the difficulty of getting gold–silver alloys into solution with aqua regia, owing to the formation of an impervious coating of silver chloride.

Of the noble metals, only platinum, palladium, and gold are completely soluble in aqua regia. Silver is attacked superficially with the formation of a film of silver chloride. Platinum is converted partly into chlorplatinic acid H_2PtCl_6, and partly into nitrosyl chlorplatinate $(NO)_2PtCl_6$.

The latter compound, as well as the excess of nitric acid used, are undesirable in subsequent work, and must always be eliminated. This is effected by evaporation of the solution on a water-bath until the syrupy residue barely smells of acid, addition of a small proportion of hot water followed by strong hydrochloric acid, and evaporation as before. The second residue is once more treated with water and hydrochloric acid, and again evaporated until all the acid is expelled. The residue from the last evaporation is normally free from nitrous compounds. Before evaporation with hydrochloric acid some operators add sodium chloride in slight excess over the quantity required to form chlorplatinate, Na_2PtCl_6. In either case, the last evaporation residue is taken up with 0·5 N hydrochloric acid.

Repeated evaporation of the aqua regia solution with hydrochloric acid to the consistency of a syrup yields a solution of pure chlorplatinic acid, which however, is readily decomposed when strongly heated ; care is therefore necessary in evaporation.

The action of aqua regia on gold is somewhat similar, resulting in the production of auric chloride $AuCl_3$, which combines easily with hydrochloric acid to give the complex chlor-auric acid, $HAuCl_4$. The latter decomposes again on being heated gently, thus: $HAuCl_4 = HCl + AuCl_3$. Further heating decomposes auric chloride, and aurous chloride AuCl is obtained:

$AuCl_3 = AuCl + Cl_2$. This salt is decomposed by hot water, auric chloride and free gold being produced, thus: $3AuCl = AuCl_3 + 2Au$. These reactions must be considered in effecting the solution of gold and its alloys in aqua regia, or other mixtures evolving chlorine, for the production of auric chloride.

As in the case of platinum, over-heating during evaporation may result in the complete decomposition of the auric chloride syrup.

The gold solution for analysis is prepared in the same way as directed for platinum.

Palladium dissolves in aqua regia with the formation of a deep-red unstable solution of palladic acid H_2PdCl_6. Upon evaporation with hydrochloric acid, as described above for platinum, the compound loses chlorine, and is converted into brown chlorpalladous acid H_2PdCl_4. No nitrosyl compound is formed, as in the case of platinum. Chlorpalladous acid has a tendency to hydrolyse in aqueous solution, with precipitation of basis compounds; it should be taken up with dilute hydrochloric acid. Iridium and rhodium in the compact form are not attacked by aqua regia, but when finely divided are to some extent soluble, depending upon the degree of subdivision. They are more or less soluble. if alloyed with a large excess of platinum. The solution of platinum–iridium in aqua regia is light- to deep-brown in colour.

Ruthenium is not acted upon by aqua regia, except to a slight extent when it is alloyed with other metals.

Dilute aqua regia is sometimes used for separating metals of the platinum group contained in the finely-divided mixtures of ignited metals obtained by precipitation in the course of analysis. Frequently the platinum metals are contained in the ignited and reduced ammonium-chloride and zinc precipitates. The usual procedure is to digest the ignited metals with dilute aqua regia (1 : 3 water) until a fresh portion of acid remains colourless.

The author is in agreement with Schoeller and Powell in their conclusions that "the results thus obtained are not sufficiently reliable for accurate work."[1] An exception may be made for ignited precipitates weighing only a few milligrammes and consisting chiefly of iridium with a little palladium and platinum. Such mixed ignited metals, upon digestion with dilute aqua regia, leave a residue of substantially pure iridium.

When igniting and reducing platinum-metal precipitates, special precautions must be taken to prevent oxidation and over-heating (see p. 390).

Fusion of Platinum-metal Alloys with Silver, Zinc, or Lead.—This procedure is adopted when alloying the sample with lead, and subsequent cupellation are undesirable.

From 0·25 to 1·0 gramme of the alloy to be assayed is mixed with the correct weight of fine silver and fine gold for parting purposes in a small crucible or annealing-cup, a little fused borax sprinkled on the surface, and the crucible covered. It is then heated in a muffle furnace at 1000°–1100° C. for one hour to complete the alloying of the constituent metals, and produce a button of uniform composition so far as possible. When cold the crucible is broken, the button removed and cleaned from all adhering slag, and then flattened into a strip for parting.

A similar procedure is adopted for alloying the sample with zinc which melts at approximately 420° C. A weighed quantity of the alloy is mixed with from 8 to 10 grammes of chemically pure zinc (electrolytic zinc is very suitable), transferred to a small carbon crucible, and covered. It is placed inside a small fireclay crucible which is then filled with charcoal of bean size, a cover placed on it, and heated in a muffle for half-an-hour at about 600° C., to allow the

[1] *Op. cit.*, p. 254.

platinum metals to dissolve in the molten zinc. The heat is then gradually raised above the boiling-point of zinc (906° C.), the vapour of which escapes, forming dense fumes of zinc oxide. When the evolution of fumes ceases, the heat is raised to over 1000° C. for half-an-hour, after which the muffle is allowed to cool, and the clay pot withdrawn.

The button of zinc is removed from the carbon crucible, and cleaned.

In view of the brittleness of the alloy it is advisable to treat it direct with nitric acid, rather than to crush it to powder and risk loss.

Fusion of the assay portion with excess of lead in a carbon crucible, and direct treatment of the lead button with nitric acid, may be resorted to in some cases with advantage, and examples of this procedure will be found in the text which follows (see pages 442, 452).

Touchstone Test for Platinum-metal Alloys.—In recent years the application of the touchstone, or streak test, for differentiating between platinum and its alloys and white base metals of similar appearance, and determining their approximate composition, has received considerable attention. Its use for this purpose has been investigated by Hoke[1] and others, and their conclusions are of interest, as the touchstone is useful as a preliminary test in dealing with unknown white alloys that resemble platinum in appearance, and as giving some indication of their nature.

The touchstone test for platinum alloys is similar to that described for gold alloys on p. 292, except that the test-needles are made of platinum, platinum–iridium and platinum–palladium, and aqua regia is used as the test acid on a hot porcelain touchstone. By using a spot plate instead of the usual touchstone, and adding a solution of stannous chloride to the acid-treated streak, gold, platinum and palladium can be detected in the presence of one another, and by comparison with standard solutions their relative proportions can be approximately determined. The use of a solution of dimethyl-glyoxime is recommended in the detection of nickel and palladium in precious-metal alloys such as white golds.

The applicability of streak analysis to the approximate determination of platinum in jewellery alloys has also been discussed by other workers,[2] who propose a mixture of 8 c.c. nitric acid (d.1·42), 24 c.c. hydrochloric acid (d.1·19), and 5·3 grammes potassium iodide as the test reagent. This mixture is stable for two or three months. For platinum–palladium alloys, acid-proof porcelain plates are recommended, and 2·5 grammes potassium chloride should be added to the acid mixture.

Attention has already been drawn to the use of iodine as a delicate test for palladium and its alloys.

(a) Assay of Jewellers' Platinum-metal Alloys.

Platinum alloyed with from 5 to 10 per cent. of iridium is used for the highest-quality jewellery. Iridium counts as platinum in the 950 parts per 1000 platinum standard now recognized by jewellers for manufacturing purposes. In the cheaper alloys the platinum may be partially replaced by palladium. Base metals may be present in amounts up to 5 per cent., two alloys in common use containing 3 per cent. and 4·5 per cent. of copper respectively.[3] These two alloys have adequate hardness and strength for most jewellery purposes, and have good working properties, and good colour. Other very high-grade alloys

[1] C. M. Hoke, "Testing Precious Metals: Gold, Silver, Platinum, Palladium," 2nd edn., 1935.

[2] *cf.* J. Kretai, *Chem. Abs.*, 1938, **32**, 3291.

[3] Atkinson and Raper, *loc. cit.*, p. 203.

for jewellery can be made with the addition of 5 per cent. of rhodium plus ruthenium, the ruthenium content being varied according to the degree of hardness required. An alloy composed of 96 per cent. palladium and 4 per cent. ruthenium is suitable for high-class jewellery. Palladium also enters into the composition of jewellery alloys, notably the palladium white golds, in which palladium is the chief bleaching agent and destroys the characteristic yellow colour of the gold with which it is alloyed.

The following procedures are typical of those which are applied to the separation and determination of the various constituents of the white alloys used in the manufacture of platinum jewellery.

The procedures described later for the assay of dental alloys and miscellaneous industrial alloys are also applicable to those used in the jewellery trade, as well as to jewellers' lemel or trade bars containing platinum metals. As these bars are rarely uniform in composition they call for special care in sampling, as described on p. 259.

Method A. The following procedure for the determination of gold, platinum, palladium, iridium, rhodium, and copper in jewellery alloys has been described by H. Holzer and E. Zaussinger.[1]

The sample is rolled out to a thin strip and, after annealing, 0·2 to 0·4 gramme is dissolved in aqua regia ; if the alloy is insoluble it is alloyed with zinc under a flux, and after extraction of the zinc with dilute hydrochloric acid, the residual black powder is dissolved in aqua regia. The solution is evaporated to dryness with hydrochloric acid, the residue dissolved in water, any siliceous matter filtered off, the filtrate boiled with an excess of sodium nitrite ($NaNO_2$), a little sodium carbonate added until alkaline to phenolphthalein, and the precipitate of gold and cupric oxide collected, washed, and leached with hydrochloric acid to remove the copper.

The nitrite filtrate is boiled with hydrochloric acid to expel NO_2, oxidised with hydrogen peroxide, and treated with a hot saturated solution of dimethyl-glyoxime to remove palladium.

The filtrate is boiled with hydrogen peroxide to destroy the glyoxime, and divided into two equal portions, both of which are boiled with $NaBrO_3$ and NaBr to separate the rhodium and iridium from the platinum ; one precipitate is ignited and weighed, after reduction and acid leaching, to give the sum of rhodium and iridium, and the other is dissolved in sulphuric acid for the separation of rhodium by Gilchrist's $TiCl_3$ method. The platinum in the bromate filtrate is recovered by boiling with $N_2H_4.2HCl$.

Method B. Another procedure for the assay of platinum alloys, developed by Beamish and Russell,[2] has been described as follows :—

When a silver bead containing 12 times as much silver as platinum metals is parted in boiling sulphuric acid, most of the palladium and a small part of the platinum, rhodium, and ruthenium dissolve, but the insoluble portion always retains some palladium.

The palladium may be recovered by neutralizing the acid[3] to pH4 with $NaHCO_3$, boiling with $NaBrO_3$, adjusting the pH to 6, and boiling for 20 minutes to precipitate PdO_2 which is collected, washed, and dissolved in hydrochloric acid for precipitation of the palladium with dimethyl-glyoxime.

The residue from the sulphuric-acid parting is digested with 1·5 aqua regia and, if gold is absent, the solution is diluted and the palladium and traces of

[1] *Z. anal. Chem.*, 1938, **111**, (9/10), 321–337 : *Met. Abs. Inst. Metals*, 1938, **5**, 307.

[2] F. E. Beamish and J. J. Russell, *Ind. and Eng. Chem.* (Anal. Edn.), 1936, **8**, (2), 141–144 : Abs. *Met. Abs. Inst. Metals*, (2), 1936, **3**, 211.

[3] $p\mathrm{H} = \log \{1/(\text{H-ion concentration})\}$.

iridium and rhodium are precipitated by neutralization and $NaBrO_3$ hydrolysis, the precipitate being added to that obtained from the sulphuric acid solution. The platinum is recovered from the hydrolysis filtrates by boiling with $H.CO_2Na$.

The insoluble residue from the aqua regia treatment is ignited and fused with Na_2O_2 in a silver crucible. The melt is dissolved in nitric acid and the solution evaporated to fumes with sulphuric acid. After dilution the rhodium and iridium are separated from the silver (derived from the crucible) by hydrolysis as described for palladium, and from one another by Gilchrist's $Ti_2(SO_4)_3$ method.

Method C. The following procedure is applicable to alloys containing gold, platinum, palladium, iridium, and rhodium, with small amounts of silver and copper.

From 0·25 to 1·0 gramme of the sample is alloyed with the required excess of fine silver, and fine gold (as determined by a preliminary assay), by melting together in an annealing-cup under a layer of fused borax.

The resulting button is cleaned and parted in dilute nitric acid (1 : 2 water), the solution decanted, and the residue subjected to two successive boils in stronger nitric acid (2 : 1 water).

The solution contains the silver, platinum, palladium, and copper. The boiling in acid should not be unduly prolonged and not too vigorous, otherwise small amounts of rhodium and iridium tend to pass into solution, or are separated from the cornet as very fine particles that remain in suspension. To collect any such particles the solution should be allowed to stand before decanting. The residue will contain the gold, rhodium and iridium.

The acid solution of nitrates is neutralized with a minimum excess of ammonia, the hot solution slightly acidified with acetic acid, and the silver precipitated with hydrochloric acid. By this procedure the silver chloride should be white and not coloured by traces of palladium and platinum. After filtering the silver chloride, the palladium in the filtrate may be precipitated with dimethyl-glyoxime, and the platinum in solution with ammonium chloride (see, however, p. 406).

The residue from nitric-acid parting is treated with hot 15 per cent. aqua regia, which dissolves the gold and leaves a residue of iridium and rhodium. The solution is freed from nitric acid by evaporation and then re-evaporation with hydrochloric acid, and the gold precipitated with ferrous chloride or sulphurous acid. The precipitate of metallic gold is washed, ignited, and weighed, and from this weight is deducted the weight of the gold added.

The iridium–rhodium insoluble residue is washed, dried, ignited, reduced, and weighed as combined rhodium and iridium and returned as such, or further treated for the separation and determination of the two metals. For this purpose fusion with bisulphate is ineffective, and fusion with lead is a more serviceable method.[1] The combined residue is fused with lead in a carbon crucible, and the resulting lead button dissolved in hot 10 per cent. nitric acid. This leaves the rhodium in a finely-divided condition and more amenable to attack by aqua regia. The *unignited* rhodium–iridium residue is washed and treated with hot 15 per cent. aqua regia, which slowly dissolves the rhodium ; a second treatment with fresh aqua regia is frequently desirable. If the proportion of rhodium in the combined residue is high, some of it will probably remain undissolved with the iridium after aqua regia treatment. In such cases it is advisable to submit the iridium residue to a second fusion with lead and to repeat the attack with aqua regia in order to effect the solution of the remaining portion of

[1] Schoeller and Powell, *op. cit.*, 278.

rhodium. This treatment should yield an insoluble residue which, after washing, igniting, and reducing, can be weighed as pure iridium.

When dealing with alloys high in rhodium, small amounts of this metal may pass into solution with the lead with the nitric-acid treatment of the lead button. To recover this, the lead in solution is precipitated with sulphuric acid (1 : 3 water) in moderate excess. The precipitate of pure lead sulphate is white, and a yellow tint indicates traces of rhodium, in which case re-precipitation is desirable. The lead sulphate is dissolved in a minimum of hot ammonium acetate solution, and then re-precipitated with a moderate excess of sulphuric acid, and set aside for the lead sulphate to settle. The precipitate is filtered off, and the filtrate is evaporated to very small bulk for the elimination of the acetic acid, diluted, and added to the main rhodium solution.

Another procedure when the rhodium content is small is to boil the unignited rhodium–iridium residue with strong sulphuric acid, thus dissolving the rhodium and leaving the iridium as an insoluble residue. The rhodium is recovered as metal by precipitation with hydrazine or other precipitant.

If it is desired to determine the silver content of the original alloy the assay portion is alloyed with zinc instead of silver, and the silver determined after solution of the zinc button in nitric acid.

A modification of the above procedure for determination of the platinum and palladium is sometimes adopted. After removal of the silver as chloride, a rapid current of hydrogen sulphide is passed through the boiling platinum, palladium filtrate, whereby both metals are precipitated as sulphides. The precipitate is filtered off, washed, dried, ignited strongly, and weighed as metallic sponge of the combined metals. If copper is present it is removed before precipitation of the platinum and palladium. The sponge is transferred to a beaker, heated with aqua regia, the nitric acid removed by evaporation with hydrochloric acid, and the solution diluted to about 50 c.c. The solution is then cooled to room temperature and glyoxime added to precipitate the palladium, which is washed, dried, ignited, reduced, and weighed as metallic palladium. The platinum may be taken by difference or determined after destroying the excess of glyoxime.

When the palladium content is relatively small it may be determined by the colour test.

Method D. A complete scheme for the separation of the six platinum metals from one another and their gravimetric determination has been devised by Gilchrist and Wichers[1].

The procedure consists, broadly, of the following series of operations :—

(1) Distillation of the osmium as OsO_4 by nitric acid from an alkaline osmate solution, absorption of the OsO_4 in hydrochloric acid containing SO_2, hydrolytic precipitation of OsO_2 at *p*H6, ignition of the precipitate in hydrogen to obtain osmium ;

(2) Distillation of RuO_4 from the residual solution by boiling with sodium bromate ($NaBrO_3$), followed by hydrolysis, etc. as for osmium, finally weighing as ruthenium ;

(3) Separation of the iridium, rhodium, and palladium from platinum by hydrolysis with sodium bromate, first at *p*H6, then at *p*H8, in chloride solutions containing a little sodium chloride, double precipitation being generally necessary ;

(4) Precipitation of the platinum in the filtrate by hydrogen sulphide, followed by ignition of the PtS_2 to platinum in air, the bromate being

[1] Raleigh Gilchrist and Edward Wichers, *J. Amer. Chem. Soc.*, 1935, **57**, 2565–2573 : Abs. *Met. Abs. Inst. Metals*, 1936, **3**, 48.

first destroyed by evaporation of the solution to dryness with hydrochloric acid;

(5) Dissolution of the iridium, rhodium, and palladium hydroxides in hydrochloric acid and precipitation of the palladium with dimethyl-glyoxime, followed by weighing of the precipitate or ignition to palladium;

(6) Separation of the rhodium from the iridium by double precipitation with titanous sulphate $Ti_2(SO_4)_3$ and subsequent precipitation of both metals with hydrogen sulphide after removal of the titanium salt with cupferron.

No details are given for obtaining the metals together in one solution starting from alloys or "sponges," etc., and the disturbing effects of base metals are not considered.

According to Schoeller and Powell,[1] the drawback to this method consists in the interference of titanium with the iridium determination. Gilchrist and Wichers find it necessary to precipitate the titanium from the combined iridium filtrates by means of cupferron. The cupferron precipitate is boiled with nitric and sulphuric acids, the resultant solution (freed from nitric acid) is re-precipitated with cupferron, and the combined filtrates are boiled down until 10 mls. of sulphuric acid remain. The iridium in the residual liquid is precipitated by hydrolysis with sodium bicarbonate and bromite, and the iridium hydroxide converted into, and weighed as, metallic iridium.

As previously stated, Schoeller and Powell give preference to the lead fusion process over the titanous sulphate method as the regular procedure for the separation of iridium from platinum, palladium, and rhodium.

Method E. Lenk[2] has devised the following procedure for the assay of alloys containing gold, palladium, and silver.

Alloys of these three metals containing less than 31 per cent. gold, that is, of about 7·5 carat quality, can be parted directly in nitric acid (d.1·3).

The solution containing the silver and palladium should be diluted to 400–900 c.c., according to the palladium content, before precipitating the silver with hydrochloric acid; after precipitation, the solution should contain 1 per cent. hydrochloric acid to reduce the solubility of the silver chloride to a minimum. The palladium in the filtrate from the silver chloride is precipitated with dimethyl-glyoxime, and the filtrate from this process is used for the determination of copper and nickel, if these metals are present.

Method F. A procedure applicable to the assay of white-gold alloys.

These substitutes for platinum are in use by jewellers, and also for dental work. They are usually of 18-carat quality containing 75 per cent. gold, in addition to 10 to 20 per cent. palladium, with smaller amounts of other metals such as copper, silver, and zinc. Two typical analyses of these alloys are given on pages 334 and 385. White gold alloys of 14-carat and 9-carat quality, containing 58·5 and 37·5 per cent. gold respectively, are also in use.

These alloys may be assayed by the parting methods previously described, preceded by cupellation to remove base metals.

Method G. Gilchrist[3] describes the following method for the separation of platinum, palladium, rhodium, and iridium from one another and for their gravimetric determination.

It is based on the complete precipitation of the hydrated dioxides of palladium, rhodium, and iridium in a boiling solution containing sodium bromate

[1] *Op. cit.*, 274.

[2] G. E. Lenk, *Metall u. Erz*, 1935, **32**, 95–98: Abs. *Met. Abs. Inst. Metals*, 1935, **2**, 242.

[3] Raleigh Gilchrist, *U.S. Bur. Stand. J. Research*, 1934, **12**, 291–303; Research Paper No. 655: *Met. Abs. Inst. Metals*, 1934, **1**, 251.

and having an acidity of pH6. Two precipitations of the hydrated dioxides were found to be sufficient to effect the quantitative isolation of platinum. The separation of palladium from rhodium and iridium by means of dimethylglyoxime, and its subsequent determination, are also described in the paper. The procedures for the separation of rhodium from iridium by titanous chloride, and for the determination of these two metals, are given in detail on p. 400.

(b) Assay of Dental Alloys containing Platinum Metals.

The chief constituents of the various precious-metal alloys now in use in dental practice, are gold, silver, platinum, palladium and copper, with sometimes minor amounts of nickel, zinc and iridium.

The composition of the alloy is varied according to whether it is to be wrought, or cast direct into shape, or whether yellow or white.

Suitable alloys for wrought dental plates generally consist of 18-carat (75 per cent.) or of 16-carat (67 per cent.) gold containing platinum, palladium and copper, but of late years white dental alloys with a palladium–silver base have been introduced to replace the more expensive gold-base alloys. A cheap alloy of this description consists of gold 25, silver 41·5, and palladium 33·5 per cent. respectively. The white palladium–silver–gold alloys are no longer regarded as substitutes in dental work, but as valuable materials superior to gold alloys in certain respects.

The high-strength wire for bands used for correcting malformations of the mouth may contain up to 30 per cent. of platinum or of palladium, or varying proportions of the two, the remainder consisting of gold, silver, and copper.[1]

The amount of platinum and palladium which may be added to dental casting-alloys is restricted, as they should melt at a temperature below 1050° C. Alloys containing 75 per cent. gold may contain up to about 15 per cent. of platinum or palladium, or varying proportions of both these metals.

Cheaper casting-alloys with a gold content of about 37·5 per cent. may contain up to 25 per cent. palladium.

These few examples give some indication of the composition of the dental alloys containing platinum metals now in use.

The methods for the analysis of dental alloys has been fully investigated by W. H. Swanger,[2] R. Gilchrist,[3] and others, whose original papers must be consulted for a detailed description of the lengthy procedure involved in the separation and determination of the numerous constituents. Some of the new methods developed as the result of research are briefly described below, and may be regarded as typical of the procedures now adopted.

Method H. This new procedure, devised by Gilchrist[4] for the analysis of dental gold alloys, provides for the separation and gravimetric determination of gold, silver, platinum, palladium, rhodium, iridium, nickel, copper, zinc, tin, and indium—a somewhat rare constituent. Briefly the method is as follows :—

Silver chloride and metallic iridium are precipitated when the alloy is decomposed by aqua regia. Tin is next separated by a new technique of controlled hydrolysis. Gold is then precipitated by sodium nitrite, and next, by adjusting the alkalinity with sodium hydrate to the end-point of thymolphthalein, copper, nickel, zinc, and indium are collectively separated from

[1] Atkinson and Raper, *loc. cit.*, 204.
[2] U.S. Bureau Stand., Sci. Paper, 532, 1926.
[3] U.S. Bureau Stand., J. Research, 1938, **20**, 745.
[4] R. Gilchrist, *Nat. Bur. Stand. J. Research*, 1938, **20**, (6), 745–771 ; and Research Paper No. 1103 : *Met. Abs. Inst. Metals*, (2), 1938, **5**, 501.

platinum and rhodium. At this alkalinity a small proportion of palladium precipitates with the base metals, but is recovered from them with dimethylglyoxime. Palladium, rhodium, and platinum are separated from one another by the usual National Bureau of Standards methods. The four base metals are separated in order by precipitating indium with sodium hydrate: copper, and subsequently zinc, with hydrogen sulphide: and nickel with dimethylglyoxime.

It was found that manganese, iron, cobalt, and chromium are also quantitatively precipitated in a solution containing nitrate at the end-point of thymolphthalein, and that it is possible to separate lead from palladium and platinum at the end-point of xylenol-blue, if the lead is precipitated as carbonate.

Method I. In this procedure, which deals primarily with the determination of gold in dental alloys,[1] the alloy is dissolved in aqua regia and the silver, iridium, and tin removed by Swanger's method.[2]

The gold is then precipitated with SO_2, washed with hot 1 per cent. hydrochloric acid, and redissolved in aqua regia. After removal of the nitric acid by boiling with hydrochloric acid and evaporating to 3 c.c., 20 c.c. of water are added, followed by 4 c.c. of 15 per cent. $N(C_2H_5)_4Cl$ for every 0·1 gramme of gold present. After 15 minutes the precipitate of $N(C_2H_4)AuCl_4$ is collected, washed four times with 5 c.c. of the precipitant, and boiled with 35 c.c. of 20 per cent. glucose solution containing 2 grammes of sodium hydroxide. The precipitated gold is collected, washed with water, then with hot hydrochloric acid, and finally with hot water; dried, ignited, and weighed.

The filtrate from the second gold precipitation is treated with hydrogen sulphide to recover any platinum or palladium co-precipitated in the SO_2 process.

Method J. Assay of dental precious-metal alloys, including the determination of indium.[3]

The solution of the alloy in appropriate acids is treated with hydrogen sulphide to remove platinum, palladium, gold, rhodium, tin, silver, and iridium, and then oxidized with nitric acid and treated with ammonia and ammonium chloride to remove nickel and zinc which remain in solution. The precipitate is dissolved in sulphuric acid and the indium–iron precipitated as sulphide. The sulphide is redissolved in acid and the indium precipitated with ammonia for ignition to In_2O_3. (In regard to this procedure, A. R. Powell points out that indium is readily separated from iron by boiling the slightly acid solution with $NaHSO_3$.)

Method K. A procedure for the assay of so-called platinized gold alloys used in dental practice.

This term is applied to gold–silver–copper alloys containing usually about 5 per cent. of platinum, but in some cases rising to 10 per cent. The alloys vary in quality from 16-carat (66·7 per cent. gold) to 20-carat (83·4 per cent. gold) and are used in dentistry as bases, clasps and wires. They are essentially jewellers' alloys to which more or less platinum is added to give greater hardness.

These alloys may be assayed for platinum and gold by the double parting method.

Method L. The following procedure is applicable to the assay of " dental alloy," a term especially applied to a white tarnish-resisting alloy containing

[1] J. Lewis Maynard, *Ind. and Eng. Chem.*, Anal. Edn., 1936, **8**, (5), 368–369: *Met. Abs. Inst. Metals*, 1936, **3**, 467.

[2] U.S. Bur. Standards, 1926, Sci. Paper 532: *Met. Abs. Inst. Metals*, 1926, **36**, 519.

[3] L. G. Wright, *Brit. J. Dental Research*, 1935, **15**, (3/4), 224–225: *Met. Abs. Inst. Metals*, 1936, **3**, 696.

70 per cent. of silver and 30 per cent. of platinum which has long been used in dental practice for swaged plates etc. The platinum content, however, varies with different makers, lower-quality alloys containing only 25 per cent. of platinum with, in some cases, the addition of about 2 per cent. of copper. In modern dental alloy the platinum may be partially or entirely replaced by the cheaper metal palladium. Thus a useful alloy consists of palladium 50 per cent. and silver 50 per cent.

The platinum in dental alloy may be determined by alloying 0·5 gramme with excess of silver, and parting in sulphuric acid as previously described. By this procedure a portion of the silver will remain undissolved in the platinum cornet, while, on the other hand, part of the platinum will be dissolved with the bulk of the silver; and for correct results this entails the determination of the amount of platinum that passes into solution, and the purification from silver of the insoluble platinum residue.

In some cases the weight of the insoluble residue from sulphuric-acid parting is sufficiently accurate for the platinum content, as the silver which it contains is counterbalanced to some extent by the loss of platinum in solution.

A more satisfactory procedure is to melt a weighed quantity of the dental alloy with excess of some base metal, such as zinc or lead, and then to treat the resulting button with hot 10 per cent. nitric acid which dissolves out the silver and base metal, and leaves an insoluble residue of platinum. The silver in solution is precipitated with a slight excess of hydrochloric acid, and the silver-chloride precipitate washed and weighed as such, or wrapped in sheet lead, cupelled, and the resulting silver bead cleaned and weighed.

The residue of platinum is collected and washed with hot water, then dried, ignited—preferably in hydrogen—and weighed.

The insoluble platinum residue from the nitric-acid treatment may be washed free from silver nitrate and then dissolved in aqua regia. The resulting solution is evaporated with hydrochloric acid to remove nitric acid, etc., and the platinum precipitated with ammonium chloride. The precipitate is washed, dried, ignited, reduced, and weighed.

(c) Assay of Miscellaneous Platinum-metal Alloys.

Under this heading are included a number of alloys containing platinum metals used in industry other than those utilized in jewellery and in dentistry.

Assay of Platinum–Iridium Alloys.—Alloys of these two metals find useful application in a number of industries whilst the iridio–platinum alloy containing from 5 to 10 per cent., of iridium is extensively used for jewellery, aeroplane and automobile magneto contacts are usually made of platinum alloyed with from 15 to 25 per cent. of iridium to give the required hardness.

Usually the iridio-platinum contact is soldered or welded to a steel or other base-metal rivet or screw, which has to be cut off before assaying. The contacts are then treated with hot hydrochloric or nitric acid to remove the base metal still adhering to them.

In view of the introduction of substitutes for contacts, such as tungsten and nickel alloys, it is advisable, before assaying, to heat all contacts to a red heat in an oxidizing atmosphere in a muffle, whereby base-metal contacts become oxidized and may thus be distinguished from iridio–platinum. Tungsten, when heated to a red heat, is converted into tungsten oxide WO_3, which is bright greenish-yellow in colour and readily distinguished from other base metals.

Iridio–platinum alloys are usually made from very pure metals, so that for identification purposes it is generally only necessary to determine one metal, and take the other by difference.

These alloys are conveniently assayed by parting, the procedure being as follows:—Take 0·20 gramme and mix with 1·0 gramme fine gold, and 3·0 grammes fine silver in an annealing-cup, cover with a little fused borax, close with a lid, and heat in a muffle at say 1100° C. for one hour until the metals are thoroughly alloyed.

When cool, break the cup, and clean the button from all adhering slag. Flatten the button, anneal, roll into a strip, and coil to a cornet. Part in nitric acid as usual, giving two boils with the stronger acid. Keep the nitric acid solutions containing the silver and platinum. Wash and anneal the cornet, reduce in hydrogen, remix with 3·0 grammes of silver as before and repeat the parting; add the second nitric parting acid to the first, and if necessary determine the platinum in solution.

Treat the cornet, after reduction in hydrogen as before, with 15 per cent. aqua regia, which dissolves the gold and leaves the iridium as an insoluble residue; this is filtered off, washed with dilute aqua regia, then ignited, reduced in hydrogen, cooled in carbon dioxide, and weighed as metal.

Treatment of the cornet with 15 per cent. aqua regia to remove the gold is slow, but it dissolves if left on the hot-plate overnight, and the determination of the iridium completed next morning.

An alternative method for the assay of iridio–platinum alloys consists of fusing 0·25 to 0·5 gramme with excess of lead in a carbon crucible, dissolving out the lead and platinum in nitric acid, and then igniting, reducing, and weighing the residual iridium, as described for the lead-fusion method. This procedure has been tested by Gilchrist[1] who confirms that the presence of palladium and rhodium does not interfere with the iridium determination.

The residual iridium should be tested for ruthenium and iron by fusion with potassium hydroxide and nitrate, distillation, and examination of the residual solution; and since this involves precipitation of the iridium as ammonium chlor-iridate, a direct iridium determination should be carried out.[2]

When both platinum and iridium are in solution together as chlorides they may be separated and determined by the following procedure:—[3]

The solution containing H_2PtCl_6 and H_2IrCl_6 is nearly neutralized with sodium carbonate, then 1 to 1·5 grammes of sodium bromate are added, followed by an excess of sodium bromide, and the solution boiled until bromine ceases to be evolved, whereby the iridium is precipitated completely as $Ir(OH)_4$. The precipitate is washed with hot, very dilute ammonium nitrate solution and heated in hydrogen at 160° to 180° C. until the filter-paper is thoroughly charred; after cooling in hydrogen the carbonized mass is ignited in the air until all carbon is burnt away, and the residue is then reduced in hydrogen, cooled in carbon dioxide, and weighed as iridium.

Extraction with dilute nitric acid is advisable to remove absorbed sodium salts. All the platinum, and any gold present, remain in the first filtrate; palladium, when present, is partly precipitated, but may be removed from the ignited iridium by extraction with dilute aqua regia.

If copper is present it is completely precipitated with the iridium by sodium bromate and sodium bromide, but it may be removed by extraction with dilute nitric acid after the preliminary carbonization in hydrogen.

For the separation of any iron present, the chloride solution is treated with $C_2H_2O_4$ until yellow, saturated with hydrogen sulphide and made ammoniacal,

[1] R. Gilchrist, U.S. Bureau, Standards, Sci. Paper, 1924, No. 19, 325.
[2] Schoeller and Powell, *op. cit.*, 296.
[3] L. Moser and H. Hackofer, *Monatsh,*. 1932, **59**, 45–60: Abs. *Met. Abs. Inst. Metals*, 1932, (3), **50**, 309.

and the precipitate of iron sulphide collected, redissolved in hydrochloric acid, and re-precipitated; the combined filtrates are then evaporated to dryness with ammonium sulphide, the residue is heated to expel ammonia and extracted with dilute ammonium nitrate solution, and the insoluble iridium sulphide is collected, ignited, reduced to metallic iridium, and weighed.

The separation and determination of platinum and iridium may also be effected by hydrolysis of the neutral chloride solution by boiling with sodium bromate and sodium bromide, iridium behaving in the same way as rhodium in the procedure described below.

Assay of Platinum-Rhodium Alloys.—A number of industrial alloys of these two metals are in use. For instance, gauzes of the alloy platinum 90 : rhodium 10 per cent. are used for the oxidation of ammonia in the production of nitric acid. Formerly this alloy was also largely used for thermocouples, but has been supplanted by the platinum 87 : rhodium 13 per cent. alloy, which gives a greater E.M.F. Experience has shown that the platinum–rhodium alloys are superior to the platinum–iridium alloys of similar composition which were formerly used for the same purpose. Another alloy containing 80 per cent. platinum and 20 per cent. rhodium has been found to be more satisfactory than pure platinum for the resistance windings of high-temperature laboratory furnaces.[1]

The procedure for the assay of platinum–rhodium alloys is the same as that described for platinum–iridium alloys above, which involves parting in nitric acid and treatment of the insoluble residue with aqua regia, whereby the rhodium is separated as a residue which is washed, ignited, reduced, and weighed as metallic rhodium.

Usually it is desirable to make a preliminary assay in order to determine the approximate proportion of rhodium in the alloy.

For some years bromate-bromide hydrolysis has been successfully applied to the separation of rhodium from platinum. The separation is effected by hydrolysis of the neutral chloride solution by boiling with sodium bromate and sodium bromide. The procedure recommended by Schoeller and Powell[2] is as follows:

The neutral chloride solution (acid solutions should be neutralized with sodium carbonate), containing normally an excess of platinum over rhodium, is diluted considerably (to 300-500 mls.), heated to boiling, and treated with a solution containing 30 grammes of sodium bromate and 100 grammes of sodium bromide per litre, the rhodium being gradually precipitated as dark-green hydroxide, with evolution of bromine. Boiling is continued for some time until the precipitate flocculates, more reagent being added and the effect noted. There should be no further evolution of bromine upon addition of reagent.

The liquid is treated with 2 per cent. sodium bicarbonate solution until neutral to bromcresol purple, and set aside on the water-bath until the precipitate has settled. This is collected, washed with 2 per cent. sodium chloride solution, dissolved in hydrochloric acid, and the precipitation repeated. If small amounts of rhodium and platinum are involved, a single precipitation may suffice.

The filtrate from the rhodium precipitate should always be tested for complete precipitation by renewed boiling with bromate reagent. If very little rhodium is present, the filtrate should be tested by boiling with a few drops of sodium bicarbonate solution, which precipitates the last of the rhodium.

[1] Atkinson and Raper, *loc. cit.*, 202.
[2] *op. cit.*, 277.

The rhodium hydroxide precipitate is returned to the beaker with hot water, the filter cleaned with a little hydrochloric acid, and the rhodium reduced to metal by boiling with a slight excess of sodium carbonate and hydrazine salt solution.

The platinum filtrate is cautiously freed from excess of the reagent by boiling with hydrochloric acid, then concentrated by evaporation, neutralized with sodium carbonate, and the platinum precipitated with hydrazine.

In both cases the precipitate is washed with 1 per cent. hydrochloric acid, ignited in a weighed crucible, reduced in hydrogen, and cooled in carbon dioxide.

If copper is present in the original platinum–rhodium solution (200 to 400 c.c.), it is rendered 0·1 N with respect to sulphuric acid, treated with 15 c.c. of 0·1 N $MgCl_2$ solution for every 100 c.c. volume, saturated with hydrogen sulphide at 100° C., and allowed to cool in the hydrogen-sulphide stream. The precipitate of copper sulphide and rhodium sulphide is collected and ignited to oxide. The residue is extracted with 1 : 1 nitric acid on the water-bath to remove copper-oxide, washed, ignited, reduced in hydrogen, and extracted twice with concentrated nitric acid; pure rhodium remains.[1]

Separation from iron is effected by hydrogen sulphide treatment of the solution after addition of $MgCl_2$.

The Assay of Platinum Solders.—Pure gold is a very satisfactory solder for platinum and its alloys, when the colour of the joint is not an objection. But for platinum jewellery, where white solders are necessary, a number of white alloys containing palladium, gold and smaller amounts of platinum, silver, and copper are used. The composition of these solders varies with different makers, but it may be remarked that gold containing from 15 to 20 per cent. of palladium is nearly white, and is ductile and workable, but is relatively hard. With moderate additions of copper the white colour of palladium is not materially affected, whilst such additions tend to lower the melting-point. A solder consisting of gold 75 and platinum 25 per cent. is used in dental work.

Platinum solders may be assayed by the methods given for jewellery and dental alloys containing gold and platinum metals.

The Assay of Platinum-metal Alloys for Spinnets.—The corrosion resistance of the platinum metals is utilized in the production of alloys for perforated nozzles or Spinnerets used in artificial silk manufacture (rayon). These vary in composition and may consist of acid-resisting alloys of platinum with small amounts of other metals, platinum–iridium, platinum–gold (80 to 90 per cent. gold), gold–palladium, or gold–platinum–rhodium. The usual methods of assaying alloys of gold and platinum-metals already described are applicable to spinnet materials.

The Assay of Platinum-metal alloys for Fountain-pen Points.—Very hard alloys usually containing varying proportions of iridium, osmium, and ruthenium (the hardest metals of the platinum group) are in use for fountain-pen points as substitutes for osmiridium, which was formerly largely used for the same purpose. Their composition varies considerably with different makers, and other metals of the platinum group may be present as well as a little nickel in some cases. The usual methods of platinum-metal assay are applicable, but it may be advisable to make a preliminary assay in order to determine what platinum-metals are present.

Assay of Platinum-Nickel Alloys.—The alloy platinum 95 per cent. and nickel 5 per cent. finds application for the filaments used in amplifying tubes in communication systems.[2]

[1] L. Moser and H. Graber, *Monatsh.*, 1932, **59**, 61–72: Abs. *Met. Abs. Inst. Metals*, 1932, (3), **50**, 310.

[2] Atkinson and Raper, *loc. cit.*, 205.

For the determination of the platinum, the alloy is dissolved in aqua regia and the nitric acid expelled by evaporation with hydrochloric acid. The platinum is then precipitated with ammonium chloride and the nickel taken by difference, or determined in the filtrate by precipitation with dimethyl-glyoxime, or other suitable reagent, in aqueous ammonia solution.

Assay of Palladium-Silver Alloys.—Silver-rich alloys containing 2·5 and 10 per cent. of palladium respectively are used as contacts in electrical systems. The alloy palladium 50 : silver 50 per cent. is used for swaged dental plates. The alloys are soluble in nitric acid. Silver is determined as chloride or by titration with N/10 ammonium thiocyanate. Palladium is determined with dimethyl-glyoxime.

When a nitrate or sulphate solution of palladium and silver is treated with hydrochloric acid, the silver chloride almost invariably carries down palladium, and is thereby coloured pink to brown.

It has, however, been shown by Schoeller and Powell[1] that silver chloride free from palladium can be obtained in one precipitation if the palladium is converted into a tetrammine-palladous salt prior to the precipitation of silver chloride. The procedure consists in treating the hot solution of the nitrates of the two metals with ammonia until the precipitate first formed redissolves and the solution becomes colourless. A small piece of litmus-paper is dropped into the solution, which is slightly acidified with acetic acid, and treated whilst hot with dilute hydrochloric acid drop by drop, with constant stirring, until no further precipitate forms, after which it is left in the dark till clear.

The precipitate is filtered by decantation and washed with very dilute nitric acid, then with water, dried, and weighed.

The filtrate is treated with a little more hydrochloric acid and the palladium precipitated with dimethyl-glyoxime.

In the case of a coloured precipitate of silver chloride containing palladium, obtained by direct precipitation in nitrate or sulphate solution, the procedure is as follows :—The precipitate is dissolved in a minimum excess of ammonia, and the hot solution slightly acidified with acetic acid. The silver is then precipitated as chloride as described above, and the small amount of palladium in the filtrate recovered by precipitation with dimethyl-glyoxime.

Assay of Palladium-Copper Alloys.[2]—The alloy containing 60 per cent. palladium and 40 per cent. copper is extensively used for electrical contacts instead of the more expensive 15 per cent. iridium–platinum alloy ; both alloys have a Vickers' pyramid hardness of 260.

The palladium in the former alloy can be determined by dissolving a weighed quantity in aqua regia and precipitating the palladium as K_2PdCl_6 by the addition of potassium chloride and C_2H_5OH, the filtrate being used for the determination of copper.

The precipitate of K_2PdCl_6 is redissolved in hot water and the palladium precipitated as metal by boiling with $H.CO_2K$; the washed palladium is ignited, reduced and, weighed.

This is a rapid method for inspection and control in works practice.

Assay of Rhodium-Copper Alloys.[3]—Alloys of copper with rhodium, especially if the rhodium exceeds 50 per cent., are very sparingly soluble in acids, even aqua regia. If the alloy is fused for some time, however, with 10

[1] *Op. cit.*, 261–262.

[2] E. E. Halls, *Indust. Chemist*, 1934, **10**, 313–314 : *Met. Abs. Inst. Metals*, 1935, **2**, 107.

[3] O. E. Zviagintzev, *Russian J. Applied Chem.*, 1932, B, **5**, (2), 217–218 : *Met. Abs. Inst. Metals*, 1933, **53**, 35.

times its weight of zinc, it dissolves completely in aqua regia. When the alloy contains more than 50 per cent. rhodium, this metal is precipitated with ammonium nitrite in excess of ammonium chloride. The precipitate (which contains some copper) is redissolved, re-precipitated, heated to redness in a stream of hydrogen, and weighed as rhodium.

If the alloy contains less than 50 per cent. rhodium, the copper is first removed as CuCNS, and the rhodium determined as above.

(d) Assay of Commercial Platinum Metals.

Assay of Commercial Platinum.—Platinum refined by modern processes is very pure, containing over 99·9 per cent. with traces of other platinum metals, and with iron and possibly copper. But various grades of commercial platinum containing small quantities of other platinum metals, together with some base metals, are also used in industry. Thus the platinum used for crucibles and chemical wares is hardened with about 3 per cent. of iridium.

For assay, two separate portions of 20 grammes each are taken, the platinum metals being determined in one, and the base metals in the other.

For determination of the platinum metals, the assay portion is dissolved in aqua regia and converted into sodium chlorplatinate by evaporation with hydrochloric acid and sodium chloride. The solution is submitted to bromate hydrolysis, and the precipitate analysed for palladium, rhodium, and iridium.

The base metals are determined[1] by converting the assay portion into chlorplatinic acid by solution in aqua regia. The platinum is then cautiously and gradually precipitated in the hot solution with a calculated deficiency of ammonium chloride (10·7 grammes). After separating the precipitate, the filtrate is evaporated to dryness and treated by the nitrite method, the precipitate containing the traces of iron and copper.

Assay of Commercial Palladium.[2]—The palladium of commerce is now of a high degree of purity, owing to improved methods of refining. The only impurity usually encountered is a minute amount of platinum.

This is determined in one or more 5-gramme portions by solution in aqua regia, evaporation with hydrochloric acid, precipitation of the solution with a calculated deficiency of potassium iodide, filtration under slight suction, and precipitation of the balance of the palladium in the filtrate with dimethylglyoxime.

The last filtrate is evaporated to dryness with aqua regia, the residue again evaporated with hydrochloric acid, and the acid chloride solution reduced with pure zinc. The combined precipitates (if more than one portion was taken) are ignited, dissolved in aqua regia, and the platinum precipitated with ammonium chloride.

Assay of Commercial Iridium.—Commercial iridium is usually in the form of sponge resulting from the refining process; small amounts of osmium and ruthenium are frequently present, with some iron and dissolved oxygen (iridium oxide). These impurities are determined in separate 2-gramme portions.[3]

Osmium and ruthenium are detected and determined by fusion of a 2-gramme portion with sodium peroxide, and subsequent distillation with bromate.

[1] Method of Schoeller and Powell, *op. cit.*, 294.
[2] Ibid.
[3] Ibid.

Another 2-gramme portion is ignited in a current of hydrogen in a weighed porcelain crucible, the iridium and crucible together being weighed before and after reduction. The difference gives oxygen. Traces of silica may also be determined in this portion, and tests made for the presence of rhodium and platinum.

A separate iron determination is made by fusion of 1 to 2 grammes of iridium with potassium hydroxide and nitrate in a gold crucible. The mass is taken up with water, the suspension boiled with alcohol, the precipitated hydroxide collected, and dissolved in aqua regia. The solution is cautiously precipitated with ammonium chloride, yielding a precipitate containing the bulk of the iridium, and a second small precipitate (mostly chlor-iridate) resulting from the treatment of the filtrate.

The filtrate from these separations is precipitated with ammonia, and the precipitate collected, dissolved in hydrochloric acid, and freed from traces of platinum metals by the nitrite process. The precipitate thus obtained is dissolved in hydrochloric acid, re-precipitated with ammonia, and ignited to ferric oxide.

Assay of Commercial Rhodium.—This metal is now produced in the form of sponge in a state of high purity for commercial purposes. It may contain traces of platinum and palladium, and possibly gold.

For assay[1] a weighed quantity of the metal is fused with lead in a carbon crucible, the button dissolved in 15 per cent. nitric acid, and the filtered extract precipitated with 40 to 50 mls. of sulphuric acid (1 : 1). The lead sulphate is filtered under suction, and the filtrate returned to the original beaker, together with the filter containing the insoluble residue from the nitric-acid treatment, and 5 grammes of ammonium sulphate added. When the paper has been destroyed, the beaker is covered and the sulphuric acid vigorously boiled for 10 to 15 minutes, the rhodium dissolving. When cool, the acid is diluted with 100 mls. of water, and the beaker is set aside in an inclined position for some hours. The clear liquid is poured off, and the deposit boiled as before with 15 mls. of sulphuric acid and a little ammonium sulphate. The cold acid is diluted, set aside overnight, and filtered, together with the first extract.

The insoluble residue is freed from lead sulphate by ammonium acetate solution, and then extracted with 15 per cent. aqua regia. The residue, unless small and obviously crystalline, should again be fused with 5 grammes of lead, and the above operations repeated.

The products of this procedure are: a combined sulphate extract; a combined aqua-regia extract; and the insoluble residue, which is analysed for iridium, ruthenium and osmium, after fusion with sodium peroxide. The extracts are worked up, after fusion with lead, for the recovery of platinum, palladium, and possibly gold. Rhodium is not determined, but is taken by difference.

For the determination of any iron present one gramme or more of the rhodium sample is fused with from 5 to 10 grammes of pure granulated zinc in a covered porcelain crucible, ammonium chloride being sprinkled on to the molten metal to maintain a bright, clean surface free from oxide crusts. The cooled zinc is washed, dissolved in hydrochloric acid, and the spongy rhodium filtered off. The filtrate is treated with excess of ammonia, and the precipitate collected, dissolved, re-precipitated, and ignited to ferric oxide.

This may have to be purified by the nitrite process.

[1] Method of Schoeller and Powell, *op. cit.*, 295.

II. ASSAY OF INDUSTRIAL PRODUCTS CONTAINING PLATINUM-GROUP METALS.

Under this head are included scrap or residual metal, sweepings from bench and floor, and lemel and other " refuse " resulting from the working of platinum and its alloys by manufacturing jewellers, in the workshops of mechanical dentists, and in the various industries in which platinum-metal alloys are used. Also included are various products containing platinum metals obtained in the metallurgical treatment of ores, etc., such as the slime or mud from electrolytic refining processes. The various products from manufacture, however, constitute by far the larger proportion of the materials containing platinum metals that are received for assay. Many of these are very complex and varied in composition, containing not only metals of the platinum group, but also gold and silver in addition to base metals, which frequently have to be separated before an accurate determination of the rarer metals is possible.

The metals most frequently present in the products from manufacture are gold and silver, platinum, palladium, iridium, and occasionally rhodium. Much less frequently they may contain small amounts of ruthenium and osmium.

The exact procedure to be adopted in the assay of these materials is not only dependent on the number of platinum metals present, but also on the amount and their ratio to one another and to any gold and silver that is present. The procedure most suitable for one class of sweep, etc. may require modification for the assay of a different type of material.

In assaying these materials it is well to recall that rhodium, ruthenium, and iridium are not attacked by aqua regia ; that platinum is dissolved by aqua regia but is not attacked by any single acid ; and that palladium is easily soluble in aqua regia, and is also attacked by hot concentrated sulphuric acid and to a less degree by nitric acid, though with these exceptions it is resistant to the common acids.

In the finely-divided state the metals are less resistant to chemical attack than in the compact form ; for example, rhodium-black is attacked by hot concentrated sulphuric acid, and osmium dissolves in nitric acid.

The following differences in the chemical behaviour of the platinum metals and their principal compounds are utilized for their separation from one another :[1]—

(*a*) Fusion of the platinum metals with excess of lead at 1000° C. converts platinum, palladium, and rhodium into an acid-soluble form, whilst iridium, ruthenium, and osmium remain in an acid-soluble form. The solubility of the platinum metals in molten lead offers the most convenient—and usually the only—means of quantitatively separating them from vastly preponderating amounts of base-metal compounds and siliceous material.

(*b*) Osmium and ruthenium are characterized by the ready volatility of their tetroxides. This permits of their quantitative separation from the remaining four platinum metals (and all other metals in general) by distillation. This operation is often preceded by a fusion with an alkaline flux.

(*c*) Fusion with lead, and fusion with alkali followed by distillation are the two fundamental operations in the analysis of the platinum group. Either operation brings about a subdivision of the six metals into two sub-groups :—

Lead Fusion	Soluble : Rhodium, Palladium, Platinum. Insoluble : Ruthenium, Osmium, Iridium.
Distillation	Volatile : Ruthenium, Osmium. Fixed : Rhodium, Palladium, Iridium, Platinum.

[1] Schoeller and Powell, *op. cit.*, 267–268.

The order in which the two operations are applied is dependent on the osmium content of the material. If the osmium content is small and the rhodium content appreciable, lead fusion should be carried out first, which leads to the following grouping :—

(i) Lead Fusion { Rhodium, Palladium, Platinum (soluble).
Ruthenium, Osmium, Iridium (insoluble).

(ii) Distillation { Ruthenium, Osmium.
Iridium (fixed).

If, on the other hand, osmium is present in substantial quantity and the rhodium content is small, fusion with alkaline flux and distillation should be carried out first, followed by lead fusion, thus reversing the order :—

(i) Distillation { Ruthenium, Osmium.
Rhodium, Palladium, Iridium, Platinum.

(ii) Lead Fusion { Rhodium, Palladium, Platinum.
Iridium.

In the resolution of the above sub-groups into their individual metals, Schoeller and Powell give the following useful information :—

(*d*) Osmium tetroxide is much more stable than ruthenium tetroxide : the former is not affected by hot hydrochloric acid (1 : 1), whereas the latter is converted into non-volatile ruthenium trichloride :

$$2RuO_4 + 16HCl = 2RuCl_3 + 5Cl_2 + 8H_2O.$$

Osmium is volatilized as tetroxide when its solutions are boiled with nitric acid (distinction from ruthenium).

For the separation of rhodium, palladium, iridium, and platinum the following reactions are available :—

(*e*) Only palladium is precipitated by dimethyl-glyoxime from acid solution.

(*f*) Iridium, rhodium, and palladium can be separated from platinum by hydrolytic precipitation from neutral chloride solution in presence of sodium bromate.

(*g*) Rhodium, palladium, and platinum are more reducible than iridium, being precipitated as metals by titanous salt solutions ; iridium tetrachloride is reduced only to the trichloride.

(*h*) Ammonium chloride precipitates quadrivalent platinum, iridium, and palladium. Quadrivalent platinum and iridium are separable from bivalent palladium and from subordinate amounts of rhodium by a judicious application of this procedure.

During recent years the separation and determination of the platinum-group metals have been the subject of considerable research in which the United States Bureau of Standards and the Platinum Institute of the U.S.S.R. Academy of Science have taken a leading part with a view to improving the methods that have long been in use. In the search for new methods the time-honoured use of ammonium chloride in the analysis of platinum metals has been discarded, and the use of hydrogen sulphide precipitation strongly recommended. This procedure will not commend itself unreservedly to many assayers in view of the readiness with which many base metals are thrown out of solution with this precipitant.

As Schoeller and Powell[1] have pointed out, " the difficulties encountered in the quantitative precipitation of the platinum metals by hydrogen sulphide

[1] *Op. cit.*, 256.

are due to the pronounced tendency of these elements to form comparatively stable complexes. Palladous sulphide (PdS) is most readily precipitated, platinic sulphide less so, whilst the precipitation of rhodium and especially of iridium by hydrogen sulphide is more erratic. The behaviour of ruthenium towards hydrogen sulphide recalls that of molybdenum, partial reduction being liable to occur, as revealed by the blue colour of the solution."

For details of the new methods developed, the various papers published by the Bureau of Standards must be consulted.

Useful information on modern methods of analysis of the platinum metals will also be found in Schoeller and Powell's well-known book (1940 edition); in the chapter on "Analysis of Platinum-group Metals" by R. E. Hickman, which appears in *Standard Methods of Chemical Analysis*, edited by W. W. Scott, and in similar works especially devoted to the subject.

Methods of Precipitation of Platinum Metals in Analysis of Sweeps.—In dealing with the somewhat complex sweeps and other materials resulting from modern manufacture, the following notes on the precipitation of the platinum metals in the course of analysis may be useful.[1]

Precipitation with Ammonium Chloride.—In spite of certain objections against its use, ammonium chloride is most usually employed as a precipitant for platinum and iridium, either alone or in admixture, palladium—if present—having been converted into chlorpalladous acid.

In order to minimize co-precipitation of other metals, it is advisable first to add a dilute solution of the reagent drop by drop in slight excess to the warm solution of iridiferous platinum in aqua regia during continuous agitation, the bulk of the platinum and iridium being precipitated. The liquid is left to cool, and diluted with an equal volume of saturated ammonium-chloride solution; if this leads to an unduly large volume of solution, a suitable quantity of solid A. R. ammonium chloride may be added instead.

The precipitate is collected, washed with half-saturated ammonium chloride, and reserved.

The filtrate is concentrated on a water-bath, with the occasional addition of a drop of sodium chlorate solution (to oxidize any chlor-iridite), until ammonium chloride begins to crystallize out at the surface. Whilst still hot, the liquid is stirred and cautiously treated with a little cold water to redissolve the crystals, and set aside overnight to cool. The clear solution is decanted through a small filter, and the small dark precipitate (mostly chlor-iridate) collected and washed as before, and added to the main precipitate.

The precipitation of platinum and iridium by the above process is quantitative within the limits of experimental error. If obtained from more or less pure solutions, the precipitates are sufficiently pure to be converted to metal by ignition. If, however, small quantities of palladium and rhodium and substantial quantities of base metals (*e.g.* iron, copper, nickel, bismuth) are present, it is advisable to purify the precipitate by Schoeller's chlorine process,[2] which consists in reconverting the precipitate into hexachloro-acids by the action of a current of chlorine upon its suspension in boiling water.

Small quantities of ammonium chlorplatinate or of chlor-iridate may be purified by simple recrystallization, the precipitate being washed on the filter with boiling water until completely dissolved, and the washings evaporated to small bulk, with subsequent addition of a little ammonium chloride after cooling. Large quantities of precipitate cannot be satisfactorily treated in

[1] Abridged from *Analysis of Minerals and Ores of the Rarer Elements* by Schoeller and Powell, who fully describe the analytical technique involved, pages 254–257.

[2] W. R. Schoeller, *Analyst*, 1930, **55**, 551.

this way owing to the large volume of water required to effect solution of the precipitate, which is sparingly soluble.

In some cases palladium is conveniently determined in chloride solutions by precipitation as chlorpalladate with ammonium chloride.

When ignited, the palladium precipitates leave a residue of spongy palladium.

Precipitation with Hydrogen Sulphide.—For the reasons given above, the precipitation of platinum-group metals as sulphides finds only limited application in their separation and determination.

Although all the platinum metals react with hydrogen sulphide, their acid solutions afford dark-brown to black precipitates, which are mostly of more or less indefinite composition. Quantitatively considered, the precipitates have no distinctive properties leading to the identification of the platinum metals or their separation from other members of the hydrogen-sulphide group. In quantitative analysis the sulphides of ruthenium and osmium are the most useful, and a modified sulphide process has been successfully applied by Schoeller and Powell[1] for the gravimetric determination of these two metals, but not for their separation from other metals.

In regard to the other four metals of the group, precipitation by hydrogen sulphide is a procedure of more questionable value. Palladium is readily precipitated by hydrogen sulphide from acid solutions, but the precipitate does not part with the whole of its sulphur on ignition, and the impure metal runs into one or several globules. The sulphide precipitate should therefore be dissolved in aqua regia, and the palladium precipitated with dimethylglyoxime or ammonium chloride.

With rhodium and iridium, Schoeller and Powell have at times achieved quantitative precipitation, whilst at others even pressure-precipitation failed to bring about the desired result. These authors, therefore, for the present "still consider hydrogen sulphide a reagent of doubtful efficiency for the recovery of these valuable metals under the arduous conditions of actual practice," and with this conclusion many who are engaged in the industrial assaying of the platinum-group metals will agree.

Precipitation with Metallic Zinc or Magnesium.—The use of zinc is a convenient means of recovering rhodium, palladium, platinum, and ruthenium from solutions, and of separating them from base metals, etc. not precipitated by zinc. The recovery of iridium by displacement with zinc is rarely complete; consequently zinc precipitation is not applicable to solutions containing iridium. Rhodium is quantitatively precipitated from chloride or sulphate solutions with zinc or magnesium. The results are usually high but are sufficiently accurate for the determination of very small amounts of rhodium in technical work.

The acid chloride solution free from iridium and nitrates is heated to about 80° C., and treated with excess of chemically pure, powdered zinc. When any ferric salt present has been reduced, the action is allowed to proceed for some time until it becomes sluggish owing to neutralization of the acid, the zinc being in excess. Strong hydrochloric acid is then added in portions to the hot solution until all excess zinc is dissolved.

The precipitate is collected on a loose filter and washed twice without delay with N. hydrochloric acid (to avoid oxidation of the finely-divided metal), the washing being completed with 0·1 N acid.

Precipitation with Sodium Nitrite.—Sodium nitrite may be described as a group reagent for the platinum metals, since the sodium-nitric salts are very soluble in water and stable in feebly alkaline solutions, in which base metals are precipitated as hydroxides or basic salts.

[1] *Op. cit.*, 246, 256, 281.

The nitrite process is a valuable method for separating large or small amounts of rhodium, iridium, platinum, and ruthenium from small quantities of base metals (lead, bismuth, copper, iron, nickel, etc.). If the base metals are present in appreciable quantity, the bulky precipitate of hydroxides (basic carbonates) absorbs a large proportion of the noble metals. Double treatment must be applied in any case. The process does not answer so well for palladium, because the double nitrite of that metal is less stable in the boiling alkaline solution; hence palladium may be found in the precipitate after double treatment. The precipitate may be tested for palladium by solution in dilute hydrochloric acid and precipitation with dimethyl-glyoxime, with or without intervening precipitation on zinc.

The chloride solution should be free from ammonium chloride. If this salt is present, it must be destroyed by boiling in a large flask with a copious excess of strong nitric acid (30 to 50 mls.). When the volume has been reduced to about 25 mls. the solution is transferred to a beaker, evaporated to dryness, and the residue evaporated twice with a little hydrochloric acid. The final residue is dissolved in hot water.

The hot, nearly neutral solution (100 mls.) is treated with 5 grammes of sodium nitrite and set aside on a covered water-bath for an hour. It is then cautiously treated with sodium hydroxide, drop by drop, until the precipitate no longer increases in bulk and flocculates readily. This is collected, washed with hot water, dissolved in hydrochloric acid, and the precipitation from the neutralized solution is repeated. The combined nitrite filtrate is acidified with hydrochloric acid and evaporated to dryness before being subjected to further manipulation.

Precipitation with Titanous Salt.—Solutions of platinum, palladium, and rhodium are reduced by titanous salt with precipitation of the metal, whilst iridium solutions are not precipitated, reduction proceeding only to the iridous state. Copper in large amounts interferes, being also precipitated; if small, it is held in solution at high acidity. Bismuth is also precipitated. The procedure is more expeditious than zinc precipitation, but the precipitate may be contaminated with titania unless the acid concentration is very high. Precipitation may be effected with titanous chloride, but Schoeller and Powell[1] favour titanous sulphate, which is prepared by electrolytic reduction at a mercury cathode of titanic-sulphate solution (obtained from a bisulphate melt of titanic oxide, or from potassium titanium oxalate by evaporation with sulphuric acid until strong white fumes are evolved).

The solution under treatment is acidified with 30 to 40 per cent. by volume of sulphuric acid, boiled, and treated with a decided excess of the titanous-salt solution. When the precipitate has coagulated, the liquid is diluted with an equal volume of water, and filtered. The precipitate is washed with 5 per cent. sulphuric acid, and returned with the filter to the beaker. It is then boiled with dilute aqua regia (1 : 3 water). The pulped filter, containing part of the rhodium free from platinum and palladium, is collected, well washed, ignited, and reserved (insoluble rhodium fraction). The filtrate is treated with dimethyl-glyoxime which precipitates the whole of the palladium.

The filtrate from the palladium precipitate is evaporated to dryness and the residue taken up with strong hydrochloric acid, which is again evaporated, with the addition of small amounts of sodium chlorate for the destruction of the dimethyl-gloxime. The final residue is taken up with hot water and a drop of hydrochloric acid, and the soluble rhodium fraction is separated from platinum by bromate hydrolysis.

[1] *Op. cit.*, 258, 275–6.

Precipitation by titanous sulphate is most useful in the separation of small quantities of platinum, palladium, and especially rhodium, from preponderating amounts of iridium when there is no need to recover and determine that element (*e.g.* in the analysis of commercial iridium).

The small precipitate produced by titanous salt is ignited, reduced, weighed, and fused with sodium chloride in a current of chlorine. The melt is dissolved in a little water and the aqueous extract filtered, any co-precipitated titania remaining insoluble.

If only rhodium is present, the aqueous extract is of a pure rose-red colour; in the presence of platinum or palladium or both, the procedure is carried out in chloride solution, the palladium being precipitated with dimethyl-glyoxime, and the rhodium separated from platinum by bromate hydrolysis.

Choice of Precipitants of Platinum Metals.[1]—The above precipitants are used for the separation of the platinum metals from iron, nickel, zinc, and base metals not precipitated by zinc or hydrogen sulphide. When the platinum metals are present in minute quantity, separation resolves itself into their concentration in lead by a crucible fusion or scorification, and cupellation of the resulting lead button with a suitable excess of silver. When the platinum metals are present in substantial amounts their separation from solution is effected by the precipitants given above. Osmium and ruthenium, being separable by distillation, need not be considered here. The separation of the other four platinum metals from base metals is effected by the following methods :—

(*a*) Ammonium-chloride precipitation is the cleanest separation method for substantial quantities of iridium or platinum, or both, especially from much iron.[2] The acid chloride solution free from nitric acid is precipitated with ammonium chloride as prescribed above. The co-precipitation of iron, etc. is of the order of 0·001 to 0·002 gramme, and is effectively counteracted by retreatment of the precipitate.

(*b*) Zinc precipitation is not applicable to solutions containing iridium.

(*c*) Hydrogen-sulphide precipitation works best in the case of palladium, for which, however, dimethyl-glyoxime precipitation is a more attractive procedure (this latter is not applicable to sulphate solutions.)

(*d*) Titanous-sulphate precipitation may be used for rhodium, palladium, and platinum.

(*e*) The nitrite process is satisfactory if the base-metal precipitate is not too bulky, and palladium quite subordinate. Gold, if present in the solution, is reduced and precipitated together with the hydroxides and basic carbonates of the base metals.

(*f*) When gold is present, the reduction of auric chloride may be effected with a freshly-prepared solution of ferrous chloride free from ferric salt. The separation of gold from iridium, rhodium, and ruthenium is quantitative in one operation.

If platinum and palladium are present, the precipitate should be ignited, dissolved in aqua regia, the solution freed from nitric acid by evaporation and re-evaporation with hydrochloric acid, and the precipitation with ferrous chloride repeated.

Ignition and Reduction of Platinum-metal Precipitates.—The ignition of the "double chlorides" of the platinum metals produced by ammonium chloride has to be effected with great care so that loss by decrepitation may be avoided. In all cases, the washed precipitate is carefully wrapped in the wet

[1] Summary by Schoeller and Powell, *op. cit.*, 267.
[2] W. R. Schoeller, *Analyst*, 1926, **51**, 392.

filter, the edges of the paper being folded over the single layer and the folds pressed against the bottom of the crucible, with the threefold layer of paper uppermost. The crucible is heated on an asbestos mat over a small flame until the paper is dry. The heat is then raised very slightly to cause a barely visible evolution of ammonium-chloride fumes. With large precipitates the procedure takes several hours, but requires very little supervision. When the ammonium salt has been expelled, the charred paper is burned away at a low temperature, after which the heat is raised to redness. The ignited precipitate is next reduced to metal.

In technical work this may be effected by coal-gas, the small luminous flame of a Bunsen burner being directed upon the hot precipitate whilst the crucible is in a slanting position, so that the escaping gas burns at the mouth of the crucible.

The reduction is, however, more satisfactorily carried out by means of hydrogen in a weighed Rose crucible ; the precipitate, after having been ignited in air, is left to cool somewhat, the crucible is covered, and a slow current of hydrogen is admitted through a tube connected by a three-way stopcock to a supply of hydrogen and carbon dioxide. The crucible is heated to dull redness, kept at this temperature for a few minutes, and allowed to cool to 200° to 300° C., when the hydrogen in the crucible is displaced by a slow current of carbon dioxide, in which the metal is allowed to cool to room temperature before weighing.

Reduction takes place with incandescence, and the blackish oxide is converted into grey metal which must be allowed to cool in the gas as a protection from oxidation before weighing.

Although platinum is very resistant to oxidation and its precipitates are converted into metal by simple ignition in air, it is advisable to observe the above precautions against oxidation when reducing large platinum precipitates.

In dealing with the ignition of precipitates of other metals of the platinum group, it must be observed that ruthenium, rhodium, palladium, and iridium precipitates yield an ignition residue consisting either wholly or partly of oxide, which must be reduced to metal by gentle ignition in hydrogen, and subsequent cooling in carbon dioxide as described above. This precaution is very necessary, since a certain amount of hydrogen may be superficially absorbed by the reduced metal and catalytically oxidised to water on subsequent contact with the air,[1] with attendant risk of positive errors in weighing the reduced metal.

Palladium when ignited in air usually becomes coated with a purple film of oxide, which may be reduced by treatment with a few drops of formic acid, instead of reduction in hydrogen.

Exclusion of air is very necessary when igniting osmium precipitates, otherwise the metal is completely volatilized as tetroxide.

Fusion with Sodium Peroxide.—Fusion with sodium peroxide forms the first step in the separation and determinatiou of osmium, ruthenium, and iridium. Materials rich in rhodium must be attacked by fusion with lead or litharge, whilst those in which platinum and palladium preponderate may be decomposed by aqua regia.

The sodium-peroxide fusion is carried out in a nickel crucible ; the material to be fused must be in the state of very fine powder which is thoroughly mixed with 15 to 20 grammes of sodium peroxide to which 2 or 3 grammes of powdered sodium hydroxide may be added. The crucible is gradually heated on an asbestos mat until the mixture sinters. It is kept in this condition for one to three hours, according to the amount of noble metals present, after which it is heated more strongly, passing from a pasty to a viscous mass, which is finally converted

[1] Schoeller and Powell, *op. cit.*, 278.

into a fluid melt by stronger heating on a triangle. The crucible is now rotated at frequent intervals, and finally heated for a short time at a bright red heat.

The second fusion period occupies about the same time as the first (sintering).

The cold melt is disintegrated with cold water in a 600-ml. beaker, the crucible removed with the aid of tongs, cleaned with a rubber-tipped glass rod, and discarded. By weighing the crucible before and after fusion, the amount of nickel introduced into the assay can be ascertained. Some prefer coating the inside of the crucible with a layer of fused sodium carbonate for the purpose of decreasing the attack on the nickel.

Distillation in one form or another being the next step, the alkaline suspension is quantitatively transferred to a distillation flask.

Fusion of Platinum Metals with Sodium Chloride in a Current of Chlorine.[1]—This is a procedure of general applicability to the platinum metals, which is most useful in the examination for purity of ignited rhodium precipitates. It introduces no impurities, but is tedious for any but small quantities of material.

A weighed quantity of the finely-divided metal is intimately mixed with two parts of sodium chloride, transferred to a Rose crucible, and heated in a current of undried chlorine, gently at first, then more strongly, and finally with the full flame of the Bunsen burner above the melting-point of the salt (800° C.) After cooling, the melt is dissolved in water, in which it should dissolve without leaving any dark insoluble residue. If a residue remains it must be filtered off, washed, ignited, reduced, then mixed with sodium chloride, and the fusion process repeated.

Distillation of Ruthenium and Osmium.[2]—These two metals can be separated from the other four platinum metals, from all other elements in general, and from one another, by distillation, a fact which greatly simplifies the analysis of the platinum group. In commercial assaying, however, determinations of ruthenium and osmium are only occasionally required. It is only necessary, therefore, to indicate briefly the general procedure and form of apparatus employed in their determination, and to refer the reader to works such as that of Schoeller and Powell dealing with platinum-metal analysis for further details of the technique involved.

The first step is the fusion of the material with sodium peroxide for the production of soluble compounds of the two metals. The resulting alkaline liquid is then distilled in a current of chlorine, or with bromate. In these operations, osmium volatilizes with the ruthenium, but is carried over by the steam into the next receiver, whilst the ruthenium tetroxide undergoes reduction to the deep reddish-brown chloride in contact with the hydrochloric acid contained in the first receiver.

The apparatus used for the separation consists of three parts, viz. the distillation train, the ruthenium absorption train, and the osmium absorption train. The distillation and absorption flasks can be used for three distinct operations :—

(*a*) Ruthenium distillations: the distillation train is connected to the ruthenium train (from absorption flasks).

(*b*) Osmium distillations: the distillation train is connected to the osmium train (two absorption flasks). A short cooler may be inserted between the distillation flask and the receiver No. 5.

(*c*) Ruthenium–osmium distillations: the distillation train is connected to the ruthenium and osmium trains in the order given above.

[1] F. Wöhler, *Pogg. Ann.*, 1834, **31**, 161. Description by Schoeller and Powell, *op. cit.*, 260.

[2] Schoeller and Powell, *op. cit.*, 269, 278, 286.

When both metals are present the complete distillation apparatus is used, and their separation effected by distillation in a current of chlorine, which is bubbled through the boiling solutions in the ruthenium and osmium trains respectively.

If ruthenium is present in appreciable quantity, orange-yellow vapours of the tetroxide are evolved, and yellow crystals deposit in the inlet tube of the receiver until fused and carried down by the rush of steam. The hydrochloric acid in the first receiver acquires a yellow colour, which soon turns a deep reddish-brown as the bulk of the ruthenium tetroxide becomes converted into trichloride by the hot acid.

The osmium and still unreduced ruthenium tetroxides are driven into the second receiver, where a further quantity of the latter is reduced to trichloride as the liquid becomes hot. Small quantities of ruthenium are also recovered in later receivers. When complete absorption of the osmium in No. 5 receiver has apparently taken place, the distillation is continued for another 15 to 20 minutes, the total time taken amounting to about $1\frac{1}{2}$ hours. The recovery of the osmium is normally complete in one operation; the osmate solution from the two receivers is transferred to a capacious beaker, and the osmium determined by precipitation as hydroxide, or as sulphide.

The gravimetric determination of osmium is always preceded by its precipitation, either as hydroxide or as sulphide, from an alkaline solution of sodium osmate as obtained by distillation. If the amount of osmium to be determined is substantial (as indicated by the more or less intense mauve colour of the osmate solution) it should be precipitated as hydroxide; if, on the other hand, the solution is barely or faintly coloured, precipitation as sulphide is to be preferred. It must be noted that the determination of osmium differs from that of the other five members of the group in that its precipitates must be ignited with the exclusion of air, as otherwise the metal is completely volatilized as tetroxide.

The reduced osmium either from the hydroxide or sulphide precipitate is allowed to cool under hydrogen, which is then displaced by dry carbon dioxide. The reduction to metal takes place below 200° C., and is frequently attended with slight decrepitation.

The results obtained by precipitation as sulphide have a tendency to be slightly high, since osmium obtained by ignition of the sulphide in hydrogen generally retains a little sulphur which has not been expelled in the subsequent ignition of the precipitate. The procedure, therefore, should be applied only to small amounts of osmium.

The ruthenium recovery by distillation is usually incomplete, due to partial decomposition of the tetroxide in the distillation flask. The contents of the four receivers forming the major ruthenium fraction are transferred to a 600-ml. beaker and evaporated to dryness on a water-bath. For the recovery of the minor ruthenium fraction, a second distillation is necessary. The distillation is carried out as before, and the contents of the receivers added to the beaker containing the major fraction. Ruthenium is obtained as chloride in the distillate, and may be determined by precipitation as sulphide. The precipitate is collected, washed with hot water, ignited wet in a weighed crucible, reduced as usual, and weighed as ruthenium.

If large quantities of ruthenium (*e.g.* more than 0·2 gramme) have to be determined, the precipitation as sulphide becomes inconvenient on account of the bulkiness of the precipitate. In such cases the bulk of the ruthenium may be precipitated from the concentrated distillate by addition of solid ammonium chloride until the liquid is saturated. The brown crystalline precipitate (ammonium chloro-ruthenite) is collected, washed twice with ammonium-chloride solution, and ignited with the usual precautions. The pale-brown filtrate and washings are precipitated with ammonium polysulphide, the balance of the ruthenium being recovered as sulphide. The double-chloride precipitate is crystalline and compact but comparatively soluble, hence the necessity for a combination process.

Osmium–Ruthenium Distillation with Bromate.[1]—This is a more recently introduced method for separating ruthenium and osmium. In this method the alkaline solution is digested with hypochlorite in the distillation flask overnight. The train is connected up and a slow current of air drawn through it. The solution is then raised to the boiling-point, and the gas-washing bottle replaced by a glass tube, through which hydrochloric acid is cautiously aspirated from a beaker until the black precipitate has dissolved. The clear, boiling solution is next treated with 25 per cent. sodium chlorate solution (10 to 20) introduced through the tube, when the tetroxides of osmium and ruthenium distil into the first receiver. The volatile compounds are then driven gradually through the ruthenium receivers, the burners being lit in succession.

When no further darkening of the reddish-brown colour can be observed in the contents of the ruthenium receivers, 10 to 20 mls. of bromite–bromide solution (120 grammes of sodium bromide and 36 grammes of sodium bromate made up with water to 1 litre) is aspirated into the distillation flask, and the distillation is continued for 10 to 15 minutes after the evolution of bromide has ceased, when the flask is disconnected. If iridium is present, incipient precipitation of its hydroxide should take place towards the end of the distillation. The recovery of osmium and ruthenium is quantitative in one operation.

Whichever distillation method has been employed, the procedure for recovering the metals other than osmium and ruthenium from the residual solution is the same. In the case of chlorine distillations, the second or minor ruthenium fraction may be included in the procedure. The hot liquid left in the distillation flask is treated with a solution of 10 grammes of sodium hydroxide and 20 mls. of alcohol and boiled for 5 minutes, thus precipitating the metals of the platinum group as hydroxides, together with the nickel resulting from the attack of the sodium peroxide on the crucible during the peroxide fusion of the original material.

The precipitate is washed several times with hot water to remove the bulk of the sodium salts, and then evaporated to dryness with 15 mls. of hydrochloric and 5 mls. of nitric acid on a water-bath. The dry residue is taken up with hot dilute hydrochloric acid, and any traces of silica filtered off.

The filtrate containing the non-volatile platinum metals, with small amounts of ruthenium, is worked up for their separation, as outlined above.

Assay of Sweep, Loose Lemel, etc., containing Platinum Metals.—The preliminary dressing and sampling of sweep and lemel, etc. containing platinum metals are conducted in the same way as described for similar materials containing gold and silver, but in view of the greater intrinsic value of the platinum-group metals every care must be taken to prevent loss.

Comparatively small lots of material rich in precious metals may be sampled for assay by fusing with six times their weight of lead, with the addition of suitable fluxes if necessary. After fusion the slag is removed from the surface of the melt, so far as possible, with a graphite stirrer, and the lead alloy granulated by pouring in a steady stream into water. After washing free from

[1] Schoeller and Powell, *op. cit.*, 271.

particles of slag, the lead granules are dried, then well mixed, and a sample taken for assay.

When the probable content of platinum metals is not known it is necessary to make a preliminary assay of the sweep in order to ascertain the best procedure to adopt in assaying it.

The assay of sweep and similar materials almost invariably involves concentration of the platinum metals in lead by crucible fusion or by scorification, with or without the addition of silver, either in the form of precipitated powder or as silver chloride, according to circumstances. As already stated, the action of molten lead on the platinum metals is of great importance. They become diffused in the excess of lead in a very finely-divided form, in which condition platinum, palladium, and rhodium are soluble in acids, whilst iridium, ruthenium, and osmium remain in a form insoluble in acids. The ratio of lead to the total content of precious metals should be about 20 to 1, but more lead is desirable in some cases.

The lead button containing precious metals may be treated by either of the following methods :—(i) it is treated direct with nitric acid ; or (ii) it is cupelled and the resulting silver button treated with either nitric acid or sulphuric acid.

Direct treatment of the lead button with acid is more tedious as solution is slow, but it must be used for the determination of osmium, appreciable quantities of iridium, rhodium and ruthenium, with or without platinum and palladium, and for any materials suspected of containing osmiridium.

When osmium is present in any form cupellation is not admissible, since this metal is partly or wholly volatilized as the tetroxide during the later stages of cupellation. If osmium is present in substantial amounts, fusion with alkali and distillation should always precede fusion with lead. If significant amounts of iridium, rhodium, or ruthenium are present in the material, cupellation involves risk of loss through their partial oxidation, some of the oxide formed adhering to the cupel. In such cases, direct solution of the lead button in nitric acid is advisable.

Materials containing much nickel or copper yield a lead button from the crucible fusion which gives trouble, as these metals cannot be completely removed by scorification. Such buttons yield, upon cupellation, a greenish-black scoria surrounding the silver bead, containing minute prills of silver and retaining rhodium and ruthenium, which would normally form a black incrustation on the bead. It is therefore advisable to treat materials rich in nickel and copper by a combination process which involves separation of the base metals with acid, and recovery of the precious metals in the residue as described below.

Schoeller and Powell[1] give preference to the lead-fusion process over the titanous-sulphate method as the regular procedure for the separation of iridium from platinum, palladium, and rhodium. The direct treatment of the lead button from crucible fusion prevents the losses of precious metal incidental to cupellation and subsequent parting operations.

The procedure for the treatment of the lead button by the two methods stated is as follows :—

(1) *Direct Treatment of Lead Button with Nitric Acid.*[2]—If considered desirable, the lead button obtained by crucible fusion may first be reduced in bulk and purified by a single, or even a double scorification to eliminate copper, nickel, and other base metals. The cleaned button is then flattened out and

[1] *Op. cit.*, 274.

[2] Procedure of Schoeller and Powell, *op. cit.*, 249.

treated with hot 15 per cent. nitric acid, the resulting liquid decanted or filtered off, and the residue washed.

The filtrate is heated to about 80° C., stirred, and the lead precipitated with sulphuric acid (1 : 3 water) in moderate excess. When cool, the liquid is filtered under slight suction and the washed lead sulphate rejected. The filtrate is evaporated to about 25 ml. and added to the beaker containing the residue from nitric-acid treatment, which is then heated until the sulphuric acid fumes strongly. After cooling somewhat, 5 grammes of solid ammonium sulphate are added, the beaker covered, and the acid liquor vigorously boiled for 10 to 15 minutes. When cold, the mass is taken up in 75 mls. of water, and the liquid boiled for a short time and set aside overnight. The solution is then filtered off.

This operation yields a sulphate solution of palladium, rhodium, and possibly silver, copper, bismuth, etc., from which the platinum metals are precipitated and are recovered either by cupellation with excess of silver and subsequent parting if the amount of rhodium and palladium present is not unduly high, or by a wet method.

The residue from the above treatment is washed with 0·5 per cent. sulphuric acid; it consists of crystalline iridium, ruthenium, and osmium (possibly alloyed with platinum and iron) with a little lead sulphate, which may be yellow to buff in colour, due to a little occluded rhodium.

To recover these traces the residue is treated with a minimum of hot ammonium acetate solution to dissolve the lead sulphate, and after filtering, the lead re-precipitated with a moderate excess of sulphuric acid. The lead sulphate, which is now white, is filtered off. The filtrate is evaporated to very small bulk for the elimination of the acetic acid, diluted, and added to the sulphate solution containing platinum metals obtained by evaporation as above.

The residue, now free from lead acetate, is treated with 100 mls. of 15 per cent. aqua regia, and digested for an hour or more. The extract, containing platinum and gold, and perhaps a little palladium and rhodium, is filtered, and the solution treated for the recovery of these metals. The residue from this operation consists of iridium, ruthenium, osmium, and possibly iron, sometimes contaminated with a little silver chloride. It is collected and washed with warm water, care being taken to ensure the complete transfer of the heavy micro-crystalline residue of iridium, etc. to the filter. The filter-paper with residue is transferred to a porcelain crucible and the paper gently charred on an asbestos mat: the carbon is then burned off at the lowest possible temperature, so as to avoid loss of any osmium present.

The gold in the aqua regia solution may be precipitated with ferrous chloride or other suitable reagent, and filtered off; and the platinum in the filtrate precipitated with ammonium chloride. The finely-divided iridium, ruthenium, and osmium in the residue are separated and determined by fusion with sodium peroxide in a nickel crucible. If the residue consists mainly of iridium it may be weighed as such, the small quantities of ruthenium and osmium being disregarded.

(2) *Cupellation of the Lead and Treatment of Resulting Silver Button.*[1]—The lead button from crucible fusion is cupelled with the precautions that are necessary when metals of the platinum group are present. The concentration of platinum and palladium (with or without gold) into a silver bead by cupellation is very satisfactory. A suitable excess of silver is necessary, at least 30 parts to 1 of platinum and palladium combined, to prevent solidification of the bead

[1] Schoeller and Powell, *op. cit.*, 251.

before all the lead has been oxidized. In the case of the other platinum metals, osmium is partly or wholly volatilized as the tetroxide during the later stages of cupellation, depending on the amount present, whilst iridium, rhodium, and ruthenium also oxidize more or less readily, the oxides staining the cupel and forming a brittle coating on the silver bead. Consequently, cupellation is limited to materials containing relatively small amounts of these metals.

A large excess of silver tends to counteract oxidation by keeping the metals in solution. Roughly speaking, the sum of these platinum metals should not exceed 0·02 gramme in a silver bead weighing at least 3 grammes.

The silver bead obtained by cupellation varies in composition according to the class of sweep, etc., but may contain five of the platinum metals (osmium excluded) with gold and excess of silver. Such beads may be treated either with (*a*) nitric, or with (*b*) sulphuric acid.

Treatment of Silver Bead—(*a*) *Nitric-acid Parting.*—This method is applicable to materials high or low in platinum and palladium, and free from or containing only very subordinate amounts of rhodium, ruthenium, and iridium; osmium is not recovered.

The most minute quantities of platinum and palladium are readily recoverable from large quantities of silver by Crooks' method, which consists of solution of the metal in nitric acid, evaporation, fusion of the silver nitrate (m.p. 209° C.), and solution of the product in hot water, and is applicable to the separation of silver from all the platinum metals, which remain insoluble whilst the silver nitrate goes into solution. A silver bead weighing at least 1 gramme should be aimed at, and there should be at least a fifty-fold excess of silver over the other noble metals: a greater excess of silver is desirable, as this gives a fluid nitrate melt, leading to a cleaner separation.

The silver bead is dissolved in nitric acid (1 : 1) and the solution evaporated to dryness; if the amount of residual silver nitrate appears too small for a satisfactory fusion, a suitable amount of strong C.P. silver-nitrate solution should be added.

The solution is evaporated to dryness, and the silver nitrate fused, set aside till cold, and dissolved in hot water. The liquid is passed through a small filter and the residue washed till free from dissolved silver, then digested with dilute aqua regia (1 : 1), when it will dissolve completely if only platinum and palladium are present. For the recovery of the latter metals the solution is evaporated with hydrochloric acid to free from nitric acid in the usual way; the palladium is separated by precipitation by dimethyl-glyoxime, and the platinum precipitated with ammonium chloride. The aqua regia solution may also contain gold which is separated by precipitation with ferrous chloride, or other suitable reagent.

If a dark residue is left after digestion with aqua regia, iridium, ruthenium, and rhodium may be present. In this case, after filtering off the solution containing platinum and palladium, the residue is collected, washed first with water, and then with dilute ammonia to remove any silver chloride, which is separated and rejected. The residue is ignited, reduced, and weighed as a check.

If rhodium preponderates, the residue should be fused with lead and the lead button treated with nitric acid for the determination of the rhodium with hydrazine. If iridium or ruthenium preponderates, fusion of the residue with sodium peroxide is applied. It is transferred to a small nickel crucible, fused with a little sodium peroxide, and the aqueous solution of the melt transferred to a 250-ml. distillation flask and distilled for ruthenium. The hot liquid left in the distillation flask is treated with a solution of 10 grammes of sodium hydroxide and 20 mls. of alcohol and boiled for 5 minutes, whereby the metals of the platinum group are precipitated as hydroxides, together with the nickel resulting from the attack of the sodium peroxide on the crucible. The precipitate is dissolved in aqua regia and the solution (after removal of silica) evaporated with a little ammonium chloride for the precipitation of iridium.

The filtrate, containing the rhodium with nickel derived from the crucible, is precipitated with zinc, and the small precipitate after ignition identified by fusion with sodium chloride in chlorine. The rhodium is then determined with hydrazine.

(*b*) *Sulphuric-acid Parting.*—Parting with sulphuric acid is confined to materials containing only gold, platinum, and palladium as payable constituents in quantities of less than about 5 oz. per ton, approximately 0·015 per cent. For richer materials nitric-acid attack should be applied.

The flattened silver bead is boiled with slightly diluted sulphuric acid (9 : 1 water), the boiling being continued for a few minutes after the evolution of sulphur dioxide has ceased. The acid, whilst still warm, is cautiously diluted with a jet of cold water introduced through the spout of the covered beaker.

The filtered solution contains, besides silver, nearly all of the palladium. The silver is precipitated with dilute hydrochloric acid, the chloride precipitate allowed to settle in the dark, the clear liquid filtered, and the precipitate re-treated—unless white—to free from traces of palladium. The palladium is then recovered from the filtrate or combined filtrates by precipitation with zinc, and is added to the residue from the sulphuric-acid parting. The combined precipitates are dissolved in aqua regia, the acid is evaporated, the residue taken up in dilute aqua regia, and the palladium precipitated with dimethylglyoxime.

This filtrate, freed from excess of the precipitant by evaporation to dryness and digestion with a little sodium chlorate, contains gold and platinum. The former is precipitated with hydrazine, and the latter with the same reagent after evaporation of the acid and neutralization with sodium carbonate.

Reporting Assay Results of Sweep, etc.—The content of platinum metals, etc., in sweeps and similar products from manufacture are usually reported in ounces and decimals of an ounce troy per hundredweight in the case of small lots, and per ton (2240 lbs. or 2000 lbs.) or per cent. in the case of large lots.

Determination of Platinum Metals in Electrolytic Slime or Mud.—This product results from the electrolytic refining of copper and other metals. The composition of anode mud varies with the composition of the anodes; thus, in the case of copper refining, the percentage of gold may range from 0·7 to 2·0, of silver from 5·0 to 40·0, of copper from 10·0 to 25·0, and of arsenic and antimony up to 10·0, with more or less bismuth, lead, and iron (Gowland).

The mud is removed from the vats, passed through a 40-mesh sieve to separate any pieces of copper, and is then boiled with sulphuric acid, care being taken to dissolve as little of the silver as possible. By this means the copper is reduced to about 1 per cent., and the iron, bismuth, and most of the arsenic are removed. The mud is then dried and sent to the gold and silver refinery.

Great care is necessary in sampling electrolytic muds, especially when platinum and its allied metals are present, as not infrequently is the case. These muds are usually assayed by scorification, from 0·25 to 0·5 gramme being taken for each assay.

In order to adjust the proper proportions of the precious metals for satisfactory parting it is usually necessary to make a preliminary assay.

Experience indicates, however, that the precious metals, including those of the platinum group, present in these muds are to a certain extent in an oxidised condition; consequently the cupellation losses are liable to be greater than usual, and there is a greater tendency for the precious metals to remain on the cupel in a subdivided state, and to spread over the face of the cupel. Therefore, in order to avoid excessive cupellation loss, it is advisable to add the requisite amount of gold and silver for parting purposes by direct fusion of the material in a small crucible. For this purpose from 0·25 to 0·5 gramme of the material, with the necessary gold and silver, is placed in an annealing-cup or small crucible containing a little powdered borax glass, and then covered with more borax glass, and a cover placed on the crucible. It is then strongly heated, and the heat maintained for about an hour to effect the thorough alloying of the platinum metals. When cold the crucible is broken, the button freed from slag, rolled into a cornet, and subjected to double parting in the usual way, or treated with acids as previously described.

III. THE ASSAY OF ORES AND CONCENTRATES.

Assay of Base-Metal Ores containing Platinum Metals.—Unrefined or "crude" platinum usually occurs in Nature as native metal alloyed with one or more of the allied metals in the form of very fine grains more or less flattened, and sometimes in the form of irregular-shaped nuggets. The grains are sometimes coated with a black layer of magnetic iron oxide, in which case it is not easily recognizable.

The alluvial deposits or sands in which crude platinum is found constitute the so-called platinum ores, which are seldom received for assay. But the rarity of the platinum-group metals, and their increasing industrial importance, has caused search to be made for them in recent years in all classes of ores and of metallurgical products in which they have become concentrated as the result of smelting and refining operations. The assayer is, therefore, frequently asked to test for the presence of the platinum-metals in ores of gold, silver, and base metals, especially those of copper and nickel. For example, there are extensive primary deposits of platinum metals in sulphide-bearing norite, a basic igneous rock occurring in the Potgietersrust and Rustenburg districts of the Transvaal.[1] These deposits contain about 5 dwt. of platinum metals per ton; platinum and palladium are present in about equal proportions, with subordinate amounts of the allied metals. The platinum metals are recovered from the concentrates that result from the dressing of the ore.

Platinum is also found, mostly in combination with arsenic, as sperrylite in the well-known copper- and nickel-bearing deposits of the Sudbury district of Ontario, Canada. Palladium, in about the same amount as platinum, and subordinate amounts of other platinum metals are also present. Although the amount of the platinum metals per ton of these sulphide deposits is very small, yet, owing to the fact that big tonnages of ore are treated annually for the recovery of the main products, viz., copper and nickel, very substantial amounts of platinum metals are produced as by-products in the final refining operations. These ores constitute the principal source of palladium.

When ores have to be tested for platinum-group metals, a crucible fusion assay is made, as described for gold ores. The resulting lead button is then treated direct with nitric acid, as described for sweep, or cupelled, and the cupel bead parted for the determination of platinum, etc.

As the platinum metals are usually present in very small quantity it is generally necessary to make several fusions and then to concentrate the precious

[1] Atkinson and Raper, *loc. cit.*, 180.

metals into one cupellation bead for parting. If any considerable amount of platinum be present, there will be a decided difference between the weighings of the two residues from the double parting, but with very small amounts of platinum the difference is very slight and is no real evidence of the presence of platinum. In fact, the weight of the nitric-acid residue may equal or possibly exceed the weight of the sulphuric-acid residue, even when traces of platinum and its allied metals are present. Dewey[1] therefore recommends parting the button in nitric acid only, when any platinum present in small amounts will readily go into solution with the silver. To this solution is added a limited quantity of dilute solution of hydrogen sulphide (1 part strong solution diluted with from 10 to 20 parts of water), sufficient to precipitate the platinum and three to five times as much silver.

After allowing the combined precipitate to settle, it is collected on a filter-paper, then dried, the paper burnt, and the residue wrapped in a small piece of lead-foil and cupelled. The resulting silver–platinum bead is then parted in strong sulphuric acid, leaving the platinum usually in the form of a sponge, which is washed, annealed, and weighed.

If other platinum metals are to be sought for and determined, the cupellation button must be dealt with as previously described.

The following method is described by Dart[2] for the assay of the platiniferous copper ores of the Rambler Mine, Wyoming.

The high-grade ores contain about 25 per cent. of copper, and low-grade ores about 2 per cent. Platinum-group metals are found to be associated with the various copper minerals present. Eight portions of the ore sample, of 1 assay-ton each, are fused with suitable fluxes. If much copper be present, the resulting lead buttons are scorified until soft, and then cupelled. The weight of the beads obtained gives the total quantity of precious metals in the ore. The beads are combined to two lots, representing 4 assay-tons each, and re-cupelled with ten times their weight of silver. The resulting beads are parted with 12 per cent. nitric acid, and finally with concentrated nitric acid.

Silver, palladium, and platinum are dissolved : gold remains, and is weighed. The bulk of the nitric acid solution should be kept as small as possible. The silver is precipitated as chloride by the addition of hydrochloric acid, and removed by filtration. The filtrate is evaporated to dryness, and a few drops of hydrochloric acid and 10 c.c. of water added. If any silver chloride separates out, it is filtered off and added to the main portion.

The filtrate is rendered ammoniacal, acidified with formic acid, and boiled for thirty minutes. Platinum and palladium are reduced, and can be filtered off, washed, ignited, and weighed. The precipitate is dissolved in aqua regia, evaporated to dryness, taken up again with hydrochloric acid and 50 c.c. of hot water, saturated with ammonium chloride crystals, and, after the addition of 10 c.c. of alcohol, is allowed to stand for twenty-four hours. The ammonium platini-chloride is filtered off, washed with ammonium chloride solution, then ignited slowly and weighed as metallic platinum.

Assay of Concentrates for Platinum-group Metals.—The concentrates resulting from the dressing of sulphide-bearing ores are frequently assayed for the presence of the platinum-group metals, which will sometimes be found to be present when they have not been detected in the original ore. The concentrates are assayed by the methods described for gold ore assays, and the resulting precious-metal button submitted to double parting in the usual way.

[1] Dewey, *Trans. Amer. Inst. Min. Eng.*, 1912, 439.
[2] *Met. and Chem. Eng.*, 1912, **10**, 219–220.

Samples of platiniferous alluvial are first concentrated by panning, and the concentrate dried and treated with a magnet to remove any magnetic particles of magnetite, etc. The heavy material and the magnetic portion are both carefully examined with a lens and particles of metallic appearance picked out. This is very necessary as some Uralian specimens are said to attract iron filings more powerfully than an ordinary magnet. The high specific gravity, the infusibility and insolubility of platinum, serve to distinguish it from other minerals.

Assay of Osmiridium (Iridosmine).—This is a native alloy of iridium and osmium in variable proportions, and may contain from 23 to 40 per cent. of osmium, 21 to 35 per cent. of iridium, ruthenium and rhodium together from 9 to 15 per cent., and from 5 to 15 per cent. of platinum. Osmiridium is the chief source of iridium and osmium, and is now recovered as a by-product of gold mining operations on the Witwatersrand, South Africa. There is also a fluctuating production of osmiridium, depending on the market price, from alluvial deposits in Tasmania.[1]

In view of the very poisonous nature of the volatile oxides of osmium the assay of materials in which osmium is present calls for special precautions. Native osmiridium when strongly ignited evolves the pungent vapour of osmium tetroxide.

For assay purposes the heavy osmiridium grains are separated from all associated minerals by ore-dressing processes.

Concentrates consisting of osmiridium intermingled with sand, etc., are subjected to preliminary treatment to remove the sand. A weighed quantity of the concentrate is mixed with granulated lead, and scorified under a layer of borax. The resulting lead button is treated with dilute nitric acid, thereby dissolving the lead and leaving a residue of osmiridium which, after washing, and drying is weighed, and is usually reported as such.

Sulphide concentrates containing much copper and nickel, as well as osmiridium, may first be boiled with strong hydrochloric acid which dissolves the sulphides. The insoluble residue containing the osmiridium and sand, etc. is washed, dried, and then scorified as before.

When both constituents, osmium and iridium, have to be separately determined, further preliminary treatment is necessary.

As osmiridium is extremely hard and defies all attempts at crushing it sufficiently fine for assay, it is first converted into fine powder by fusion with zinc, which converts the hard, crystalline native alloy into an amorphous mass, which is easily reduced to an impalpable powder by grinding. Thus prepared, the mineral is opened up by fusion with sodium peroxide.

From 2 to 3 grammes[2] of the osmiridium concentrate are accurately weighed and mixed with from 8 to 10 grammes of chemically pure zinc, and fused in a small carbon crucible, as previously described. When cold, the contents of the crucible are accurately weighed, the porous metallic mass showing a slight increase in weight owing to retention of a little zinc. It is ground to a very fine powder in an agate mortar, and a quantity of material equivalent to 1 gramme of the original material weighed into a nickel crucible. The sodium-peroxide fusion and lixiviation are carried out as described on page 448. The alkaline suspension thus obtained is transferred to the distillation flask forming part of the train described on page 449, then digested overnight with electrolytic hypochlorite solution, and distilled in a current of chlorine or with bromate.

[1] Atkinson and Raper, *loc. cit.*, 181.
[2] Schoeller and Powell, *op. cit.*, 293.

The distillation yields three products, viz. an osmium distillate, a ruthenium distillate, and the residual solution, which is worked up to give an ammonium-chloride and a zinc precipitate.

The combined ignited and reduced precipitates are fused with lead, and the resultant lead button is extracted with acids, giving a residue of pure iridium (or iridium containing the minor ruthenium fraction if chlorine distillation has been used) and two acid extracts containing platinum, rhodium, and usually gold; these latter are treated for the determination of the three metals.

Palladium does not normally occur in osmiridium, and the only base metals found in the mineral are iron and copper in minute quantities.

The South African Platinum Concentrates containing osmiridium (in minute grains mostly finer than 200-mesh) differ from all other platinum ores in that they contain platinum and palladium minerals insoluble in aqua regia, but soluble after roasting in air, which removes arsenic, antimony, and sulphur. The whole of the material can normally be crushed to 100-mesh size, hence portions of 1 gramme can be taken for analysis.

CHAPTER XXV

LABORATORY WORK IN A CYANIDE MILL

The daily work of an assayer in a cyanide mill varies considerably, especially in regard to the testing of the cyanide solutions, which has now become a part of the routine work. In some plants simple titrations to determine the strength of the cyanide solutions, with an occasional determination of the amount of gold in solution and assays of " headings " and " tailings," are all that is called for. On the other hand, in some modern mills a careful check is kept on all the ore and solutions at every stage of the treatment, and the number of ore assays and other determinations on control samples, " headings," " tailings," slimes, solutions, bullions, and various products is necessarily very large.

The assay work in a cyanide mill laboratory may be classified under six heads, viz.[1] :—

(1) Mine Samples.
(2) Reduction-works Samples.
(3) Cyanide Solutions.
(4) Cyanide Bullion.
(5) Cyanide Mill-products.
(6) Testing of Ores for Cyanide Treatment.

(1) Assay of Mine Samples.

Mine samples consist of stope and development samples from the mine, and in the majority of cases come to the assayers in a prepared state.

For the degree of accuracy required of them they should be passed through a sieve of at least 80-mesh, although with some ores a 60-mesh may be satisfactory. Before being assayed, the samples should be thoroughly mixed by rolling on a sheet of rubber or paper. On some mines it is customary for the sampler to pan a portion of every sample before handing it to the assayer. This, in practised hands, gives an idea of the value of the ore, and also enables the assayer to ascertain the pyritic content, etc., of the ore for fluxing purposes. The ores are assayed as described in Chapter XIII. Development or prospecting samples are assayed in duplicate. The gold content of mine samples is usually reported in pennyweight and decimals of a pennyweight per ton to the first place of decimals.

(2) Assay of Reduction-works Samples.[2]

The term " reduction-works samples " is applied to samples of solids taken from a running plant, and may consist of :—

(*a*) Screen samples (or heads) and tailings (tails).
(*b*) Cyanide pulp samples.
(*c*) Spitz concentrates and residues.
(*d*) Sand charges and residues.
(*e*) Slime charges and residues.

The sampling and assaying of these various products are dealt with at the end of this chapter.

[1] Whitby, *J. Chem. Met. and Min. Soc. of S. Africa,* 1906, **6**, 269.
[2] For fuller details consult *Rand Metallurgical Practice,* 3rd edn., vol. I, 320–21, 446–49.

(3) Assay of Cyanide Solutions.

Much of the information contained in this section has been derived from the authoratitive work on cyanide solutions by Clennell.[1] Most of the methods used in cyanide mills for the assay of the various cyanide solutions are technical methods in which rapidity is aimed at rather than extreme accuracy. In the hands of competent assayers the methods give, as a general rule, good comparative results, sufficiently accurate to be of asssitance to the cyanide manager in the control of his plant; but there is little doubt that in many mills the assaying of solutions, with a view to their correct valuation, is not given the consideration that its importance deserves.

One of the most important functions of the cyanide manager is to reduce the consumption of potassium cyanide to the minimum and raise the percentage of extraction to the maximum at least cost: consequently the testing of the solutions in the laboratory is absolutely necessary if economy in working is to be effected. The ordinary works tests for this purpose usually consist in determining—

(*a*) The free cyanide present;
(*b*) The total cyanide; and
(*c*) The total or protective alkali.

Other analytical determinations in the assay of mill cyanide solutions are described subsequently to these.

These tests and some few others will be used daily, whilst other tests are used only at long intervals. Complete analysis of working cyanide solutions are made in many mills at varying intervals.

The strength of the cyanide solution used in practice varies with the nature of the ore and the conditions of working, but it usually contains less than 1 per cent. of alkaline cyanide. Thus the maximum strength used in the Witwatersrand mills is 0·06 per cent. KCN, and the alkalinity 0·005 per cent. CaO.[2] In ordinary working conditions solutions are not saturated with oxygen; the rate of attack on the gold is thus diminished and the cyanide strength can be correspondingly reduced. From considerations based on experiment and on diffusion rates, it is deduced that in works practice the best cyanide strength for rapid action is between 0·02 and 0·03 per cent. KCN, but the exact strength can be determined only by actual trials for each case involved.[3] For instance, although the strong solutions on the Rand contain only 0·05 to 0·1 per cent. KCN, a strength of 0.5 per cent. or more may be necessary in the treatment of some ores, especially silver ores. Still stronger solutions were used in the early days of the process, but were found to be unnecessary.

In the case of silver ores stronger solutions are generally used. Silver sulphide is more slowly dissolved than metallic gold, and strong solutions containing up to 1 per cent. KCN are required.

(a) Determination of Free Cyanide.—" The term 'free cyanide' is used to indicate the equivalent, in terms of potassium cyanide, of all the cyanogen which is present in the solution as simple cyanides of alkalis and alkaline earth metals, such as potassium, sodium, ammonium, calcium, and barium cyanides. It does not include cyanogen present in the form of double cyanides or hydrocyanic acid."[4]

[1] Consult Clennell, *Chemistry of Cyanide Solutions*; M'A. Johnston, *Rand Metallurgical Practice* (chapters on Assaying and Testing); Argall, *Western Methods of Smelter Analysis.*
[2] Rose & Newman, *Metallurgy of Gold*, 7th edn., p. 427.
[3] *Ibid.*, pp. 305, 307, 339.
[4] J. E. Clennell, *Chemistry of Cyanide Solutions.*

The determination of the cyanogen is of importance, since this is the active solvent of the gold, and so long as the cyanogen present is readily soluble it is a matter of indifference whether the base be sodium or potassium (see remarks on commercial cyanide, p. 473). The percentage of cyanide in mill solutions is invariably determined by Liebig's[1] well-known method, which is based upon the fact that when a solution of silver nitrate is added to an alkaline solution containing cyanogen, no permanent precipitate of silver cyanide occurs until all the cyanogen has combined with the alkali and silver to form a double salt; thus in the presence of potassium the reactions are:—

$$(a)\ AgNO_3 + KCN = AgCN + KNO_3,$$
$$(b)\ AgCN + KCN = KAg(CN)_2.$$

As soon as the whole of the potassium cyanide has been converted into the double salt, the addition of more silver solution results in its decomposition and the precipitation of silver cyanide; thus:—

$$(c)\ AgNO_3 + KAg(CN)_2 = 2AgCN + KNO_3.$$

Hence the completion of the first stage can be taken as the end of the reaction, which will be indicated by the first appearance of a permanent precipitate (AgCN) causing a white turbidity or opalescence in the solution. If, therefore, the silver solution be of known strength, the quantity of cyanogen present is easily found. It will be noted from the above reactions that one molecule of silver nitrate is equivalent to two molecules of potassium cyanide.

The first stage in the reaction between silver nitrate and potassium cyanide is not interfered with by the presence of soluble iodides (and other haloids). If, therefore, potassium iodide is added to the solution, no silver iodide is formed until the cyanide has been entirely converted into the double salt. When that stage is complete, the further addition of silver nitrate results in the precipitation of silver iodide, which produces a yellowish turbidity and is easily recognized. The end of the reaction is more clearly defined when potassium iodide is added, and both methods of determining the finishing-point are largely used in cyanide mills.

Standard Solutions.—Two standard silver-nitrate solutions of different strengths are usually employed in testing cyanide solutions: one for testing strong solutions, and the other for testing dilute solutions. The standard silver nitrate for strong solutions is prepared by dissolving 13·046 grammes of pure crystallized silver nitrate in distilled water, and diluting to 1000 c.c. Each c.c. of this solution is equivalent to 0·01 gramme of potassium cyanide in the mill solution tested.

For testing dilute cyanide solutions it is more convenient to use a standard silver-nitrate solution of half this strength, *i.e.* containing 6·523 grammes of silver nitrate per litre. Each c.c. of this solution is equivalent to 0·005 gramme potassium cyanide (or 1 lb. KCN per ton) in the mill solution tested.

Standardizing Silver Solutions.—The silver-nitrate solution may be standardized by titration with sodium chloride. Weigh out 1 gramme of chemically pure dry sodium chloride, and dissolve in 100 c.c. of distilled water. Take 10 c.c. of this solution by means of a pipette and transfer to a small flask, dilute to 100 c.c., add a little sodium bicarbonate, and 2 c.c. of a solution of neutral potassium chromate as an indicator. Fill a burette with the standard solution of silver nitrate and run into the salt solution until a permanent faint blood-red colour is produced, due to the formation of silver chromate. One c.c. of N/10 sodium chloride = 0·0058378 gramme of sodium chloride, equivalent to 0·016966

[1] *Ann. der Chem. und Pharm.*, **77**, 102.

gramme of silver nitrate. From these values it is easy to calculate the amount of silver nitrate in the number of cubic centimetres of the solution required by a given quantity of sodium chloride, and hence the strength of the silver-nitrate solution is ascertained.

Indicator Solution.—The strength of the potassium-iodide solution used as an indicator in testing cyanide solutions varies from a 1 per cent. to a 10 per cent. solution ; the latter strength is more usual. Three or four drops of this solution are added to the cyanide solution before titration.

Titration.—Practice varies considerably in different mills, both with regard to the strength of the silver-nitrate solution, and also the amount of the working cyanide solution taken for the test ; but, as a general rule, silver solutions of the strength given above are used for the titrations and 50 c.c. or 100 c.c. of the weak cyanide solutions, and 5 c.c. or 10 c.c. of the strong solutions used for the tests. The weak solutions are generally titrated without dilution, but some chemists dilute all solutions containing, say, over 0·10 per cent. KCN, using 50 c.c. of solution diluted with 50 c.c. of water. Below this strength it is not advisable to dilute with water. Strong solutions (using 5 c.c. or 10 c.c.) are diluted to about 100 c.c. with water.

For very strong solutions containing, say, over 1·0 per cent. of potassium cyanide more dilution is necessary ; otherwise the silver cyanide may be precipitated in a granular form not readily soluble in excess of cyanide. In this case 10 c.c. of the solution are diluted with water to a total volume of 100 c.c. Then 10 c.c. of this diluted solution are used for the titration ; but in calculating the result, it must be borne in mind that the diluted solution contains only one-tenth of the original strong cyanide solution.

The cyanide solution is measured in a burette or pipette, then transferred to a small conical flask, the indicator solution added, and the silver-nitrate solution run in from a burette until it gives a precipitate of silver cyanide which does not redissolve on thoroughly agitating the solution in the flask. If the strength of the cyanide solution is approximately known, the silver solution may be added rapidly at first, and then drop by drop when finishing. The titration should be performed in a good light and the flask placed upon a dead-black surface, so that the first indications of a faint permanent precipitate may be more readily observed. If the cyanide solution used for the determination is measured in a pipette, it is advisable to place about 9 inches of rubber tubing over the end of the pipette as a safeguard against drawing cyanide solution into the mouth.

Titration of Turbid Cyanide Solutions.—There is no difficulty in observing the finishing-point with clear cyanide solutions, but when the solutions are at all turbid, they must be filtered : otherwise the end-point cannot be observed with any approach to accuracy. "Solutions occurring in ore treatment sometimes contain very finely-divided matter in suspension, which will pass through all ordinary filter-papers, and cannot be removed even by repeated filtration. In such cases the liquid may generally be clarified by the addition of lime,[1] which causes flocculation and settlement of the suspended matter, and allows the liquid to be filtered perfectly clear." This treatment is only permissible, however, with solutions free from zinc, and otherwise comparatively pure ; and when a perfectly correct titration of the free cyanide is required, it is necessary to add potassium iodide indicator after lime has been used.

Liebig's method works admirably with pure cyanide solutions, but gives uncertain and inaccurate results with ordinary working solutions, particularly in the presence of zinc. As, however, it is generally only necessary to obtain

[1] Clennell, *op. cit.*, p. 7.

relative commercial results, and as a knowledge of the exact strength of the working solution in actual free KCN or its equivalent is not essential, this method is in general use.[1]

Titration of Cyanide Solutions containing Zinc.—Zinc occurs in cyanide solutions chiefly as double cyanide of zinc and potassium or other metal. When solutions containing much zinc are titrated as described above, the finishing-point is almost always obscure and indefinite owing to the formation of a white, flocculent precipitate, which occurs long before the whole of the potassium cyanide has been converted into the soluble double salt of potassium and silver $KAg(CN)_2$. When it is desired to determine the "free cyanide" in solutions containing zinc, it is necessary to determine first the so-called "total cyanide" by the method given below, in which the cyanogen found represents not only the potassium cyanide, but also the double zinc compound. A determination of the zinc (p. 471) is also made, and from the results obtained, the amount of "free cyanide" may be readily calculated, as 1 part of zinc corresponds to 4 parts of potassium cyanide. A similar allowance must be made if small quantities of copper are present.

The "free cyanide" in solutions containing zinc therefore equals: total cyanide less zinc $\times$ 4. The presence of cyanogen, in the form of ferrocyanides or thiocyanates, does not affect this result.

(b) **Determination of Total Cyanide.**—Strictly speaking, the term "total cyanide" should indicate the equivalent (as potassium cyanide) of all the cyanogen contained in the solution. In practice, however, it is generally taken to mean "the equivalent, in terms of potassium cyanide, of all the cyanogen existing in the form of simple cyanides, hydrocyanic acid, and certain readily decomposable double cyanides, notably that of zinc." Some other double cyanides, such as those of silver and copper, are generally excluded, together with ferro- and ferricyanides, thiocyanate, and similar bodies.[2] The total cyanide is generally determined in practice by titration with silver nitrate after the addition of an excess of caustic alkali. To obtain accurate results, it is necessary to use the potassium-iodide indicator, and it is convenient to add the caustic alkali to the indicator. An alkaline-iodide indicator is therefore prepared by dissolving 40 grammes of caustic soda and 10 grammes of potassium iodide in water and making up to 1000 c.c.

Fifty c.c. of the cyanide solution to be tested are taken and from 5 c.c. to 10 c.c. of the alkaline indicator added. The solution is then titrated with standard silver nitrate (6·523 grammes per litre) until a distinct yellow coloration is obtained, disregarding any white turbidity. The finishing-point is generally quite sharp and definite, and if a white cloudiness occurs before the true finishing-point, it may in most cases be removed by adding a little ammonium hydrate, which in moderate amounts does not affect the accuracy of the test. Each c.c. of silver nitrate added will represent 0·005 per cent. of potassium cyanide (equivalent to total cyanide).

On the Rand,[3] 5 c.c. of a 10-per-cent. solution of caustic soda are added to 100 c.c. of the working solution with the addition of a few drops of a 10-per-cent. solution of potassium-iodide indicator. In ordinary practice the 100 c.c. used for the determination of "free cyanide" are also employed for the determination of the total cyanide, after the addition of alkali and indicator.

(c) **Determination of Protective Alkali.**—The term "protective alkali" merely means the real alkalinity, or alkalinity due to hydrates in working cyanide solutions, and is explained by Clennell[4] as follows:—

[1] Argall, *Mill and Smelter Methods of Analysis*, p. 59.
[2] Argall, *loc. cit.*
[3] *Rand Metallurgical Practice*, 3rd edn., vol. I, 324.
[4] *Chemistry of Cyanide Solutions.*

"The term is based on the assumption that certain constituents of an ordinary working solution will be wholly or partially neutralized, on addition of a dilute mineral acid or carbonic acid, before any decomposition of cyanide occurs. In the treatment of ores, the addition of lime or caustic soda is made with the object of protecting the cyanide from the decomposing effect of various substances contained in the ore, and of the carbonic acid in the air. Since a cyanide solution undergoes gradual decomposition with evolution of hydrocyanic acid, even when an excess of caustic alkali is present, the protection afforded is only partial and temporary. There is, however, a fairly definite point in the neutralization of an ordinary solution (containing free cyanide, hydrates, carbonates, etc.) at which the decomposition, with formation of hydrocyanic acid, begins to take place with marked rapidity. In the absence of zinc this point corresponds with tolerable exactness to the alkalinity of the hydrates and carbonates present towards phenolphthalein, and if the amount of acid corresponding to this alkalinity be added, with agitation, to such a solution, the cyanide strength, as shown by *immediate* titration with silver nitrate, will remain apparently unchanged." In this case, therefore:

$$\text{Protective alkali} = \text{alkalinity of hydrates} + \tfrac{1}{2}\ \text{alkalinity of carbonates.}$$

"For practical purposes it is most important to know the alkalinity exclusive of cyanide, as it is this alkali which is chiefly of use in preventing the unnecessary waste of cyanide by reactions due to the presence of base-metal compounds, termed 'cyanicides,' and to the carbonic acid of the air."

The protective alkali may be determined (accurately in the absence of zinc) by the method devised by Clennell, which is based on the following facts:—

(*a*) That double cyanides of silver and potassium are neutral to phenolphthalein;

(*b*) That the amount of acid required to neutralize the hydrates and carbonates towards phenolphthalein is a measure of the protective alkali as defined;

(*c*) That the presence of cyanide or double cyanide of silver and potassium does not interfere with the titration of alkali, the double salt not being decomposed until the whole of the alkali is neutralized.

The method consists in titration with standard acid and phenolphthalein indicator, after addition of silver nitrate.

The method as conducted on the Rand is described by Rose and Newman as follows:—

The solutions required are standard silver nitrate; phenolphthalein, prepared by dissolving 1 gramme of the powder in 100 c.c. of 60-per-cent. alcohol or methylated spirits; decinormal oxalic acid, N/10 $C_2H_2O_4$, prepared by dissolving 6·3 grammes of oxalic acid crystals, $C_2H_2O_4$, $2H_2O$, in distilled water and diluting to 1000 c.c.; potassium ferrocyanide, $K_4Fe(CN)_6$, prepared by dissolving 10 grammes of $K_4Fe(CN)_6$ in 100 c.c. of distilled water.

METHOD.—Measure into a conical flask 100 c.c. of cyanide solution. Add just sufficient silver nitrate solution to give a permanent precipitate, and then 5 c.c. of the ferrocyanide solution and two drops of phenolphthalein. Titrate with decinormal oxalic acid till the phenolphthalein colour is discharged. Then each c.c. of oxalic acid solution used multiplied by 0·004 gives the percentage of alkalinity in terms of caustic soda. The portion of solution used to determine free cyanide may be used afterwards to estimate the protective alkali.

The use of potassium ferrocyanide in this determination is desirable, so as neither to include $K_2Zn(CN)_4$ as free alkali, nor to neutralize free alkali present by excess of $AgNO_3$.

Bicarbonates are neutral to phenolphthalein, hence are not determined. They have no protective influence in this case.

Influence of Zinc.—When solutions containing zinc are titrated with standard acid and phenolphthalein, only a small portion of the double cyanide of zinc and potassium, K_2Zn (CN), is shown as alkali. Moreover, the addition of excess of silver nitrate after the first turbidity causes the $K_2Zn(CN)_4$ solution to became quite neutral to phenolphthalein at some point before the complete precipitation of the cyanide. Green[1] has pointed out that when large quantities of zinc are present (in the absence of much hydrate or carbonate of the alkali), the addition of silver nitrate causes the solution to become acid to phenolphthalein.

To obviate the difficulties due to the presence of zinc, Green has devised the following method by which the protective alkali is determined after the addition of silver nitrate and potassium ferrocyanide. In this method the total cyanide is first determined in the ordinary way by titration with silver nitrate, using the alkaline-iodide indicator.

Another portion, say 50 c.c. of the original solution, is now taken, an excess of ferrocyanide solution added, and then a little more silver solution than was used in the previous test, to ensure the complete conversion of all the cyanide into silver salts. Phenolphthalein is then added, and the liquid titrated with N/10 standard acid, as described above.

Determination of Free Oxygen.—Several methods for the determination of free oxygen in cyanide solutions have been introduced, but the colorimetric method devised by H. A. White[2] appears to be that in most general use.

This depends on the darkening of pyrogallic acid (" pyro ") by oxygen, and comparing the colour so produced with a standard coloured solution. Two solutions are required, viz. (i) water saturated with oxygen, and (ii) stock dye solution.

The solution saturated with atmospheric oxygen is prepared by half-filling a Winchester quart-bottle with distilled water and shaking well until it is saturated. The temperature of the water is then taken, from which the number of milligrammes of oxygen per litre is ascertained by reference to any of the well-known published tables on the solution of gases in liquids.

The dye solution is made by dissolving, in distilled water, Diamond Brown, Caramel, or other suitable brown dye. From this stock solution a set of colour standards is prepared by simple dilution. To standardize the stock solution, water saturated with oxygen is placed in a 250-c.c. stoppered bottle, and 0·1 gramme of solid " pyro " and 1 c.c. of 8-per-cent. NaOH solution added, thus obtaining a brown-coloured solution. Suppose this colour corresponds to an oxygen content of 8 mg. per litre; then by diluting the stock dye solution, colour-standards corresponding to 7, 6, 5, etc. down to ½ mgm. oxygen per litre can be prepared. Such standards keep well if the bottles are completely filled, the stopper sealed with paraffin wax, and stored in the dark.

To determine the oxygen in a mill cyanide solution, 250 c.c. are treated with " pyro " and caustic soda as described, every precaution being taken to exclude air. The colour produced is then matched with the standard set, and the oxygen content thus determined reported in milligrammes and half-milligrammes per litre.

" For perfect matching it is necessary to add a little acid, methyl orange and potassium chromate. . . . The colour in the mill solution fades slowly. A difficulty in the method is that it gives a violet or purple shade with foul solutions

[1] L. M. Green, *Trans. Inst. Min. and Met.*, 1901, **10**, 29–37.

[2] *J. Chem. Met. and Min., Soc. S. Africa*, 1918, **18**, 292, and 1919, **19**, 177.

containing complex ferrous or ferric salts, which is difficult to match. Strengthening the solution with NaCN improves matters" (Rose and Newman). According to Tromp and Schilz[1], a minimum amount of calcium salt must be present in White's method in order to develop the full colour of the pyrogallol.

Determination of Ferrocyanides.—Any substance which is capable of absorbing oxygen will obviously affect the efficiency of mill cyanide solutions; hence it is desirable to determine the amount of such de-oxidizing or reducing agents.

The chief reducing agents present in cyanide solutions are ferrocyanides and thiocyanates (sulphocyanides).

Many methods have been devised for the determination of ferrocyanides in mill solutions, but those involving titration with permanganate appear to be most used. Solutions of pure potassium cyanide and other simple cyanides acidified with sulphuric acid give hardly any reaction with permanganate, a single drop of N/10 $KMnO_4$ being usually sufficient to produce a permanent pink tint, owing to the fact that the liberated hydrocyanic acid scarcely affects the permanganate at all. When, however, ferrocyanides, thiocyanates, sulphides, nitrites, formates, and some other substances are present, the pink colour at first disappears rapidly as each drop of permanganate is added, but with continued additions of permanganate the pink colour eventually becomes permanent.

A method used on the Rand is described by A. M'A. Johnson,[2] and consists in titrating the cyanide solution with N/100 potassium permanganate solution prepared by dissolving 0·316 gramme of the crystals in a litre of distilled water.

The determination is made on 50 c.c. of the cyanide solution, which is first treated with an excess of a solution of ferric sulphate made acid with sulphuric acid to precipitate the ferrocyanides. The filtrate is used for the determination of sulphocyanide, as described below. The precipitate is treated with a slight excess of caustic soda and again filtered. The filtrate is acidified with sulphuric acid and then titrated with permanganate until a permanent pink is obtained. One c.c. of N/100 potassium permanganate solution equals 0·00368 gramme potassium ferrocyanide $K_4Fe(CN)_6$.

Determination of Sulphocyanide.[3] This may be determined by titration with permanganate, after removal of ferrocyanides. The filtrate from the ferrocyanide determination as described above is titrated with N/100 permanganate solution. The red colour of the filtrate is gradually destroyed and a permanent pink obtained. The amount of sulphocyanide in the 50 c.c. of cyanide solution originally taken is then calculated. One c.c. N/100 $KMnO_4$ solution equals 0·000162 potassium sulphocyanide (KCNS).

The method is inaccurate in the presence of thiosulphates.

Determination of Sulphates.—Sulphates are usually determined by the ordinary gravimetric method, by precipitating with barium chloride and weighing the resultant barium sulphate.

Sharwood[4] recommends adding excess of hydrochloric acid to 100 c.c. of the cyanide solution to decompose cyanides, then boiling until all odour of hydrocyanic acid has disappeared and filtering off any precipitate of zinc or copper ferrocyanides, Prussian blue, or silver chloride that may be formed.

The clear filtrate is heated nearly to boiling, after the addition of a little hydrochloric acid if necessary, and then a slight excess of barium-chloride

[1] *J. Chem. Met. and Min. Soc. S. Africa*, 1935, **35**, 313.
[2] *Rand Metallurgical Practice*, vol. I, p. 327.
[3] *Op. cit.*
[4] *Eng. and Min. J.*, 1898, **66**, 216.

solution added. After standing several hours, the precipitate of barium sulphate is washed by decantation several times with hot water, and finally collected on a small filter, then washed with hot water until free from chlorides, dried, heated to dull redness, and weighed as barium sulphate. The dry precipitate should, if not too small, be detached as completely as possible from the paper, which should be burnt separately. The weight of $BaSO_4$ multiplied by 0·61 equals Na_2SO_4.

Determination of Thiosulphates and Sulphites.—These constituents are determined by difference in two separate quantities of the mill solution.

A usual procedure is to take 100 c.c. of cyanide solution for each determination and add one drop of methyl-orange indicator and a few drops of starch solution to each. Both solutions are then titrated with normal sulphuric acid until red, and then a further 1 c.c. of acid added. To one solution 10 c.c. of a 10-per-cent. solution of formaldehyde are added, then both solutions are titrated with N/10 iodine solution to a distinct permanent blue.

Formaldehyde destroys the sulphite, hence the smaller result gives the thiosulphate. The difference between the two results gives the sulphite. The factors for calculation are :—

$$1 \text{ c.c. N/10 iodine} = \begin{cases} 0{\cdot}0158 \text{ gramme thiosulphate } (Na_2S_2O_3). \\ 0{\cdot}0063 \text{ gramme sulphite } (Na_2SO_3). \end{cases}$$

Determination of Gold and Silver in Mill Cyanide Solutions.—The amount of gold in cyanide solutions obtained in the cyanide process varies considerably. The solutions entering the precipitators may contain up to 10 dwt. of gold per ton on a rich mine, whilst the solution leaving the precipitation boxes may contain merely a trace of gold, less than 0·005 dwt.[1] For " rich " solutions 1 or 2 A.T. of the liquid may be taken, and for " poor " solutions 10 to 20 A.T. or more could be used. In general mine assay-office practice, however, it is advisable to adopt a fixed quantity, and it will generally be found that when using a balance turning to $\frac{1}{100}$ mg., 20 A.T. (584 c.c. for the ton of 2000 lbs.) of solution is a convenient amount.

Since the specific gravity of ordinary gold-bearing cyanide solutions is very little above that of water, exactly 10 A.T., 20 A.T., or even more, of solution may be measured out for assay. It may be well to point out that where liquids are being assayed, cubic centimetres are held to be equivalent to grammes, and the results may be stated as " so many parts by weight in so many parts by measure." By using measured quantities of solution corresponding to the assay-ton system, the weight of gold obtained from the assay gives at once, or with little calculation, the amount of gold per ton. The results of assays of cyanide solutions are reported in pennyweight and decimals of a pennyweight per ton of 2000 lbs. (*i.e.* short ton).

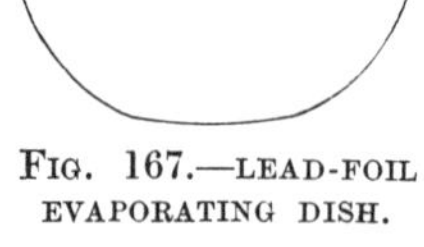

FIG. 167.—LEAD-FOIL EVAPORATING DISH.

Many methods have been proposed for the assay of gold in cyanide solutions of which the following may be given :—

(*a*) *Evaporation Method.*—This method, is accurate, but entails much tedious work, and other methods, in which rapidity is aimed at rather than extreme accuracy, have been devised and are in use in many mines where a large number of determinations have to be made daily.

A convenient volume of solution (preferably 10 or 20 A.T.) is measured out, poured into an oiled porcelain dish, about 10 grammes of litharge added, with some carbon and the solution carefully evaporated to almost complete dryness on a hot plate. It is desirable to evaporate the solution somewhat before

[1] M'A. Johnston, *Rand. Met. Pract.*, vol. I, p. 311.

adding the litharge. The heat must be reduced towards the end of the operation, otherwise loss may result owing to "spurting," or from caking of the dried residue on the dish. Should caking take place, the hard scale may be removed by moistening with dilute nitric acid. Assays that spurt must be repeated. When dry, the residue is removed by means of a spatula, and transferred to a glazed paper. Any litharge adhering to the dish is removed by adding a little borax, and wiping out with a small piece of filter-paper. The residue is then mixed with about 30 grammes of a flux[1] consisting of—litharge, 100 parts ; sand, 25 parts ; charcoal, 3 parts. Fuse the charge, cupel the resulting lead, and part as usual.

In the case of solutions containing over 0·5 dwt. of gold, evaporation may be performed in dishes or trays made of lead-foil, about 4 × 2 inches, and 1 inch high, and weighing from 25 to 35 grammes (Fig. 167). Such a dish is easily extemporized out of a piece of lead-foil. From 1 to 2 A.T. (30 to 60 c.c.) of the solution are measured out, placed in the lead dish, and evaporated to dryness. The dish with the dried residue is rolled up and then scorified and cupelled, the resulting bead being parted and weighed in the usual manner. This method has the disadvantage of permitting the use of comparatively small quantities of solution only. When the gold and silver are not present in sufficient quantity to give a bead of convenient size for weighing with 1 to 2 A.T. of solution, a larger quantity must be used. If 20 or 30 A.T. are taken for assay, evaporation must be begun in glass or porcelain vessels, and finished in a lead tray after having been very carefully transferred when the solution has been sufficiently reduced in bulk.

(*b*) *Lead Acetate Method* (Chiddey).[2]—In this method, zinc shavings and a lead salt are used to produce a lead sponge on which the gold is precipitated and subsequently separated by cupellation. The quantities of the several reagents used, and the method of procedure adopted, vary somewhat with different assayers.

An adaption of the method which, according to M'A. Johnston,[3] gives more uniform results, comparable indeed with the evaporation assay, is to measure 584 c.c. (20 A.T.) of solution into a beaker of 500 to 700 c.c. capacity, boil, add 25 to 30 c.c. of saturated lead-acetate solution and about 3 grammes of zinc shavings rolled into a loose ball, and again boil for half-an-hour. Add 25 c.c. of commercial hydrochloric acid, and keep at nearly boiling point until the zinc is dissolved. The heating is continued until all effervescence has ceased, and the reduced lead has collected into a spongy mass. The spongy lead is washed by several decantations in water and then pressed into a ball with the aid of a flattened glass rod, and then with the finger, care being taken to include any small loose pieces. The ball is dried slightly in a cupel tray over a hot-plate, and then cupelled. The silver for inquartation is added as silver nitrate solution or as metallic silver.

A. J. Clark[4] recommends the addition of 10 to 20 c.c. of lead-acetate solution containing 10 to 20 per cent. of the salt, and then adds zinc dust (3 to 4 grammes) in the form of an emulsion or suspended in water. This latter is prepared by mixing 100 grammes of zinc dust with 300 c.c. of water in a bottle fitted with a ¼-inch glass tube passing through the cork. This mixture is shaken into a capsule of the proper size used as a measure. The lead sponge, after treatment with 20

[1] M'A. Johnston, *op. cit.*
[2] A. Chiddey, *Eng. and Min. J.*, 1921, **111**, 629.
[3] *Op. cit.*, vol. I, p. 312.
[4] Allan J. Clark, Homestake Mining Co., described by Fulton & Sharwood, *Fire Assaying.*

c.c. of strong hydrochloric acid as described, is filtered on a " quick " paper and washed twice. The sponge and filter-paper are removed and as much water as possible squeezed out. It is then placed in a 2-inch scorifier with 10 to 15 grammes of test lead and 3 to 5 grammes of borax glass. Transfer at once to the muffle, burn the paper, scorify for only a few minutes, pour, cupel the lead button, part, and weigh. Unless silver is to be determined, silver-foil should be added to the scorifier for inquartation, or a measured volume of dilute silver-nitrate solution may be added to the cyanide solution from a burette.

Fulton states that " Clark's method gives somewhat better results on low-grade solutions than evaporation methods with litharge or litharge-bearing flux." It is also more rapid than evaporation. Some assayers use aluminium plates instead of zinc for precipitation.

According to Wogrinz,[1] Chiddey's method is not applicable to the determination of gold in cyanide solutions containing ferrocyanide. In this case the solution is evaporated to fumes with sulphuric acid in a Kjeldahl flask, cooled, diluted with water, boiled and filtered, the precipitate dissolved in aqua regia, the nitric acid removed by evaporation with hydrochloric acid, and the gold precipitated by adding lead acetate and boiling with pure zinc powder. The washed precipitate is then extracted with boiling concentrated nitric acid and the residual gold collected, washed, ignited, and weighed.

(*c*) *Precipitation with Cuprous Chloride.*—This method of determining the gold in cyanide solutions is much favoured in many mills. From 10 to 20 A.T. of the solution are transferred to a conical flask, then 10 c.c. of a well-reduced solution of cuprous chloride added, and the liquid allowed to stand, with occasional shaking, for 30 minutes. A few drops (in all not more than 1 c.c.) of a 5-per-cent. solution of potassium ferrocyanide are then added, and the liquid well shaken for a few minutes. The ferrocyanide gives a brown precipitate of copper ferrocyanide which helps to collect the finely-divided gold and facilitates filtration. Excess of ferrocyanide must, however, be avoided as it dissolves gold and leads to loss. The precipitate is filtered off, then mixed with fluxes, and fused, and the resulting lead button cupelled in the usual way for the recovery of the gold. This method gives low results if much sulphocyanide is present in the cyanide solution tested.

The cuprous chloride solution may conveniently be prepared by placing 300 grammes of crushed copper sulphate into a Winchester quart-bottle, which is then three-parts filled with strong commercial hydrochloric acid, space being left for the addition of as much copper wire (about 0·05 inch diameter, previously flattened) as can be inserted. After standing for a few days the copper wire reduces the mixture to the cuprous state, which is indicated by its becoming nearly colourless and clear ; it rapidly oxidizes, however to a brown solution on exposure to air.

As cuprous chloride is somewhat troublesome to prepare and to keep in the cuprous state, many assayers prefer to use copper sulphate, sodium sulphite, and sulphuric acid for the gold determination instead of cuprous chloride.

The following solutions are necessary :—

Sodium cyanide	Saturated solution.
Copper sulphate	15 or 30 per cent. solution.
Sodium sulphite	10 or 20 ,, ,,
Sulphuric acid	20 or 40 ,, ,,
Potassium ferrocyanide ...	2 or 4 ,, ,,

[1] A. Wogrinz, *Zeits. anal. Chem.*, 1937, **108**, (6/7), 266–267 : Abs., *Met. Abs. Inst. Met.*, 1937, **4**, 201.

In carrying out the assay, 20 A.T. of the cyanide solution to be tested are transferred to a conical flask, and 10 c.c. or 20 c.c. (according to their strength) of the above reagents added in the order given, the flask being well shaken after each addition. When SO_2 is freely evolved the precipitate is filtered off and allowed to drain, no washing being necessary. When thoroughly drained the precipitate is fused with 2 A.T. of a suitable flux, with the addition of carbon to give a lead button of about 20 grammes. It is usual to put part of the flux in the crucible, add the precipitate, and then cover with the remainder of the flux. The silver necessary for inquartation is generally added in the form of a measured quantity of silver nitrate of known strength.

(*d*) *Electrolytic Method* (Crichton).[1]—In this method the gold is deposited on a lead cathode which is cupelled to obtain the gold. The beakers, each containing 10 or 20 A.T. of cyanide solution, are arranged in a row on a special stand. The anodes consist of $\frac{5}{16}$-inch carbon rods, held in position in the centre of each beaker by means of clamps fitted to a horizontal copper bar which runs parallel to and 6 inches above the line of beakers. Thus all the anodes can be lowered or lifted in one operation. The cathodes consist of cylinders of assay lead-foil which fit just inside the beakers and have the lower edges serrated to allow better circulation. The cathodes are connected by means of insulated flexible wires with a copper rod which runs along the front side of the row of beakers. The terminals are made by almost severing narrow strips of the cathodes and folding the strips over the ends of the flexible wires.

The current for deposition of the gold is obtained from three 2-volt accumulator cells in series, and varies in amperage according to the strength of the cyanide solutions used. Thus, when the strength is 0·3 per cent. KCN, 0·1 ampere is required ; and when it is 0·02 per cent. KCN, the current is about 0·04 ampere. The time required for complete deposition of the gold is from three to six hours, the quantity of solution used being immaterial. With 10 or 20 A.T. of solution complete deposition of the gold is effected in four hours.

When deposition is complete, the carbons are raised clear of the beakers, the current switched off, and the lead cathodes disconnected and removed to a hot-plate to dry. When dry, these are folded into a small compass and cupelled with a little silver, and the buttons parted and weighed. To assist the deposition of the gold in weak solutions a little KCN should be added, and in impure solutions a little ammonium hydrate is advantageous.

Reporting the Results.—In reporting assay values of cyanide and other solutions, the results should be given in parts by weight in a stated volume of the solution. Usually the report is made in pennyweight per ton of solution. The use of the " fluid ton of 32 cubic feet " is recommended. It closely approximates to 2000 lbs., and is in common use.

Determination of *Zinc* in Cyanide Solutions.—" Zinc is generally supposed to occur in cyanide solutions as a double cyanide of the type of $K_2Zn(CN)_4$, though there is some evidence to show that this is partially dissociated in dilute solutions, which probably contain a portion of the zinc as the simple cyanide $Zn(CN)_2$. In certain cases potassium zincate, $Zn(OK)_2$, or some similar compound may be present " (Clennell).

Zinc[2] in cyanide solutions is conveniently determined by precipitation with sodium sulphide, then adding a slight excess of standard iodine solution to decompose the zinc sulphide, and determining the excess of iodine with standard sodium thiosulphate. 100 c.c. of the working solution are heated to about 70° C. and an excess of sodium sulphide added. The precipitate of zinc sulphide

[1] C. Crichton, *J. Chem. Met. and Min. Soc. of S. Africa*, 1911, **12**, 90–92.
[2] A. M'A. Johnston, *Rand Met. Pract.*, vol. I, p. 328.

is filtered off and washed with hot water until free from soluble sulphide. The precipitate and filter-paper are then transferred to a small stoppered bottle (about 150 c.c. capacity), and sufficient N/10 iodine solution is run into the flask to leave an excess of not more than 5 c.c. (30 to 35 c.c. N/10 iodine is usually sufficient). After shaking, the excess iodine is determined by titrating with N/10 sodium thiosulphate. Then—

1 c.c. of N/10 iodine = 0·00325 gramme of zinc.

The method is quick and accurate.

Determination of Copper in Cyanide Solutions.[1]—The presence of copper in the ore under treatment by the cyanide process, especially when the metal occurs as carbonate, has a very injurious effect, giving rise to high consumptions of cyanide. Hence in some cases it becomes a matter of importance to determine the quantity of copper in the solution.

The presence of copper, even in very small amount, may be detected by acidulating the cyanide solution with any mineral acid, and adding a few drops of dilute ferrocyanide solution, which gives the characteristic reddish-brown colour. Ammonium hydrate, of course, gives no blue colour in solutions containing soluble cyanides, so that complete decomposition (*e.g.* by boiling with sulphuric and nitric acids) is necessary before this reagent can be applied for the detection of copper.

Copper may in most cases be rapidly determined by the colorimetric method with ammonium hydrate as follows :—

100 c.c. of the solution are strongly acidulated with HCl, heated to boiling, then 0·5 gramme potassium chlorate added, and the boiling continued until most of the chlorous gases are expelled. The liquid is then made alkaline with ammonium hydrate and filtered. The tint of the filtrate is compared with that of a similar volume of water, to which HCl and ammonium hydrate have been added in about the same quantities as used in the test, and to which standard copper solution is added until the colours of the two liquids are alike.

For this purpose a standard solution of copper nitrate is prepared, containing 0·1 per cent. of copper and about 1 per cent. of sulphuric acid. The ammoniacal copper solution from the cyanide liquid, diluted to a suitable volume (according to the amount of copper present), say 10 c.c. for every milligramme of copper, is placed in a clear glass cylinder, and a nearly equal volume of distilled water in a similar cylinder. To the latter is added the same amount of hydrochloric acid and ammonium hydrate as used in the test, and then the standard copper solution run in from a burette, with constant shaking, until the tint in the two cylinders is the same.

Each c.c. of solution run in represents 0·001 gramme of copper. The best results are obtained when the amount of copper is between 5 and 15 milligrammes. With practice, the amount of copper present may be determined to within 0·0002 gramme.

The treatment of the solution with potassium chlorate described above is generally sufficient for the complete oxidation of all cyanogen compounds, but in a few cases it may be necessary to add a little bromine.

When much ferrocyanide is present, however, the decomposition should be effected by evaporation with nitro-hydrochloric acid, followed by sulphuric acid and boiling until white fumes of SO_3 are freely given off.

In cases where the quantity of copper present is sufficient, it may be determined accurately in the liquid by taking 100 c.c. or more of the solution,

[1] Clennell, *op. cit.*, 133.

adding 5 c.c. of concentrated nitric acid and 5 c.c. of concentrated sulphuric acid, and evaporating until white fumes are given off. This liquid is diluted to about 100 c.c., then 10 c.c. of concentrated sulphuric acid are added, one or more sheets of aluminium foil introduced, and the solution boiled for ten minutes. The metallic copper thus precipitated is dissolved, and the copper determined by the ordinary analytical methods, such as titration with standard sodium thiosulphate.

Remarks on Commercial "Cyanide."—The " cyanide " used in cyanide mills and known in commerce as " potassium cyanide " actually consists of the mixed cyanides of sodium and potassium, containing varying amounts of impurities such as alkaline carbonates, sulphates, etc. It has been found that potassium cyanide carrying sodium cyanide is more soluble and more efficient than pure potassium cyanide, weight for weight, owing to the larger percentage of cyanogen it contains, as shown below. The combined cyanide can be made and sold more cheaply than the pure potassium cyanide, and is therefore invariably used. What is wanted in most cases is merely a soluble cyanide, and it is a matter of indifference whether the base be sodium or potassium.

It is customary to express the quality of a sample of commercial cyanide by stating that it contains so much per cent. of potassium cyanide ; or in other words, the quality is expressed by the cyanogen content, in terms of potassium cyanide. Since, however, sodium cyanide contains more cyanogen than is contained in an equal weight of potassium cyanide, this method of expressing the quality gives rise to expressions such as " 130 per cent. cyanide." This is explained as follows :—

The molecular weight of sodium cyanide (NaCN) is 49, and of potassium cyanide (KCN) 65 ; and since 49 parts of sodium cyanide are equivalent to 65 parts of potassium cyanide, it is evident that a pure sample of sodium cyanide (= 100) would contain cyanide equivalent to 132·7 per cent. of potassium cyanide.

The molecular weight of cyanogen (CN) is 26, therefore the percentage of cyanogen as sodium cyanide is :—

$$\frac{\text{CN} \times 100}{\text{NaCN}} = \frac{26 \times 100}{49} = 53{\cdot}1 \text{ per cent.},$$

and the percentage of cyanogen in potassium cyanide is :—

$$\frac{\text{CN} \times 100}{\text{KCN}} = \frac{26 \times 100}{65} = 40{\cdot}0 \text{ per cent.}$$

Pure potassium cyanide containing 40 per cent. of cyanogen corresponds in the trade to " 100 per cent. cyanide," and the percentage of cyanogen in commercial samples is calculated on this basis. Thus " 98 per cent. cyanide," which is in common use, contains only 39·2 per cent. cyanogen, and " 125 per cent. cyanide " contains 50 per cent. cyanogen.

Assay of Commercial Cyanide.—Since the " potassium " cyanide of commerce may contain impurities such as alkaline carbonates, and small quantities of alkaline chlorides, sulphates, and sulphides, it is usual to test it for its cyanide contents.

Determination of Free Cyanide.—Break off 20 grammes of the sample of cyanide to be tested, in clean fresh pieces, and weigh accurately to the nearest centigramme. The portion selected should be taken through the thickness of the cake so as to obtain a representative sample. Dissolve in water containing a little sodium hydrate, transfer the solution to a 2-litre flask, and dilute with distilled water to 2 litres. After thorough mixing, a measured portion, 50 or 100 c.c., is titrated, as previously described on page 464, with standard silver-nitrate solution and a few drops of potassium iodide as an indicator. It is well to bear in mind that the result does not give the actual potassium cyanide present, but the cyanogen, which may in reality be present as a sodium, calcium, ammonium, or other salt. This cyanogen is, however, reported in terms of its equivalent in potassium cyanide, as stated above. The percentage may be calculated by multiplying the number of c.c. of standard silver nitrate (10 c.c. equivalent to 0·1 gramme KCN) used by 40 (50 c.c. being one-fortieth of 2 litres) and dividing by the weight of commercial cyanide originally taken.

Detection of Impurities in Commercial Cyanide.—It is sometimes desirable to test for impurities in commercial cyanide, although it is seldom necessary to make exact determinations. The commonest impurity in commercial cyanide is carbonate of sodium or of potassium.

The various impurities may be detected in the following manner:—

Detection of Carbonates.—These may be tested for by dissolving, say, 2 grammes of the commercial cyanide in a little water and adding a solution of barium chloride. If carbonates are present, a white precipitate of barium carbonate will be formed which, if filtered off, washed, and treated with hydrochloric acid, will dissolve with effervescence.

The amount of carbonic acid representing the carbonates present may be determined[1] by dissolving 10 or 20 grammes of the commercial cyanide in water, adding calcium nitrate, then shaking, and allowing the solution to stand. The precipitate of calcium carbonate is then filtered off, washed, ignited, and weighed as CaO. The weight of CaO multiplied by 0·786 gives the weight of CO_2 in the 20 grammes of cyanide taken.

Detection of Chlorides.—Chlorides may be detected by adding an excess of silver-nitrate solution, which gives a white curdy precipitate of silver chloride. Chlorides may be determined volumetrically as follows[2]:—

L. Siebold has shown "that chlorides, when present, may be conveniently determined in the same liquid in which the cyanide has been determined by neutralizing the excess of free alkali (which should not be ammonia) by the cautious addition of dilute nitric acid, adding a few drops of a solution of neutral potassium chromate, and continuing the addition of the silver solution until the red tint due to the formation of silver chromate remains permanent. If cyanide only be present. the volume of silver solution now required will be exactly equal to that previously employed to obtain a permanent turbidity, whereas any excess over this amount represents the silver solution corresponding to the chlorides present."

Using a standard solution containing 13·039 grammes of silver nitrate per litre, the silver equivalents for complete precipitation will be as follows[3]:—

$$1 \text{ c.c. standard } AgNO_3 = \begin{cases} 0{\cdot}005 \text{ gramme potassium cyanide.} \\ 0{\cdot}002723 \text{ gramme chlorine.} \\ 0{\cdot}004489 \quad \text{,,} \quad \text{sodium chloride.} \\ 0{\cdot}005726 \quad \text{,,} \quad \text{potassium chloride.} \end{cases}$$

Detection of Sulphates.—Sulphates are detected by the formation of a white precipitate of barium sulphate obtained on adding a solution of barium chloride

[1] Seamon, *Chem. News*, 1910, p. 18, from *The Chemical Engineer*, 1910, **10.**
[2] Quoted by Argall, *Mill and Smelter Methods*, p. 81.
[3] Clennell, *op. cit.*, p. 109.

to the cyanide solution which has been previously acidulated with hydrochloric acid to effect the decomposition of the cyanides. The determination of sulphates is usually made by the ordinary gravimetric method with barium chloride. The cyanide solution may be completely decomposed by evaporating several times to dryness with nitric and hydrochloric acids, and finally with the latter alone. The residue is taken up with water and a little hydrochloric acid, the solution heated nearly to boiling, and barium chloride added in slight excess.

Detection of Sulphides.—Soluble sulphides produce a marked effect in checking the solution of the gold ; consequently in South Africa the amount of sulphide which a commercial cyanide should contain is limited to the second place of decimals per cent.[1] A very small percentage of alkaline sulphides frequently occurs, however, in samples of commercial cyanide, and it is sometimes desirable to test for them, although it is seldom necessary to make exact determinations.

The presence of sulphides may be detected by adding a drop or two of a solution of lead tartrate to 4 or 5 c.c. of caustic-soda solution and adding this alkaline lead salt to a clear solution of the commercial cyanide. If any sulphide is present, a brown coloration or a black precipitate will be formed according to the amount present.

" A solution of lead tartrate in excess of soda is found to be most suitable as a test for the presence of soluble sulphides, especially in the presence of cyanides, since ordinary lead salts yield a white precipitate of lead cyanide which may obscure the black or brown coloration " (Clennell).

The most delicate test for sulphides is to add a solution of nitro-prusside to the cyanide solution. The most minute trace of alkaline sulphide is shown by the appearance of a brilliant purple colour. Nitro-prussides are formed by adding nitric acid to a solution of potassium ferrocyanide, as described below.

Dr Loewy's method[2] for the determination of sulphides in cyanide solutions is to add sodium nitro-prusside, which gives the violet colour. He dissolves 2 grammes of the sample of cyanide to be tested in 100 c.c. of water, and compares it with a solution of cyanide of similar strength free from sulphides. He then adds 1 c.c. of a solution containing 5 per cent. of sodium nitro-prusside and a very little cyanide to each solution, and then adds a solution of sodium sulphide of known strength to the pure cyanide until the violet colour is of the same depth in each. The calculation of the result is then simple. The method is applicable to solutions in which the amount of sulphur lies between 0·0005 and 0·0015 per cent. The solutions are adjusted to bring the sulphur within these limits. The sodium nitro-prusside solution is prepared thus :—

Preparation of Sodium Nitro-prusside.[3]—Concentrated nitric acid is diluted with its own volume of water. This diluted acid is mixed with powdered potassium ferrocyanide in the proportion of :—

2 parts $K_4Fe(CN)_63H_2O$.
5 parts dilute HNO_3.

The salt dissolves to a coffee-coloured liquid evolving CO_2, N, CN, and HCN. It is warmed on a water-bath until the liquid gives a dark green or slate-coloured precipitate instead of a blue precipitate with ferrous sulphate. It is then cooled, neutralized with sodium carbonate, and filtered.

[1] Salamon, *J. Soc. Chem. Ind.*, 1910, **29**, 1426.
[2] Described by Rose, *Met. of Gold*, 5th edn., p. 371.
[3] Clennell, *Cyanide Solutions*, p. 93.

It may be pointed out that the presence of sulphides is shown at once when a sample of commercial cyanide is being tested for its strength in cyanide, inasmuch as the first few drops of silver-nitrate solution produce at once a brown coloration due to the formation of silver sulphide. This coloration obscures and entirely vitiates the result of the titration. When sulphides are present they are removed by precipitation before titrating, by adding freshly-precipitated lead carbonate and shaking the solution.

Detection of Silicates.—The presence of silicates can be detected and the amount determined by evaporating a solution of the commercial cyanide to dryness with hydrochloric acid. The residue is treated with water acidulated with hydrochloric acid and the insoluble residue of silica filtered off, washed, dried, ignited, and weighed.

(4) Assay of Cyanide Bullion.

Bullion samples consist of (*a*) mill bars, (*b*) cyanide bars, (*c*) lead bars. Zinc–gold slimes are also included under this head.

(*a*) *Mill Bars.*—The term "mill bars" is given to the bars of gold resulting from the treatment of the ores by amalgamation prior to cyanide treatment. The standard of mill bars is higher than that of bars obtained by the cyanide process. The bars are sampled and assayed as previously described in Chapter XVII. As a rule, 0·5 gramme is taken for assay and the assay made in duplicate.

(*b*) *Cyanide Bars.*—As previously pointed out, these bars vary greatly in composition and present considerable difficulty in sampling owing to the liquation which takes place on solidification. The methods of sampling and of assaying cyanide bars have been described in Chapter XVII.

The assays are usually made on 0·5-gramme samples, and on account of the irregularity of composition it is advisable to make the assays in triplicate and to use checks.

Scorification, before cupellation, is sometimes recommended for cyanide bars of very low quality, say below 700 fine.

(*c*) *Lead Bars.*—These consist of bars of lead bullion obtained by melting the zinc-box precipitate (often called "gold slimes") with litharge and fluxes in a reverberatory furnace. The lead bars are usually soft, and contain up to about 8 per cent. of gold. The bars are subject to more or less segregation during solidification, and care is required in sampling, as discussed on p. 323.

The methods of sampling in use on the Rand, are as follows[1] :—

(1) Dip sample taken from the furnace before tapping.
(2) Dip sample taken from the kettle that receives the molten metal before casting into moulds (if such is used).
(3) Sampling the cast lead bars by drilling or cutting chips.

All three methods are said to give equally good results if the lead does not contain more than from 4 to 5 per cent. of gold and silver.

The usual method of assay on the Rand is to scorify from 0·5 to 1 A.T. with from 40 to 50 grammes of lead and a little borax and then to cupel and part in the ordinary way. If the approximate fineness of the lead bullion is known, the silver necessary for parting purposes can be added direct during the scorification or cupellation. The assays should be made in duplicate, and in cases where extreme accuracy is desired, checks should be used.

Details of the methods of sampling and assaying lead bullion are given in Chapter XX.

[1] *Rand Met. Pract.*, vol. I, pp. 280 and 319.

Assay of Zinc-Gold Slimes.—This is the precipitate formed by zinc acting on cyanide solutions of gold. The precipitate varies considerably in composition, and gold may be present in any proportion up to about 50 per cent. according to the treatment given before melting.

The average composition of slime obtained on the Rand after acid-treatment and calcination is approximately as follows[1] :—

Gold	34 to 37 per cent.
Silver	3·5 to 4·5 ,,
Base metals, silica, alumina, iron oxides, etc.	remainder.

The amount of impurity, especially of lead, contained in the precipitate is largely dependent on the preliminary treatment to which it has been subjected. The sample is received in a dry condition, but before being assayed the moisture must be very carefully determined : it is preferable to make the same sample serve for both moisture and assay.

The precipitate may be assayed either by crucible fusion or by scorification ; both methods are in general use. The following is a suitable stock flux :—

Litharge	600 grammes.
Sodium carbonate	350 ,,
Borax glass	100 ,,
Sand	100 ,,
Charcoal	8 ,,

In some charges recommended for the assay of gold slime the sand is omitted, but in this case there is a greater tendency for the slag to retain gold and silver.

For one-tenth assay ton (3 grammes) of precipitate, take 80 grammes of the flux.

For one-fifth assay ton (6 grammes) of precipitate, take 120 grammes of the flux.

The weighed quantity of precipitate is mixed with two-thirds of the flux and the remainder used as a cover. The charge is fused in a rather hot fire for thirty minutes. The resultant lead button is freed from slag and then cupelled and parted as usual. The cupel and slag absorption is determined and added.

Since the composition of the precipitate is so variable, several assays should be made on each sample, and the average of the results taken.

If the scorification method is used, from 2 to 3 grammes of the precipitate are scorified with 40 grammes of lead and a cover of borax, and the resultant lead cupelled, etc.

Scorification is usually adopted for this class of material on the Rand,[2] a preliminary assay being first made to determine the approximate percentage of fine gold. From 0·25 to 1·0 gramme is then scorified with fifteen times its weight of lead-foil or shot, with the addition of the silver necessary for parting, as ascertained by the preliminary assay. At least five separate scorifications should be made.

Fulton and Crawford[3] have investigated the different methods for the determination of gold and silver in zinciferous precipitates.

Experimental assays were made of a well-mixed sample of gold and silver precipitate from the zinc boxes of the cyanide process, to obtain a rapid and accurate method of analysis. Scorification gave low results due to volatilization,

[1] Rose and Newman, *Metallurgy of Gold*, 7th edn., p. 401.
[2] *Rand Metallurgical Practice*, vol. I, p. 320.
[3] *School of Mines Quart.*, 1901, **22**, 153–162 (abs. *J. Soc. Chem. Ind.*, 1901, **20**, 749).

and so was discarded. The crucible method yielded variable results, and the combination wet and dry method gave results high in silver and low in gold, due probably to its solubility in nitrous acid as a result of the action of the nitric acid on the zinc present. To overcome this difficulty a combination method was tried, using sulphuric acid in place of nitric acid. One-tenth assay ton of precipitate was treated with 20 c.c. of concentrated sulphuric acid and 60 c.c. of water and boiled for one hour, after which the solution was cooled and diluted to 100 c.c. Normal sodium-chloride solution 75 c.c., and lead acetate solution 20 c.c. were added, and the precipitate settled, filtered, washed, and dried. The paper was burnt off, and the residue scorified until half covered over in the scorifier, then poured and the button cupelled. The results were good and uniform, the slag corrections being small. The gold is lower than in the crucible assay, while the silver is higher, but the results are not so erratic, and it is not necessary to make more than two assays to get a reliable figure, as is the case with the crucible method. The original sample should be assayed at once or kept from the air as the zinc oxidizes quickly, and sufficiently to alter the percentage of gold and silver present. Care should be taken that the sulphuric acid is free from nitric acid.

(5) Assay of Cyanide Mill-Products.

Under the head of mill-products are included :—

(*a*) Black sands (mill concentrates).
(*b*) Concentrates (Fruevanner concentrates).
(*c*) Screen samples.
(*d*) Slime samples.
(*e*) Crushed plumbago crucibles and liners.
(*f*) Slags and sweepings.
(*g*) Litharge dross.

(*a*) *Assay of "Black Sands" (Mill Concentrates)*.—This material consists largely of magnetite (Fe_3O_4), which is a heavy mineral and on concentration separates from the lighter gangue materials. The sands should be passed through a sieve of 100-mesh before being assayed, the sieve being watched carefully for metallics.

The assay of black sands is made by crucible fusion in the same way as basic ores, as described in Chapter XII.

Whitby gives the following charges for the assay of 1 A.T. of black sand obtained in gold mills on the Rand [1]:—

Ore	1·0 assay ton.	1·0 assay ton.
Litharge	3·0 ,,	—
Red lead	—	2·5 ,,
Sodium carbonate	2·0 ,,	2·0 ,,
Borax	0·5 ,,	0·5 ,,
Weight of lead button obtained	50 to 58 grammes.	35 to 40 grammes.

If one-half assay ton is taken for assay, half the above quantities of flux are used, in which case the lead button will weigh about 25 grammes.

(*b*) *Assay of Concentrates (Fruevanner)*.—These consist of the heavy metallic minerals, mainly sulphides, that have been separated from the accompanying gangue or rock by concentration. Samples of concentrates should be crushed to pass 100-mesh sieve ; metallics may be present. Concentrates are assayed by the methods described for pyritic material in Chapter XIII. In many

[1] *J. Chem. Met. and Min. Soc. of South Africa*, 1906, p. 271.

gold mills, samples of concentrates are roasted before being assayed, but with proper precautions they may be satisfactorily assayed by direct fusion.

The following flux is recommended by Whitby for the direct fusion of concentrates obtained on the Rand[1] :—

Concentrates	0·5 assay ton.
Red lead	2·0 ,,
Sodium carbonate	1·5 ,,
Borax	0·5 ,,

The sulphur present will act as a reducing agent and give a button of lead weighing about 50 grammes.

(c) *Assay of Screen Samples.*—These represent all the finely-crushed ore or pulp charged into the leaching-tanks and all the tailings discharged from the plant. The samples are taken by means of a bucket or some mechanical sampling device, as described on page 87. The preparation of these samples for assay requires great care, as they are the most difficult of all the mill samples from which to obtain accurate results. Each bucket must be kept separate and the contents dried and quartered down; but as a general rule, before quartering, the sample should be passed through a sieve of the same aperture as, or slightly larger than, that used in the battery, so that any coarse particles due to damaged screens may be detected and crushed to the same fineness as the remainder of the sample. When the bulk of the contents of each bucket has been sufficiently reduced by quartering, the final portion selected for assay is crushed to pass an 80- or 100-mesh sieve, and then well mixed and assayed in the same way as described for gold ores.

The most reliable results are obtained by keeping the assay portions from each bucket quite distinct and making separate assays, and then averaging all the results obtained.

In some mills, however, it is the practice to mix together the assay portions from several buckets and assay the sample thus obtained. This, of course, effects a saving of time, but in making a larger number of separate assays the errors which might arise from the presence of coarse gold are probably eliminated. Other reduction-works samples are quartered down in the manner described above.

A greater degree of accuracy is required in the results of reduction-works samples than with mine samples; the assays are therefore made in triplicate or even in quadruplicate.

The following charges are given by Whitby as typical of those used on the Rand.

General charge for screen and pulp samples :—

Ore (or pulp)	2·5 A.T. or	75 grammes.
Litharge	3·0 ,,	90 ,,
Sodium carbonate	2·5 ,,	75 ,,
Borax	0·5 ,,	15 ,,
Charcoal, a sufficiency.		

For sand charges (*i.e.* originals) and residues for treatment by a concentrating plant, and also slime charges and residues for slime plant :—

Ore	2·5 A.T. or	75 grammes.
Litharge	2·0 ,,	60 ,,
Sodium carbonate	2·5 ,,	75 ,,
Borax	0·5 ,,	15 ,,
Charcoal, a sufficiency.		

[1] Whitby, *loc. cit.*, p. 271.

(*d*) *Assay of Slime Samples.*—When no spitzlutte, fruevanner, or other concentrate is separately treated, the charge for sand originals and residues is necessarily altered, as they will then contain the bulk of the pyrites. In this case the following charge is used :—

For sands, originals, and residues containing pyrites :—

Ore	2·5 A.T. or	75 grammes.
Litharge	3·0 ,,	90 ,,
Sodium carbonate	2·5 ,,	75 ,,
Borax	0·75 to 1·0 ,,	22·5 to 30 ,,
No charcoal.		

When much pyrites is present, iron nails should be added.

In assaying residues some assayers use as much as 5 A.T., or from 100 to 150 grammes.

The treatment of slime samples necessarily involves evaporation to dryness. With residues this means a loss of more or less gold, which is deposited on the side of the evaporating vessel, and not recoverable. The discrepancies in slime-residue assays and the assays of slimes undergoing treatment are attributed by Whitby to this cause. He states that the following method of preparing slimes residues for assay gives satisfactory results. To the slimes residues is added first a little cyanide and then small quantities of solutions of copper sulphate, dilute sulphuric acid, and sodium sulphite. After shaking vigorously, the sample is turned out into a shallow, enamelled drying-pan and dried.

Residues (containing undissolved gold) resulting from the treatment of sand and slimes are transferred to three-gallon buckets filled with water and settled with lime-water for six hours, the water is then syphoned off, and the residues dried and assayed.

The results of reduction-works samples are reported in pennyweight and decimals of a pennyweight per ton to the second place of decimals ; consequently a fine assay balance must be used for the final weighings. This should be sensitive to $\frac{1}{200}$ milligramme with a 0·5 milligramme rider for beam work.

(*e*) *Assay of Graphite Crucibles and Liners.*—Graphite (plumbago) crucibles are extensively used in cyanide mills for smelting the zinc–gold precipitate. They often contain a considerable amount of gold and silver, which filters into the pores of the crucible and also adheres to the surface. A large proportion of this may be separated as " metallics " by crushing the sample and passing it through a sieve of 100- or 150-mesh. The metallics are collected and assayed separately. The fine material presents difficulty in assaying on account of the graphite and zinc it contains : it is well mixed and assayed as described on page 379 where typical charges used in South African cyanide mills for this material are given. The method of calculating the value of material containing metallics is given on page 205.

Detachable clay liners are sometimes used inside the graphite crucibles. These are crushed and sampled and assayed in the same way as graphite pots, but since they contain very little graphite, due allowance must be made for this fact in making up the charge for the assay, and the material fluxed as an ordinary siliceous ore.

(*f*) *Assay of Slags and Sweepings.*—The slags and sweepings which result from the smelting and casting of bullion always contain more or less precious metal. The methods of assay for these materials are described on page 141.

(*g*) *Assay of Litharge Dross.*—This consists of impure litharge obtained from the cupellation of the auriferous lead resulting from the melting of zinc–gold slime. Litharge dross is assayed as described on page 131.

Number of Control Assays made in a Cyanide Mill.—The actual number of assays made monthly in a cyanide mill varies considerably and is dependent upon the size of the mill and the amount of ore being treated annually. Some indication of the assay work involved in controlling production is given in the following figures, which show the number of assays made on the various products each month at a producing mine in Rhodesia, South Africa, crushing 7000 tons of ore per month.

Development samples	200 assays.	Amount taken 2 A.T.
Stope samples	400 ,,	,, 2 ,,
Screen-pulp samples	200 ,,	,, 2 ,,
Battery tailing samples	100 ,,	,, 2 ,,
Cyanide mill :—		
Sand charges	30 ,,	,, 4 ,,
Slime charges	30 ,,	,, 4 ,,
Sand residues	30 ,,	,, 4 ,,
Slime residues	30 ,,	,, 4 ,,
Sand solutions (entering precipitating boxes)	30 ,,	,, 10 ,,
Sand solutions (leaving precipitating boxes)	30 ,,	,, 10 ,,
Slime solutions (entering precipitating boxes)	30 ,,	,, 10 ,,
Slime solutions (leaving precipitating boxes)	30 ,,	,, 10 ,,
Samples of bullion	8 ,,	,, 0·5 gramme.
Total assays per month ...	1148	

With the exception of the bullion samples, only one assay of each sample is made as a general rule.

(6) Testing of Ores for Cyanide Treatment.

The assayer is sometimes called upon to make preliminary tests on ore samples to ascertain their suitability for cyanide treatment, and to determine the maximum gold extraction that can be looked for, together with the probable consumption of cyanide and best strength of solution to use. In large Customs works, where ores from all sources are treated, it is usual to make such tests on each lot of ore received. A complete analysis of the ore should be made before treatment, special search being made for any elements, such as tellurium and sulphides, which may seriously affect the economic treatment of the ore by the cyanide process. Preliminary tests are conveniently made on ore crushed to pass a 20- or 30-mesh sieve.

Microscopical examination.—Examination of the ore samples with a good lens or microscope frequently affords very useful information on the associated minerals and their amount, as well as the mode of occurrence of the gold. Similar examination should be applied to the concentrates obtained by washing a suitable quantity of the ore. This concentration will bring into prominence any native gold, and also any sulphides, arsenides, tellurides, etc. that may influence cyanide treatment.

Determination of the Acidity of an Ore.—The acidity of an ore is due mainly to the presence of free sulphuric acid and soluble sulphates (chiefly iron) resulting from the decomposition of sulphides. These products are destructive to cyanide solutions, destroying more or less their alkalinity and in a proportionate degree reducing their efficiency. The soluble salts and acids that consume cyanide are termed *cyanicides.* To prevent the waste of cyanides due to the action of cyanicides it is necessary to neutralize the latter before treating the ore with cyanide solution.

In practice the acidity of the ores is neutralized by the addition of either lime or caustic soda, and a preliminary test is made to determine how much lime or soda must be added for each ton of the ore in order to counteract the acidity. " Whether this acidity should be reported in terms of the lime or of the caustic soda required to neutralize it will depend on which of these reagents is to be used in the actual practice."

In most cyanide mills lime is used in preference to caustic soda, except in cases where tests have proved lime to be unsuitable. The amount of alkali or lime to be mixed with a charge of ore is determined by adding, little by little, an alkaline solution of known strength to a given weight of ore until the whole is neutral to litmus or other indicator.

The strength of the standard solution of caustic soda and the quantity of ore used for this determination vary considerably in different cyanide mills, but in many mills the standard solutions are so adjusted that each c.c. corresponds to 1 lb. of caustic soda or of lime per ton of ore, and the quantity of ore taken for the test is calculated accordingly.

The following may be taken as examples of the method of calculation :—

Example 1.—A normal solution of caustic soda contains 40 grammes NaOH per litre, but the solutions used in practice are usually weaker.

Suppose the strength of the solution is one-fourth normal (N/4), it will contain 10 grammes of NaOH per litre, and 1 c.c. will contain 0·01 gramme of NaOH. Then, since 1 c.c. of the standard solution is equivalent to 0·01 gramme of caustic soda, if we take 2000 times this weight of ore for the test (*i.e.* 0·01 gramme $\times$ 2000 = 20·0 grammes), each c.c. of standard solution will be equivalent to 1 lb. of caustic soda per ton (2000 lbs.) of ore. It is frequently more satisfactory to take a larger quantity of ore for the determination, and if 200 grammes is taken then each c.c. of the standard solution will be equivalent to one-tenth (0·1) of a pound of caustic soda per ton of ore. If the acidity is to be reported in terms of lime, the result obtained for caustic soda (expressed in pounds per ton) is multiplied by 0·7, since 1 part of caustic soda is equivalent to 0·7 part of lime ; thus (NaOH = 40, CaO = $\frac{56}{2}$[1] = 28, and $\frac{28 \cdot 0}{40 \cdot 0} = 0 \cdot 7$).

Therefore if the result of the determination of the acidity of an ore showed that it required the addition of 2·5 lbs. of caustic soda per ton, then the equivalent amount of lime required would be 2·5 $\times$ 0·7 = 1·75 lbs. per ton of ore.

Example 2.—As shown above, a normal solution of caustic soda contains 40 grammes of NaOH, which is equivalent to 28·0 grammes of lime. One c.c. of normal solution is therefore equivalent to 0·04 gramme NaOH or 0·028 gramme CaO, and 1 c.c. of a soda solution one-fifth normal (N/5) will be equivalent to $\frac{0 \cdot 04}{5} = 0 \cdot 008$ gramme NaOH or $\frac{0 \cdot 028}{5} = 0 \cdot 0056$ gramme CaO.

Then, since 1 c.c. of the N/5 standard solution is equivalent to 0·0056 gramme of lime, if we take 2000 times this weight of ore for the determination, viz. 0·0056 gramme $\times$ 2000 = 11·2 grammes, each c.c. of the standard solution will be equivalent to 1 lb. of lime per ton (2000 lbs.) of ore.

In practice ten times this weight of ore would be taken (*i.e.* 112·0 grammes), and 1 c.c. of the standard solution would in this case be equivalent to one-tenth of a lb. of lime per ton of ore.

If a normal solution of caustic soda is used for the titration and 56 grammes (the calculated quantity) of ore is taken, then 1 c.c. of the standard solution will be equivalent to 1 lb. of lime per ton of 2000 lbs.

[1] Calcium being divalent, one-half of the molecular weight is taken.

The test for acidity may be made by placing 200 grammes of the pulverized ore into a tall, wide-mouthed stoppered bottle (of colourless glass), adding 1000 c.c. of water, and shaking well for five or ten minutes. A N/4 standard solution of caustic soda (10 grammes NaOH in 1000 c.c. distilled water) is then run in from a burette until the liquid in the bottle is neutral. A piece of litmus paper or a small quantity of litmus or phenolphthalein solution is used as an indicator.

As stated above, each c.c. of the standard soda solution will indicate that 0·1 lb. of caustic soda must be added to each ton of ore (2000 lbs.) to neutralize the cyanicides present, provided that 200 grammes of ore are used for the determination. For example: If 13·0 c.c. of standard solution of caustic soda were required, then $0 \cdot 1 \times 13 = 1 \cdot 30$ lbs. of caustic soda must be added to each ton of ore or $1 \cdot 30 \times 0 \cdot 7 = 0 \cdot 91$ lb. of lime per ton of ore.

It must be remembered that the results obtained refer to pure caustic soda and pure lime, and since the soda and lime of commerce are not pure, a larger quantity will be required in practice than that indicated by the laboratory test.

It is also the practice in some mills in determining the acidity of an ore to add an excess of standard caustic soda solution, the additions of soda solution being continued until the indicator shows that the alkaline reaction is permanent. The amount of soda solution added is carefully noted and the excess of soda then determined by titrating back with standard acid until the solution is neutral. Thus:

$$HCl + NaOH = NaCl + H_2O;$$

or, if the return is to be made in terms of lime,

$$2HCl + CaO = CaCl_2 + H_2O.$$

Standard hydrochloric acid or sulphuric acid is usually employed.

If the ore is strictly neutral (*i.e.* free from cyanicides) the quantity of standard acid required will be the same as the quantity of standard soda used. If, as is frequently the case, the ore is acid, less standard acid will be used.

For example: If 14 c.c. of soda solution were used and only 10 c.c. of standard acid were required, then the ore will have done the work of the remaining 4 c.c. of acid. Therefore, to neutralize this acidity in the ore an addition of 0·28 lb. of lime per ton of ore will be required.

When the result of the test for acidity indicates that the ore requires the addition of more than about 3 lbs. of caustic soda per ton, it is usual to wash the ore first with water to remove the free sulphuric acid and soluble salts and then to test the acidity of the washed ore. An approximate determination of the acidity of an ore may be made[1] by taking 200 grammes of the crushed ore along with 200 c.c. of fresh water, and adding, little by little, small quantities of powdered lime until an alkaline reaction, using phenolphthalein as indicator, is obtained. The pulp is shaken after each addition of lime. From the amount of lime added is calculated the quantity necessary to be added to the ore bins. This test gives a figure which represents a quantity of lime slightly in excess of that actually required in practice, but it is advisable to err on the side of excess rather than deficiency in the addition of lime. A more accurate figure would, of course, be obtained by titrating a portion of the water in contact with the ore against standard alkali, as described above.

Acidity after Washing.—Take 200 grammes of ore and shake well with 1000 c.c. of water in a stoppered bottle as before, allow the ore to settle, and decant the clear liquid as far as possible. This wash-water is tested for cyanide consumption, etc., as directed below under " tests for extraction." The residue

[1] *Rand Met. Pract.*, vol. I, p. 346.

is then treated immediately, without drying, with an excess of a standard solution of caustic soda and well shaken for ten or fifteen minutes. About 500 c.c. of water are then added, the whole again shaken, and then the excess of soda determined by titrating back with a standard solution of acid, as described above. The acidity of the washed ore is due to insoluble basic salts, which are converted by the action of the caustic soda into ferric hydrate and soluble sulphates, thus :

$$Fe_2O_3 . 2SO_3 + 4NaOH + H_2O = Fe_2(HO)_6 + 2Na_2SO_4$$
$$2Fe_2O_3 . SO_3 + 4NaOH + 4H_2O = 2Fe_2(HO)_6 + 2Na_2SO_4.$$

The water filtrate may be mixed with a measured quantity of cyanide solution and subsequently titrated with standard silver nitrate to determine how much cyanide is decomposed by it. In this way the value of the water-wash on the particular ore is determined.

Test for Cyanide Consumption.—To determine the amount of cyanide consumed, place 200 grammes of the ore (ground to 30-mesh) with 200 c.c. of a cyanide solution of known strength (say 0·1 or 0·2 per cent.) in a stoppered bottle and shake for fifteen minutes, then leave undisturbed for twelve hours, and finally shake again for five minutes. Most of the decomposition of cyanide (due to the presence of cyanicides) takes place after shaking for the first few minutes, but not the whole, hence it is necessary to let the solution stand as described.

If power is obtainable, agitation is carried on in suitable jars. Allow the ore to settle, filter off some of the liquor, and assay 20 c.c. for cyanide by titration with standard silver nitrate in the usual way.

The difference in the strength of the cyanide solution before and after extraction gives the percentage of cyanide consumed in treating 200 grammes of ore, and from this result the consumption of cyanide per ton of ore can be calculated. Thus, if the original solution contains 0·2 per cent. of cyanide, and the solution after treatment contains 0·17 per cent., showing a consumption of $0{\cdot}2 - 0{\cdot}17 = 0{\cdot}03$ per cent.; this will be equivalent to a consumption of $\frac{0{\cdot}03 \times 2000}{100} = 0{\cdot}6$ lb. of cyanide per ton of ore (2000 lbs.) treated.

The filtrate may be tested for soluble sulphides, ferrocyanides, sulphocyanides (thiocyanates), copper, zinc, etc., and for its reducing power.

The tests recommended by Clennell[1] are as follows :—

(1) Add hydrochloric acid. A white turbidity may indicate the presence of soluble sulphides or of zinc.

(2) After adding a slight excess of hydrochloric acid, add a few drops of potassium ferrocyanide $K_4Fe(CN)_6$. A brown-coloured precipitate indicates copper, a white flocculent precipitate may indicate zinc.

(3) Add lead tartrate dissolved in excess of caustic soda : a brown or black coloration indicates soluble sulphides.

(4) Acidulate with sulphuric acid, and add dilute potassium permanganate drop by drop. If it is decolorized, it indicates that the solution has *reducing power*, the permanganate being affected by ferrocyanides and sulphocyanides, sulphides, organic matter, etc. If this reaction be strong, oxidation (*i.e.* aeration) will probably be necessary before or during the practical cyanide extraction.

Further Tests on Ores.—After making the above preliminary tests, experiments with 1 or 2 lbs. (0·5 to 1 kilogramme) of the ore are made to obtain information on the following points :—

[1] *J. Chem. Met. and Min. Soc. of South Africa*, 1898.

(1) The most economical degree of fineness to which the ore may be crushed for subsequent cyanide treatment.
(2) The best strength of cyanide solution to use.
(3) The best proportion of this solution to the ore.
(4) The length of time necessary for satisfactory results.
(5) The necessity for using alkalis.
(6) The necessity for oxidation.
(7) The necessity for any preparatory treatment.

Extraction Tests.—These tests are preferably made in wide-mouthed, glass-stoppered bottles. To determine the best fineness of ore the tests are carried out by first crushing portions of the ore to pass a 10-, 20-, 30-, 40-, and 60-mesh sieve respectively, and subjecting each to treatment with cyanide solution a trifle above the probable working strength, under exactly similar conditions of quantity, time, etc. If the preliminary test for "acidity" has shown that the ore contains cyanicides, add to each test the amount of lime found to be necessary to neutralize acidity. If the ore contains much free acid, it should first be subjected to a preliminary wash with water, as described above. The wash-water is carefully decanted off, and then the cyanide treatment is begun.

The tests should be made in duplicate, one kilogramme (1000 grammes or 2 lbs.) of the pulverized ore and the required quantity of pulverized lime are well mixed, transferred to the stoppered bottle, 1 litre of cyanide solution (strength say 0·2 per cent. KCN) added, and the bottle placed on an agitator, frequently a revolving shaft, and agitated for six to twelve hours.

After agitation, allow the ore to settle, then decant the solution on to a filter. Test 10 c.c. of the filtrate for cyanide consumption by titration with standard silver nitrate. Now wash out the ore from the bottle on to the filter, wash twice with water, then dry and carefully sample and assay the residue (*i.e.* tailings) for gold in the usual manner. The ore before treatment (*i.e.* the heads) having been carefully sampled and assayed, the difference between this assay result and that obtained by assaying the residue gives the quantity of precious metals that have been removed by cyanide treatment. The results may be checked by recovering the gold from the filtrate by evaporation, as described on p. 468.

To determine the Best Strength of Solution.—The ore of the fineness yielding the best result is then tried with cyanide solutions of different strengths.

Clennell recommends tests with solutions of 0·01, 0·05, 0·1, 0·2, and 0·5 per cent. potassium cyanide. The tests are carried out as described above. The results of these tests will indicate the best strength of cyanide solution to use.

To determine the Best Proportions of Solution to Ore.—The best strength of solution having been ascertained, tests are then made to ascertain the best proportion of this solution to the ore. For this purpose the ore of suitable mesh is treated with this solution in the proportions of 1, $1\frac{1}{4}$, $1\frac{1}{2}$, 2, 3, and 4 assay tons of solution to 1 assay ton of ore.

The ore before and after treatment is assayed and the difference between the results noted. A comparison of the extraction results obtained for each test will show the best proportions of solution to ore to use.

Time of Contact with Solution.—The best proportion of solution to ore having been ascertained, the time of contact required to attain the best results on the ore is now determined.

A number of tests of ore of suitable mesh are treated with the proper proportion of solution of correct strength, in stoppered bottles as before. The solution is allowed to remain in contact with the ore for 1, 2, 3, 4, 7, and 14 days. A comparison of the assay results from each test will give the time required for extraction.

Percolation Tests.—The rate at which a cyanide solution will percolate through a mass of ore is largely dependent on the nature of the ore and the size of the particles. With coarse ore and clean sand the percolation is always quicker and more uniform than with finely pulverized material, clayey ore, or mixtures of sand and slime. "A physical examination of the ore will give a good idea of the screen aperture through which it must be passed in order to obtain a good extraction. For example, if the ore is a porous or cellular oxidized product, perhaps crushing through a 0·44-inch screen aperture will suffice; if of dense and solid structure it should be crushed to pass screen apertures varying from 0·024 to 0·018 inch. Flinty material, pyritic and telluric ores, silver ores, etc., may have to be reduced to impalpable powder to obtain the desired extraction."[1]

The method of ascertaining the fineness of the ore that gives the best extraction has been described above.

Percolation tests are made in duplicate in glass percolators (such as that described and illustrated in Fig. 120 on page 232,) holding about 4 lbs. of ore. A false bottom is made of sand as described, or it may be constructed of a perforated board with a piece of canvas stretched across it to act as a filter. The strength of the cyanide solution to be used for the percolation tests on a particular ore is previously determined by bottle tests.

The dry ore, after being thoroughly incorporated with the required amount of lime, is placed evenly and carefully on the false bottom of the percolator and gently pressed around the sides to prevent channelling.

The cyanide solution of correct strength is then carefully added on top until the ore is thoroughly submerged. The solution is allowed to stand on the ore for four hours, and then allowed to percolate through by opening the stop-cock. Allow the ore to drain dry and then let it stand to aerate it: this will probably occupy a day. Then treat again with cyanide solution and allow to drain dry as before; after the second day's treatment turn the ore out from one of the percolators when drained dry, and cut out a sample for assay, to determine the extraction, and return the remainder to the percolator. Keep up this treatment and sampling daily, until the final extraction is reached so far as one percolator is concerned.

Allow the duplicate to run on without being disturbed, but it must be drained dry and allowed to aerate daily.

The percolation tests (as just outlined) should follow the usual cyanide practice, as follows:—

(1) Treat with water, or alkaline wash, if required.
(2) Treat with weak cyanide solution.
(3) Treat with strong cyanide solution.
(4) Treat again with weak cyanide solution.
(5) Treat with water to wash. Finish.

If tests have shown that the preliminary wash before cyanide treatment is unnecessary, the strong solution is added at once, followed by weak solutions and finally water to wash. When the solutions are run in on top of the ore and allowed to percolate downwards there is a tendency, especially with fine ore, for packing to take place, and to obviate this the solution may be run in from below and allowed to rise through the ore. This must be performed slowly to prevent the formation of holes or channels in the charge.

[1] Argall, *Mill and Smelter Methods*, p. 93.

In reporting the results of cyanide tests on ores the following information should be given :—

Character of the ore, with analysis if necessary.
Size of mesh to which ore is crushed.
Weight of ore taken for the test.
Assay value before treatment.
" " after (say 4) hours treatment.
" " " (" 8) " "
" " " (" 12) " "
" " " (" 16) " "
Lime used per ton of ore.
Strength of potassium cyanide solution used.
Amount of solution used per ton of ore treated.
Potassium cyanide consumed per ton.
Total time of treatment, percolation, and washing.
Percentage of gold and silver extracted.

Although preliminary tests such as those described afford some indication of the probable treatment for a particular ore, in mill practice it is essential that they should be followed by large-scale tests in some suitable plant of, say, 10 tons per day capacity. "Each ore-deposit is a problem in itself, and the information derived from a series of tests should bring into prominence the characteristics to which most attention must be paid in plant operation."

APPENDIX

ATOMIC WEIGHTS AND OTHER CONSTANTS OF METALS

Metal.	Symbol.	Atomic Weight 0 = 16.	Atomic Number.	Specific Gravity.	Melting Point ° C.
Aluminium	Al	26·97	13	2·56	657
Antimony	Sb	121·76	51	6·71	632
Arsenic	As	74·91	33	5·67	450
Barium	Ba	137·36	56	3·75	850
Beryllium	Be	9·02	4	1·93	Below 960
Bismuth	Bi	209·00	83	9·80	266
Cadmium	Cd	112·41	48	8·60	322
Calcium	Ca	40·08	20	1·57	780
Chromium	Cr	52·01	24	6·80	1482
Cobalt	Co	58·94	27	8·50	1464
Copper	Cu	63·57	29	8·82	1084
Gold	Au	197·20	79	19·32	1064
Indium	In	114·76	49	7·42	155
Iridium	Ir	193·1	77	22·42	2454
Iron	Fe	55·84	26	7·86	1505
Lead	Pb	207·22	82	11·37	327
Lithium	Li	6·94	3	0·59	186
Magnesium	Mg	24·32	12	1·74	633
Manganese	Mn	54·93	25	8·00	1207
Mercury	Hg	200·61	80	13·59	−39
Molybdenum	Mo	96·0	42	8·60	2500
Nickel	Ni	58·69	28	8·80	1427
Osmium	Os	191·5	76	22·48	Higher than ruthenium
Palladium	Pd	106·7	46	12·16	1553
Platinum	Pt	195·23	78	21·40	1773·5
Potassium	K	39·096	19	0·87	62
Rhodium	Rh	102·91	45	12·41	1966
Rubidium	Rb	85·44	37	1·52	38
Ruthenium	Ru	101·7	44	12·20	Higher than indium
Silver	Ag	107·88	47	10·53	962
Sodium	Na	22·997	11	0·97	95
Strontium	Sr	87·63	38	2·54	800
Tantalum	Ta	181·4	73	12·8	2910
Tellurium	Te	127·61	52	6·25	440
Thallium	Tl	204·39	81	11·85	303
Thorium	Th	232·12	90	11·10	1450
Tin	Sn	118·70	50	7·29	232
Titanium	Ti	47·90	22	3·54	2000
Tungsten	W	184·0	74	19·10	3100
Uranium	U	238·14	92	18·70	...
Vanadium	V	50·95	23	5·50	1680
Zinc	Zn	65·38	30	7·15	419
Zirconium	Zr	91·22	40	4·15	1500

THE METRIC SYSTEM OF WEIGHTS AND MEASURES

On this system the multiples and sub-multiples are arranged on a decimal basis. The multipliers are designated by the Greek prefixes :—*deka* = 10, *hecto* = 100, *kilo* = 1000. For the subdivisions Latin prefixes are employed :—*deci* = $\frac{1}{10}$, *centi* = $\frac{1}{100}$, *milli* = $\frac{1}{1000}$.

Length.—The unit of length is the metre.

1 kilometre	=	1000 metres	=	0·6214 mile
1 hectometre	=	100 ,,	=	109·361 yards
1 dekametre	=	10 ,,	=	32·8 feet
1 metre	=	1 ,,	=	39·37 inches
1 decimetre	=	0·1 ,,	=	3·937 ,,
1 centimetre	=	0·01 ,,	=	0·3937 ,,
1 millimetre	=	0·001 ,,	=	0·0394 ,,

Mass.—The unit of mass, the gramme (or gram.), is derived from the metre, and represents very nearly the mass of one cubic centimetre of water at its temperature of maximum density, 4° C.

1 kilogramme	=	1000 grammes	=	2·2046 lbs.
1 hectogramme	=	100 ,,	=	3·5274 ozs. (avoir).
1 dekagramme	=	10 ,,	=	154·3236 grains
1 gramme	=	1 ,,	=	15·4324 ,,
1 decigramme	=	0·1 ,,	=	1·5432 ,,
1 centigramme	=	0·01 ,,	=	0·1543 ,,
1 milligramme	=	0·001 ,,	=	0·0154 ,,

Volume.—The unit of volume is the litre (1000 c.c.) (one cubic centimetre of water at 4° C. equals 1 gramme).

1 kilolitre	=	1000 litres	=	220·4 imp gallons.
1 hectolitre	=	100 ,,	=	22·04 ,, ,,
1 dekalitre	=	10 ,,	=	2·20 ,, ,,
1 litre	=	1 ,,	=	1·76 imp. pints.
1 decilitre	=	0·1 ,,	=	3·52 Brit. fluid ounces.
1 centilitre	=	0·01 ,,	=	0·352 ,, ,,
1 millilitre	=	0·001 ,,	=	16·894 ,, minims.

RELATION OF IMPERIAL TO METRIC STANDARDS

Standards of Mass.

1 grain	=	0·0648	grammes.	
1 ounce troy	=	31·1035	,,	480 grains. 1·0971 ounces avoir.
1 ounce avoir.	=	28·350	,,	437·5 grains. 0·9114 ounce troy.
1 pennyweight	=	1·5552	,,	(24 grains).
1 pound troy	=	373·25	,,	5760 grains. 12 ounces troy. 0·822857 lb. avoir.
1 pound avoir.	=	453·593	,,	7000 grains. 16 ounces avoir. 1·215277 lbs. troy.

1 hundredweight (cwt.) = 50·8 kilogrammes (= 112 pounds).
1 British ton (2240 lbs. Long Ton) = 1016 kilogrammes.
1 American ton (2000 lbs. Short Ton) = 908 kilogrammes.

Standards of Capacity.

1 fluid ounce = 0·0284123 litre = 28·417 cubic centimetres.
1 pint = 0·5682454 ,, = 568·336 ,, ,,
1 gallon = 4·545963 litres.
1 U.S.A. gallon = 0·83254 imperial gallon.

1 British fluid ounce = volume of a weight of 437·5 grains (*i.e.* 1 oz. Av.) of water = 1·73 cub. ins.
1 U.S.A. fluid ounce = volume of a weight of 455·6 grains of water = 1·8 cub. ins.
1 Imperial pint = 20 fluid ounces = 567 c.c.
1 U.S.A. ,, = 16 ,, ,, = 473·15 c.c.
1 Imperial Gallon = 277·274 cub ins.
1 U.S.A. ,, = 231 cub. ins.

RELATION OF METRIC TO IMPERIAL STANDARDS

Standards of Mass.

1 milligramme = 0·015 grain.
1 gramme = 15·432 grains. { 0·032151 ounce troy. 0·0352736 ounce avoir. 0·0022046 pound.
1 kilogramme = 2·204622 pounds.
100 kilos (1 quintal) = 1·968 cwt.
1000 kilos (1 tonne) = 0·9842 British ton.

Standards of Capacity.

1 cubic centimetre (water at 4° C.) = 1 gramme.
1 litre (1000 c.c.) = 1·7598 pints.

To convert cubic centimetres into (British) fluid ounces multiply by 0·0352
,, litres ,, ,, ,, ,, ,, ,, 35·2
,, British fluid ounces into cubic centimetres ,, ,, 28·42
,, pints into litres into cubic centimetres ,, ,, 0·567

USEFUL DATA

Table to Convert Metric Weights into Avoirdupois and Troy Weights.

As 1 gramme is equal to 15·432 grains, or ·03527 avoirdupois ounce, or ·03215 troy ounce; to convert:

Grammes .	. into grains . . .	multiply by	15·432
Centigrammes	. ,, grains	,,	0·15432
Milligrammes	. ,, grains	,,	0·01543
Kilogrammes	. ,, avoirdupois ounces .	,,	35·2739
Grammes .	. ,, avoirdupois ounces .	,,	·03527
Kilogrammes	. ,, avoirdupois pounds .	,,	2·2046
Kilogrammes	. ,, troy ounces . .	,,	32·1507
Grammes .	. ,, troy ounces . .	,,	·03215

Table to Convert Avoirdupois and Troy Weights into Metric Weights.

As 1 grain is equal to 0·0648 gramme, and 1 avoirdupois ounce is equal to 28·3495 grammes, and 1 troy ounce is equal to 31·1035 grammes ; to convert :

Grains	into grammes	multiply by	0·0648
Grains	,, centigrammes	,,	6·4799
Grains	,, milligrammes	,,	64·799
Avoirdupois ounces	,, kilogrammes	,,	0·2835
Avoirdupois ounces	,, grammes	,,	28·3495
Avoirdupois lbs.	,, kilogrammes	,,	0·4536
Troy ounces	,, kilogrammes	,,	0·0311
Troy ounces	,, grammes	,,	31·1035

Table to Convert English Linear Measure into Metric Linear Measure.

As one inch is equal to 0·0254 metres ; to convert :

Inches	into metres	multiply by	0·0254
Inches	,, centimetres	,,	2·5399
Inches	,, millimetres	,,	25·3997

Table to Convert Metric Linear Measure into English Linear Measure.

As one metre is equal to 39·370 inches ; to convert :

Metres	into inches	multiply by	39·370
Centimetres	,, inches	,,	0·3937
Millimetres	,, inches	,,	0·03937

To convert Metric tons (2204·6 lb.) into English long ton (2240 lb.) multiply by 1·016

,,	Long tons	,, metric tons	,,	·9842
,,	Short tons (2000 lb.)	,, Long tons (2240 lb.)	,,	·950
,,	Long tons	,, Short tons	,,	1·120

Area of circle	= square of diameter multiplied by 0·7854.
Area of sphere	= square of diameter multiplied by 3·14159.
Solid contents of sphere	= one-sixth of the cube of the diameter multiplied by 3·14159.
Area of curved surface of cylinder	= diameter × 3·14159 × length.
Solid contents of cylinder	= square of diameter × length × 0·7854.

Comparison of Thermometer Scales.

Centigrade.	Fahrenheit.	Réaumur.
Water freezes at 0°.	Water freezes at 32°.	Water freezes at 0°.
Water boils at 100°.	Water boils at 212°.	Water boils at 80°.

To convert Centigrade degrees into degrees of Fahrenheit, multiply by 9, divide by 5, and add 32.

To convert Fahrenheit degrees into degrees of Centigrade, subtract 32, multiply by 5, and divide by 9.

To convert Réaumur degrees into degrees of Fahrenheit, multiply by 9, divide by 4, and add 32.

To convert Fahrenheit degrees into degrees of Réaumur, subtract 32, multiply by 4, and divide by 9.

To convert Réaumur degrees into degrees of Centigrade, multiply by 5, and divide by 4.

To convert Centigrade degrees into degrees of Réaumur, multiply by 4 and divide by 5.

HYDROMETERS

A quick and convenient method of determining the density or specific gravity of a liquid is by the use of the hydrometer. The method is based upon the fact that a floating body is buoyed up more by a heavy liquid than by a light one. The hydrometer consists of a hollow glass bulb or float, carrying above it a thin tube, with a scale marked on it, or on a paper enclosed in it. A smaller bulb is also blown beneath, into which mercury or lead shot is put, so as to adjust the weight and at the same time cause the instrument to float in a vertical position. The liquid to be examined is poured into a tall glass jar, and the hydrometer immersed so that it floats freely; the specific gravity may then be read off from the stem, which is so graduated that the number level with the surface of the liquid shows the specific gravity of the liquid. It is manifest that the denser the liquid, the higher the instrument will float in it, the weight of the liquid displaced being always equal to that of the hydrometer.

The temperature of the liquid must be adjusted to the temperature required by the hydrometer, which is usually 15·5° C. The graduation of hydrometers is not made to any uniform system. Those marked in degrees Baumé or Twaddell, or according to specific gravity, are most commonly used. Hydrometers are only accurate within comparatively small ranges of density, hence they are usually supplied in sets as given below. Two different sets are used, one for liquids lighter than water, the other for liquids such as acids, that are heavier.

Baumé's Hydrometer.—The degrees on Baumé's hydrometer agree among themselves in being at equal distances along the stem, but they are proportional neither to the specific gravity nor to the percentage of salt in the solution. They may be converted into an ordinary statement of specific gravity by the use of the following formulæ:—

(1) For liquids heavier than water,

$$\text{Specific gravity} = \frac{144{\cdot}3}{144{\cdot}3 - \text{degrees Baumé}}.$$

For example, 30° Baumé corresponds to a specific gravity of 1·262, thus

$$\frac{144{\cdot}3}{144{\cdot}3 - 30^\circ} = \frac{144{\cdot}3}{114{\cdot}3} = 1{\cdot}262.$$

(2) For liquids lighter than water,

$$\text{Specific gravity} = \frac{146}{136 + \text{degrees Baumé}}.$$

For general purposes a hydrometer graduated according to specific gravity is the most convenient. In these instruments the distances between the divisions become less as the densities increase. With the aid of the above formulæ a Baumé hydrometer may readily be converted into one showing the actual specific gravity.

To reduce specific gravity to degrees Baumé the formula is :—

$$\frac{144 \cdot 3}{\text{Specific gravity}} - 136 = \text{degrees Baumé.}$$

Twaddell's Hydrometer.—The graduations of this instrument correspond to equal differences of density. Each degree is 0·005 in excess of unity (water = 1·000).

Thus,

1° Twaddell = 1·005 specific gravity.
2° ,, = 1·010 ,,
7° ,, = 1·035 ,,

The relation between the specific gravity of a liquid and the reading of the instrument may be expressed as follows :—

To find the specific gravity, multiply the reading in degrees Twaddell by 0·005, and add 1. For example 31° Twaddell corresponds to a specific gravity of 1·155, thus

$$31 \times 0{\cdot}005 = 0{\cdot}155 + 1{\cdot}000 = 1{\cdot}155.$$

The degrees Twaddell may also be converted into specific gravity by multiplying the reading by 5, adding 1000 (water = 1000) and cutting off three decimal places.

For example,

25° Twaddell = 1·125 specific gravity.

To reduce specific gravity (water 1000) to Twaddell, deduct 1000 and divide the remainder by 5.

The following sets of hydrometers are those most generally supplied :—

Twaddell.	Baumé, for heavy liquids.	Baumé, for light liquids.
No. 1 0 to 24 = 1·00 to 1·12 sp. gr.	0 = 1·000 sp. gr.	60 = ·706 sp. gr.
2 24 to 48 = 1·12 to 1·24 ,,	10 = 1·075 ,,	50 = ·761 ,,
3 48 to 74 = 1·24 to 1·37 ,,	20 = 1·161 ,,	40 = ·817 ,,
4 74 to 102 = 1·37 to 1·51 ,,	30 = 1·263 ,,	36 = ·837 ,,
5 102 to 138 = 1·51 to 1·69 ,,	40 = 1·385 ,,	30 = ·871 ,,
6 138 to 170 = 1·69 to 1·85 ,,	50 = 1·532 ,,	26 = ·892 ,,
	60 = 1·714 ,,	20 = ·928 ,,
	70 = 1·946 ,,	10 = 1·000 ,,

REAGENTS OF NORMAL STRENGTH

The following normal reagents are prepared by dissolving the equivalent weight in grammes of the various salts in water, and diluting to 1 litre (1000 c.c.).

Ammonium oxalate $(NH_4)_2C_2O_42H_2O$,	equivalent	weight . .	80·0
Ammonium thiocyanate NH_4CNS	,,	,, . .	76·0
Potassium cyanide, KCN	,,	,, . .	65·0
Potassium hydroxide, KOH	,,	,, . .	56·0
Potassium iodide, KI	,,	,, . .	166·0
Sodium carbonate, $Na_2CO_310H_2O$	,,	,, . .	143·0
Sodium chloride, NaCl	,,	,, . .	58·5
Sodium hydroxide, NaOH	,,	,, . .	40·0
Silver nitrate, $AgNO_3$	,,	,, . .	170·0
Sodium thiosulphate, $Na_2S_2O_35H_2O$	,,	,, . .	124·0

NITRIC ACID

Percentage, by Weight, of Concentrated Acid in Aqueous Solutions of Nitric Acid of different Specific Gravities. Temperature, 15° C. (60° F.).

Specific gravity.	HNO_3 per cent.	Specific gravity.	HNO_3 per cent.	Specific gravity.	HNO_3 per cent.
1·530	100·0	1·405	66·0	1·205	33·0
1·527	99·0	1·400	65·0	1·198	32·0
1·524	98·0	1·395	64·0	1·192	31·0
1·520	97·0	1·390	63·0	1·185	30·0
1·516	96·0	1·386	62·0	1·179	29·0
1·513	95·0	1·380	61·0	1·172	28·0
1·509	94·0	1·374	60·0	1·166	27·0
1·506	93·0	1·368	59·0	1·159	26·0
1·503	92·0	1·363	58·0	1·152	25·0
1·499	91·0	1·358	57·0	1·145	24·0
1·495	90·0	1·353	56·0	1·138	23·0
1·492	89·0	1·346	55·0	1·132	22·0
1·488	88·0	1·341	54·0	1·126	21·0
1·485	87·0	1·335	53·0	1·120	20·0
1·482	86·0	1·329	52·0	1·114	19·0
1·478	85·0	1·323	51·0	1·108	18·0
1·474	84·0	1·317	50·0	1·102	17·0
1·470	83·0	1·311	49·0	1·096	16·0
1·467	82·0	1·304	48·0	1·089	15·0
1·463	81·0	1·298	47·0	1·083	14·0
1·460	80·0	1·291	46·0	1·077	13·0
1·456	79·0	1·284	45·0	1·071	12·0
1·452	78·0	1·277	44·0	1·065	11·0
1·449	77·0	1·270	43·0	1·060	10·0
1·445	76·0	1·264	42·0	1·053	9·0
1·442	75·0	1·257	41·0	1·047	8·0
1·438	74·0	1·251	40·0	1·041	7·0
1·435	73·0	1·244	39·0	1·034	6·0
1·431	72·0	1·238	38·0	1·028	5·0
1·427	71·0	1·232	37·0	1·022	4·0
1·423	70·0	1·225	36·0	1·016	3·0
1·418	69·0	1·218	35·0	1·010	2·0
1·414	68·0	1·212	34·0	1·004	1·0
1·410	67·0				

SULPHURIC ACID

Percentage, by Weight, of Concentrated Acid in Aqueous Solutions of Sulphuric Acid of varying Specific Gravity. Temperature 15° C. (60° F.).

Specific gravity.	H_2SO_4 per cent.	Specific gravity.	H_2SO_4 per cent.	Specific gravity.	H_2SO_4 per cent.
1·843	100·0	1·568	66·0	1·247	33·0
1·842	99·0	1·557	65·0	1·239	32·0
1·841	98·0	1·545	64·0	1·231	31·0
1·841	97·0	1·534	63·0	1·223	30·0
1·840	96·0	1·523	62·0	1·215	29·0
1·838	95·0	1·512	61·0	1·206	28·0
1·836	94·0	1·501	60·0	1·198	27·0
1·834	93·0	1·490	59·0	1·190	26·0
1·831	92·0	1·480	58·0	1·182	25·0
1·827	91·0	1·469	57·0	1·174	24·0
1·822	90·0	1·458	56·0	1·167	23·0
1·816	89·0	1·448	55·0	1·159	22·0
1·809	88·0	1·438	54·0	1·151	21·0
1·802	87·0	1·428	53·0	1·144	20·0
1·794	86·0	1·418	52·0	1·136	19·0
1·786	85·0	1·408	51·0	1·129	18·0
1·777	84·0	1·398	50·0	1·121	17·0
1·767	83·0	1·388	49·0	1·113	16·0
1·756	82·0	1·379	48·0	1·106	15·0
1·745	81·0	1·370	47·0	1·098	14·0
1·734	80·0	1·361	46·0	1·091	13·0
1·722	79·0	1·351	45·0	1·083	12·0
1·710	78·0	1·342	44·0	1·075	11·0
1·698	77·0	1·333	43·0	1·068	10·0
1·686	76·0	1·324	42·0	1·061	9·0
1·675	75·0	1·315	41·0	1·053	8·0
1·663	74·0	1·306	40·0	1·046	7·0
1·651	73·0	1·297	39·0	1·039	6·0
1·639	72·0	1·289	38·0	1·032	5·0
1·627	71·0	1·281	37·0	1·025	4·0
1·615	70·0	1·272	36·0	1·019	3·0
1·604	69·0	1·264	35·0	1·013	2·0
1·592	68·0	1·256	34·0	1·006	1·0
1·580	67·0				

HYDROCHLORIC ACID

Percentage, by Weight, of Concentrated Acid in Aqueous Solutions of Hydrochloric Acid of different Specific Gravities. Temperature, 15° C. (60° F.).

Specific gravity.	HCl per cent.	Specific gravity.	HCl per cent.	Specific gravity.	HCl per cent.
1·2000	40·78	1·1410	28·54	1·0798	16·31
1·1982	40·37	1·1389	28·13	1·0778	15·90
1·1964	39·96	1·1369	27·72	1·0758	15·49
1·1946	39·55	1·1349	27·32	1·0738	15·08
1·1928	39·14	1·1328	26·91	1·0718	14·68
1·1910	38·74	1·1308	26·50	1·0697	14·27
1·1893	38·33	1·1287	26·10	1·0677	13·86
1·1875	37·92	1·1267	25·69	1·0657	13·45
1·1857	37·51	1·1247	25·28	1·0637	13·05
1·1846	37·11	1·1226	24·87	1·0617	12·64
1·1822	36·70	1·1206	24·46	1·0597	12·23
1·1802	36·29	1·1185	24·06	1·0577	11·82
1·1782	35·88	1·1164	23·65	1·0557	11·41
1·1762	35·47	1·1143	23·24	1·0537	11·01
1·1741	35·07	1·1123	22·83	1·0517	10·60
1·1721	34·66	1·1102	22·43	1·0497	10·19
1·1701	34·25	1·1082	22·02	1·0477	9·78
1·1681	33·84	1·1061	21·61	1·0457	9·38
1·1661	33·43	1·1041	21·20	1·0437	8·97
1·1641	33·03	1·1020	20·79	1·0417	8·56
1·1620	32·62	1·1000	20·39	1·0397	8·15
1·1599	32·21	1·0980	19·98	1·0377	7·75
1·1578	31·80	1·0960	19·57	1·0357	7·34
1·1557	31·40	1·0939	19·16	1·0337	6·93
1·1536	30·99	1·0919	18·76	1·0318	6·52
1·1515	30·58	1·0899	18·35	1·0298	6·11
1·1494	30·17	1·0879	17·94	1·0279	5·51
1·1473	29·76	1·0859	17·53	1·0259	5·30
1·1452	29·36	1·0838	17·12	1·0239	4·89
1·1431	28·95	1·0818	16·72	1·0200	4·01

NORMAL SOLUTIONS

In volumetric analysis standard solutions are designated *normal solutions* (distinguished by the letter N) when they are of such strength that 1 litre contains 1 gramme of replaceable hydrogen or a weight of the reagent in grammes, equal to the *chemical equivalent* of that reagent. Therefore, a normal solution of sulphuric acid, which is a dibasic acid, will contain :—

$$H_2 + S + O_4$$
$$2 + 32 + 64 = 98;\ 98 \div 2 = 49 \text{ grammes per litre,}$$

or 0·049 gramme per cubic centimetre.

In the case of alkalis, etc., the normal solution must contain the chemical equivalent (weighed in grammes) of the metal or group of elements that replace the hydrogen. Therefore a normal solution of sodium hydroxide (caustic soda) will contain :—

$$Na + H + O$$
$$23 + 1 + 16 = 40 \text{ grammes per litre,}$$

or 0·04 gramme per cubic centimetre since the valency (or combining power) of sodium and hydrogen are the same.

A normal solution of barium hydroxide (barium hydrate) would contain :—

$$Ba + H_2 + O_2$$
$$137 + 2 + 32 = 171 ; 171 \div 2 = 85{\cdot}5 \text{ grammes per litre,}$$

or 0·085 gramme per cubic centimetre since the valency of barium is twice that of hydrogen.

From the above remarks it will be obvious that equal quantities of any normal acid solution, and of a normal alkali solution, will neutralise each other. Also the quantity of any solution of a single acid or an alkali that is neutralised by the addition of 1 c.c. of a normal solution will contain 0·001 gramme of the hydrogen equivalent of its active constituent.

It is frequently convenient to use standard solutions the strength of which bears a definite ratio to that of the normal solution. Thus solutions of *one-half, one-tenth,* and *one-hundredth* of the strength of the normal solution are called respectively *semi-normal, deci-normal,* and *centi-normal,* and are distinguished by the signatures $\frac{N}{2}$, $\frac{N}{10}$, and $\frac{N}{100}$.

Solutions that are stronger than the normal solution are designated by prefixing the numeral denoting the number of times it is stronger than normal; thus a solution twice the normal strength is designated 2N, and a solution five times normal strength 5N.

NORMAL ACIDS

Sulphuric Acid.—H_2SO_4 equivalent = 49. Normal sulphuric acid contains 49 grammes of H_2SO_4 per litre. Ordinary concentrated acid, sp. gr. 1·84, has a strength of only 36 times normal, and may be designated 36N. Therefore one volume of strong acid diluted with 30 volumes of water = normal strength.

Nitric Acid.—HNO_3equivalent = 63. Concentrated acid, sp. gr. 1·42 = 16N. One volume of strong acid diluted with 11 volumes of water = normal strength.

Hydrochloric Acid.—HCl equivalent = 36·5. Concentrated acid, sp. gr. 1·16 — 10N. One volume of strong acid diluted with 5 volumes of water = normal strength.

NITRO-HYDROCHLORIC ACID, OR AQUA REGIA

This mixture termed aqua regia (because it dissolves the noble metals) usually consists of 1 part by measure of nitric acid and 3 parts of hydrochloric acid. Its solvent action depends upon the fact that it contains free chlorine, liberated by the oxidizing action of nitric acid on the hydrogen of the hydrochloric acid, thus :—

$$3HCl + HNO_3 = Cl_2 + NOCl + 2H_2O.$$

The mixture is best prepared when required, and is chiefly used for the solution of gold and platinum, and certain sulphides, which do not dissolve in either nitric or hydrochloric acid separately. When solutions in aqua regia are evaporated, chlorides are left.

METHOD OF TESTING WEIGHTS[1]

"As a rule, the weights of the best makers possess greater accuracy than is required of them in ordinary quantitative processes; nevertheless their examination should never be omitted. The readiest method of detecting errors in the values of the denominations is to place one of the gramme weights on the pan of a delicate balance, adjusted to perfect equilibrium, and equipoise it with pieces of brass or small shot, and finally tin foil. The weight is then removed and replaced by the second gramme weight, and the balance caused to oscillate. If the excursions on either side of the zero are of equal amplitude, the weights are equivalent; if not, the deviation must be noted. It should not exceed 1 division of the scale from the zero-point. The third gramme weight is next tried in the same way, and it is then replaced by platinum weights of the smaller denominations to make up 1 gramme, and the balance again caused to oscillate, every deviation from the equilibrium being carefully noted. In the same way the 2-gramme piece is compared with two of the single grammes, the 5-gramme piece with the $2 + 1 + 1 + 1$ gramme pieces, and each of the 10-gramme pieces with the $5 + 2 + 1 + 1 + 1$ gramme pieces. The larger pieces also should not show greater variations than 1 division from the zero, since the value of 1-scale division with a heavy load on the pans is almost invariably greater than with a diminished load."

[1] T. E. Thorpe, *Quantitative Chemical Analysis.*

INDEX